SECOND
EDITION

ELECTRONICS FUNDAMENTALS
Circuits, Devices, and Applications

Thomas L. Floyd

Merrill, an imprint of Macmillan Publishing Company
New York

Collier Macmillan Canada, Inc.
Toronto

Maxwell Macmillan International Publishing Group
New York   Oxford   Singapore   Sydney

To the electronics instructor, in the hope that this book will make your job a little easier

Administrative Editor: David Garza
Developmental Editor: Carol Hinklin Thomas
Production Editor: Rex Davidson
Art Coordinator: Ruth Kimpel
Photo Editor: Gail Meese
Cover Art: Hans & Cassady, Inc.
Cover Designer: Brian Deep
Page Layout: Ellen Robison

This book was set in Times Roman.

Credits: Part 1 opening photo by Merrill Publishing/Bruce Johnson; Part 2 opening photo and Chapters 2, 3, 8, 10, 17, and 21 opening photos courtesy of Tektronix, Inc.; Part 3 opening and Chapter 5 opening photos courtesy of Motorola Semiconductor Products, Inc.; Chapters 1 and 6 opening photos by Merrill Publishing; Chapters 4, 15, 16 and 20 photos courtesy of DeVry, Inc.; Chapters 7, 9, 14, 18, and 19 by Merrill Publishing/Jo Hall; Chapter 11 opening photo by Merrill Publishing/Larry Hamill; Chapter 12 opening photo courtesy of Tandy Corp.; Chapter 13 opening photo courtesy of Hewlett-Packard; four-color insert photos by Larry Hamill, and four-color insert illustrations by Hans & Cassady, Inc.

Collier Macmillan Canada, Inc.

Library of Congress Cataloging-in-Publication Data

Floyd, Thomas L.
    Electronics fundamentals : circuits, devices, and applications /
  Thomas L. Floyd. -- 2nd ed.
       p.   cm.
  Includes index.
  ISBN 0-675-21310-X
   1. Electronics. 2. Electronic circuits. 3. Electronic apparatus
and appliances.   I. Title.
TK7816.F57   1991
621.381--dc20
                                                  90-43627
                                                  CIP

Printing: 4 5 6 7 8 9   Year: 2 3 4

P. Tang

MERRILL'S INTERNATIONAL SERIES
IN ENGINEERING TECHNOLOGY

ADAMSON	*Applied Pascal for Technology*, 20771-1
	Structured BASIC Applied to Technology, 20772-X
	Structured C for Technology, 20993-5
	Structured C for Technology (w/disks), 21289-8
ANTONAKOS	*The 68000 Microprocessor: Hardware and Software, Principles and Applications*, 21043-7
ASSER/STIGLIANO/ BAHRENBURG	*Microcomputer Servicing: Practical Systems and Troubleshooting*, 20907-2
	Microcomputer Theory and Servicing, 20659-6
	Lab Manual to accompany Microcomputer Theory and Servicing, 21109-3
ASTON	*Principles of Biomedical Instrumentation and Measurement*, 20943-9
BATESON	*Introduction to Control System Technology, Third Edition*, 21010-0
BEACH/JUSTICE	*DC/AC Circuit Essentials*, 20193-4
BERLIN	*Experiments in Electronic Devices, Second Edition*, 20881-5
	The Illustrated Electronics Dictionary, 20451-8
BERLIN/GETZ	*Experiments in Instrumentation and Measurement*, 20450-X
	Fundamentals of Operational Amplifiers and Linear Integrated Circuits, 21002-X
	Principles of Electronic Instrumentation and Measurement, 20449-6
BOGART	*Electronic Devices and Circuits, Second Edition*, 21150-6
BOGART/BROWN	*Experiments in Electronic Devices and Circuits, Second Edition*, 21151-4
BOYLESTAD	*DC/AC: The Basics*, 20918-8
	Introductory Circuit Analysis, Sixth Edition, 21181-6
BOYLESTAD/KOUSOUROU	*Experiments in Circuit Analysis, Sixth Edition*, 21182-4
	Experiments in DC/AC Basics, 21131-X
BREY	*Microprocessors and Peripherals: Hardware, Software, Interfacing, and Applications, Second Edition*, 20884-X
	The 8086/8088 Microprocessor Family, Second Edition, 21309-6
BROBERG	*Lab Manual to accompany Electronic Communication Techniques, Second Edition*, 21257-X
BUCHLA	*Digital Experiments: Emphasizing Systems and Design, Second Edition*, 21180-8
	Experiments in Electric Circuits Fundamentals, Second Edition, 21409-2
	Experiments in Electronics Fundamentals: Circuits, Devices, and Applications, Second Edition, 21407-6
COOPER	*Introduction to VersaCAD*, 21164-6
COX	*Digital Experiments: Emphasizing Troubleshooting, Second Edition*, 21196-4
CROFT	*Getting a Job: Resume Writing, Job Application Letters, and Interview Strategies*, 20917-X
DAVIS	*Technical Mathematics*, 20338-4
	Technical Mathematics with Calculus, 20965-X
	Study Guide to accompany Technical Mathematics, 20966-8
	Study Guide to accompany Technical Mathematics with Calculus, 20964-1

Preface

This second edition of *Electronics Fundamentals: Circuits, Devices, and Applications* covers the essential topics in dc circuits, ac circuits, and electronic devices with emphasis on applications and troubleshooting. The coverage provides a thorough foundation upon which to develop skills in electronics and related fields.

This book is written particularly for students in technician-level programs in technical institutes, community colleges, and vocational/technical schools. Emphasis is on the understanding and application of basic concepts. Mathematics is held to a minimum and is used only in a supportive role where it is essential for the development of a complete and meaningful coverage of a topic or for providing a means of analyzing basic circuit operation.

FEATURES

- Chapter objectives
- Application assignments
- Section reviews with answers at the end of the chapter
- Numerous examples (over 275)
- Calculator sequences
- Chapter summaries
- Chapter glossaries with a comprehensive glossary at the end of the book
- Chapter formula list
- Two problem sets in each chapter (one easier, the other more difficult)
- Functional two-color format
- Two functional full-color sections
- Extensively illustrated (over 1300 figures)
- Test bank
- Transparency package with 200 transparency masters
- Instructor's manual
- Coordinated lab manual available
- Study Guide available

For those who used the first edition of this text, you will find a number of major improvements in this edition:

☐ More end-of-chapter problems
☐ Increased use of second color in the illustrations
☐ A glossary at the end of each chapter in addition to the comprehensive glossary at the end of this book
☐ A completely revised functional full-color section for the dc chapters and a new one for the ac chapters
☐ References to color sections marked by the margin logo COLOR INSERT
☐ Use of test instruments expanded and integrated throughout the text
☐ Increased and improved troubleshooting coverage
☐ Troubleshooting problems marked by the margin logo TBLS + −
☐ A coverage of batteries included in the appendix
☐ Multiple-choice self-tests
☐ Many areas reworked to update or expand coverage and to improve the clarity of the material
☐ Expanded test bank
☐ Improved lab manual
☐ Videotape on current, voltage, and resistance free with adoption

CHAPTER ORGANIZATION

Chapter Opener   The chapter opener provides an overview of the material to be covered and a set of performance-based objectives that lists the specific things a student will be able to do upon completion of the chapter.

Application Assignment   The application assignment at the beginning of each chapter places students in a practical on-the-job situation and requires them to complete the assignment based on knowledge gained during the study of the chapter. These application assignments are designed to be both educational and motivational. They put the students in a situation to which they can relate but which they are generally unable to handle before studying the chapter. After completing the material in the chapter, however, they can successfully carry out the assignment and thereby derive a sense of accomplishment. Solutions or suggested approaches to each application assignment are provided in the Instructor's Manual.

Calculator Sequences   Calculator sequences are included in selected examples to show the students how to use a typical calculator to obtain a solution. Generally, these are placed where a new procedure is first introduced. Keep in mind that there may be differences from one type of calculator to another in the labeling of certain keys and in the keystroke sequences required for certain functions.

Section Reviews   Section reviews are found at the end of each section and consist of questions and problems that focus on the key concepts presented in that section. These provide the students with frequent feedback on their comprehension of the material. Answers are given at the end of the chapter.

Summary   The summary at the end of the chapter provides a concise review of the material covered in the chapter.

Formula List   A list of all numbered equations that are used in the chapter is at the end of the chapter.

Self-Test   A multiple-choice self-test appears at the end of the chapter. This test allows students to conveniently check their comprehension of the material covered in the chapter before proceeding to the problem sets. A complete set of answers is at the end of the book.

Problem Sets   In each of the first fifteen chapters, there are *two* problem sets. Problem Set A contains relatively straightforward exercises, and Problem Set B generally contains more demanding and thought-provoking problems. Troubleshooting problems and problems related to the color sections are indicated by logos placed in the margin. Answers to most of the odd-numbered problems are at the end of the book.

Color Sections   A number of examples, section reviews, and problems relate to either the dc circuits color section (following page 240) or to the ac circuits color section (following page 416). References to these functional full-color sections are indicated by a margin logo. In these referenced exercises, students are required to identify simple component relationships in actual breadboard assemblies or on printed circuit boards, to determine resistance values from the color band, and to take readings from instruments connected to operating circuits. The color sections utilize computer-generated art to insure distinct color codes and clarity of circuit connections on the component boards.

OVERALL ORGANIZATION OF THE TEXT

The book is divided into three parts: DC Circuits, consisting of Chapters 1 through 7; AC Circuits, consisting of Chapters 8 through 15; and Electronic Devices, consisting of Chapters 16 through 21.

For those wishing to introduce capacitance and inductance prior to Chapter 8 (Introduction to Alternating Current and Voltage), all of Chapter 9 except Sections 9–6 and 9–7 and all of Chapter 10 except Sections 10–6 and 10–7 can be added to the dc coverage. Then Sections 9–6, 9–7, 10–6, and 10–7 can be included in the ac coverage.

Those wishing to cover capacitive circuit topics before moving into inductive circuits may choose to cover Chapter 12 immediately after Chapter 9 and then cover Chapters 10, 11, and 13.

The organization and content of this book are suited particularly to those programs that cover dc, ac, and devices during the first year in a two-term or a three-term sequence.

ACKNOWLEDGMENTS

The efforts of many people were required to produce this book. In particular, I want to express my appreciation to Carol Thomas, Rex Davidson, Ruth Kimpel, Gail Meese, Brian Deep, and Dave Garza at Merrill Publishing Company for their dedication and hard work. Also, I want to thank the following reviewers for their many valuable comments and suggestions: Dave Buchla, Yuba College (CA); Phillip Chiarelli, Electronics Institute (MD); Linda Hartman, ITT Technical Institute (WA); Floyd Hastings, Francis Tuttle High Tech Center (OK); Mike Merchant, Microcomputer Technical Institute

(TX); Bill Rice, Catonsville C. C. (MD); Tom Stultz, GTE, Grafton (MA); and Davis Wilson, Arizona Western College (AZ).

A special thanks to Hamid Yousefzadeh at DeVry Atlanta for his painstaking accuracy check of the manuscript, to Gary Farmer, Omar Adham, and Dave Brandt at DeVry Columbus for their help in producing the new full-color photographs, and to my wife Sheila for her support during this revision.

Tom Floyd

Contents

12   FREQUENCY RESPONSE OF *RC* CIRCUITS   458

13   FREQUENCY RESPONSE OF *RL* CIRCUITS   512

14   *RLC* CIRCUITS AND RESONANCE   550

15   PULSE RESPONSE OF *RC* AND *RL* CIRCUITS   606

PART THREE   **DEVICES**

16   INTRODUCTION TO SEMICONDUCTOR DEVICES   652

17   DIODES AND APPLICATIONS   680

18   TRANSISTORS AND THYRISTORS   732

PART ONE DC CIRCUITS

1 Introduction

This chapter presents a brief history of the fields of electricity and electronics and discusses some of the many areas of application. Also, to aid you throughout the book, the basics of scientific notation and metric prefixes are introduced, and the quantities and units commonly used in electronics are presented.

After completing this chapter, you will be able to

☐ Recognize some common electrical components and measuring instruments.
☐ State the electrical quantities and their units.
☐ Use scientific notation (powers of ten).
☐ Use metric prefixes to express large and small quantities.

**APPLICATION
ASSIGNMENT**

At the beginning of each chapter starting with Chapter 2, you will find an Application Assignment that relates to that chapter. These assignments present a variety of practical job situations that a technician might encounter in industry.

As you study each chapter, think about how to approach the Application Assignment. When you have completed each chapter, you should have a sufficient knowledge of the topics covered to enable you to carry out the assignment.

1–1   A BRIEF HISTORY OF ELECTRICITY AND ELECTRONICS

One of the first important discoveries about static electricity is attributed to William Gilbert (1540–1603). Gilbert was an English physician who, in a book published in 1600, described how amber differs from magnetic loadstones in its attraction of certain materials. He found that when amber was rubbed with a cloth, it attracted only light-weight objects, whereas loadstones attracted only iron. Gilbert also discovered that other substances, such as sulfur, glass, and resin, behave as amber does. He used the Latin word *elektron* for amber and originated the word *electrica* for the other substances that acted similarly to amber. The word *electricity* was used for the first time by Sir Thomas Browne (1605–82), an English physician.

Another Englishman, Stephen Gray (1696–1736), discovered that some substances conduct electricity and some do not. Following Gray's lead, a Frenchman named Charles du Fay experimented with the conduction of electricity. These experiments led him to believe that there were two kinds of electricity. He called one type *vitreous electricity* and the other type *resinous electricity*. He found that objects having vitreous electricity repelled each other and those having resinous electricity attracted each other. It is known today that two types of electrical *charge* exist, *positive* and *negative*. Negative charge results from an excess of electrons in a material and positive charge results from a deficiency of electrons.

Benjamin Franklin (1706–90) conducted studies in electricity in the mid-1700s. He theorized that electricity consisted of a single *fluid,* and he was the first to use the terms *positive* and *negative*. In his famous kite experiment, Franklin showed that lightning is electricity.

Charles Augustin de Coulomb (1736–1806), a French physicist, in 1785 proposed the laws that govern the attraction and repulsion between electrically charged bodies. Today, the unit of electrical charge is called the *coulomb*.

Luigi Galvani (1737–98) experimented with current electricity in 1786. Galvani was a professor of anatomy at the University of Bologna in Italy. Electrical current was once known as *galvanism* in his honor.

In 1800, Alessandro Volta (1745–1827), an Italian professor of physics, discovered that the chemical action between moisture and two different metals produced electricity. Volta constructed the first battery, using copper and zinc plates separated by paper that had been moistened with a salt solution. This battery, called the *voltaic pile,* was the first source of steady electric current. Today, the unit of electrical potential energy is called the *volt* in honor of Professor Volta.

A Danish scientist, Hans Christian Oersted (1777–1851), is credited with the discovery of electromagnetism, in 1820. He found that electrical current flowing through a wire caused the needle of a compass to move. This finding showed that a magnetic field exists around a current-carrying conductor and that the field is produced by the current.

The modern unit of electrical current is the *ampere* (also called *amp*) in honor of the French physicist André Ampère (1775–1836). In 1820, Ampère measured the magnetic effect of an electrical current. He found that two wires carrying current can attract and repel each other, just as magnets can. By 1822, Ampère had developed the fundamental laws that are basic to the study of electricity.

One of the most well known and widely used laws in electrical circuits today is *Ohm's law*. It was formulated by Georg Simon Ohm (1787–1854), a German teacher, in 1826. Ohm's law gives us the relationship among the three important electrical quantities of *resistance, voltage,* and *current.*

Although it was Oersted who discovered electromagnetism, it was Michael Faraday (1791–1867) who carried the study further. Faraday was an English physicist who believed that if electricity could produce magnetic effects, then magnetism could produce electricity. In 1831 he found that a moving magnet caused an electric current in a coil of wire placed within the field of the magnet. This effect, known today as *electromagnetic induction,* is the basic principle of electric generators and transformers.

Joseph Henry (1797–1878), an American physicist, independently discovered the same principle in 1831, and it is in his honor that the unit of inductance is called the *henry.* The unit of capacitance, the *farad,* is named in honor of Michael Faraday.

In the 1860s, James Clerk Maxwell (1831–79), a Scottish physicist, produced a set of mathematical equations that expressed the laws governing electricity and magnetism. These formulas are known as *Maxwell's equations.* Maxwell also predicted that electromagnetic waves (radio waves) that travel at the speed of light in space could be produced.

It was left to Heinrich Rudolph Hertz (1857–94), a German physicist, to actually produce these waves that Maxwell predicted. Hertz performed this work in the late 1880s. Today, the unit of frequency is called the *hertz.*

The Beginning of Electronics

The early experiments in electronics involved electric currents in glass tubes. One of the first to conduct such experiments was a German named Heinrich Geissler (1814–79). Geissler found that when he removed most of the air from a glass tube, the tube glowed when an electrical potential was placed across it.

Around 1878, Sir William Crookes (1832–1919), a British scientist, experimented with tubes similar to those of Geissler. In his experiments, Crookes found that the current in the tubes seemed to consist of particles.

Thomas Edison (1847–1931), experimenting with the carbon-filament light bulb that he had invented, made another important finding. He inserted a small metal plate in the bulb. When the plate was positively charged, a current flowed from the filament to the plate. This device was the first *thermionic diode.* Edison patented it but never used it.

The electron was discovered in the 1890s. The French physicist Jean Baptiste Perrin (1870–1942) demonstrated that the current in a vacuum tube consists of negatively charged particles. Some of the properties of these particles were measured by Sir Joseph Thomson (1856–1940), a British physicist, in experiments he performed between 1895 and 1897. These negatively charged particles later became known as *electrons.* The charge on the electron was accurately measured by an American physicist, Robert A. Millikan (1868–1953), in 1909. As a result of these discoveries, electrons could be controlled, and the electronic age was ushered in.

Putting the Electron to Work

A vacuum tube that allowed electrical current in only one direction was constructed in 1904 by John A. Fleming, a British scientist. The tube was used to detect electromagnetic waves. Called the *Fleming valve,* it was the forerunner of the more recent vacuum diode tubes.

Major progress in electronics, however, awaited the development of a device that could boost, or *amplify,* a weak electromagnetic wave or radio signal. This device was the *audion,* patented in 1907 by Lee de Forest, an American. It was a triode vacuum tube capable of amplifying small electrical signals.

Two other Americans, Harold Arnold and Irving Langmuir, made great improvements in the triode tube between 1912 and 1914. About the same time, de Forest and Edwin Armstrong, an electrical engineer, used the triode tube in an *oscillator circuit.* In 1914, the triode was incorporated in the telephone system and made the transcontinental telephone network possible.

The tetrode tube was invented in 1916 by Walter Schottky, a German. The tetrode, along with the pentode (invented in 1926 by Tellegen, a Dutch engineer), provided great improvements over the triode. The first television picture tube, called the *kinescope,* was developed in the 1920s by Vladimir Zworykin, an American researcher.

During World War II, several types of microwave tubes were developed that made possible modern microwave radar and other communications systems. In 1939, the *magnetron* was invented in Britain by Henry Boot and John Randall. In the same year, the *klystron* microwave tube was developed by two Americans, Russell Varian and his brother Sigurd Varian. The *traveling-wave* tube was invented in 1943 by Rudolf Komphner, an Austrian-American.

The Computer

The computer probably has had more impact on modern technology than any other single type of electronic system. The first electronic digital computer was completed in 1946 at the University of Pennsylvania. It was called the Electronic Numerical Integrator and Computer (ENIAC). One of the most significant developments in computers was the *stored program* concept, developed in the 1940s by John von Neumann, an American mathematician.

Solid State Electronics

The crystal detectors used in the early radios were the forerunners of modern solid state devices. However, the era of solid state electronics began with the invention of the *transistor* in 1947 at Bell Labs. The inventors were Walter Brattain, John Bardeen, and William Shockley. Figure 1–1 shows these three men, along with the notebook entry describing the historic discovery.

In the early 1960s, the integrated circuit was developed. It incorporated many transistors and other components on a single small *chip* of semiconductor material. Integrated circuit technology continues to be developed and improved, allowing more complex circuits to be built on smaller chips. The introduction of the microprocessor in the early 1970s created another electronics revolution: the entire processing portion of a computer placed on a single, small, silicon chip. Continued development brought about complete computers on a single chip by the late 1970s.

(a)

(b)

FIGURE 1–1

(a) Nobel Prize winners Drs. John Bardeen, William Shockley, and Walter Brattain, shown left to right, with apparatus used in their first investigations that led to the invention of the transistor. The trio received the 1956 Nobel Physics award for their invention of the transistor, which was announced by Bell Laboratories in 1948. (b) The laboratory notebook entry of scientist Walter H. Brattain recorded the events of December 23, 1947, when the transistor effect was discovered at Bell Telephone Laboratories. The notebook entry describes the event and adds, "This circuit was actually spoken over and by switching the device in and out a distinct gain in speech level could be heard and seen on the scope presentation with no noticeable change in quality." [(a) and (b) courtesy of Bell Laboratories]

**SECTION REVIEW
1–1**

1. Who developed the first battery?
2. The unit of what electrical quantity is named after André Ampère?
3. What contribution did Georg Simon Ohm make to the study of electricity?
4. In what year was the transistor invented?
5. What major development followed the invention of the transistor?

1-2

CAREERS IN ELECTRONICS

The field of electronics is very diverse, and career opportunities are available in many areas. Because electronics is currently found in so many different applications and new technology is being developed at a fast rate, its future is limitless. There is hardly an area of our lives that is not enhanced to some degree by electronics technology. Those who acquire a sound, basic knowledge of electrical and electronic principles and are willing to continue learning will always be in demand.

The importance of obtaining a thorough understanding of the basic principles contained in this text cannot be overemphasized. Most employers prefer to hire people who have both a thorough grounding in the basics and the ability and eagerness to grasp new concepts and techniques. If you have a good training in the basics, an employer will train you in the specifics of the job to which you are assigned.

There are many types of job classifications for which a person with training in electronics technology may qualify. A few of the most common job functions are discussed briefly in the following paragraphs.

Service Shop Technician

Technical personnel in this category are involved in the repair or adjustment of both commercial and consumer electronic equipment that is returned to the dealer or manufacturer for service. Specific areas include TVs, VCRs, stereo equipment, CB radios, computer equipment, and so on. This area also offers opportunities for self-employment.

Manufacturing Technician

Manufacturing personnel are involved either in the testing of electronic products at the assembly-line level or in the maintenance and troubleshooting of electronic and electromechanical systems used in the testing and manufacturing of products. Virtually every type of manufacturing plant, regardless of its product, uses automated equipment that is electronically controlled.

Laboratory Technician

These technicians are involved in breadboarding, prototyping, and testing new or modified electronic systems in research and development laboratories. They generally work closely with engineers during the development phase of a product.

Field Service Technician

Field service personnel service and repair electronic equipment—for example, computer systems, radar installations, automatic banking equipment, and security systems—at the user's location.

Engineering Assistant/Associate Engineer

Personnel in this category work closely with engineers in the implementation of a concept and in the basic design and development of electronic systems. Engineering assistants are frequently involved in a project from its initial design through the early manufacturing stages.

Technical Writer

Technical writers compile technical information and then use the information to write and produce manuals and audiovisual materials. A broad knowledge of a particular system and the ability to clearly explain its principles and operation are essential.

Technical Sales

Technically trained people are in demand as sales representatives for high-technology products. The ability both to understand technical concepts and to communicate the technical aspects of a product to a potential customer is very valuable. In this area, as in technical writing, competency in expressing yourself orally and in writing is essential. Actually, being able to communicate well is very important in any technical job category because you must be able to record data clearly and explain procedures, conclusions, and actions taken so that others can readily understand what you are doing.

**SECTION REVIEW
1–2**

1. For which of the following reasons have you selected or are you considering selecting electronics as a career?
   (a) An interesting and exciting field.
   (b) Good opportunities for advancement.
   (c) Good income.
   (d) Clean and comfortable working environment.
   (e) Very promising future.
   (f) Like to work with electronic equipment.
   (g) Like to understand how things work.
2. Name several ways in which electronics affects your everyday life.

1–3   APPLICATIONS OF ELECTRICITY AND ELECTRONICS

Electricity and electronics are diverse technological fields with very broad applications. There is hardly an area of our lives that is not dependent to some extent on electricity and electronics. Some of the applications are discussed here in a general way to give you an idea of the scope of the fields.

Computers

One of the most important electronic systems is the digital computer; its applications are broad and diverse. For example, computers have applications in business for record keeping, accounting, payrolls, inventory control, market analysis, and statistics, to name but a few.

Scientific fields utilize the computer to process huge amounts of data and to perform complex and lengthy calculations. In industry, the computer is used for controlling and monitoring intricate manufacturing processes. Communications, navigation, medical, military, and home uses are a few of the other areas in which the computer is used extensively.

The computer's success is based on its ability to perform mathematical operations extremely fast and to process and store large amounts of information.

Computers vary in complexity and capability, ranging from very large systems with vast capabilities down to a computer on a chip with much more limited performance. Figure 1–2 shows typical computers.

(a)   (b)

FIGURE 1–2
Typical computer systems. (a) Large computer system (courtesy of Unisys Corp.). (b) Personal computer (courtesy of Tandy Corp.).

Communications

Electronic communications encompasses a wide range of specialized fields. Included are space and satellite communications, commercial radio and television, citizens' band and amateur radio, data communications, navigation systems, radar, telephone systems, military applications, and specialized radio applications such as police, aircraft, and so on. Computers are used to a great extent in many communications systems. Figure 1–3 shows a telephone switching system as an example of electronic communications.

FIGURE 1–3
A telephone switching system (courtesy of AG Communication Systems).

Automation

Electronic systems are employed extensively in the control of manufacturing processes. Computers and specialized electronic systems are used in industry for various purposes, for example, control of ingredient mixes, operation of machine tools, product inspection, and control and distribution of power. Figure 1–4 shows an example of automation in a manufacturing facility using robots.

FIGURE 1–4
Robots on an automobile assembly line (courtesy of Ford Motor Co.).

Medicine

Electronic devices and systems are finding ever-increasing applications in the medical field. The familiar *electrocardiograph* (ECG), used for the diagnosis of heart and other circulatory ailments, is a widely used medical electronic instrument. A closely related instrument is the *electromyograph,* which uses a cathode ray tube display rather than an ink trace.

The *diagnostic sounder* uses ultrasonic sound waves for various diagnostic procedures in neurology, for heart chamber measurement, and for detection of certain types of tumors. The *electroencephalograph* (EEG) is similar to the electrocardiograph. It records the electrical activity of the brain rather than heart activity. Another electronic instrument used in medical procedures is the *coagulograph.* This instrument is used in analysis of blood clots.

Electronic instrumentation is also used extensively in intensive-care facilities. Heart rate, pulse, body temperature, respiration, and blood pressure can be monitored on a continuous basis. Monitoring equipment is also used a great deal in operating rooms. Some typical medical electronic equipment is pictured in Figure 1–5.

(a)

(b)

(c)

FIGURE 1–5
Typical medical instrumentation (courtesy of Hewlett-Packard Co.).

Consumer Products

Electronic products used directly by the consumers for information, entertainment, recreation, or work around the home are an important segment of the total electronics market. For example, the electronic calculator and digital watch are popular examples of consumer electronics. The small personal computer is used widely by hobbyists and is also becoming a common household appliance.

Electronic systems are used in automobiles to control and monitor engine functions, control braking, provide entertainment, and display useful information to the driver.

Most appliances such as microwave ovens, washers, and dryers are available with electronic controls. Home entertainment, of course, is largely electronic. Examples are television, radio, stereo, and recorders. Also, many new games for adults and children incorporate electronic devices.

SECTION REVIEW
1–3

1. Name some of the areas in which electronics is used.
2. The computerization of manufacturing processes is an example of _____ .

1–4

CIRCUIT COMPONENTS AND MEASURING INSTRUMENTS

In this text you will study many types of electrical components and measuring instruments. This coverage of dc/ac circuit fundamentals provides the foundation for understanding electronic devices and circuits. Several types of electrical and electronic components and instruments that you will be studying in detail in this and in other courses are introduced briefly in this section.

Resistors

Resistors *resist,* or limit, electrical current in a circuit. Several common types of resistors are shown in Figure 1–6.

(a)   (b)   (c)

(d)

FIGURE 1–6
Typical fixed and variable resistors. (a) Carbon-composition resistors with standard power ratings of 1/8 W, 1/4 W, 1/2 W, 1 W, and 2 W (courtesy of Allen-Bradley Co.). (b) Wirewound resistors (courtesy of Dale Electronics, Inc.). (c) Variable resistors or potentiometers (courtesy of Bourns, Inc.). (d) Resistor networks (courtesy of Bourns, Inc.).

Capacitors

Capacitors *store* electrical charge; they are found in a variety of applications. Figure 1–7 shows several typical capacitors.

FIGURE 1–7

Typical capacitors. [(a,d,e,h) courtesy of Phillips Components. (b) courtesy of Murata Erie North America. (c,f,g) courtesy of Sprague Electric Co.].

Inductors

Inductors, also known as *coils,* are used to store energy in an electromagnetic field; they serve many useful functions in an electrical circuit. Figure 1–8 shows several typical inductors.

FIGURE 1–8

Typical inductors. [(a) and (b) courtesy of Dale Electronics, Inc. (c) courtesy of Delevan.].

Transformers

Transformers are sometimes used to couple ac voltages from one point in a circuit to another, or to increase or decrease the ac voltage. Several types of transformers are shown in Figure 1–9.

FIGURE 1–9
Typical transformers (courtesy of Dale Electronics, Inc.).

Semiconductor Devices

Several varieties of diodes, transistors, and integrated circuits are shown in Figure 1–10.

(a) Discrete devices   (b) Integrated circuits

FIGURE 1–10
A grouping of typical semiconductor devices (courtesy of Motorola Semiconductor Products, Inc.).

Electronic Instruments

Figure 1–11 shows a variety of instruments that are discussed throughout the text. Typical instruments include the power supply, for providing voltage and current; the voltmeter, for measuring voltage; the ammeter, for measuring current; the ohmmeter, for measuring resistance; the wattmeter, for measuring power; and the oscilloscope for observing and measuring ac voltages.

(a)   (b)   (c)   (d)

FIGURE 1–11
Typical instruments. (a) dc power supply (courtesy of B&K Precision). (b) Analog multimeter (courtesy of B&K Precision). (c) Digital multimeter (courtesy of B&K Precision). (d) Oscilloscope (courtesy of Tektronix, Inc.).

**SECTION REVIEW
1–4**

1. Name three types of common electrical components.
2. What instrument is used for measuring electrical current?

1–5   ELECTRICAL UNITS

In electronics work, you must deal with measurable quantities. For example, you must be able to express how many volts are measured at a certain test point, how much current is flowing through a wire, or how much power a certain amplifier produces.

In this section you will learn the units and symbols for many of the electrical quantities that are used throughout the book. Definitions of the quantities are presented as they are needed in later chapters.

Symbols are used in electronics to represent both quantities and their units. One symbol is used to represent the name of the quantity, and another is used to represent the unit of measurement of that quantity. For example, *P* stands for *power,* and W stands for *watts,* which is the unit of power. Table 1–1 lists the most important quantities, along with their SI units and symbols. The term *SI* is the French abbreviation for *International System* (*Système International* in French).

TABLE 1–1
Electrical quantities and units with SI symbols.

Quantity	Symbol	Unit	Symbol
Capacitance	C	farad	F
Charge	Q	coulomb	C
Conductance	G	siemen	S
Current	I	ampere	A
Energy	W	joule	J
Frequency	f	hertz	Hz
Impedance	Z	ohm	Ω
Inductance	L	henry	H
Power	P	watt	W
Reactance	X	ohm	Ω
Resistance	R	ohm	Ω
Time	t	second	s
Voltage	V	volt	V

**SECTION REVIEW
1–5**

1. What does *SI* stand for?
2. Without referring to Table 1–1, list as many electrical quantities as possible, including their symbols, units, and unit symbols.

1–6

SCIENTIFIC NOTATION

In electronics work, you will encounter both very small and very large numbers. For example, it is common to have electrical current values of only a few thousandths or even a few millionths of an ampere. On the other hand, you will find resistance values of several thousand or several million ohms. This range of values is typical of many other electrical quantities also.

Powers of Ten

Scientific notation uses *powers of ten*, a method that makes it much easier to express large and small numbers and to do calculations involving such numbers.

Table 1–2 lists some powers of ten, both positive and negative. The power of ten is expressed as an exponent of the base 10 in each case. The exponent indicates the

TABLE 1–2
Some positive and negative powers of ten.

$1{,}000{,}000 = 10^6$	$0.000001 = 10^{-6}$
$100{,}000 = 10^5$	$0.00001 = 10^{-5}$
$10{,}000 = 10^4$	$0.0001 = 10^{-4}$
$1{,}000 = 10^3$	$0.001 = 10^{-3}$
$100 = 10^2$	$0.01 = 10^{-2}$
$10 = 10^1$	$0.1 = 10^{-1}$
$1 = 10^0$	

number of decimal places to the right or left of the decimal point in the expanded number. If the power is *positive*, the decimal point is moved to the *right*. For example,

$$10^4 = 1 \times 10^4 = 1.0000. = 10,000.$$

If the power is *negative*, the decimal point is moved to the *left*. For example,

$$10^{-4} = 1 \times 10^{-4} = .0001. = 0.0001$$

EXAMPLE 1–1

Express each number as a positive power of ten.
(a) 200   (b) 5000   (c) 85,000   (d) 3,000,000

Solution
In each case there are many possibilities for expressing the number in powers of ten. We do not show all possibilities in this example but include the most common powers of ten used in electrical work.
(a) $200 = 0.0002 \times 10^6 = 0.2 \times 10^3$
(b) $5000 = 0.005 \times 10^6 = 5 \times 10^3$
(c) $85,000 = 0.085 \times 10^6 = 8.5 \times 10^4 = 85 \times 10^3$
(d) $3,000,000 = 3 \times 10^6 = 3000 \times 10^3$

EXAMPLE 1–2

Express each number as a negative power of ten.
(a) 0.2   (b) 0.005   (c) 0.00063   (d) 0.000015

Solution
Again, all the possible ways to express each number as a power of ten are not given. The most commonly used powers are included, however.
(a) $0.2 = 2 \times 10^{-1} = 200 \times 10^{-3} = 200,000 \times 10^{-6}$
(b) $0.005 = 5 \times 10^{-3} = 5000 \times 10^{-6}$
(c) $0.00063 = 0.63 \times 10^{-3} = 6.3 \times 10^{-4} = 630 \times 10^{-6}$
(d) $0.000015 = 0.015 \times 10^{-3} = 1.5 \times 10^{-5} = 15 \times 10^{-6}$

EXAMPLE 1–3

Express each of the following powers of ten as a regular decimal number:
(a) 10^5 (b) 2×10^3 (c) 3.2×10^{-2} (d) 250×10^{-6}

Solution
(a) $10^5 = 1 \times 10^5 = 100,000$ (b) $2 \times 10^3 = 2000$
(c) $3.2 \times 10^{-2} = 0.032$ (d) $250 \times 10^{-6} = 0.000250$

Calculating with Powers of Ten

The advantage of scientific notation is in addition, subtraction, multiplication, and division of very small or very large numbers.

Rules for Addition

The rules for adding numbers in powers of ten are as follows:

1. Convert the numbers to be added to the *same* power of ten.
2. Add the numbers directly to get the sum.
3. Bring down the common power of ten, which is the power of ten of the sum.

EXAMPLE 1–4

Add 2×10^6 and 5×10^7.

Solution
1. Convert both numbers to the same power of ten:

$$(2 \times 10^6) + (50 \times 10^6)$$

2. Add $2 + 50 = 52$.
3. Bring down the common power of ten (10^6); the sum is 52×10^6.

Rules for Subtraction

The rules for subtracting numbers in powers of ten are as follows:

1. Convert the numbers to be subtracted to the *same* power of ten.
2. Subtract the numbers directly to get the difference.
3. Bring down the common power of ten, which is the power of ten of the difference.

EXAMPLE 1–5

Subtract 25×10^{-12} from 75×10^{-11}.

Solution
1. Convert each number to the same power of ten:

$$(75 \times 10^{-11}) - (2.5 \times 10^{-11})$$

2. Subtract $75 - 2.5 = 72.5$.
3. Bring down the common power of ten (10^{-11}); the difference is 72.5×10^{-11}.

Rules for Multiplication

The rules for multiplying numbers in powers of ten are as follows:

1. Multiply the numbers directly.
2. Add the powers of ten algebraically (the powers do not have to be the same).

EXAMPLE 1–6

Multiply 5×10^{12} and 3×10^{-6}.

Solution
Multiply the numbers, and algebraically add the powers:

$$(5 \times 10^{12})(3 \times 10^{-6}) = 15 \times 10^{12+(-6)} = 15 \times 10^6$$

Rules for Division

The rules for dividing numbers in powers of ten are as follows:

1. Divide the numbers directly.
2. Subtract the power of ten in the denominator from the power of ten in the numerator.

EXAMPLE 1–7

Divide 50×10^8 by 25×10^3.

Solution

The division problem is written with a numerator and denominator as

$$\frac{50 \times 10^8}{25 \times 10^3}$$

Dividing the numbers and subtracting 3 from 8, we get

$$\frac{50 \times 10^8}{25 \times 10^3} = 2 \times 10^{8-3} = 2 \times 10^5$$

**SECTION REVIEW
1–6**

1. Scientific notation uses powers of ten (T or F).
2. Express 100 as a power of ten.
3. Do the following operations:
    (a) $(1 \times 10^5) + (2 \times 10^5)$ (b) $(3 \times 10^6)(2 \times 10^4)$
    (c) $(8 \times 10^3) \div (4 \times 10^2)$ (d) $(25 \times 10^{-6}) - (130 \times 10^{-7})$

1–7   METRIC PREFIXES

In electrical and electronics work, certain powers of ten are used more often than others. The most frequently used powers of ten are 10^9, 10^6, 10^3, 10^{-3}, 10^{-6}, 10^{-9}, and 10^{-12}.

It is common practice to use *metric prefixes* to represent these quantities. Table 1–3 lists the metric prefix for each of the commonly used powers of ten.

TABLE 1–3
Metric prefixes and their symbols.

Power of Ten	Value	Metric Prefix	Metric Symbol	Power of Ten	Value	Metric Prefix	Metric Symbol
10^9	one billion	giga	G	10^{-6}	one-millionth	micro	μ
10^6	one million	mega	M	10^{-9}	one-billionth	nano	n
10^3	one thousand	kilo	k	10^{-12}	one-trillionth	pico	p
10^{-3}	one-thousandth	milli	m				

Use of Metric Prefixes

Now we use examples to illustrate metric prefixes. The number 2000 can be expressed in scientific notation as 2×10^3. Suppose we wish to represent 2000 W with a metric prefix. Since $2000 = 2 \times 10^3$, the metric prefix *kilo* (k) is used for 10^3. So we can express 2000 W as 2 kW.

As another example, 0.015 A can be expressed as 15×10^{-3} A. The metric prefix *milli* (m) is used for 10^{-3}. So 0.015 becomes 15 mA.

EXAMPLE 1–8

Express each quantity using a metric prefix.
(a) 50,000 V   **(b)** 25,000,000 Ω **(c)** 0.000036 A

Solution
(a) 50,000 V $= 50 \times 10^3$ V $= 50$ kV
(b) 25,000,000 $\Omega = 25 \times 10^6 \Omega = 25$ MΩ
(c) 0.000036 A $= 36 \times 10^{-6}$ A $= 36 \mu$A

Entering Numbers with Metric Prefixes on the Calculator

To enter a number expressed in scientific notation on the calculator, use the EXP key (the EE key on some calculators). The following example shows how to enter numbers with metric prefixes on a typical scientific calculator.

EXAMPLE 1–9

(a)   Enter 3.3 kΩ (3.3×10^3 Ω) on the calculator.
(b)   Enter 450 μA (450×10^{-6} A) on the calculator.

Solution
(a) **Step 1:** Enter ③ ⟨·⟩ ③. The display shows 3.3.
    Step 2: Press ⟨EXP⟩. The display shows 3.3 00.
    Step 3: Enter ③. The display shows 3.3 03.
(b) **Step 1:** Enter ④ ⑤ ⓪. The display shows 450.
    Step 2: Press ⟨EXP⟩. The display shows 450 00.
    Step 3: Press ⟨+/−⟩. Enter ⑥. The display shows 450 − 06.

Engineering Notation

Engineering notation is a term that refers to the application of scientific notation in which the powers of ten are limited to multiples of three, such as 10^3, 10^{-3}, 10^6, 10^{-6}, 10^9, 10^{-9}, 10^{12}, and 10^{-12}. The reason for this is that all units used in the fields of engineering and technology, such as ohms, amps, volts, watts, and so on, are expressed with prefixes of k (kilo, 10^3), m (milli, 10^{-3}), M (mega, 10^6), μ (micro, 10^{-6}), n (nano, 10^{-9}), p (pico, 10^{-12}), and so on.

**SECTION REVIEW
1–7**

1. List the metric prefix for each of the following powers of ten: 10^6, 10^3, 10^{-3}, 10^{-6}, 10^{-9}, and 10^{-12}.
2. Use an appropriate metric prefix to express 0.000001 A.

SUMMARY

☐ Electronics is a varied field with many opportunities.

☐ Areas of application include computers, communications, automation, medicine, military, and consumer products.

☐ Resistors limit electrical current.

☐ Capacitors store electrical charge.

☐ Inductors store energy electromagnetically.

☐ Inductors are also known as *coils*.

☐ Transformers magnetically couple ac voltages.

☐ Semiconductor devices include diodes, transistors, and integrated circuits.

☐ Power supplies provide current and voltage.

☐ Voltmeters measure voltage.

☐ Ammeters measure current.

☐ Ohmmeters measure resistance.

☐ Coulomb (C) is the unit of charge *(Q)*.

☐ Farad (F) is the unit of capacitance *(C)*.

☐ Henry (H) is the unit of inductance *(L)*.

☐ Hertz (Hz) is the unit of frequency *(f)*.

☐ Siemen (S) is the unit of conductance *(G)*.

☐ Watt (W) is the unit of power *(P)*.

SELF-TEST

1. Which of the following is not an electrical quantity?
   (a) current   (b) voltage   (c) time   (d) power

2. The unit of current is
   (a) volt   (b) watt   (c) ampere   (d) joule

3. The unit of voltage is
   (a) ohm   (b) watt   (c) volt   (d) farad

4. The unit of resistance is
   (a) ampere   (b) henry   (c) hertz   (d) ohm

5. 15,000 W is the same as
   (a) 15 mW   (b) 15 kW   (c) 15 MW   (d) 15 μW

6. The quantity 4.7×10^3 is the same as
   (a) 470   (b) 4700   (c) 47,000   (d) 0.0047

7. The quantity 56×10^{-3} is the same as
   (a) 0.056   (b) 0.560   (c) 560   (d) 56,000

8. The number 3,300,000 can be expressed as
   (a) 3300×10^3 (b) 3.3×10^{-6}
   (c) 3.3×10^6 (d) either (a) or (c)

9. Ten milliamperes can be expressed as
   (a) 10 MA   (b) 10 μA   (c) 10 kA   (d) 10 mA

10. Five thousand volts can be expressed as
    (a) 5000 V   (b) 5 MV   (c) 5 kV   (d) either (a) or (c)

11. Twenty million ohms can be expressed as
    (a) 20 mΩ (b) 20 MW (c) 20 MΩ (d) 20 μΩ
12. Hertz is the unit of
    (a) power (b) inductance (c) frequency (d) time

PROBLEMS

Section 1–6   Scientific Notation

1. Express each of the following numbers as a power of ten:
   (a) 3000 (b) 75,000 (c) 2,000,000

2. Express each number as a power of ten.
   (a) 1/500 (b) 1/2000 (c) 1/5,000,000

3. Express each of the following numbers in three ways, using 10^3, 10^4, and 10^5:
   (a) 8400 (b) 99,000 (c) 0.2×10^6

4. Express each of the following numbers in three ways, using 10^{-3}, 10^{-4}, and 10^{-5}:
   (a) 0.0002 (b) 0.6 (c) 7.8×10^{-2}

5. Express each power of ten in regular decimal form.
   (a) 2.5×10^{-6} (b) 50×10^2 (c) 3.9×10^{-1}

6. Express each power of ten in regular decimal form.
   (a) 45×10^{-6} (b) 8×10^{-9} (c) 40×10^{-12}

7. Add the following numbers:
   (a) $(92 \times 10^6) + (3.4 \times 10^7)$ (b) $(5 \times 10^3) + (85 \times 10^{-2})$
   (c) $(560 \times 10^{-8}) + (460 \times 10^{-9})$

8. Perform the following subtractions:
   (a) $(3.2 \times 10^{12}) - (1.1 \times 10^{12})$ (b) $(26 \times 10^8) - (1.3 \times 10^9)$
   (c) $(150 \times 10^{-12}) - (8 \times 10^{-11})$

9. Perform the following multiplications:
   (a) $(5 \times 10^3)(4 \times 10^5)$ (b) $(12 \times 10^{12})(3 \times 10^2)$
   (c) $(2.2 \times 10^{-9})(7 \times 10^{-6})$

10. Divide the following:
    (a) $(10 \times 10^3) \div (2.5 \times 10^2)$ (b) $(250 \times 10^{-6}) \div (50 \times 10^{-8})$
    (c) $(4.2 \times 10^8) \div (2 \times 10^{-5})$

Section 1–7   Metric Prefixes

11. Express each of the following as a quantity having a metric prefix:
    (a) 31×10^{-3} A (b) 5.5×10^3 V (c) 200×10^{-12} F

12. Express the following using metric prefixes:
    (a) 3×10^{-6} F (b) 3.3×10^6 Ω (c) 350×10^{-9} A

13. Express each quantity as a power of ten.
    (a) 5 μA (b) 43 mV (c) 275 kΩ (d) 10 MW

14. Add the following quantities:
    (a) 6 mA + 3 μA
    (b) 550 mV + 3.2 V
    (c) 12 kΩ + 6800 Ω
    (d) 15 MW + 750 kW

ANSWERS TO SECTION REVIEWS

Section 1–1

1. Volta.   2. Current.
3. He established the relationship among current, voltage, and resistance as expressed in Ohm's law.
4. 1947.   5. Integrated circuits.

Section 1–3

1. Computers, communications, automation, medicine, and consumer products.
2. Automation.

Section 1–4

1. Resistors, capacitors, inductors, and transformers.
2. Ammeter.

Section 1–5

1. It is the abbreviation for Système International.
2. Refer to Table 1–1 after you have compiled your list.

Section 1–6

1. T.   2. 10^2.
3. **(a)** 3×10^5;   **(b)** 6×10^{10};   **(c)** 2×10^1;   **(d)** 12×10^{-6}.

Section 1–7

1. Mega (M), kilo (k), milli (m), micro (μ), nano (n), and pico (p).
2. 1 μA (one microampere).

2 The Basic Electrical Quantities

The three basic electrical quantities presented in this chapter are voltage, current, and resistance. No matter what type of electrical or electronic equipment you may work with, these quantities will always be of primary importance.

To help you understand voltage, current, and resistance, the basic structure of the atom is discussed and the concept of charge is introduced. The basic electric circuit is studied, along with techniques for measuring voltage, current, and resistance.

After completing this chapter, you will be able to

☐ Describe the structure of an atom.
☐ State what an electron is.
☐ Define voltage, current, and resistance.
☐ Discuss how voltage produces current.
☐ Discuss how resistance restricts current.
☐ Identify various types of voltage sources.
☐ Identify various types of fixed and variable resistors.
☐ Determine resistance value by color code.
☐ Describe a basic electrical circuit.
☐ Identify closed and open circuits.
☐ Identify various types of switches.
☐ Identify various types of fuses and circuit breakers.
☐ Use both analog and digital multimeters.
☐ Measure voltage, current, and resistance.

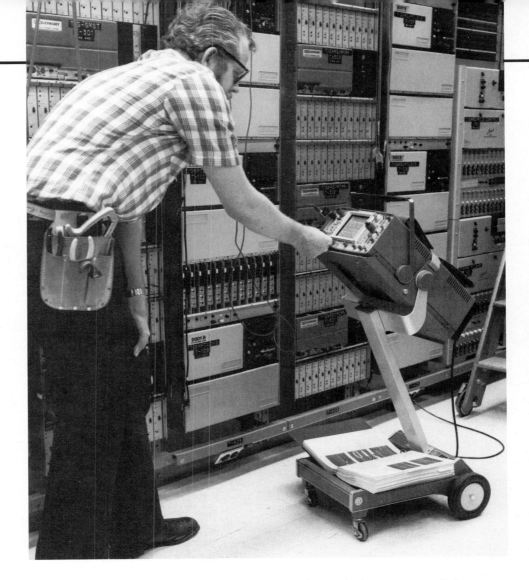

APPLICATION ASSIGNMENT

As head technician for special effects, you are asked to devise a system of six lamps for use in the stage lighting for a concert. There are three requirements for the system:

1. Each of the lamps can be turned on one at a time in any sequence.
2. Only one lamp can be turned on at a time.
3. A single control can brighten or dim any lamp that is on.

After studying this chapter, you should be able to solve this problem.

2–1

ATOMS

An atom is the smallest particle of an element that still retains the characteristics of that element. Different elements have different types of atoms. In fact, every element has a unique atomic structure.

Atoms have a planetary type of structure, consisting of a central nucleus surrounded by orbiting electrons. The nucleus consists of positively charged particles called *protons* and uncharged particles called *neutrons*. The electrons are the basic particles of negative charge.

Each type of atom has a certain number of electrons and protons that distinguishes the atom from all other atoms of other elements. For example, the simplest atom is that of hydrogen. It has one proton and one electron, as pictured in Figure 2–1(a). The helium atom, shown in Figure 2–1(b), has two protons and two neutrons in the nucleus, which is orbited by two electrons.

(a) Hydrogen

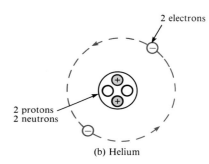

(b) Helium

FIGURE 2–1
Hydrogen and helium atoms.

Atomic Weight and Number

All elements are arranged in the periodic table of the elements in order according to their atomic number, which is the number of protons in the nucleus. The elements can also be arranged by their atomic weight, which is approximately the number of protons and neutrons in the nucleus. For example, hydrogen has an atomic number of one and an atomic weight of one. The atomic number of helium is two, and its atomic weight is four.

In their normal, or neutral, state, all atoms of a given element have the same number of electrons as protons. So the positive charges cancel the negative charges, and the atom has a net charge of zero.

The Copper Atom

Since copper is the most commonly used metal in electrical applications, let us examine its atomic structure. The copper atom has 29 electrons in orbit around the nucleus. They do not all occupy the same orbit, however. They move in orbits at varying distances from the nucleus. The orbits in which the electrons revolve are called *shells*. The number of electrons in each shell follows a predictable pattern according to the formula, $2N^2$, where N is the number of the shell.

The first shell of any atom can have up to 2 electrons, the second shell up to 8 electrons, the third shell up to 18 electrons, and the fourth shell up to 32 electrons. A copper atom is shown in Figure 2–2. Notice that the outermost shell, called the *valence shell*, has only one electron, called the *valence electron*.

FIGURE 2–2
The copper atom.

4th shell: 1 electron

3rd shell: 18 electrons

29 protons +

1st shell: 2 electrons

2nd shell: 8 electrons

Free Electrons

When the electron in the outer shell of the copper atom gains sufficient energy from the surrounding media, it can break away from the parent atom and become what is called a *free electron*. Materials with fewer valence electrons in their atoms have more free electrons because the valence electrons are more loosely bound to the atoms. The free electrons in the copper material are capable of moving from one atom to another in the material. In other words, they drift randomly from atom to atom within the copper. As you will see, the free electrons make electrical *current* possible.

Three categories of materials are used in electronics: *conductors, semiconductors,* and *insulators*.

Conductors

Conductors are materials that allow current to flow easily. They have a large number of free electrons and are characterized by one to three valence electrons in their atomic structures. Most metals are good conductors. Silver is the best conductor, and copper is next. Copper is the most widely used conductive material because it is less expensive than silver. Copper wire is commonly used as a conductor in electric circuits.

Semiconductors

Semiconductors are classed below the conductors in their ability to carry current because they have fewer free electrons than do conductors. Semiconductors are characterized by four valence electrons in their atomic structures. However, because of their unique characteristics, certain semiconductor materials are the basis for modern electronic devices such as the diode, transistor, and integrated circuit. Silicon and germanium are common semiconductor materials.

Insulators

Insulating materials are poor conductors of electric current. In fact, they are used to *prevent* current where it is not wanted. Compared to conductive materials, insulators have very few free electrons and are characterized by more than four valence electrons in their atomic structures.

**SECTION REVIEW
2–1**

1. What is the basic particle of negative charge?
2. Define *atom*.
3. What does a typical atom consist of?
4. Do all elements have the same types of atoms?
5. What is a free electron?

2–2   ELECTRICAL CHARGE

There are two types of charge: *positive* and *negative*. The electron is the smallest particle that exhibits negative electrical charge. When an excess of electrons exists in a material, there is a net negative electrical charge. When a deficiency of electrons exists, there is a net positive electrical charge. The charge of an electron and that of a proton are equal in magnitude. Electrical charge is symbolized by Q.

Static electricity is the presence of a net positive or negative charge in a material. Everyone has experienced the effects of static electricity from time to time, for example, when attempting to touch a metal surface or another person or when the clothes in a dryer cling together.

Materials with charges of opposite polarity are attracted to each other, and materials with charges of the same polarity are repelled, as indicated in Figure 2–3. There is a force acting between charges, as evidenced by the attraction or repulsion. This force, called an *electric field,* consists of invisible lines of force as shown in Figure 2–4.

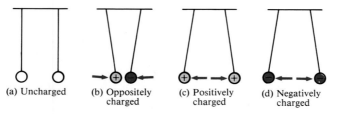

(a) Uncharged   (b) Oppositely charged   (c) Positively charged   (d) Negatively charged

FIGURE 2–3
Attraction and repulsion of electrical charges.

FIGURE 2–4
Electric field between oppositely charged surfaces.

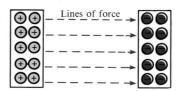

Coulomb: The Unit of Charge

Electrical charge is measured in *coulombs,* abbreviated C. *One coulomb is the total charge possessed by 6.25 × 10^18 electrons.* A single electron has a charge of 1.6×10^{-19} C. This unit is named for Charles Coulomb (1736–1806), a French scientist.

How Positive and Negative Charges Are Created

Consider a neutral atom—that is, one that has the same number of electrons and protons and thus has no net charge. If a valence electron is pulled away from the atom by the application of energy, the atom is left with a net positive charge (more protons than electrons) and becomes a *positive ion.* If an atom acquires an extra electron in its outer shell, it has a net negative charge and becomes a *negative ion.*

The amount of energy required to free a valence electron is related to the number of electrons in the outer shell. An atom can have up to eight valence electrons. The more complete the outer shell, the more stable the atom and thus the more energy is required to release an electron. Figure 2–5 illustrates the creation of a positive and a negative ion when sodium chloride dissolves and the sodium atom gives up its single valence electron to the chloride atom.

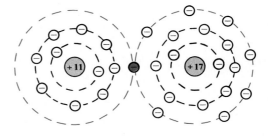

Sodium atom
(11 protons, 11 electrons)

Chlorine atom
(17 protons, 17 electrons)

(a) The sodium atom has a single
valence electron.

(b) The atoms combine by sharing
the valence electron to form
sodium chloride (table salt).

Positive sodium ion
(11 protons, 10 electrons)

Negative chlorine ion
(17 protons, 18 electrons)

(c) When dissolved, the sodium atom gives up the valence electron to become a positive
ion, and the chlorine atom retains the extra valence electron to become a negative ion.

FIGURE 2–5
Example of the formation of positive and negative ions.

EXAMPLE 2–1

How many coulombs do 93.75×10^{16} electrons represent?

Solution

$$Q = \frac{\text{number of electrons}}{\text{number of electrons in one coulomb}}$$

$$= \frac{93.75 \times 10^{16} \text{ electrons}}{6.25 \times 10^{18} \text{ electrons/C}}$$

$$= 0.15 \text{ C}$$

**SECTION REVIEW
2–2**

1. What is the symbol for charge?
2. What is the unit of charge, and what is its symbol?
3. How much charge, in coulombs, is there in 10×10^{12} electrons?

2–3

VOLTAGE

As you have seen, there is a force of attraction between a positive and a negative charge. A certain amount of energy (work) is required to overcome the force and move the charges a given distance apart. All opposite charges possess a certain potential energy because of the separation between them. The difference in potential energy of the charges is the *potential difference*. For example, consider a water tank that is supported several feet above the ground. A given amount of energy (work) is required to pump water up to fill the tank. Once the water is stored in the tank, it has a certain potential energy which, if released, can be used to perform work. For example, the water can be allowed to fall down a chute to turn a water wheel.

   Potential difference in electrical terms is more commonly called *voltage (V)* and is expressed as energy *(W)* per unit charge *(Q)*:

$$V = \frac{W}{Q} \qquad\qquad (2\text{--}1)$$

where *W* is expressed in joules (J) and *Q* is in coulombs (C).

Volt: The Unit of Voltage

The unit of voltage is the *volt,* symbolized by V.

One volt is the potential difference (voltage) between two points when one joule of energy is used to move one coulomb of charge from one point to the other.

EXAMPLE 2–2

If 50 J of energy are available for every 10 C of charge, what is the voltage?

Solution

$$V = \frac{W}{Q} = \frac{50 \text{ J}}{10 \text{ C}} = 5 \text{ V}$$

Sources of Voltage

The Battery

A voltage source is a source of potential energy that is also called *electromotive force* (emf). The battery is one type of voltage source that converts chemical energy into electrical energy. A voltage exists between the electrodes (terminals) of a battery, as shown by a voltaic cell in Figure 2–6. One electrode is positive and the other negative as a result of the separation of charges caused by the chemical action when two different conducting materials are dissolved in the electrolyte.

FIGURE 2–6
A voltaic cell converts chemical energy into electrical energy.

Battery cells are generally classified as *primary cells,* which cannot be recharged, and *secondary cells,* which can be recharged by reversal of the chemical action (Figure 2.7). The amount of voltage provided by a cell varies depending on the materials used

FIGURE 2–7
Typical batteries.

in the construction of the cell. For example, a carbon-zinc cell (flashlight battery) produces 1.5 V, whereas a lead-acid cell used in automobile batteries produces 2 V. Individual cells are combined to produce higher voltages. A 9-V battery is made up of six 1.5-V cells and a 12-V car battery consists of six 2-V cells. Some typical cells and multicell batteries are shown in Figure 2–7. For additional discussion of batteries, see Appendix B.

The Electronic Power Supply

These voltage sources convert the ac voltage from the wall outlet to a constant (dc) voltage which is available across two terminals, as indicated in Figure 2–8(a). Typical commercial power supplies are shown in part (b).

(a)

(b)

FIGURE 2–8
Electronic power supplies. (b) courtesy of B&K Precision.

The Solar Cell

The operation of solar cells is based on the principle of *photovoltaic action,* which is the process whereby light energy is converted directly into electrical energy. A basic solar cell consists of two layers of different semiconductive materials joined together to form a junction. When one layer is exposed to light, many electrons acquire enough energy to break away from their parent atoms and cross the junction. This process forms negative ions on one side of the junction and positive ions on the other, and thus a potential difference (voltage) is developed. Basic construction of a solar cell is shown in Figure 2–9.

FIGURE 2–9
Basic construction of solar cell.

The Generator

Generators convert mechanical energy into electrical energy using a principle called *electromagnetic induction* (to be studied later). A conductor is rotated through a magnetic field, and a voltage is produced across the conductor. A typical generator is pictured in Figure 2–10.

FIGURE 2–10
Cutaway view of dc generator.

**SECTION REVIEW
2–3**

1. Define *voltage*.
2. How much is the voltage when there are 24 J of energy for 10 C of charge?
3. List four sources of voltage.

2–4

CURRENT

As you have seen, there are free electrons available in all conductive and semiconductive materials. These electrons drift randomly in all directions, from atom to atom, within the structure of the material, as indicated in Figure 2–11.

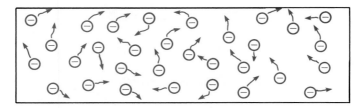

FIGURE 2–11
Random motion of free electrons in a material.

Now, if a voltage is placed across the conductive or semiconductive material, one end becomes positive and the other negative, as indicated in Figure 2–12. The repulsive force between the negative voltage at the left end causes the free electrons (negative charges) to move toward the right. The attractive force between the positive voltage at the right end pulls the free electrons to the right. The result is a net movement of the free electrons from the negative end of the material to the positive end, as shown in Figure 2–12.

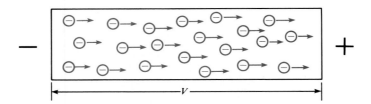

FIGURE 2–12
Electrons flow from negative to positive when a voltage is applied across a conductive or semiconductive material.

The movement of these free electrons from the negative end of the material to the positive end is the *electrical current,* symbolized by *I*.

Electrical current is defined as the rate of flow of electrons in a conductive or semiconductive material.

Current is measured by the number of electrons (amount of charge, Q) that flow past a point in a unit of time:

$$I = \frac{Q}{t} \tag{2–2}$$

where *I* is current, *Q* is the charge of the electrons, and *t* is the time.

Ampere: The Unit of Current

Current is measured in a unit called the *ampere,* or *amp* for short, symbolized by A. It is named after André Ampère (1775–1836), a French physicist whose work contributed to the understanding of electrical current and its effects.

One ampere (1 A) is the amount of current when a number of electrons having one coulomb (1 C) of charge move past a given point in one second (1 s).

See Figure 2–13. One coulomb is the charge carried by 6.25×10^{18} electrons.

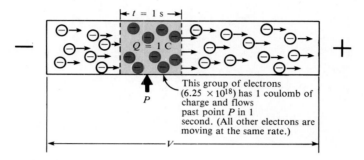

FIGURE 2–13
Illustration of 1 A of current in a material (1 C/s).

EXAMPLE 2–3

Ten coulombs of charge flow past a given point in a wire in 2 s. How many amperes of current are flowing?

Solution

$$I = \frac{Q}{t} = \frac{10\ \text{C}}{2\ \text{s}} = 5\ \text{A}$$

**SECTION REVIEW
2–4**

1. Define *current* and state its unit.
2. How many electrons make up one coulomb of charge?
3. What is the current in amperes when 20 C flow past a point in a wire in 4 s?

2–5

RESISTANCE

When current flows in a material, the free electrons move through the material and occasionally collide with atoms. These collisions cause the electrons to lose some of their energy, and thus their movement is restricted. The more collisions, the more the flow of electrons is restricted. This restriction varies and is determined by the type of material. The property of a material that restricts the flow of electrons is called *resistance,* designated R.

R

FIGURE 2–14
Resistance/resistor symbol.

Resistance is the opposition to current.

The schematic symbol for resistance is shown in Figure 2–14.

When current flows through any material that has resistance, heat is produced by the collisions of electrons and atoms. Therefore, wire, which typically has a very small resistance, becomes warm when there is current through it.

(a)

(b)

(c)

FIGURE 2–15
Typical fixed resistors [(a) and (b) courtesy of Stackpole Carbon Co. (c) resistor networks courtesy of Bourns, Inc.].

(a) Cutaway view of a carbon-composition resistor (Courtesy of Allen-Bradley Co.)

(b) Cutaway view of a chip resistor

FIGURE 2–16
Two types of fixed resistors.

Ohm: The Unit of Resistance

Resistance, R, is expressed in the unit of *ohms,* named after Georg Simon Ohm and is symbolized by the Greek letter omega (Ω).

> **There is one ohm (1 Ω) of resistance when there is one ampere (1 A) of current in a material with one volt (1 V) applied.**

Resistors

Components that are specifically designed to have a certain amount of resistance are called *resistors.* The principal applications of resistors are to limit current, divide voltage, and, in certain cases, generate heat. Although there are a variety of different types of resistors that come in many shapes and sizes, they can all be placed in one of two main categories: *fixed* or *variable.*

Fixed Resistors

The fixed resistor is available with a large selection of ohmic values that are set during manufacturing and cannot be changed easily. Fixed resistors are constructed using various methods and materials. Several common types are shown in Figure 2–15.

One common fixed resistor is the *carbon-composition* type, which is made with a mixture of finely ground carbon, insulating filler, and a resin binder. The ratio of carbon to insulating filler sets the resistance value. The mixture is formed into rods, and lead connections are made. The entire resistor is then encapsulated in an insulated coating for protection. The chip resistor is another type of fixed resistor and is in the category of SMT (surface mount technology) components. It has the advantage of a very small size for compact assemblies. Figure 2–16(a) shows the construction of a typical carbon-composition resistor and Figure 2-16(b) shows the construction of a chip resistor.

Other types of fixed resistors include carbon film, metal oxide, metal film, metal glaze, and wire-wound. In film resistors, a resistive material is deposited evenly onto a high-grade ceramic rod. The resistive film may be carbon (carbon film), nickel chromium (metal film), a mixture of metals and glass (metal glaze), or metal and insulating oxide (metal oxide). In these types of resistors, the desired resistance value is obtained by removing part of the resistive material in a helical pattern along the rod using a *spiraling* technique. Very close tolerance can be achieved with this method.

Wire-wound resistors are constructed with resistive wire wound around an insulating rod and then sealed. Normally, wire-wound resistors are used because of their relatively high power ratings. Some typical fixed resistors are shown in Figure 2–17.

1. Coloring bands 4. Substrates
2. Helixing   5. Insulation
3. Film   6. Terminations

(a) Carbon film

2. Metal film   4. Epoxy coating

1. End cap   3. Ceramic   5. Leads

(b) Metal film

Resistive element   Housing

Termination

(c) Resistor network

FIGURE 2–17
Construction views of typical fixed resistors [(a) and (b) courtesy of Stackpole Carbon Co. (c) courtesy of Bourns, Inc.]

Resistor Color Codes

Fixed resistors with value tolerances of 5%, 10%, or 20% are color coded with four bands to indicate the resistance value and the tolerance. This color-code band system is shown in Figure 2–18, and the color code is listed in Table 2–1.

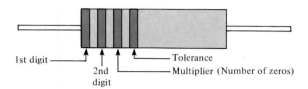

1st digit ——
2nd digit
— Tolerance
— Multiplier (Number of zeros)

FIGURE 2–18
Color-code bands on a resistor.

TABLE 2–1
Resistor color code.

		Digit	Color
		0	Black
		1	Brown
Resistance value,		2	Red
first three bands.		3	Orange
1st band—1st digit		4	Yellow
2nd band—2nd digit		5	Green
3rd band—number		6	Blue
of zeros		7	Violet
		8	Gray
		9	White
Tolerance, fourth band		5%	Gold
		10%	Silver
		20%	No band

The color code is read as follows:

1. Beginning at the banded end, the first band is the first digit of the resistance value. If it is not clear which is the banded end, start from the end that does not begin with a gold or silver band.
2. The second band is the second digit.
3. The third band is the number of zeros, or the *multiplier*.
4. The fourth band indicates the tolerance and is usually gold or silver.

For example, a 5% tolerance means that the *actual* resistance value is within $\pm 5\%$ of the color-coded value. Thus, a 100-Ω resistor with a tolerance of $\pm 5\%$ can have acceptable values as low as 95 Ω and as high as 105 Ω.

For resistance values less than 10 Ω, the third band is either gold or silver. Gold represents a multiplier of 0.1, and silver represents 0.01. For example, a color code of red, violet, gold, and silver represents 2.7 Ω with a tolerance of $\pm 10\%$.

EXAMPLE 2–4

COLOR
INSERT

Find the resistance values in ohms and the percent tolerance for each of the color-coded resistors shown in Figure 2–19 (a full-color photo appears following page 240).

FIGURE 2–19

(a)

(b)

(c)

Solution

(a) First band is red = 2, second band is violet = 7, third band is orange = 3 zeros, fourth band is silver = 10% tolerance.

$$R = 27,000 \ \Omega \ \pm 10\%$$

(b) First band is brown = 1, second band is black = 0, third band is brown = 1 zero, fourth band is silver = 10% tolerance.

$$R = 100 \ \Omega \ \pm 10\%$$

(c) First band is green = 5, second band is blue = 6, third band is green = 5 zeros, fourth band is gold = 5% tolerance.

$$R = 5,600,000 \ \Omega \ \pm 5\%$$

Certain precision resistors with tolerances of 1% or 2% are color coded with five bands. Beginning at the banded end, the first band is the first digit of the resistance value, the second band is the second digit, the third band is the third digit, the fourth band is the multiplier, and the fifth band indicates the tolerance. Table 2–1 applies, except that gold indicates 1% and silver indicates 2%.

Numerical labels are also commonly used on certain types of resistors where the resistance value and tolerance are stamped on the body of the resistor. For example, a common system uses R to designate the decimal point and letters to indicate tolerance as follows:

$$F = \pm 1\%, \quad G = \pm 2\%, \quad J = \pm 5\%, \quad K = \pm 10\%, \quad M = \pm 20\%$$

For values above 100 Ω, three digits are used to indicate resistance value, followed by a fourth digit that specifies the number of zeros. For values less than 100 Ω, R indicates the decimal point.

Some examples are as follows: 6R8M is a 6.8-Ω $\pm 20\%$ resistor; 3301F is a 3300-Ω $\pm 1\%$ resistor; and 2202J is a 22,000-Ω $\pm 5\%$ resistor.

Resistor Reliability Band

The fifth band on some color-coded resistors indicates the resistor's reliability in percent of failures per 1000 hours (1000 h) of use. The fifth-band reliability color code is listed in Table 2–2. For example, a brown fifth band means that if a group of like resistors are operated under standard conditions for 1000 h, 1% of the resistors in that group will fail.

TABLE 2–2
Fifth-band reliability color code.

Color	Failures (%) during 1000 hours of operation
Brown	1.0%
Red	0.1%
Orange	0.01%
Yellow	0.001%

Variable Resistors

Variable resistors are designed so that their resistance values can be changed easily with a manual or an automatic adjustment.

Two basic uses for variable resistors are to divide voltage and to control current. The variable resistor used to divide voltage is called a *potentiometer*. The variable resistor used to control current is called a *rheostat*. Schematic symbols for these types are shown in Figure 2–20. The potentiometer is a *three-terminal device*, as indicated in part (a). Terminals 1 and 2 have a fixed resistance between them, which is the total resis-

(a) Potentiometer   (b) Rheostat   (c) Potentiometer connected
                                                as a rheostat

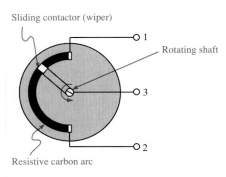

(d) Basic potentiometer construction

FIGURE 2–20
Potentiometer and rheostat symbols.

tance. Terminal 3 is connected to a moving contact (wiper). We can vary the resistance between 3 and 1 or between 3 and 2 by moving the contact up or down.

Figure 2–20(b) shows the rheostat as a *two-terminal* variable resistor. Part (c) shows how we can use a potentiometer as a rheostat by connecting terminal 3 to either terminal 1 or terminal 2. Part (d) shows a basic simplified construction diagram of a potentiometer. Some typical potentiometers are pictured in Figure 2–21.

(a)

(b)

FIGURE 2–21
(a) Typical potentiometers (courtesy of Allen-Bradley Co.). (b) Trimmer potentiometers with construction views (courtesy of Bourns Trimpot).

Potentiometers and rheostats can be classified as *linear* or *tapered,* as shown in Figure 2–22, where a potentiometer with a total resistance of 100 Ω is used as an example. As shown in part (a), in a linear potentiometer, the resistance between either terminal and the moving contact varies linearly with the position of the moving contact. For example, one-half of a turn results in one-half the total resistance. Three-quarters of a turn results in three-quarters of the total resistance between the moving contact and one terminal, or one-quarter of the total resistance between the other terminal and the moving contact.

1. Quarter turn   2. Half turn   3. Three-quarter turn

(a) Linear

1. Quarter turn   2. Half turn   3. Three-quarter turn

(b) Tapered (nonlinear)

FIGURE 2–22
Examples of (a) linear and (b) tapered potentiometers.

In the tapered potentiometer, the resistance varies nonlinearly with the position of the moving contact, so that one-half of a turn does not necessarily result in one-half the total resistance. This concept is illustrated in Figure 2–22(b), where the nonlinear values are arbitrary.

**SECTION REVIEW
2–5**

1. Define *resistance* and name its unit.
2. What are the two main categories of resistors? Briefly explain the difference between them.
3. In the resistor color code, what does each band represent?
4. Determine the resistance and tolerance for each of the following color codes:
    (a) yellow, violet, red, gold
    (b) blue, red, orange, silver
    (c) brown, gray, black, gold
5. What is the basic difference between a rheostat and a potentiometer?

2–6

(a)

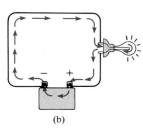

(b)

FIGURE 2–23
A simple electric circuit.

FIGURE 2–24
Schematic diagram for
the circuit in Figure 2–23.

THE ELECTRIC CIRCUIT

Basically, an electric circuit consists of a *voltage source*, a *load*, and a *path for current* between the source and the load. Figure 2–23 shows an example of a simple electric circuit: a battery connected to a lamp with two wires. The battery is the voltage source, the lamp is the load on the battery because it draws current from the battery, and the two wires provide the current path from the negative terminal of the battery to the lamp and back to the positive terminal of the battery, as shown in part (b). Current flows through the filament of the lamp (which has a resistance), causing it to emit visible light. Current flows through the battery by chemical action. In many practical cases, one terminal of the battery is connected to a *ground* point. For example, in automobiles, the negative battery terminal is connected to the metal chassis of the car. The chassis is the ground for the automobile electrical system. The concept of circuit ground is covered in detail in a later chapter.

The Electrical Schematic

An electric circuit can be represented by a *schematic diagram* using standard symbols for each element, as shown in Figure 2–24 for the simple circuit in Figure 2–23. The purpose of a schematic diagram is to show in an organized manner how the various components in a given circuit are interconnected so that the operation of the circuit can be determined.

Closed and Open Circuits

The example circuit in Figure 2–23 illustrated a *closed circuit*—that is, a circuit in which the current has a complete path. When the current path is broken so that current cannot flow, the circuit is called an *open circuit*.

Switches

Switches are commonly used for controlling the opening or closing of circuits by either mechanical or electronic means. For example, a switch is used to turn a lamp on or off as illustrated in Figure 2–25. Each circuit pictorial is shown with its associated schematic diagram. The type of switch indicated is a *single-pole–single-throw* (SPST) toggle switch.

Figure 2–26 shows a somewhat more complicated circuit using *single-pole–double-throw* (SPDT) type of switch to control the current to two different lamps. When one lamp is on, the other is off, and vice versa, as illustrated by the two schematic diagrams which represent each of the switch positions.

The term *pole* refers to the movable arm in a switch, and the term *throw* indicates the number of contacts that are affected (either opened or closed) by a single switch action (a single movement of a pole).

In addition to the SPST and the SPDT switches already introduced, several other types are of importance:

☐ *Double-pole–single-throw (DPST).* The DPST switch permits simultaneous opening or closing of two sets of contacts. The symbol is shown in Figure 2–27(a). The dashed line indicates that the contact arms are mechanically linked so that both move with a single switch action.

(a) Current flows in a *closed* circuit (switch is ON or in the *closed* position).

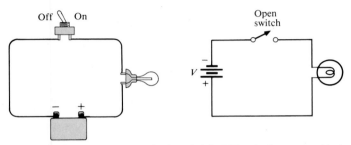

(b) No current flows in an *open* circuit (switch is OFF or in the *open* position).

FIGURE 2–25
Basic closed and open circuits using an SPST switch for control.

(a) Pictorial

(b) Schematic diagram showing L_1 on and L_2 off

(c) Schematic diagram showing L_2 on and L_1 off

FIGURE 2–26
An example of an SPDT switch controlling two lamps.

☐ *Double-pole–double-throw (DPDT).*   The DPDT switch provides connection from one set of contacts to either of two other sets. The schematic symbol is shown in Figure 2–27(b).

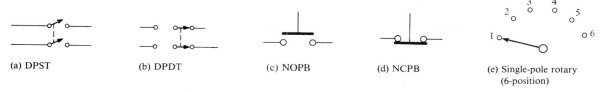

(a) DPST (b) DPDT (c) NOPB (d) NCPB (e) Single-pole rotary (6-position)

FIGURE 2–27
Switch symbols.

☐ *Push button (PB).* In the normally open push-button switch (NOPB), shown in Figure 2–27(c), connection is made between two contacts when the button is depressed, and connection is broken when the button is released. In the normally closed push-button switch (NCPB), shown in Figure 2–27(d), connection between the two contacts is broken when the button is depressed.

☐ *Rotary.* In a rotary switch, a knob is turned to make connection between one contact and any one of several others. A symbol for a simple six-position rotary switch is shown in Figure 2–27(e).

Several varieties of switches are shown in Figure 2–28.

(a)

(b)

(c)

(d)

(e)

FIGURE 2–28
Switches. (a) Typical toggle-lever switches (courtesy of Eaton Corporation). (b) Rocker switches (courtesy of Eaton Corporation). (c) Rocker DIP (dual-in-line package) switches (courtesy of Grayhill, Inc.). (d) Push-button switches (courtesy of Eaton Corporation). (e) Rotary-position switches (courtesy of Grayhill, Inc.).

Protective Devices

Fuses and *circuit breakers* are placed in the current path and are used to deliberately create an open circuit when the current exceeds a specified number of amperes due to a malfunction or other abnormal condition in a circuit. For example, a 20-A fuse or circuit breaker will open a circuit when the current exceeds 20 A.

The major difference between a fuse and a circuit breaker is that when a fuse is "blown," it must be replaced, but when a circuit breaker opens, it can be reset and reused repeatedly. The purpose of both of these devices is to protect against damage to a circuit due to excess current or to prevent a hazardous condition created by the overheating of wires and other components when the current is too great. Because fuses cut off excess current more quickly than circuit breakers, fuses are used whenever delicate electronic equipment needs to be protected. Several typical fuses and circuit breakers, along with their schematic symbols, are shown in Figure 2–29.

(a)

(b)

(c)   (d)   (e)   (f)

FIGURE 2–29
Fuses and circuit breakers. (a) Power fuses (courtesy of Bussmann, Cooper Industries.). (b) Fuses and fuse holders (courtesy of Bussmann, Cooper Industries). (c) Fuse symbol. (d) Circuit breakers (courtesy of Bussmann, Cooper Industries). (e) Remote control circuit breaker (courtesy of Eaton Corp.) and (f) Circuit breaker symbol.

Wires

Wires are a common form of conductive material used in electrical circuits. They vary in diameter size and are arranged according to standard *gage numbers,* called *American Wire Gage* (AWG) sizes. The larger the gage number is, the smaller the wire diameter is. The AWG sizes are listed in Table 2–3.

TABLE 2–3
American Wire Gage (AWG) sizes and resistances for solid round copper.

AWG #	Area (CM)	Ω/1000 ft at 20°C	AWG #	Area (CM)	Ω/1000 ft at 20°C
0000	211,600	0.0490	19	1,288.1	8.051
000	167,810	0.0618	20	1,021.5	10.15
00	133,080	0.0780	21	810.10	12.80
0	105,530	0.0983	22	642.40	16.14
1	83,694	0.1240	23	509.45	20.36
2	66,373	0.1563	24	404.01	25.67
3	52,634	0.1970	25	320.40	32.37
4	41,742	0.2485	26	254.10	40.81
5	33,102	0.3133	27	201.50	51.47
6	26,250	0.3951	28	159.79	64.90
7	20,816	0.4982	29	126.72	81.83
8	16,509	0.6282	30	100.50	103.2
9	13,094	0.7921	31	79.70	130.1
10	10,381	0.9989	32	63.21	164.1
11	8,234.0	1.260	33	50.13	206.9
12	6,529.0	1.588	34	39.75	260.9
13	5,178.4	2.003	35	31.52	329.0
14	4,106.8	2.525	36	25.00	414.8
15	3,256.7	3.184	37	19.83	523.1
16	2,582.9	4.016	38	15.72	659.6
17	2,048.2	5.064	39	12.47	831.8
18	1,624.3	6.385	40	9.89	1049.0

As the table shows, the size of a wire is also specified in terms of its *cross-sectional area,* as illustrated also in Figure 2–30. The unit of cross-sectional area is the *circular mil,* abbreviated CM. One circular mil is the area of a wire with a diameter of 0.001 inch (0.001 in., or 1 mil). We find the cross-sectional area by expressing the diameter in thousandths of an inch (mils) and squaring it, as follows:

$$A = d^2$$

where A is the cross-sectional area in circular mils and d is the diameter in mils.

FIGURE 2–30
Cross-sectional area of a wire.

0.001 in.

Cross-sectional area, A $A = 1$ CM

EXAMPLE 2–5

What is the cross-sectional area of a wire with a diameter of 0.005 in.?

Solution

$$d = 0.005 \text{ in.} = 5 \text{ mils}$$
$$A = d^2 = 5^2 = 25 \text{ CM}$$

Wire Resistance

Although copper wire conducts electricity extremely well, it still has some resistance, as do all conductors. The resistance of a wire depends on four factors: (1) type of material, (2) length of wire, (3) cross-sectional area, and (4) temperature.

Each type of conductive material has a characteristic called its *resistivity*, ρ. For each material, ρ is a constant value at a given temperature. The formula for the resistance of a wire of length l and cross-sectional area A is

$$R = \frac{\rho l}{A}$$

This formula tells us that resistance increases with resistivity and length and decreases with cross-sectional area. For resistance to be calculated in ohms, the length must be in feet (ft), the cross-sectional area in circular mils, and the resistivity in CM-Ω/ft.

EXAMPLE 2–6

Find the resistance of a 100-ft length of copper wire with a cross-sectional area of 810.1 CM. The resistivity of copper is 10.4 CM-Ω/ft.

Solution

$$R = \frac{\rho l}{A} = \frac{(10.4 \text{ CM-}\Omega\text{/ft})(100 \text{ ft})}{810.1 \text{ CM}} = 1.284 \ \Omega$$

Table 2–3 lists the resistance of the various standard wire sizes in ohms per 1000 ft at 20°C. For example, a 1000-ft length of 14-gage copper wire has a resistance of 2.525 Ω. A 1000-ft length of 22-gage wire has a resistance of 16.14 Ω. For a given length, the smaller wire has more resistance. Thus, for a given voltage, larger wires can carry more current than smaller ones.

SECTION REVIEW 2–6

1. What are the basic elements of an electric circuit?
2. Define *open circuit*.
3. Define *closed circuit*.
4. What is the purpose of a fuse?
5. Which wire is larger in diameter, AWG #3 or AWG #22?

2–7

BASIC CIRCUIT MEASUREMENTS

Voltage, current, and resistance measurements are commonly required in electronics work. Special types of instruments are used to measure these basic electrical quantities.

The instrument used to measure voltage is a *voltmeter*, the instrument used to measure current is an *ammeter*, and the instrument used to measure resistance is an *ohmmeter*. Commonly, all three instruments are combined into a single instrument known as a *multimeter*, or VOM (volt-ohm-milliammeter), in which you can choose what specific quantity to measure by selecting the switch setting.

(a)   (b)

FIGURE 2–31
Typical multimeters. (a) Analog and (b) digital (courtesy of B&K Precision).

Typical multimeters are shown in Figure 2–31. Part (a) shows an analog meter, that is, with a needle pointer, and part (b) shows a digital multimeter (DMM), which provides a digital readout of the measured quantity.

Meter Symbols

Throughout this book, certain symbols will be used to represent the different meters, as shown in Figure 2–32. You may see any of three types of symbols for voltmeters, ammeters, and ohmmeters, depending on which symbol most effectively conveys the

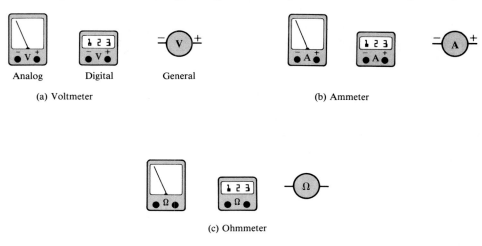

FIGURE 2–32
Meter symbols used in this book.

information required. Although the digital meter is much more widely used than the analog meter in industry, we will use the analog meter symbol in certain situations to illustrate better the operation of a circuit when relative measurements or changes in quantities need to be depicted by the position or movement of the needle. The digital meter symbol is used when fixed values are to be indicated in a circuit. The general schematic symbol is used to indicate placement of meters in a circuit when no values or value changes need to be shown.

How to Measure Current with an Ammeter

Figure 2–33 illustrates how to measure current with an ammeter. Part (a) shows the simple circuit in which the current through the resistor is to be measured. Connect the ammeter *in the current path* by first opening the circuit, as shown in part (b). Then insert the meter as shown in part (c). As you will learn later, such a connection is a *series* connection. The polarity of the meter must be such that the current flows in at the negative terminal and out at the positive.

(a) Circuit in which the current is to be measured

(b) Open the circuit either between the resistor and the negative terminal or between the resistor and the positive terminal of source.

(c) Install the ammeter with polarity as shown (negative to negative–positive to positive).

FIGURE 2–33
Example of an ammeter connection.

How to Measure Voltage with a Voltmeter

To measure voltage, connect the voltmeter *across the component* for which the voltage is to be found. As you will learn later, such a connection is a *parallel* connection. The negative terminal of the meter must be connected to the negative side of the circuit, and the positive terminal of the meter to the positive side of the circuit. Figure 2–34 shows a voltmeter connected to measure the voltage across the resistor.

FIGURE 2–34
Example of a voltmeter connection.

How to Measure Resistance with an Ohmmeter

To measure resistance, connect the ohmmeter across the resistor. *The resistor must first be removed or disconnected from the circuit*. This procedure is shown in Figure 2–35.

(a) Disconnect the resistor from the
circuit to avoid damage to
the meter and/or incorrect
measurement.

(b) Measure the resistance.
(Polarity is not important.)

FIGURE 2–35
Example of using an ohmmeter.

Reading Analog Multimeters

The face of a typical analog meter is represented in Figure 2–36. This particular instrument can be used to measure both direct current (dc) and alternating current (ac) quantities as well as resistance values. It has four selectable functions: dc volts (DC VOLTS), dc milliamperes (DC MA), ac volts (AC VOLTS), and OHMS. Most analog multimeters are similar to this one, although range selections and scales may vary.

Within each function there are several ranges, as indicated by the brackets around the selector switch. For example, the DC VOLTS function has 0.3-V, 3-V, 12-V,

FIGURE 2–36
A typical analog multimeter.

60-V, 300-V, and 600-V ranges. Thus, dc voltages from 0.3 V full-scale to 600 V full-scale can be measured. On the DC MA function, direct currents from 0.06 mA full-scale to 120 mA full-scale can be measured. On the ohm scale, the settings are $\times 1$, $\times 10$, $\times 100$, $\times 1000$, and $\times 100000$.

The Ohm Scale

Ohms are read on the top scale of the meter. This scale is *nonlinear;* that is, the values represented by each division (large or small) vary as you go across the scale. In Figure 2–36, notice how the scale becomes more compressed as you go from right to left.

To read the actual value in ohms, multiply the number on the scale as indicated by the pointer by the factor selected by the switch. For example, when the switch is set at $\times 100$ and the pointer is at 20, the reading is $20 \times 100 = 2000 \ \Omega$.

As another example, assume that the switch is at $\times 10$ and the pointer is at the seventh small division between the 1 and 2 marks, indicating 17 Ω (1.7 $\times$ 10). Now, if the meter remains connected to the same resistance and the switch setting is changed to $\times 1$, the pointer will move to the second small division between the 15 and 20 marks. This, of course, is also a 17-Ω reading, illustrating that a given resistance value can often be read at more than one switch setting.

The ac-dc Scales

The second, third, and fourth scales from the top, labeled ''AC'' and ''DC,'' are used in conjunction with the DC VOLTS, DC MA, and AC VOLTS functions. The upper ac-dc scale ends at the 300 mark and is used with the range settings that are multiples of three, such as 0.3, 3, and 300. For example, when the switch is at 3 on the DC VOLTS function, the 300 scale has a full-scale value of 3 V. At the range setting of 300, the full-scale value is 300 V, and so on.

The middle ac-dc scale ends at 60. This scale is used in conjunction with range settings that are multiples of 6, such as 0.06, 60, and 600. For example, when the switch is at 60 on the DC VOLTS function, the full-scale value is 60 V.

The lower ac-dc scale ends at 12 and is used in conjunction with switch settings that are multiples of 12, such as 1.2, 12, and 120.

The remaining scales are for ac current and for decibels, which are discussed later in the book.

EXAMPLE 2-7   In Figure 2–37, determine the quantity that is being measured and its value.

FIGURE 2–37

Solution

(a) The switch is set on the DC VOLTS function and the 60-V range. The reading taken from the middle dc scale is 18 V.

(b) The switch is set on the DC mA function and the 12-mA range. The reading taken from the upper dc scale is 7.2 mA.

(c) The switch is set on the ohm (OHMS) function and the × 1000 range. The reading taken from the ohm scale (top scale) is 7 kΩ.

Digital Multimeters (DMMs)

DMMs are the most widely used type of electronic measuring instrument. Generally, DMMs provide more functions, better accuracy, greater ease of reading, and greater

FIGURE 2–38
Typical digital multimeters (DMMs) [(top left) courtesy of Triplett Corp. (top right and bottom) reproduced with permission from the John Fluke Manufacturing Co. Inc.].

reliability than do many analog meters. Analog meters have at least one advantage over DMMs, however: They can track short-term variations and trends in a measured quantity that many DMMs are too slow to respond to. Several typical DMMs are shown in Figure 2–38.

DMM Functions

The basic functions found on most DMMs include the following:

☐ ohms
☐ dc voltage and current
☐ ac voltage and current

Some DMMs provide special functions such as transistor or diode tests, power measurement, and decibel measurement for audio amplifier tests.

DMM Displays

DMMs are available with either LCD (liquid-crystal display) or LED (light-emitting diode) readouts. The LCD is the most commonly used readout in battery-powered instruments because it requires only very small amounts of current. A typical battery-powered DMM with an LCD readout operates on a 9-V battery that will last from a few hundred hours to 2000 hours and more. The disadvantages of LCD readouts are that (a) they are difficult or impossible to see in low-light conditions and (b) they are relatively slow to respond to measurement changes. LEDs, on the other hand, can be seen in the dark and respond quickly to changes in measured values. LED displays require much more current than LCDs, and, therefore, battery life is shortened when they are used in portable equipment.

Both LCD and LED DMM displays are in a *seven-segment* format. Each digit in the display consists of seven separate segments, as shown in Figure 2–39(a). Each of the ten decimal digits is formed by activation of appropriate segments, as illustrated in Figure 2–39(b). In addition to the seven segments, there is also a decimal point.

(a)   (b)

FIGURE 2–39
Seven-segment display.

Resolution

The resolution of a meter is the smallest increment of a quantity that the meter can measure. The smaller the increment, the better the resolution. One factor that determines the resolution of a meter is the number of digits in the display.

Because many meters have 3½ digits in their display, we will use this case for illustration. A 3½-digit multimeter has three digit positions that can indicate from 0 through 9, and one digit position that can indicate only a value of 1. This latter digit, called the *half-digit,* is always the most significant digit in the display. For example, suppose that a DMM is reading 0.999 V, as shown in Figure 2–40(a). If the voltage increases by 0.001 V to 1 V, the display correctly shows 1.000 V, as shown in part (b). The "1" is the half-digit. Thus, with 3½ digits, a variation of 0.001 V, which is the resolution, can be observed.

(a) Resolution: 0.001 V

(b) Resolution: 0.001 V

(c) Resolution: 0.001 V

(d) Resolution: 0.01 V

FIGURE 2–40

A 3½-digit DMM illustrates how the resolution changes with the number of digits in use.

Now, suppose that the voltage increases to 1.999 V. This value is indicated on the meter as shown in part (c) of the figure. If the voltage increases by 0.001 V to 2 V, the half-digit cannot display the ''2,'' so the display shows 2.00. The half-digit is blanked and only three digits are active, as indicated in part (d). With only three digits active, the resolution is 0.01 V rather than 0.001 V as it is with 3½ active digits. The resolution remains 0.01 V up to 19.99 V. The resolution goes to 0.1 V for readings of 20.0 V to 199.9 V. At 200 V, the resolution goes to 1 V, and so on.

The resolution capability of a DMM is also determined by the internal circuitry and the rate at which the measured quantity is *sampled*. DMMs with displays of 4½ through 8½ digits are also available.

Accuracy

The accuracy of a DMM is established strictly by its internal circuitry. For typical meters, accuracies range from 0.01% to 0.5%, with some precision laboratory-grade meters going to 0.002%.

SECTION REVIEW 2–7

1. Name the meters for measurement of (a) current, (b) voltage, and (c) resistance.
2. Place two ammeters in the circuit of Figure 2–26 to measure the current through either lamp (be sure to observe the polarities). How can the same measurements be accomplished with only one ammeter?
3. Show how to place a voltmeter to measure the voltage across lamp 2 in Figure 2–26.
4. The multimeter in Figure 2–36 is set on the 3-V range to measure dc voltage. The pointer is at 150 on the upper ac-dc scale. What voltage is being measured?
5. How do you set up the meter to measure 275 V dc, and on what scale do you read the voltage?
6. If you expect to measure a resistance in excess of 20 kΩ, where do you set the switch?
7. List two common types of DMM displays, and discuss the advantages and disadvantages of each.
8. Define *resolution* in a DMM.

SUMMARY

- ☐ An atom is the smallest particle of an element that retains the characteristics of that element.
- ☐ The electron is the basic particle of negative electrical charge.
- ☐ The proton is the basic particle of positive charge.
- ☐ An ion is an atom that has gained or lost an electron and is no longer neutral.
- ☐ When electrons in the outer orbit of an atom (valence electrons) break away, they become free electrons.
- ☐ Free electrons make current possible.
- ☐ Like charges repel each other, and opposite charges attract each other.
- ☐ Voltage must be applied to a circuit before current can flow.
- ☐ Resistance limits the current.
- ☐ Basically, an electric circuit consists of a source, a load, and a current path.
- ☐ An open circuit is one in which the current path is broken.
- ☐ A closed circuit is one which has a complete current path.

☐ An ammeter is connected in line with the current path.

☐ A voltmeter is connected across the current path.

☐ An ohmmeter is connected across a resistor after removal from circuit.

☐ Figure 2–41 shows the electrical symbols introduced in this chapter.

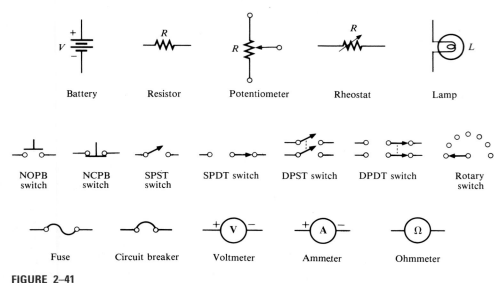

| Battery | Resistor | Potentiometer | Rheostat | Lamp |

| NOPB switch | NCPB switch | SPST switch | SPDT switch | DPST switch | DPDT switch | Rotary switch |

| Fuse | Circuit breaker | Voltmeter | Ammeter | Ohmmeter |

FIGURE 2–41

GLOSSARY

Ammeter An electrical instrument used to measure current.

Ampere (A) The unit of electrical current.

Atom The smallest particle of an element possessing the unique characteristics of that element.

Atomic number The number of electrons in a neutral atom.

Atom weight The number of protons and neutrons in the nucleus of an atom.

AWG American wire gage. A standardization based on wire diameter.

Battery An energy source that uses a chemical reaction to convert chemical energy into electrical energy.

Capacitor An electrical device possessing the property of capacitance.

Charge An electrical property of matter that exists because of an excess or a deficiency of electrons. Charge can be either positive or negative.

Choke An inductor. The term is used more commonly in connection with inductors used to block or choke off high frequencies.

Circuit An interconnection of electrical components designed to produce a desired result. A basic circuit consists of a source, a load, and an interconnecting current path.

Circuit breaker A resettable protective device used for interrupting excessive current in an electric circuit.

Circular mil (CM) The unit of the cross-sectional area of a wire.

Closed circuit A circuit with a complete current path.

Coil A common term for an inductor for the primary or secondary winding of a transformer.

Conductance The ability of a circuit to allow current. The unit is the siemen (S).

Conductor   A material in which electrical current can flow with relative ease. An example is copper.

Coulomb (C)   The unit of electrical charge.

Covalent   Related to the bonding of two or more atoms by the interaction of their valence electrons.

Current   The rate of flow of electrons.

Electrical   Related to the use of electrical voltage and current to achieve desired results.

Electron   The basic particle of electrical charge in matter. The electron possesses negative charge.

Electronic   Related to the movement and control of free electrons in semiconductors or vacuum devices.

Element   One of the unique substances that make up the known universe. Each element is characterized by a unique atomic structure.

Energy   The ability to do work.

Free electron   A valence electron that has broken away from its parent atom and is free to move from atom to atom within the atomic structure of a material.

Fuse   A protective device that burns open when excessive current flows in a circuit.

Generator   An energy source that produces electrical signals.

Inductor   An electrical device that has the capability to store energy in its magnetic field created by current.

Insulator   A material that does not allow current under normal conditions.

Joule (J)   The unit of energy.

Load   The device in a circuit upon which work is done.

Neutron   An atomic particle having no electrical charge.

Ohm (Ω)   The unit of resistance.

Ohmmeter   An instrument for measuring resistance.

Open circuit   A circuit in which there is not a complete current path.

Potentiometer   A three-terminal variable resistor.

Power supply   An electronic instrument that produces voltage, current, and power from the ac power line or batteries in a form suitable for use in various applications to power electronic equipment.

Proton   A positively charged atomic particle.

Resistance   Opposition to current. The unit is the ohm (Ω).

Resistor   An electrical component designed specifically to provide resistance.

Rheostat   A two-terminal variable resistor.

Schematic   A symbolized diagram of an electrical or electronic circuit.

Semiconductor   A material that has a conductance value between that of a conductor and insulator. Silicon and germanium are examples.

Source   Any device that produces energy.

Switch   An electrical device for opening and closing a current path.

Tapered   Nonlinear, such as a tapered potentiometer.

Tolerance   The limits of variation in the value of a component.

Valence   Related to the outer shell or orbit of an atom.

Volt   The unit of voltage or electromotive force.

Voltage   The amount of energy available to move a certain number of electrons from one point to another in an electrical circuit.

Wiper   The sliding contact in a potentiometer.

FORMULAS

$$V = \frac{W}{Q}$$ Voltage equals energy divided by charge.   (2–1)

$$I = \frac{Q}{t}$$ Current equals charge divided by time.   (2–2)

SELF-TEST

1. A neutral atom with an atomic number of three has how many electrons?
   (a) 1   (b) 3   (c) none   (d) depends on the type of atom

2. Electron orbits are called
   (a) shells   (b) nuclei   (c) waves   (d) valences

3. Materials in which no current can flow are called
   (a) filters   (b) conductors   (c) insulators   (d) semiconductors

4. When placed close together, a positively charged material and a negatively charged material will
   (a) repel   (b) become neutral   (c) attract   (d) exchange charges

5. The charge on a single electron is
   (a) 6.25×10^{-18} C   (b) 1.6×10^{-19} C   (c) 1.6×10^{-19} J
   (d) 3.14×10^{-6} C

6. *Potential difference* is another term for
   (a) energy   (b) voltage   (c) distance of an electron from the nucleus
   (d) charge

7. The unit of energy is the
   (a) watt   (b) coulomb   (c) joule   (d) volt

8. Which one of the following is not a type of energy source?
   (a) battery   (b) solar cell   (c) generator   (d) potentiometer

9. Which one of the following is not a possible condition in an electric circuit?
   (a) voltage and no current   (b) current and no voltage
   (c) voltage and current   (d) no voltage and no current

10. Electrical current is defined as
    (a) free electrons   (b) the rate of flow of free electrons
    (c) the energy required to move electrons   (d) the charge on free electrons

11. Current will not flow in a circuit when
    (a) a switch is closed   (b) a switch is open   (c) there is no voltage
    (d) (a) and (c)   (e) (b) and (c)

12. The primary purpose of a resistor is to
    (a) increase current   (b) limit current   (c) produce heat
    (d) resist current change

13. Potentiometers and rheostats are types of
    (a) voltage sources   (b) variable resistors   (c) fixed resistors
    (d) circuit breakers

14. The current in a given circuit is not to exceed 22 A. Which value of fuse is best?
    (a) 10 A   (b) 25 A   (c) 20 A   (d) a fuse is not necessary

PROBLEM SET A   **Section 2–2   Electrical Charge**

1. How many coulombs of charge do 50×10^{31} electrons possess?

2. How many electrons does it take to make 80 µC of charge?

Section 2–3   Voltage

3. Determine the voltage in each of the following cases:
   (a) 10 J/C   (b) 5 J/2 C   (c) 100 J/25 C

4. Five hundred joules of energy are used to move 100 C of charge through a resistor. What is the voltage across the resistor?

5. What is the voltage of a battery that uses 800 J of energy to move 40 C of charge through a resistor?

6. How much energy does a 12-V battery use to move 2.5 C through a circuit?

Section 2–4   Current

7. Determine the current in each of the following cases:
   (a) 75 C in 1 s   (b) 10 C in 0.5 s   (c) 5 C in 2 s

8. Six-tenths coulomb passes a point in 3 s. What is the current in amperes?

9. How long does it take 10 C to flow past a point if the current is 5 A?

10. How many coulombs pass a point in 0.1 s when the current is 1.5 A?

Section 2–5   Resistance

COLOR
INSERT
11. Figure 2–42(a) shows color-coded resistors (a full-color photo appears following page 240). Determine the resistance value and the tolerance of each.

12. Find the minimum and the maximum resistance within the tolerance limits for each resistor in Figure 2–42(a).

COLOR
INSERT
13. (a) If you need a 270-Ω resistor, what color bands should you look for?
    (b) From the selection of resistors in Figure 2–42(b) (a full-color photo appears following page 240), choose the following values: 330 Ω, 2.2 kΩ, 56 kΩ, 100 kΩ, and 39 kΩ.

14. The adjustable contact of a linear potentiometer is set at the mechanical center of its adjustment. If the total resistance is 1000 Ω, what is the resistance between each end terminal and the adjustable contact?

Section 2–6   The Electric Circuit

15. Trace the current path in Figure 2–26(a) with the switch in position 2.

16. With the switch in either position, redraw the circuit in Figure 2–26(b) with a fuse connected to protect the circuit against excessive current.

Section 2–7   Basic Circuit Measurements

17. Show the placement of an ammeter and a voltmeter to measure the current and the source voltage in Figure 2–43.

18. Show how you would measure the resistance of R_2 in Figure 2–43.

FIGURE 2–43

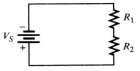

19. In Figure 2–44, how much voltage does each meter indicate when the switch is in position 1? In position 2?

20. In Figure 2–44, indicate how to connect an ammeter to measure the current from the voltage source regardless of the switch position.

FIGURE 2–44

21. What are the voltage readings in Figures 2–45(a) and 2–45(b)?

(a)

(b)

FIGURE 2–45

22. How much resistance is the ohmmeter in Figure 2–46 measuring?

23. Determine the resistance indicated by each of the following ohmmeter readings and range settings:

(a) pointer at 2, range setting at R × 100

(b) pointer at 15, range setting at R × 10M

(c) pointer at 45, range setting at R × 100

24. A multimeter has the following ranges: 1 mA, 10 mA; 100 mA, 100 mV, 1 V, 10 V; R × 1, R × 10, R × 100. Indicate schematically how you would connect the multimeter in Figure 2–47 for the following:

(a) I_1

(b) V_1

(c) R_1

In each case indicate the *function* to which you would set the meter and the *range* that you would use.

FIGURE 2–46

FIGURE 2–47

PROBLEM SET B

25. A resistor with a current of 2 A through it converts 1000 J of electrical energy to heat energy in 15 s. What is the voltage across the resistor?

26. If 574×10^{15} electrons flow through a wire in 250 ms, what is the current in amperes?

27. A 120-V source is to be connected to a 1500-Ω resistive load by two lengths of wire as shown in Figure 2–48. The voltage source is to be located 50 ft from the load. Using Table 2–3, determine the gage number of the *smallest* wire that can be used if the *total* resistance of the two lengths of wire is not to exceed 6 Ω.

FIGURE 2–48

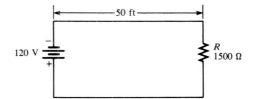

28. Determine the resistance and tolerance of each resistor labeled as follows:
    (a) 4R7J   (b) 5602M   (c) 1501F

29. There is only one circuit in Figure 2–49 in which it is possible to have all lamps on at the same time. Determine which circuit it is.

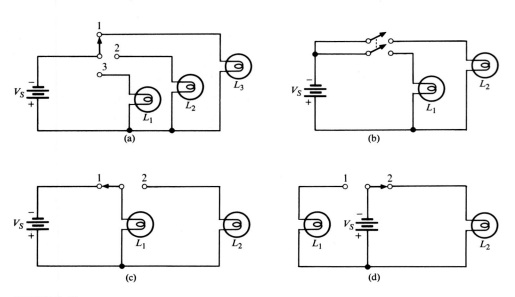

FIGURE 2–49

30. Through which resistor in Figure 2–50 does current always flow, regardless of the position of the switches?

31. In Figure 2–50, show the proper placement of ammeters to measure the current through each resistor and the current out of the battery.

32. Show the proper placement of voltmeters to measure the voltage across each resistor in Figure 2–50.

33. Devise a switch arrangement whereby two voltage sources (V_{S1} and V_{S2}) can be connected simultaneously to either of two resistors (R_1 and R_2) as follows:

V_{S1} connected to R_1 and V_{S2} connected to R_2 or

V_{S1} connected to R_2 and V_{S2} connected to R_1

FIGURE 2–50

34. The different sections of a stereo system are represented by the blocks in Figure 2–51. Show how a single switch can be used to connect the phonograph, the CD (compact disk)

player, the tape deck, the AM tuner, or the FM tuner to the amplifier by a single knob control. Only one section can be connected to the amplifier at any time.

FIGURE 2-51

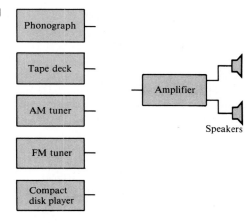

Phonograph

Tape deck

AM tuner

Amplifier

Speakers

FM tuner

Compact disk player

ANSWERS TO SECTION REVIEWS

Section 2-1

1. Electron.
2. The smallest particle of an element that retains the unique characteristics of the element.
3. A positively charged nucleus surrounded by orbiting electrons.
4. No.
5. An outer-shell electron that has drifted away from the parent atom.

Section 2-2

1. Q.   2. Coulomb, C.   3. 1.6×10^{-6} C.

Section 2-3

1. Energy per unit charge.   2. 2.4 V.
3. Battery, power supply, solar cell, and generator.

Section 2-4

1. Rate of flow of electrons, ampere (A).   2. 6.25×10^{18}.
3. 5 A.

Section 2-5

1. Opposition to current, Ω.
2. Fixed and variable. The value of a fixed resistor cannot be changed, but that of a variable resistor can.
3. *First band:* first digit of resistance value. *Second band:* second digit of resistance value. *Third band:* number of zeros. *Fourth band:* tolerance.
4. **(a)** 4700 Ω $\pm 5\%$;   **(b)** 62,000 Ω $\pm 10\%$;   **(c)** 18 Ω $\pm 5\%$.
5. A rheostat has two terminals; a potentiometer has three terminals.

Section 2–6

1.   Source, load, and current path between source and load.
2.   A circuit that has no path for current.
3.   A circuit that has a complete path for current.
4.   Protect a circuit against excessive current.
5.   AWG #3.

Section 2–7

1.   **(a)** Ammeter;   **(b)** voltmeter;   **(c)** ohmmeter.
2.   See Figure 2–52.

FIGURE 2–52

3.   See Figure 2–53.
4.   1.5 V.
5.   DC volts, 300 setting; 300 scale.
6.   Ohms × 1000.
7.   LED, LCD.
8.   Smallest increment that the meter can measure.

FIGURE 2–53

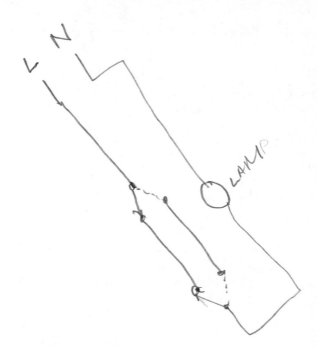

3

Ohm's Law and Power

Georg Simon Ohm (1787–1854) experimentally found that voltage, current, and resistance are all related in a specific way. This basic relationship, known as *Ohm's law,* is one of the most fundamental and important laws in the fields of electricity and electronics. In this chapter, the meaning of Ohm's law is examined, and its use in practical circuit applications is discussed and demonstrated by numerous examples.

In addition to Ohm's law, the concept and definition of power in electric circuits are introduced. Practical considerations are emphasized.

After completing this chapter, you will be able to

☐ Apply Ohm's law to determine voltage, current, and resistance in electric circuits.
☐ Work with voltage, current, and resistance in both large and small units.
☐ Describe the linear relationship between current and voltage.
☐ Define power and how it relates to voltage, current, and resistance.
☐ Explain the difference between power and energy.
☐ Determine the amount of power in a resistive circuit.
☐ Select resistors that will handle the amount of power in a given circuit.
☐ Explain how energy is measured in kilowatthours.

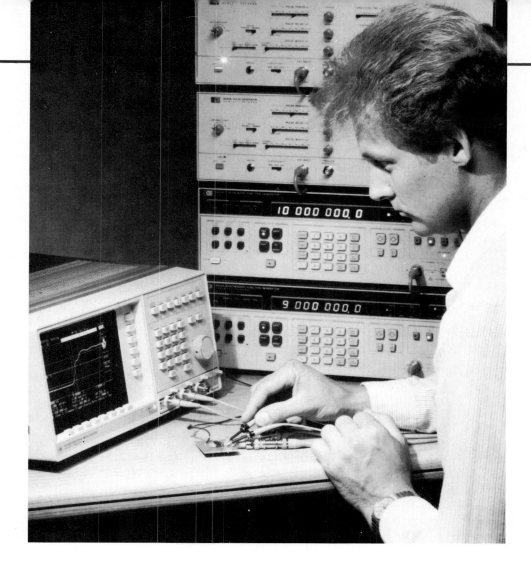

APPLICATION ASSIGNMENT

As a technician in the development lab for a large manufacturer of electronic equipment, you are given three specific assignments during your first week on the job:

1. By specifying the necessary current capacity, choose a 5-V power supply that is capable of providing current to a 75-Ω load.
2. Measure the current through a certain device when 25 V are applied, and determine the resistance value and power rating of a "dummy" load resistor that can be substituted for the device for test purposes.
3. Specify the source voltage necessary for a certain display element that requires 20 mA of current to achieve a desired illumination. The internal resistance of the element is known to be 100 Ω.

After studying this chapter, you should be able to complete this assignment.

3–1

OHM'S LAW

Ohm's law describes how *current, voltage,* and *resistance* are related and provides a formula that you can use to determine any one of the quantities when you know the other two. We will now examine Ohm's law step-by-step using the simple circuit in Figure 3–1 to demonstrate its meaning. Then you will see how to use the formula in practical circuit calculations.

Variable voltage source

(a) Pictorial   (b) Schematic

FIGURE 3–1
Circuit used for the demonstration of Ohm's law.

In Figure 3–1 there is a variable voltage source with which the voltage can be adjusted over a range of values. A voltmeter is connected across the terminals of the voltage source, and an ammeter is connected in the current path with the resistor.

How Voltage Affects Current

Figure 3–2 demonstrates how a change in the voltage affects the current. Specific values are used for illustration. In part (a), the voltage is set at 10 V, and the resulting current of 1 A is measured. Notice that the resistance is kept at 10 Ω. In part (b), the voltage is reduced to 5 V, and a resulting current of 0.5 A is measured. Notice that when the voltage is halved, the current is halved. In part (c), the voltage is increased to 20 V (double its initial value), and a resulting current of 2 A is measured. Thus, as you can see, doubling the voltage also doubles the current, tripling the voltage triples the current, and so on.

Linearly Proportional Relationship between Current and Voltage

The previous demonstration showed that the voltage and current in a resistive circuit are directly related and *linearly proportional;* that is, they have a straight-line relationship. If the voltage goes up by a certain amount, the current will go up by a proportional amount. If the voltage goes down by a certain amount, the current will go down by a proportional amount. This relationship is illustrated by the straight-line graph of Figure 3–3 for a resistance of 10 Ω. For any constant resistance, a straight-line (linear) relationship between current and voltage will always exist.

(a) Initial setting, V = 10 V

(b) Voltage decreased to 5 V

(c) Voltage increased to 20 V

FIGURE 3–2
Demonstration of Ohm's law: How voltage affects current with resistance constant.

FIGURE 3–3
Graph of current versus voltage for Figure 3–2 with resistance constant.

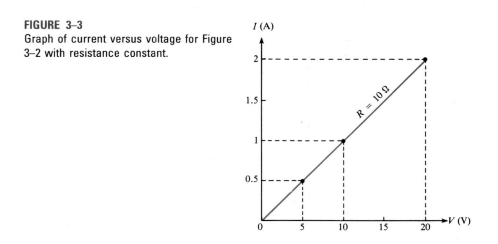

How Resistance Affects Current

Figure 3–4 demonstrates how a change in resistance affects the current when the voltage is kept at a constant value. Again, specific values are used for illustration. In part (a), the voltage is set at 10 V and remains there for each step. A 10-Ω resistor is connected in the circuit, and the resulting current is 1 A as indicated. In part (b), the resistor is changed to 5 Ω, and the resulting current is 2 A. As you can see, halving the resistance doubles the current. In part (c), the resistor is changed to 20 Ω, and the resulting current is 0.5 A. In this case, doubling the resistance reduces the current to half its original value.

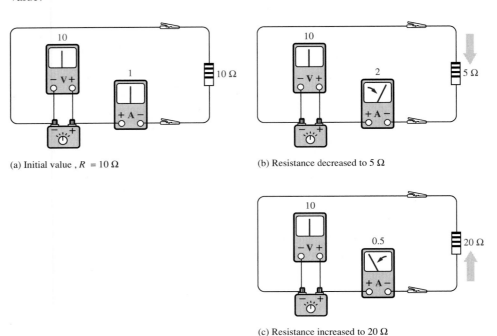

(a) Initial value , $R = 10\ \Omega$

(b) Resistance decreased to 5 Ω

(c) Resistance increased to 20 Ω

FIGURE 3–4
Demonstration of Ohm's law: How resistance affects current with voltage constant.

This demonstration shows that current is *inversely* proportional to resistance. When resistance goes up by a certain amount, the current goes down by a proportional amount. When resistance goes down by a certain amount, the current goes up by a proportional amount.

The Ohm's Law Formula

Figures 3–2 and 3–4 demonstrate Ohm's law, which can be stated by a formula as follows:

$$I = \frac{V}{R} \qquad\qquad (3\text{--}1)$$

This formula describes what was indicated by the demonstration:

For a constant resistance, if the voltage in a circuit is increased, more current will flow, and if the voltage is decreased, less current will flow. Also, for a constant voltage, if the resistance in a circuit is increased, less current will flow, and if the resistance is decreased, more current will flow.

Ohm's law is illustrated as follows.

$I = V / R$ Increase V, I increases $I = V / R$ Decrease V, I decreases R constant

$I = V / R$ Increase R, I decreases $I = V / R$ Decrease R, I increases V constant

---

EXAMPLE 3–1

Verify that the Ohm's law formula in Equation (3–1) gives results that agree with the values in the circuits of Figures 3–2 and 3–4.

Solution

In Figure 3–2, $R = 10\ \Omega$. For $V = 10$ V:

$$I = \frac{V}{R} = \frac{10\ \text{V}}{10\ \Omega} = 1\ \text{A}$$

For $V = 5$ V:

$$I = \frac{V}{R} = \frac{5\ \text{V}}{10\ \Omega} = 0.5\ \text{A}$$

For $V = 20$ V:

$$I = \frac{V}{R} = \frac{20\ \text{V}}{10\ \Omega} = 2\ \text{A}$$

In Figure 3–4, $V = 10$ V. For $R = 5\ \Omega$:

$$I = \frac{V}{R} = \frac{10\ \text{V}}{5\ \Omega} = 2\ \text{A}$$

For $R = 20\ \Omega$:

$$I = \frac{V}{R} = \frac{10\ \text{V}}{20\ \Omega} = 0.5\ \text{A}$$

Thus, the formula verifies the values of both figures.

The calculator sequence for $V = 20$ V is

Other Forms of the Ohm's Law Formula

As you have seen, you can use the Ohm's law formula in Equation (3–1) to find the current in a circuit when the voltage and resistance are known. By a simple algebraic change, you can use the formula to find the voltage when the current and resistance are known. By multiplying both sides of Equation (3–1) by R, we get:

$$V = IR \qquad\qquad (3-2)$$

We obtain another simple change in Equation (3–1) by transposing the R and the I:

$$R = \frac{V}{I} \qquad\qquad (3-3)$$

With this formula, you can find the resistance when the voltage and current are known.
    Keep in mind that Equations (3–1), (3–2), and (3–3) are all equivalent expressions of Ohm's law.

A Graphic Aid for Ohm's Law

You may find the graphic aid in Figure 3–5(a) of value for applying any of the three forms of Ohm's law. It is a way to help you remember the formulas, as illustrated in Figure 3–5(b).

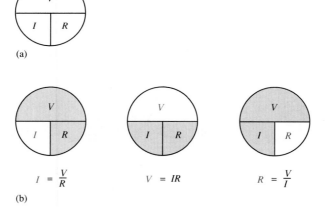

(a)

(b)

$$I = \frac{V}{R} \qquad\qquad V = IR \qquad\qquad R = \frac{V}{I}$$

FIGURE 3–5
A graphic aid for the Ohm's law formulas.

SECTION REVIEW 3–1

1. Briefly state Ohm's law in words.
2. If the voltage across a resistor is tripled, does the current increase or decrease? By how much?
3. There is a fixed voltage across a variable resistor, and you measure a current of 10 mA. If you double the resistance, how much current will you measure?
4. Will you observe any change in current in a circuit where both the voltage and the resistance are doubled?

3–2   APPLICATION OF OHM'S LAW

How to Determine *I* When *V* and *R* Are Known

In this section you will learn to determine current values when you know the values of voltage and resistance. In these problems, the formula $I = V/R$ is used. In order to get current in *amperes,* you must express the value of *V* in *volts* and the value of *R* in *ohms*.

EXAMPLE 3–2

How many amperes of current are flowing in the circuit of Figure 3–6?

FIGURE 3–6

100 V
I
R
20 Ω

Solution
Into the formula $I = V/R$, substitute 100 V for *V* and 20 Ω for *R*. Divide 20 Ω into 100 V as follows:

$$I = \frac{V}{R} = \frac{100 \text{ V}}{20 \text{ } \Omega} = 5 \text{ A}$$

There are 5 A of current in this circuit.

In electronics work, resistance values of thousands of ohms or even millions of ohms are common. As you learned in Chapter 1, large values of resistance are indicated by the metric system prefixes *kilo* (k) and *mega* (M). Thus, thousands of ohms are expressed in *kilohms* (kΩ), and millions of ohms in *megohms* (MΩ). The following examples illustrate how to use kilohms and megohms when using Ohm's law to find the current.

EXAMPLE 3–3

Calculate the current in Figure 3–7.

1 kΩ

50 V

I

FIGURE 3–7

Solution
Remember that 1 kΩ is the same as 1000 Ω. Substituting 50 V for *V* and 1000 Ω for *R* gives the current in amperes as follows:

$$I = \frac{V}{R} = \frac{50 \text{ V}}{1000 \text{ } \Omega} = 0.05 \text{ A}$$

The calculator sequence is

⑤ ⓪ ÷ ① EXP ③ =

In Example 3–3, 0.05 A can be expressed as 50 mA. Thus, when *volts* are divided by *kilohms*, the current will be in *milliamperes*.

If *volts* are used when resistance values are in *megohms*, the current is in *microamperes* (μA).

EXAMPLE 3–4

Determine the amount of current in the circuit of Figure 3–8.

FIGURE 3–8

Solution
Recall that 5 MΩ equals 5,000,000 Ω. Substituting 25 V for V and 5,000,000 Ω for R gives the following result:

$$I = \frac{V}{R} = \frac{25 \text{ V}}{5,000,000 \ \Omega} = 0.000005 \text{ A}$$

Notice that 0.000005 A equals 5 μA, which can also be expressed as 5×10^{-6} A.
The calculator sequence is

$\boxed{2}$ $\boxed{5}$ $\boxed{\div}$ $\boxed{5}$ $\boxed{\text{EXP}}$ $\boxed{6}$ $\boxed{=}$

Small voltages, usually less than 50 V, are common in electronic circuits. Occasionally, however, large voltages are encountered. For example, the high-voltage supply in a television receiver is around 20,000 V (20 kV), and transmission voltages generated by the power companies may be as high as 345,000 V (345 kV).

EXAMPLE 3–5

How much current will flow through 100 MΩ when 50 kV are applied?

Solution
In this case, we divide 50 kV by 100 MΩ to get the current. Using 50,000 V for 50 kV and 100,000,000 Ω for 100 MΩ, we obtain the current as follows:

$$I = \frac{V}{R} = \frac{50 \text{ kV}}{100 \text{ M}\Omega} = \frac{50,000 \text{ V}}{100,000,000 \ \Omega}$$
$$= 0.0005 \text{ A} = 0.5 \text{ mA}$$

Remember to move the decimal point three places to the right to convert from amperes to milliamperes.
The calculator sequence is

$\boxed{5}$ $\boxed{0}$ $\boxed{\text{EXP}}$ $\boxed{3}$ $\boxed{\div}$ $\boxed{1}$ $\boxed{0}$ $\boxed{0}$ $\boxed{\text{EXP}}$ $\boxed{6}$ $\boxed{=}$

How to Determine *V* When *I* and *R* Are Known

In this section you will learn to determine voltage values when the current and resistance are known. In these problems, the formula $V = IR$ is used. To obtain voltage in *volts*, you must express the value of *I* in *amperes* and the value of *R* in *ohms*.

EXAMPLE 3–6

In the circuit of Figure 3–9, how much voltage is needed to produce 5 A of current?

FIGURE 3–9

Solution

Substitute 5 A for *I* and 100 Ω for *R* into the formula $V = IR$ as follows:

$$V = IR = (5\ A)(100\ \Omega) = 500\ V$$

Thus, 500 V are required to produce 5 A of current through a 100-Ω resistor.

EXAMPLE 3–7

How much voltage will be measured across the resistor in Figure 3–10?

FIGURE 3–10

Solution

Note that 5 mA equals 0.005 A. Substituting the values for *I* and *R* into the formula $V = IR$, we get the following result:

$$V = IR = (5\ mA)(50\ \Omega)$$
$$= (0.005\ A)(50\ \Omega) = 0.25\ V = 250\ mV$$

When *milliamperes* are multiplied by *ohms*, we get *millivolts*.

The calculator sequence is

⑤ EXP +/− ③ × ⑤ ⓪ =

EXAMPLE 3–8

The circuit in Figure 3–11 has a current of 10 mA. What is the voltage?

FIGURE 3–11

DC CIRCUITS

Solution

Note that 10 mA equals 0.010 A and that 3.3 kΩ equals 3300 Ω. Substituting these values into the formula $V = IR$, we get

$$V = IR = (10 \text{ mA})(3.3 \text{ k}\Omega)$$
$$= (0.010 \text{ A})(3300 \text{ }\Omega) = 33 \text{ V}$$

When *milliamperes* and *kilohms* are multiplied, the result is *volts*.
The calculator sequence is

How to Determine *R* When *V* and *I* Are Known

In this section you will learn to determine resistance values when the current and voltage are known. In these problems, the formula $R = V/I$ is used. To find resistance in *ohms*, you must express the value of I in *amperes* and the value of V in *volts*.

EXAMPLE 3–9

In the circuit of Figure 3–12, how much resistance is needed to draw 3 A of current from the battery?

FIGURE 3–12

Solution

Substitute 12 V for V and 3 A for I into the formula $R = V/I$:

$$R = \frac{V}{I} = \frac{12 \text{ V}}{3 \text{ A}} = 4 \text{ }\Omega$$

EXAMPLE 3–10

Suppose that the ammeter in Figure 3–13 indicates 5 mA of current and the voltmeter reads 150 V. What is the value of R?

FIGURE 3–13

Solution

Note that 5 mA equals 0.005 A. Substituting the voltage and current values into the formula $R = V/I$, we get

$$R = \frac{V}{I} = \frac{150 \text{ V}}{5 \text{ mA}} = \frac{150 \text{ V}}{0.005 \text{ A}} = 30,000 \ \Omega = 30 \text{ k}\Omega$$

Thus, if *volts* are divided by *milliamperes*, the resistance will be in *kilohms*.

SECTION REVIEW 3-2

1. $V = 10$ V and $R = 5 \ \Omega$. Find I.
2. If a 5-MΩ resistor has 20 kV across it, how much current flows?
3. How much current will 10 kV across 2 kΩ produce?
4. $I = 1$ A and $R = 10 \ \Omega$. Find V.
5. What voltage do you need to produce 3 mA of current in a 3-kΩ resistance?
6. A battery produces 2 A of current into a 6-Ω resistive load. What is the battery voltage?
7. $V = 10$ V and $I = 2$ A. Find R.
8. You have a resistor across which you measure 25 V, and your ammeter indicates 50 mA of current. What is the resistor's value in kilohms? In ohms?

3-3

POWER AND ENERGY

Power is the rate at which energy is used.

In other words, power, symbolized by P, is a certain amount of energy used in a certain length of time, expressed as follows:

$$\text{Power} = \frac{\text{energy}}{\text{time}}$$

$$P = \frac{W}{t} \qquad (3\text{--}4)$$

Energy is measured in *joules* (J), time is measured in *seconds* (s), and power is measured in *watts* (W). Energy in *joules* divided by time in *seconds* gives power in *watts*. For example, if 50 J of energy are used in 2 s, the power is 50 J/2 s = 25 W. By definition,

one watt is the amount of power when one joule of energy is consumed in one second.

Thus, the number of joules consumed in 1 s is always equal to the number of watts. For example, if 75 J are used in 1 s, the power is 75 W. The following example illustrates use of these units.

EXAMPLE 3-11

An amount of energy equal to 100 J is used in 5 s. What is the power in watts?

Solution

$$P = \frac{\text{energy}}{\text{time}} = \frac{100 \text{ J}}{5 \text{ s}} = 20 \text{ W}$$

Amounts of power much less than one watt are common in certain areas of electronics. As with small current and voltage values, metric prefixes are used to designate small amounts of power. Thus, *milliwatts* (mW) and *microwatts* (μW) are commonly found in some applications.

In the electrical utilities field, *kilowatts* (kW) and *megawatts* (MW) are common units. Radio and television stations also use large amounts of power to transmit signals.

EXAMPLE 3–12

Express the following powers using appropriate metric prefixes:
(a) 0.045 W   (b) 0.000012 W   (c) 3500 W   (d) 10,000,000 W

Solution
(a) 0.045 W = 45 mW   (b) 0.000012 W = 12 μW
(c) 3500 W = 3.5 kW   (d) 10,000,000 W = 10 MW

The Kilowatthour (kWh) Unit of Energy

Since power is the rate of energy usage, power utilized over a period of time represents energy consumption. If we multiply *power* and *time*, we have *energy*, symbolized by *W*:

$$\text{Energy} = \text{power} \times \text{time}$$
$$W = Pt \tag{3–5}$$

Earlier, the *joule* was defined as a unit of energy. However, there is another way of expressing energy. Since power is expressed in *watts* and time in *seconds*, we can use units of energy called the *wattsecond* (Ws), *watthour* (Wh), and *kilowatthour* (kWh).

When you pay your electric bill, you are charged on the basis of the amount of *energy* you use. Because power companies deal in huge amounts of energy, the most practical unit is the *kilowatthour. You use a kilowatthour of energy when you use the equivalent of 1000 W of power for 1 h.* For example, a 100-W light bulb burning for 10 h uses 1 kWh of energy.

EXAMPLE 3–13

Determine the number of kilowatthours for each of the following energy consumptions:
(a) 1400 W for 1 h   (b) 2500 W for 2 h   (c) 100,000 W for 5 h

Solution
(a) 1400 W = 1.4 kW
    Energy = (1.4 kW)(1 h) = 1.4 kWh
(b) 2500 W = 2.5 kW
    Energy = (2.5 kW)(2 h) = 5 kWh
(c) 100,000 W = 100 kW
    Energy = (100 kW)(5 h) = 500 kWh

1. Define *power*.
2. Write the formula for power in terms of energy and time.
3. Define *watt*.
4. Express each of the following values of power in the most appropriate units:
    (a) 68,000 W   (b)   0.005 W   (c)   0.000025 W
5. If you use 100 W of power for 10 h, how much energy (in kilowatthours) have you consumed?
6. Convert 2000 Wh to kilowatthours.
7. Convert 360,000 Ws to kilowatthours.

3–4

POWER IN AN ELECTRIC CIRCUIT

When current flows through resistance, the collisions of the electrons give off heat, resulting in a loss of energy as indicated in Figure 3–14. There is always a certain

FIGURE 3–14
Power in an electric circuit is seen as heat given off by the resistance.

amount of power in an electric circuit, and it is dependent on the amount of *resistance* and on the amount of *current*, expressed as follows:

$$P = I^2R \qquad (3–6)$$

We can produce an *equivalent* expression for power by substituting V for IR (I^2 is $I \times I$).

$$P = I^2R = (I \times I)R = I(IR) = (IR)I$$
$$P = VI \qquad (3–7)$$

We obtain another *equivalent* expression by substituting V/R for I (Ohm's law) as follows:

$$P = VI = V\left(\frac{V}{R}\right)$$
$$P = \frac{V^2}{R} \qquad (3–8)$$

The three power formulas in Equations (3–6), (3–7), and (3–8) are also known as the *Watt's law formulas*.

How to Use the Appropriate Power Formula

To calculate the power in a resistance, you can use any one of the three power formulas, depending on what information you have. For example, assume that you know the values of current and voltage. In this case you calculate the power with the formula $P = VI$. If you know I and R, use the formula $P = I^2R$. If you know V and R, use the formula $P = V^2/R$.

EXAMPLE 3–14

Calculate the power in each of the three circuits of Figure 3–15.

(a)   (b)   (c)

FIGURE 3–15

Solution

In circuit (a), V and I are known. The power is determined as follows:

$$P = VI = (10 \text{ V})(2 \text{ A}) = 20 \text{ W}$$

In circuit (b), I and R are known. The power is determined as follows:

$$P = I^2R = (2 \text{ A})^2(50 \text{ }\Omega) = 200 \text{ W}$$

In circuit (c), V and R are known. The power is determined as follows:

$$P = \frac{V^2}{R} = \frac{(5 \text{ V})^2}{10 \text{ }\Omega} = 2.5 \text{ W}$$

The calculator sequence for circuit (c) is

EXAMPLE 3–15

A 100-W light bulb operates on 120 V. How much current does it require?

Solution

Use the formula $P = VI$ and solve for I as follows: Transpose the terms to get I on the left of the equation:

$$VI = P$$

Divide both sides of equation by V to get I by itself:

$$\frac{\cancel{V}I}{\cancel{V}} = \frac{P}{V}$$

The V's cancel on the left, leaving

$$I = \frac{P}{V}$$

Substituting 100 W for P and 120 V for V yields

$$I = \frac{P}{V} = \frac{100 \text{ W}}{120 \text{ V}} = 0.833 \text{ A}$$

SECTION REVIEW
3–4

1. If there are 10 V across a resistor and a current of 3 A through it, what is the power?
2. How much power does the source in Figure 3–16 generate? What is the power in the resistor? Are the two values the same? Why?

FIGURE 3–16

3. If a current of 5 A is flowing through a 50-Ω resistor, what is the power?
4. How much power is produced by 20 mA through a 5-kΩ resistor?
5. Five volts are applied to a 10-Ω resistor. What is the power?
6. How much power does a 2-kΩ resistor with 8 V across it produce?
7. What is the resistance of a 75-W bulb that takes 0.5 A?

3–5

THE POWER RATING OF RESISTORS

As you know, a resistor gives off heat when current flows through it. There is a limit to the amount of heat that a resistor can give off, which is specified by its power rating.

The *power rating* is the maximum amount of power that a resistor can dissipate without being damaged by excessive heat buildup. The power rating is not related to the ohmic value (resistance) but rather is determined mainly by the physical size and shape of the resistor. The larger the *surface area* of a resistor, the more power it can dissipate. *The surface area of a cylindrically shaped resistor is equal to the length (l) times the circumference (c),* as indicated in Figure 3–17.

FIGURE 3–17
The power rating of a resistor is directly related to its surface area.

FIGURE 3–18
Carbon-composition resistors with standard
power ratings of 1/8 W, 1/4 W, 1/2 W, 1 W,
and 2 W (courtesy of Allen-Bradley Co.).

Carbon-composition resistors are available in standard power ratings from 1/8 W to 2 W, as shown in Figure 3–18. Available power ratings for other types of resistors vary. For example, carbon-film and metal-film resistors have ratings up to 10 W, and wire-wound resistors have ratings up to 225 W or greater. Figure 3–19 shows some of these resistors.

(a) Vitreous-enameled resistor

(b) "Corrib" corrugated ribbon resistor

(c) Molded vitreous-enameled wire-wound axial lead resistor

(d) Thin resistors

(e) Metal-film precision resistors

FIGURE 3–19
Typical resistors with high power ratings (courtesy of Ohmite Manufacturing Co.)

How to Select the Proper Power Rating for an Application

When a resistor is used in a circuit, its power rating must be greater than the maximum power that it will have to handle. For example, if a carbon-composition resistor is to dissipate 0.75 W in a circuit application, its rating should be at least the next higher standard value which is 1 W.

EXAMPLE 3–16

Choose an adequate power rating for each of the carbon-composition resistors in Figure 3–20.

FIGURE 3–20

(a)   (b)

Solution

(a) The actual power is

$$P = \frac{V^2}{R} = \frac{(10 \text{ V})^2}{62 \text{ }\Omega} = \frac{100 \text{ V}^2}{62 \text{ }\Omega} = 1.6 \text{ W}$$

Select a resistor with a power rating higher than the actual power. In this case, a 2-W resistor should be used.

(b) The actual power is

$$P = I^2R = (10 \text{ mA})^2(1000 \text{ }\Omega) = 0.1 \text{ W}$$

A 1/8-W (0.125-W) resistor should be used in this case.

Resistor Failures

When the power in a resistor is greater than its rating, the resistor will become excessively hot. As a result, either the resistor will burn open or its resistance value will be greatly altered.

A resistor that has been damaged because of overheating can often be detected by the charred or altered appearance of its surface. If there is no visual evidence, a resistor that is suspected of being damaged can be checked with an ohmmeter for an open or increased resistance value.

EXAMPLE 3–17

Determine whether the resistor in each circuit of Figure 3–21 has possibly been damaged by overheating.

(a)   (b)   (c)

FIGURE 3–21

Solution

(a)

$$P = \frac{V^2}{R} = \frac{(9\ \text{V})^2}{100\ \Omega} = 0.81\ \text{W}$$

The rating of the resistor is 1/4 W (0.25 W), which is insufficient to handle the power. The resistor may be burned out.

(b)

$$P = \frac{V^2}{R} = \frac{(24\ \text{V})^2}{1.5\ \text{k}\Omega} = 0.384\ \text{W}$$

The rating of the resistor is 1/2 W (0.5 W), which is sufficient to handle the power.

(c)

$$P = \frac{V^2}{R} = \frac{(5\ \text{V})^2}{10\ \Omega} = 2.5\ \text{W}$$

The rating of the resistor is 1 W, which is insufficient to handle the power. The resistor may be burned out.

How to Check a Resistor with an Ohmmeter

A typical analog multimeter (VOM) and a digital multimeter are shown in Figures 3–22(a) and 3–22(b), respectively. The large switch on the analog meter is called a *range switch*. Notice the resistance (OHMS) settings on both meters. For the analog meter in part (a), each setting indicates the amount by which the *ohms scale* (top scale) on the meter is to be *multiplied*. For example, if the pointer is at 50 on the ohms scale and the range switch is set at ×100, the resistance being measured is 50 × 100 Ω = 5000 Ω. *If the resistor is open, the pointer will stay at full left scale (∞ means infinite) regardless of the range switch setting.* For the digital meter, you use the range switch to select the appropriate setting for the value of resistance being measured. You do not have to multiply to get the correct reading because you have a direct digital readout of the resistance value.

The Basic Analog Ohmmeter

A basic ohmmeter, shown in Figure 3–23(a), contains a battery and a variable resistor in series with the meter movement. To measure resistance, connect the leads across the external resistor to be measured, as shown in part (b). This connection completes the circuit, allowing the internal battery to produce current through the movement, in turn causing a deflection of the pointer proportional to the value of the external resistance being measured.

(a)

(b)

FIGURE 3–22
Typical multimeters (courtesy of Triplett Corp.).

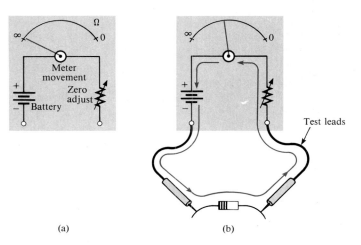

(a)   (b)

FIGURE 3–23
Basic analog ohmmeter circuit.

Zero Adjustment of an Analog Ohmmeter

When the ohmmeter leads are open, as in Figure 3–24(a), the pointer is at full left scale, indicating *infinite* (∞) resistance (open circuit). When the leads are *shorted*, as in Figure 3–24(b), the pointer is at full right scale, indicating *zero* resistance.

The purpose of the variable resistor is to adjust the current so that the pointer is at exactly zero when the leads are shorted. It is used to compensate for changes in the internal battery voltage due to aging. Only analog ohmmeters need to be zeroed and they should be rezeroed each time the ohms scale is changed.

FIGURE 3–24
Zero adjustment.

Adjust pointer to zero.

(a) Open (infinite resistance)   (b) Short (zero resistance)

Because of the ohmmeter's internal battery, any other voltage source must be disconnected from a circuit before a resistance measurement is made. If the voltage is left on, there is a risk of damaging the meter.

The following steps should be used in measuring resistance:

1. Disconnect the voltage from the circuit.
2. To prevent other parts of the circuit from altering the measurement, disconnect the resistor to be checked from its circuit.
3. Select the proper range switch setting. On the analog meter, it is best to select a setting that produces a reading on the right half of the scale for better accuracy. On the digital meter, set the switch at the lowest setting that is greater than the value you expect to measure.
4. Touch the test leads together, and use the *ohms adjust* knob to set the pointer at 0 Ω (analog meter only).
5. Connect the meter leads across the resistor. Polarity is not important in resistance measurements. Be careful not to touch the test lead contacts with your fingers because your body resistance will affect the reading.

**SECTION REVIEW
3–5**

1. Name two important values associated with a resistor.
2. How does the physical size of a resistor determine the amount of power that it can handle?
3. List the standard power ratings of carbon-composition resistors.
4. A resistor must handle 0.3 W. What standard size carbon resistor should be used to dissipate the energy properly?

5. Why is an internal battery required in an ohmmeter?

6. If the pointer indicates 8 on the ohmmeter scale and the range switch is set at R × 1k, what resistance is being measured?

SUMMARY

☐ Voltage and current are linearly proportional.

☐ Ohm's law gives the relationship of voltage, current, and resistance.

☐ Current is directly proportional to voltage.

☐ Current is inversely proportional to resistance.

☐ A kilohm (kΩ) is one thousand ohms.

☐ A megohm (MΩ) is one million ohms.

☐ A microampere (μA) is one-millionth of an ampere.

☐ A milliampere (mA) is one-thousandth of an ampere.

☐ The amount of heat that a resistor can dissipate is specified by its power rating.

☐ The power rating of a resistor is determined by its physical size and shape.

☐ Power rating of a resistor is not related to resistance value.

☐ A resistor normally opens when it burns out.

☐ The symbol for infinity is ∞.

☐ The formula wheel in figure 3–25 gives the Ohm's law and Watt's law relationships.

FIGURE 3–25

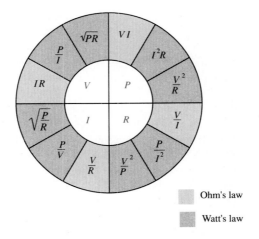

Ohm's law

Watt's law

GLOSSARY

Joule (J) The unit of energy.

Kilowatthour (kWh) A common unit of energy used mainly by utility companies.

Linear Characterized by a straight-line relationship.

Ohm (Ω) The unit of resistance.

Ohm's law A law stating that current is directly proportional to voltage and inversely proportional to resistance.

Power The rate of energy consumption.

Watt (W) The unit of power.

Watt's law A law that states the relationships of power to current, voltage, and resistance.

FORMULAS

$$I = \frac{V}{R} \qquad \text{Ohm's law} \tag{3–1}$$

$$V = IR \qquad \text{Ohm's law} \tag{3–2}$$

$$R = \frac{V}{I} \qquad \text{Ohm's law} \tag{3–3}$$

$$P = \frac{W}{t} \qquad \text{Power equals energy divided by time.} \tag{3–4}$$

$$W = Pt \qquad \text{Energy equals power multiplied by time.} \tag{3–5}$$

$$P = I^2 R \qquad \text{Power equals current squared times resistance.} \tag{3–6}$$

$$P = VI \qquad \text{Power equals voltage times current.} \tag{3–7}$$

$$P = \frac{V^2}{R} \qquad \text{Power equals voltage squared divided by resistance.} \tag{3–8}$$

SELF-TEST

1. Ohm's law states that
   (a) current equals voltage times resistance
   (b) voltage equals current times resistance
   (c) resistance equals current divided by voltage
   (d) voltage equals current squared times resistance

2. When the voltage across a resistor is doubled, the current will
   (a) triple   (b) halve   (c) double   (d) not change

3. When 10 V are applied across a 20-Ω resistor, the current is
   (a) 10 A   (b) 0.5 A   (c) 200 A   (d) 2 A

4. When 10 mA flow through a 1-kΩ resistor, the voltage across the resistor is
   (a) 100 V   (b) 0.1 V   (c) 10 kV   (d) 10 V

5. If 20 V are applied across a resistor and there are 6.06 mA of current, the resistance is
   (a) 3.3 kΩ (b) 33 kΩ (c) 330 Ω (d) 3.03 kΩ

6. A current of 250 μA through a 4.7-kΩ resistor produces a voltage drop of
   (a) 53.2 V   (b) 1.175 mV   (c) 18.8 V   (d) 1.175 V

7. A resistance of 2.2 MΩ is connected across a 1-kV source. The resulting current is approximately
   (a) 2.2 mA   (b) 0.455 mA   (c) 45.5 μA   (d) 0.455 A

8. Power can be defined as
   (a) energy   (b) heat   (c) the rate at which energy is used
   (d) the time required to use energy

9. For 10 V and 50 mA, the power is
   (a) 500 mW   (b) 0.5 W   (c) 500,000 μW   (d) (a), (b), and (c)

10. When the current through a 10-kΩ resistor is 10 mA, the power is
    (a) 1 W   (b) 10 W   (c) 100 mW   (d) 1000 μW

11. A 2.2-kΩ resistor dissipates 0.5 W. The current is
    (a) 15.1 mA   (b) 0.227 mA   (c) 1.1 mA   (d) 4.4 mA

12. A 330-Ω resistor dissipates 2 W. The voltage is
    (a) 2.57 V   (b) 660 V   (c) 6.6 V   (d) 25.7 V

13. The power rating of a carbon-composition resistor that is to handle up to 1.1 W should be
    (a) 0.25 W   (b) 1 W   (c) 2 W   (d) 5 W

14. A 22-Ω half-watt resistor and a 220-Ω half-watt resistor are connected across a 10-V source. Which one(s) will overheat?
    (a) 22 Ω (b) 220 Ω (c) both   (d) neither

15. When the needle of an analog ohmmeter indicates infinity, the resistor being measured is
    (a) overheated   (b) shorted   (c) open   (d) reversed

PROBLEM SET A

Section 3–1   Ohm's Law

1. The current in a circuit is 1 A. Determine what the current will be when
   (a) the voltage is tripled
   (b) the voltage is reduced by 80%
   (c) the voltage is increased by 50%

2. The current in a circuit is 100 mA. Determine what the current will be when
   (a) the resistance is increased by 100%
   (b) the resistance is reduced by 30%
   (c) the resistance is quadrupled

3. The current in a circuit is 10 mA. What will the current be if the voltage is tripled and the resistance is doubled?

Section 3–2   Application of Ohm's Law

4. Determine the current in each case.
   (a) $V = 5$ V, $R = 1$ Ω (b) $V = 15$ V, $R = 10$ Ω
   (c) $V = 50$ V, $R = 100$ Ω (d) $V = 30$ V, $R = 15$ kΩ
   (e) $V = 250$ V, $R = 5$ MΩ

5. Determine the current in each case.
   (a) $V = 9$ V, $R = 2.7$ kΩ (b) $V = 5.5$ V, $R = 10$ kΩ
   (c) $V = 40$ V, $R = 68$ kΩ (d) $V = 1$ kV, $R = 2$ kΩ
   (e) $V = 66$ kV, $R = 10$ MΩ

6. A 10-Ω resistor is connected across a 12-V battery. How much current flows through the resistor?

7. A resistor is connected across the terminals of a dc voltage source in each part of Figure 3–26. Determine the current in each resistor. (Full-color photos appear following page 240.)

      (a)   (b)   (c)

FIGURE 3–26

8. Calculate the voltage for each value of I and R.
   (a) $I = 2$ A, $R = 18$ Ω   (b) $I = 5$ A, $R = 50$ Ω
   (c) $I = 2.5$ A, $R = 600$ Ω   (d) $I = 0.6$ A, $R = 47$ Ω
   (e) $I = 0.1$ A, $R = 500$ Ω

9. Calculate the voltage for each value of I and R.
   (a) $I = 1$ mA, $R = 10$ Ω   (b) $I = 50$ mA, $R = 33$ Ω
   (c) $I = 3$ A, $R = 5$ kΩ   (d) $I = 1.6$ mA, $R = 2.2$ kΩ
   (e) $I = 250$ μA, $R = 1$ kΩ   (f) $I = 500$ mA, $R = 1.5$ MΩ
   (g) $I = 850$ μA, $R = 10$ MΩ   (h) $I = 75$ μA, $R = 50$ Ω

10. Three amperes of current are measured through a 27-Ω resistor connected across a voltage source. How much voltage does the source produce?

11. Assign a voltage value to each source in the circuits of Figure 3–27 to obtain the indicated amounts of current.

27 kΩ

V

3 mA

(a)

100 MΩ

V

5 μA

(b)

50 Ω

V

2.5 A

(c)

FIGURE 3–27

12. Calculate the resistance for each value of V and I.
    (a) $V = 10$ V, $I = 2$ A   (b) $V = 90$ V, $I = 45$ A
    (c) $V = 50$ V, $I = 5$ A   (d) $V = 5.5$ V, $I = 10$ A
    (e) $V = 150$ V, $I = 0.5$ A

13. Calculate R for each set of V and I values.
    (a) $V = 10$ kV, $I = 5$ A   (b) $V = 7$ V, $I = 2$ mA
    (c) $V = 500$ V, $I = 250$ mA   (d) $V = 50$ V, $I = 500$ μA
    (e) $V = 1$ kV, $I = 1$ mA

14. Six volts are applied across a resistor. A current of 2 mA is measured. What is the value of the resistor?

15. Choose the correct value of resistance to get the current values indicated in each circuit of Figure 3–28.

R

8 V

2 A

(a)

R

12 V

4 mA

(b)

R

30 V

150 μA

(c)

FIGURE 3–28

Section 3–3   Power and Energy

16. What is the power when energy is consumed at the rate of 350 J/s?

17. How many watts are used when 7500 J of energy are consumed in 5 h?

18. Convert the following to kilowatts:
    (a) 1000 W   (b) 3750 W   (c) 160 W   (d) 50,000 W

19. Convert the following to megawatts:
    (a) 1,000,000 W   (b) 3×10^6 W   (c) 15×10^7 W
    (d) 8700 kW

20. Convert the following to milliwatts:
    (a) 1 W   (b) 0.4 W   (c) 0.002 W   (d) 0.0125 W

21. Convert the following to microwatts:
    (a) 2 W   (b) 0.0005 W   (c) 0.25 mW   (d) 0.00667 mW

22. Convert the following to watts:
    (a) 1.5 kW   (b) 0.5 MW   (c) 350 mW   (d) 9000 μW

Section 3–4   Power in an Electric Circuit

23. If a resistor has 5.5 V across it and 3 mA flowing through it, what is the power?

24. An electric heater works on 115 V and draws 3 A of current. How much power does it use?

25. How much power is produced by 500 mA of current through a 4.7-kΩ resistor?

26. Calculate the power handled by a 10-kΩ resistor carrying 100 μA.

27. If there are 60 V across a 600-Ω resistor, what is the power?

28. A 50-Ω resistor is connected across the terminals of a 1.5-V battery. What is the power dissipation in the resistor?

29. If a resistor is to carry 2 A of current and handle 100 W of power, how many ohms must it be? Assume that the voltage can be adjusted to any required value.

30. Convert 5×10^6 Wmin to kWh.

31. Convert 6700 Ws to kWh.

32. How many watthours does 50 W used for 12 h equal? How many kilowatthours?

Section 3–5   The Power Rating of Resistors

33. A 6.8-kΩ resistor has burned out in a circuit. You must replace it with another resistor with the same resistance value. If the resistor carries 10 mA, what should its power rating be? Assume that you have available carbon-composition resistors in all the standard power ratings.

34. A certain type of power resistor comes in the following ratings: 3 W, 5 W, 8 W, 12 W, 20 W. Your particular application requires a resistor that can handle approximately 8 W. Which rating would you use? Why?

PROBLEM SET B

35. The filament of a light bulb in the circuit of Figure 3–29(a) has a certain amount of resistance, represented by an equivalent resistance in Figure 3–29(b). If the bulb operates with 120 V and 0.8 A of current, what is the resistance of its filament?

36. A certain electrical device has an unknown resistance. You have available a 12-V battery and an ammeter. How would you determine the value of the unknown resistance? Draw the necessary circuit connections.

FIGURE 3-29

37. A variable voltage source is connected to the circuit of Figure 3–30. Start at 0 V and increase the voltage in 10-V steps up to 100 V. Determine the current at each voltage value, and plot a graph of V versus I. Is the graph a straight line? What does the graph indicate?

FIGURE 3-30

38. In a certain circuit, $V = 1$ V and $I = 5$ mA. Determine the current for each of the following voltages in the same circuit:
   (a) $V = 1.5$ V   **(b)** $V = 2$ V   **(c)** $V = 3$ V
   (d) $V = 4$ V   **(e)** $V = 10$ V

39. Figure 3–31 is a graph of voltage versus current for three resistance values. Determine R_1, R_2, and R_3.

FIGURE 3-31

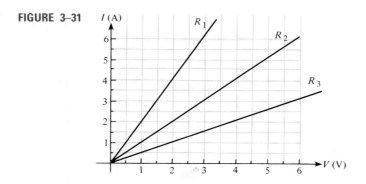

40. You are measuring the current in a circuit that is operated on a 10-V battery. The ammeter reads 50 mA. Later, you notice that the current has dropped to 30 mA. Eliminating the possibility of a resistance change, you must conclude that the voltage has changed. How much has the voltage of the battery changed, and what is its new value?

41. If you wish to increase the amount of current in a resistor from 100 mA to 150 mA by changing the 20-V source, by how many volts should you change the source? To what new value should you set it?

42. By varying the rheostat (variable resistor) in the circuit of Figure 3–32, you can change the amount of current. The setting of the rheostat is such that the current is 750 mA. What is the ohmic value of this setting? To adjust the current to 1 A, to what resistance value must you set the rheostat?

FIGURE 3–32

43. A certain resistor has the following color code: orange, orange, red, gold. Determine the maximum and minimum currents you should expect to measure when a 12-V source is connected across the resistor.

44. A 6-V source is connected to a 100-Ω resistor by two 12-ft lengths of 18-gage copper wire. Refer to Table 2–3 to determine the following:
    (a) current   **(b)** resistor voltage
    (c) voltage across each length of wire

45. A certain appliance uses 300 W. If it is allowed to run continuously for 30 days, how many kilowatthours of energy does it consume?

46. At the end of a 31-day period, your utility bill shows that you have used 1500 kWh. What is your average daily power?

47. A certain type of power resistor comes in the following ratings: 3 W, 5 W, 8 W, 12 W, 20 W. Your particular application requires a resistor that can handle approximately 10 W. Which rating would you use? Why?

48. A 12-V source is connected across a 10-Ω resistor for 2 min.
    (a) What is the power?
    (b) How much energy is used?
    (c) If the resistor remains connected for an additional minute, does the power increase or decrease?

49. The rheostat in Figure 3–33 is used to control the current to a heating element. When the rheostat is adjusted to a value of 8 Ω or less, the heating element can burn out. What is the rated value of the fuse needed to protect the circuit if the voltage across the heating element at the point of maximum current is 100 V?

FIGURE 3–33

ANSWERS TO
SECTION REVIEWS

Section 3–1

1. Current varies directly with voltage and inversely with resistance.
2. Increases, three times.   3. 5 mA.   4. No.

Section 3–2

1. 2 A.   2. 4 mA.   3. 5 A.
4. 10 V.   5. 9 V.   6. 12 V.
7. 5 Ω.   8. 0.5 kΩ, 500 Ω.

Section 3–3

1. The rate at which energy is used.   2. $P = W/t$.
3. The unit of power. One watt is the power when 1 J of energy is used in 1 s.
4. (a) 68 kW;   (b) 5 mW;   (c) 25 μW.
5. 1 kWh.   6. 2 kWh.   7. 0.1 kWh.

Section 3–4

1. 30 W.
2. 1.2 W, 1.2 W. Yes, because all energy produced by the source is dissipated by the resistance.
3. 1250 W.   4. 2 W.   5. 2.5 W.
6. 32 mW.   7. 300 Ω.

Section 3–5

1. Ohmic value, power rating.   2. A larger surface area dissipates more energy.
3. 0.125 W, 0.25 W, 0.5 W, 1 W, 2 W.   4. At least 0.5 W.
5. To produce current through the resistance being measured.
6. 8 kΩ.

4

Series Circuits

Resistive circuits can be of two basic forms: *series* or *parallel*. In this chapter we discuss series circuits. Parallel circuits are covered in Chapter 5, and combinations of series and parallel circuits are examined in Chapter 6. In this chapter you will see how Ohm's law is used in series circuits, and you will study another very important law, called *Kirchhoff's voltage law*. Also, several important applications of series circuits are given.

After completing this chapter, you will be able to

☐ Identify a series circuit.
☐ Determine the current in a series circuit.
☐ Determine the total resistance of a series circuit.
☐ Apply Ohm's law to find the current, voltages, and resistances in a series circuit.
☐ Connect voltage sources to achieve a higher voltage.
☐ Apply Kirchhoff's voltage law.
☐ Devise and use voltage dividers.
☐ Determine the total power in a series circuit.
☐ Measure voltages with respect to ground.
☐ Identify and troubleshoot common faults in series circuits.

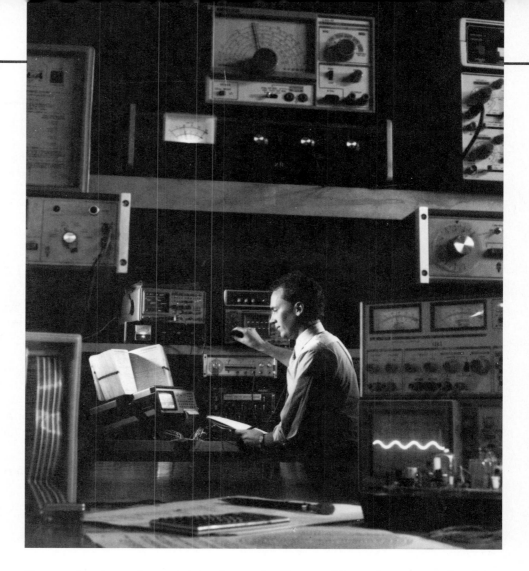

APPLICATION ASSIGNMENT

You are the electronics technician for a scientific expedition to investigate volcanic activity. The scientists will use several specialized electronic sensing devices to make measurements on or near the volcanoes and to send the information to a central processing unit.

You are assigned to develop a circuit that will be part of the central unit. The circuit will reduce the signals transmitted from the remote sensing devices and will receive only one signal at a time. The required circuit will be a voltage divider with a *single* switch-selectable output. The output voltage is to be either ½, ¼, ⅛, or 1/16 of the input amplitude, depending on the setting of the switch. The total resistance of the voltage divider is to be 100 kΩ.

After completing this chapter, you will have the basic knowledge to complete this assignment.

4–1

RESISTORS IN SERIES

Resistors in series are connected end-to-end, or in a "string," as shown in Figure 4–1. Part (a) of the figure shows two resistors connected in series between point A and point B. Part (b) shows three in series, and part (c) shows four in series. Of course, there can be any number of resistors in a series circuit.

FIGURE 4–1
Resistors in series.

R_1 R_2
A o———/\/\/———/\/\/———o B

(a)

R_1 R_2 R_3
A o———/\/\/———/\/\/———/\/\/———o B

(b)

R_1 R_2 R_3 R_4
A o———/\/\/—/\/\/—/\/\/—/\/\/———o B

(c)

The only way for electrons to get from point A to point B in any of the connections of Figure 4–1 is to go through *each* of the resistors. The following is an important way to identify a series circuit:

> **A series circuit provides only one path for current between two points in a circuit so that the same current flows through each series resistor.**

Identifying Series Circuits

In an actual circuit diagram, a series circuit may not always be as easy to identify as those in Figure 4–1. For example, Figure 4–2 shows series resistors drawn in other ways. Remember, *if there is only one current path between two points, the resistors between those two points are in series,* no matter how they appear in a diagram.

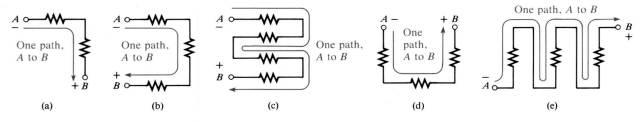

(a)   (b)   (c)   (d)   (e)

FIGURE 4–2
Some examples of series circuits. Notice that the current must be the same at all points.

EXAMPLE 4–1

Five resistors are positioned on a circuit board as shown in Figure 4–3. Wire them together in series so that, starting from the negative ($-$) terminal, R_1 is first, R_2 is second, R_3 is third, and so on. Draw a schematic diagram showing this connection.

FIGURE 4-3

Solution

The wires are connected as shown in Figure 4–4(a), which is the *assembly diagram*. The *schematic diagram* is shown in Figure 4–4(b). Note that the schematic diagram does not necessarily show the actual physical arrangement of the resistors as does the assembly diagram. The purpose of the *schematic* is to show how components are connected *electrically*. The purpose of the *assembly* diagram is to show how components are arranged *physically*.

FIGURE 4-4

(a) Assembly

(b) Schematic

EXAMPLE 4-2

Describe how the resistors on the printed circuit (PC) board in Figure 4–5 are related electrically.

FIGURE 4-5

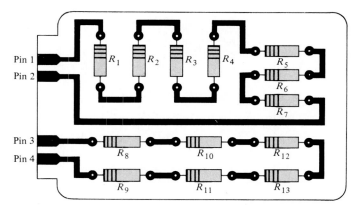

Solution

Resistors R_1 through R_7 are in series with each other. This series combination is connected between pins 1 and 2 on the PC board.

Resistors R_8 through R_{13} are in series with each other. This series combination is connected between pins 3 and 4 on the PC board.

1. How can you identify a series circuit?
2. Complete the schematic diagrams for the circuits in each part of Figure 4–6 by connecting the resistors in series in numerical order from *A* to *B*.
3. Now connect each *group* of series resistors in Figure 4–6 in series.

FIGURE 4–6

(a)   (b)   (c)

4–2

CURRENT IN A SERIES CIRCUIT

Equal Current through All Points

Figure 4–7 shows three resistors connected in series to a voltage source. *At any point in this circuit, the current into that point must equal the current out of that point*, as illustrated by the directional arrows. Notice that the current out of each of the resistors must equal the current in, because it is essentially forced through the resistor and out the other end. Since there are no paths other than the single series path into which the current can branch off, the current in each section of the circuit must be the same as the current in all other sections. *Although resistance reduces the current, the reduction is felt at all points equally.*

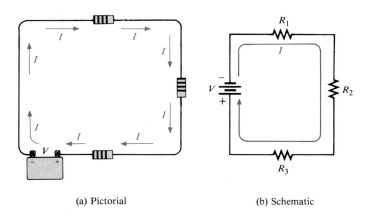

(a) Pictorial   (b) Schematic

FIGURE 4–7
Current entering any point in a series circuit is the same as the current leaving that point.

In Figure 4–8, the battery supplies 1 A of current to the series resistors. One ampere is flowing out of the negative terminal of the source. When ammeters are connected in series at several points as shown, *each* meter reads 1 A.

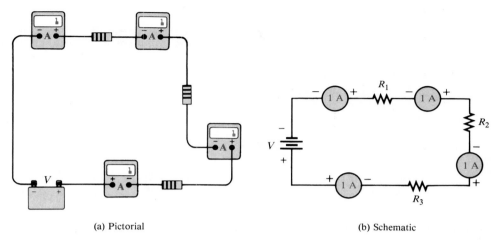

<table>
<tr><td>(a) Pictorial</td><td>(b) Schematic</td></tr>
</table>

FIGURE 4–8
Current is the same at all points in a series circuit.

**SECTION REVIEW
4–2**

1. In a series circuit with a 10-Ω and a 5-Ω resistor in series, 2 A flow through the 10-Ω resistor. How much current flows through the 5-Ω resistor?

2. A milliammeter is connected between points A and B in Figure 4–9. It measures 50 mA. If you move the meter and connect it between points C and D, how much current will it indicate? Between E and F?

FIGURE 4–9

3. In Figure 4–10, how much current does ammeter 1 indicate? Ammeter 2?

FIGURE 4–10

4–3

TOTAL SERIES RESISTANCE

The total resistance of a series circuit is equal to the sum of the resistances of each individual resistor.

This fact is understandable because each of the resistors in series offers opposition to the current in direct proportion to its ohmic value. A greater number of resistors connected in series creates *more opposition* to current. More opposition to current implies a higher value of resistance. So every time a resistor is added in series, the total resistance increases.

How Series Resistor Values Add

Figure 4–11 illustrates how series resistances add to *increase* the total resistance (R_T). Part (a) of the figure has a single 10-Ω resistor. Part (b) shows another 10-Ω resistor connected in series with the first one, making a total resistance of 20 Ω. If a third 10-Ω resistor is connected in series with the first two, as shown in part (c), the total resistance becomes 30 Ω.

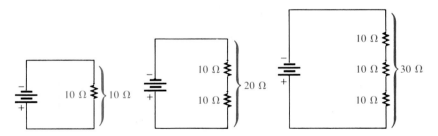

FIGURE 4–11
Total resistance increases with each additional series resistor.

Series Resistance Formula

For *any number* of individual resistors connected in series, the total resistance is the sum of each of the individual values:

$$R_T = R_1 + R_2 + R_3 + \cdots + R_n \qquad (4\text{–}1)$$

where R_T is the total resistance and R_n is the last resistor in the series string (n can be any positive integer equal to the number of resistors in series). For example, if there are four resistors in series ($n = 4$), the total resistance formula is

$$R_T = R_1 + R_2 + R_3 + R_4$$

If there are six resistors in series ($n = 6$), the total resistance formula is

$$R_T = R_1 + R_2 + R_3 + R_4 + R_5 + R_6$$

To illustrate the calculation of total series resistance, let us determine R_T of the circuit of Figure 4–12, where V_S is the source voltage. This circuit has five resistors in series. To find the total resistance, we simply add the values as follows:

$$R_T = 50\ \Omega + 100\ \Omega + 25\ \Omega + 10\ \Omega + 5\ \Omega = 190\ \Omega$$

This equation illustrates an important point: In Figure 4–12, the order in which the resistances are added does not matter; we still get the same total. Also, we can physically change the positions of the resistors in the circuit without affecting the total resistance.

FIGURE 4–12
Example of five resistors in series.

EXAMPLE 4–3

Connect the resistors in Figure 4–13 (a full-color photo appears following page 240) in series, and determine the total resistance, R_T.

FIGURE 4–13

(a) Pictorial

FIGURE 4–14

(b) Schematic

Solution
The resistors are connected as shown in Figure 4–14.
We find the total resistance by adding all the values, as follows:

$$R_T = 33\ \Omega + 68\ \Omega + 100\ \Omega + 47\ \Omega + 10\ \Omega = 258\ \Omega$$

EXAMPLE 4–4

Calculate R_T for each circuit in Figure 4–15.

Solution
(a) Add the resistor values:

$$R_T = 1\ \text{k}\Omega + 2.2\ \text{k}\Omega + 4.7\ \text{k}\Omega + 10\ \text{k}\Omega = 17.9\ \text{k}\Omega$$

FIGURE 4–15

(a)   (b)

(b)   Add the resistor values:

$$R_T = 4.7 \text{ k}\Omega + 10 \text{ k}\Omega + 1 \text{ k}\Omega + 2.2 \text{ k}\Omega = 17.9 \text{ k}\Omega$$

Notice that the total resistance does not depend on the position of the resistors. Both circuits are identical in terms of total resistance.

The calculator sequence for part (a) is

[1] [EXP] [3] [+] [2] [.] [2] [EXP] [3] [+]
[4] [.] [7] [EXP] [3] [+] [1] [0] [EXP] [3] [=]

EXAMPLE 4–5   Determine R_4 in the circuit of Figure 4–16.

FIGURE 4–16

Solution
From the ohmmeter reading, $R_T = 17.9 \text{ k}\Omega$.

$$R_T = R_1 + R_2 + R_3 + R_4$$

Solving for R_4 yields

$$R_4 = R_T - (R_1 + R_2 + R_3)$$
$$= 17.9 \text{ k}\Omega - (1 \text{ k}\Omega + 2.2 \text{ k}\Omega + 4.7 \text{ k}\Omega)$$
$$= 10 \text{ k}\Omega$$

Equal-Value Series Resistors

When a circuit has more than one resistor of the *same* value in series, there is a shortcut method to obtain the total resistance: Simply multiply the *ohmic value* of the resistors having the same value by the *number* of equal-value resistors that are in series. This method is essentially the same as adding the values. For example, five 100-Ω resistors in series have an R_T of 5(100 Ω) = 500 Ω. In general, the formula is expressed as

$$R_T = nR \tag{4–2}$$

where n is the number of equal-value resistors and R is the value.

EXAMPLE 4–6   Find the R_T of eight 22-Ω resistors in series.

Solution
We find R_T by adding the values as follows:

$$R_T = 22 \ \Omega + 22 \ \Omega + 22 \ \Omega + 22 \ \Omega + 22 \ \Omega + 22 \ \Omega + 22 \ \Omega + 22 \ \Omega$$
$$= 176 \ \Omega$$

However, it is much easier to multiply:

$$R_T = 8(22 \ \Omega) = 176 \ \Omega$$

**SECTION REVIEW
4–3**

COLOR
INSERT

1. Calculate R_T between points A and B for each circuit in Figure 4–17. (A full-color photo appears following page 240.)
2. The following resistors are in series: one 100 Ω, two 47 Ω, four 12 Ω, and one 330 Ω. What is the total resistance?

(a)   (b)   (c)

FIGURE 4–17

3. Suppose that you have one resistor each of the following values: 1 kΩ, 2.7 kΩ, 3.3 kΩ, and 1.8 kΩ. To get a total resistance of 10 kΩ, you need one more resistor. What should its value be?

4. What is the R_T for twelve 47-Ω resistors in series?

4–4

OHM'S LAW IN SERIES CIRCUITS

In this section, we use Ohm's law and the basic concepts of series circuits to analyze several circuits.

EXAMPLE 4–7

As you have seen, some resistors are not color coded with bands but have the values stamped on the resistor body. When the circuit board shown in Figure 4–18 was assembled, someone mounted the resistors with the labels turned down, and there is no documentation showing the resistor values (inexcusable!). Assume that there are a voltmeter, ammeter, and power supply immediately available, but no ohmmeter. Without removing the resistors from the board, use Ohm's law to determine the resistance of each one.

FIGURE 4–18

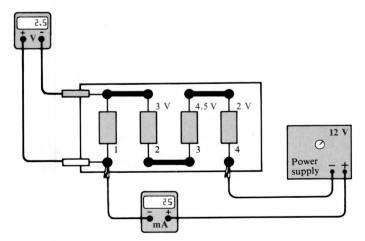

FIGURE 4–19
The voltmeter readings for each resistor are indicated.

Solution

The resistors are all in series, so the current is the same through each one. Measure the current by connecting a 12-V source (arbitrary value) and an ammeter as shown in Figure 4–19. Measure the voltage across each resistor by placing the voltmeter across the first resistor. Then repeat this measurement for the other three resistors. For illustration, the voltage values indicated in color are assumed to be the measured values.

Determine the resistance of each resistor by substituting the measured values of current and voltage into the Ohm's law formula as follows:

$$R_1 = \frac{V_1}{I} = \frac{2.5 \text{ V}}{25 \text{ mA}} = 100 \ \Omega$$

$$R_2 = \frac{V_2}{I} = \frac{3 \text{ V}}{25 \text{ mA}} = 120 \ \Omega$$

$$R_3 = \frac{V_3}{I} = \frac{4.5 \text{ V}}{25 \text{ mA}} = 180 \ \Omega$$

$$R_4 = \frac{V_4}{I} = \frac{2 \text{ V}}{25 \text{ mA}} = 80 \ \Omega$$

The calculator sequence for R_1 is

[2] [.] [5] [÷] [2] [5] [EXP] [+/−] [3] [=]

EXAMPLE 4–8

Find the current in the circuit of Figure 4–20.

FIGURE 4–20

Solution

The current is determined by the voltage and the *total resistance*. First, calculate the total resistance as follows:

$$R_T = R_1 + R_2 + R_3 + R_4$$
$$= 82 \ \Omega + 18 \ \Omega + 15 \ \Omega + 10 \ \Omega = 125 \ \Omega$$

Next, using Ohm's law, calculate the current as follows:

$$I = \frac{V_S}{R_T} = \frac{25 \text{ V}}{125 \ \Omega} = 0.2 \text{ A}$$

Remember, the *same* current flows at all points in the circuit. Thus, *each* resistor has 0.2 A through it.

EXAMPLE 4–9

In the circuit of Figure 4–21, 1 mA of current flows. For this amount of current, what must the source voltage V_S be?

FIGURE 4–21

Solution

In order to calculate V_S, we must determine R_T as follows:

$$R_T = 1.2 \text{ k}\Omega + 5.6 \text{ k}\Omega + 1.2 \text{ k}\Omega + 1.5 \text{ k}\Omega = 9.5 \text{ k}\Omega$$

Now use Ohm's law to get V_S:

$$V_S = IR_T = (1 \text{ mA})(9.5 \text{ k}\Omega) = 9.5 \text{ V}$$

EXAMPLE 4–10

Calculate the voltage across each resistor in Figure 4–22, and find the value of V_S.

FIGURE 4–22

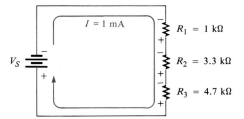

Solution

By Ohm's law, the voltage across each resistor is equal to its resistance multiplied by the current through it. Using the Ohm's law formula $V = IR$, we determine the voltage across each of the resistors. Keep in mind that the same current flows through each series resistor.

The voltage across R_1 is

$$V_1 = IR_1 = (1 \text{ mA})(1 \text{ k}\Omega) = 1 \text{ V}$$

The voltage across R_2 is

$$V_2 = IR_2 = (1 \text{ mA})(3.3 \text{ k}\Omega) = 3.3 \text{ V}$$

The voltage across R_3 is

$$V_3 = IR_3 = (1 \text{ mA})(4.7 \text{ k}\Omega) = 4.7 \text{ V}$$

The source voltage V_S is equal to the current times the *total resistance:*

$$R_T = 1 \text{ k}\Omega + 3.3 \text{ k}\Omega + 4.7 \text{ k}\Omega = 9 \text{ k}\Omega$$
$$V_S = (1 \text{ mA})(9 \text{ k}\Omega) = 9 \text{ V}$$

Notice that if you add the voltage drops of the resistors, they total 9 V, which is the same as the source voltage.

The calculator sequence for V_S is

**SECTION REVIEW
4–4**

1. A 10-V battery is connected across three 100-Ω resistors in series. What is the current through each resistor?

2. How much voltage is required to produce 5 A through the circuit of Figure 4–23?

FIGURE 4–23

R_1
$10 \ \Omega$

V_S

$R_2 \gtrless 15 \ \Omega$

$18 \ \Omega$
R_3

3. How much voltage is dropped across each resistor in Figure 4–23 with 5 A?

4. Four equal-value resistors are connected in series with a 5-V source. Five milliamperes of current are measured. What is the value of each resistor?

4–5

VOLTAGE SOURCES IN SERIES

When cells are placed in a flashlight, they are connected series-aiding to produce a larger voltage, as illustrated in Figure 4–24. In this example, three 1.5-V cells are placed in series to produce a total voltage (V_T) as follows:

$$V_T = V_{S1} + V_{S2} + V_{S3} = 1.5 \text{ V} + 1.5 \text{ V} + 1.5 \text{ V} = 4.5 \text{ V}$$

(a) Flashlight with series batteries

4.5 V

1.5 V   1.5 V   1.5 V

V_{S1} V_{S2} V_{S3}

(b) Schematic of flashlight circuit

FIGURE 4–24
Example of series-aiding voltage sources.

Series voltage sources are *added* when their polarities are in the same direction and are *subtracted* when their polarities are in opposite directions, or series-opposing. For example, if one of the batteries in the flashlight is turned around, as indicated in the schematic of Figure 4–25, its voltage subtracts, because it has a negative value, and reduces the total voltage:

$$V_T = V_{S1} - V_{S2} + V_{S3} = 1.5 \text{ V} - 1.5 \text{ V} + 1.5 \text{ V} = 1.5 \text{ V}$$

FIGURE 4–25
Opposite polarities subtract.

1.5 V

1.5 V   1.5 V   1.5 V

V_{S1} V_{S2} V_{S3}

EXAMPLE 4–11   What is the total source voltage (V_T) in Figure 4–26?

FIGURE 4–26

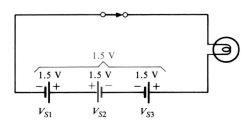

10 V V_{S1}

5 V V_{S2}

3 V V_{S3}

R

Solution

The polarity of each source is the same—that is, they are series-aiding (the sources are connected in the same direction in the circuit). So we sum the three voltages to get the total:

$$V_T = V_{S1} + V_{S2} + V_{S3} = 10 \text{ V} + 5 \text{ V} + 3 \text{ V} = 18 \text{ V}$$

The three individual sources can be replaced by a single equivalent source of 18 V with its polarity as shown in Figure 4–27.

FIGURE 4–27

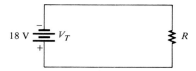

EXAMPLE 4–12   Determine V_T in Figure 4–28.

FIGURE 4–28

Solution

These sources are connected in *opposite* directions, or are series-opposing. If you go clockwise around the circuit, you go from plus to minus through V_{S1}, and minus to plus through V_{S2}. The total voltage is the *difference* of the two source voltages. The total voltage has the same polarity as the larger-value source. Here we will choose V_{S2} to be positive:

$$V_T = V_{S2} - V_{S1} = 25 \text{ V} - 15 \text{ V} = 10 \text{ V}$$

The two sources in Figure 4–28 can be replaced by a 10-V equivalent one with polarity, as shown in Figure 4–29.

FIGURE 4–29

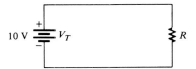

SECTION REVIEW 4–5

1. How many 12-V batteries must be connected in series to produce 60 V? Sketch a schematic that shows the battery connections.

2. The resistive circuit in Figure 4–30 is used to bias a transistor amplifier. Show how to connect the resistors to two 15-V power supplies in order to get 30 V across the series circuit.

FIGURE 4–30

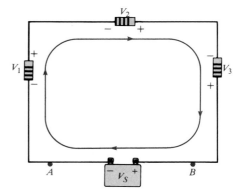

(a)

(b)

FIGURE 4–31

3. Determine the total source voltage in each circuit of Figure 4–31.
4. Sketch the *equivalent single source* circuit for each circuit of Figure 4–31.

4–6   KIRCHHOFF'S VOLTAGE LAW

In an electric circuit, the voltages across the resistors (called *voltage drops*) *always* have polarities *opposite* to the source voltage polarity. For example, in Figure 4–32, follow a clockwise loop around the circuit and note that the source polarity is *plus-to-minus* and each voltage drop is *minus-to-plus*.

FIGURE 4–32
Kirchhoff's voltage law: The sum of the voltage drops equals the source voltage.

Also notice in Figure 4–32 that the *current flows out of the negative side of the source* and through the resistors as indicated by the arrows. The *current flows into the negative side of each resistor and out the positive side*. The reason is that as the electrons flow through a resistor, they lose energy and are therefore at a lower energy level when they emerge. The lower energy side is less negative (more positive) than the higher energy side. The drop in energy level across a resistor creates a potential difference, or *voltage drop,* with a minus-to-plus polarity in the direction of the current.

Notice that the voltage from point A to point B in the circuit of Figure 4–32 equals the source voltage, V_S. Also, the voltage from A to B is the *sum* of the series resistor voltage drops. Therefore, the source voltage is equal to the sum of the three voltage drops.

FIGURE 4–33
Sum of *n* voltage drops equals the source voltage.

This discussion is an example of *Kirchhoff's voltage law*, which is generally stated as follows:

The sum of all the voltage drops around a single closed loop in a circuit is equal to the total source voltage in that loop.

The general concept of Kirchhoff's voltage law is illustrated in Figure 4–33 and expressed by Equation (4–3):

$$V_S = V_1 + V_2 + V_3 + \cdots + V_n \tag{4–3}$$

where *n* represents the number of voltage drops.

Another Way to State Kirchhoff's Voltage Law

If all the voltage drops around a closed loop are added and then this total is subtracted from the source voltage, the result is zero. This result occurs because the sum of the voltage drops equals the source voltage.

The algebraic sum of all voltages (both source and drops) around a closed path is zero.

You can verify Kirchhoff's voltage law by connecting a circuit and measuring each resistor voltage and the source voltage as illustrated in Figure 4–34. When the resistor voltages are added together, their sum will equal the source voltage. Any number of resistors can be added.

FIGURE 4–34
Experimental verification of Kirchhoff's voltage law.

EXAMPLE 4–13

Determine the source voltage V_S in Figure 4–35 where the two voltage drops are given.

FIGURE 4–35

Solution

By Kirchhoff's voltage law, the source voltage (applied voltage) must equal the sum of the voltage drops. Adding the voltage drops gives us the value of the source voltage:

$$V_S = 5 \text{ V} + 10 \text{ V} = 15 \text{ V}$$

EXAMPLE 4–14

Determine the unknown voltage drop, V_3, in Figure 4–36.

FIGURE 4–36

Solution

By Kirchhoff's voltage law, the sum of all the voltages around the circuit is zero (the signs of the voltage drops are opposite the sign of the source).

$$V_S - V_1 - V_2 - V_3 = 0$$

The value of each voltage drop except V_3 is known. Substitute these values into the equation as follows:

$$50 \text{ V} - 12 \text{ V} - 25 \text{ V} - V_3 = 0$$

Next combine the known values:

$$13 \text{ V} - V_3 = 0$$

Transpose 13 V to the right side of the equation, and cancel the minus signs:

$$-V_3 = -13 \text{ V}$$
$$V_3 = 13 \text{ V}$$

The voltage drop across R_3 is 13 V, and its polarity is as shown in Figure 4–36.

EXAMPLE 4–15

Find the value of R_4 in Figure 4–37.

Solution

In this problem we must use both Ohm's law *and* Kirchhoff's voltage law. Follow this procedure carefully.

First use Ohm's law to find the voltage drop across each of the *known* resistors:

$$V_1 = IR_1 = (200 \text{ mA})(10 \text{ }\Omega) = 2 \text{ V}$$
$$V_2 = IR_2 = (200 \text{ mA})(50 \text{ }\Omega) = 10 \text{ V}$$
$$V_3 = IR_3 = (200 \text{ mA})(100 \text{ }\Omega) = 20 \text{ V}$$

FIGURE 4-37

$R_2 = 50\ \Omega$ $R_3 = 100\ \Omega$

$I = 200$ mA

$R_1 = 10\ \Omega$ $R_4 = ?$

100 V

Next, use Kirchhoff's voltage law to find V_4, the voltage drop across the *unknown* resistor:

$$V_S - V_1 - V_2 - V_3 - V_4 = 0$$
$$100\ \text{V} - 2\ \text{V} - 10\ \text{V} - 20\ \text{V} - V_4 = 0$$
$$68\ \text{V} - V_4 = 0$$
$$V_4 = 68\ \text{V}$$

Now that we know V_4, we can use Ohm's law to calculate R_4 as follows:

$$R_4 = \frac{V_4}{I} = \frac{68\ \text{V}}{200\ \text{mA}} = 340\ \Omega$$

SECTION REVIEW
4-6

1. State Kirchhoff's voltage law in two ways.
2. A 50-V source is connected to a series resistive circuit. What is the total of the voltage drops in this circuit?
3. Two equal-value resistors are connected in series across a 10-V battery. What is the voltage drop across each resistor?
4. In a series circuit with a 25-V source, there are three resistors. One voltage drop is 5 V, and the other is 10 V. What is the value of the third voltage drop?

4-7

VOLTAGE DIVIDERS

A series circuit acts as a *voltage divider*. In this section you will see what this term means and why voltage dividers are an important application of series circuits.

To illustrate how a series string of resistors acts as a voltage divider, we will examine Figure 4-38(a), where there are two resistors in series. As you already know, there are two voltage drops: one across R_1 and one across R_2. These voltage drops are V_1 and V_2, respectively, as indicated in the diagram.

Since the same current flows through each resistor, the voltage drops are proportional to the values of the resistors. For example, if the value of R_2 is twice that of R_1, then the value of V_2 is twice that of V_1. In other words, the total voltage drop *divides* among the series resistors in amounts directly proportional to the resistance values. The

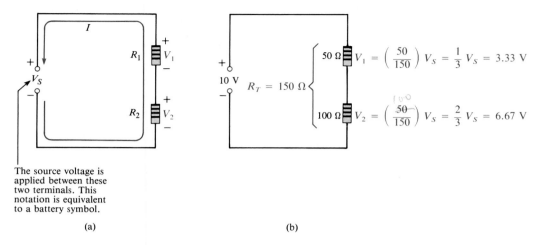

The source voltage is
applied between these
two terminals. This
notation is equivalent
to a battery symbol.

(a)   (b)

FIGURE 4–38
A two-resistor voltage divider.

smallest resistance has the least voltage, and the largest resistance has the most voltage
(V = IR). For example, in Figure 4–38(b), if V_S is 10 V, R_1 is 50 Ω, and R_2 is 100
Ω, then V_1 is one-third the total voltage, or 3.33 V, because R_1 is one-third the *total*
resistance. Likewise, V_2 is two-thirds V_S, or 6.67 V, because R_2 is two-thirds the total
resistance.

Voltage Divider Formula

Although you can use Ohm's law to determine the voltage drop across any resistor in a
series circuit, you must first know the current. It is often easier to use resistance ratios
to find the voltage across a particular resistor as was demonstrated in Figure 4–38(b).
The general voltage divider formula is

$$V_x = \left(\frac{R_x}{R_T}\right) V_S \qquad (4\text{--}4)$$

where V_x is the voltage across any resistor or combination of resistors, R_x is the resistance across which the voltage is to be determined, and R_T is the total series resistance.

FIGURE 4–39
Five-resistor voltage divider.

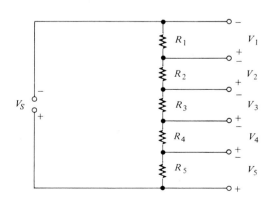

For example, in Figure 4–39, $V_1 = (R_1/R_T)V_S$, $V_2 = (R_2/R_T)V_S$, $V_3 = (R_3/R_T)V_S$, and so on.

The following general statement about voltage dividers applies to series resistive circuits and describes Equation (4–4):

> **The voltage drop across any resistor or combination of resistors in a series circuit is equal to the ratio of that resistance value to the total resistance, multiplied by the source voltage.**

The following examples illustrate how the voltage divider formula is used.

EXAMPLE 4–16

Determine the voltage across R_1 and the voltage across R_2 in the voltage divider in Figure 4–40.

FIGURE 4–40

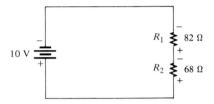

Solution
Use the voltage divider formula, $V_x = (R_x/R_T)V_S$. In this problem we are looking for V_1; so $V_x = V_1$ and $R_x = R_1$. The total resistance is

$$R_T = R_1 + R_2 = 82 \ \Omega + 68 \ \Omega = 150 \ \Omega$$

R_1 is 82 Ω and V_S is 10 V. Substituting these values into the voltage divider formula, we have

$$V_1 = \left(\frac{R_1}{R_T}\right)V_S = \left(\frac{82 \ \Omega}{150 \ \Omega}\right)10 \text{ V} = 5.47 \text{ V}$$

There are two ways to find the value of V_2 in this problem: Kirchhoff's voltage law or the voltage divider formula.

First, using Kirchhoff's voltage law, we know that $V_S = V_1 + V_2$. By substituting the values for V_S and V_1, we can solve for V_2 as follows:

$$V_2 = 10 \text{ V} - 5.47 \text{ V} = 4.53 \text{ V}$$

A second way is to use the voltage divider formula to find V_2 as follows:

$$V_2 = \left(\frac{R_2}{R_T}\right)V_S = \left(\frac{68 \ \Omega}{150 \ \Omega}\right)10 \text{ V} = 4.53 \text{ V}$$

We get the same result either way.
The calculator sequence for V_2 is

⑥ ⑧ ⊞ ① ⑤ ⓪ ⊟ ⊠ ① ⓪ ⊟

EXAMPLE 4–17   Calculate the voltage drop across each resistor in the voltage divider of Figure 4–41.

FIGURE 4–41

Solution
Look at the circuit for a moment and consider the following: The total resistance is 1000 Ω. We can examine the circuit and determine that 10% of the total voltage is across R_1 because it is 10% of the total resistance (100 Ω is 10% of 1000 Ω). Likewise, we see that 22% of the total voltage is dropped across R_2 because it is 22% of the total resistance (220 Ω is 22% of 1000 Ω). Finally, R_3 drops 68% of the total voltage (680 Ω is 68% of 1000 Ω).

Because of the convenient values in this problem, it is easy to figure the voltages mentally. Such is not always the case, but sometimes a little thinking will produce a result more efficiently and eliminate some calculating.

Although we have already reasoned through this problem, the calculations will verify our results:

$$V_1 = \left(\frac{R_1}{R_T}\right)V_S = \left(\frac{100\ \Omega}{1000\ \Omega}\right)100\ \text{V} = 10\ \text{V}$$

$$V_2 = \left(\frac{R_2}{R_T}\right)V_S = \left(\frac{220\ \Omega}{1000\ \Omega}\right)100\ \text{V} = 22\ \text{V}$$

$$V_3 = \left(\frac{R_3}{R_T}\right)V_S = \left(\frac{680\ \Omega}{1000\ \Omega}\right)100\ \text{V} = 68\ \text{V}$$

Notice that the sum of the voltage drops is equal to the source voltage, in accordance with Kirchhoff's voltage law. This check is a good way to verify your results.

The Potentiometer as an Adjustable Voltage Divider

Recall from Chapter 2 that a potentiometer is a variable resistor with three terminals. A potentiometer connected to a voltage source is shown in Figure 4–42. Notice that the two end terminals are labeled 1 and 2. The adjustable terminal or wiper is labeled 3. The potentiometer acts as a voltage divider. We can illustrate this concept better by separating the total resistance into two parts, as shown in Figure 4–42(c). The resistance between terminal 1 and terminal 3 (R_{13}) is one part, and the resistance between terminal 3 and terminal 2 (R_{32}) is the other part. So this potentiometer actually is a two-resistor voltage divider that can be manually adjusted.

Figure 4–43 shows what happens when the wiper terminal (3) is moved. In part (a), the wiper is exactly centered, making the two resistances equal. If we measure the

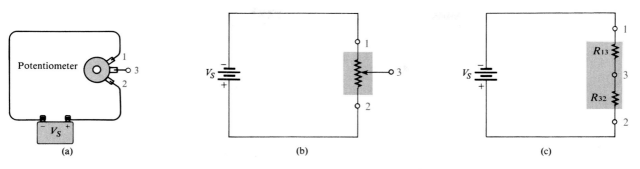

FIGURE 4–42
The potentiometer as a voltage divider.

FIGURE 4–43
Adjusting the voltage divider.

voltage across terminals 3 to 2 as indicated by the voltmeter symbol, we have one-half of the total source voltage. When the wiper is moved up, as in part (b), the resistance between terminals 3 and 2 increases, and the voltage across it increases proportionally. When the wiper is moved down, as in part (c), the resistance between terminals 3 and 2 decreases, and the voltage decreases proportionally.

Applications of Voltage Dividers

The volume control of radio or TV receivers is a common application of a potentiometer used as a voltage divider. Since the loudness of the sound is dependent on the amount of voltage associated with the audio signal, you can increase or decrease the volume by adjusting the potentiometer, that is, by turning the knob of the volume control on the set. The block diagram in Figure 4–44 shows how a potentiometer can be used for volume control in a typical receiver.

FIGURE 4–44
A voltage divider used for volume control.

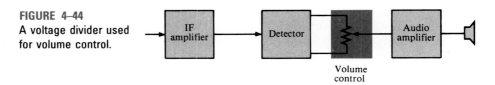

Another application of a voltage divider is illustrated in Figure 4–45, which depicts a potentiometer voltage divider as a fuel-level sensor in an automobile gas tank. As shown in part (a), the float moves up as the tank is filled and moves down as the tank empties. The float is mechanically linked to the wiper arm of a potentiometer, as shown in part (b). The output voltage varies proportionally with the position of the wiper arm. As the fuel in the tank decreases, the sensor output voltage also decreases. The output voltage goes to the indicator circuitry, which controls the fuel gauge or digital readout to show the fuel level. The schematic of this system is shown in part (c).

FIGURE 4–45

A potentiometer voltage divider used as an automotive fuel-level sensor.

Still another application for voltage dividers is in setting the dc operating voltage *(bias)* in transistor amplifiers. Figure 4–46 shows a voltage divider used for this purpose. You will study transistor amplifiers and biasing later, so it is important that you understand the basics of voltage dividers at this point.

FIGURE 4–46
The voltage divider as a bias circuit for a transistor amplifier.

These examples are only three out of an almost endless number of applications of voltage dividers.

SECTION REVIEW 4–7

1. If two series resistors of equal value are connected across a 20-V source, how much voltage is there across each resistor?
2. A 56-Ω resistor and an 82-Ω resistor are connected as a voltage divider. The source voltage is 100 V. Sketch the circuit, and determine the voltage across each of the resistors.
3. The circuit of Figure 4–47 is an adjustable voltage divider. If the potentiometer is linear, where would you set the wiper in order to get 5 V from B to A and 5 V from C to B?

FIGURE 4–47

4–8   POWER IN A SERIES CIRCUIT

The total amount of power in a series resistive circuit is equal to the sum of the powers in each resistor in series:

$$P_T = P_1 + P_2 + P_3 + \cdots + P_n \qquad (4\text{--}5)$$

where n is the number of resistors in series, P_T is the total power, and P_n is the power in the last resistor in series. In other words, the powers are additive.

The power formulas that you learned in Chapter 3 are, of course, directly applicable to series circuits. Since the same current flows through each resistor in series, the following formulas are used to calculate the total power:

$$P_T = V_S I$$
$$P_T = I^2 R_T$$
$$P_T = \frac{V_S^2}{R_T}$$

where V_S is the total source voltage across the series circuit and R_T is the total resistance. Example 4–18 illustrates how to calculate total power in a series circuit.

EXAMPLE 4–18   Determine the total amount of power in the series circuit in Figure 4–48.

FIGURE 4–48

Solution
We know that the source voltage is 15 V. The total resistance is

$$R_T = 10 \ \Omega + 12 \ \Omega + 56 \ \Omega + 22 \ \Omega = 100 \ \Omega$$

The easiest formula to use is $P_T = V_S^2/R_T$, since we know both V_S and R_T:

$$P_T = \frac{V_S^2}{R_T} = \frac{(15 \ \text{V})^2}{100 \ \Omega} = \frac{225 \ \text{V}^2}{100 \ \Omega} = 2.25 \ \text{W}$$

If the power of each resistor is determined separately and all these powers are added, the same result is obtained. We will work through another calculation to illustrate.
    First, find the current as follows:

$$I = \frac{V_S}{R_T} = \frac{15 \ \text{V}}{100 \ \Omega} = 0.15 \ \text{A}$$

Next, calculate the power for each resistor using $P = I^2R$:

$$P_1 = (0.15 \ \text{A})^2(10 \ \Omega) = 0.225 \ \text{W}$$
$$P_2 = (0.15 \ \text{A})^2(12 \ \Omega) = 0.270 \ \text{W}$$
$$P_3 = (0.15 \ \text{A})^2(56 \ \Omega) = 1.260 \ \text{W}$$
$$P_4 = (0.15 \ \text{A})^2(22 \ \Omega) = 0.495 \ \text{W}$$

Now, add these powers to get the total power:

$$P_T = 0.225 \ \text{W} + 0.270 \ \text{W} + 1.260 \ \text{W} + 0.495 \ \text{W}$$
$$= 2.25 \ \text{W}$$

This result shows that the sum of the individual powers is equal to the total power as determined by one of the power formulas.

1. If you know the power in each resistor in a series circuit, how can you find the total power?
2. The resistors in a series circuit have the following powers: 2 W, 5 W, 1 W, and 8 W. What is the total power in the circuit?
3. A circuit has a 100-Ω, a 330-Ω, and a 680-Ω resistor in series. A current of 1 A flows through the circuit. What is the total power?

4–9

CIRCUIT GROUND

Voltage is relative. That is, the voltage at one point in a circuit is always measured relative to another point. For example, if we say that there are +100 V at a certain point in a circuit, we mean that the point is 100 V more positive than some *reference point* in the circuit. This *reference point* in a circuit is usually called *ground*.

The term *ground* derives from the method used in ac power lines, in which one side of the line is neutralized by connecting it to a water pipe or a metal rod driven into the ground. This method of grounding is called *earth ground*.

In most electronic equipment, the metal chassis that houses the assembly or a large conductive area on a printed circuit board is used as the *common* or *reference point*, called the *chassis ground* or *circuit ground*, as illustrated in Figure 4–49. This ground provides a convenient way of connecting all common points within the circuit back to one side of the battery or other energy source. The chassis or circuit ground does not necessarily have to be connected to earth ground. However, in many cases it is earth grounded in order to prevent a shock hazard due to a potential difference between chassis and earth ground.

(a) Physical circuit   (b) Schematic diagram

FIGURE 4–49
Illustration of circuit ground.

In summary, ground is the reference point in electronic circuits. It has a potential of zero volts (0 V) *with respect to all other points in the circuit that are referenced to it,* as illustrated in Figure 4–50. In part (a), the negative side of the source is grounded, and all voltages indicated are positive with respect to ground. In part (b), the positive side of the source is ground. The voltages at all other points are therefore negative with respect to ground.

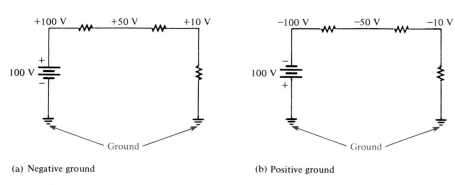

(a) Negative ground   (b) Positive ground

FIGURE 4–50
Examples of negative and positive grounds.

Measurement of Voltages with Respect to Ground

When voltages are measured in a circuit, one meter lead is connected to the circuit ground, and the other to the point at which the voltage is to be measured. In a negative ground circuit, the negative meter terminal is connected to the circuit ground. The positive terminal of the voltmeter is then connected to the positive voltage point. Measurement of positive voltage is illustrated in Figure 4–51, where the meter reads the voltage at point *A* with respect to ground.

FIGURE 4–51
Measuring a voltage with respect to negative ground.

For a circuit with a positive ground, the positive voltmeter lead is connected to ground, and the negative lead is connected to the negative voltage point, as indicated in Figure 4–52. Here the meter reads the voltage at point *A* with respect to ground.

When voltages must be measured at several points in a circuit, the ground lead can be clipped to ground at one point in the circuit and left there. The other lead is then moved from point to point as the voltages are measured. This method is illustrated in Figure 4–53.

FIGURE 4–52
Measuring a voltage with respect to positive ground.

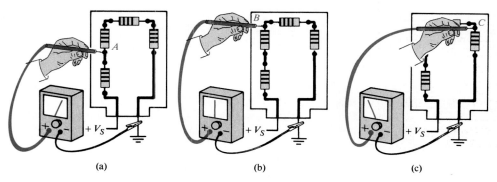

(a)   (b)   (c)

FIGURE 4–53
Measuring voltages at several points in a circuit.

Measurement of Voltage across an Ungrounded Resistor

Voltage can normally be measured across a resistor, as shown in Figure 4–54, even though neither side of the resistor is connected to circuit ground. In some cases, when the meter is not isolated from power line ground, the negative lead of the meter will ground one side of the resistor and alter the operation of the circuit. In this situation,

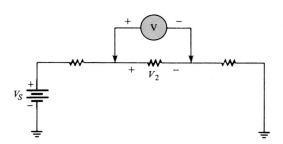

FIGURE 4–54
Measuring voltage across a resistor.

another method must be used, as illustrated in Figure 4–55. The voltages on each side of the resistor are measured *with respect to ground*. The *difference* of these two measurements is the voltage drop across the resistor.

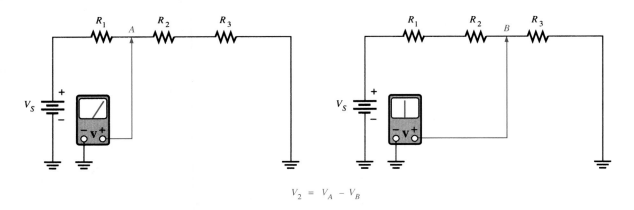

$$V_2 = V_A - V_B$$

FIGURE 4–55
Measuring voltage across a resistor with two separate measurements to ground.

EXAMPLE 4–19

Determine the voltages at each of the indicated points in each circuit of Figure 4–56. Assume that 25 V are dropped across each resistor.

FIGURE 4–56

Solution
In circuit (a), the voltage polarities are as shown. Point E is ground. Single-letter subscripts denote voltage at a point with respect to ground. The voltages with respect to ground are as follows:

$$V_E = 0 \text{ V}$$
$$V_D = +25 \text{ V}$$
$$V_C = +50 \text{ V}$$
$$V_B = +75 \text{ V}$$
$$V_A = +100 \text{ V}$$

In circuit (b), the voltage polarities are as shown. Point D is ground. The voltages with respect to ground are as follows:

$$V_E = -25 \text{ V}$$
$$V_D = 0 \text{ V}$$
$$V_C = +25 \text{ V}$$
$$V_B = +50 \text{ V}$$
$$V_A = +75 \text{ V}$$

In circuit (c), the voltage polarities are as shown. Point C is ground. The voltages with respect to ground are as follows:

$$V_E = -50 \text{ V}$$
$$V_D = -25 \text{ V}$$
$$V_C = 0 \text{ V}$$
$$V_B = +25 \text{ V}$$
$$V_A = +50 \text{ V}$$

SECTION REVIEW
4–9

1. The common point in a circuit is called _____.
2. Most voltages in a circuit are referenced to ground (T or F).
3. The housing or chassis is often used as circuit ground (T or F).
4. What is the symbol for ground?
5. What does *earth ground* mean?

4–10

TROUBLESHOOTING SERIES CIRCUITS

Open Circuit

The most common failure in a series circuit is an *open*. For example, when a resistor or a lamp burns out, it creates a break in the current path, as illustrated in Figure 4–57.

An open in a series circuit prevents current.

How to Check for an Open Element in a Series Circuit

Sometimes a visual check will reveal a charred resistor or an open lamp filament. However, it is possible for a resistor to open without showing visible signs of damage. In this situation, a *voltage check* of the series circuit is required. The general procedure is as follows: Measure the voltage across each resistor in series. *The voltage across all of the good resistors will be zero. The voltage across the open resistor will equal the total voltage across the series combination.*

(a) Complete circuit

Open filament

(b) Open circuit

FIGURE 4–57
An open circuit prevents current.

The above condition occurs because an open resistor will prevent current through the series circuit. With no current, there can be no voltage drop across any of the good resistors. Since $IR = 0$, in accordance with *Ohm's law*, the voltage on each side of a good resistor is the same. The total voltage must then appear across the open resistor in accordance with *Kirchhoff's voltage law*, as illustrated in Figure 4–58. To fix the circuit, replace the open resistor.

(a) Voltage check shows that R_3 is open.

$V_T = V_1 + V_2 + V_3 + V_4$
$V_3 = V_T - V_1 - V_2 - V_4$
$V_3 = 10\text{ V} - 0\text{ V} - 0\text{ V} - 0\text{ V}$
$V_3 = 10\text{ V}$

Kirchhoff's voltage law requires that total voltage appear across an open.

(b)

FIGURE 4–58
Troubleshooting a series circuit for an open element.

Short Circuit

Sometimes a short occurs when two conductors touch or a foreign object such as solder or a wire clipping accidentally connects two sections of a circuit together. This situation is particularly common in circuits with a high component density. Several potential causes of short circuits are illustrated on the PC board in Figure 4–59.

FIGURE 4–59
Examples of shorts on a PC board.

When there is a short, a portion of the series resistance is bypassed (all of the current goes through the short), thus reducing the total resistance as illustrated in Figure 4–60. Notice that the current increases as a result of the short.

A short in a series circuit causes more current.

FIGURE 4–60
The effect of a short in a series circuit.

$$I = \frac{100\ V}{100\ \Omega} = 1\ A$$

(a) Before short

$$I = \frac{100\ V}{28\ \Omega} = 3.57\ A$$

(b) After short

SECTION REVIEW 4–10

1. When a resistor fails, it will normally open (T or F).

2. The total voltage across a string of series resistors is 24 V. If one of the resistors is open, how much voltage is there across it? How much is there across each of the good resistors?

SUMMARY
- ☐ A series circuit provides only one path for current.
- ☐ The same amount of current flows at all points in a series circuit.
- ☐ The total series resistance is the sum of all resistors in the series circuit.
- ☐ Series voltage sources add when their polarities are the same and subtract when their polarities are opposite.
- ☐ The voltage across a resistor is called a *voltage drop*.
- ☐ The polarity of a voltage drop is opposite to that of the source voltage.
- ☐ The voltage across each series resistor equals the ratio of the resistance value to the total resistance, multiplied by the source voltage.
- ☐ The highest resistance in series has the greatest voltage drop, and the lowest resistance has the least voltage drop.
- ☐ Total power dissipation is the sum of powers dissipated by the individual resistors.
- ☐ There is no current in an open circuit.
- ☐ All of the voltage appears across an open series element and none across the other series elements.
- ☐ A short circuit reduces the total resistance.

GLOSSARY

Ground   In electric circuits, the common or reference point.

Kirchhoff's voltage law   a law stating that (1) the sum of the voltage drops around a closed loop equals the source voltage or (2) the sum of all the voltages (drops and sources) around a closed loop is zero.

Open circuit   A circuit in which the current path is broken.

Series   In an electrical circuit, a relationship of components in which the components are connected such that they provide a single current path between two points.

Short circuit   A zero or abnormally low resistance between two points. Usually an inadvertant condition.

FORMULAS

$$R_T = R_1 + R_2 + R_3 + \cdots + R_n \qquad \text{Total resistance of } n \text{ resistors in series} \qquad (4\text{--}1)$$

$$R_T = nR \qquad \text{Total resistance of } n \text{ equal-value resistors in series} \qquad (4\text{--}2)$$

$$V_S = V_1 + V_2 + V_3 + \cdots + V_n \qquad \text{Kirchhoff's voltage law} \qquad (4\text{--}3)$$

$$V_x = \left(\frac{R_x}{R_T}\right) V_S \qquad \text{Voltage divider formula} \qquad (4\text{--}4)$$

$$P_T = P_1 + P_2 + P_3 + \cdots + P_n \qquad \text{Total power} \qquad (4\text{--}5)$$

SELF-TEST

1. Five resistors are connected in series and there is a current of 2 A into the first resistor. The amount of current out of the second resistor is
   (a) 2 A   (b) 1 A   (c) 4 A   (d) 0.4 A

2. To measure the current out of the third resistor in a circuit consisting of four series resistors, an ammeter can be placed
   (a) between the third and fourth resistors   (b) between the second and third resistors
   (c) at the positive terminal of the source   (d) at any point in the circuit

3. When a third resistor is connected in series with two series resistors, the total resistance
   (a) remains the same   (b) increases   (c) decreases
   (d) increases by one-third

4. When one of four series resistors is removed from a circuit and the circuit reconnected, the current
   (a) decreases by the amount of current through the removed resistor
   (b) decreases by one-fourth
   (c) quadruples
   (d) increases

5. A series circuit consists of three resistors with values of 100 Ω, 220 Ω, and 330 Ω. The total resistance is
   (a) less than 100 Ω   (b) the average of the values   (c) 550 Ω   (d) 650 Ω

6. A 9-V battery is connected across a series combination of 68-Ω, 33-Ω, 100-Ω, and 47-Ω resistors. The amount of current is
   (a) 36.3 mA   (b) 27.56 A   (c) 22.32 mA   (d) 326.6 mA

7. While putting four 1.5-V batteries in a flashlight, you accidentally put one of them in backward. The light will be
   (a) brighter than normal   (b) dimmer than normal   (c) off   (d) the same

8. If you measure all the voltage drops and the source voltage in a series circuit and add them together, taking into consideration the polarities, you will get a result equal to
   (a) the source voltage   (b) the total of the voltage drops   (c) zero
   (d) the total of the source voltage and the voltage drops

9. There are six resistors in a given series circuit and each resistor has 5 V dropped across it. The source voltage is
   (a) 5 V   (b) 30 V   (c) dependent on the resistor values
   (d) dependent on the current

10. A series circuit consists of a 4.7-kΩ, a 5.6-kΩ, and a 10-kΩ resistor. The resistor that has the most voltage across it is
    (a) the 4.7-kΩ   (b) the 5.6-kΩ   (c) the 10-kΩ
    (d) impossible to determine from the given information

11. Which of the following series combinations dissipates the most power when connected across a 100-V source?
    (a) one 100-Ω resistor   (b) two 100-Ω resistors   (c) three 100-Ω resistors
    (d) four 100-Ω resistors

12. The total power in a certain circuit is 10 W. Each of the five equal-value series resistors making up the circuit dissipates
    (a) 10 W   (b) 50 W   (c) 5 W   (d) 2 W

13. When you connect an ammeter in a series resistive circuit and turn on the source voltage, the meter reads zero. You should check for
    (a) a broken wire   (b) a shorted resistor   (c) an open resistor   (d) (a) and (c)

14. While checking out a series resistive circuit, you find that the current is higher than it should be. You should look for
    (a) an open circuit   (b) a short   (c) a low resistor value   (d) (b) and (c)

PROBLEM SET A

Section 4–1   Resistors in Series

1. Connect each set of resistors in Figure 4–61 in series between points A and B.

FIGURE 4–61

2. Determine which resistors in Figure 4–62 are in series. Show how to interconnect the pins to put all the resistors in series.

FIGURE 4–62

Section 4–2   Current in a Series Circuit

3. What is the current through each of four resistors in a series circuit if the source voltage is 12 V and the total resistance is 120 Ω?

4. The current from the source in Figure 4–63 is 5 mA. How much current does each milliammeter in the circuit indicate?

FIGURE 4–63

Section 4–3   Total Series Resistance

5. An 82-Ω resistor and a 56-Ω resistor are connected in series. What is the total resistance?

6. Find the total resistance of each group of series resistors shown in Figure 4–64 (a full-color photo appears following page 240).

(a)   (b)   (c)   (d)

FIGURE 4–64

7. Determine R_T for each circuit in Figure 4–65. Show how to measure R_T with an ohmmeter.

FIGURE 4–65

(a)   (b)   (c)

8. What is the total resistance of twelve 5.6-kΩ resistors in series?

9. Six 47-Ω resistors, eight 100-Ω resistors, and two 22-Ω resistors are in series. What is the total resistance?

10. The total resistance in Figure 4–66 is 20 kΩ. What is the value of R_5?

FIGURE 4–66

Section 4–4   Ohm's Law in Series Circuits

11. What is the current in each circuit of Figure 4–67? Show how to connect an ammeter in each case.

FIGURE 4–67

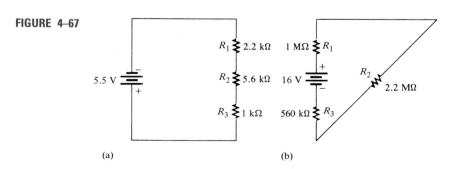

(a)   (b)

12. Determine the voltage across each resistor in Figure 4–67.

13. Three 470-Ω resistors are in series with a 500-V source. How much current is there?

14. Four equal-value resistors are in series with a 5-V source, and a current of 1 mA is measured. What is the value of each resistor?

Section 4–5   Voltage Sources in Series

15. A 5-V battery and a 9-V battery are connected in series with their polarities in the same direction. What is the total voltage?

16. Determine the total source voltage in each circuit of Figure 4–68.

FIGURE 4–68

(a)

(b)

(c)

Section 4–6   Kirchhoff's Voltage Law

17. The following voltage drops are measured across each of three resistors in series: 5.5 V, 8.2 V, and 12.3 V. What is the value of the source voltage to which these resistors are connected?

18. Five resistors are in series with a 20-V source. The voltage drops across four of the resistors are 1.5 V, 5.5 V, 3 V, and 6 V. How much voltage is across the fifth resistor?

19. Determine the unspecified voltage drop(s) in each circuit of Figure 4–69. Show how to connect a voltmeter to measure each unknown voltage drop.

(a)

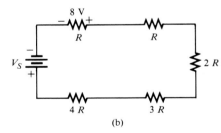
(b)

FIGURE 4–69

Section 4–7   Voltage Dividers

20. The total resistance of a series circuit is 500 Ω. What percentage of the total voltage appears across a 22-Ω resistor in the series circuit?

21. Find the voltage between A and B in each voltage divider of Figure 4–70.

22. What is the voltage across each resistor in Figure 4–71? R is the lowest value and all others are multiples of that value as indicated.

23. What is the voltage across each resistor in Figure 4–72 (a full-color photo appears following page 240)?

FIGURE 4–70

               (a)   (b)

FIGURE 4–71

FIGURE 4–72

Section 4–8   Power in a Series Circuit

24. Five series resistors each dissipate 50 mW of power. What is the total power?

25. Use the results of Problem 23 to find the total power in Figure 4–72.

Section 4–9   Circuit Ground

26. Determine the voltage at each point with respect to ground in Figure 4–73.

FIGURE 4–73

27. In Figure 4–74, how would you determine the voltage across R_2 by measurement, without connecting a meter directly across the resistor?

28. Determine the voltage at each point with respect to ground in Figure 4–74.

FIGURE 4–74

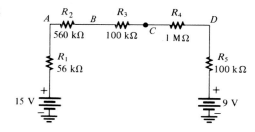

Section 4–10   Troubleshooting* Series Circuits

29. By observing the meters in Figure 4–75, determine the types of failures in the circuits and which components have failed.

FIGURE 4–75

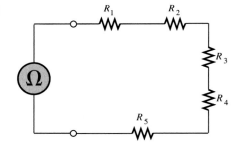

30. Is the ohmmeter reading in Figure 4–76 correct? (A full-color photo appears following page 240). If not, what is wrong?

FIGURE 4–76

PROBLEM SET B

31. Determine the unknown resistance (R_3) in the circuit of Figure 4–77.

32. You have the following resistor values available to you in the lab in unlimited quantities: 10 Ω, 100 Ω, 470 Ω, 560 Ω, 680 Ω, 1 kΩ, 2.2 kΩ, and 5.6 kΩ. All of the other standard values are out of stock. A project that you are working on requires an 18-kΩ resistance. What combination of available values can you use to obtain the needed value?

33. Determine the voltage at each point in Figure 4–78 with respect to the negative side of the battery.

34. Find all the unknown quantities (in color) in Figure 4–79.

---

*(TBLS) indicates troubleshooting.

FIGURE 4-77

FIGURE 4-78

FIGURE 4-79

$R_3 = R_4$

$P_6 = 112$ mW

35. There are 250 mA in a series circuit with a total resistance of 1.5 kΩ. The current must be reduced by 25%. Determine how much resistance must be added in order to accomplish this reduction in current.

36. Four ½-W resistors are in series: 47 Ω, 68 Ω, 100 Ω, and 120 Ω. To what maximum value can the current be raised before the power rating of one of the resistors is exceeded? Which resistor will burn out first if the current is increased above the maximum?

37. A certain series circuit is made up of a ⅛-W resistor, a ¼-W resistor, and a ½-W resistor. The total resistance is 2400 Ω. If each of the resistors is operating at its maximum power level, determine the following:
    (a) I (b) V_S (c) the value of each resistor

38. Using 1.5-V batteries, a switch, and three lamps, devise a circuit to apply 4.5 V across one lamp, two lamps in series, or three lamps in series with a single-control switch. Draw the schematic diagram.

39. Develop a variable voltage divider to provide output voltages ranging from a minimum of 10 V to a maximum of 100 V using a 120-V source. The maximum voltage must be at the maximum resistance setting of the potentiometer. The minimum voltage must be at the minimum resistance (zero ohms) setting. The maximum current is to be 10 mA.

40. Using the standard resistor values given in Appendix D, build a voltage divider to provide the following approximate voltages with respect to the negative terminal of a 30-V source:

8.18 V, 14.73 V, and 24.55 V. The current drain on the source must be limited to no more than 1 mA. The number of resistors, their resistance values, and their power ratings must be specified. Draw a schematic diagram showing the circuit with all resistor values indicated.

41. On the double-sided PC board in Figure 4–80, identify each group of series resistors. Note that many of the interconnections feed through the board from the top side to the bottom side.

FIGURE 4–80

Top

Bottom

42. The table shows the results of resistance measurements on the PC circuit board in Figure 4–80 (a full-color photo appears following page 240). Are these results correct? If not, identify the possible problems.

Between Pins	Resistance
1 and 2	∞
1 and 3	∞
1 and 4	4.23 kΩ
1 and 5	∞
1 and 6	∞
2 and 3	23.6 kΩ
2 and 4	∞
2 and 5	∞
2 and 6	∞
3 and 4	∞
3 and 5	∞
3 and 6	∞
4 and 5	∞
4 and 6	∞
5 and 6	19.9 kΩ

43. You measure 15 kΩ between pins 5 and 6 on the PC board in Figure 4–80. Does this indicate a problem? If so, identify it.

44. In checking out the PC board in Figure 4–80, you measure 17.83 kΩ between pins 1 and 2. Also, you measure 13.6 kΩ between pins 2 and 4. Does this indicate a problem on the PC board? If so, identify the fault.

45. The three groups of series resistors on the PC board in Figure 4–80 are connected in series with each other to form a single series circuit by connecting pin 2 to pin 4 and pin 3 to pin 5. A voltage source is connected across pins 1 and 6 and an ammeter is placed in series. As you increase the source voltage, you observe the corresponding increase in current. Suddenly, the current drops to zero and you smell smoke. All resistors are ½ W.
  (a) What has happened?
  (b) Specifically, what must you do to fix the problem?
  (c) At what voltage did the failure occur?

ANSWERS TO SECTION REVIEWS

Section 4–1

1. There is a single current path.   **2.** See Figure 4–81.   **3.** See Figure 4–82.

FIGURE 4–81

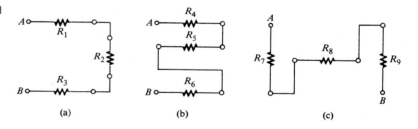

(a)   (b)   (c)

FIGURE 4–82

(a)   (b)   (c)

Section 4–2
1. 2 A.   **2.** 50 mA between C and D, 50 mA between E and F.   **3.** 2 A, 2 A.

Section 4–3
1. **(a)** 143 Ω;   **(b)** 105 Ω;   **(c)** 4020 Ω.   **2.** 572 Ω.   **3.** 1.2 kΩ.
4. 564 Ω.

Section 4–4
1. 0.033 A or 33 mA.   **2.** 215 V.   **3.** $V_1 = 50$ V, $V_2 = 75$ V, $V_3 = 90$ V.
4. 250 Ω.

Section 4–5
1. Five; see Figure 4–83.   2. See Figure 4–84.
3. (a) 75 V;   (b) 15 V.   4. See Figure 4–85.

FIGURE 4–83

FIGURE 4–84   FIGURE 4–85

Section 4–6
1. (a) The sum of the voltages around a closed path is zero.
    (b) The sum of the voltage drops equals the total source voltage.
2. 50 V.   3. 5 V.   4. 10 V.

Section 4–7
1. 10 V.   2. 40.6 V across the 56-Ω, 59.4 V across the 82-Ω. See Figure 4–86.
3. At the midpoint.

FIGURE 4–86

Section 4–8
1. Add the powers in each resistor.   2. 16 W.   3. 1110 W.

Section 4–10
1. T.   2. 24 V, 0 V.

5

Parallel Circuits

In this chapter, parallel circuits are introduced. Parallel circuits are found in many applications, for example, lighting systems, sound systems, appliances, and most electronic equipment.

After completing this chapter, you will be able to

☐ Identify a parallel circuit.
☐ Apply Kirchhoff's current law.
☐ Determine total resistance of a parallel circuit.
☐ Apply Ohm's law to find voltage, currents, and resistances in a parallel circuit.
☐ Describe how parallel circuits act as current dividers.
☐ Apply the current divider principle.
☐ Determine the total power in a parallel circuit.
☐ Locate an open path in a parallel circuit.

APPLICATION ASSIGNMENT

As the technician in charge of testing a particular PC board assembly, you are required to devise a systematic test procedure for checking out each printed circuit (PC) board for open resistors as it comes off the assembly line. Each board consists of three groups of parallel resistors. Each group has several 1/2-W resistors in parallel and is connected to pins on the PC board as follows:

☐   Parallel group 1(R_1 through R_4): 1 kΩ, 1.5 kΩ, 1.8 kΩ, 2.2 kΩ. Pins 1 and 2.
☐   Parallel group 2 (R_5 through R_8): 10 kΩ, 22 kΩ, 27 kΩ, 39 kΩ. Pins 3 and 4.
☐   Parallel group 3 (R_9 through R_{12}): 47 kΩ, 56 kΩ, 68 kΩ, 82 kΩ. Pins 5 and 6.

Each parallel group is separate and not interconnected with another group. Each group must be checked individually for a single open resistor, and that resistor must be identified for replacement. The PC board layout is similar to the one shown in Figure 5–75 (a full-color photo appears following page 240) except for the number and values of resistors. The material that you will cover in this chapter will enable you to develop the required test procedure.

5–1

RESISTORS IN PARALLEL

When two or more components, each forming a separate currrent path, are connected across the same voltage source, they are in parallel. A parallel circuit provides more than one path for current between two given points.

Each parallel path is called a *branch*. Two resistors connected in parallel are shown in Figure 5–1(a). As shown in part (b), the current out of the source divides when it gets to point *A*. Part of it goes through R_1 and part through R_2. The two currents come back together at point *B*. If additional resistors are connected in parallel with the first two, more current paths are provided, as shown in Figure 5–1(c).

FIGURE 5–1
Resistors in parallel.

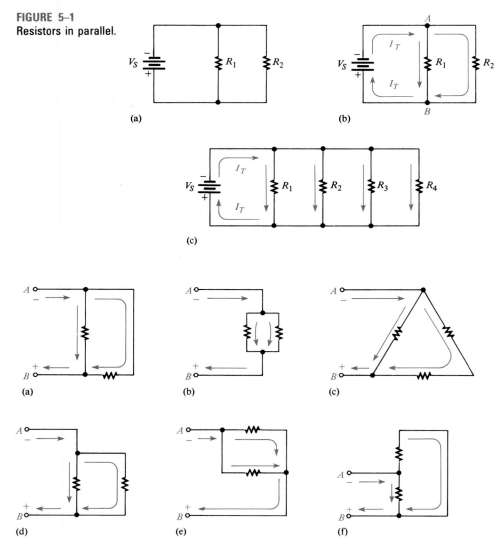

(a)

(b)

(c)

FIGURE 5–2
Examples of circuits with two parallel paths.

(a)

(b)

(c)

(d)

(e)

(f)

Identifying Parallel Circuits

It is obvious that the resistors in Figure 5–1 are connected in parallel. However, in actual circuit diagrams, the parallel relationship often is not so clear. It is important that you learn to recognize parallel circuits regardless of how they may be drawn.

A way to identify a parallel resistive circuit is as follows:

If there is more than one resistor connected from point *A* to point *B* and if the voltage across each of the resistors is the same, then the resistors are in parallel.

Figure 5–2 shows parallel resistors drawn in different ways between two points labeled *A* and *B*. Notice that in each case, the current "travels" two paths going from *A* to *B*, and the voltage across each branch is the same. Although these figures show only two parallel paths, there can be any number of resistors in parallel.

EXAMPLE 5–1

Five resistors are positioned on a circuit board as shown in Figure 5–3. Show the wiring required to connect all the resistors in parallel, and draw a schematic diagram showing this circuit.

FIGURE 5–3

Solution

Wires are connected as shown in the assembly wiring diagram of Figure 5–4(a). The schematic diagram is shown in Figure 5–4(b). Again, note that the schematic does not necessarily have to show the actual physical arrangement of the resistors. The purpose of the schematic is to show how components are connected *electrically*.

(a) Assembly wiring diagram

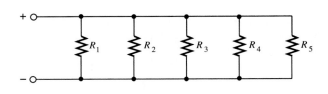

(b) Schematic diagram

FIGURE 5–4

EXAMPLE 5–2   Determine the parallel groupings in Figure 5–5.

FIGURE 5–5

Solution
Resistors R_1 through R_4 and R_{11} and R_{12} are all in parallel. This parallel combination is connected to pins 1 and 4.

Resistors R_5 through R_{10} are all in parallel. This combination is connected to pins 2 and 3.

**SECTION REVIEW
5–1**

1. How do you identify a parallel circuit?
2. Complete the schematic diagrams for the circuits in each part of Figure 5–6 by connecting the resistors in parallel between points A and B.
3. Now connect each *group* of parallel resistors in Figure 5–6 in parallel with each other.

FIGURE 5–6

(a)   (b)   (c)

5–2   VOLTAGE IN A PARALLEL CIRCUIT

As mentioned, each path in a parallel circuit is sometimes called a *branch*.

> **The voltage across any branch of a parallel combination is equal to the voltage across each of the other branches in parallel.**

To illustrate voltage drop in a parallel circuit, let us examine Figure 5–7(a). Points A, B, C, and D along the left side of the parallel circuit are *electrically the same point* because the voltage is the same along this line. You can think of all of these points as being connected by a single wire to the negative terminal of the battery. The points E,

FIGURE 5–7
Voltage across parallel
branches is the same.

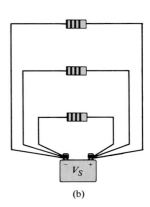

(a)

(b)

F, G, and *H* along the right side of the circuit are all at a voltage equal to that of the positive terminal of the source. Thus, voltage across each parallel resistor is the same, and is equal to the source voltage.

Figure 5–7(b) is the same circuit as in part (a), drawn in a slightly different way. Here the left side of each resistor is connected to a single point, which is the negative battery terminal. The right side of each resistor is connected to a single point, which is the positive battery terminal. The resistors are still all in parallel across the source.

In Figure 5–8, a 12-V battery is connected across three parallel resistors. When the voltage is measured across the battery and then across each of the resistors, the readings are the same. Thus, as you can see, the same voltage appears across each branch in a parallel circuit.

FIGURE 5–8
The same voltage
appears across each
resistor in parallel.

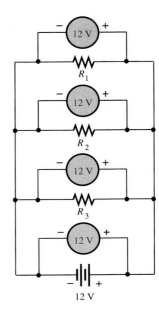

(a) Pictorial

(b) Schematic

SECTION REVIEW
5–2

1. A 10-Ω and a 22-Ω resistor are connected in parallel with a 5-V source. What is the voltage across each of the resistors?
2. A voltmeter connected across R_1 in Figure 5–9 measures 118 V. If you move the meter and connect it across R_2, how much voltage will it indicate? What is the source voltage?
3. In Figure 5–10, how much voltage does voltmeter 1 indicate? Voltmeter 2?

FIGURE 5–9   **FIGURE 5–10**

5–3

KIRCHHOFF'S CURRENT LAW

Kirchhoff's current law is stated as follows:

> **The sum of the currents into a junction is equal to the sum of the currents out of that junction.**

A *junction* is any point in a circuit where two or more circuit paths come together. In a parallel circuit, a junction is a point where the parallel branches connect together. The following is another way to state Kirchhoff's current law:

> **The total current into a junction is equal to the total current out of that junction.**

For example, in the circuit of Figure 5–11, point A is one junction and point B is another. Let us start at the negative terminal of the source and follow the current. The total current I_T flows from the source and *into* the junction at point A. At this point, the

FIGURE 5–11
Kirchhoff's current law: The current into a junction equals the current out of that junction.

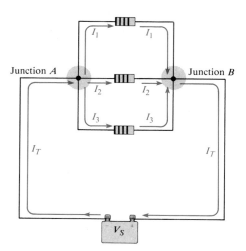

current splits up among the three branches as indicated. Each of the three branch currents (I_1, I_2, and I_3) flows *out of* junction A. Kirchhoff's current law says that the total current into junction A is equal to the total current out of junction A; that is,

$$I_T = I_1 + I_2 + I_3$$

Now, following the currents in Figure 5–11 through the three branches, you see that they come back together at point B. Currents I_1, I_2, and I_3 flow into junction B, and I_T flows out. Kirchhoff's current law formula at this junction is therefore the same as at junction A:

$$I_T = I_1 + I_2 + I_3$$

General Formula for Kirchhoff's Current Law

The previous discussion was a specific case to illustrate Kirchhoff's current law, often abbreviated KCL. Now let us look at the general case. Figure 5–12 shows a *generalized* circuit junction where a number of branches are connected to a point in the circuit.

FIGURE 5–12
Generalized circuit junction to illustrate Kirchhoff's current law.

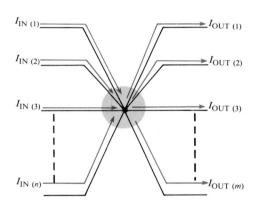

$$I_{IN\,(1)} + I_{IN\,(2)} + I_{IN\,(3)} + \cdots + I_{IN\,(n)}$$
$$= I_{OUT\,(1)} + I_{OUT\,(2)} + I_{OUT\,(3)} + \cdots + I_{OUT\,(m)}$$

Currents $I_{IN(1)}$ through $I_{IN(n)}$ flow into the junction (n can be any number). Currents $I_{OUT(1)}$ through $I_{OUT(m)}$ flow out of the junction (m can be any number but not necessarily equal to n). By Kirchhoff's current law, the sum of the currents into a junction must equal the sum of the currents out of the junction. With reference to Figure 5–12, the general formula for Kirchhoff's current law is

$$I_{IN(1)} + I_{IN(2)} + \cdots + I_{IN(n)} = I_{OUT(1)} + I_{OUT(2)} + \cdots + I_{OUT(m)} \quad \textbf{(5–1)}$$

If all of the terms on the right side of Equation (5–1) are brought over to the left side, their signs change to negative, and a zero is left on the right side. Kirchhoff's current law is sometimes stated in this way:

The algebraic sum of all the currents entering and leaving a junction is equal to zero.

This statement is just another equivalent way of stating what we have just discussed.

You can verify Kirchhoff's current law by connecting a circuit and measuring each branch current and the total current from the source, as illustrated in Figure 5–13. When the branch currents are added together, their sum will equal the total current. This rule applies for any number of branches.

The following three examples illustrate use of Kirchhoff's current law.

FIGURE 5–13
Experimental verification of Kirchhoff's current law.

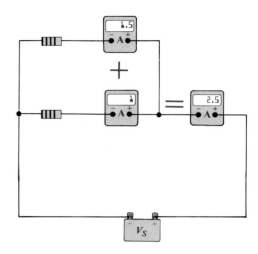

EXAMPLE 5–3   The branch currents in the circuit of Figure 5–14 are known. Determine the total current entering junction A and the total current leaving junction B.

FIGURE 5–14

Solution
The total current out of junction A is the sum of the two branch currents. So the total current into A is

$$I_T = I_1 + I_2 = 5 \text{ mA} + 12 \text{ mA} = 17 \text{ mA}$$

The total current entering point B is the sum of the two branch currents. So the total current out of B is

$$I_T = I_1 + I_2 = 5 \text{ mA} + 12 \text{ mA} = 17 \text{ mA}$$

The calculator sequence is

⑤ (EXP) (+/−) ③ (+) ① ② (EXP) (+/−) ③ (=)

EXAMPLE 5–4   Determine the current through R_2 in Figure 5–15.

FIGURE 5–15

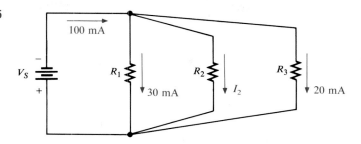

Solution
The total current into the junction of the three branches is known, and two of the branch currents are known. The current equation at this junction is

$$I_T = I_1 + I_2 + I_3$$

Solving for I_2, we get

$$I_2 = I_T - I_1 - I_3 = 100 \text{ mA} - 30 \text{ mA} - 20 \text{ mA} = 50 \text{ mA}$$

EXAMPLE 5–5   Use Kirchhoff's current law to find the current measured by ammeters A_1 and A_2 in Figure 5–16.

FIGURE 5–16

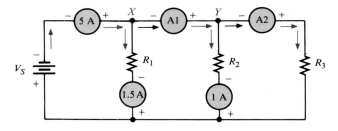

Solution
The total current into junction X is 5 A. Two currents flow out of junction X: 1.5 A through resistor R_1 and the current through A1. Kirchhoff's current law at junction X is

$$5 \text{ A} = 1.5 \text{ A} + I_{A1}$$

Solving for I_{A1} yields

$$I_{A1} = 5 \text{ A} - 1.5 \text{ A} = 3.5 \text{ A}$$

The total current into junction Y is $I_{A1} = 3.5$ A. Two currents flow out of junction Y: 1 A through resistor R_2 and the current through A_2 and R_3. Kirchhoff's current law applied at junction Y gives

$$3.5 \text{ A} = 1 \text{ A} + I_{A2}$$

Solving for I_{A2} yields

$$I_{A2} = 3.5 \text{ A} - 1 \text{ A} = 2.5 \text{ A}$$

SECTION REVIEW
5–3

1. A total current of 2.5 A flows into the junction of three parallel branches. What is the sum of all three branch currents?
2. In Figure 5–17, 100 mA and 300 mA flow into the junction. What is the amount of current out of the junction?
3. Determine I_1 in the circuit of Figure 5–18.

FIGURE 5–17   **FIGURE 5–18**

4. Two branch currents enter a junction, and two branch currents leave the same junction. One of the currents entering the junction is 1 A, and one of the currents leaving the junction is 3 A. The total current entering and leaving the junction is 8 A. Determine the value of the unknown current entering the junction and the value of the unknown current leaving the junction.

5–4

TOTAL PARALLEL RESISTANCE

When resistors are connected in parallel, the total resistance of the circuit decreases. The total resistance of a parallel combination is always less than the value of the smallest resistor.

For example, if a 10-Ω resistor and a 100-Ω resistor are connected in parallel, the total resistance is less than 10 Ω. The exact value must be calculated, and you will learn how to do so later in this section.

How the Number of Current Paths Affects Resistance

As you know, when resistors are connected in parallel, the current has more than one path. The number of current paths is equal to the number of parallel branches.

For example, in Figure 5–19(a), there is only one current path, since it is a series circuit. A certain amount of current, I_1, flows through R_1. If resistor R_2 is connected in parallel with R_1, as shown in Figure 5–19(b), an *additional* amount of current, I_2, flows through R_2. The total current coming from the source has *increased* with the addition of the parallel branch. An increase in the *total current* from the source means that the total resistance has decreased, in accordance with Ohm's law. Additional resistors connected in parallel will further reduce the resistance and increase the total current.

(a)   (b)

FIGURE 5–19
Adding resistors in parallel reduces total resistance and increases total current.

Formula for Total Parallel Resistance

The circuit in Figure 5–20 shows a general case of n resistors in parallel (n can be any number). From Kirchhoff's current law, the current equation is

$$I_T = I_1 + I_2 + I_3 + \cdots + I_n$$

FIGURE 5–20
Circuit with n resistors in parallel.

Since V_S is the voltage across each of the parallel resistors, by Ohm's law, $I_1 = V_S/R_1$, $I_2 = V_S/R_2$, and so on. By substituting into the current equation, we get

$$\frac{V_S}{R_T} = \frac{V_S}{R_1} + \frac{V_S}{R_2} + \frac{V_S}{R_3} + \cdots + \frac{V_S}{R_n}$$

We factor V_S out of the right side of the equation and cancel it with V_S on the left side, leaving only the resistance terms:

$$\frac{1}{R_T} = \frac{1}{R_1} + \frac{1}{R_2} + \frac{1}{R_3} + \cdots + \frac{1}{R_n} \tag{5–2}$$

We get another useful form of Equation (5–2) by taking the reciprocal of (that is, by inverting) both sides of the equation:

$$R_T = \cfrac{1}{\left(\dfrac{1}{R_1}\right) + \left(\dfrac{1}{R_2}\right) + \left(\dfrac{1}{R_3}\right) + \cdots + \left(\dfrac{1}{R_n}\right)} \qquad (5\text{–}3)$$

Equation (5–3) shows that to find the total parallel resistance, add all the $1/R$ terms and then take the reciprocal of the sum. Example 5–6 shows how to use this formula in a specific case, and Example 5–7 shows a simple method of using the calculator in determining parallel resistances. The reciprocal of resistance ($1/R$) is called *conductance* and is symbolized by G. We will, however, continue to use $1/R$.

EXAMPLE 5–6

Calculate the total parallel resistance between points A and B of the circuit in Figure 5–21.

FIGURE 5–21

Solution

First, find the reciprocal of each of the three resistors as follows:

$$\frac{1}{R_1} = \frac{1}{100 \ \Omega} = 0.01$$

$$\frac{1}{R_2} = \frac{1}{47 \ \Omega} = 0.0213$$

$$\frac{1}{R_3} = \frac{1}{22 \ \Omega} = 0.0455$$

Next, calculate R_T by adding $1/R_1$, $1/R_2$, and $1/R_3$ and taking the reciprocal of the sum, as follows:

$$R_T = \frac{1}{0.01 + 0.0213 + 0.0455} = \frac{1}{0.0768} = 13.02 \ \Omega$$

Notice that the value of R_T (13.02 Ω) is smaller than the smallest value in parallel, which is R_3 (22 Ω).

Calculator Solution

The parallel resistance formula is easily solved on an electronic calculator. The general procedure is to enter the value of R_1 and then take its reciprocal by pressing the [1/x] key. Next press the $\boxed{+}$ key; then enter the value of R_2 and take its reciprocal. Repeat this procedure until all of the resistor values have been entered and the reciprocal of each has been added. The final step is to press the $\boxed{1/x}$ key to convert $1/R_T$ to R_T. The total parallel resistance is now on the display. This calculator procedure is illustrated in Example 5–7.

EXAMPLE 5–7

Show the steps required for a calculator solution of Example 5–6.

Solution
Step 1: Enter 100. Display shows 100.
Step 2: Press $\boxed{1/x}$ key. Display shows 0.01.
Step 3: Press $\boxed{+}$ key. Display shows 0.01.
Step 4: Enter 47. Display shows 47.
Step 5: Press $\boxed{1/x}$ key. Display shows 0.021276595.
Step 6: Press $\boxed{+}$ key. Display shows 0.031276595.
Step 7: Enter 22. Display shows 22.
Step 8: Press $\boxed{1/x}$ key. Display shows 0.045454545.
Step 9: Press $\boxed{=}$ key. Display shows 0.076731141.
Step 10: Press $\boxed{1/x}$ key. Display shows 13.03251828.
The number displayed in Step 10 is the total *resistance* in ohms.

Two Resistors in Parallel

Equation (5–3) is a general formula for finding the total resistance of any number of resistors in parallel. It is often useful to consider only two resistors in parallel because this situation occurs commonly in practice.

Derived from Equation (5–3), the formula for two resistors in parallel is

$$R_T = \frac{R_1 R_2}{R_1 + R_2} \qquad (5–4)$$

Equation (5–4) states that

> **the total resistance of two resistors in parallel is equal to the product of the two resistors divided by the sum of the two resistors.**

This equation is sometimes referred to as the "product-over-the-sum" formula. Example 5–8 illustrates how to use it.

EXAMPLE 5–8

Calculate the total resistance between the positive and negative terminals of the source of the circuit in Figure 5–22.

FIGURE 5–22

Solution

Use Equation (5–4) as follows:

$$R_T = \frac{R_1 R_2}{R_1 + R_2} = \frac{(680\ \Omega)(330\ \Omega)}{680\ \Omega + 330\ \Omega}$$

$$= \frac{224{,}400\ \Omega^2}{1{,}010\ \Omega} = 222.18\ \Omega$$

The calculator sequence is

⑥ ⑧ ⓪ ✕ ③ ③ ⓪ ÷ ((⑥ ⑧ ⓪ + ③ ③ ⓪)) =

Resistors of Equal Value in Parallel

Another special case of parallel circuits is the parallel connection of several resistors having the same value. The following is a shortcut method of calculating R_T when this case occurs:

$$R_T = \frac{R}{n} \qquad\qquad (5\text{–}5)$$

Equation (5–5) says that when any number of resistors (n), all having the same resistance (R), are connected in parallel, R_T is equal to the resistance divided by the number of resistors in parallel. Examples 5–9 and 5–10 show how to use this formula.

EXAMPLE 5–9

Find the total resistance between points A and B in Figure 5–23.

FIGURE 5–23

Solution

There are five 100-Ω resistors in parallel. Use Equation (5–5) as follows:

$$R_T = \frac{R}{n} = \frac{100\ \Omega}{5} = 20\ \Omega$$

EXAMPLE 5–10

A stereo amplifier drives two 8-Ω speakers in parallel as shown in Figure 5–24. What is the total resistance across the output terminals of the amplifier?

Solution

The total parallel resistance of the two 8-Ω speakers is

$$R_T = \frac{R}{n} = \frac{8\ \Omega}{2} = 4\ \Omega$$

FIGURE 5–24

Notation for Parallel Resistors

Sometimes, for convenience, parallel resistors are designated by two parallel vertical marks. For example, R_1 in parallel with R_2 can be written as $R_1 \parallel R_2$. Also, when several resistors are in parallel with each other, this notation can be used. For example,

$$R_1 \parallel R_2 \parallel R_3 \parallel R_4 \parallel R_5$$

indicates that R_1 through R_5 are all in parallel.

This notation is also used with resistance values. For example,

$$10 \text{ k}\Omega \parallel 5 \text{ k}\Omega$$

means that a 10-kΩ resistor is in parallel with a 5-kΩ resistor.

**SECTION REVIEW
5–4**

COLOR
INSERT

1. Does the total resistance increase or decrease as more resistors are connected in parallel?
2. The total parallel resistance is always less than _____ .
3. Determine R_T for each circuit in Figures 5–25, 5–26, and 5–27. (Full-color photos appear following page 240.)

FIGURE 5–25

FIGURE 5–26

FIGURE 5–27

5–5

OHM'S LAW IN PARALLEL CIRCUITS

In this section, several circuits are analyzed using Ohm's law and the basic concepts of parallel circuits.

EXAMPLE 5–11

The circuit board in Figure 5–28 has three resistors in parallel used for bias modification in an instrumentation amplifier. The values of two of the resistors are known from the color codes, but the third resistor is unmarked (maybe the bands are worn off from handling). Determine the value of the unknown resistor without disconnecting it from the circuit board or without using an ohmmeter.

FIGURE 5–28

Unknown

1.8 kΩ

1 kΩ

Solution

If we can determine the total resistance of the three resistors in parallel, then we can use the parallel resistance formula to calculate the unknown resistance. We can use Ohm's law to find the total resistance if voltage and total current are known.

In Figure 5–29, a 10-V source (arbitrary value) is connected across the resistors and the total current is measured. Using these measured values, we find that the total resistance is

$$R_T = \frac{V}{I_T} = \frac{10 \text{ V}}{20.1 \text{ mA}} = 498 \ \Omega$$

Use the reciprocal resistance formula to find the unknown resistance as follows:

$$\frac{1}{R_T} = \frac{1}{R_1} + \frac{1}{R_2} + \frac{1}{R_3}$$

$$\frac{1}{R_3} = \frac{1}{R_T} - \frac{1}{R_1} - \frac{1}{R_2} = \frac{1}{498 \ \Omega} - \frac{1}{1 \text{ k}\Omega} - \frac{1}{1.8 \text{ k}\Omega} = 0.000453$$

$$R_3 = \frac{1}{0.000453} = 2.2 \text{ k}\Omega$$

The calculator sequence for R_3 is

[4] [9] [8] [1/x] [−] [1] [EXP] [3] [1/x] [−] [1] [·] [8] [EXP] [3] [1/x] [=] [1/x]

FIGURE 5-29

EXAMPLE 5-12   Find the total current produced by the battery in Figure 5–30.

FIGURE 5-30

Solution
Step 1: The battery "sees" a total parallel resistance that determines the amount of current generated. First, calculate R_T:

$$R_T = \frac{R_1 R_2}{R_1 + R_2} = \frac{(100\ \Omega)(56\ \Omega)}{100\ \Omega + 56\ \Omega}$$

$$= \frac{5600\ \Omega^2}{156\ \Omega} = 35.9\ \Omega$$

Step 2: The battery voltage is 100 V. Use Ohm's law to find I_T:

$$I_T = \frac{100\ V}{35.9\ \Omega} = 2.79\ A$$

The calculator sequence for R_T is

$\boxed{1}\ \boxed{0}\ \boxed{0}\ \boxed{\times}\ \boxed{5}\ \boxed{6}\ \boxed{\div}\ \boxed{(}\ \boxed{(}\ \boxed{1}\ \boxed{0}\ \boxed{0}\ \boxed{+}\ \boxed{5}\ \boxed{6}\ \boxed{)}\ \boxed{)}\ \boxed{=}$

and the calculator sequence for I_T is

$\boxed{1}\ \boxed{0}\ \boxed{0}\ \boxed{\div}\ \boxed{3}\ \boxed{5}\ \boxed{.}\ \boxed{9}\ \boxed{=}$

EXAMPLE 5–13

Determine the current through each resistor in the parallel circuit of Figure 5–31.

FIGURE 5–31

Solution

The voltage across each resistor (branch) is equal to the source voltage. That is, the voltage across R_1 is 20 V, the voltage across R_2 is 20 V, and the voltage across R_3 is 20 V. The current through each resistor is determined as follows:

$$I_1 = \frac{V_S}{R_1} = \frac{20 \text{ V}}{1 \text{ k}\Omega} = 20 \text{ mA}$$

$$I_2 = \frac{V_S}{R_2} = \frac{20 \text{ V}}{2.2 \text{ k}\Omega} = 9.1 \text{ mA}$$

$$I_3 = \frac{V_S}{R_3} = \frac{20 \text{ V}}{560 \text{ }\Omega} = 35.7 \text{ mA}$$

EXAMPLE 5–14

Find the voltage V_S across the parallel circuit in Figure 5–32.

FIGURE 5–32

Solution

We know the total current into the parallel circuit. We need to know the total resistance, and then we can apply Ohm's law to get the voltage. The total resistance is

$$R_T = \frac{1}{\left(\dfrac{1}{R_1}\right) + \left(\dfrac{1}{R_2}\right) + \left(\dfrac{1}{R_3}\right)} = \frac{1}{0.00455 + 0.00179 + 0.001}$$

$$= \frac{1}{0.00734} = 136.24 \text{ }\Omega$$

$$V_S = I_T R_T = (10 \text{ mA})(136.24 \text{ }\Omega) = 1.3624 \text{ V}$$

1. A 10-V battery is connected across three 68-Ω resistors that are in parallel. What is the total current from the battery?

2. How much voltage is required to produce 2 A of current through the circuit of Figure 5–33?

3. How much current is there through each resistor of Figure 5–33?

4. There are four equal-value resistors in parallel with a 12-V source, and 6 mA of current from the source. What is the value of each resistor?

5. A 1-kΩ and a 2.2-kΩ resistor are connected in parallel. A total of 100 mA flows through the parallel combination. How much voltage is dropped across the resistors?

FIGURE 5–33

5–6

THE CURRENT DIVIDER PRINCIPLE

In a parallel circuit, when the total current flows into the junction of the parallel branches, it *divides* among the branches. Thus, a parallel circuit acts as a current divider. This current divider principle is illustrated in Figure 5–34 for a two-branch parallel circuit in which part of the total current I_T goes through R_1 and part through R_2.

FIGURE 5–34
Total current divides between two branches.

Since the same voltage is across each of the resistors in parallel, the branch currents are inversely proportional to the values of the resistors. For example, if the value of R_2 is twice that of R_1, then the value of I_2 is one-half that of I_1. In other words,

the total current divides among parallel resistors in a manner inversely proportional to the resistance values.

The branches with higher resistance have less current, and the branches with lower resistance have more current, in accordance with Ohm's law. If all the branches have the same resistance, the branch currents are all equal.

Figure 5–35 shows specific values to demonstrate how the currents divide according to the branch resistances. Notice that in this case the resistance of the upper branch

FIGURE 5–35
The branch with the lowest resistance has the most current, and the branch with the highest resistance has the least current.

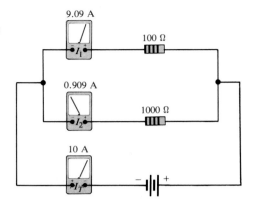

is one-tenth the resistance of the lower branch, but the upper branch current is *ten times* the lower branch current.

Formulas for Two Branches

You already know how to use Ohm's law $(I = V/R)$ to determine the current in any parallel branch when the voltage and resistance are known. When the voltage is not known but the total current is, you can find both of the branch currents (I_1 and I_2) by using the following formulas:

$$I_1 = \left(\frac{R_2}{R_1 + R_2} \right) I_T \qquad (5\text{–}6)$$

$$I_2 = \left(\frac{R_1}{R_1 + R_2} \right) I_T \qquad (5\text{–}7)$$

The current in either branch is equal to the *opposite* branch resistance divided by the sum of the two resistors and then multiplied by the total current.

EXAMPLE 5–15

Find I_1 and I_2 in Figure 5–36.

FIGURE 5–36

Solution
Using Equation (5–6), we determine I_1 as follows:

$$I_1 = \left(\frac{R_2}{R_1 + R_2} \right) I_T = \left(\frac{47\ \Omega}{147\ \Omega} \right) 100 \text{ mA}$$

$$= 31.97 \text{ mA}$$

Using Equation (5–7), we determine I_2 as follows:

$$I_2 = \left(\frac{R_1}{R_1 + R_2}\right)I_T = \left(\frac{100 \ \Omega}{147 \ \Omega}\right)100 \ \text{mA}$$

$$= 68.03 \ \text{mA}$$

The calculator sequence for I_2 is

[1] [0] [0] [÷] [(] [(] [1] [0] [0] [+] [4] [7] [)] [×] [1] [0] [0] [EXP] [+/−] [3] [=]

General Formula for Any Number of Branches

A generalized parallel circuit with n branches is shown in Figure 5–37 (n represents any number).

FIGURE 5–37
Generalized parallel circuit with n branches.

The current in any branch can be determined with the following formula:

$$I_x = \left(\frac{R_T}{R_x}\right)I_T \tag{5–8}$$

where I_x represents any branch current (I_1, I_2, and so on) and R_x represents any resistance (R_1, R_2, and so on). For example, the formula for current in the second branch is

$$I_2 = \left(\frac{R_T}{R_2}\right)I_T$$

Equation (5–8) can be used for a parallel circuit with any number of branches. The formulas for the special case of two branches are actually derived from this general formula. Notice that you must determine R_T in order to use Equation (5–8) to find a branch current.

EXAMPLE 5–16

Determine the current through each resistor in the circuit of Figure 5–38.

FIGURE 5–38

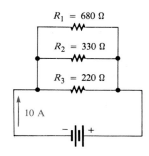

Solution

First calculate the total parallel resistance:

$$R_T = \cfrac{1}{\left(\cfrac{1}{R_1}\right) + \left(\cfrac{1}{R_2}\right) + \left(\cfrac{1}{R_3}\right)}$$

$$= \cfrac{1}{\left(\cfrac{1}{680 \ \Omega}\right) + \left(\cfrac{1}{330 \ \Omega}\right) + \left(\cfrac{1}{220 \ \Omega}\right)} = 110.5 \ \Omega$$

The total current is 10 A. Using Equation (5–8), we calculate each branch current as follows:

$$I_1 = \left(\frac{R_T}{R_1}\right)I_T = \left(\frac{110.5 \ \Omega}{680 \ \Omega}\right)10 \ \text{A} = 1.63 \ \text{A}$$

$$I_2 = \left(\frac{R_T}{R_2}\right)I_T = \left(\frac{110.5 \ \Omega}{330 \ \Omega}\right)10 \ \text{A} = 3.35 \ \text{A}$$

$$I_3 = \left(\frac{R_T}{R_3}\right)I_T = \left(\frac{110.5 \ \Omega}{220 \ \Omega}\right)10 \ \text{A} = 5.02 \ \text{A}$$

SECTION REVIEW
5–6

COLOR INSERT

COLOR INSERT

1. A circuit has the following resistors in parallel: 220-Ω, 100-Ω, 68-Ω, 56-Ω, and 22-Ω. Which resistor has the most current through it? The least current?

2. Determine the current through the right-most resistor in Figure 5–39. (A full-color photo appears following page 240.)

3. Find the currents through each resistor in the circuit of Figure 5–40. (A full-color photo appears following page 240.)

FIGURE 5–39

FIGURE 5–40

5–7 POWER IN A PARALLEL CIRCUIT

The total amount of power in a parallel resistive circuit is equal to the sum of the powers in each resistor in parallel. Equation (5–9) states this formula in a concise way for any number of resistors in parallel:

$$P_T = P_1 + P_2 + P_3 + \cdots + P_n \qquad (5\text{--}9)$$

where P_T is the total power and P_n is the power in the last resistor in parallel. As you can see, the powers are additive, just as in the series circuit.

EXAMPLE 5–17

The amplifier in the stereo system of Figure 5–41 drives four speakers as shown. If the maximum voltage to the speakers is 15 V, how much power must the amplifier be able to deliver to the speakers?

FIGURE 5–41

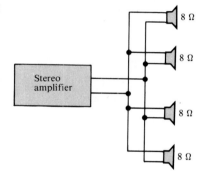

Solution

The speakers are connected in parallel to the amplifier output, so the voltage across each is the same. The maximum power to each speaker is

$$P = \frac{V^2}{R} = \frac{(15 \text{ V})^2}{8 \text{ } \Omega} = 28.125 \text{ W}$$

The total power that the amplifier must be capable of delivering to the speaker system is four times the power in an individual speaker because the total power is the sum of the individual powers:

$$P_T = P + P + P + P = 4P = 4(28.125 \text{ W}) = 112.5 \text{ W}$$

SECTION REVIEW 5–7

1. The resistors in a parallel circuit have the following powers: 2 W, 5 W, 1 W, and 8 W. What is the total power in the circuit?

2. A circuit has a 1-kΩ, a 2.2-kΩ, and a 3.9-kΩ resistor in parallel. A total current of 1 A flows into the parallel circuit. What is the total power?

5-8   EXAMPLES OF PARALLEL CIRCUIT APPLICATIONS

Parallel circuits are found in some form in virtually every electronic device. In many of these applications, the parallel relationship of components may not be obvious until you have covered some advanced topics that will come later. Therefore, in this section, some common and familiar applications of parallel circuits are discussed.

Automotive

One advantage of a parallel circuit over a series circuit is that when one branch opens, the other branches are not affected. For example, Figure 5–42 shows a simplified diagram of an automobile lighting system. When one headlight on your car goes out, it does not cause the other lights to go out, because they are all in parallel.

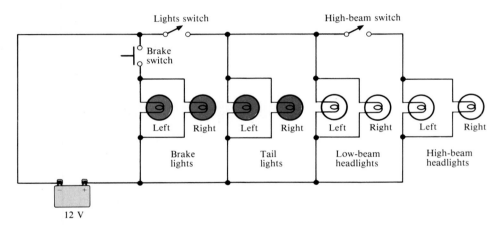

FIGURE 5–42
Simplified diagram of the exterior light system of an automobile. All lights are *off* with switches in the positions shown.

Notice that the brake lights are switched on independently of the head and tail-lights. They come on only when the driver closes the brake light switch by depressing the brake pedal.

When the *lights* switch is closed, both low-beam headlights and both taillights are on. The high-beam headlights are on only when both the *lights* switch and the *high-beam* switch are closed. If any one of the lights burns out (opens), there is still current in each of the other lights.

Residential

Another common use of parallel circuits is in residential electrical systems. All the lights and appliances in a home are wired in parallel. Figure 5–43(a) shows a typical room wiring arrangement with two switch-controlled lights and three wall outlets in parallel.

Figure 5–43(b) shows a simplified parallel arrangement of four heating elements on an electric range. The four-position switches in each branch allow the user to control the amount of current through the heating elements by selecting the appropriate limiting resistor. The lowest resistor value (H setting) allows the highest amount of current for

FIGURE 5–43
Examples of parallel
circuits in residential
wiring and appliances.

(a) Simplified diagram of room wiring

(b) Simplified diagram of a four-burner range

maximum heat. The highest resistor value (L setting) allows the least amount of current for minimum heat; M designates the medium settings.

Ammeters

Another example in which parallel circuits are used is the familiar analog (needle-type) ammeter or milliammeter. Parallel circuits are an important part of the operation of the ammeter because they allow the user to select various ranges in order to measure many different current values.

The mechanism in an ammeter that causes the pointer to move in proportion to the current is called the *meter movement,* which is based on a magnetic principle that you will learn later. Right now, all you need to know is that a meter movement has a certain resistance and a maximum current. This maximum current, called the *full-scale deflection* current, causes the pointer to go all the way to the end of the scale. For example, a certain meter movement has a 50-Ω resistance and a full-scale deflection current of 1 mA. A meter with this particular movement can measure currents of 1 mA or less. Currents greater than 1 mA will cause the pointer to "peg" (or stop) at full scale. Figure 5–44 illustrates a 1-mA meter.

(a) Half-scale deflection

(b) Full-scale deflection

(c) Pegged

FIGURE 5–44
A 1-mA meter.

FIGURE 5–45
A 10-mA meter.

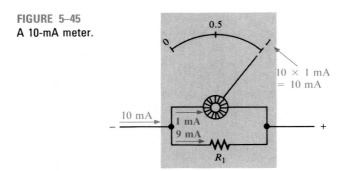

Figure 5–45 shows a simple ammeter with a resistor in parallel with the meter movement; this resistor is called a *shunt* resistor. Its purpose is to bypass any current in excess of 1 mA around the meter movement. The figure specifically shows 9 mA through the shunt resistor and 1 mA through the meter movement. Thus, up to 10 mA can be measured. To find the actual current value, simply multiply the reading on the scale by 10.

A practical ammeter has a range switch that permits the selection of several full-scale current settings. In each switch position, a certain amount of current is bypassed through a parallel resistor as determined by the resistance value. In our example, the current through the movement is never greater than 1 mA.

Figure 5–46 illustrates a meter with three ranges: 1 mA, 10 mA, and 100 mA. When the range switch is in the 1-mA position, all of the current coming into the meter goes through the meter movement. In the 10-mA setting, up to 9 mA goes through R_1 and up to 1 mA through the movement. In the 100-mA setting, up to 99 mA goes through R_2, and the movement can still have only 1 mA for full-scale.

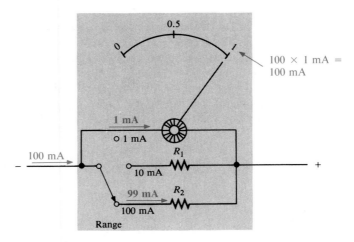

FIGURE 5–46
A milliammeter with three ranges.

For example, in Figure 5–46, if 50 mA of current are being measured, the needle points to the 0.5 mark on the scale; you must multiply 0.5 by 100 to find the current value. In this situation, 0.5 mA flows through the movement (half-scale deflection), and 49.5 mA flows through R_2.

Effect of the Ammeter on a Circuit

As you know, an ammeter is connected in *series* to measure the current in a circuit. Ideally, the meter should not alter the current that it is intended to measure. In practice, however, the meter unavoidably has some effect on the circuit, because its *internal resistance* is connected in series with the circuit resistance. However, in most cases, the meter's internal resistance is so small compared to the circuit resistance that it can be neglected.

For example, if the meter has a 50-Ω movement (R_M) and a 100-μA full-scale current (I_M), the voltage dropped across the movement is

$$V_M = I_M R_M = (100 \ \mu A)(50 \ \Omega) = 5 \ mV$$

The shunt resistance (R_{SH}) for the 10-mA range, for example, is

$$R_{SH} = \frac{V_M}{I_{SH}} = \frac{5 \ mV}{9.9 \ mA} = 0.505 \ \Omega$$

As you can see, the total resistance of the ammeter on the 10-mA range is the resistance of the movement in parallel with the shunt resistance:

$$R_M \parallel R_{SH} = 50 \ \Omega \parallel 0.505 \ \Omega = 0.5 \ \Omega$$

EXAMPLE 5–18

How much does an ammeter with a 100-μA, 50-Ω movement affect the current in the circuit of Figure 5–47?

FIGURE 5–47

(a) Circuit

(b) Circuit with ammeter connected

Solution

The true current in the circuit is

$$I = \frac{10 \ V}{1200 \ \Omega} = 8.3333 \ mA$$

The meter is set on the 10-mA range in order to measure this particular amount of current. It was found that the meter's resistance on the 10-mA range is 0.5 Ω. When the meter is connected in the circuit, its resistance is in series with the 1200-Ω resistor. Thus, there is a total of 1200.5 Ω.

The current actually measured by the ammeter is

$$I = \frac{10 \text{ V}}{1200.5 \ \Omega} = 8.3299 \text{ mA}$$

This current differs from the true circuit current by only 0.04%. Therefore, the meter does not significantly alter the current value, a situation which, of course, is necessary because the measuring instrument should not change the quantity that is to be measured accurately.

SECTION REVIEW 5–8

1. For the ammeter in Figure 5–46, what is the maximum resistance that the meter will have when connected in a circuit? What is the maximum current that can be measured at the setting?
2. Do the shunt resistors have resistance values considerably less than or more than that of the movement? Why?

5–9

TROUBLESHOOTING PARALLEL CIRCUITS

Open Branches

Recall that an open circuit is one in which the current path is interrupted and no current flows. In this section we examine what happens when a branch of a parallel circuit opens.

If a switch is connected in a branch of a parallel circuit, as shown in Figure 5–48, an open or a closed path can be made by the switch. When the switch is closed, as in Figure 5–48(a), R_1 and R_2 are in parallel. The total resistance is 50 Ω (two 100-Ω resistors in parallel). Current flows through both resistors. If the switch is

(a)   (b)

FIGURE 5–48
When switch opens, total current decreases and current through R_2 remains unchanged.

opened, as in Figure 5–48(b), R_1 is effectively removed from the circuit, and the total resistance is 100 Ω. Current now flows only through R_2.

In general,

when an open circuit occurs in a parallel branch, the total resistance increases, the total current decreases, and the same current continues through each of the remaining parallel paths.

The decrease in total current equals the amount of current that was previously in the open branch. The other branch currents remain the same.

Consider the lamp circuit in Figure 5–49. There are four bulbs in parallel with a 120-V source. In part (a), current is flowing through each bulb. Now suppose that one of the bulbs burns out, creating an open path as shown in Figure 5–49(b). This light will go out because no current is flowing through the open path. Notice, however, that current continues to flow through all the other parallel bulbs, and they continue to glow. The open branch does not change the voltage across the parallel branches; it remains at 120 V.

(a)   (b)

FIGURE 5–49
When lamp opens, total current decreases and other branch currents remain unchanged.

You can see that a parallel circuit has an advantage over a series connection in lighting systems because if one or more of the parallel bulbs burn out, the others will stay on. In a *series* circuit, when one bulb goes out, all of the others go out also because the current path is *completely* interrupted.

When a resistor in a parallel circuit opens, the open resistor cannot be located by measurement of the voltage across the branches, because the same voltage exists across all the branches. Thus, there is no way to tell which resistor is open by simply measuring voltage. The good resistors will always have the same voltage as the open one, as illustrated in Figure 5–50 (note that the middle resistor is open).

FIGURE 5–50
Parallel branches (open or not) have the same voltage.

If a visual inspection does not reveal the open resistor, it must be located by current measurements. In practice, measuring current is more difficult than measuring voltage because you must insert the ammeter in *series* to measure the current. Thus, a wire or a printed circuit connection must be cut or disconnected, or one end of a component must be lifted off the circuit board, in order to connect the ammeter in series.

This procedure, of course, is not required when voltage measurements are made, because the meter leads are simply connected *across* a component.

Where to Measure the Current

In a parallel circuit, the total current should be measured. When a parallel resistor opens, I_T is less than its normal value. Once I_T and the voltage across the branches are known, a few calculations will determine the open resistor when all the resistors are of *different* values.

Consider the two-branch circuit in Figure 5–51(a). If one of the resistors opens, the total current will equal the current in the good resistor. Ohm's law quickly tells us what the current in each resistor should be:

$$I_1 = \frac{50 \text{ V}}{560 \text{ }\Omega} = 0.089 \text{ A}$$

$$I_2 = \frac{50 \text{ V}}{100 \text{ }\Omega} = 0.5 \text{ A}$$

(a)

(b)

(c)

FIGURE 5–51
Finding an open path by current measurement.

If R_2 is open, the total current is 0.089 A, as indicated in Figure 5–51(b). If R_1 is open, the total current is 0.5, as indicated in Figure 5–51(c). This procedure can be extended to any number of branches having *unequal* resistances.

If the parallel resistances are all equal, the current in each branch must be checked until a branch is found with no current. This is the open resistor.

EXAMPLE 5–19   In Figure 5–52, there is a total current of 31.09 mA, and the voltage across the parallel branches is 20 V. Is there an open resistor, and, if so, which one is it?

FIGURE 5-52

Solution

Calculate the current in each branch:

$$I_1 = \frac{V}{R_1} = \frac{20 \text{ V}}{10 \text{ k}\Omega} = 2 \text{ mA}$$

$$I_2 = \frac{V}{R_2} = \frac{20 \text{ V}}{4.7 \text{ k}\Omega} = 4.26 \text{ mA}$$

$$I_3 = \frac{V}{R_3} = \frac{20 \text{ V}}{2.2 \text{ k}\Omega} = 9.09 \text{ mA}$$

$$I_4 = \frac{V}{R_4} = \frac{20 \text{ V}}{1 \text{ k}\Omega} = 20 \text{ mA}$$

The total current *should* be

$$I_T = I_1 + I_2 + I_3 + I_4 = 2 \text{ mA} + 4.26 \text{ mA} + 9.09 \text{ mA} + 20 \text{ mA}$$
$$= 35.35 \text{ mA}$$

The actual *measured* current is 31.09 mA, as stated, which is 4.26 mA less than normal, indicating that the branch carrying 4.26 mA is open. Thus, R_2 must be open.

When two or more branches have the same normal current, the open branch can be isolated only by direct measurements of the branch current or resistance. When a zero current measurement or an infinite resistance measurement occurs, you have found the open branch. Remember, however, that the resistor must be disconnected for a resistance measurement.

SECTION REVIEW 5–9

1. If a parallel branch opens, what changes can be detected in the circuit's voltage and the currents, assuming that the parallel circuit is across a constant voltage source?

2. What happens to the total resistance if one branch opens?

3. If several light bulbs are connected in parallel and one of the bulbs opens (burns out), will the others continue to glow?

4. One ampere of current flows in each branch of a parallel circuit. If one branch opens, what is the current in each of the remaining branches?

5. A three-branch circuit normally has the following branch currents: 1 A, 2.5 A, and 1.2 A. If the total current measure 3.5 A, which branch is open?

SUMMARY
- ☐ The same voltage appears across each branch in a parallel circuit.
- ☐ The total resistance of a parallel circuit is less than the smallest branch resistance.
- ☐ In a parallel circuit, the total current divides into branch currents. According to Ohm's law, each branch current is inversely proportional to the branch resistance.
- ☐ The largest branch resistance has the smallest current, and the smallest branch resistance has the largest current.
- ☐ The total power in a parallel circuit is the sum of the powers in each resistor.
- ☐ When a parallel branch opens, the total resistance increases and the total current decreases. The other branch currents are not affected.

GLOSSARY

Branch   One current path in a parallel circuit.

Kirchhoff's current law   A law stating that the total current into a junction equals the total current out of the junction.

Parallel   The relationship in electric circuits in which two or more current paths are connected between the same two points.

Junction   A point at which two or more branches are connected.

FORMULAS

$$I_{IN(1)} + I_{IN(2)} + I_{IN(3)} + \cdots + I_{IN(n)}$$
$$= I_{OUT(1)} + I_{OUT(2)} + I_{OUT(3)} + \cdots + I_{OUT(m)} \qquad \text{Kirchhoff's current law} \qquad (5\text{--}1)$$

$$\frac{1}{R_T} = \frac{1}{R_1} + \frac{1}{R_2} + \frac{1}{R_3} + \cdots + \frac{1}{R_n} \qquad \begin{array}{l}\text{Reciprocal for total parallel}\\ \text{resistance}\end{array} \qquad (5\text{--}2)$$

$$R_T = \frac{1}{\left(\dfrac{1}{R_1}\right) + \left(\dfrac{1}{R_2}\right) + \left(\dfrac{1}{R_3}\right) + \cdots + \left(\dfrac{1}{R_n}\right)} \qquad \text{Total parallel resistance} \qquad (5\text{--}3)$$

$$R_T = \frac{R_1 R_2}{R_1 + R_2} \qquad \text{Special case for two resistors in parallel} \qquad (5\text{--}4)$$

$$R_T = \frac{R}{n} \qquad \text{Special case for } n \text{ equal-value resistors in parallel} \qquad (5\text{--}5)$$

$$I_1 = \left(\frac{R_2}{R_1 + R_2}\right) I_T \qquad \text{Two-branch current divider formula} \qquad (5\text{--}6)$$

$$I_2 = \left(\frac{R_1}{R_1 + R_2}\right) I_T \qquad \text{Two-branch current divider formula} \qquad (5\text{--}7)$$

$$I_x = \left(\frac{R_T}{R_x}\right) I_T \qquad \text{General current divider formula} \qquad (5\text{--}8)$$

$$P_T = P_1 + P_2 + P_3 + \cdots + P_n \qquad \text{Total power} \qquad (5\text{--}9)$$

SELF-TEST

1. In a parallel circuit, each resistor has
   (a) the same current   (b) the same voltage   (c) the same power
   (d) a, b, and c

2. When a 1.2-kΩ resistor and a 100-Ω resistor are connected in parallel, the total resistance is
   (a) greater than 1.2 kΩ   (b) greater than 100 Ω but less than 1.2 kΩ
   (c) less than 100 Ω but greater than 90 Ω   (d) less than 90 Ω

3. A 330-Ω resistor, a 270-Ω resistor, and a 68-Ω resistor are all in parallel. The total resistance is approximately
   (a) 668 Ω   (b) 47 Ω   (c) 68 Ω   (d) 22 Ω

4. Eight resistors are in parallel. The two lowest-value resistors are both 1 kΩ. The total resistance
   (a) cannot be determined   (b) is greater than 1 kΩ   (c) is less than 1 kΩ
   (d) is less than 500 Ω

5. When an additional resistor is connected across an existing parallel circuit, the total resistance
   (a) decreases   (b) increases   (c) remains the same
   (d) increases by the value of the added resistor

6. If one of the resistors in a parallel circuit is removed, the total resistance
   (a) decreases by the value of the removed resistor   (b) remains the same
   (c) increases   (d) doubles

7. The currents into a junction flow along two paths. One current is 5 A and the other is 3 A. The total current out of the junction is
   (a) 2 A   (b) unknown   (c) 8 A   (d) the larger of the two

8. The following resistors are in parallel across a voltage source: 390-Ω, 560-Ω, and 820-Ω. The resistor with the least current is
   (a) 390 Ω   (b) 560 Ω   (c) 820 Ω
   (d) impossible to determine without knowing the voltage

9. A sudden decrease in the total current into a parallel circuit may indicate
   (a) a short   (b) an open resistor   (c) a drop in source voltage
   (d) either b or c

10. In a four-branch parallel circuit, there are 10 mA of current in each branch. If one of the branches opens, the current in each of the other three branches is
    (a) 13.33 mA   (b) 10 mA   (c) 0 A   (d) 30 mA

11. In a certain three-branch parallel circuit, R_1 has 10 mA through it, R_2 has 15 mA through it, and R_3 has 20 mA through it. After measuring a total current of 35 mA, you can say that
    (a) R_1 is open   (b) R_2 is open   (c) R_3 is open
    (d) the circuit is operating properly

12. If there are a total of 100 mA into a parallel circuit consisting of three branches and two of the branch currents are 40 mA and 20 mA, the third branch current is
    (a) 60 mA   (b) 20 mA   (c) 160 mA   (d) 40 mA

13. A complete short develops across one of five parallel resistors on a PC board. The most likely result is
    (a) the shorted resistor will burn out
    (b) one or more of the other resistors will burn out
    (c) the fuse in the power supply will blow
    (d) the resistance values will be altered

14. The power dissipation in each of four parallel branches is 1 W. The total power dissipation is
    (a) 1 W   (b) 4 W   (c) 0.25 W   (d) 16 W

PROBLEM SET A
Section 5–1   Resistors in Parallel
   1.   Connect the resistors in Figure 5–53 in parallel across the battery.

FIGURE 5–53

   2.   Determine whether or not all the resistors in Figure 5–54 are connected in parallel on the printed circuit board. Sketch the schematic.

FIGURE 5–54

Section 5–2   Voltage in a Parallel Circuit
   3.   Determine the voltage across and the current through each parallel resistor if the total voltage is 12 V and the total resistance is 600 Ω. There are four resistors, all of equal value.
   4.   The source voltage in Figure 5–55 is 100 V. How much voltage does each of the meters V1–V3 read?

FIGURE 5–55

Section 5–3   Kirchhoff's Current Law
   5.   The following currents are measured in the same direction in a three-branch parallel circuit: 250-mA, 300-mA, and 800-mA. What is the value of the current into the junction of these three branches?
   6.   Five hundred milliamperes (500 mA) flow into five parallel resistors. The currents through four of the resistors are 50 mA, 150 mA, 25 mA, and 100 mA. How much current flows through the fifth resistor?

7. How much current flows through R_2 and R_3 in Figure 5–56 if R_2 and R_3 have the same resistance? Show how to connect ammeters to measure these currents.

FIGURE 5–56

Section 5–4   Total Parallel Resistance

8. The following resistors are connected in parallel: 1-MΩ, 2.2-MΩ, 4.7-MΩ, 12-MΩ, and 22-MΩ. Determine the total resistance.

COLOR
INSERT

9. Find the total resistance for each group of parallel resistors in Figure 5–57. (A full-color photo appears following page 240.)

10. Calculate R_T for each circuit in Figure 5–58.

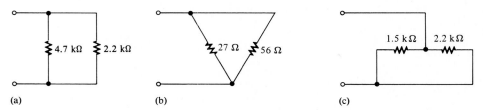

(a)   (b)   (c)

FIGURE 5–58

11. What is the total resistance of eleven 22-kΩ resistors in parallel?

12. Five 15-Ω, ten 100-Ω, and two 10-Ω resistors are all connected in parallel. What is the total resistance?

Section 5–5   Ohm's Law in Parallel Circuits

13. What is the total current I_T in each circuit of Figure 5–59?

FIGURE 5–59

(a)   (b)

14. Three 33-Ω resistors are connected in parallel with a 110-V source. How much current is flowing from the source?

FIGURE 5–60

(a)   (b)

15. Which circuit of Figure 5–60 has more total current?

16. Four equal-value resistors are connected in parallel. Five volts are applied across the parallel circuit, and 2.5 mA are measured from the source. What is the value of each resistor?

Section 5–6   The Current Divider Principle

17. How much branch current should each meter in Figure 5–61 indicate?

FIGURE 5–61

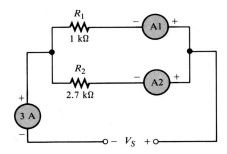

18. Using the current divider formula, determine the current in each branch of the circuits of Figure 5–62.

FIGURE 5–62

(a)   (b)

Section 5–7   Power in a Parallel Circuit

19. Five parallel resistors each handle 40 mW. What is the total power?

20. Determine the total power in each circuit of Figure 5–62.

21. Six light bulbs are connected in parallel across 110 V. Each bulb is rated at 75 W. How much current flows through each bulb, and what is the total current?

Section 5–9   Troubleshooting Parallel Circuits

22. If one of the bulbs burns out in Problem 21, how much current will flow through each of the remaining bulbs? What will the total current be?

TBLS
+ −

23. In Figure 5–63, the current and voltage measurements are indicated. Has a resistor opened, and, if so, which one?

FIGURE 5–63

TBLS
+ −

24. What is wrong with the circuit in Figure 5–64?

FIGURE 5–64

PROBLEM SET B

25. In the circuit of Figure 5–65, determine resistances R_2, R_3, and R_4.

FIGURE 5–65

26. The total resistance of a parallel circuit is 25 Ω. If the total current is 100 mA, how much current flows through a 220-Ω resistor that makes up part of the parallel circuit?

27. What is the current through each resistor in Figure 5–66? R is the lowest-value resistor, and all others are multiples of that value as indicated.

FIGURE 5–66

28. A certain parallel network consists of only ½-W resistors. The total resistance is 1 kΩ, and the total current is 50 mA. If each resistor is operating at one-half its maximum power level, determine the following:
   (a) the number of resistors   **(b)** the value of each resistor
   (c) the current in each branch   **(d)** the applied voltage

29. Find the values of the unspecified labeled quantities in each circuit of Figure 5–67.

FIGURE 5–67

30. What is the total resistance between point A and ground in Figure 5–68 for the following conditions?
   (a) S_1 and S_2 open   **(b)** S_1 closed, S_2 open
   (c) S_1 open, S_2 closed   **(d)** S_1 and S_2 closed

FIGURE 5–68

31. To what minimum value can the 100-Ω rheostat in Figure 5–69 be adjusted before the 0.5-A fuse blows?

FIGURE 5–69

32. Determine the total current from the source and the current through each resistor for each switch position in Figure 5–70.

FIGURE 5–70

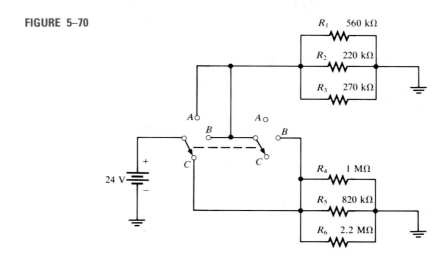

33. The electrical circuit in a room has a ceiling lamp that draws 1.25 A and four wall outlets. Two table lamps that each draw 0.833 A are plugged into two outlets, and a TV set that draws 1 A is connected to the third outlet. When all of these items are in use, how much current is in the main line serving the room? If the main line is protected by a 5-A circuit breaker, how much current can be drawn from the fourth outlet? Draw a schematic of this wiring.

34. The total resistance of a parallel circuit is 25 Ω. How much current flows through a 220-Ω resistor that makes up part of the parallel circuit if the total current is 100 mA?

35. Identify which groups of resistors are in parallel on the double-sided PC board in Figure 5–71.

FIGURE 5–71

Side 1

Side 2

FIGURE 5–72

FIGURE 5–73

36. If the total resistance in Figure 5–72 is 200 Ω, what is the value of R_2?

37. Determine the unknown resistances in Figure 5–73.

38. A total of 250 mA flows into a parallel circuit with a total resistance of 1.5 kΩ. The current must be increased by 25%. Determine how much resistance to add in parallel to accomplish this increase in current.

39. What is wrong with the circuit in Figure 5–74 (a full-color photo appears following page 240) if 25 V are applied across the red and black leads?

FIGURE 5–74

40. Develop a test procedure to check the circuit in Figure 5–75 (a full-color photo appears following page 240) to make sure that there are no open components. You must do this test without removing any component from the board. List the procedure in a detailed step-by-step format.

FIGURE 5–75

41. A certain parallel circuit consists of five ½-W resistors with the following values: 1.8 kΩ, 2.2 kΩ, 3.3 kΩ, 3.9 kΩ, and 4.7 kΩ. As you slowly increase the voltage across the parallel circuit, the total current slowly increases. Suddenly, the total current drops to a lower value.
(a) Excluding a power supply failure, what happened?
(b) What is the maximum voltage you should have applied?
(c) Specifically, what must be done to repair the fault?

42. For the circuit board shown in Figure 5–76 (a full-color photo appears following page 240), determine the resistance between the following pins if there is a short between pins 2 and 4.
  (a) pins 1 and 2
  (b) pins 2 and 3
  (c) pins 3 and 4
  (d) pins 1 and 4

43. For the circuit board shown in Figure 5–76 (a full-color photo appears following page 240), determine the resistance between the following pins if there is a short between pins 3 and 4.
  (a) pins 1 and 2
  (b) pins 2 and 3
  (c) pins 2 and 4
  (d) pins 1 and 4

FIGURE 5–76

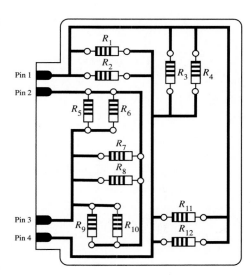

ANSWERS TO SECTION REVIEWS

Section 5–1

1. A parallel connection has more than one current path between two given points.

2. See Figure 5–77.   4. See Figure 5–78.

FIGURE 5–77

(a)   (b)   (c)

FIGURE 5-78

Section 5-2

1. 5 V.   **2.** 118 V, 118 V.   **3.** 50 V, 50 V.

Section 5-3

1. 2.5 A.   **2.** 400 mA.   **3.** 3 μA.   **4.** 7 A entering, 5 A leaving.

Section 5-4

1. Decrease.   **2.** The smallest resistance value.   **3.** 688 Ω, 250 Ω, 29.93 Ω.

Section 5-5

1. 0.44 A.   **2.** 444.36 V
3. 0.653 A through the 680-Ω; 1.35 A through the 330-Ω.   **4.** 8 kΩ.
5. 68.75 V.

Section 5-6

1. The 22-Ω has most current; the 220-Ω has least current.   **2.** 0.97 mA.
3. $I_1 = 3.26$ mA, $I_2 = 6.73$ mA.

Section 5-7

1. 16 W.   **2.** 584.47 W.

Section 5-8

1. 50 Ω, 1 mA.

2. Less than, because the shunt resistors must allow currents much greater than the current through the meter movement.

Section 5-9

1. No change in voltage; total current decreases.
2. When the total resistance is greater than it should be, a branch is open.
3. Yes.
4. 1 A.
5. The 1.2-A branch is open.

6

Series-Parallel Circuits

Various combinations of both series and parallel resistors are often used in electronic circuits. In this chapter, examples of such series-parallel arrangements are examined and analyzed. An important circuit called the *Wheatstone bridge* is introduced, and circuits with more than one voltage source are analyzed in simple steps. Also, you will learn how complex circuits can be simplified using *Thevenin's theorem*. Troubleshooting series-parallel circuits for shorts and opens is also covered.

After completing this chapter, you will be able to

☐ Identify series and parallel portions of a series-parallel circuit and recognize the relationships of all the resistors.
☐ Determine the total resistance of a series-parallel circuit.
☐ Determine the currents and voltages in a series-parallel circuit.
☐ Determine how resistive loads will affect voltage divider circuits.
☐ Recognize and determine the effect of a voltmeter on the circuit voltages that it is measuring.
☐ Understand and use a Wheatstone bridge.
☐ Apply the superposition method to evaluate circuits with more than one source.
☐ Apply Thevenin's theorem to simplify complex circuits.
☐ Use troubleshooting techniques to identify opens and shorts in a series-parallel circuit.

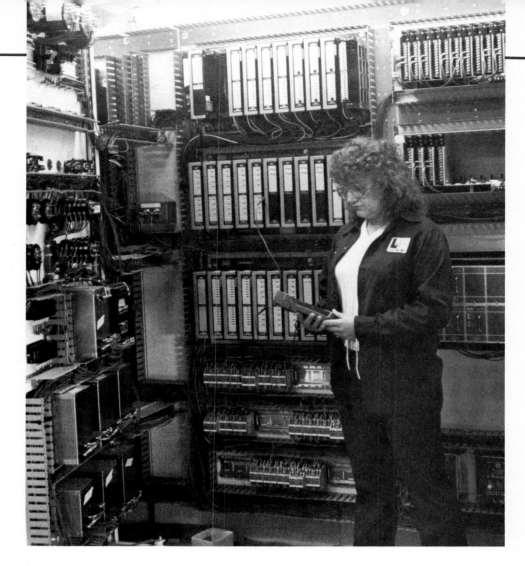

APPLICATION ASSIGNMENT

Recall that the Application Assignment in Chapter 4 required a voltage divider circuit to provide reduction (attenuation) of signals from several electronic sensing devices.

Although you did a great job on that assignment, it turns out that an additional requirement is necessary. As a result, you are asked to review the circuit to make sure that it meets the following specification: The output signal voltage from the voltage divider for any of the switch positions must not change more than 5% when an instrument with an internal input resistance of 1 MΩ is connected to the output.

Your assignment is to analyze the voltage divider that was developed in Chapter 4 under the loading condition specified and determine if redesign is required. After studying this chapter, you will be able to complete this assignment.

6–1

IDENTIFYING SERIES-PARALLEL RELATIONSHIPS

A series-parallel circuit consists of combinations of both series and parallel current paths. It is important to be able to identify how the components in a circuit are arranged in terms of their series and parallel relationships.

Figure 6–1(a) shows a simple series-parallel combination of resistors. Notice that the resistance from point A to point B is R_1. The resistance from point B to point C is R_2 and R_3 in parallel ($R_2 \parallel R_3$). The resistance from point A to point C is R_1 in series with the parallel combination of R_2 and R_3, as indicated in Figure 6–1(b).

When the circuit of Figure 6–1(a) is connected to a voltage source as shown in part (c), the *total* current flows through R_1 and divides at point B into the two parallel paths. These two branch currents then recombine, and the total current flows into the positive source terminal as shown.

(a)   (b)   (c)

FIGURE 6–1
A simple series-parallel circuit.

Now, to illustrate series-parallel relationships, we will increase the complexity of the circuit in Figure 6–1(a) step-by-step. In Figure 6–2(a), another resistor (R_4) is connected in series with R_1. The resistance between points A and B is now $R_1 + R_4$, and this combination is in series with the parallel combination of R_2 and R_3, as illustrated in part (b).

(a)

(b)

FIGURE 6–2
R_4 is added to the circuit in series with R_1.

In Figure 6–3(a), R_5 is connected in series with R_2. The series combination of R_2 and R_5 is in parallel with R_3. This entire series-parallel combination is in series with the $R_1 + R_4$ combination, as illustrated in part (b).

(a)   (b)

FIGURE 6–3
R_5 is added to the circuit in series with R_2.

In Figure 6–4(a), R_6 is connected in parallel with the series combination of R_1 and R_4. The series-parallel combination of R_1, R_4, and R_6 is in series with the series-parallel combination of R_2, R_3, and R_5, as indicated in part (b).

(a)   (b)

FIGURE 6–4
R_6 is added to the circuit in parallel with the series combination of R_1 and R_4.

EXAMPLE 6–1

Identify the series-parallel relationships in Figure 6–5.

FIGURE 6–5

Solution
Starting at the negative terminal of the source, follow the current paths. All the current produced by the source must go through R_1, which is in series with the rest of the circuit.

The total current takes two paths when it gets to point A. Part of it flows through R_2, and part of it through R_3. R_2 and R_3 are in parallel with each other, and this parallel combination is in series with R_1.

At point B, the currents through R_2 and R_3 come together again. Thus, the total current flows through R_4. R_4 is in series with R_1 and the parallel combination of R_2 and R_3. The currents are shown in Figure 6–6, where I_T is the total current.

In summary, R_1 and R_4 are in series with the parallel combination of R_2 and R_3.

FIGURE 6–6

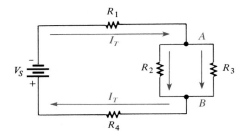

EXAMPLE 6–2

Describe the series-parallel combination between points A and D in Figure 6–7.

FIGURE 6–7

Solution
Between points B and C, there are two parallel paths. The lower path consists of R_4, and the upper path consists of a *series* combination of R_2 and R_3. This parallel combination is in series with both R_1 and R_5.

In summary, R_1 and R_5 are in series with the parallel combination of R_4 and $R_2 + R_3$.

EXAMPLE 6–3

Describe the total resistance between each pair of points in Figure 6–8.

FIGURE 6–8

Solution
1. From point *A* to *B*: R_1 is in *parallel* with the *series* combination of R_2 and R_3.
2. From point *A* to *C*: R_3 is in *parallel* with the *series* combination of R_1 and R_2.
3. From point *B* to *C*: R_2 is in *parallel* with the *series* combination of R_1 and R_3.

Redrawing a Schematic to Determine the Series-Parallel Relationships

Sometimes it is difficult to see the series-parallel relationships on a schematic diagram because of the way in which it is drawn. In such a situation, it helps to redraw the diagram so that the relationships become clear. The following example gives a simple illustration.

EXAMPLE 6-4   Identify the series-parallel relationships in Figure 6–9.

FIGURE 6-9

Solution
Sometimes it is easier to see a particular circuit arrangement if it is drawn in a different way. In this example, the circuit schematic is redrawn in Figure 6–10, which better illustrates the series-parallel relationships. Now you can see that R_2 and R_3 are in parallel with each other and also that R_4 and R_5 are in parallel with each other. Both parallel combinations are in series with each other and with R_1.

FIGURE 6-10

Determining Relationships on a Printed Circuit Board

Usually, the physical arrangement of components on a PC board bears no resemblance to the actual circuit relationships. By tracing out the circuit on the PC board and rearranging the components on paper into a recognizable form, you can determine the series-parallel relationships. An example will illustrate.

EXAMPLE 6–5   Determine the relationships of the resistors on the PC board in Figure 6–11.

FIGURE 6–11

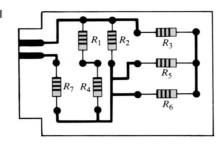

Solution

In Figure 6–12(a), the schematic is drawn approximately in the same arrangement as that of the resistors on the board. In part (b), the resistors are reoriented so that the series-parallel relationships are obvious. R_1 and R_4 are in series; $R_1 + R_4$ is in parallel with R_2; R_5 and R_6 are in parallel and this combination is in series with R_3. The R_3, R_5, and R_6 series-parallel combination is in parallel with both R_2 and the $R_1 + R_4$ combination. This entire series-parallel combination is in series with R_7, as Figure 6–12(c) illustrates.

(a)

(b)

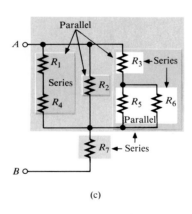

(c)

FIGURE 6–12

SECTION REVIEW
6–1

1. A certain series-parallel circuit is described as follows: R_1 and R_2 are in parallel. This parallel combination is in series with another parallel combination of R_3 and R_4. Sketch the circuit.
2. In the circuit of Figure 6–13, describe the series-parallel relationships of the resistors.

FIGURE 6–13

3. Which resistors are in parallel in Figure 6–14?
4. Identify the parallel arrangements in Figure 6–15.
5. Are the parallel combinations in Figure 6–15 in series?

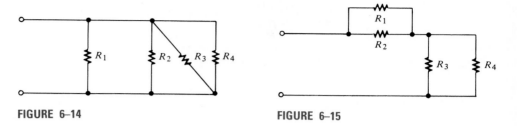

FIGURE 6–14   **FIGURE 6–15**

6–2

ANALYSIS OF SERIES-PARALLEL CIRCUITS

Several quantities are important in a circuit that is a series-parallel configuration of
resistors. In this section you will learn how to find total resistance, total current, branch
currents, and the voltage across any portion of a circuit.

Total Resistance

In Chapter 4, you learned how to determine total series resistance. In Chapter 5, you
learned how to determine total parallel resistance.

   To find the total resistance R_T of a series-parallel combination, first identify the
series and parallel relationships, and then apply what you have previously learned. The
following two examples illustrate this general approach.

EXAMPLE 6–6   Determine R_T between points A and B of the circuit in Figure 6–16.

FIGURE 6–16

Solution

R_2 and R_3 are in parallel, and this parallel combination is in series with R_1. First find
the parallel resistance of R_2 and R_3. Since R_2 and R_3 are equal in value, divide the value
by 2:

$$R_{2\text{-}3} = \frac{R}{n} = \frac{100\ \Omega}{2} = 50\ \Omega$$

Now, since R_1 is in series with $R_{2\text{-}3}$, add their values as follows:

$$R_T = R_1 + R_{2\text{-}3} = 10\ \Omega + 50\ \Omega = 60\ \Omega$$

The calculator sequence is

$$\boxed{1}\ \boxed{0}\ \boxed{0}\ \boxed{\div}\ \boxed{2}\ \boxed{=}\ \boxed{+}\ \boxed{1}\ \boxed{0}\ \boxed{=}$$

EXAMPLE 6–7   Find R_T of the circuit in Figure 6–17.

FIGURE 6–17

Solution

In the *upper branch*, R_2 is in series with R_3. We will call the series combination $R_{2\text{-}3}$. It is equal to $R_2 + R_3$:

$$R_{2\text{-}3} = R_2 + R_3 = 82\ \Omega + 18\ \Omega = 100\ \Omega$$

In the *lower branch*, R_4 and R_5 are in parallel with each other. We will call this parallel combination $R_{4\text{-}5}$. It is calculated as follows:

$$R_{4\text{-}5} = \frac{R_4\,R_5}{R_4 + R_5} = \frac{(60\ \Omega)(40\ \Omega)}{60\ \Omega + 40\ \Omega} = 24\ \Omega$$

Also in the *lower branch*, the parallel combination of R_4 and R_5 is in series with R_6. This series-parallel combination is designated $R_{4\text{-}5\text{-}6}$ and is calculated as follows:

$$R_{4\text{-}5\text{-}6} = R_6 + R_{4\text{-}5} = 76\ \Omega + 24\ \Omega = 100\ \Omega$$

Figure 6–18 shows the original circuit in a simplified *equivalent* form.

FIGURE 6–18

Now we can find the resistance between points A and B. It is $R_{2\text{-}3}$ in parallel with $R_{4\text{-}5\text{-}6}$. Since these resistances are equal, the equivalent resistance is calculated as follows:

$$R_{AB} = \frac{100 \ \Omega}{2} = 50 \ \Omega$$

Finally, the total resistance is R_1 in series with R_{AB}:

$$R_T = R_1 + R_{AB} = 100 \ \Omega + 50 \ \Omega = 150 \ \Omega$$

Total Current

Once the total resistance and the source voltage are known, you can find total current in a circuit by applying Ohm's law. Total current is the total source voltage divided by the total resistance:

$$I_T = \frac{V_S}{R_T}$$

For example, let us find the total current in the circuit of Example 6–7 (Figure 6–17). Assume that the source voltage is 30 V. The calculation is as follows:

$$I_T = \frac{V_S}{R_T} = \frac{30 \ \text{V}}{150 \ \Omega} = 0.2 \ \text{A}$$

Branch Currents

Using the *current divider formula, Kirchhoff's current law,* or *Ohm's law,* or combinations of these, you can find the current in any branch of a series-parallel circuit. In some cases it may take repeated application of the formula to find a given current.

EXAMPLE 6–8

Determine the current through R_4 in Figure 6–19 if $V_S = 50$ V.

FIGURE 6–19

Solution

First the current (I_2) into the junction of R_3 and R_4 must be found. Once we know this current, we can use the current divider formula to find I_4.

Notice that there are two main branches in the circuit. The left-most branch consists of only R_1. The right-most branch has R_2 in series with the parallel combination of R_3 and R_4. The voltage across both of these main branches is the same and equal to 50 V. We can find the current (I_2) into the junction of R_3 and R_4 by calculating the resistance of the right-most main branch and then applying Ohm's law as follows:

$$R_{2\text{-}3\text{-}4} = R_2 + \frac{R_3 R_4}{R_3 + R_4} = 330\ \Omega + \frac{(330\ \Omega)(560\ \Omega)}{890\ \Omega}$$

$$= 537.6\ \Omega$$

$$I_2 = \frac{V_S}{R_{2\text{-}3\text{-}4}} = \frac{50\ \text{V}}{537.6\ \Omega}$$

$$= 0.093\ \text{A}$$

Using the current divider formula, calculate I_4 as follows:

$$I_4 = \left(\frac{R_3}{R_3 + R_4}\right)I_2 = \left(\frac{330\ \Omega}{890\ \Omega}\right)0.093\ \text{A}$$

$$= 0.035\ \text{A}$$

You can find the current in R_3 by applying Kirchhoff's law ($I_2 = I_3 + I_4$) and subtracting I_4 from I_2 ($I_3 = I_2 - I_4$).

You can find the current in R_1 by using Ohm's law ($I_1 = V_S/R_1$).

The calculator sequence is

3	3	0	×	5	6	0	÷	(	3	3	0
+	5	6	0	)	+	3	3	0	=	STO	
5	0	÷	RCL	=	STO	3	3	0	÷		
(	3	3	0	+	5	6	0	)	×	RCL	=

Voltage Relationships

The circuit in Figure 6–20 will be used to illustrate voltage relationships in a series-parallel circuit. Voltmeters are connected to measure each of the resistor voltages, and the readings are as indicated.

Some general observations about Figure 6–20 follow:

1. V_1 and V_2 are equal because R_1 and R_2 are in parallel. (Recall that voltages across parallel branches are the same.) V_1 and V_2 are the same as the voltage from A to B.
2. V_3 is equal to $V_4 + V_5$ because R_3 is in parallel with the series combination of R_4 and R_5. (V_3 is the same as the voltage from B to C.)
3. V_4 is about one-third of the voltage from B to C because R_4 is about one-third of the resistance $R_4 + R_5$ (by the voltage divider principle).
4. V_5 is about two-thirds of the voltage from B to C because R_5 is about two-thirds of $R_4 + R_5$.
5. $V_1 + V_3$ equals V_S because, by Kirchhoff's voltage law, the sum of the voltage drops must equal the source voltage.

The following example will verify the meter readings in Figure 6–20.

FIGURE 6–20
Illustration of voltage relationships.

EXAMPLE 6–9

Verify that the voltmeter readings in Figure 6–20 are correct. The circuit is redrawn as a schematic in Figure 6–21.

FIGURE 6–21

Solution

The resistance from point A to point B is the parallel combination of R_1 and R_2:

$$R_{AB} = \frac{R_1 R_2}{R_1 + R_2} = \frac{(1 \text{ k}\Omega)(2.2 \text{ k}\Omega)}{3.2 \text{ k}\Omega} = 688 \ \Omega$$

The resistance from point B to point C is R_3 in parallel with the series combination of R_4 and R_5:

$$R_4 + R_5 = 330\ \Omega + 680\ \Omega = 1010\ \Omega = 1.01\ k\Omega$$

$$R_{BC} = \frac{R_3(R_4 + R_5)}{R_3 + R_4 + R_5} = \frac{(1.8\ k\Omega)(1.01\ k\Omega)}{2.81\ k\Omega} = 647\ \Omega$$

The resistance from A to B is in series with the resistance from B to C, so the total circuit resistance is

$$R_T = R_{AB} + R_{BC} = 688\ \Omega + 647\ \Omega = 1335\ \Omega$$

Using the voltage divider principle, we find the voltages as follows:

$$V_{AB} = \left(\frac{R_{AB}}{R_T}\right)V_S = \left(\frac{688\ \Omega}{1335\ \Omega}\right)100\ V = 51.54\ V$$

$$V_{BC} = \left(\frac{R_{BC}}{R_T}\right)V_S = \left(\frac{647\ \Omega}{1335\ \Omega}\right)100\ V = 48.46\ V$$

$$V_1 = V_2 = V_{AB} = 51.54\ V$$

$$V_3 = V_{BC} = 48.46\ V$$

$$V_4 = \left(\frac{R_4}{R_4 + R_5}\right)V_{BC} = \left(\frac{330\ \Omega}{1010\ \Omega}\right)48.46\ V = 15.83\ V$$

$$V_5 = \left(\frac{R_5}{R_4 + R_5}\right)V_{BC} = \left(\frac{680\ \Omega}{1010\ \Omega}\right)48.46\ V = 32.63\ V$$

SECTION REVIEW 6–2

1. Find the total resistance between A and B in the circuit of Figure 6–22.
2. Find I_3 in Figure 6–22.
3. Find V_2 in Figure 6–22.
4. Determine R_T and I_T in Figure 6–23.

FIGURE 6–22

FIGURE 6–23

6–3

LOADED VOLTAGE DIVIDERS

Voltage dividers were introduced in Chapter 4. In this section we will discuss the effects of resistive loads on voltage dividers. For example, the voltage divider in Figure 6–24(a) produces an output voltage (V_{OUT}) of 5 V because the two resistors are of equal value. This voltage is the *unloaded output voltage. When a load resistor, R_L, is con-*

(a) Unloaded   (b) Loaded

FIGURE 6–24
A voltage divider with both unloaded and loaded outputs.

FIGURE 6–25
The effect of a load resistor.

(a) No load

(b) R_L not significantly greater than R_2

(c) R_L much greater than R_2

nected from the output to ground as shown in Figure 6–24(b), the output voltage is reduced by an amount that depends on the value of R_L. The load resistor is in parallel with R_2, reducing the resistance from point A to ground and, as a result, also reducing the voltage across the parallel combination. This is one effect of *loading* a voltage divider. Another effect of a loaded condition is that more current is drawn from the source because the total resistance of the circuit is reduced.

The larger R_L is compared to R_2, the less the output voltage is reduced from its unloaded value, as illustrated in Figure 6–25. The reason is that when two resistors are connected in parallel and one of the resistors is much greater than the other, the total resistance is close to the value of the smaller resistance.

EXAMPLE 6–10

Determine both the unloaded and the loaded output voltages of the voltage divider in Figure 6–26 for the following two values of load resistance:

(a) $R_L = 10 \text{ k}\Omega$
(b) $R_L = 100 \text{ k}\Omega$

FIGURE 6–26

Solution
(a) The *unloaded* output voltage is

$$V_{\text{OUT}} = \left(\frac{10 \text{ k}\Omega}{14.7 \text{ k}\Omega} \right) 5 \text{ V} = 3.4 \text{ V}$$

With the 10-kΩ load resistor connected, R_L is in parallel with R_2, which gives 5 kΩ, as shown by the equivalent circuit in Figure 6–27(a).
The *loaded* output voltage is

$$V_{\text{OUT}} = \left(\frac{5 \text{ k}\Omega}{9.7 \text{ k}\Omega} \right) 5 \text{ V} = 2.6 \text{ V}$$

FIGURE 6–27

(a) $R_L = 10 \text{ k}\Omega$

(b) $R_L = 100 \text{ k}\Omega$

(b) With the 100-kΩ load, the resistance from output to ground is

$$\frac{R_2 R_L}{R_2 + R_L} = \frac{(10\ \text{k}\Omega)(100\ \text{k}\Omega)}{110\ \text{k}\Omega} = 9.1\ \text{k}\Omega$$

as shown in Figure 6–27(b).

The *loaded* output voltage is

$$V_{\text{OUT}} = \left(\frac{9.1\ \text{k}\Omega}{13.8\ \text{k}\Omega}\right) 5\ \text{V} = 3.3\ \text{V}$$

Notice that with the larger value of R_L, the output is reduced from its unloaded value by much less than it is with the smaller R_L. This problem illustrates the loading effect of R_L on the voltage divider.

Load Current and Bleeder Current

In a loaded voltage divider circuit, the total current drawn from the source consists of currents through the load resistors and the divider resistors. Figure 6–28 shows a voltage divider with two voltage outputs or two *taps*. Notice that the total current, I_T, flows through R_1 to point A where the current divides in I_{RL1}, which flows through R_{L1} and into I_2, which flows through R_2. At point B, the current I_2 divides into I_{RL2}, which flows through R_{L2} and into I_3, which flows through R_3. I_3 is called the *bleeder current*, which is the current left after the total load current is subtracted from the total current into the circuit:

$$I_{\text{BLEEDER}} = I_T - I_{RL1} - I_{RL2}$$

FIGURE 6–28
Currents in a two-tap loaded voltage divider.

EXAMPLE 6–11

Determine the load currents and the bleeder current in the two-tap loaded voltage divider in Figure 6–28.

Solution

The equivalent resistance from point A to ground is the 100-kΩ load resistor R_{L1} in parallel with the combination of R_2 in series with the parallel combination of R_3 and R_{L2}. This solution is as follows.

R_3 in parallel with R_{L2} is R_B:

$$R_B = \frac{(6 \text{ k}\Omega)(100 \text{ k}\Omega)}{106 \text{ k}\Omega} = 5.66 \text{ k}\Omega$$

R_2 in series with R_B is $R_{2\text{-}B}$:

$$R_{2\text{-}B} = 6 \text{ k}\Omega + 5.66 \text{ k}\Omega = 11.66 \text{ k}\Omega$$

R_{L1} in parallel with $R_{2\text{-}B}$ is R_A:

$$R_A = \frac{(100 \text{ k}\Omega)(11.66 \text{ k}\Omega)}{111.66 \text{ k}\Omega} = 10.44 \text{ k}\Omega$$

$$R_T = R_A + R_1 = 10.44 \text{ k}\Omega + 12 \text{ k}\Omega = 22.44 \text{ k}\Omega$$

R_A is the total resistance from point A to ground.

The voltage across R_{L1} is determined as follows:

$$V_{RL1} = V_A = \left(\frac{R_A}{R_T}\right)V_S = \left(\frac{10.44 \text{ k}\Omega}{22.44 \text{ k}\Omega}\right)24 \text{ V} = 11.17 \text{ V}$$

$$I_{RL1} = \frac{V_{RL1}}{R_{L1}} = \frac{11.17 \text{ V}}{100 \text{ k}\Omega} = 0.112 \text{ mA}$$

The voltage at point B is determined by the parallel combination of R_{L2} and R_3 (R_B), R_2, and the voltage at point A:

$$V_B = \left(\frac{R_B}{R_2 + R_B}\right)V_A = \left(\frac{5.66 \text{ k}\Omega}{11.66 \text{ k}\Omega}\right)11.17 \text{ V} = 5.42 \text{ V}$$

The load current through R_{L2} is

$$I_{RL2} = \frac{V_{RL2}}{R_{L2}} = \frac{V_B}{R_{L2}} = \frac{5.42 \text{ V}}{100 \text{ k}\Omega} = 0.054 \text{ mA}$$

The bleeder current is

$$I_3 = \frac{V_B}{R_3} = \frac{5.42 \text{ V}}{6 \text{ k}\Omega} = 0.903 \text{ mA}$$

Bipolar Voltage Dividers

An example of a voltage divider that produces both positive and negative voltages from a single source is shown in Figure 6–29. Notice that neither the positive nor the nega-

FIGURE 6-29
A bipolar voltage divider. The positive and negative voltages are with respect to ground.

tive terminal of the source is connected to ground. The voltages at points A and B are *positive with respect to ground,* and the voltages at points C and D are *negative with respect to ground.*

SECTION REVIEW 6–3

1. A load resistor is connected to an output on a voltage divider. What effect does the load resistor have on the output voltage?

2. A larger-value load resistor will cause the output voltage to change less than a smaller-value one will (T or F).

3. For the voltage divider in Figure 6–30, determine the unloaded output voltage. Also determine the output voltage with a 10-kΩ load resistor connected from the output to ground.

FIGURE 6-30

6–4

THE LOADING EFFECT OF A VOLTMETER

As you have learned, voltmeters must be connected in parallel with a resistor in order to measure the voltage across the resistor. Because of its internal resistance, a voltmeter puts a load on the circuit and will affect, to a certain extent, the voltage that is being measured. Until now, we have ignored the loading effect. Why?

The internal resistance of a voltmeter is very high, and normally it has negligible effect on the circuit that is being measured. However, if the internal resistance of the voltmeter is not sufficiently greater than the circuit resistance across which it is connected, the loading effect will cause the measured voltage to be less than its actual value. *You should always be aware of this effect.*

Why a Voltmeter Can Load a Circuit

When a voltmeter is connected to a circuit as shown, for example, in Figure 6–31(a), its internal resistance appears in parallel with R_3, as shown in part (b). The resistance from point A to point B is altered by the loading effect of the voltmeter's internal resistance R_M and is equal to $R_3 \parallel R_M$, as indicated in part (c).

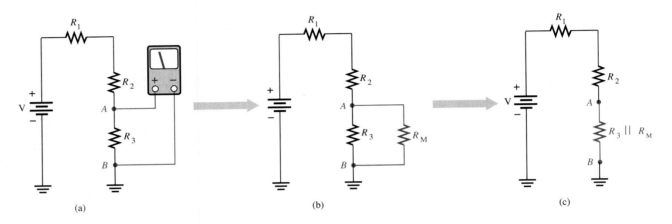

FIGURE 6–31
The loading effect of a voltmeter.

If R_M is much greater than R_3, the resistance from A to B changes very little, and the meter indicates the actual voltage. If R_M is not sufficiently greater than R_3, the resistance from A to B is reduced significantly, and the voltage across R_3 is altered by the loading effect of the meter. A good rule of thumb is that if the meter resistance is at least ten times greater than the resistance across which it is connected, the loading effect can be neglected.

Internal Resistance of a Voltmeter

There are three basic categories of voltmeters: the electromagnetic analog voltmeter, whose internal resistance is determined by its *sensitivity factor*, the electronic analog voltmeter, whose internal resistance is set by an input amplifier and is typically at least 10 MΩ, and the digital voltmeter, whose internal resistance is also typically at least 10 MΩ. The electronic analog voltmeter and the digital voltmeter present fewer loading problems than the electromagnetic type because their internal resistances are higher.

As mentioned, the internal resistance of an electromagnetic voltmeter is specified by the sensitivity factor. A common sensitivity value is 20,000 Ω/V. The internal resistance depends on the dc voltage range switch setting of the meter and is determined by multiplying the sensitivity by the range setting. For example, if the dc voltage range switch is set at 10 V, the internal resistance is

$$R_M = 20,000 \ \Omega/V \times 10 \ V = 200,000 \ \Omega$$

The lower the range setting, the less the internal resistance and the greater the loading effect of the voltmeter on a circuit. Always use the highest possible range setting that will allow an accurate reading from the meter scale. However, you would not use a 60-V setting to measure 1 V because you could not read 1 V accurately on the 60-V meter scale. The following example shows how the range setting affects the loading.

EXAMPLE 6–12

How much does the voltmeter affect the voltage being measured for each range switch setting indicated in Figure 6–32? Assume the meter has a sensitivity of 20,000 Ω/V.

FIGURE 6–32

(a)

(b)

(c)

Solution

The unloaded voltage across R_2 in the voltage divider circuit is

$$V_{R2} = \left(\frac{10 \text{ k}\Omega}{28 \text{ k}\Omega}\right) 15 \text{ V} = 5.357 \text{ V}$$

(a) The internal resistance of the meter in part (a) of the figure is

$$R_M = 20,000 \ \Omega/V \times 12 \ V = 240 \ k\Omega$$

The meter's resistance in parallel with R_2 is

$$\frac{R_2 R_M}{R_2 + R_M} = \frac{(10 \ k\Omega)(240 \ k\Omega)}{250 \ k\Omega} = 9.6 \ k\Omega$$

The voltage actually measured by the meter is

$$V_{R2} = \left(\frac{9.6 \ k\Omega}{27.6 \ k\Omega}\right) 15 \ V = 5.22 \ V$$

(b) The internal resistance of the meter in part (b) of the figure is

$$R_M = 20,000 \ \Omega/V \times 60 \ V = 1.2 \ M\Omega$$

and

$$\frac{R_2 R_M}{R_2 + R_M} = \frac{(10 \ k\Omega)(1.2 \ M\Omega)}{1.21 \ M\Omega} = 9917 \ \Omega$$

The voltage actually measured is

$$V_{R2} = \left(\frac{9917 \ \Omega}{27,917 \ \Omega}\right) 15 \ V = 5.33 \ V$$

(c) The internal resistance of the meter in part (c) of the figure is

$$R_M = 20,000 \ \Omega/V \times 300 \ V = 6 \ M\Omega$$

and

$$\frac{R_2 R_M}{R_2 + R_M} = \frac{(10 \ k\Omega)(6 \ M\Omega)}{6.01 \ M\Omega} = 9983 \ \Omega$$

The voltage actually measured is

$$V_{R2} = \left(\frac{9983 \ \Omega)}{27,983 \ \Omega}\right) 15 \ V = 5.35 \ V$$

As you can see, the loading decreases as you increase the range setting on the voltmeter. However, the 12-V setting is the best choice in this case because, although the loading effect is greater, you can read the expected voltage much more accurately than you can on the 60-V range. Of course, using the 300-V range is out of the question in this case.

**SECTION REVIEW
6–4**

1. What does it mean when we say a voltmeter loads a circuit?
2. If a voltmeter with a 10-MΩ internal resistance is measuring the voltage across a 1-kΩ resistor, should you normally be concerned about the loading effect?

3. A certain electromagnetic-type voltmeter is set on the 12-V range to measure 2.5 V. If you switch it down to the 3-V range, will there be an increase or a decrease in the measured voltage? What is the advantage of using the 3-V range instead of the 12-V range in this situation?

4. What is the internal resistance of a 30,000-Ω/V voltmeter on its 50-V range setting?

6–5   THE WHEATSTONE BRIDGE

The *bridge circuit* is widely used in measurement devices and other applications that you will learn later. For now, we will consider the *balanced bridge,* which can be used to measure unknown resistance values. This circuit, shown in Figure 6–33(a), is known as a *Wheatstone bridge.* Part (b) is the same circuit electrically, but it is drawn in a different way.

(a)

(b)

FIGURE 6–33
Wheatstone bridge.

A bridge is said to be *balanced* when the voltage (V_{OUT}) across the output terminals A and B is *zero;* that is, $V_A = V_B$. If V_A equals V_B, then $V_{R1} = V_{R2}$, since the top sides of both R_1 and R_2 are connected to the same point. Also, $V_{R3} = V_{R4}$, since the bottom sides of both R_3 and R_4 connect to the same point. The voltage ratios can be written as

$$\frac{V_1}{V_3} = \frac{V_2}{V_4}$$

Substituting by Ohm's law yields

$$\frac{I_1 R_1}{I_1 R_3} = \frac{I_2 R_2}{I_2 R_4}$$

The currents cancel, leaving

$$\frac{R_1}{R_3} = \frac{R_2}{R_4}$$

Solving for R_1, we get

$$R_1 = R_3 \left(\frac{R_2}{R_4} \right)$$

How can we use this formula to determine an unknown resistance? First, let us make R_3 a *variable* resistor and call it R_V. Also, we set the ratio R_2/R_4 to a known value. If R_V is adjusted until the bridge is balanced, the product of R_V and the ratio R_2/R_4 is equal to R_1, which is our unknown resistor (R_{UNK}). The formula for unknown resistance is stated as follows:

$$R_{UNK} = R_V\left(\frac{R_2}{R_4}\right) \tag{6--1}$$

The bridge is balanced when the voltage across the output terminals equals zero ($V_A = V_B$). A *galvanometer* (a meter that measures small currents in either direction and is zero at center scale) is connected between the output terminals. Then R_V is adjusted until the galvanometer shows zero current ($V_A = V_B$), indicating a balanced condition. The setting of R_V multiplied by the ratio R_2/R_4 gives the value of R_{UNK}. Figure 6--34 shows this arrangement. For example, if $R_2/R_4 = \frac{1}{10}$ and $R_V = 680\ \Omega$, then $R_{UNK} = (680\ \Omega)(\frac{1}{10}) = 68\ \Omega$.

FIGURE 6--34
Balanced Wheatstone bridge.

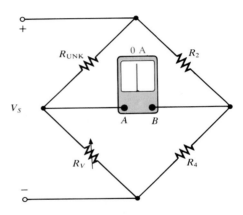

EXAMPLE 6--13   What is R_{UNK} under the balanced bridge conditions shown in Figure 6--35?

FIGURE 6--35

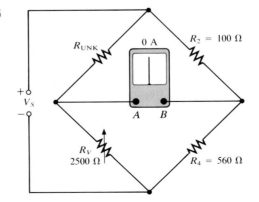

Solution

$$R_{\text{UNK}} = R_V \left(\frac{R_2}{R_4} \right) = 2500 \ \Omega \left(\frac{100 \ \Omega}{560 \ \Omega} \right)$$

$$= 446.43 \ \Omega$$

A Bridge Application

Bridge circuits are used for many measurements other than that of unknown resistance. One application of the Wheatstone bridge is in accurate temperature measurement. A temperature-sensitive element such as a thermistor is connected in a Wheatstone bridge as shown in Figure 6–36. An amplifier is connected across the output from A to B in order to increase the output voltage from the bridge to a usable value. The bridge is calibrated so that it is balanced at a specified reference temperature. As the temperature changes, the resistance of the sensing element changes proportionately, and the bridge becomes unbalanced. As a result, V_{AB} changes and is amplified (increased) and converted to a form for direct temperature readout on a gauge or a digital-type display.

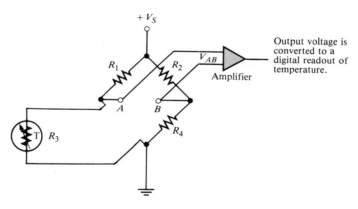

FIGURE 6–36
A simplified circuit for temperature measurement.

SECTION REVIEW 6–5

1. Under what condition is a bridge balanced?
2. What is the unknown resistance in Figure 6–34 when $R_V = 3.3 \ \text{k}\Omega$, $R_2 = 10 \ \text{k}\Omega$, and $R_4 = 2.2 \ \text{k}\Omega$?

6–6

CIRCUITS WITH MORE THAN ONE VOLTAGE SOURCE

Some circuits require more than one voltage source. For example, certain types of amplifiers require both a positive and a negative voltage source for proper operation.

In this section, a general method is presented for evaluating series-parallel circuits having multiple voltage sources. The method is illustrated with two-source circuits.

However, once you understand the basic principle, you can easily extend this method to circuits with any number of sources.

The Superposition Method

The superposition method is a way to determine currents and voltages in a circuit that has multiple sources by taking *one source at a time*. The other sources are replaced by their internal resistances. Recall from Chapter 4 that the *ideal* voltage source has a *zero* internal resistance. In this section, all voltage sources will be treated as ideal in order to simplify the coverage.

The steps in applying the superposition method are as follows:

Step 1:   Take one voltage source at a time and replace each of the other voltage sources with a short (a short represents zero resistance).

(a) Problem: Find I_2.

(b) Replace V_{S2} with zero resistance (short).

(c) Find R_T and I_T looking from V_{S1}:
$R_{T1} = R_1 + R_2 \| R_3$
$I_{T1} = V_{S1}/R_{T1}$

(d) Find I_2 due to V_{S1} (current divider):
$I_{2(1)} = \left(\dfrac{R_3}{R_2 + R_3}\right) I_{T1}$

(e) Replace V_{S1} with zero resistance (short).

(f) Find R_T and I_T looking from V_{S2}: $R_{T2} = R_3 + R_1 \| R_2$
$I_{T2} = V_{S2}/R_{T2}$

(g) Find I_2 due to V_{S2}:
$I_{2(2)} = \left(\dfrac{R_1}{R_1 + R_2}\right) I_{T2}$

(h) Add $I_{2(1)}$ and $I_{2(2)}$ to get the actual I_2 (they are in same direction):
$I_2 = I_{2(1)} + I_{2(2)}$

FIGURE 6–37
Demonstration of the superposition method.

Step 2: Determine the current or voltage that you need just as if there were only that *one* source in the circuit.

Step 3: Repeat Steps 1 and 2 for each source in turn.

Step 4: To find an actual current or voltage, add or subtract the currents or voltages due to each individual source. If the currents are in the same direction or the voltages of the same polarity, add them. If the currents are in opposite directions or the voltages of opposite polarities, subtract them.

An example of the approach to superposition is demonstrated in Figure 6–37 for a series-parallel circuit with two voltage sources. Study the steps in this figure.

EXAMPLE 6–14   Find the total current through R_3 in Figure 6–38.

FIGURE 6–38
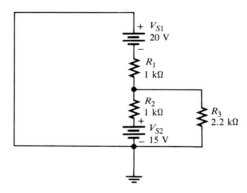

Solution

Step 1: Find the current through R_3 due to source V_{S1} by replacing source V_{S2} with a short, as shown in Figure 6–39(a).
Looking from V_{S1},

$$R_{T1} = R_1 + \frac{R_2 R_3}{R_2 + R_3}$$

$$= 1\ \text{k}\Omega + \frac{(1\ \text{k}\Omega)(2.2\ \text{k}\Omega)}{3.2\ \text{k}\Omega} = 1.69\ \text{k}\Omega$$

$$I_{T1} = \frac{V_{S1}}{R_{T1}} = \frac{20\ \text{V}}{1.69\ \text{k}\Omega} = 11.83\ \text{mA}$$

Now apply the current divider formula to find the current through R_3 due to source V_{S1} as follows:

$$I_{3(1)} = \left(\frac{R_2}{R_2 + R_3}\right)I_{T1} = \left(\frac{1\ \text{k}\Omega}{3.2\ \text{k}\Omega}\right)11.83\ \text{mA} = 3.7\ \text{mA}$$

Notice that this current flows through R_3 in the direction shown in Figure 6–39(a).

Step 2: Next find I_3 due to source V_{S2} by replacing source V_{S1} with a short, as shown in Figure 6–39(b).

FIGURE 6–39

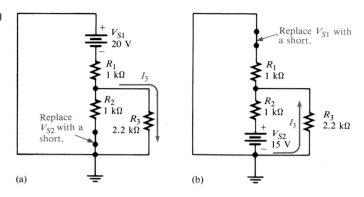

(a)   (b)

Looking from V_{S2},

$$R_{T2} = R_2 + \frac{R_1 R_3}{R_1 + R_3}$$

$$= 1 \text{ k}\Omega + \frac{(2.2 \text{ k}\Omega)(1 \text{ k}\Omega)}{3.2 \text{ k}\Omega} = 1.69 \text{ k}\Omega$$

$$I_{T2} = \frac{V_{S2}}{R_{T2}} = \frac{15 \text{ V}}{1.69 \text{ k}\Omega} = 8.88 \text{ mA}$$

Now apply the current divider formula to find the current through R_3 due to source V_{S2} as follows:

$$I_{3(2)} = \left(\frac{R_1}{R_1 + R_3} \right) I_T = \left(\frac{1 \text{ k}\Omega}{3.2 \text{ k}\Omega} \right) 8.88 \text{ mA} = 2.78 \text{ mA}$$

Notice that this current flows through R_3 in the direction shown in Figure 6–39(b).

Step 3: Calculation of the total current through R_3 is as follows:

$$I_{3(T)} = I_{3(1)} - I_{3(2)} = 3.7 \text{ mA} - 2.78 \text{ mA} = 0.92 \text{ mA}$$

This current flows through R_3 in the same direction as the larger component current, $I_{3(1)}$.

**SECTION REVIEW
6–6**

1. Using the superposition theorem, find the current through R_1 in Figure 6–40.
2. If two currents are in opposing directions through a branch of a circuit, in what direction is the net current?

FIGURE 6–40

6–7

THEVENIN'S THEOREM

Thevenin's theorem gives us a method for simplifying a circuit to a standard equivalent form. The Thevenin equivalent form of any resistive circuit consists of an *equivalent voltage source* (V_{TH}) and an *equivalent resistance* (R_{TH}), arranged as shown in Figure 6–41. The values of the equivalent voltage and resistance depend on the values in the original circuit. Any resistive circuit can be simplified regardless of its complexity.

FIGURE 6–41
The general form of a Thevenin equivalent circuit. Any resistive circuit can be reduced to this form.

The Meaning of Equivalency in Thevenin's Theorem

Although a Thevenin equivalent circuit is not the same as its original circuit, it *acts* the same in terms of the output voltage and current. For example, as shown in Figure 6–42, place a resistive circuit of any complexity in a box with only the output terminals exposed. Then place the Thevenin equivalent of that circuit in an *identical* box with, again, only the output terminals exposed. Connect identical load resistors across the output terminals of each box. Next connect a voltmeter and an ammeter to measure the voltage and current for each load as shown in the figure. The measured values will be identical (neglecting tolerance variations), and you will not be able to determine which box contains the original circuit and which contains the Thevenin equivalent. That is, in terms of your observations, both circuits are the same. This condition is sometimes known as *terminal equivalency,* because both circuits look the same from the "viewpoint" of the two output terminals.

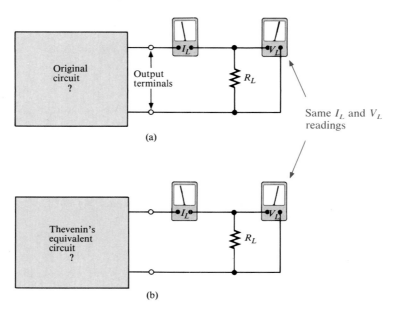

FIGURE 6–42
Which box contains the original circuit and which contains the Thevenin equivalent circuit?

How to Find the Thevenin Equivalent of a Circuit

To find the Thevenin equivalent of any circuit, determine the equivalent voltage, V_{TH}, and the equivalent resistance, R_{TH}. For example, in Figure 6–43, the Thevenin equivalent for the circuit between points A and B is found as follows.

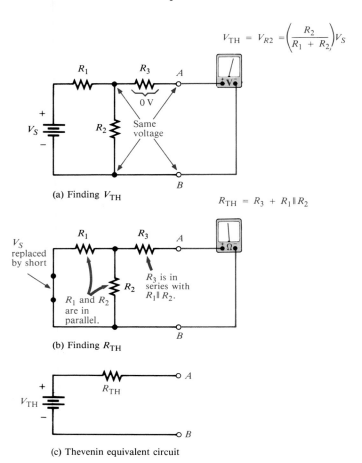

$$V_{TH} = V_{R2} = \left(\frac{R_2}{R_1 + R_2}\right)V_S$$

(a) Finding V_{TH}

$$R_{TH} = R_3 + R_1 \| R_2$$

(b) Finding R_{TH}

(c) Thevenin equivalent circuit

FIGURE 6–43
Example of the simplification of a circuit by Thevenin's theorem.

In part (a), the voltage across the designated points A and B is the Thevenin equivalent voltage. In this particular circuit, the voltage from A to B is the same as the voltage across R_2 because there is no current through R_3 and, therefore, no voltage drop across it. V_{TH} is expressed as follows for this particular example:

$$V_{TH} = \left(\frac{R_2}{R_1 + R_2}\right)V_S$$

In part (b), the resistance between points A and B with the source replaced by a short (zero internal resistance) is the Thevenin equivalent resistance. In this particular circuit, the resistance from A to B is R_3 in series with the parallel combination of R_1 and R_2. Therefore, R_{TH} is expressed as follows:

$$R_{TH} = R_3 + \frac{R_1 R_2}{R_1 + R_2}$$

The Thevenin equivalent circuit is shown in part (c).

EXAMPLE 6–15

Find the Thevenin equivalent between the output terminals of the circuit in Figure 6–44.

FIGURE 6–44

Solution

V_{TH} equals the voltage across $R_2 + R_3$, as shown in Figure 6–45(a). Use the voltage divider principle to find V_{TH}:

$$V_{TH} = \left(\frac{R_2 + R_3}{R_1 + R_2 + R_3}\right) V_S = \left(\frac{69\ \Omega}{169\ \Omega}\right) 10\ V = 4.08\ V$$

(a) The voltage from A to B is V_{TH} and equals $V_{2\text{-}3}$.

(b) Looking from terminals A and B, R_4 appears in series with the combination of R_1 in parallel with $R_2 + R_3$.

(c) Thevenin equivalent circuit

FIGURE 6–45

To find R_{TH}, first replace the source with a short to simulate a zero internal resistance. Then R_1 appears in parallel with $R_2 + R_3$, and R_4 is in series with the series-parallel combination of R_1, R_2, and R_3 as indicated in Figure 6–45(b):

$$R_{TH} = R_4 + \frac{R_1(R_2 + R_3)}{R_1 + R_2 + R_3} = 100\ \Omega + \frac{(100\ \Omega)(69\ \Omega)}{169\ \Omega} = 140.8\ \Omega$$

The resulting Thevenin equivalent circuit is shown in part (c).

How the Thevenin Equivalency Depends on the Viewpoint

The Thevenin equivalent for any circuit depends on the location of the two points from between which the circuit is "viewed." In Figure 6–44, we viewed the circuit from between the two points labeled A and B. Any given circuit can have more than one Thevenin equivalent, depending on how the viewpoints are designated. For example, if you view the circuit in Figure 6–46 from between points A and C, you obtain a completely different result than if you viewed it from between points A and B or from between points B and C.

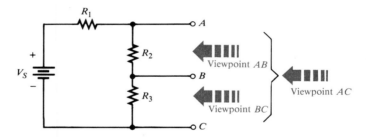

FIGURE 6–46
Thevenin's equivalent depends on viewpoint.

In Figure 6–47(a), when viewed from points A and C, V_{TH} is the voltage across $R_2 + R_3$ and can be expressed using the voltage divider formula as follows:

$$V_{TH} = \left(\frac{R_2 + R_3}{R_1 + R_2 + R_3} \right) V_S$$

Also, as shown in part (b), the resistance between points A and C is $R_2 + R_3$ in parallel with R_1 (the source is replaced by a short) and can be expressed as follows:

$$R_{TH} = \frac{R_1(R_2 + R_3)}{R_1 + R_2 + R_3}$$

The resulting Thevenin equivalent circuit is shown in part (c).

When viewed from points B and C, as indicated in Figure 6–47(d), V_{TH} is the voltage across R_3 and can be expressed as follows:

$$V_{TH} = \left(\frac{R_3}{R_1 + R_2 + R_3}\right)V_S$$

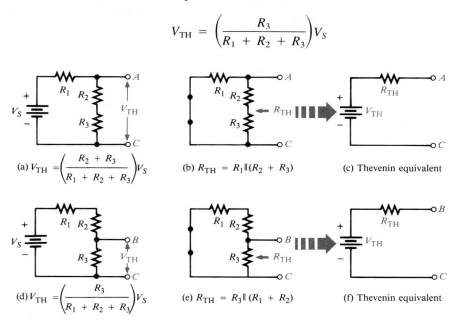

(a) $V_{TH} = \left(\dfrac{R_2 + R_3}{R_1 + R_2 + R_3}\right)V_S$ (b) $R_{TH} = R_1 \| (R_2 + R_3)$ (c) Thevenin equivalent

(d) $V_{TH} = \left(\dfrac{R_3}{R_1 + R_2 + R_3}\right)V_S$ (e) $R_{TH} = R_3 \| (R_1 + R_2)$ (f) Thevenin equivalent

FIGURE 6–47
Example of circuit Thevenized from two viewpoints, resulting in two different equivalent circuits. (The V_{TH} and R_{TH} values are different.)

As shown in part (e), the resistance between points B and C is R_3 in parallel with the series combination of R_1 and R_2:

$$R_{TH} = \frac{R_3(R_1 + R_2)}{R_1 + R_2 + R_3}$$

The resulting Thevenin equivalent is shown in part (f).

Thevenizing a Portion of a Circuit

In many cases it helps to Thevenize only a portion of a circuit. For example, when we need to know the equivalent circuit as viewed by one particular resistor in the circuit, we remove that resistor and apply Thevenin's theorem to the remaining part of the circuit as viewed from the points between which that resistor was connected. Figure 6–48 illustrates the Thevenizing of part of a circuit.

Using this type of approach, you can easily find the voltage and current for a specified resistor for any number of resistor values using only Ohm's law. This method eliminates the necessity of reanalyzing the original circuit for each different resistance value.

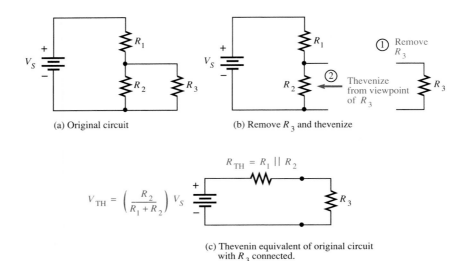

(a) Original circuit

(b) Remove R_3 and thevenize

$$V_{TH} = \left(\frac{R_2}{R_1 + R_2}\right) V_S$$

$$R_{TH} = R_1 \| R_2$$

(c) Thevenin equivalent of original circuit with R_3 connected.

FIGURE 6–48
Example of Thevenizing a portion of a circuit. In this case, the circuit is Thevenized from the viewpoint of R_3.

Thevenizing a Bridge Circuit

The usefulness of Thevenin's theorem is perhaps best illustrated when it is applied to a Wheatstone bridge circuit. For example, when a load resistor is connected to the output terminals of a Wheatstone bridge, as shown in Figure 6–49, the circuit is very difficult to analyze because it is not a straightforward series-parallel arrangement. If you doubt that this analysis is difficult, try to identify which resistors are in parallel and which are in series.

FIGURE 6–49
Wheatstone bridge with load resistor is not a series-parallel circuit.

Using Thevenin's theorem, we can simplify the bridge circuit to an equivalent circuit viewed from the load resistor as shown step-by-step in Figure 6–50. Study carefully the steps in this figure. Once the equivalent circuit for the bridge is found, the voltage and current for any value of load resistor can easily be determined.

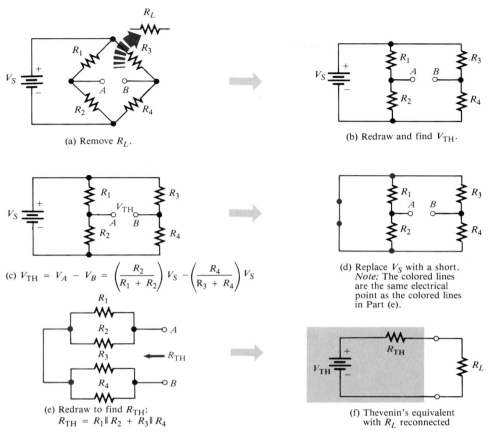

(a) Remove R_L.

(b) Redraw and find V_{TH}.

(c) $V_{TH} = V_A - V_B = \left(\dfrac{R_2}{R_1 + R_2}\right) V_S - \left(\dfrac{R_4}{R_3 + R_4}\right) V_S$

(d) Replace V_S with a short.
Note: The colored lines
are the same electrical
point as the colored lines
in Part (e).

(e) Redraw to find R_{TH}:
$R_{TH} = R_1 \| R_2 + R_3 \| R_4$

(f) Thevenin's equivalent
with R_L reconnected

FIGURE 6–50
Simplifying a Wheatstone bridge with Thevenin's theorem.

EXAMPLE 6–16   Determine the voltage and current for the load resistor, R_L, in the bridge circuit of
Figure 6–51.

FIGURE 6–51

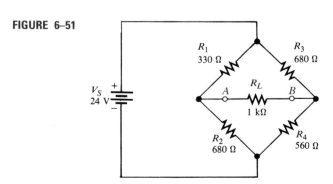

Solution
Step 1: Remove R_L.
Step 2: Thevenize the bridge as viewed from between points A and B, as was shown in Figure 6–50.

$$V_{TH} = V_A - V_B = \left(\frac{R_2}{R_1 + R_2}\right)V_S - \left(\frac{R_4}{R_3 + R_4}\right)V_S$$

$$= \left(\frac{680\ \Omega}{1010\ \Omega}\right)24\ V - \left(\frac{560\ \Omega}{1240\ \Omega}\right)24\ V$$

$$= 16.16\ V - 10.84\ V = 5.32\ V$$

$$R_{TH} = \frac{R_1 R_2}{R_1 + R_2} + \frac{R_3 R_4}{R_3 + R_4} = \frac{(330\ \Omega)(680\ \Omega)}{1010\ \Omega} + \frac{(680\ \Omega)(560\ \Omega)}{1240\ \Omega}$$

$$= 222.18\ \Omega + 307.1\ \Omega = 529.28\ \Omega$$

Step 3: Connect the load resistor to points A to B of the equivalent circuit, and determine the load voltage and current as illustrated in Figure 6–52.

FIGURE 6–52 Thevenin's equivalent for the Wheatstone bridge

$$V_{RL} = \left(\frac{R_L}{R_L + R_{TH}}\right)V_{TH} = \left(\frac{1\ k\Omega}{1.52928\ \Omega}\right)5.32\ V = 3.48\ V$$

$$I_{RL} = \frac{V_{RL}}{R_L} = \frac{3.48\ V}{1\ k\Omega} = 3.48\ mA$$

SECTION REVIEW 6–7

1. What are the two components of a Thevenin equivalent circuit?
2. Draw the general form of a Thevenin equivalent circuit.
3. How is V_{TH} determined?
4. How is R_{TH} determined?
5. For the original circuit in Figure 6–53, draw the Thevenin equivalent circuit as viewed by R_L.

FIGURE 6–53

6–8 TROUBLESHOOTING SERIES-PARALLEL CIRCUITS

Troubleshooting is the process of identifying and locating a failure or problem in a circuit. Some troubleshooting techniques have already been discussed in relation to both series circuits and parallel circuits. Now we will extend these methods to the series-parallel networks.

Opens and *shorts* are typical problems that occur in electric circuits. As mentioned before, if a resistor burns out, it will normally produce an *open circuit*. Bad solder connections, broken wires, and poor contacts can also be causes of open paths. Pieces of foreign material, such as solder splashes, broken insulation on wires, and so on, can often lead to shorts in a circuit. A *short* is a zero resistance path between two points.

In addition to complete opens or shorts, partial opens or partial shorts can develop in a circuit. A partial open would be a much higher than normal resistance, but not infinitely large. A partial short would be a much lower than normal resistance, but not zero.

The following examples illustrate troubleshooting series-parallel circuits.

EXAMPLE 6–17   From the indicated voltmeter reading in Figure 6–54, determine if there is a fault. If there is a fault, identify it as either a short or an open.

FIGURE 6–54

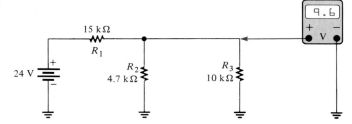

Solution

First determine what the voltmeter *should* be indicating. Since R_2 and R_3 are in parallel, their combined resistance is

$$R_{2\text{-}3} = \frac{R_2 R_3}{R_2 + R_3} = \frac{(4.7 \text{ k}\Omega)(10 \text{ k}\Omega)}{14.7 \text{ k}\Omega} = 3.2 \text{ k}\Omega$$

The voltage across the parallel combination is determined by the voltage divider formula as follows:

$$V_{2\text{-}3} = \left(\frac{R_{2\text{-}3}}{R_1 + R_{2\text{-}3}}\right)V_S = \left(\frac{3.2\ \text{k}\Omega}{18.2\ \text{k}\Omega}\right)24\ \text{V} = 4.22\ \text{V}$$

This calculation shows that 4.22 V is the voltage reading that you *should* get on the meter. But the meter reads 9.6 V instead. This value is incorrect, and, because it is higher than it should be, R_2 or R_3 is probably open. Why? Because if either of these two resistors is open, the resistance across which the meter is connected is larger than expected. A higher resistance will drop a higher voltage in this circuit, which is, in effect, a voltage divider.

Let us start by assuming that R_2 is open. If it is, the voltage across R_3 is as follows:

$$V_3 = \left(\frac{R_3}{R_1 + R_3}\right)V_S = \left(\frac{10\ \text{k}\Omega}{25\ \text{k}\Omega}\right)24\ \text{V} = 9.6\ \text{V}$$

This calculation shows that R_2 is open. Replace R_2 with a new resistor.

EXAMPLE 6–18   Suppose that you measure 24 V with the voltmeter in Figure 6–55. Determine if there is a fault, and, if there is, identify it.

FIGURE 6–55

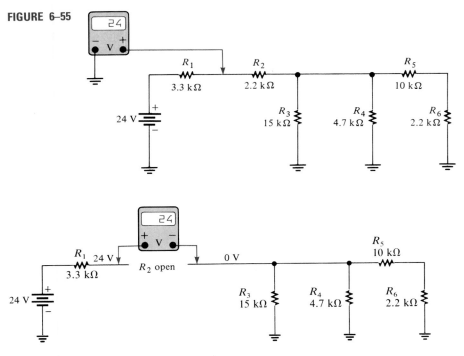

FIGURE 6–56

Solution

There is no voltage drop across R_1 because both sides of the resistor are at $+24$ V. Either no current is flowing through R_1 from the source, which tells us that R_2 is open in the circuit, or R_1 is shorted.

If R_1 were open, the meter would not read 24 V. The most logical failure is an open R_2. If it *is* open, then no current will flow from the source. To verify this, measure across R_2 with the voltmeter as shown in Figure 6–56. If R_2 is open, the meter will indicate 24 V. The right side of R_2 will be at zero volts because no current is flowing through any of the other resistors to cause a voltage drop.

EXAMPLE 6–19

The two voltmeters in Figure 6–57 indicate the voltages shown. Determine if there are any opens or shorts in the circuit and, if so, where they are located.

FIGURE 6–57

Solution

First let us see if the voltmeter readings are correct. R_1, R_2, and R_3 act as a voltage divider. The voltage across R_3 (V_A) is calculated as follows:

$$V_A = \left(\frac{R_3}{R_1 + R_2 + R_3} \right) V_S$$

$$= \left(\frac{3.3 \text{ k}\Omega}{21.6 \text{ k}\Omega} \right) 24 \text{ V} = 3.67 \text{ V}$$

The voltmeter A reading (V_A) is correct.

Now let us see if the voltmeter B reading (V_B) is correct. $R_6 + R_7$ is in parallel with R_5. The series-parallel combination of R_5, R_6, and R_7 is in series with R_4. The resistance of the R_5, R_6, and R_7 combination is calculated as follows:

$$R_{5\text{-}6\text{-}7} = \frac{(R_6 + R_7)R_5}{R_5 + R_6 + R_7}$$

$$= \frac{(17.2 \text{ k}\Omega)(10 \text{ k}\Omega)}{27.2 \text{ k}\Omega} = 6.3 \text{ k}\Omega$$

$R_{5\text{-}6\text{-}7}$ and R_4 form a voltage divider, and voltmeter B is measuring the voltage across $R_{5\text{-}6\text{-}7}$. Is it correct? We check as follows:

$$V_B = \left(\frac{R_{5\text{-}6\text{-}7}}{R_4 + R_{5\text{-}6\text{-}7}} \right) V_S$$

$$= \left(\frac{6.3 \text{ k}\Omega}{11 \text{ k}\Omega} \right) 24 \text{ V} = 13.75 \text{ V}$$

Thus, the actual measured voltage (6.65 V) at this point is incorrect. Some further thought will help to isolate the problem.

We know that R_4 is not open, because if it were, the meter would read 0 V. If there were a short across it, the meter would read 24 V. Since the actual voltage is much less than it should be, R_{5-6-7} must be less than the calculated value of 6.3 kΩ. The most likely problem is a short across R_7. If there is a short from the top of R_7 to ground, R_6 is effectively in parallel with R_5. In this case,

$$R_5 \parallel R_6 = \frac{R_5 R_6}{R_5 + R_6}$$

$$= \frac{(10 \text{ k}\Omega)(2.2 \text{ k}\Omega)}{12.2 \text{ k}\Omega} = 1.8 \text{ k}\Omega$$

Then V_B is

$$V_B = \left(\frac{1.8 \text{ k}\Omega}{6.5 \text{ k}\Omega}\right) 24 \text{ V} = 6.65 \text{ V}$$

This value for V_B agrees with the voltmeter B reading. So there is a short across R_7. If this were an actual circuit, you would try to find the physical cause of the short.

SECTION REVIEW
6–8

1. Name two types of common circuit faults.
2. For the following faults in Figure 6–58, determine what voltage would be measured at point A:
   (a) no faults   (b) R_1 open   (c) short across R_5 (d) R_3 and R_4 open
   (e) R_2 open

FIGURE 6–58

3. In Figure 6–59 one of the resistors in the circuit is open. Based on the meter reading, determine which is the bad resistor.

FIGURE 6–59

SUMMARY

☐ A series-parallel circuit is a combination of both series components and parallel components.

☐ When a load resistor is connected across a voltage divider output, the output voltage decreases. This effect is called *loading.*

☐ To minimize the loading effect, the load resistance should be much larger than the resistance across which it is connected.

☐ Ground is considered at zero volts with respect to other points in a circuit.

☐ A Wheatstone bridge is used to measure an unknown resistance.

☐ When a Wheatstone bridge is *balanced,* the output voltage is zero.

☐ To find any current or voltage in a circuit with two or more voltage sources, take the sources one at a time using the superposition method.

☐ Any resistive circuit, no matter how complex, can be replaced by its Thevenin equivalent.

☐ The Thevenin equivalent circuit is made up of an equivalent resistance (R_{TH}) in series with an equivalent voltage source (V_{TH}).

GLOSSARY

Bleeder current   The current left after the total load current is subtracted from the total current into the circuit.

Load   An element (resistor or other component) connected across the output terminals of a circuit that draws current from the circuit.

Sensitivity factor   The ohms-per-volt rating of a voltmeter.

Superposition   A method for analyzing circuits with two or more sources by examining the effects of each source by itself and then combining the effects.

Terminal equivalency   A condition that occurs when two circuits produce the same load voltage and load current when the same value of load resistance is connected to either circuit.

Thevenin's theorem   A circuit theorem that provides for reducing any resistive circuit to a single equivalent voltage source in series with an equivalent resistance.

FORMULA

$$R_{UNK} = R_V\left(\frac{R_2}{R_4}\right) \qquad \text{Unknown resistance in a Wheatstone bridge} \qquad (6\text{--}1)$$

SELF-TEST

1. Which of the following statements are true concerning Figure 6–60?
   (a) R_1 and R_2 are in series with R_3, R_4, and R_5
   (b) R_1 and R_2 are in series
   (c) R_3, R_4, and R_5 are in parallel
   (d) The series combination of R_1 and R_2 is in parallel with the series combination of R_3, R_4, and R_5
   (e) b and d

2. The total resistance of Figure 6–60 can be found with which of the following formulas?
   (a) $R_1 + R_2 + R_3 \parallel R_4 \parallel R_5$
   (b) $R_1 \parallel R_2 + R_3 \parallel R_4 \parallel R_5$
   (c) $(R_1 + R_2) \parallel R_3 + R_4 + R_5)$
   (d) none of the above

FIGURE 6–60

3. If all of the resistors in Figure 6–60 have the same value, when voltage is applied across terminals *A* and *B*, the current is
   (a) greatest in R_5
   (b) greatest in R_3, R_4, and R_5
   (c) greatest in R_1 and R_2
   (d) the same in all the resistors

4. Two 1-kΩ resistors are in series and this series combination is in parallel with a 2.2-kΩ resistor. The voltage across one of the 1-kΩ resistors is 6 V. The voltage across the 2.2-kΩ resistor is
   (a) 6 V   (b) 3 V   (c) 12 V   (d) 13.2 V

5. The parallel combination of a 330-Ω resistor and a 470-Ω resistor is in series with the parallel combination of four 1-kΩ resistors. A 100-V source is connected across the circuit. The resistor with the most current has a value of
   (a) 1 kΩ   (b) 330 Ω   (c) 470 Ω

6. In the circuit described in Question 5, the resistor(s) with the most voltage has a value of
   (a) 1 kΩ   (b) 470 Ω   (c) 330 Ω

7. In the circuit of Question 5, the percentage of the total current through any single 1-kΩ resistor is
   (a) 100%   (b) 25%   (c) 50%   (d) 31.25%

8. The output of a certain voltage divider is 9 V with no load. When a load is connected, the output voltage
   (a) increases   (b) decreases   (c) remains the same   (d) becomes zero

9. A certain voltage divider consists of two 10-kΩ resistors in series. Which of the following load resistors will have the most effect on the output voltage?
   (a) 1 MΩ   (b) 20 kΩ   (c) 100 kΩ   (d) 10 kΩ

10. When a load resistance is connected to the output of a voltage divider circuit, the current drawn from the source
    (a) decreases   (b) increases   (c) remains the same   (d) is cut off

11. The output voltage of a balanced Wheatstone bridge is
    (a) equal to the source voltage
    (b) equal to zero
    (c) dependent on all of the resistor values in the bridge
    (d) dependent on the value of the unknown resistor

12. The *primary* method of analyzing a circuit with two or more voltage sources is usually
    (a) Thevenin's theorem
    (b) Ohm's law
    (c) superposition
    (d) Kirchhoff's law

13. In a certain two-source circuit, one source acting alone produces 10 mA through a given branch. The other source acting alone produces 8 mA in the opposite direction through the same branch. The total current through the branch is
    (a) 10 mA   (b) 8 mA   (c) 18 mA   (d) 2 mA

14. A Thevenin equivalent circuit consists of
    (a) A voltage source in series with a resistance
    (b) A voltage source in parallel with a resistance
    (c) A current source in parallel with a resistance
    (d) Two voltage sources and a resistance

15. You are measuring the voltage at a given point in a circuit that has very high resistance values and the measured voltage is a little lower than it should be. This is possibly because of
    (a) one or more of the resistance values being off
    (b) the loading effect of the voltmeter
    (c) the source voltage is too low
    (e) all the above

PROBLEM SET A

Section 6–1   Identifying Series-Parallel Relationships

1. Identify the series-parallel relationships in Figure 6–61 as seen from the source terminals.

FIGURE 6–61

2. Visualize and sketch the following series-parallel combinations
   (a) R_1 in series with the parallel combination of R_2 and R_3
   (b) R_1 in parallel with the series combination of R_2 and R_3
   (c) R_1 in parallel with a branch containing R_2 in series with a parallel combination of four other resistors

3. Visualize and sketch the following series-parallel circuits:
   (a) a parallel combination of three branches, each containing two series resistors
   (b) a series combination of three parallel circuits, each containing two parallel resistors

4. In each circuit of Figure 6–62, identify the series and parallel relationships of the resistors viewed from the source.

FIGURE 6–62

(a)   (b)

DC CIRCUITS

Section 6–2 Analysis of Series-Parallel Circuits

5. A certain circuit is composed of two parallel resistors. The total resistance is 667 Ω. One of the resistors is 1 kΩ. What is the other resistor?

6. For the circuit in Figure 6–63, determine the total resistance between points A and B. (A full-color photo appears following page 240.)

7. Determine the total resistance for each circuit in Figure 6–62.

8. Determine the current through each resistor in Figure 6–61; then calculate each voltage drop.

9. Determine the current through each resistor in both circuits of Figure 6–62; then calculate each voltage drop.

10. In Figure 6–64, find the following:
    (a) total resistance between terminals A and B
    (b) total current drawn from a 6-V source connected from A to B
    (c) current through R_5
    (d) voltage across R_2

FIGURE 6–64

Section 6–3 Loaded Voltage Dividers

11. A voltage divider consists of two 56-kΩ resistors and a 15-V source. Calculate the unloaded output voltage. What will the output voltage be if a load resistor of 1 MΩ is connected to the output?

12. A 12-V battery output is divided down to obtain two output voltages. Three 3.3-kΩ resistors are used to provide the two outputs. Only one output at a time is loaded. Determine the output voltages when both load resistances are 10 kΩ.

13. Which will cause a smaller decrease in output voltage for a given voltage divider, a 10-kΩ load or a 56-kΩ load?

14. In Figure 6–65, determine the current drain on the battery with no load across the output terminals. With a 10-kΩ load, what is the battery current?

FIGURE 6–65

Section 6–4 The Loading Effect of a Voltmeter

15. On which one of the following voltage range settings will a voltmeter present the minimum load on a circuit?
    (a) 1 V   (b) 10 V   (c) 100 V   (d) 1000 V

16. Determine the internal resistance of a 20,000-Ω/V voltmeter on each of the following range settings.
    (a) 0.5 V (b) 1 V (c) 5 V (d) 50 V (e) 100 V
    (f) 1000 V

17. The voltmeter described in Problem 16 is used to measure the voltage across R_4 in Figure 6-62(a).
    (a) What range should be used?
    (b) How much less is the voltage measured by the meter than the actual voltage?

18. Repeat Problem 17 if the voltmeter is used to measure the voltage across R_4 in the circuit of Figure 6-62(b).

Section 6-5   The Wheatstone Bridge

19. A resistor of unknown value is connected to a Wheatstone bridge circuit. The bridge parameters for a balanced condition are set as follows: $R_V = 18$ kΩ and $R_2/R_4 = 0.02$. What is R_{UNK}?

20. A bridge network is shown in Figure 6-66. To what value must R_V be set in order to balance the bridge?

FIGURE 6-66

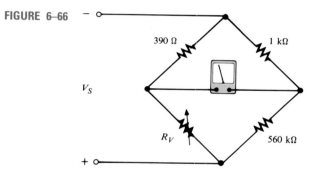

Section 6-6   Circuits with More than One Voltage Source

21. In Figure 6-67, use the superposition method to find the total current in R_3.

22. In Figure 6-67, what is the total current through R_2?

FIGURE 6-67

Section 6-7   Thevenin's Theorem

23. Reduce the circuit in Figure 6-68 to its Thevenin equivalent as viewed from terminals A and B.

FIGURE 6-68

24. For each circuit in Figure 6–69, determine the Thevenin equivalent as seen by R_L.

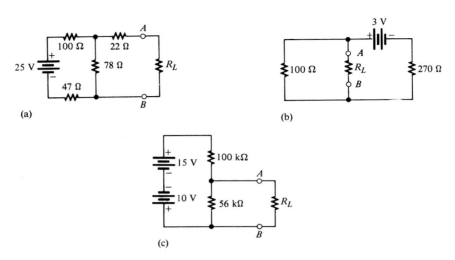

(a)   (b)

(c)

FIGURE 6–69

Section 6–8   Troubleshooting Series-Parallel Circuits

25. Is the voltmeter reading in Figure 6–70 correct? (A full-color photo appears following page 240.) If not, what is the problem?

26. If R_2 in Figure 6–71 opens, what voltages will be read at points A, B, and C?

FIGURE 6–71

27. Check the meter readings in Figure 6–72 and locate any fault that may exist.

FIGURE 6–72

PROBLEM SET B

28. In each circuit of Figure 6–73, identify the series and parallel relationships of the resistors viewed from the source.

FIGURE 6–73

(a)

(b)

COLOR
INSERT

29. Draw the schematic of the PC board layout in Figure 6–74 showing resistor values (a full-color photo appears following page 240) and identify the series-parallel relationships. Which resistors can be removed with no effect on R_T?

FIGURE 6–74

30. For the circuit shown in Figure 6–75, calculate the following:
    (a) total resistance across the source. — $250\ \Omega$
    (b) total current from the source. — $240\ mA$.
    (c) current through the 900-Ω resistor. $60\ mA$
    (d) voltage from point A to point B. $12\ V$

FIGURE 6–75

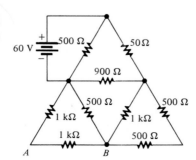

31. Determine the total resistance and the voltage at points A, B, and C in the circuit of Figure 6–76.

FIGURE 6–76

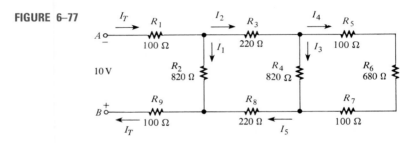

32. Determine the total resistance between terminals A and B of the circuit in Figure 6–77. Also calculate the current in each branch with 10 V between A and B.

33. What is the voltage across each resistor in Figure 6–77?

FIGURE 6–77

34. Determine the voltage, V_{AB}, in Figure 6–78.

35. Find the value of R_2 in Figure 6–79.

FIGURE 6–78   **FIGURE 6–79**

FIGURE 6–80

36. Find I_T and V_{OUT} in Figure 6–80.

37. Design a voltage divider to provide a 6-V output with no load and a minimum of 5.5 V across a 1-kΩ load. The source voltage is 24 V, and the unloaded current is not to exceed 100 mA.

38. Determine the resistance values for a voltage divider that must meet the following specifications: The current under an unloaded condition is not to exceed 5 mA. The source voltage is to be 10 V. A 5-V output and a 2.5-V output are required. Sketch the circuit. Determine the effect on the output voltages if a 1-kΩ load is connected to each output.

39. Using the superposition method, calculate the current in the right-most branch of Figure 6–81.

FIGURE 6–81

40. Find the current through R_L in Figure 6–82.

FIGURE 6–82

41. Using Thevenin's theorem, find the voltage across R_4 in Figure 6–83.

FIGURE 6–83

42. Find the resistance between point A and each of the other points (R_{AB}, R_{AC}, R_{AD}, etc.) in Figure 6–84.

FIGURE 6–84

FIGURE 6–85

B   A
Side 1

A   B
Side 2

43. Develop a schematic for the double-sided PC board in Figure 6–85.

44. Lay out a PC board for the circuit in Figure 6–73(b). The battery is to be connected external to the board.

45. Look at the meters in Figure 6–86 and determine if there is a fault in the circuit. If there is a fault, identify it.

FIGURE 6–86

46. Are the meter readings in Figure 6–87 correct?

FIGURE 6–87

47. There is one fault in Figure 6–88. Based on meter indications, determine what the fault is.

FIGURE 6–88

48. Look at the meters in Figure 6–89 and determine if there is a fault in the circuit. If there is a fault, identify it.

FIGURE 6–89

ANSWERS TO SECTION REVIEWS

Section 6–1

1. See Figure 6–90.

FIGURE 6–90

2. R_1 and R_2 are in series with the parallel combination of R_3 and R_4.
3. All resistors.
4. R_1 and R_2 are in parallel; R_3 and R_4 are in parallel. 5. Yes.

Section 6–2
1. 599 Ω.   **2.** 11.2 mA.   **3.** 3.7 V.   **4.** 89 Ω, 11.2 mA.

Section 6–3
1. It decreases the output voltage.   **2.** T.   **3.** 19.2 V, 4.19 V.

Section 6–4
1. The internal resistance of the meter appears in parallel with the circuit resistance across which it is connected, reducing the resistance between those two points of the circuit and drawing current from the circuit.

2. No.

3. Decrease; you can read the scale more accurately.

4. 1.5 MΩ.

Section 6–5
1. Zero output.   **2.** 15 kΩ.

Section 6–6
1. 6.67 mA.   **2.** In the direction of the larger.

Section 6–7
1. V_{TH} and R_{TH}.   **2.** See Figure 6–91.

FIGURE 6–91

3. V_{TH} is the open circuit voltage between two terminals in a circuit.

4. R_{TH} is the resistance as viewed from two terminals in a circuit, with all sources replaced by their internal resistances.

5. See Figure 6–92.

FIGURE 6–92

Section 6–8
1. Opens and shorts.

2. **(a)** 62.8 V;   **(b)** 62.8 V;   **(c)** 62 V;   **(d)** 100 V;   **(e)** 0 V.

3. 10 kΩ

KEY TO RESISTOR COLORS

DC Color Insert: Actual Circuits for Selected Examples, Section Reviews, and Problems

The experimenter boards shown in this full-color section are commonly found in many school laboratories and are used in industry, however, a brief explanation of the board layout may be useful to those unfamiliar with them.

Notice that each row perpendicular to the center slot consists of five connection points. Each of these five points is internally connected to each other to form a common electrical tie point. Each row, however, is electrically separate from all the other rows. The groups of points running parallel with the top of the board are connected together and the groups of points running parallel with the bottom of the board are connected together. See Appendix A for further explanation.

FIGURE 2–19
For Example 2–4.

FIGURE 2–42 (a)
For problems 11 and 12.

FIGURE 2–42 (b)
For Problem 13 (b).

FIGURE 3–26 (a)
For Problem 7.

Closeup

Closeup

FIGURE 3–26 (b)

FIGURE 3–26 (c)

Closeup

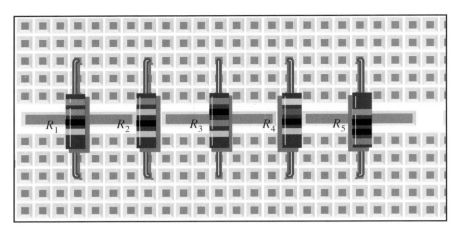

FIGURE 4–13
For Example 4–3.

FIGURE 4–17
For Section Review 4–3, Question 1.

(a)

(b)

(c)

(a)

(b)

(c)

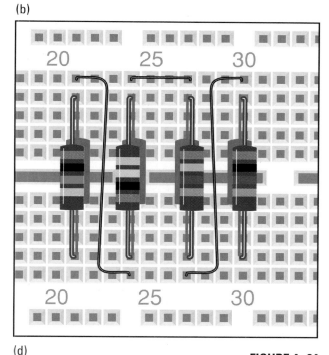

(d)

FIGURE 4–64
For Problem 6.

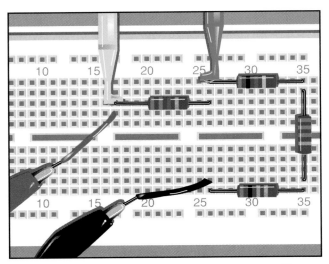

FIGURE 4–72
For Problem 23.

Closeup

FIGURE 4–76
For Problem 30.

Closeup

FIGURE 4-80
For Problems 42, 43, 44 and 45

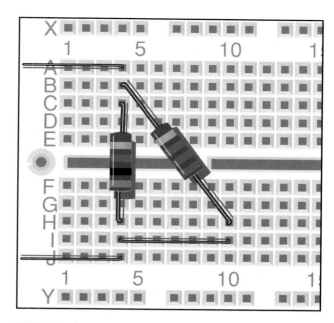

FIGURE 5–25
For Section Review 5–4, Question 3.

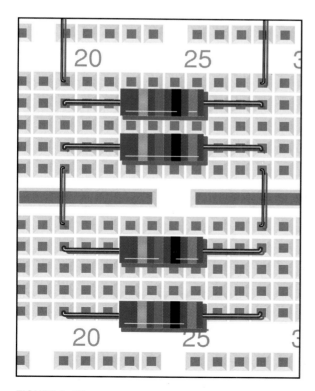

FIGURE 5–26
For Section Review 5–4, Question 3.

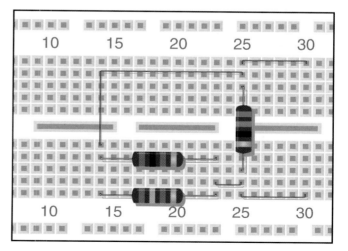

FIGURE 5–27
For Section Review 5–4,
Question 3.

FIGURE 5–39
For Section Review 5–6, Question 2.

Closeup

FIGURE 5–40
For Section Review 5–6, Question 3.

Closeup

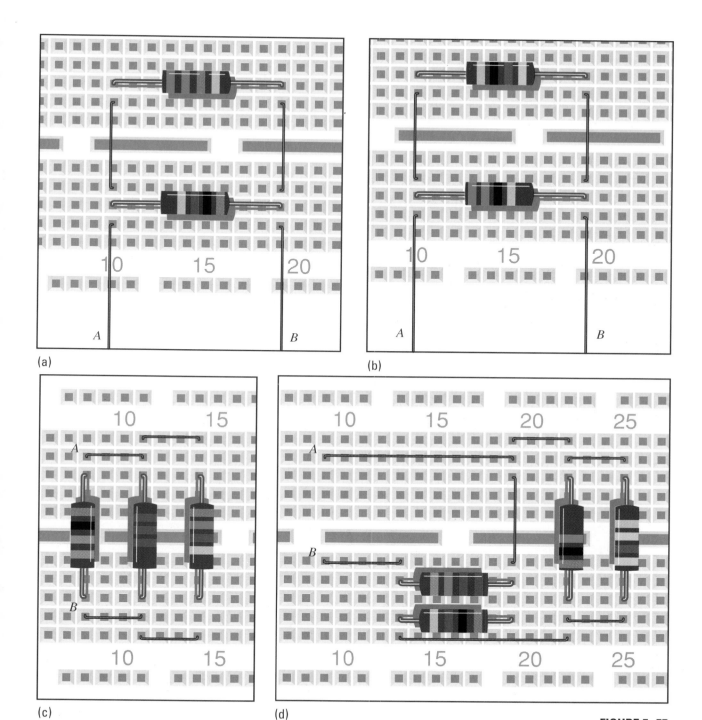

(a)

(b)

(c)

(d)

FIGURE 5–57
For Problem 9.

FIGURE 5–74
For Problem 39.

Closeup

FIGURE 5–75
For Problem 40.

FIGURE 5–76
For problems 42 and 43.

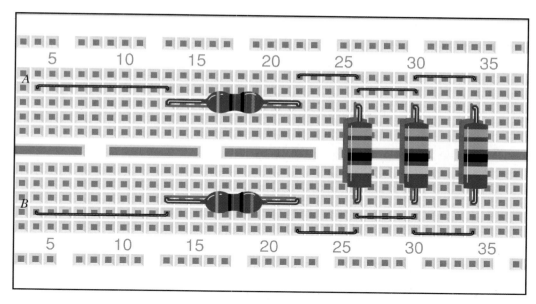

FIGURE 6–63
For Problem 6.

FIGURE 6–70
For Problem 25.

Closeup

FIGURE 6–74
For Problem 29.

A   B

7 Magnetism and Electromagnetism

This chapter is somewhat of a departure from the previous six chapters because two new concepts are introduced: magnetism and electromagnetism. The operation of many types of electrical devices is based partially on magnetic or electromagnetic principles.

After completing this chapter, you will be able to

☐ Describe the characteristics of magnetic fields.
☐ Explain how materials are magnetized.
☐ Show how a magnetic switch operates.
☐ Apply magnetic switches to alarm systems and other applications.
☐ Discuss electromagnetism.
☐ Explain how electromagnetism is used to record on magnetic tape.
☐ Describe how solenoids, relays, speakers, and dc generators operate.
☐ State Faraday's law.

In this chapter, you will become familiar with several devices that can be used in security systems. Using only magnetic and electromagnetic devices, devise a system that will detect illegal entry into your house or apartment through any window or door. When someone breaks into a room, the system will turn on a light in that room and open the door of a cage inside the house and in which there is a vicious guard dog.

7–1

THE MAGNETIC FIELD

A permanent magnet, such as the bar magnet shown in Figure 7–1, has a magnetic field surrounding it. The magnetic field consists of *lines of force*, or flux lines, that radiate from the north pole (N) to the south pole (S) and back to the north pole through the magnetic material. For clarity, only a few lines of force are shown in the figure. Imagine, however, that many lines surround the magnet in three dimensions. The lines shrink to the smallest possible size and blend together, although they do not touch. This effectively forms a continuous magnetic field surrounding the magnet.

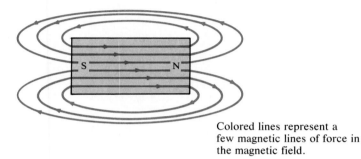

Colored lines represent a few magnetic lines of force in the magnetic field.

FIGURE 7–1
Magnetic lines of force around a bar magnet.

Attraction and Repulsion of Magnetic Poles

When *unlike* poles of two permanent magnets are placed close together, an attractive force is produced by the magnetic fields, as indicated in Figure 7–2(a). When two *like* poles are brought close together, they repel each other, as shown in part (b).

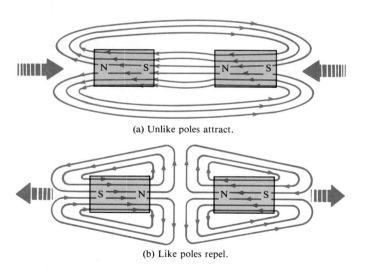

(a) Unlike poles attract.

(b) Like poles repel.

FIGURE 7–2
Magnetic attraction and repulsion.

Altering a Magnetic Field

When a nonmagnetic material, such as paper, glass, wood, or plastic, is placed in a magnetic field, the lines of force are unaltered, as shown in Figure 7–3(a). However, when a *magnetic* material such as iron is placed in the magnetic field, the lines of force tend to change course and pass through the iron rather than through the surrounding air. They do so because the iron provides a magnetic path that is more easily established than that of air. Figure 7–3(b) illustrates this principle.

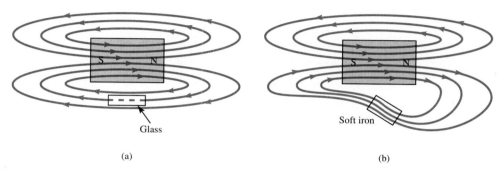

(a)   (b)

FIGURE 7–3
Effect of (a) nonmagnetic and (b) magnetic materials on a magnetic field.

Magnetic Flux

The group of force lines going from the north pole to the south pole of a magnet is called the *magnetic flux*, symbolized by ϕ (the lowercase Greek letter phi). The number of lines of force in a magnetic field determines the value of the flux. The more lines of force, the greater the flux and the stronger the magnetic field.

The unit of magnetic flux is the *weber* (Wb). One weber equals 10^8 lines. In most practical situations, the weber is a very large unit; thus, the microweber (μWb) is more common. One microweber equals 100 lines of magnetic flux.

Magnetic Flux Density

The *flux density* is the amount of flux per unit area in the magnetic field. Its symbol is B, and its unit is the tesla (T). One tesla equals one weber per square meter (Wb/m^2). The following formula expresses the flux density:

$$B = \frac{\phi}{A}$$
(7–1)

where ϕ is the flux and A is the cross-sectional area of the magnetic field.

EXAMPLE 7–1   Find the flux density in a magnetic field in which the flux in 0.1 m^2 is 800 μWb.

Solution

$$B = \frac{\phi}{A} = \frac{800 \ \mu\text{Wb}}{0.1 \ \text{m}^2} = 8000 \times 10^{-6} \ \text{T}$$

How Materials Become Magnetized

Ferromagnetic materials, such as iron, nickel, and cobalt become magnetized when placed in the magnetic field of a magnet. We have all seen a permanent magnet pick up paper clips, nails, iron filings, and so on. In these cases, the object becomes magnetized (that is, it actually becomes a magnet itself) under the influence of the permanent magnetic field and becomes attracted to the magnet. When removed from the magnetic field, the object tends to lose its magnetism.

Ferromagnetic materials have minute *magnetic domains* created within their atomic structure by the orbital motion and spin of electrons. These domains can be viewed as very small bar magnets with north and south poles. When the material is not exposed to an external magnetic field, the magnetic domains are randomly oriented, as shown in Figure 7–4(a). When the material is placed in a magnetic field, the domains align themselves, as shown in part (b). Thus, the object itself effectively becomes a magnet.

FIGURE 7–4
Ferromagnetic domains in (a) an unmagnetized and (b) a magnetized material.

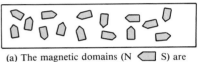

(a) The magnetic domains (N S) are randomly oriented in the unmagnetized material.

(b) The magnetic domains become aligned when the material is magnetized.

An Application

Permanent magnets have almost endless applications, one of which is presented here as an illustration. Figure 7–5 shows a typical magnetically operated, normally closed (NC) switch. When the magnet is near the switch mechanism, the metallic arm is held in its NC position. When the magnet is moved away, the spring pulls the arm up, breaking the contact as shown in Figure 7–6.

FIGURE 7–5
Magnet and switch set (courtesy of Tandy Corp.).

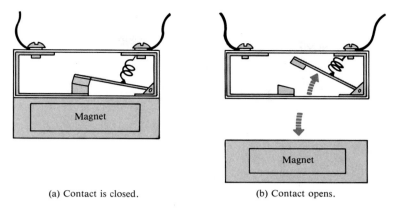

(a) Contact is closed.   (b) Contact opens.

FIGURE 7–6
Operation of a magnetic switch.

Switches of this type are commonly used in perimeter alarm systems to detect entry into a building through windows or doors. As Figure 7–7 shows, several openings can be protected by magnetic switches wired to a common transmitter. When any one of the switches opens, the transmitter is activated and sends a signal to a central receiver and alarm unit.

FIGURE 7–7
Connection of a typical perimeter alarm system.

SECTION REVIEW 7–1

1. When the north poles of two magnets are placed close together, do they repel or attract each other?
2. What is magnetic flux?

7–2

ELECTROMAGNETISM

Current produces a magnetic field, called an *electromagnetic field,* around a conductor, as illustrated in Figure 7–8. The invisible lines of force of the magnetic field form a concentric circular pattern around the conductor and are continuous along its length.

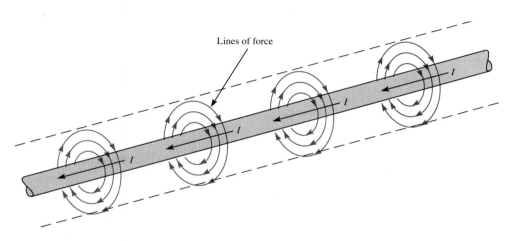

Lines of force

FIGURE 7–8
Magnetic field around a current-carrying conductor.

Although the magnetic field cannot be seen, it is capable of producing visible effects. For example, if a current-carrying wire is inserted through a sheet of paper in a perpendicular direction, iron filings placed on the surface of the paper arrange themselves along the magnetic lines of force in concentric rings, as illustrated in Figure 7–9(a). Part (b) of the figure illustrates that the north pole of a compass placed in the electromagnetic field will point in the direction of the lines of force. The field is stronger closer to the conductor and becomes weaker with increasing distance from the conductor.

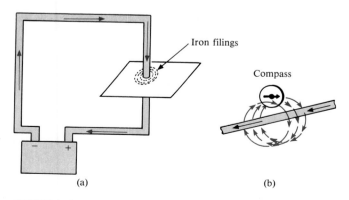

Iron filings

Compass

(a)   (b)

FIGURE 7–9
Visible effects of an electromagnetic field.

Direction of the Lines of Force

The direction of the lines of force surrounding the conductor are indicated in Figure 7–10. When the direction of current is right to left, as in part (a), the lines are in a *clockwise* direction. When current is left to right, as in part (b), the lines are in a *counterclockwise* direction.

(a)   (b)

FIGURE 7–10
Magnetic lines of force around a current-carrying conductor.

Left-Hand Rule

An aid to remembering the direction of the lines of force is illustrated in Figure 7–11. Imagine that you are grasping the conductor with your left hand, with your thumb pointing in the direction of current. Your fingers indicate the direction of the magnetic lines of force.

FIGURE 7–11
Illustration of left-hand rule.

Electromagnetic Properties

Several important properties relating to electromagnetic fields are now presented.

Permeability (μ)

The ease with which a magnetic field can be established in a given material is measured by the *permeability* of that material. The higher the permeability, the more easily a magnetic field can be established.

The symbol of permeability is μ, and its value varies depending on the type of material. The permeability of a vacuum (μ_0) is $4\pi \times 10^{-7}$ Wb/At·m (weber/ampere-turn·meter) and is used as a reference. Ferromagnetic materials typically have permeabilities hundreds of times larger than that of a vacuum, indicating that a magnetic field can be set up with relative ease in these materials. Ferromagnetic materials include iron, steel, nickel, cobalt, and their alloys.

The *relative permeability* (μ_r) of a material is the ratio of its absolute permeability (μ) to the permeability of a vacuum. (μ_0). Since μ_r is a ratio, it has no units.

$$\mu_r = \frac{\mu}{\mu_0} \tag{7-2}$$

Reluctance ($\mathcal{R}$)

Reluctance ($\mathcal{R}$) is the opposition to the establishment of a magnetic field in a material. The value of reluctance is directly proportional to the length (l) of the magnetic path, and inversely proportional to the permeability (μ) and to the cross-sectional area (A) of the material as expressed by the following equation.

$$\mathcal{R} = \frac{l}{\mu A} \tag{7-3}$$

Reluctance in magnetic circuits is analogous to resistance in electric circuits. The unit of reluctance can be derived using l in meters, A (area) in square meters, and μ in Wb/At·m as follows.

$$\mathcal{R} = \frac{l}{\mu A} = \frac{\cancel{m}}{(\text{Wb/At·}\cancel{m})(\cancel{m^2})} = \frac{\text{At}}{\text{Wb}}$$

At/Wb is ampere-turns/weber.

EXAMPLE 7-2

What is the reluctance of a material that has a length of 0.05 m, a cross-sectional area of 0.012 m², and a permeability of 3500×10^{-6} Wb/At·m?

Solution

$$\mathcal{R} = \frac{l}{\mu A} = \frac{0.05 \text{ m}}{(3500 \times 10^{-6} \text{ Wb/At·m})(0.012 \text{ m}^2)}$$

$$= 1190.48 \text{ At/Wb}$$

Magnetomotive Force (mmf)

As you have learned, current in a conductor produces a magnetic field. The force that produces the magnetic field is called the *magnetomotive force* (mmf). The unit of mmf, the ampere-turn (At), is established on the basis of the current in a single loop (turn) of wire. The formula for mmf is as follows:

$$F_m = NI \tag{7-4}$$

where F_m is the magnetomotive force, N is the number of turns of wire, and I is the current.

Figure 7-12 illustrates that a number of turns of wire carrying a current around a magnetic material create a force that sets up flux lines through the magnetic path. The

FIGURE 7–12
A basic magnetic circuit.

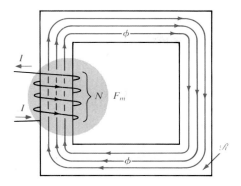

amount of flux depends on the magnitude of the mmf and on the reluctance of the material, as expressed by the following equation:

$$\phi = \frac{F_m}{\mathcal{R}} \qquad (7\text{–}5)$$

This is known as the *Ohm's law for magnetic circuits* because the flux (ϕ) is analogous to current, the mmf (F_m) is analogous to voltage, and the reluctance ($\mathcal{R}$) is analogous to resistance.

EXAMPLE 7–3

How much flux is established in the magnetic path of Figure 7–13 if the reluctance of the material is 0.28×10^5 At/Wb?

FIGURE 7–13

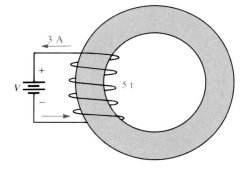

Solution

$$\phi = \frac{F_m}{\mathcal{R}} = \frac{NI}{\mathcal{R}} = \frac{(5)(3\ \text{A})}{0.28 \times 10^5\ \text{At/Wb}} = 5.36 \times 10^{-4}\ \text{Wb}$$

EXAMPLE 7–4

There are two amperes of current through a wire with 5 turns.
(a) What is the mmf?
(b) What is the reluctance of the circuit if the flux is 250 μWb?

Solution

(a) $N = 5$ and $I = 2$ A

$\qquad F_m = NI = (5)(2 \text{ A}) = 10$ At

(b) $\mathscr{R} = \dfrac{F_m}{\phi} = \dfrac{10 \text{ At}}{250 \ \mu\text{Wb}} = 0.04 \times 10^6$ At/Wb

The Electromagnet

An electromagnet is based on the properties that you have just learned. A basic electromagnet is simply a coil of wire wound around a core material that can be easily magnetized. When a length of wire is formed into a coil, as shown in Figure 7–14(a), current through the coil produces a magnetic field, as illustrated in Figure 7–14(b). The magnetic lines of force around each loop (turn) in the coil effectively add to the lines of force around the adjoining loops, forming a strong magnetic field within and around the coil, as shown. The net direction of the total magnetic field creates a north and a south pole, as shown in part (b).

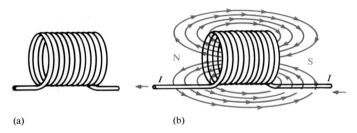

(a)   (b)

FIGURE 7–14
A coil of wire forms an electromagnet. When current flows through it, a three-dimensional electromagnetic field is created, surrounding the coil in all directions.

To understand the formation of the total magnetic field in a coil, let's discuss the interaction of the magnetic fields around two adjacent loops. The magnetic lines of force around adjacent loops are deflected into an outer path when the loops are brought close together. This effect occurs because the magnetic lines of force are in *opposing* directions between adjacent loops, as illustrated in Figure 7–15(a). The total magnetic field for the two loops is depicted in part (b) of the figure. For simplicity, only single lines

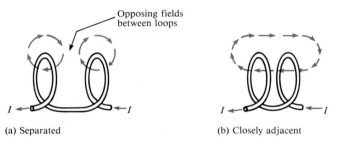

(a) Separated   (b) Closely adjacent

FIGURE 7–15
Interaction of magnetic lines of force in two adjacent loops of a coil.

of force are shown. This effect is additive for many closely adjacent loops in a coil; that is, each additional loop adds to the strength of the electromagnetic field.

The shape of the electromagnet can be designed for various applications. For example, Figure 7–16 shows a U-shaped magnetic core. When the coil of wire is connected to a battery and current flows, as shown in part (a), a magnetic field is established as indicated. If the current is reversed, as shown in part (b), the direction of the magnetic field is also reversed. The closer the north and south poles are brought together, the smaller the air gap between them becomes, and the easier it becomes to establish a magnetic field, because the reluctance is lessened.

FIGURE 7–16
Reversing the current in the coil causes the electromagnetic field to reverse.

A good example of one application of an electromagnet is the process of recording on magnetic tape. In this situation, the *recording head* is an electromagnet with a narrow air gap, as shown in Figure 7–17. Current sets up a magnetic field across the air gap, and as the recording head passes over the magnetic tape, the tape is permanently magnetized. In digital recording, for example, the tape is magnetized in one direction for a binary 1 and in the other direction for a binary 0, as illustrated in the figure. This

FIGURE 7–17
A recording head recording information on magnetic tape by magnetizing the tape as it passes by.

magnetization in different directions is accomplished by reversing the coil current in the recording head.

1. Explain the difference between magnetism and electromagnetism.
2. What happens to the magnetic field in an electromagnet when the current through the coil is reversed?

7–3

ELECTROMAGNETIC DEVICES

The recording head is one example of an electromagnetic device. Several other examples are now presented.

FIGURE 7–18
Basic solenoid structure.

(a) Solenoid

(b) Basic construction

(c) Cutaway view

The Solenoid

The solenoid is a type of electromagnetic device that has a *movable* iron core called a *plunger*. The movement of this iron core depends on both an electromagnetic field and a mechanical spring force. The basic structure of a solenoid is shown in Figure 7–18. It consists of a cylindrical coil of wire wound around a nonmagnetic hollow form. A stationary iron core is fixed in position at the end of the shaft and a sliding iron core is attached to the stationary core with a spring.

In the at-rest, or unenergized, state, the plunger is extended as shown in Figure 7–19(a). The solenoid is energized by current through the coil, as shown in part (b).

(a) Unenergized – plunger extented

(b) Energized – plunger retracted

FIGURE 7–19
Basic solenoid operation.

The current sets up an electromagnetic field that magnetizes both iron cores as indicated. The south pole of the stationary core attracts the north pole of the movable core causing it to slide inward, thus retracting the plunger and compressing the spring. As long as the coil current flows, the plunger remains retracted by the attractive force of the magnetic fields. When the current is cut off, the magnetic fields collapse and the force of the compressed spring pushes the plunger back out. The solenoid is used for many applications, such as opening and closing valves, automobile door locks, and so on.

The Relay

Relays differ from solenoids in that the electromagnetic action is used to open or close electrical contacts rather than to provide mechanical movement. Figure 7–20 shows the

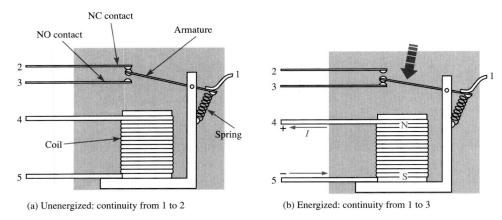

(a) Unenergized: continuity from 1 to 2   (b) Energized: continuity from 1 to 3

FIGURE 7-20
Basic structure of a single-pole–double-throw relay.

basic operation of a relay with one normally open (NO) contact and one normally closed (NC) contact (single-pole–double-throw). When there is no coil current, the armature is held against the upper contact by the spring, thus providing continuity from terminal 1 to terminal 2, as shown in part (a) of the figure. When energized with coil current, the armature is pulled down by the attractive force of the electromagnetic field and makes connection with the lower contact to provide continuity from terminal 1 to terminal 3, as shown in part (b).

A typical relay and its schematic symbol are shown in Figure 7–21.

FIGURE 7-21
A typical relay.

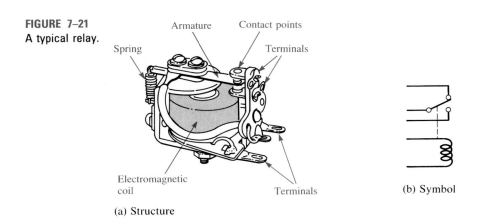

(a) Structure

(b) Symbol

The Speaker

Permanent-magnet speakers are commonly used in stereos, radios, and TV, and their operation is based on the principle of electromagnetism. A typical speaker is constructed with a permanent magnet and an electromagnet, as shown in Figure 7–22(a). The cone of the speaker consists of a paper-like diaphragm to which is attached a hollow cylinder with a coil around it, forming an electromagnet. One of the poles of the permanent magnet is positioned within the cylindrical coil. When current flows through the coil in one direction, the interaction of the permanent magnetic field with the electromagnetic field causes the cylinder to move to the right, as indicated in Figure 7–22(b). Current through the coil in the other direction causes the cylinder to move to the left, as shown in part (c).

FIGURE 7–22
Basic speaker operation.

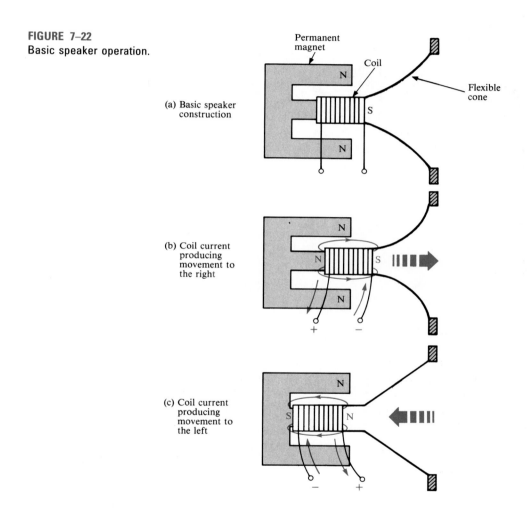

(a) Basic speaker construction

(b) Coil current producing movement to the right

(c) Coil current producing movement to the left

The movement of the coil cylinder causes the flexible diaphragm also to move in or out, depending on the direction of the coil current. The amount of coil current determines the intensity of the magnetic field, which controls the amount that the diaphragm moves.

As shown in Figure 7–23, when an audio signal (voice or music) is applied to the coil, the current varies in both direction and amount. In response, the diaphragm will *vibrate* in and out by varying amounts and at varying rates. Vibration in the diaphragm

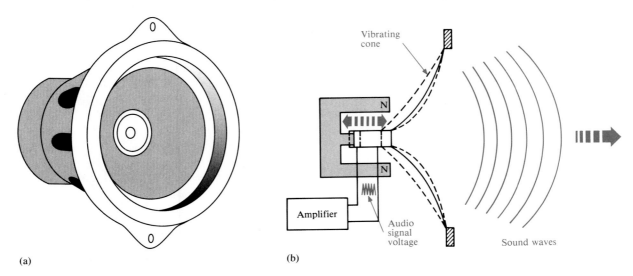

(a)   (b)

FIGURE 7–23
The speaker converts audio signal voltages into sound waves.

causes the air that is in contact with it to vibrate in the same manner. These air vibrations move through the air as sound waves.

Analog Meter Movement

The d'Arsonval movement is the most common type used in analog multimeters. In this type of meter movement, the pointer is deflected in proportion to the amount of current through a coil. Figure 7–24 shows a basic d'Arsonval meter movement. It consists of a

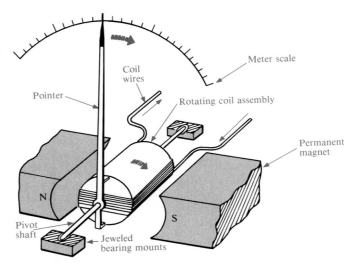

FIGURE 7–24
The basic d'Arsonval meter movement.

coil of wire wound on a bearing-mounted assembly that is placed between the poles of a permanent magnet. A pointer is attached to the moving assembly. With no current through the coil, a spring mechanism keeps the pointer at its left-most (zero) position. When current flows through the coil, electromagnetic forces act on the coil, causing a rotation to the right. The amount of rotation depends on the amount of current.

Figure 7–25 illustrates how the interaction of magnetic fields produces rotation of the coil assembly. The current flows inward at the "cross" and outward at the "dot" in the single winding shown. The inward current produces a counterclockwise electromagnetic field that reinforces the permanent magnetic field below it. The result is an upward force on the left side of the coil as shown. A downward force is developed on the right side of the coil, where the current is outward. These forces produce a clockwise rotation of the coil assembly and are opposed by a spring mechanism. The indicated forces and the spring force are balanced at the value of the current. When current is removed, the spring force returns the pointer to its zero position.

FIGURE 7–25
When the electromagnetic field interacts with the permanent magnetic field, forces are exerted on the rotating coil assembly, causing it to move clockwise and thus deflecting the pointer.

⊕ Current away from you
⊙ Current toward you

**SECTION REVIEW
7–3**

1. Explain the difference between a solenoid and a relay.
2. What is the movable part of a solenoid called?
3. What is the movable part of a relay called?
4. Upon what basic principle is the d'Arsonval meter movement based?

7–4

MAGNETIC HYSTERESIS

When a *magnetizing force* is applied to a material, the flux density in the material changes in a certain manner that we will now examine.

Magnetizing Force (*H*)

The magnetizing force in a material is defined to be the magnetomotive force (F_m) per unit length (*l*) of the material, as expressed by the following equation. The unit of magnetizing force (*H*) is ampere-turns per meter (At/m).

$$H = \frac{F_m}{l} \tag{7–6}$$

where $F_m = NI$. Note that the magnetizing force depends on the number of turns (N) of the coil of wire, the current (I) through the coil, and the length (l) of the material. It does not depend on the type of material.

Since $\phi = F_m/\mathcal{R}$, as F_m increases, the flux increases. Also, the magnetizing force (H) increases. Recall that the flux density (B) is the flux per unit cross-sectional area ($B = \phi/A$), so B is also proportional to H. The curve showing how these two quantities (B and H) are related is called the *B-H* curve, or the hysteresis curve. The parameters that influence both B and H are illustrated in Figure 7–26.

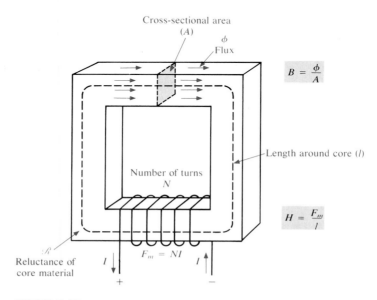

FIGURE 7–26
Parameters that determine the magnetizing force (H) and the flux density (B).

The Hysteresis Curve

H can be readily increased or decreased by varying the current through the coil of wire, and it can be reversed by reversing the voltage polarity across the coil.

We start by assuming a magnetic core is unmagnetized so that $B = 0$. As the magnetizing force H is increased from zero, the flux density B increases proportionally as indicated by the curve in Figure 7–27(a). When H reaches a certain value H_1, the value of B begins to level off and, as H continues to increase, B reaches a *saturation* value when H reaches a value H_{sat} as illustrated in Figure 7–27(b). Once saturation is reached, a further increase in H will not increase B.

Now, if H is decreased to zero, B will fall back along a different path to a *residual value* (B_R), as shown in Figure 7–27(c). This indicates that the material continues to be magnetized even with the magnetizing force removed ($H = 0$). The ability of a material, once magnetized, to maintain a magnetized state without the presence of a magnetizing force is called *retentivity*.

Reversal of the magnetizing force is represented by negative values of H on the curve and is achieved by reversing the current in the coil of wire. An increase in H in

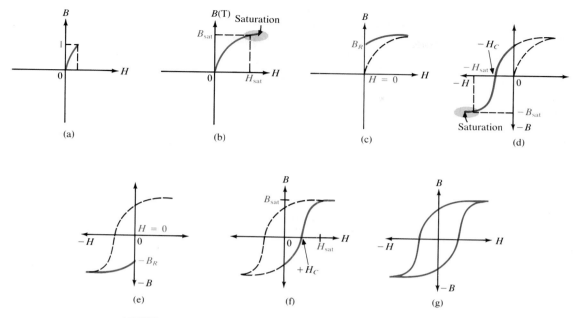

FIGURE 7–27
Development of a magnetic hysteresis curve.

the negative direction causes saturation to occur at a value $-H_{sat}$ where the flux density is at its maximum negative value as indicated in Figure 7–27(d). When the magnetizing force is removed ($H = 0$), the flux density goes to its negative *residual* value ($-B_R$) as shown in part (e). From the $-B_R$ value, the flux density follows the curve indicated in part (f) back to its maximum positive value when the magnetizing force equals H_{sat} in the positive direction. The complete *B-H* curve is shown in part (g) and is called the *hysteresis curve*. The magnetizing force required to make the flux density zero is called the *coercive force, H_C*.

Retentivity

The retentivity of a material is indicated by the ratio of B_R to B_{sat}. Materials with a low retentivity do not retain a magnetic field very well while those with high retentivities exhibit values of B_R very close to the saturation value of B. Depending on the application, retentivity in a magnetic material can be an advantage or a disadvantage. In permanent magnets and memory cores, for example, high retentivity is required. In ac motors, retentivity is undesirable because the residual magnetic field must be overcome each time the current reverses, thus wasting energy.

**SECTION REVIEW
7–4**

1. For a given wirewound core, how does an increase in current through the coil affect the flux density?
2. Define *retentivity*.

7–5   ELECTROMAGNETIC INDUCTION

When a conductor is moved through a magnetic field, a voltage is produced across the conductor. This principle is known as *electromagnetic induction,* and the resulting voltage is an *induced voltage*.

    The principle of electromagnetic induction is widely applied in electrical circuits, as you will learn in this chapter and later in the study of transformers. The operation of electrical motors and generators is also based on this principle.

Relative Motion

When a wire is moved across a magnetic field, there is a relative motion between the wire and the magnetic field. Likewise, when a magnetic field is moved past a stationary wire, there is also relative motion. In either case, there is an *induced voltage* in the wire as a result of this motion, as Figure 7–28 indicates.

    The amount of the induced voltage depends on the *rate* at which the wire and the magnetic field move with respect to each other: The faster the relative speed, the greater the induced voltage.

FIGURE 7–28
Relative motion between a wire and a magnetic field.

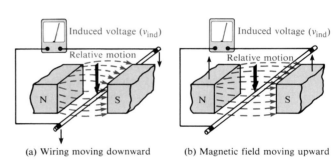

(a) Wiring moving downward   (b) Magnetic field moving upward

Polarity of the Induced Voltage

If the conductor in Figure 7–28 is moved first one way and then another in the magnetic field, a reversal of the polarity of the induced voltage will be observed. As the wire is moved downward, a voltage is induced with the polarity indicated in Figure 7–29(a). As the wire is moved upward, the polarity is as indicated in part (b) of the figure. The lowercase v stands for instantaneous voltage.

FIGURE 7–29
Polarity of induced voltage depends on direction of motion.

 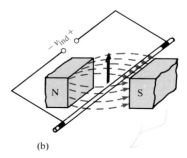

(a)   (b)

Induced Current

When a load resistor is connected to the wire in Figure 7–29, the voltage induced by the relative motion in the magnetic field will cause a current in the load, as shown in Figure 7–30. This current is called the *induced current*.

FIGURE 7–30
Induced current in a load as the wire moves through the magnetic field.

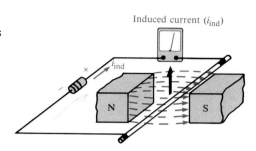

The principle of producing a voltage and a current in a load by moving a conductor across a magnetic field is the basis for electrical generators. The concept of a conductor existing in a moving magnetic field is fundamental to inductance in an electrical circuit. The lowercase i stands for instantaneous current.

Forces on a Current-Carrying Conductor in a Magnetic Field (Motor Action)

Figure 7–31(a) shows current inward through a wire in a magnetic field. The electromagnetic field set up by the current interacts with the permanent magnetic field; as a result, the permanent lines of force *above* the wire tend to be deflected down under the wire, because they are opposite in direction to the electromagnetic lines of force. Therefore, the flux density above is reduced, and the magnetic field is weakened. The flux density below the conductor is increased, and the magnetic field is strengthened. An upward force on the conductor results, and the conductor tends to move toward the weaker magnetic field.

Figure 7–31(b) shows the current outward, resulting in a force on the conductor in the downward direction. This principle is the basis for electrical motors.

FIGURE 7–31
Forces on a current-carrying conductor in a magnetic field.

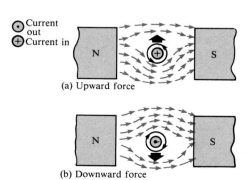

Faraday's Law

Michael Faraday discovered the principle of electromagnetic induction in 1831. He found that moving a magnet through a coil of wire induced a voltage across the coil, and that when a complete path was provided, the induced voltage caused an induced current, as you have seen. Faraday's observations are as follows:

1. The amount of voltage induced in a coil is directly proportional to the rate of change of the magnetic field with respect to the coil.
2. The amount of voltage induced in a coil is directly proportional to the number of turns of wire in the coil.
3. The voltage induced across a coil equals the number of turns in the coil times the rate of change of the magnetic flux.

Part 1 of Faraday's law is demonstrated in Figure 7–32, where a bar magnet is moved through a coil, thus creating a changing magnetic field. In part (a) of the figure, the magnet is moved at a certain rate, and a certain induced voltage is produced as indicated. In part (b), the magnet is moved at a faster rate through the coil, creating a greater induced voltage.

$$V = N \times \frac{\Delta \phi}{\Delta t}$$

(a) As magnet moves to the right, magnetic field is changing with respect to coil, and a voltage is induced.

(b) As magnet moves more rapidly to the right, magnetic field is changing more rapidly with respect to coil, and a greater voltage is induced.

FIGURE 7–32
A demonstration of the first part of Faraday's law: The amount of induced voltage is directly proportional to the rate of change of the magnetic field with respect to the coil.

Part 2 of Faraday's law is demonstrated in Figure 7–33. In part (a), the magnet is moved through the coil and a voltage is induced as shown. In part (b), the magnet is moved at the same speed through a coil that has a greater number of turns. The greater number of turns creates a greater induced voltage.

(a) Magnet moves through coil and induces a voltage.

(b) Magnet moves at same rate through a coil with more turns (loops) and induces a greater voltage.

FIGURE 7–33
A demonstration of the second part of Faraday's law: The amount of induced voltage is directly proportional to the number of turns in the coil.

**SECTION REVIEW
7–5**

1. What is the induced voltage across a stationary conductor in a stationary magnetic field?
2. When the speed at which a conductor is moved through a magnetic field is increased, does the induced voltage increase, decrease, or remain the same?
3. When there is current through a conductor in a magnetic field, what happens?

7–6

APPLICATIONS OF ELECTROMAGNETIC INDUCTION

In this section, two interesting applications of electromagnetic induction are discussed: an automotive crankshaft position sensor, and a dc generator. Although there are many varied applications, these two are representative.

Automotive Crankshaft Position Sensor

An interesting automotive application is a type of engine sensor that detects the crankshaft position directly using electromagnetic induction. The electronic engine controller in many automobiles uses the position of the crankshaft to set ignition timing and, sometimes, to adjust the fuel control system. Figure 7–34 shows the basic concept. A steel disk is attached to the engine's crankshaft by an extension rod; the protruding tabs on the disk represent specific crankshaft positions.

As illustrated in Figure 7–34, as the disk rotates with the crankshaft, the tabs periodically pass through the air gap of the permanent magnet. Since steel has a much

FIGURE 7–34
A crankshaft position sensor that produces a voltage when a tab passes through the air gap of the magnet.

lower reluctance than does air (a magnetic field can be established in steel much more easily than in air), the magnetic flux suddenly increases as a tab comes into the air gap, causing a voltage to be induced across the coil. This process is illustrated in Figure 7–35. The electronic engine control circuit uses the induced voltage as an indicator of the crankshaft position.

(a) There is no changing magnetic field, so there is no induced voltage.

(b) Insertion of the steel tap reduces the reluctance of the air gap, causing the magnetic flux to increase and thus inducing a voltage.

FIGURE 7–35
As the tab passes through the air gap of the magnet, the coil senses a change in the magnetic field, and a voltage is induced.

A dc Generator

Figure 7–36 shows a simplified dc generator consisting of a single loop of wire in a permanent magnetic field. Notice that each end of the loop is connected to a *split-ring* arrangement. This conductive metal ring is called a *commutator*. As the loop is rotated in the magnetic field, the split commutator ring also rotates. Each half of the split ring rubs against the fixed contacts, called *brushes*, and connects the loop to an external circuit.

FIGURE 7–36
A basic dc generator.

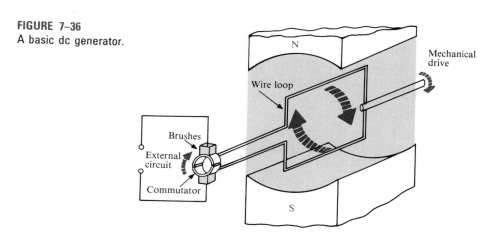

As the loop rotates through the magnetic field, it "cuts" through the flux lines at varying angles, as illustrated in Figure 7–37. At position *A* in its rotation, the loop of wire is effectively moving *parallel* with the magnetic field. Therefore, at this instant, the rate at which it is cutting through the magnetic flux lines is *zero*.

FIGURE 7–37
End view of loop cutting through the magnetic field.

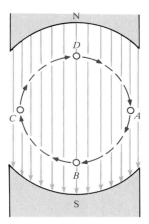

As the loop moves from position *A* to position *B*, it cuts through the flux lines at an increasing rate. At position *B*, it is moving effectively *perpendicular* to the magnetic field and thus is cutting through a maximum number of lines. As the loop rotates from position *B* to position *C*, the rate at which it cuts the flux lines decreases to minimum (zero) at *C*. From position *C* to position *D*, the rate at which the loop cuts the flux lines increases to a maximum at *D* and then back to a minimum again at *A*.

As you previously learned, when a wire moves through a magnetic field, a voltage is induced, and by Faraday's law, the amount of induced voltage is proportional to the number of loops (turns) in the wire and the rate at which it is moving with respect to the magnetic field. Now you know that the *angle* at which the wire moves with respect to the magnetic flux lines determines the amount of induced voltage, because the rate at which the wire cuts through the flux lines depends on the angle of motion.

Figure 7–38 illustrates how a voltage is induced in the external circuit as the single loop rotates in the magnetic field. Assume that the loop is in its instantaneous horizontal position, so the induced voltage is zero. As the loop continues in its rotation, the induced voltage builds up to a maximum at position *B*, as shown in part (a) of the figure. Then, as the loop continues from *B* to *C*, the voltage decreases to zero at *C*, as shown in part (b).

During the second half of the revolution, shown in parts (c) and (d), the brushes switch to opposite commutator sections, so the polarity of the voltage remains the same

(a) Position *B*: Loop is moving perpendicular to flux lines, and voltage is maximum.

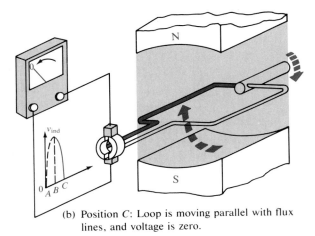

(b) Position *C*: Loop is moving parallel with flux lines, and voltage is zero.

(c) Position *D*: Loop is moving perpendicular to flux lines, and voltage is maximum.

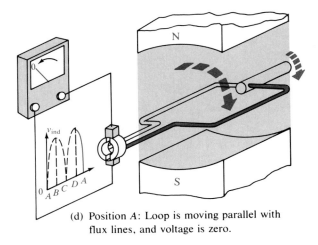

(d) Position *A*: Loop is moving parallel with flux lines, and voltage is zero.

FIGURE 7–38
Operation of a basic dc generator.

across the output. Thus, as the loop rotates from position C to position D and then back to A, the voltage increases from zero at C to a maximum at D and back to zero at A.

Figure 7–39 shows how the induced voltage varies as the loop goes through several rotations (three in this case). This voltage is a dc voltage because its polarity does not change. However, the voltage is pulsating between zero and its maximum value.

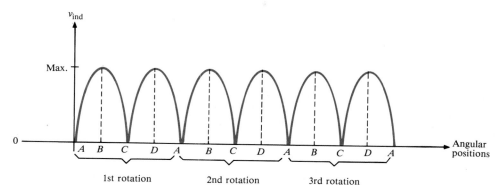

FIGURE 7–39
Induced voltage over three rotations of the loop.

When more loops are added, the voltages induced across each loop are combined across the output. Since the voltages are offset from each other, they do not reach their maximum or zero values at the same time. A smoother dc voltage results, as shown in Figure 7–40 for two loops. The variations can be further smoothed out by *filters* to achieve a nearly constant dc voltage. (Filters are discussed in a later chapter.)

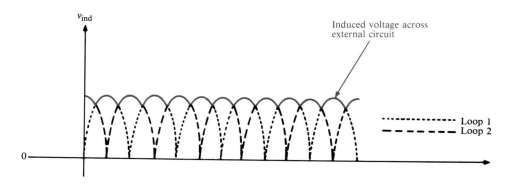

FIGURE 7–40
The induced voltage for a two-loop generator. There is much less variation in the induced voltage.

**SECTION REVIEW
7–6**

1. If the steel disk in the crankshaft position sensor has stopped with a tab in the magnet's air gap, what is the induced voltage?

2. What happens to the induced voltage if the loop in the basic dc generator suddenly begins rotating at a faster speed?

TABLE 7–1

Symbol	Quantity	Unit
B	Magnetic flux density	Tesla (T)
ϕ	Flux	Weber (Wb)
μ	Permeability	Weber/ampere-turn·meter (Wb/At·m)
$\mathcal{R}$	Reluctance	At/Wb
F_m	Magnetomotive force (mmf)	Ampere-turn (At)
H	Magnetizing force	At/m

SUMMARY

☐ Unlike magnetic poles attract each other, and like poles repel each other.

☐ Materials that can be magnetized are called *ferromagnetic*.

☐ When current flows through a conductor, it produces an electromagnetic field around the conductor.

☐ You can use the left-hand rule to establish the direction of the electromagnetic lines of force around a conductor.

☐ An electromagnet is basically a coil of wire around a magnetic core.

☐ When a conductor moves within a magnetic field, or when a magnetic field moves relative to a conductor, a voltage is induced across the conductor.

☐ The faster the relative motion between a conductor and a magnetic field, the greater is the induced voltage.

☐ Table 7–1 summarizes the magnetic quantities.

GLOSSARY

Ampere-turn   The unit of magnetomotive force (mmf).

Electromagnetic field   A formation of a group of magnetic lines of force surrounding a conductor created by electrical current in the conductor.

Electromagnetic induction   The phenomenon or process by which a voltage is produced in a conductor when there is relative motion between the conductor and a magnetic or electromagnetic field.

Faraday's law   A law stating that the voltage induced across a coil of wire equals the number of turns in the coil times the rate of change of the magnetic flux.

Flux density   The number of lines of force per unit area in a magnetic field.

Hysteresis   A lagging in the values of resulting magnetization in a magnetic material due to a changing magnetizing force.

Magnetic field intensity   The amount of mmf per unit length of magnetic material.

Magnetic flux   The lines of force between the north and south poles of a permanent magnet or an electromagnet.

Permeability   The measure of ease with which a magnetic field can be established in a material.

Relay   An electromagnetically controlled mechanical device in which electrical contacts are opened or closed by a magnetizing current.

Reluctance   The opposition to the establishment of a magnetic field.

Retentivity   The ability of a material, once magnetized, to maintain a magnetized state without the presence of a magnetizing force.

Solenoid An electromagnetically controlled device in which the mechanical movement of a shaft or plunger is activated by a magnetizing current.

Tesla The unit of flux density.

Weber The unit of magnetic flux.

FORMULAS

$$B = \frac{\phi}{A}$$ Magnetic flux density   (7–1)

$$V = N A \frac{\Delta \phi}{\Delta t}$$

$$\mu_r = \frac{\mu}{\mu_0}$$ Relative permeability   (7–2)

$$V = L \frac{\Delta I}{\Delta t}$$

$$\mathcal{R} = \frac{l}{\mu A}$$ Reluctance   (7–3)

$$F_m = NI$$ Magnetomotive force   (7–4)

$$\phi = \frac{F_m}{\mathcal{R}}$$ Magnetic flux   (7–5)

$$H = \frac{F_m}{l}$$ Magnetic flux intensity   (7–6)

SELF-TEST

1. When the south poles of two bar magnets are brought close together, there will be
   (a) a force of attraction
   (b) a force of repulsion
   (c) an upward force
   (d) no force

2. A magnetic field is made up of
   (a) positive and negative charges
   (b) magnetic domains
   (c) flux lines
   (d) magnetic poles

3. The direction of a magnetic field is from
   (a) north pole to south pole
   (b) south pole to north pole
   (c) inside to outside the magnet
   (d) front to back

4. Reluctance in a magnetic circuit is analogous to
   (a) voltage in an electric circuit
   (b) current in an electric circuit
   (c) power in an electric circuit
   (d) resistance in an electric circuit

5. The unit of magnetic flux is the
   (a) tesla   (b) weber   (c) electron-volt   (d) ampere-turn

6. The unit of magnetomotive force is the
   (a) tesla   (b) weber   (c) ampere-turn   (d) electron-volt

7. The unit of flux density is the
   (a) tesla   **(b)** weber   **(c)** ampere-turn   **(d)** electron-volt
8. The electromagnetic activation of a movable shaft is the basis for
   (a) relays   **(b)** circuit breakers   **(c)** magnetic switches   **(d)** solenoids
9. When current flows through a wire placed in a magnetic field,
   (a) the wire will overheat
   (b) the wire will become magnetized
   (c) a force is exerted on the wire
   (d) the magnetic field will be cancelled
10. A coil of wire is placed in a changing magnetic field. If the number of turns in the coil is increased, the voltage induced across the coil will
   (a) remained unchanged
   (b) decrease
   (c) increase
   (d) be excessive
11. If a conductor is moved back and forth at a constant rate in a constant magnetic field, the voltage induced in the conductor will
   (a) remain constant
   (b) reverse polarity
   (c) be reduced
   (d) be increased
12. In the crankshaft position sensor in Figure 7–34, the induced voltage across the coil is caused by
   (a) current in the coil
   (b) rotation of the disk
   (c) a tab passing through the magnetic field
   (d) acceleration of the disk's rotational speed

PROBLEM SET A

Section 7–1   The Magnetic Field

1. The cross-sectional area of a magnetic field is increased, but the flux remains the same. Does the flux density increase or decrease?
2. In a certain magnetic field the cross-sectional area is 0.5 m² and the flux is 1500 μWb. What is the flux density?
3. What is the flux in a magnetic material when the flux density is 2500×10^{-6} T and the cross-sectional area is 150 cm²?

Section 7–2   Electromagnetism

4. What happens to the compass needle in Figure 7–9 when the current through the conductor is reversed?
5. What is the relative permeability of a ferromagnetic material whose absolute permeability is 750×10^{-6} Wb/At·m?
6. Determine the reluctance of a material with a length of 0.28 m and a cross-sectional area of 0.08 m² if the absolute permeability is 150×10^{-7} Wb/At·m.
7. What is the magnetomotive force in a 50-turn coil of wire with 3 A through it?

Section 7–3   Electromagnetic Devices

8. Typically, when a solenoid is activated, is the plunger extended or retracted?
9. **(a)** What force moves the plunger when a solenoid is activated?
   (b) What force causes the plunger to return to its at-rest position?

FIGURE 7–41

10. Explain the sequence of events in the circuit of Figure 7–41 starting when switch 1 (SW1) is closed.

11. What causes the pointer in a d'Arsonval movement to deflect when there is current through the coil?

Section 7–4   Magnetic Hysteresis

12. What is the magnetizing force in Problem 7 if the length of the core is 0.2 m?

13. How can the flux density in Figure 7–42 be changed without altering the physical characteristics of the core?

14. In Figure 7–42, determine the following:
  (a) H **(b)** ϕ **(c)** B

FIGURE 7–42

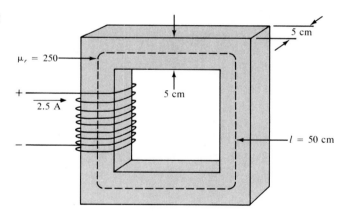

15. Determine from the hysteresis curves in Figure 7–43 which material has the most retentivity.

FIGURE 7–43

Section 7–5   Electromagnetic Induction

16. According to Faraday's law, what happens to the induced voltage across a given coil if the rate of change of magnetic flux doubles?

17. The voltage induced across a certain coil is 100 mV. A 100-Ω resistor is connected to the coil terminals. What is the induced current?

Section 7–6   Applications of Electromagnetic Induction

18. In Figure 7–34, why is there no induced voltage when the steel disk is not rotating?

19. Explain the purpose of the commutator and brushes in Figure 7–36.

PROBLEM SET B

20. A basic one-loop dc generator is rotated at 60 rev/s. How many times each second does the dc output voltage peak (reach a maximum)?

21. Assume that another loop, 90° from the first loop, is added to the dc generator in Problem 20. Make a graph of voltage versus time to show how the output voltage appears. Let the maximum voltage be 10 V.

ANSWERS TO SECTION REVIEWS

Section 7–1

1. Repel.   2.   The group of lines of force that make up a magnetic field.

Section 7–2

1. Electromagnetism is produced by current through a conductor. An electromagnetic field exists only when current flows. A magnetic field exists independently of current.

2. The direction of the magnetic field also reverses.

Section 7–3

1. A solenoid provides mechanical movement of a shaft. A relay provides contact closure.

2. Plunger.

3. Armature.

4. Interaction of magnetic fields.

Section 7–4

1. An increase in current increases the flux density.

2. The ability of a material to remain magnetized after removal of the magnetizing force.

Section 7–5

1. Zero.   2.   Increase.   3.   A force is exerted on the conductor.

Section 7–6

1. Zero.   2.   Increases.

PART TWO

AC CIRCUITS

8 Introduction to Alternating Current and Voltage

This chapter provides an introduction to alternating current (ac) circuits. Alternating voltages and currents fluctuate with time and periodically change polarity and direction according to certain patterns called *waveforms*. Particular emphasis is given to the sine wave because of its basic importance in ac circuits. Other types of waveforms are also introduced, including pulse, triangular, and sawtooth. The use of the oscilloscope for displaying and measuring waveforms is discussed.

The concept of phasors is introduced to help you understand ac circuits. Phasors provide a concise format for representing sine wave voltages and currents and their relationships with each other.

After completing this chapter, you will be able to

- Identify sine waves and measure their characteristics.
- Explain how frequency and period are related.
- Measure the following voltage or current values of a sine wave: instantaneous, peak, peak-to-peak, rms, and average.
- Describe how sine waves are generated.
- Measure points on a sine wave in terms of angular units.
- Determine the phase angle between two or more sine waves.
- Discuss the meaning of phase lead and phase lag.
- Express sine waves with a mathematical formula.
- Discuss phasors and how they can be used to represent sine waves.
- Apply Ohm's law and Kirchhoff's laws to ac circuits as well as to dc circuits.
- Describe the characteristics of pulse, triangular, and sawtooth waveforms.
- Discuss how nonsinusoidal waveforms are made up of various harmonics
- Use the oscilloscope to measure waveform parameters.

APPLICATION ASSIGNMENT

As an electronic technician working for a company that designs and manufactures laboratory instruments, you have been assigned to a team that is responsible for the design and testing of a new signal generator that produces various types of time-varying voltages with adjustable frequency and amplitude ranges. The project engineer has assigned you to test the prototype model for the proper output voltages.

You must measure the frequency, period, and amplitude for all output waveforms to assure compliance with the design specifications. In addition, you must measure the rise time, fall time, and duty cycle for the specified ranges of adjustment for the pulse output. The specifications are as follows:

1. *Sine wave output* Frequency adjustable from 30 Hz to 30 kHz; amplitude adjustable from 0.5 V peak to 25 V peak.
2. *Pulse output* Frequency adjustable from 1 kHz to 50 kHz; amplitude adjustable from 0.1 V to 15 V; duty cycle adjustable from 10% to 75%; rise and fall times to be no greater than 0.1 μs.
3. *Sawtooth output* Frequency adjustable from 100 Hz to 10 kHz; amplitude adjustable from 1 V to 30 V.

Determine the settings on the oscilloscope for the minimum and maximum of each range and the points on the waveforms where you will take the measurements. After completing this chapter, you will be able to carry out this assignment.

8–1

THE SINE WAVE

The sine wave is one very common type of alternating current (ac) and alternating voltage. It is also referred to as a *sinusoidal wave* or, simply, *sinusoid*. The electrical service provided by the power companies is in the form of sinusoidal voltage and current. In addition, other types of waveforms are composites of many individual sine waves called *harmonics,* discussed in Section 8–9.

Sine waves are produced by two types of sources: rotating electrical machines (ac generators) or electronic oscillator circuits, which are in instruments known as electronic signal generators. A typical electronic signal generator is shown in Figure 8–1(a) and (b). The schematic symbol for an ac voltage source is shown in part (c).

(a) ac generator, cutaway view (courtesy of General Electric)

(b) Electronic signal generator (courtesy of B & K Precision)

FIGURE 8–1
Sine wave sources.

(c) Symbol for an ac
voltage source

Figure 8–2 is a graph showing the general shape of a sine wave, which can be either current or voltage. Notice how the voltage (or current) varies with time. Starting at zero, it *increases* to a positive maximum (peak), returns to zero, and then increases to a *negative* maximum (peak) before returning again to zero.

FIGURE 8–2
Sine wave.

The Polarity of a Sine Wave

As you have seen, a sine wave changes *polarity* at its zero value; that is, it *alternates* between positive and negative values. When a sine wave voltage is applied to a resistive circuit, as in Figure 8–3, an alternating sine wave current results. When the voltage changes polarity, the current correspondingly changes direction as indicated.

FIGURE 8–3
Alternating current and voltage.

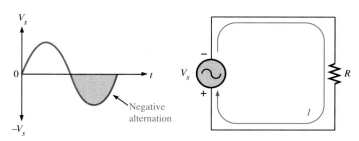

(a) Positive voltage: current direction as shown.

(b) Negative voltage: current reverses direction.

During the *positive alternation* of the applied voltage V_s, the current is in the direction shown in Figure 8–3(a). During a *negative alternation* of the applied voltage, the current is in the opposite direction, as shown in part (b). The combined positive and negative alternations make up one *cycle* of a sine wave.

The Period of a Sine Wave

As you have seen, a sine wave varies with time in a definable manner. Time is designated by t. The time required for a sine wave to complete one full cycle is called the

AC CIRCUITS

period (T), as indicated in Figure 8–4(a). Typically, a sine wave continues to repeat itself in identical cycles, as shown in part (b). Since all cycles of a repetitive sine wave are the same, the period is always a fixed value for a given sine wave.

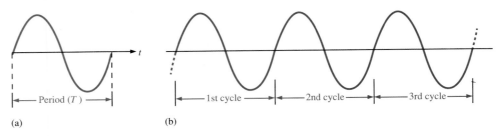

(a)   (b)

FIGURE 8–4
The period of a sine wave is the same for each cycle.

The period of a sine wave does not necessarily have to be measured between the zero crossings at the beginning and end of a cycle. It can be measured from any point in a given cycle to the *corresponding* point in the next cycle.

EXAMPLE 8–1

What is the period of the sine wave in Figure 8–5?

FIGURE 8–5

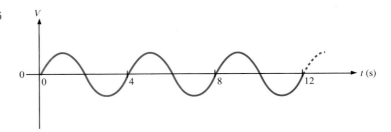

Solution
As you can see, it takes four seconds (4 s) to complete each cycle. Therefore, the period is 4 s.

$$T = 4 \text{ s}$$

EXAMPLE 8–2

Show three possible ways to measure the period of the sine wave in Figure 8–6. How many cycles are shown?

FIGURE 8–6

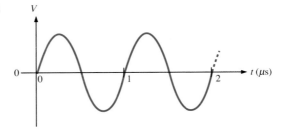

Solution

Method 1: The period can be measured from one zero crossing to the corresponding zero crossing in the next cycle.

Method 2: The period can be measured from the positive peak in one cycle to the positive peak in the next cycle.

Method 3: The period can be measured from the negative peak in one cycle to the negative peak in the next cycle.

These measurements are indicated in Figure 8–7, where two cycles of the sine wave are shown. Keep in mind that you obtain the same value for the period no matter which of these points on the waveform you use.

FIGURE 8–7
Measurement of the period.

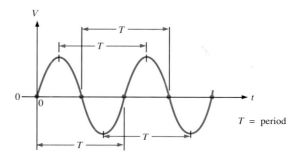

T = period

The Frequency of a Sine Wave

Frequency is the number of cycles that a sine wave completes in 1 s.

The more cycles completed in one second, the higher the frequency.

Figure 8–8 shows two sine waves. The sine wave in part (a) completes two full cycles in one second. The one in part (b) completes four cycles in one second. Therefore, the sine wave in part (b) has twice the frequency of the one in part (a).

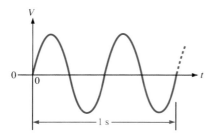

(a) Lower frequency: fewer cycles per second

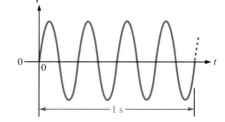

(b) Higher frequency: more cycles per second

FIGURE 8–8
Illustration of frequency.

The Unit of Frequency

Frequency is measured in units of *hertz*, abbreviated Hz. One hertz is equivalent to one cycle per second; 60 Hz is 60 cycles per second; and so on. The symbol for frequency is f.

Relationship of Frequency and Period

The relationship between frequency and period is very important. The formulas for this relationship are as follows:

$$f = \frac{1}{T} \qquad\qquad (8\text{--}1)$$

$$T = \frac{1}{f} \qquad\qquad (8\text{--}2)$$

There is a *reciprocal* relationship between f and T. Knowing one, you can calculate the other with the $\boxed{1/x}$ key on your calculator. This inverse relationship is logical because a sine wave with a longer period goes through fewer cycles in one second than one with a shorter period.

EXAMPLE 8–3   Which sine wave in Figure 8–9 has the higher frequency? Determine the period and the frequency of both waveforms.

(a)

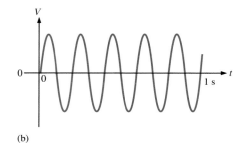

(b)

FIGURE 8–9

Solution

In part (a) of the figure, three cycles take 1 s. Therefore, one cycle takes 0.333 s (one-third second), and this is the period.

$$T = 0.333 \text{ s}$$

$$f = \frac{1}{T} = \frac{1}{0.333 \text{ s}} = 3 \text{ Hz}$$

In part (b) of the figure, five cycles take 1 s. Therefore, one cycle takes 0.2 s (one-fifth second), and this is the period.

$$T = 0.2 \text{ s}$$

$$f = \frac{1}{T} = \frac{1}{0.2 \text{ s}} = 5 \text{ Hz}$$

EXAMPLE 8–4

The period of a certain sine wave is 10 ms. What is the frequency?

Solution
Using Equation (8–1), we obtain the following result:

$$f = \frac{1}{T} = \frac{1}{10 \text{ ms}} = \frac{1}{10 \times 10^{-3} \text{ s}}$$
$$= 0.1 \times 10^3 \text{ Hz} = 100 \text{ Hz}$$

EXAMPLE 8–5

The frequency of a sine wave is 60 Hz. What is the period?

Solution
Using Equation (8–2), we determine the period as follows:

$$T = \frac{1}{f} = \frac{1}{60 \text{ Hz}} = 0.01667 \text{ s} = 16.67 \text{ ms}$$

SECTION REVIEW 8–1

1. Describe one cycle of a sine wave.
2. At what point does a sine wave change polarity?
3. How many maximum points does a sine wave have during one cycle?
4. How is the period of a sine wave measured?
5. Define *frequency*, and state its unit.
6. Determine f when $T = 5 \; \mu$s.
7. Determine T when $f = 120$ Hz.

8–2

VOLTAGE AND CURRENT VALUES OF A SINE WAVE

There are several ways to measure the value of a sine wave in terms of its voltage or its current magnitude. These are *instantaneous, peak, peak-to-peak, rms,* and *average* values.

Instantaneous Value

Figure 8–10 illustrates that at any point in time on a sine wave, the voltage (or current) has an *instantaneous* value. This instantaneous value is different at different points along the curve. Instantaneous values are positive during the positive alternation and negative during the negative alternation. Instantaneous values of voltage and current are symbolized by lowercase v and i, respectively, as shown in part (a). The curve is shown for voltage only, but applies equally for current when v's are replaced with i's. An example is shown in part (b), where the instantaneous voltage is 3.1 V at 1 μs, 7.07 V at 2.5 μs, 10 V at 5 μs, 0 V at 10 μs, -3.1 V at 11 μs, and so on.

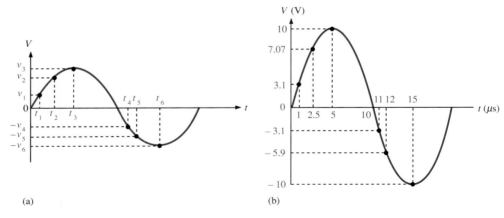

FIGURE 8–10
Example of instantaneous values of a sine wave voltage.

Peak Value

The peak value of a sine wave is the value of voltage (or current) at the positive or the negative maximum (peaks) with respect to zero. Since the peaks are equal in magnitude, a sine wave is characterized by a single peak value, as is illustrated in Figure 8–11. For a given sine wave, the peak value is constant and is represented by V_p or I_p. In this case, the peak value is 8 V.

FIGURE 8–11
Example of the peak value of a sine wave voltage.

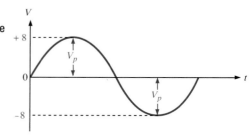

Peak-to-Peak Value

The peak-to-peak value of a sine wave, as illustrated in Figure 8–12, is the voltage (or current) from the positive peak to the negative peak. Of course, it is always *twice the peak value* as expressed in the following equations:

$$V_{pp} = 2V_p \tag{8–3}$$

$$I_{pp} = 2I_p \tag{8–4}$$

where peak-to-peak values are represented by the symbols V_{pp} or I_{pp}. In Figure 8–12, the peak-to-peak value is 16 V.

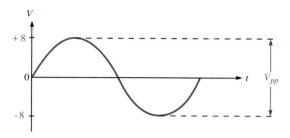

FIGURE 8–12
Example of the peak-to-peak value of a sine wave voltage.

rms Value

The term *rms* stands for *root mean square*. It refers to the mathematical process by which this value is derived. The rms value is also referred to as the *effective value*. Most ac voltmeters display rms voltage. The 110 V at your wall outlet is an rms value.

    The rms value of a sine wave is actually a measure of the *heating effect* of the sine wave. For example, when a resistor is connected across an ac (sine wave) voltage source, as shown in Figure 8–13(a), a certain amount of heat is generated by the power in the resistor. Part (b) shows the *same* resistor connected across a dc voltage source.

FIGURE 8–13
When the same amount of heat is being produced in both cases, the sine wave has an rms value equal to the dc voltage.

The value of the ac voltage can be adjusted so that the resistor gives off the same amount of heat as it does when connected to the dc source.

> **The rms value of a sine wave is equal to the dc voltage that produces the same amount of heat in a resistance as the sinusoidal voltage.**

The peak value of a sine wave can be converted to the corresponding rms value using the following relationships for either voltage or current:

$$V_{rms} = \sqrt{0.5}\ V_p \cong 0.707V_p \qquad (8\text{--}5)$$

$$I_{rms} = \sqrt{0.5}\ I_p \cong 0.707I_p \qquad (8\text{--}6)$$

Using these formulas, we can also determine the peak value knowing the rms value as follows:

$$V_p = \left(\frac{1}{0.707}\right)V_{rms}$$

$$V_p = \sqrt{2}\ V_{rms} \cong 1.414V_{rms} \qquad (8\text{--}7)$$

Similarly,

$$I_p = \sqrt{2}\ I_{rms} \cong 1.414I_{rms} \qquad (8\text{--}8)$$

To find the peak-to-peak value, simply double the peak value:

$$V_{pp} = 2.828V_{rms} \qquad (8\text{--}9)$$

and

$$I_{pp} = 2.828I_{rms} \qquad (8\text{--}10)$$

Average Value

The average value of a sine wave when taken over *one complete cycle* is always zero, because the positive values (above the zero crossing) offset the negative values (below the zero crossing).

To be useful for comparison purposes, the average value of a sine wave is defined over a *half-cycle* rather than over a full cycle. The average value is the total area under the half-cycle curve divided by the distance of the curve along the horizontal axis. The result is expressed in terms of the peak value as follows for both voltage and current sine waves:

$$V_{avg} = \left(\frac{2}{\pi}\right)V_p \cong 0.637V_p \qquad (8\text{--}11)$$

$$I_{avg} = \left(\frac{2}{\pi}\right)I_p \cong 0.637I_p \qquad (8\text{--}12)$$

The half-cycle average value of a sine wave voltage is illustrated in Figure 8–14.

FIGURE 8–14
Half-cycle average value.

EXAMPLE 8–6 Determine V_p, V_{pp}, V_{rms}, and V_{avg} for the sine wave in Figure 8–15.

FIGURE 8–15

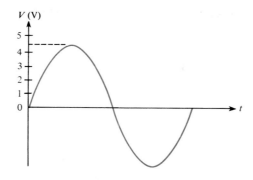

Solution

As taken directly from the graph, $V_p = 4.5$ V. From this value we calculate the other values:

$$V_{pp} = 2V_p = 2(4.5 \text{ V}) = 9 \text{ V}$$

$$V_{rms} = 0.707V_p = 0.707(4.5 \text{ V}) = 3.182 \text{ V}$$

$$V_{avg} = 0.637V_p = 0.637(4.5 \text{ V}) = 2.867 \text{ V}$$

SECTION REVIEW 8–2

1. Determine V_{pp} when
   (a) $V_p = 1$ V   (b) $V_{rms} = 1.414$ V   (c) $V_{avg} = 3$ V
2. Determine V_{rms} when
   (a) $V_p = 2.5$ V   (b) $V_{pp} = 10$ V   (c) $V_{avg} = 1.5$ V $(\frac{1.5}{.637}) \times .707 = 1.66 \text{ V} \checkmark$

8–3

SINE WAVE VOLTAGE SOURCES

There are two basic methods of generating sine wave voltages: *electromagnetic* and *electronic*. Sine waves are produced electromagnetically by ac generators and electronically by oscillator circuits.

An ac Generator

Figure 8–16 shows a basic ac generator consisting of a single loop of wire in a permanent magnetic field. Notice that each end of the loop is connected to a separate solid conductive ring called a *slip ring*. As the loop rotates in the magnetic field, the slip rings also rotate and rub against the brushes which connect the loop to an external load. Compare this generator to the basic dc generator in Chapter 7, and note the difference in the ring and brush arrangements.

As you learned in Chapter 7, when a conductor moves through a magnetic field, a voltage is induced. Figure 8–17 illustrates how a sine wave voltage is produced by the basic ac generator as the loop rotates. An oscilloscope is used to display the voltage waveform.

FIGURE 8–16
A basic ac generator.

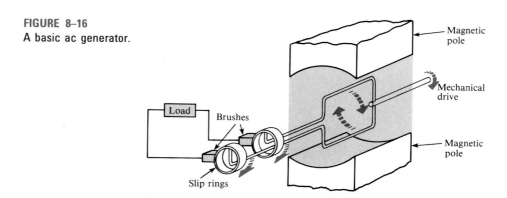

To begin, Figure 8–17(a) shows the loop rotating through the first quarter of a revolution. It goes from an instantaneous horizontal position, where the induced voltage is zero, to an instantaneous vertical position, where the induced voltage is maximum. As shown on the display, this part of the rotation produces the first quarter of the sine wave cycle as the voltage builds up from zero to its positive maximum. Part (b) of the figure shows the loop completing the first half of a revolution. During this part of the rotation, the voltage decreases from its positive maximum back to zero.

During the second half of the revolution, illustrated in Figures 8–17(c) and 8–17(d), the loop is cutting through the magnetic field in the opposite direction, so the

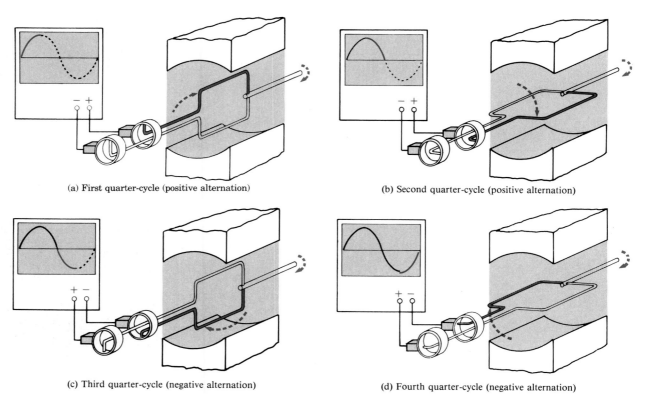

(a) First quarter-cycle (positive alternation)

(b) Second quarter-cycle (positive alternation)

(c) Third quarter-cycle (negative alternation)

(d) Fourth quarter-cycle (negative alternation)

FIGURE 8–17
One revolution of the loop generates one cycle of the sine wave voltage.

voltage produced has a polarity opposite to that produced during the first half of the revolution. After one complete revolution of the loop, one full cycle of the sine wave voltage has been produced. As the loop continues to rotate, repetitive cycles of the sine wave are generated.

Factors That Affect Frequency in ac Generators

You have seen that one revolution of the conductor through the magnetic field in the basic ac generator (also called an *alternator*) produces one cycle of induced sinusoidal voltage. It is obvious that the rate at which the conductor is rotated determines the time for completion of one cycle. For example, if the conductor completes 60 revolutions in 1 s, the period of the resulting sine wave is 1/60 s, corresponding to a frequency of 60 Hz. Thus, the faster the conductor rotates, the higher the resulting frequency of the induced voltage, as illustrated in Figure 8–18.

FIGURE 8–18
Frequency is directly proportional
to the rate of rotation of the loop.

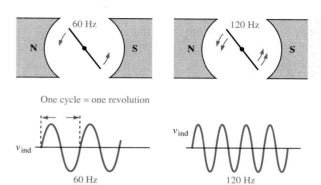

Another way of achieving a higher frequency is to increase the number of magnetic poles. In the previous discussion, two magnetic poles were used to illustrate the ac generator principle. During one revolution, the conductor passes under a north pole and a south pole, thus producing one cycle of a sine wave. When four magnetic poles are used instead of two, as shown in Figure 8–19, one cycle is generated during *one-half* a revolution. This doubles the frequency for the same rate of rotation.

An expression for frequency in terms of the number of poles and the number of revolutions per second (rev/s) is as follows:

$$f = \text{number of pole pairs} \times (\text{rev/s}) \tag{8–13}$$

FIGURE 8–19
Four poles achieve a
higher frequency than
two for the same rev/s.

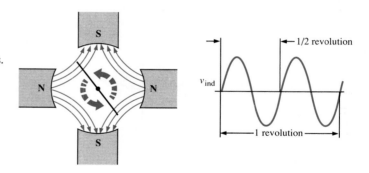

EXAMPLE 8–7

A four-pole generator has a rotation speed of 100 rev/s. Determine the frequency of the output voltage.

Solution

$$f = \text{number of pole pairs} \times (\text{rev/s}) = 2(100) = 200 \text{ Hz}$$

Factors That Affect Voltage Amplitude in ac Generators

Recall from Chapter 7 that the voltage induced in a conductor depends on the number of turns (N) and the rate of change with respect to the magnetic field. Therefore, when the speed of rotation of the conductor is increased, not only does the frequency of the induced voltage increase—so also does the amplitude. Since the frequency value normally is fixed, the most practical method of increasing the amount of induced voltage is to increase the number of loops.

Signal Generators

The signal generator is an instrument that electronically produces sine waves for use in testing or controlling electronic circuits and systems. There are a variety of signal generators, ranging from special-purpose instruments that produce only one type of waveform in a limited frequency range, to programmable instruments that produce a wide range of frequencies and a variety of waveforms. All signal generators consist basically of an oscillator, which is an electronic circuit that produces sine wave voltages whose amplitude and frequency can be adjusted. Typical signal generators are shown in Figure 8–20.

(a)

(b)

FIGURE 8–20
Typical signal generators [(a) courtesy of B&K Precision. (b) courtesy of Hewlett-Packard].

1. How are sine waves generated electromagnetically?
2. List two types of sine wave sources.

8–4

ANGULAR RELATIONSHIPS OF A SINE WAVE

As you have seen, sine waves can be measured along the horizontal axis on a time basis. However, since the time for completion of one full cycle or any portion of a cycle is frequency-dependent, it is often useful to specify points on the sine wave in terms of an *angular measurement* expressed in degrees or radians. Angular measurement is independent of frequency.

As previously discussed, sine wave voltage can be produced by rotating electromechanical machines. As the rotor of the ac generator goes through a full 360° of rotation, the resulting voltage output is one full cycle of a sine wave. Thus, the angular measurement of a sine wave can be related to the angular rotation of a generator, as shown in Figure 8–21.

FIGURE 8–21

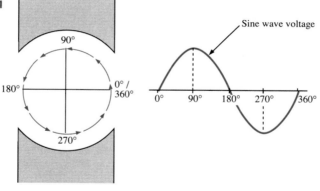

Angular Measurement

A *radian* (rad) is defined as the angular distance along the circumference of a circle equal to the radius of the circle. One radian is equivalent to 57.3°, as illustrated in Figure 8–22. In a 360° revolution, there are 2π radians.

> π **is the ratio of the circumference of any circle to its diameter and has a constant value of approximately 3.1416.**

FIGURE 8–22
Angular measurement showing relationship of radian to degrees.

AC CIRCUITS

TABLE 8–1

Degrees	Radians (rad)
0	0
45	$\pi/4$
90	$\pi/2$
135	$3\pi/4$
180	π
225	$5\pi/4$
270	$3\pi/2$
315	$7\pi/4$
360	2π

Scientific calculators have a π key so that the actual numerical value does not have to be entered.

Table 8–1 lists several values of degrees and the corresponding radian values. These angular measurements are illustrated in Figure 8–23.

FIGURE 8–23
Angular measurements.

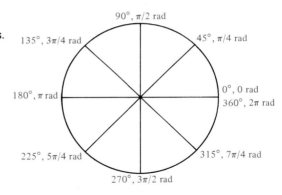

Radian-Degree Conversion

Radians can be converted to degrees using Equation (8–14).

$$\text{rad} = \left(\frac{\pi \text{ rad}}{180°}\right) \times \text{degrees} \qquad (8\text{–}14)$$

Similarly, degrees can be converted to radians with Equation (8–15).

$$\text{degrees} = \left(\frac{180°}{\pi \text{ rad}}\right) \times \text{rad} \qquad (8\text{–}15)$$

EXAMPLE 8–8   (a) Convert 60° to radians.
(b) Convert $\pi/6$ rad to degrees.

Solution

(a) $\text{Rad} = \left(\dfrac{\pi \text{ rad}}{180°}\right)60° = \dfrac{\pi}{3} \text{ rad} = 1.047 \text{ rad}$

(b) $\text{Degrees} = \left(\dfrac{180°}{\pi \text{ rad}}\right)\left(\dfrac{\pi}{6} \text{ rad}\right) = 30°$

Sine Wave Angles

The angular measurement of a sine wave is based on 360°, or 2π radians, for a complete cycle. A half-cycle is 180° or π radians; a quarter-cycle is 90°, or $\pi/2$ radians; and so on. Figure 8–24(a) shows angles in degrees for a full cycle of a sine wave; part (b) shows the same points in radians.

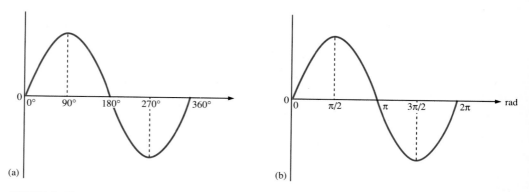

(a)

(b)

FIGURE 8–24
Sine wave angles.

Phase of a Sine Wave

The phase of a sine wave is an angular measurement that specifies the position of that sine wave *relative* to a reference. Figure 8–25 shows one cycle of a sine wave to be used as the *reference*. Note that the first positive-going crossing of the horizontal axis (zero crossing) is at 0° (0 rad), and the positive peak is at 90° ($\pi/2$ rad). The negative-

FIGURE 8–25
Phase reference.

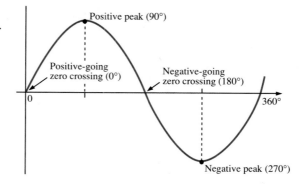

going zero crossing is at 180° (π rad), and the negative peak is at 270° ($3\pi/2$ rad). The cycle is completed at 360° (2π rad). When the sine wave is shifted left or right with respect to this reference, there is a *phase shift*.

Figure 8–26 illustrates phase shifts of a sine wave. In part (a) of the figure, sine wave B is shifted to the right by 90° ($\pi/2$ rad). Thus, there is a *phase angle* of 90° between sine wave A and sine wave B. In terms of time, the positive peak of sine wave B occurs *later* than the positive peak of sine wave A, because time increases to the right along the horizontal axis. In this case, sine wave B is said to *lag* sine wave A by 90° or $\pi/2$ radians. In other words, sine wave A *leads* sine wave B by 90°.

In Figure 8–26(b), sine wave B is shown shifted left by 90°. Thus, again there is a phase angle of 90° between sine wave A and sine wave B. In this case, the positive peak of sine wave B occurs earlier in time than that of sine wave A; therefore, sine wave B is said to *lead* by 90°. In both cases there is a 90° *phase angle* between the two waveforms.

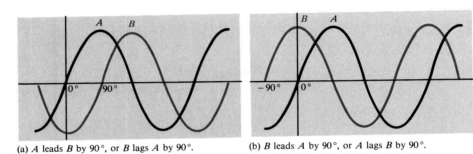

(a) A leads B by 90°, or B lags A by 90°.   (b) B leads A by 90°, or A lags B by 90°.

FIGURE 8–26
Illustration of a phase shift.

EXAMPLE 8–9   What are the phase angles between the two sine waves in Figures 8–27(a) and 8–27(b)?

FIGURE 8–27

(a)

(b)

Solution

(a) The zero crossing of sine wave A is at 0°, and the corresponding zero crossing of sine wave B is at 45°. There is a 45° phase angle between the two waveforms (45° − 0° = 45°) with sine wave A leading.

(b) The zero crossing of sine wave B is at 30°, and the corresponding zero crossing of sine wave A is at 60 °. There is a 30° phase angle between the two waveforms (60° − 30° = 30°) with sine wave B leading.

SECTION REVIEW 8–4

1. When the positive-going zero crossing of a sine wave occurs at 0°, at what angle does each of the following points occur?
   (a) positive peak   **(b)** negative-going zero crossing
   (c) negative peak   **(d)** end of first complete cycle
2. A half cycle is completed in _____ degrees, or _____ radians.
3. A full cycle is completed in _____ degrees, or _____ radians.
4. Determine the phase angle between the two sine waves in Figure 8–28.

FIGURE 8–28

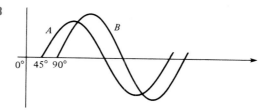

8–5

THE SINE WAVE EQUATION

As you have seen, a sine wave can be graphically represented by voltage (or current) values on the vertical axis and by angular measurement (in degrees or radians) along the horizontal axis. A generalized graph of one cycle of a sine wave is shown in Figure 8–29. The *amplitude, A,* is the maximum value of the voltage or current on the vertical axis, and angular values run along the horizontal axis. The variable y is an instantaneous value representing either voltage or current at a given angle, θ.

All sine wave curves follow a specific mathematical formula. The general expression for the sine wave curve in Figure 8–29 is

$$y = A \sin \theta \tag{8–16}$$

FIGURE 8–29
One cycle of a generalized sine wave showing amplitude and phase.

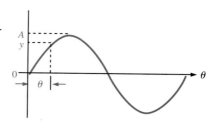

AC CIRCUITS

This expression states that any point on the sine wave, represented by an instantaneous value y, is equal to the maximum value times the sine (sin) of the angle θ at that point. For example, if a certain sine wave voltage has a peak value of 10 V, the instantaneous voltage at a point 60° along the horizontal axis can be calculated as follows:

$$v = V_p \sin \theta = 10 \sin 60° = 10(0.866) = 8.66 \text{ V}$$

Figure 8–30 shows this instantaneous value on the curve. You can find the sine of any angle on your calculator by first entering the value of the angle and then pressing the $\boxed{\sin}$ key. Verify that your calculator is in the degree mode, and, if not, use the $\boxed{\text{DRG}}$ key.

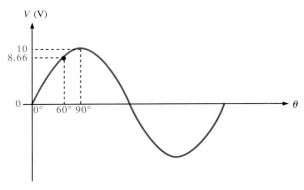

FIGURE 8–30
Illustration of the instantaneous value at $\theta = 60°$.

Expressions for Phase-Shifted Sine Waves

When a sine wave is shifted to the right of the reference (lagging) by a certain angle, ϕ, as illustrated in Figure 8–31(a), the general expression is

$$y = A \sin(\theta - \phi) \tag{8–17}$$

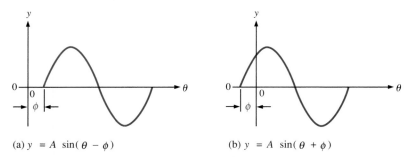

(a) $y = A \sin(\theta - \phi)$ (b) $y = A \sin(\theta + \phi)$

FIGURE 8–31
Shifted sine waves.

When a sine wave is shifted to the left of the reference (leading) by a certain angle, ϕ, as shown in Figure 8–31(b), the general expression is

$$y = A \sin(\theta + \phi) \qquad (8\text{–}18)$$

EXAMPLE 8–10

Determine the instantaneous value at the 90° reference point on the horizontal axis for each sine wave voltage in Figure 8–32.

FIGURE 8–32

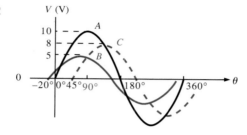

Solution
Sine wave A is the reference. Sine wave B is shifted left 20° with respect to A, so it leads. Sine wave C is shifted right 45° with respect to A, so it lags.

$$v_A = V_p \sin \theta$$
$$= 10 \sin 90° = 10(1) = 10 \text{ V}$$
$$v_B = V_p \sin(\theta + \phi_B)$$
$$= 5 \sin(90° + 20°) = 5 \sin 110° = 5(0.9397) = 4.7 \text{ V}$$
$$v_C = V_p \sin(\theta - \phi_C)$$
$$= 8 \sin(90° - 45°) = 8 \sin 45° = 8(0.7071) = 5.66 \text{ V}$$

SECTION REVIEW 8–5

1. Calculate the instantaneous value at 120° for the sine wave in Figure 8–30.
2. Determine the instantaneous value at the 45° point on the reference axis of a sine wave shifted 10° to the left of the zero reference ($V_p = 10$ V).

8–6

INTRODUCTION TO PHASORS

Phasors can be used to represent time-varying quantities, such as sine waves, in terms of their magnitude and angular position (phase angle). Examples of phasors are shown in Figure 8–33. The *length* of the phasor "arrow" represents the magnitude. The angle, θ (relative to 0°), represents the angular position, as shown in part (a). The specific phasor example in part (b) has a magnitude of 2 and a phase angle of 60°. The phasor in part (c) has a magnitude of 3 and a phase angle of 180°. The phasor in part (d) has a magnitude of 1 and a phase angle of $-45°$.

FIGURE 8–33
Examples of phasors.

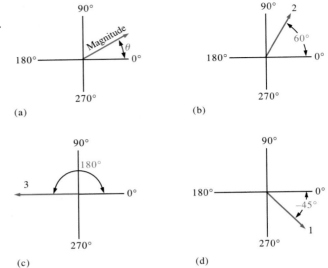

(a)

(b)

(c)

(d)

Phasor Representation of a Sine Wave

A full cycle of a sine wave can be represented by rotation of a phasor through 360°.

> **The instantaneous value of the sine wave at any point is equal to the vertical distance from the tip of the phasor to the horizontal axis.**

Figure 8–34 shows how the phasor "traces out" the sine wave as it goes from 0° to 360°. You can relate this concept to the rotation in an ac generator.

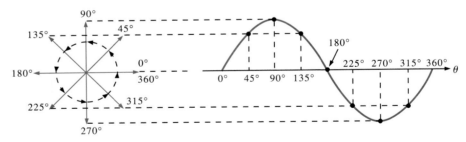

FIGURE 8–34
Sine wave represented by rotational phasor motion.

Notice in Figure 8–34 that the length of the phasor is equal to the *peak* value of the sine wave (observe the 90° and the 270° points). The angle of the phasor measured from 0° is the corresponding angular point on the sine wave.

Derivation of the Sine Wave Formula

Let's examine a phasor representation at one specific angle. Figure 8–35 shows a voltage phasor at an angular position of 45° and the corresponding point on the sine wave.

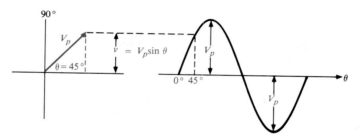

FIGURE 8–35
Right triangle derivation of sine wave formula, $v = V_p \sin \theta$.

The instantaneous value of the sine wave at this point is related to both the position and the length of the phasor. As previously mentioned, the vertical distance from the phasor tip down to the horizontal axis represents the instantaneous value of the sine wave at that point.

Notice that when a vertical line is drawn from the phasor tip down to the horizontal axis, a *right triangle* is formed, as shown in the figure. The length of the phasor is the *hypotenuse* of the triangle, and the vertical projection is the *opposite side*. From trigonometry, *the opposite side of a right triangle is equal to the hypotenuse times the sine of the angle θ*. In this case, the length of the phasor is the *peak* value of the sine wave voltage, V_p. Thus, the opposite side of the triangle, which is the instantaneous value, can be expressed as $v = V_p \sin \theta$. Recall that this formula is the one stated earlier for calculating instantaneous sine wave values. Of course, this formula also applies to a current sine wave.

Positive and Negative Phasor Angles

The position of a phasor at any instant can be expressed as a *positive* angle, as you have seen, or as an *equivalent negative angle*. Positive angles are measured counterclockwise from 0°. Negative angles are measured clockwise from 0°. For a given positive angle θ, the corresponding negative angle is $\theta - 360°$, as illustrated in Figure 8–36(a). In

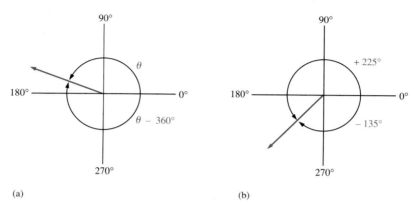

(a)   (b)

FIGURE 8–36
Positive and negative phasor angles.

AC CIRCUITS

part (b), a specific example is shown. The angle of the phasor in this case can be expressed as $+225°$ or $-135°$.

EXAMPLE 8–11   For each phasor in Figure 8–37, determine the instantaneous sine wave value. Also express each positive angle shown as an equivalent negative angle. The length of each phasor represents the peak value of the sine wave.

FIGURE 8–37

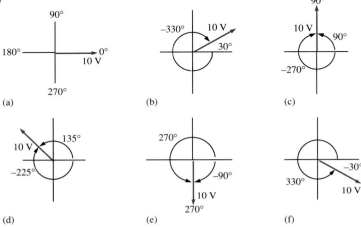

Solution
(a) $\theta = 0°$

$\quad v = 10 \sin 0° = 10(0) = 0 \text{ V}$

(b) $\theta = 30° = -330°$

$\quad v = 10 \sin 30° = 10(0.5) = 5 \text{ V}$

(c) $\theta = 90° = -270°$

$\quad v = 10 \sin 90° = 10(1) = 10 \text{ V}$

(d) $\theta = 135° = -225°$

$\quad v = 10 \sin 135° = 10(0.707) = 7.07 \text{ V}$

(e) $\theta = 270° = -90°$

$\quad v = 10 \sin 270° = 10(-1) = -10 \text{ V}$

(f) $\theta = 330° = -30°$

$\quad v = 10 \sin 330° = 10(-0.5) = -5 \text{ V}$

The equivalent negative angles are shown in Figure 8–37.

Phasor Diagrams

A phasor diagram shows the *relative relationship* of two or more sine waves of the same frequency. A phasor in a *fixed* position represents a *complete* sine wave, because once the phase angle between two or more sine waves of the same frequency is established, it remains constant throughout the cycles. For example, the two sine waves in Figure 8–38(a) can be represented by a phasor diagram, as shown in part (b). As you can see, sine wave B leads sine wave A by 30°. The length of the phasors can be used to represent peak, rms, or average values as long as the representation is consistent.

FIGURE 8–38
Example of a phasor diagram.

EXAMPLE 8–12

Use a phasor diagram to represent the sine waves in Figure 8–39.

FIGURE 8–39

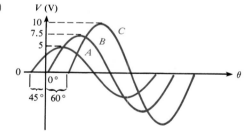

Solution
The phasor diagram representing the sine waves is shown in Figure 8–40. In this case, the length of each phasor represents the peak value of the sine wave.

FIGURE 8–40

SECTION REVIEW
8-6

1. What is a phasor?
2. Sketch a phasor diagram to represent the two sine waves in Figure 8–41. Use peak values.

FIGURE 8–41

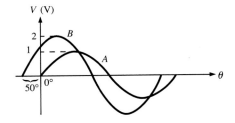

8-7   OHM'S LAW AND KIRCHHOFF'S LAWS IN AC CIRCUITS

When time-varying ac voltages such as the sine wave are applied to resistive circuits, the circuit laws that you studied earlier still apply. Ohm's law applies to resistive ac circuits in the same way that it applies to dc circuits. If a sine wave voltage is applied across a resistor as shown in Figure 8–42, a sine wave current flows. The current is zero when the voltage is zero and is maximum when the voltage is maximum. When the voltage changes polarity, the current reverses direction. As a result, the voltage and current are said to be *in phase* with each other.

When using Ohm's law in ac circuits, remember that both the voltage and the current must be expressed consistently, that is, both as peak values, both as rms values, both as average values, and so on.

FIGURE 8–42
A sine wave voltage produces a sine wave current.

EXAMPLE 8–13

Determine the rms voltage across each resistor and the rms current in Figure 8–43. The source voltage is given as an rms value.

FIGURE 8–43

Solution
The total resistance of the circuit is

$$R_T = R_1 + R_2 = 1 \text{ k}\Omega + 560 \text{ }\Omega = 1.56 \text{ k}\Omega$$

Use Ohm's law to find the rms current:

$$I_{rms} = \frac{V_{s(rms)}}{R_T} = \frac{110 \text{ V}}{1.56 \text{ k}\Omega} = 70.5 \text{ mA}$$

The rms voltage drop across each resistor is

$$V_{1(rms)} = I_{rms}R_1 = (70.5 \text{ mA})(1 \text{ k}\Omega) = 70.5 \text{ V}$$
$$V_{2(rms)} = I_{rms}R_2 = (70.5 \text{ mA})(560 \text{ }\Omega) = 39.5 \text{ V}$$

Kirchhoff's voltage and current laws apply to ac circuits as well as to dc circuits. Figure 8–44 illustrates Kirchhoff's voltage law in a resistive circuit that has a sine wave voltage source. As you can see, the source voltage is the sum of all the voltage drops across the resistors, just as in a dc circuit.

FIGURE 8–44
Illustration of Kirchhoff's voltage law in an ac circuit.

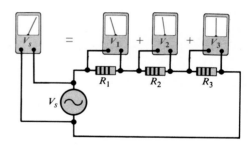

EXAMPLE 8–14

(a) Find the unknown peak voltage drop in Figure 8–45(a). All values are given in rms.

(b) Find the total rms current in Figure 8–45(b). All values are given in rms.

FIGURE 8–45

(a)

(b)

Solution
(a) Use Kirchhoff's voltage law to find V_3:

$$V_s = V_1 + V_2 + V_3$$

Solve for V_3:

$$V_{3(\text{rms})} = V_{s(\text{rms})} - V_{1(\text{rms})} - V_{2(\text{rms})}$$
$$= 24 \text{ V} - 12 \text{ V} - 8 \text{ V} = 4 \text{ V}$$

Convert rms to peak:

$$V_{3(\text{peak})} = 1.414 V_{3(\text{rms})} = 1.414(4 \text{ V}) = 5.66 \text{ V}$$

(b) Use Kirchhoff's current law to find I_T:

$$I_{T(\text{rms})} = I_{1(\text{rms})} + I_{2(\text{rms})}$$
$$= 10 \text{ A} + 3 \text{ A} = 13 \text{ A}$$

**SECTION REVIEW
8–7**

1. A sine wave voltage with an average value of 12.5 V is applied to a circuit with a resistance of 330 Ω. What is the average current in the circuit?

2. The peak voltage drops in a series resistive circuit are 6.2 V, 11.3 V, and 7.8 V. What is the rms value of the source voltage?

8–8

SUPERIMPOSED DC AND AC VOLTAGES

In many practical circuits, you will find both dc and ac voltages combined. An example of this is in amplifier circuits where ac signal voltages are superimposed on dc operating voltages. Figure 8–46 shows a dc source and an ac source in series. These two voltages will add algebraically to produce an ac voltage "riding" on a dc level, as measured across the resistor.

FIGURE 8–46
Superimposed dc and ac voltages.

If V_{dc} is greater than the peak value of the sine wave, the combined voltage is a sine wave that never reverses polarity. That is, we have a sine wave riding on a dc level, as shown in Figure 8–47(a). If V_{dc} is less than the peak value of the sine wave, the sine wave will be negative during a portion of its lower half-cycle, as illustrated in

Figure 8–47(b). In either case the sine wave will reach a maximum voltage equal to $V_{dc} + V_p$, and it will reach a minimum voltage equal to $V_{dc} - V_p$.

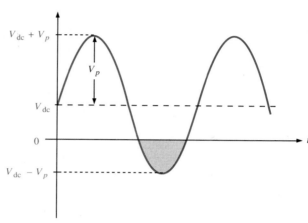

(a) $V_{dc} > V_p$. The sine wave never goes negative.

(b) $V_{dc} < V_p$. The sine wave reverses polarity during a portion of its cycle.

FIGURE 8–47
Sine waves with dc levels.

EXAMPLE 8–15   Determine the voltage across the resistor in each circuit of Figure 8–48.

(a)   (b)

FIGURE 8–48

Solution

(a)   The maximum voltage across R is

$$V_{max} = V_{dc} + V_p = 12 \text{ V} + 10 \text{ V} = 22 \text{ V}$$

The minimum voltage across R is

$$V_{min} = V_{dc} - V_p = 12 \text{ V} - 10 \text{ V} = 2 \text{ V}$$

Therefore, V_R is a nonalternating sine wave that varies from $+22$ V to $+2$ V, as shown in Figure 8–49(a).

(b)   The maximum voltage across R is

$$V_{max} = V_{dc} + V_p = 6 \text{ V} + 10 \text{ V} = 16 \text{ V}$$

The minimum voltage across R is

$$V_{min} = V_{dc} - V_p = -4 \text{ V}$$

Therefore, V_R is an alternating sine wave that varies from $+16$ V to -4 V, as shown in Figure 8–49(b).

FIGURE 8–49

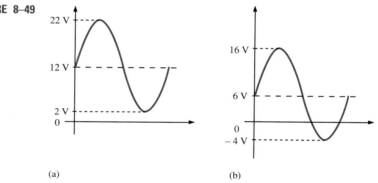

(a)   (b)

---

1.   What is the maximum peak value of the resulting voltage when a sine wave with $V_p = 5$ V is added to a dc voltage of $+2.5$ V?
2.   Will the resulting voltage in Problem 1 alternate polarity?

8–9

NONSINUSOIDAL WAVEFORMS

Sine waves are very important in electronics, but they are by no means the only type of ac or time-varying waveform. Two other categories are discussed next: the *pulse waveform* and the *triangular waveform*.

Pulse Waveforms

Basically, a pulse can be described as a very rapid transition from one voltage or current level (baseline) to another, and then, after an interval of time, a very rapid transition back to the original baseline level. The transitions in level are called *steps*. An *ideal* pulse consists of two opposite-going steps of equal amplitude. Figure 8–50(a) shows an

FIGURE 8–50
Ideal pulses.

(a) Positive-going pulse

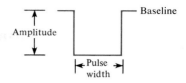

(b) Negative-going pulse

ideal positive-going pulse consisting of two equal but opposite instantaneous steps separated by an interval of time called the *pulse width*. Part (b) of the figure shows an ideal negative-going pulse. The height of the pulse measured from the baseline is its amplitude.

In many applications, analysis is simplified by treating all pulses as ideal (composed of instantaneous steps and perfectly rectangular in shape). Actual pulses, however, are never ideal. All pulses possess certain characteristics that cause them to be different from the ideal.

In practice, pulses cannot change from one level to another instantaneously. Time is always required for a transition (step), as illustrated in Figure 8–51(a). As you can see, there is an interval of time during which the pulse is rising from its lower value to its higher value. This interval is called the *rise time, t_r.*

The accepted definition of rise time is the time required for the pulse to go from 10% of its full amplitude to 90% of its full amplitude.

The interval of time during which the pulse is falling from its higher value to its lower value is called the *fall time, t_f.*

The accepted definition of fall time is the time required for the pulse to go from 90% of its full amplitude to 10% of its full amplitude.

Pulse width also requires a precise definition for the nonideal pulse, because the rising and falling edges are not vertical.

A widely accepted definition of pulse width (t_w) is the time between the point on the rising edge, where the value is 50% of full amplitude, to the point on the falling edge, where the value is 50% of full amplitude.

Pulse width is shown in Figure 8–51(b).

(a)

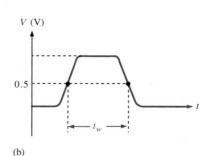

(b)

FIGURE 8–51
Nonideal pulse.

Repetitive Pulses

Any waveform that repeats itself at fixed intervals is *periodic*. Some examples of periodic pulse waveforms are shown in Figure 8–52. Notice that in each case, the pulses repeat at regular intervals. The rate at which the pulses repeat is the pulse repetition rate, which is the *fundamental frequency* of the waveform. The frequency can be expressed in *hertz* or in *pulses per second*. The time from one pulse to the corresponding point on the next pulse is the period T.

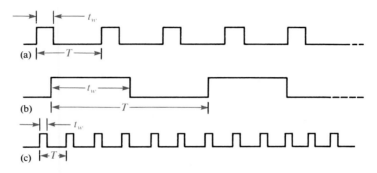

FIGURE 8–52
Examples of repetitive pulse waveforms.

Another very important characteristic of periodic pulse waveforms is the *duty cycle*.

The duty cycle is defined as the ratio of the pulse width (t_w) to the period T and is usually expressed as a percentage:

$$\text{Percent duty cycle} = \left(\frac{t_w}{T}\right)100\% \qquad (8\text{--}19)$$

EXAMPLE 8–16

Determine the period, frequency, and duty cycle for the pulse waveform in Figure 8–53.

FIGURE 8–53

Solution

$$T = 10 \ \mu s$$

$$f = \frac{1}{T} = \frac{1}{10 \ \mu s} = 0.1 \ \text{MHz}$$

$$\text{Percent duty cycle} = \left(\frac{1 \ \mu s}{10 \ \mu s}\right)100\% = 10\%$$

Square Waves

A square wave is a pulse waveform with a duty cycle of 50%. Thus, the pulse width is equal to one-half of the period. A square wave is shown in Figure 8–54.

FIGURE 8–54
A square wave has a duty cycle of 50%.

The Average Value of a Pulse Waveform

An alternating square wave that has equal positive and negative amplitudes has a zero average value. If its positive amplitude is greater than its negative amplitude, it has a positive nonzero average value. If its negative amplitude is greater, it has a negative nonzero average value. For periodic pulse waveforms that are not square waves, the duty cycle also affects the average value, as we will demonstrate.

> **The average value (V_{avg}) of a pulse waveform is equal to its baseline value plus the product of its duty cycle and its amplitude.**

The lower level of the waveform is taken as the baseline. The formula is as follows:

$$V_{avg} = \text{baseline} + (\text{duty cycle})(\text{amplitude}) \qquad (8\text{–}20)$$

The following example illustrates the calculation of the average value of pulse waveforms.

EXAMPLE 8–17   Determine the average value of each of the waveforms in Figure 8–55.

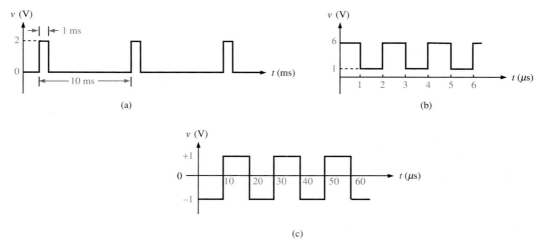

FIGURE 8–55

Solution

(a) Here the baseline is at 0 V, the amplitude is 2 V, and the duty cycle is 10%. The average value is as follows:

$$V_{avg} = \text{baseline} + (\text{duty cycle})(\text{amplitude})$$
$$= 0 \text{ V} + (0.1)(2 \text{ V}) = 0.2 \text{ V}$$

(b) This waveform has a baseline of $+1$ V, an amplitude of 5 V, and a duty cycle of 50%. The average value is

$$V_{avg} = \text{baseline} + (\text{duty cycle})(\text{amplitude})$$
$$= 1 \text{ V} + (0.5)(5 \text{ V}) = 1 \text{ V} + 2.5 \text{ V} = 3.5 \text{ V}$$

(c) This waveform is a square wave with a baseline of -1 V and an amplitude of 2 V. The average value is

$$V_{avg} = \text{baseline} + (\text{duty cycle})(\text{amplitude})$$
$$= -1 \text{ V} + (0.5)(2 \text{ V}) = -1 \text{ V} + 1 \text{ V} = 0 \text{ V}$$

This is an *alternating* square wave, and, like an alternating sine wave, it has an average of zero.

Triangular and Sawtooth Waveforms

Triangular and sawtooth waveforms are formed by voltage or current *ramps*. A ramp is a *linear* increase or decrease in the voltage or current. Figure 8–56 shows both positive- and negative-going ramps. In part (a) of the figure, the ramp has a positive slope; in part (b), the ramp has a negative slope. The *slope* of a voltage ramp is $\pm V/t$ and is expressed in units of V/s. The slope of a current ramp is $\pm I/t$ and is expressed in units of A/s.

FIGURE 8–56
Ramps.

(a) Positive ramp

(b) Negative ramp

EXAMPLE 8–18

What are the slopes of the voltage ramps in Figure 8–57?

Solution

(a) The voltage increases from 0 V to $+10$ V in 5 ms. Thus, $V = 10$ V and $t = 5$ ms. The slope is

$$\frac{V}{t} = \frac{10 \text{ V}}{5 \text{ ms}} = 2 \text{ V/ms} = 2 \text{ kV/s}$$

FIGURE 8–57

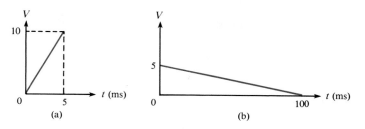

(b)   The voltage decreases from $+5$ V to 0 V in 100 ms. Thus, $V = -5$ V and $t = 100$ ms. The slope is

$$\frac{V}{t} = \frac{-5 \text{ V}}{100 \text{ ms}} = -0.05 \text{ V/ms} = -50 \text{ V/s}$$

Triangular Waveforms

Figure 8–58 shows that a triangular waveform is composed of positive- and negative-going ramps having equal slopes. The period of this waveform is measured from one peak to the next corresponding peak, as illustrated. This particular triangular waveform is alternating and has an average value of zero.

FIGURE 8–58
Alternating triangular waveform.

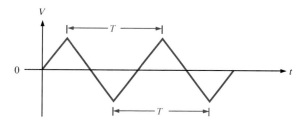

Figure 8–59 depicts a triangular waveform with a nonzero average value. The frequency for triangular waves is determined in the same way as for sine waves, that is, $f = 1/T$.

FIGURE 8–59
Nonalternating triangular waveform.

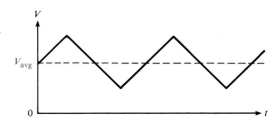

Sawtooth Waveforms

The sawtooth wave is actually a special case of the triangular wave consisting of two ramps, one of much longer duration than the other. Sawtooth waveforms are commonly used in many electronic systems, as you will see in the case of the oscilloscope. Another example is the electron beam that sweeps across the screen of your TV receiver, creating the picture; it also is controlled by sawtooth voltages and currents. One sawtooth wave produces the horizontal beam movement, and the other produces the vertical beam movement. The sawtooth is sometimes called a *sweep*.

Figure 8–60 is an example of a sawtooth wave. Notice that it consists of a positive-going ramp of relatively long duration, followed by a negative-going ramp of relatively short duration.

FIGURE 8–60
Alternating sawtooth waveform.

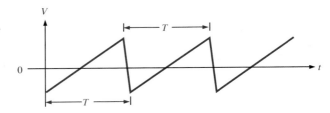

Harmonics

A repetitive nonsinusoidal waveform is composed of a *fundamental frequency* and *harmonic frequencies*. The fundamental frequency is the repetition rate of the waveform, and the harmonics are higher-frequency sine waves that are multiples of the fundamental.

Odd Harmonics

Odd harmonics are frequencies that are *odd multiples* of the fundamental frequency of a waveform. For example, a 1-kHz square wave consists of a fundamental of 1 kHz and odd harmonics of 3 kHz, 5 kHz, 7 kHz, and so on. The 3-kHz frequency in this case is called the *third harmonic;* the 5-kHz frequency is the *fifth harmonic;* and so on.

Even Harmonics

Even harmonics are frequencies that are *even multiples* of the fundamental frequency. For example, if a certain wave has a fundamental of 200 Hz, the second harmonic is 400 Hz, the fourth harmonic is 800 Hz, the sixth harmonic is 1200 Hz, and so on. These are even harmonics.

Composite Waveform

Any variation from a pure sine wave produces harmonics. A nonsinusoidal wave is a composite of the fundamental and the harmonics. Some types of waveforms have only odd harmonics, some have only even harmonics, and some contain both. The shape of the wave is determined by its harmonic content. Generally, only the fundamental and the first few harmonics are of significant importance in determining the waveshape.

A *square wave* is an example of a waveform that consists of a fundamental and only odd harmonics. When the instantaneous values of the fundamental and each

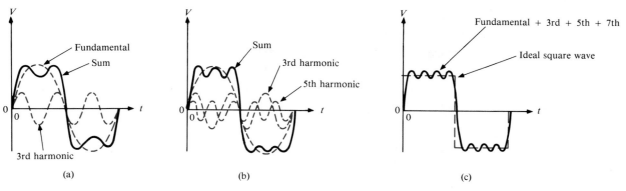

FIGURE 8–61
Odd harmonics combine to produce a square wave.

odd harmonic are added algebraically at each point, the resulting curve will have the shape of a square wave, as illustrated in Figure 8–61. In part (a) of the figure, the fundamental and the third harmonic produce a waveshape that begins to resemble a square wave. In part (b), the fundamental, third, and fifth harmonics produce a closer resemblance. When the seventh harmonic is included, as in part (c), the resulting wave-shape becomes even more like a square wave. As more harmonics are included, a square wave is approached.

**SECTION REVIEW
8–9**

1. Define the following parameters:
    (a) rise time   **(b)** fall time   **(c)** pulse width
2. In a certain repetitive pulse waveform, the pulses are 200 μs wide and occur once every millisecond. What is the frequency of this waveform?
3. Determine the duty cycle, amplitude, and average value of the waveform in Problem 2 if the pulses go from -2 V to $+6$ V.
4. What is the second harmonic of a fundamental frequency of 1 kHz?

8–10

OSCILLOSCOPE MEASUREMENTS

The oscilloscope, or *scope* for short, is one of the most widely used and versatile test instruments. It displays on a screen the graph of a voltage as it changes with time so that various measurements can be made.

Figure 8–62 shows two typical oscilloscopes. The one in part (a) is a simpler and less expensive model. The one in part (b) has better performance characteristics and plug-in modules that provide a variety of specialized functions.

You can think of the oscilloscope as essentially a "graphing machine" that graphs out a voltage and shows you how it varies with time. The shape of the sine wave that you have seen throughout this chapter is simply a graph of voltage versus time. Of

(a)   (b)

FIGURE 8–62
Oscilloscopes [(a) courtesy of B&K Precision. (b) courtesy of Tektronix, Inc.].

course, oscilloscopes can display any type of waveform, but we will use the sine wave to illustrate some basic concepts.

The Oscilloscope Screen

Generally, the screen of an oscilloscope is divided into ten horizontal divisions and eight vertical divisions, as shown in Figure 8–63. Each major division is divided into five small or minor divisions on both the horizontal and the vertical axes. The vertical axis on the scope screen is the voltage scale and the horizontal axis is the time scale. Values of voltage are read vertically and values of time are read horizontally.

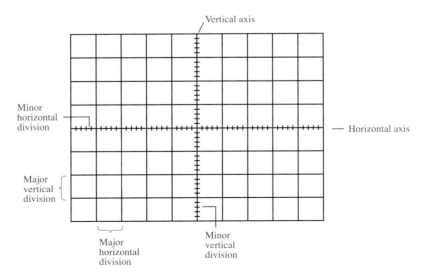

FIGURE 8–63
The oscilloscope screen.

Cathode-Ray Tube (CRT)

The oscilloscope is built around the cathode-ray tube (CRT), which is the device that displays the waveforms. The screen of the scope is the front of the CRT.

The CRT is a vacuum tube device containing an *electron gun* that emits a narrow, focused *beam* of electrons. A phosphorescent coating on the face of the tube forms the *screen*. The beam is electronically focused and accelerated so that it strikes the screen, causing light to be emitted at the point of impact.

Figure 8–64 shows the basic construction of a CRT. The *electron gun* assembly contains a *heater*, a *cathode*, a *control grid*, and *accelerating* and *focusing grids*. The heater carries current that indirectly heats the cathode. The heated cathode emits electrons. The amount of voltage on the control grid determines the flow of electrons and thus the *intensity* of the beam. The electrons are accelerated by the accelerating grid and are focused by the focusing grid into a narrow beam that converges at the screen. The beam is further accelerated to a high speed after it leaves the electron gun by a high voltage on the *anode* surfaces of the CRT.

FIGURE 8–64
Basic construction of a CRT.

The purpose of the *deflection plates* in the CRT is to produce a "bending," or deflection, of the electron beam. This deflection allows the position of the point of impact on the screen to be varied. There are two sets of deflection plates: one set for *vertical deflection,* and the other set for *horizontal deflection*.

Figure 8–65 illustrates how a waveform (a sine wave in this case) is "plotted out" on the screen. A sawtooth voltage called the *horizontal sweep signal* is internally generated and applied to the horizontal deflection plates. Each cycle of the sawtooth causes the electron beam to sweep across the screen from left to right and then quickly return for another sweep. The external signal that is to be displayed on the screen is applied to the vertical deflection plates. This causes the electron beam to move up or down vertically as it is swept across the screen, tracing out a pattern that follows the variation of the external signal voltage over an interval of time.

Operation of the Oscilloscope

Figure 8–66 shows a typical oscilloscope with its screen and controls. The basic functions of the primary controls are indicated. This is a dual-channel instrument, as most

Path of beam traces out variation of external signal with time

Vertical deflection plates

Horizontal deflection plates

External signal deflects beam vertically

Internal sawtooth horizontal sweep voltage deflects beam left to right

FIGURE 8–65
How a waveform is traced out on the scope screen.

Vertical position

Trigger controls

Horizontal position

Time/division selector switch

Screen

External trigger input

Power-On Switch

Input coupling switch

Channel 1 input

Channel 2 input

Volts/division switch CH1

Volts/division switch CH2

FIGURE 8–66
A typical dual-channel oscilloscope (courtesy of B & K Precision).

scopes are. *Dual-channel* means that you can display two waveforms on the screen at the same time.

The Volts/Division Control

Notice that in Figure 8–66 there are two VOLTS/DIV switches, one for channel 1 (CH1) and one for channel 2 (CH2). The volts/division selector switch sets the number of volts to be represented by each major division on the *vertical* scale. For example, Figure 8–67(a) shows one cycle of a sine wave on the scope screen with the VOLTS/DIV switch set at 1 V. This means that each of the major vertical divisions is 1 V. The peak of the sine wave is three divisions high, and, since each division is 1 V, the peak value of the sine wave is 3 V (3 divisions × 1 V/division) = 3 V.

For the *same* sine wave, if the VOLTS/DIV setting is changed to 2 V, the sine wave will appear as shown in Figure 8–67(b). Notice that now the peak is 1½ major divisions high. Even though each division is now 2 V, the peak value is still 3 V because 1.5 divisions × 2 V/division = 3 V.

So, as you can see, it is relatively easy to measure the peak value by counting the number of vertical divisions from the zero crossing to the peak of the waveform and multiplying by the VOLTS/DIV setting.

FIGURE 8–67
Using the VOLTS/DIV control.

The Time/Division Control

The TIME/DIV selector switch sets the number of seconds, milliseconds, or microseconds to be represented by each major division on the *horizontal* scale. It actually controls how fast the electron beam sweeps horizontally across the screen.

For example, Figure 8–68(a) shows one cycle of a sine wave displayed on the screen. In this case, the entire cycle covers ten horizontal divisions. The TIME/DIV switch is set at 10 μs. This means that each major horizontal division is 10 μs. Since a full cycle of the sine wave covers 10 divisions, the period of the sine wave is 100 μs (10 divisions × 10 μs/division) = 100 μs. From this, the frequency is $f = 1/100 \ \mu$s = 10 kHz.

For the *same* sine wave, if the TIME/DIV setting is changed to 20 μs, the sine wave will appear as shown in Figure 8–68(b). Notice that now a full cycle covers only five horizontal divisions and there are two cycles on the screen. Even though each division is now 20 μs, the period is still 100 μs because 5 divisions × 20 μs/division = 100 μs.

So, as you can see, it is easy to measure the period and get the frequency by counting the number of divisions covered by one cycle and then multiplying by the TIME/DIV setting.

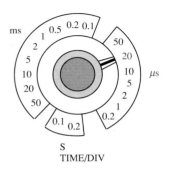

FIGURE 8–68
Using the TIME/DIV control.

Power Switch and Light

This switch turns the power to the scope on and off. The light indicates when the power is on.

Intensity

The intensity control knob varies the brightness of the trace on the screen. Caution should be used so that the intensity is not left too high for an extended period of time, especially when the beam forms a motionless dot on the screen. Damage to the screen can result from excessive intensity.

Focus

This control focuses the beam so that it converges to a tiny point at the screen. An out-of-focus condition results in a fuzzy trace.

Horizontal Position

This control knob adjusts the neutral horizontal position of the beam. It is used to reposition horizontally a waveform display for more convenient viewing or measurement.

Vertical Position

The two vertical position controls move the traces up or down for easier measurement or observation.

AC-GND-DC Switch

This switch, located below the volts/div control, allows the input signal to be ac coupled, dc coupled, or grounded. The ac coupling eliminates any dc component on the input signal. The dc coupling permits dc values to be displayed. The ground position allows a 0-V reference to be established on the screen.

Signal Inputs

The signals to be displayed are connected into the channel 1 and channel 2 input connectors. This connection is normally done with a special *probe* that minimizes the loading effect of the scope's input resistance on the circuit being measured.

Trigger Control

The trigger controls allow the beam to be triggered from various selected sources. The triggering of the beam causes it to begin its sweep across the screen. It can be triggered from an internally generated signal derived from an input signal, or from the line voltage, or from an externally applied trigger signal. The *modes* of triggering are auto, normal, and TV. In the auto mode, sweep occurs in the absence of an adequate trigger signal. In the normal mode, a trigger signal must be present for the sweep to occur. The TV mode provides triggering on the TV field or TV line signals. The *slope* switch allows the triggering to occur on either the positive-going slope or the negative-going slope of the trigger waveform. The *level* control selects the voltage level on the trigger signal at which the triggering occurs.

Basically, the trigger controls provide for synchronization of the horizontal sweep waveform and the input signal waveform. As a result, the display of the input signal is stable on the screen, rather than appearing to drift across the screen.

EXAMPLE 8–19   Determine the peak value and period of each sine wave in Figure 8–69 from the scope displays and the indicated settings for VOLTS/DIV and TIME/DIV.

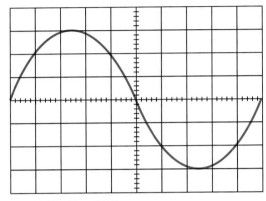

(a) VOLTS/DIV = 0.5 V
    TIME/DIV   = 2 ms

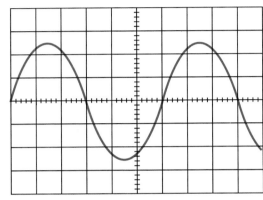

(b) VOLTS/DIV = 50 mV
    TIME/DIV   = 0.1 ms

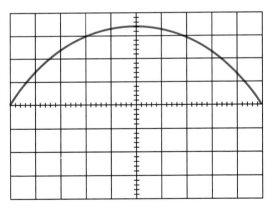

(c) VOLTS/DIV = 2 V
    TIME/DIV   = 10 μs
    Sine wave is centered
    vertically on the screen.

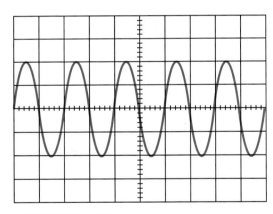

(d) VOLTS/DIV = 5 V
    TIME/DIV   = 2 μs

FIGURE 8–69

Solution
(a)   Vertical scale:

$$V_p = 3 \text{ divisions} \times 0.5 \text{ V/division} = 1.5 \text{ V}$$

Horizontal scale (one cycle covers ten divisions):

$$T = 10 \text{ divisions} \times 2 \text{ ms/division} = 20 \text{ ms}$$

(b) Vertical scale:

$$V_p = 2.5 \text{ divisions} \times 50 \text{ mV/division} = 125 \text{ mV}$$

Horizontal scale (one cycle covers six divisions):

$$T = 6 \text{ divisions} \times 0.1 \text{ ms/division} = 0.6 \text{ ms} = 600 \text{ } \mu s$$

(c) Vertical scale:

$$V_p = 3.4 \text{ divisions} \times 2 \text{ V/division} = 6.8 \text{ V}$$

Horizontal scale (one half-cycle covers ten divisions):

$$T = 20 \text{ divisions} \times 10 \text{ } \mu s/\text{division} = 200 \text{ } \mu s$$

(d) Vertical scale:

$$V_p = 2 \text{ divisions} \times 5 \text{ V/division} = 10 \text{ V}$$

Horizontal scale (one cycle covers two divisions):

$$T = 2 \text{ divisions} \times 2 \text{ } \mu s/\text{division} = 4 \text{ } \mu s$$

**SECTION REVIEW
8–10**

1. What does *CRT* stand for?
2. On an oscilloscope, voltage is measured (horizontally, vertically) on the screen and time is measured (horizontally, vertically).
3. What can an oscilloscope do that a multimeter cannot?

SUMMARY

□ The sine wave is a time-varying, periodic waveform.
□ The sine wave is a common form of ac (alternating current).
□ Alternating current changes direction in response to changes in the polarity of the source voltage.
□ One cycle of an alternating sine wave consists of a positive alternation and a negative alternation.
□ Two common sources of sine waves are the electromagnetic ac generator and the electronic oscillator circuit.
□ A full cycle of a sine wave is 360°, or 2π radians. A half-cycle is 180°, or π radian. A quarter-cycle is 90°, or $\pi/2$ radians.
□ A sine wave voltage can be generated by a conductor rotating in a magnetic field.
□ Phase angle is the difference in degrees or radians between two sine waves.

☐ The angular position of a phasor represents the angle of a sine wave, and the length of the phasor represents the amplitude.

☐ A pulse consists of a transition from a baseline level to an amplitude level, followed by a transition back to the baseline level.

☐ A triangle or sawtooth wave consists of positive- and negative-going ramps.

☐ Harmonic frequencies are odd or even multiples of the repetition rate of a nonsinusoidal waveform.

☐ Conversions of sine wave values are summarized in Table 8–2.

TABLE 8–2

To Change From	To	Multiply By	Divide By
Peak	rms	0.707	
Peak	Peak-to-peak	2	
Peak	Average	0.637	
rms	Peak		0.707
Peak-to-peak	Peak		2
Average	Peak		0.637

GLOSSARY

Alternating current (ac)   Current that reverses direction in response to a change in source voltage polarity.

Average value   The average of a sine wave over one half-cycle. It is 0.637 times the peak value.

Baseline   The normal level of a pulse waveform. The voltage level in the absence of a pulse.

CRT   Cathode-ray tube.

Cycle   One repetition of a periodic waveform.

Degree   The unit of angular measure corresponding to 1/360 of a complete revolution.

Duty cycle   A characteristic of a pulse waveform that indicates the percentage of time that a pulse is present during a cycle. The ratio of pulse width to period.

Effective value   A measure of the heating effect of a sine wave. Also known as the rms (root mean square) value.

Falling edge   The negative-going transition of a pulse.

Fall time   The time interval required for a pulse to change from 90% to 10% of its full amplitude.

Frequency   A measure of the rate of change of a periodic function. The number of cycles completed in 1 s. The unit of frequency is the hertz.

Generator   An energy source that produces electrical signals.

Harmonics   The frequencies contained in a composite waveform, which are integer multiples of the repetition frequency (fundamental).

Hertz (Hz)   The unit of frequency. One hertz equals one cycle per second.

Hypotenuse   The longest side of a right triangle.

Input   The voltage, current, or power applied to an electrical circuit to produce a desired result.

Instantaneous value   The voltage or current value of a waveform at a given instant in time.

Lag   A condition of the phase or time relationship of waveforms in which one waveform is behind the other in phase or time.

Lead   A condition of the phase or time relationship of waveforms in which one waveform is ahead of the other in phase or time. Also, a wire or cable connection to a device or instrument.

Leading edge   The first step or transition of a pulse.

Magnitude   The value of a quantity, such as the number of volts of voltage or the number of amperes of current.

Oscillator   An electronic circuit that produces a time-varying signal without an external input signal using positive feedback.

Oscilloscope   A measurement instrument that displays signal waveforms on a screen.

Peak value   The voltage or current value of a waveform at its maximum positive or negative points.

Peak-to-peak value   The voltage or current value of a waveform measured from its minimum to its maximum points.

Period   The time interval of one complete cycle of a periodic waveform.

Periodic   Characterized by a repetition at fixed time intervals.

Phase   The relative displacement of a time-varying waveform in terms of its occurrence with respect to a reference.

Phasor   A representation of a sine wave in terms of both magnitude and phase angle.

Pulse   A type of waveform that consists of two equal and opposite steps in voltage or current separated by a time interval.

Pulse width   The time interval between the opposite steps of an ideal pulse. For a nonideal pulse, the time between the 50% points on the leading and trailing edges.

Radian   A unit of angular measurement. There are 2π rads in one complete revolution. One radian equals 57.3°.

Ramp   A type of waveform characterized by a linear increase or decrease in voltage or current.

Right angle   A 90° angle.

Rise time   The time interval required for a pulse to change from 10% to 90% of its amplitude.

Rising edge   The positive-going transition of a pulse.

Root mean square (rms)   The value of a sine wave that indicates its heating effect, also known as the effective value. It is equal to 0.707 times the peak value.

Sawtooth   A type of electrical waveform composed of ramps. A special case of a triangular wave.

Trailing edge   The second step or transition of a pulse.

Triangular wave   A type of electrical waveform that consists of two ramps.

Trigger   The activating unit of some electronic devices or instruments.

Waveform   The pattern of variations of a voltage or current showing how the quantity changes with time.

FORMULAS

$$f = \frac{1}{T} \quad \text{Frequency} \tag{8-1}$$

$$T = \frac{1}{f} \quad \text{Period} \tag{8-2}$$

$$V_{pp} = 2V_p \quad \text{Peak-to-peak voltage (sine wave)} \tag{8-3}$$

$$I_{pp} = 2I_p \quad \text{Peak-to-peak current (sine wave)} \tag{8-4}$$

$$V_{rms} \cong 0.707V_p \qquad \text{Root-mean-square voltage (sine wave)} \qquad (8\text{--}5)$$

$$I_{rms} \cong 0.707I_p \qquad \text{Root-mean-square current (sine wave)} \qquad (8\text{--}6)$$

$$V_p \cong 1.414V_{rms} \qquad \text{Peak voltage (sine wave)} \qquad (8\text{--}7)$$

$$I_p \cong 1.414I_{rms} \qquad \text{Peak current (sine wave)} \qquad (8\text{--}8)$$

$$V_{pp} = 2.828V_{rms} \qquad \text{Peak-to-peak voltage (sine wave)} \qquad (8\text{--}9)$$

$$I_{pp} = 2.828I_{rms} \qquad \text{Peak-to-peak current (sine wave)} \qquad (8\text{--}10)$$

$$V_{avg} \cong 0.637V_p \qquad \text{Average voltage (sine wave)} \qquad (8\text{--}11)$$

$$I_{avg} \cong 0.637I_p \qquad \text{Average current (sine wave)} \qquad (8\text{--}12)$$

$$f = \text{number of pole pairs} \times \text{(rev/s)} \qquad \text{Frequency} \qquad (8\text{--}13)$$

$$\text{rad} = \left(\frac{\pi \text{ rad}}{180°}\right) \times \text{degrees} \qquad (8\text{--}14)$$

$$\text{degrees} = \left(\frac{180°}{\pi \text{ rad}}\right) \times \text{rad} \qquad (8\text{--}15)$$

$$y = A \sin \theta \qquad \text{General equation for a sine wave} \qquad (8\text{--}16)$$

$$y = A \sin(\theta - \phi) \qquad \text{Lagging sine wave} \qquad (8\text{--}17)$$

$$y = A \sin(\theta + \phi) \qquad \text{Leading sine wave} \qquad (8\text{--}18)$$

$$\text{Percent duty cycle} = \left(\frac{t_w}{T}\right)100\% \qquad (8\text{--}19)$$

$$V_{avg} = \text{baseline} + \text{(duty cycle)(amplitude)} \qquad \text{Average value of pulse waveform} \qquad (8\text{--}20)$$

SELF-TEST

1. The difference between alternating current (ac) and direct current (dc) is
   (a) ac changes value and dc does not
   (b) ac changes direction and dc does not
   (c) both a and b
   (d) neither a nor b

2. During each cycle, a sine wave reaches a peak value
   (a) one time   (b) two times   (c) four times
   (d) a number of times depending on the frequency

3. A sine wave with a frequency of 12 kHz is changing at a faster rate than a sine wave with a frequency of
   (a) 20 kHz   (b) 15,000 Hz   (c) 10,000 Hz   (d) 1.25 MHz

4. A sine wave with a period of 2 ms is changing at a faster rate than a sine wave with a period of
   (a) 1 ms   (b) 0.0025 s   (c) 1.5 ms   (d) 1000 μs

5. When a sine wave has a frequency of 60 Hz, in 10 s it goes through
   (a) 6 cycles   (b) 10 cycles   (c) 1/16 cycle   (d) 600 cycles

6. If the peak value of a sine wave is 10 V, the peak-to-peak value is
    (a) 20 V   **(b)** 5 V   **(c)** 100 V   **(d)** none of these

7. If the peak value of a sine wave is 20 V, the rms value is
    (a) 14.14 V   **(b)** 6.37 V   **(c)** 7.07 V   **(d)** 0.707 V

8. The average value of a 10-V peak sine wave over one complete cycle is
    (a) 0 V   **(b)** 6.37 V   **(c)** 7.07 V   **(d)** 5 V

9. The average half-cycle value of a sine wave with a 20 V peak is
    (a) 0 V   **(b)** 6.37 V   **(c)** 12.74 V   **(d)** 14.14 V

10. One sine wave has a positive-going zero crossing at 10° and another sine wave has a positive-going zero crossing at 45°. The phase angle between the two waveforms is
    (a) 55°   **(b)** 35°   **(c)** 0°   **(d)** none of these

11. The instantaneous value of a 15-A peak sine wave at a point 32° from its positive-going zero crossing is
    (a) 7.95 A   **(b)** 7.5 A   **(c)** 2.13 A   **(d)** 7.95 V

12. If the rms current through a 10-kΩ resistor is 5 mA, the rms voltage drop across the resistor is
    (a) 70.7 V   **(b)** 7.07 V   **(c)** 5 V   **(d)** 50 V

13. Two series resistors are connected to an ac source. If there is 6.5 V rms across one resistor and 3.2 V rms across the other, the peak source voltage is
    (a) 9.7 V   **(b)** 9.19 V   **(c)** 13.72 V   **(d)** 4.53 V

14. A 10-kHz pulse waveform consists of pulses that are 10 μs wide. Its duty cycle is
    (a) 100%   **(b)** 10%   **(c)** 1%   **(d)** not determinable

15. The duty cycle of a square wave
    (a) varies with the frequency
    (b) varies with the pulse width
    (c) both a and b
    (d) is 50%

PROBLEM SET A

Section 8–1   The Sine Wave

1. Calculate the frequency for each of the following periods:
    (a) 1 s   **(b)** 0.2 s   **(c)** 50 ms
    (d) 1 ms   **(e)** 500 μs   **(f)** 10 μs

2. Calculate the period for each of the following frequencies:
    (a) 1 Hz   **(b)** 60 Hz   **(c)** 500 Hz
    (d) 1 kHz   **(e)** 200 kHz   **(f)** 5 MHz

3. A sine wave goes through 5 cycles in 10 μs. What is its period?

4. A sine wave has a frequency of 50 kHz. How many cycles does it complete in 10 ms?

Section 8–2   Voltage and Current Values of a Sine Wave

5. A sine wave has a peak value of 12 V. Determine the following values:
    (a) rms   **(b)** peak-to-peak   **(c)** half-cycle average

6. A sinusoidal current has an rms value of 5 mA. Determine the following values:
    (a) peak   **(b)** half-cycle average   **(c)** peak-to-peak

7. For the sine wave in Figure 8–70, determine the peak, peak-to-peak, rms, and average values.

FIGURE 8–70

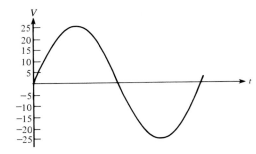

Section 8–3   Sine Wave Voltage Sources

8. The conductive loop on the rotor of a simple two-pole, single-phase generator rotates at a rate of 250 rev/s. What is the frequency of the induced output voltage?

9. A certain four-pole generator has a speed of rotation of 3600 rev/min. What is the frequency of the voltage produced by this generator?

10. At what speed of rotation must a four-pole generator be operated to produce a 400-Hz sine wave voltage?

Section 8–4   Angular Relationships of a Sine Wave

11. Sine wave A has a positive-going zero crossing at 30°. Sine wave B has a positive-going zero crossing at 45°. Determine the phase angle between the two signals. Which signal leads?

12. One sine wave has a positive peak at 75°, and another has a positive peak at 100°. How much is each sine wave shifted in phase from the 0° reference? What is the phase angle between them?

13. Make a sketch of two sine waves as follows: Sine wave A is the reference, and sine wave B lags A by 90°. Both have equal amplitudes.

14. Convert the following angular values from degrees to radians:
    (a) 30°   (b) 45°   (c) 78°
    (d) 135°   (e) 200°   (f) 300°

15. Convert the following angular values from radians to degrees:
    (a) $\pi/8$ (b) $\pi/3$ (c) $\pi/2$
    (d) $3\pi/5$ (e) $6\pi/5$ (f) 1.8π

Section 8–5   The Sine Wave Equation

16. A certain sine wave has a positive-going zero crossing at 0° and an rms value of 20 V. Calculate its instantaneous value at each of the following angles:
    (a) 15°   (b) 33°   (c) 50°
    (d) 110°   (e) 70°   (f) 145°
    (g) 250°   (h) 325°

17. For a particular 0° reference sinusoidal current, the peak value is 100 mA. Determine the instantaneous value at each of the following points:
    (a) 35°   (b) 95°   (c) 190°
    (d) 215°   (e) 275°   (f) 360°

18. For a 0° reference sine wave with an rms value of 6.37 V, determine its instantaneous value at each of the following points:
    (a) $\pi/8$ radians   (b) $\pi/4$ radians   (c) $\pi/2$ radians
    (d) $3\pi/4$ radians   (e) π radians   (f) $3\pi/2$ radians
    (g) 2π radians

19. Sine wave *A* lags sine wave *B* by 30°. Both have peak values of 15 V. Sine wave *A* is the reference with a positive-going crossing at 0°. Determine the instantaneous value of sine wave *B* at 30°, 45°, 90°, 180°, 200°, and 300°.

20. Repeat Problem 19 for the case when sine wave *A* *leads* sine wave *B* by 30°.

Section 8–6   Introduction to Phasors

21. Draw a phasor diagram to represent the sine waves in Figure 8–71.

FIGURE 8–71

22. Sketch the sine waves represented by the phasor diagram in Figure 8–72. The phasor lengths represent peak values.

FIGURE 8–72

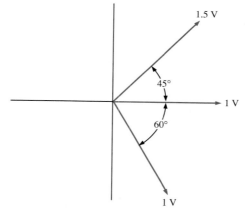

Section 8–7   Ohm's Law and Kirchhoff's Laws in ac Circuits

23. A sinusoidal voltage is applied to the resistive circuit in Figure 8–73. Determine the following:

   (a) I_{rms} (b) I_{avg} (c) I_p (d) I_{pp} (e) *i* at the positive peak

FIGURE 8–73

FIGURE 8–74   **FIGURE 8–75**

24. Find the average values of the voltages across R_1 and R_2 in Figure 8–74. Values shown are rms.

25. Determine the rms voltage across R_3 in Figure 8–75.

Section 8–8   Superimposed dc and ac Voltages

26. A sine wave with an rms value of 10.6 V is riding on a dc level of 24 V. What are the maximum and minimum values of the resulting waveform?

27. How much dc voltage must be added to a 3-V rms sine wave in order to make the resulting voltage nonalternating (no negative values)?

28. A 6-V peak sine wave is riding on a dc voltage of 8 V. If the dc voltage is lowered to 5 V, how far negative will the sine wave go?

Section 8–9   Nonsinusoidal Waveforms

29. From the graph in Figure 8–76, determine the approximate values of t_r, t_f, t_w, and amplitude.

FIGURE 8–76

30. Determine the duty cycle for each pulse waveform in Figure 8–77.

31. Find the average value of each pulse waveform in Figure 8–77.

32. What is the frequency of each waveform in Figure 8–77?

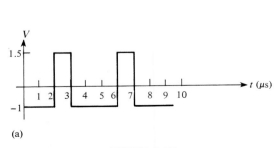

(a)   (b)

FIGURE 8–77

FIGURE 8–78

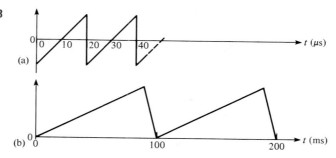

(a)

(b)

33. What is the frequency of each sawtooth waveform in Figure 8–78?

34. A square wave has a period of 40 μs. List the first six odd harmonics.

35. What is the fundamental frequency of the square wave mentioned in Problem 34?

Section 8–10   Oscilloscope Measurements

36. Determine the peak value and the period of the sine wave displayed on the scope screen in Figure 8–79.

37. Determine the rms value and the frequency of the sine wave displayed on the scope screen in Figure 8–80.

FIGURE 8–79

FIGURE 8–80

38. Find the amplitude, pulse width, and duty cycle for the pulse waveform displayed on the scope screen in Figure 8–81.

FIGURE 8–81

PROBLEM SET B

39. A certain sine wave has a frequency of 2.2 kHz and an rms value of 25 V. Assuming a given cycle begins (zero crossing) at $t = 0$ s, what is the change in voltage from 0.12 ms to 0.2 ms?

40. Figure 8–82 shows a sinusoidal voltage source in series with a dc source. Effectively, the two voltages are superimposed. Sketch the voltage across R_L. Determine the maximum current through R_L and the average voltage across R_L.

FIGURE 8–82

FIGURE 8–83

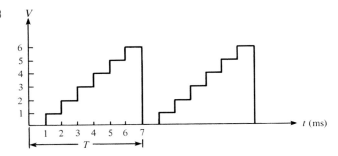

41. A nonsinusoidal waveform called a *stairstep* is shown in Figure 8–83. Determine its average value.

COLOR INSERT

42. Refer to the oscilloscope in Figure 8–84. (A full-color photo appears following page 416.)
 (a) How many cycles are displayed?
 (b) What is the rms value of the sine wave?
 (c) What is the frequency of the sine wave?

COLOR INSERT

43. Accurately sketch on a grid representing the scope screen in Figure 8–84 how the sine wave will appear if the volts/division control is moved to the next higher setting.

COLOR INSERT

44. Accurately sketch on a grid representing the scope screen in Figure 8–84 how the sine wave will appear if the time/division control is moved to the next lower setting.

45. The signal generator in Figure 8–85 is supplying a 20-V rms sine wave with a frequency of 500 Hz to the circuit board. Sketch the output of the circuit board as it would appear on the scope.

FIGURE 8–85

46. Examine the circuit board and the oscilloscope display in Figure 8–86 and determine the peak value and the frequency of the sine wave produced by the signal generator.

FIGURE 8–86

ANSWERS TO SECTION REVIEWS

Section 8–1

1. From the zero crossing through a positive peak, then through zero to a negative peak and back to the zero crossing.
2. At the zero crossings.   **3.** 2.
4. From one zero crossing to the next *corresponding* zero crossing, or from one peak to the next *corresponding* peak.
5. The number of cycles completed in one second, hertz.
6. 200 kHz.   **7.** 8.33 ms.

Section 8–2

1. **(a)** 2 V; **(b)** 4 V; **(c)** 9.42 V.

2. **(a)** 1.77 V; **(b)** 3.54 V; **(c)** 1.66 V.

Section 8–3

1. By rotating a conductor in a magnetic field.

2. AC generators and signal generators.

Section 8–4

1. **(a)** 90°; **(b)** 180°; **(c)** 270°; **(d)** 360°. **2.** 180, π.

3. 360, 2π. **4.** 45°.

Section 8–5

1. 8.66 V. **2.** 8.19 V.

Section 8–6

1. A graphic representation of the magnitude and angular position of a time-varying quantity.

2. See Figure 8–87.

FIGURE 8–87

Section 8–7

1. 37.9 mA. **2.** 17.89 V.

Section 8–8

1. 7.5 V. **2.** Yes.

Section 8–9

1. **(a)** The time interval from 10% to 90% of the rising pulse edge;
(b) The time interval from 90% to 10% of the falling pulse edge;
(c) The time interval from 50% of the leading pulse edge to 50% of the trailing pulse edge.

2. 1 kHz. **3.** 20%, 8 V, −0.4 V. **4.** 2 kHz.

Section 8–10

1. Cathode-ray tube. **2.** Vertically, horizontally.

3. Display the pattern of the variation of a voltage as a function of time.

9 Capacitors

The capacitor is an electrical device that can store electrical charge, thereby creating an electric field which in turn stores energy; the measure of that energy-storing ability is called *capacitance.* In this chapter, the basic capacitor is introduced and its characteristics are studied. The physical construction and electric properties of various types of capacitors are discussed.

The basic behavior of capacitors in both dc and ac circuits is studied, and series and parallel combinations are analyzed. Representative applications and methods of testing capacitors also are discussed.

After completing this chapter, you will be able to

☐ Describe the basic construction of a capacitor.
☐ Define capacitance and tell how it is measured
☐ Explain how a capacitor stores energy.
☐ State Coulomb's law and discuss how it relates to an electric field and the storage of energy.
☐ Illustrate the charging and discharging of a capacitor.
☐ Relate various physical parameters to capacitance value.
☐ Identify types of capacitors according to the dielectric material.
☐ Determine total series capacitance.
☐ Determine total parallel capacitance.
☐ Define *time constant* as related to a capacitive circuit.
☐ Relate the charging and discharging of a capacitor to the time constant.
☐ Explain why a capacitor blocks dc.
☐ Define *phase shift* and explain how a capacitor causes a phase shift between current and voltage.
☐ Define *capacitive reactance* and determine its value in a circuit.
☐ Explain why a capacitor produces no energy loss.
☐ Explain the significance of *reactive power* in a capacitive circuit.
☐ Describe some common capacitor applications.
☐ Check out a capacitor with an ohmmeter.

APPLICATION ASSIGNMENT

A defective stereo amplifier is returned for repair. Your supervisor thinks that a coupling or a bypass capacitor in the amplifier circuitry may be faulty. Your assignment is to check all points in the circuit to which a capacitor is connected for the proper ac and/or dc voltage levels, which are specified on the schematic.

After checking the voltage levels, you must remove from the circuit board any capacitor that you suspect of being bad and test it for proper functioning and value. After completing this chapter, you will be able to carry out this assignment.

9-1

THE CAPACITOR

Basic Construction

In its simplest form, a capacitor is an electrical device constructed of two parallel conductive plates separated by an insulating material called the *dielectric*. Connecting leads are attached to the parallel plates. A basic capacitor is shown in Figure 9–1(a), and the schematic symbol is shown in part (b).

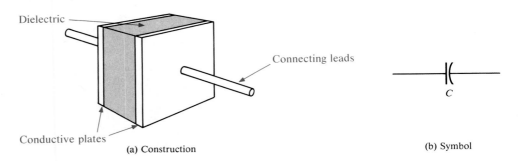

Dielectric

Connecting leads

Conductive plates

(a) Construction

(b) Symbol

C

FIGURE 9-1
The basic capacitor.

How a Capacitor Stores Charge

In the neutral state, both plates of a capacitor have an equal number of free electrons, as indicated in Figure 9–2(a). When the capacitor is connected to a voltage source through a resistor, as shown in part (b), electrons (negative charge) are removed from plate *A*, and an equal number are deposited on plate *B*. As plate *A* loses electrons and plate *B* gains electrons, plate *A* becomes positive with respect to plate *B*. During this charging process, electrons flow only through the connecting leads and the source. No electrons flow through the dielectric of the capacitor because it is an *insulator*. The movement of electrons ceases when the voltage across the capacitor equals the source voltage, as indicated in part (c). If the capacitor is disconnected from the source, it retains the stored charge for a long period of time (the length of time depends on the type of capacitor) and still has the voltage across it, as shown in part (d). Actually, the charged capacitor can be considered as a temporary battery.

Capacitance

The amount of charge per unit of voltage that a capacitor can store is its *capacitance*, designated *C*. That is, capacitance is a measure of a capacitor's ability to store charge. The more charge per unit of voltage that a capacitor can store, the greater its capacitance, as expressed by the following formula:

$$C = \frac{Q}{V}$$

(9–1)

where *C* is capacitance, *Q* is charge, and *V* is voltage.

FIGURE 9–2
Illustration of a capacitor storing charge.

The Unit of Capacitance

The *farad* (F) is the basic unit of capacitance. By definition,

> *one farad* **is the amount of capacitance when** *one coulomb* **of charge is stored with** *one volt* **across the plates.**

Most capacitors that you will use in electronics work have capacitance values in microfarads (μF) and picofarads (pF). A microfarad is one-millionth of a farad (1 μF = 1 $\times$ 10^{-6} F), and a picofarad is one-trillionth of a farad (1 pF = 1 $\times$ 10^{-12} F). Conversions for farads, microfarads, and picofarads are given in Table 9–1.

TABLE 9–1

Conversions for farads, microfarads, and picofarads.

To convert from	To	Multiply
Farads	Microfarads	Farads by 10^6
Farads	Picofarads	Farads by 10^{12}
Microfarads	Farads	Microfarads by 10^{-6}
Microfarads	Picofarads	Microfarads by 10^6
Picofarads	Farads	Picofarads by 10^{-12}
Picofarads	Microfarads	Picofarads by 10^{-6}

EXAMPLE 9–1
(a) A certain capacitor stores 50 μC when 10 V are applied across its plates. What is its capacitance?
(b) A 2-μF capacitor has 100 V across its plates. How much charge does it store?
(c) Determine the voltage across a 100-pF capacitor that is storing 2 μC of charge.

Solution
(a) Use Equation (9–1):

$$C = \frac{Q}{V} = \frac{50\ \mu C}{10\ V} = 5\ \mu F$$

The calculator sequence is

$\boxed{5}\ \boxed{0}\ \boxed{EXP}\ \boxed{+/-}\ \boxed{6}\ \boxed{\div}\ \boxed{1}\ \boxed{0}\ \boxed{=}$

(b) Solve Equation (9–1) for Q as follows:

$$\frac{Q}{V} = C$$

$$\frac{Q}{\cancel{V}}(\cancel{V}) = C(V)$$

$$Q = CV = (2\ \mu F)(100\ V) = 200\ \mu C$$

(c) Solve Equation (9–1) for V as follows:

$$\cancel{C}\left(\frac{V}{\cancel{C}}\right) = \frac{Q}{\cancel{V}}\left(\frac{\cancel{V}}{C}\right)$$

$$V = \frac{Q}{C} = \frac{2\ \mu C}{100\ pF} = 20 \times 10^3\ V = 20\ kV$$

EXAMPLE 9–2
Convert the following values to microfarads:
(a) 0.00001 F   (b) 0.005 F   (c) 1000 pF   (d) 200 pF

Solution
(a) 0.00001 F $\times$ 10^6 = 10 μF   (b) 0.005 F $\times$ 10^6 = 5000 μF
(c) 1000 pF $\times$ 10^{-6} = 0.001 μF   (d) 200 pF $\times$ 10^{-6} = 0.0002 μF

EXAMPLE 9–3
Convert the following values to picofarads:
(a) 0.1 $\times$ 10^{-8} F   (b) 0.000025 F   (c) 0.01 μF   (d) 0.005 μF

Solution
(a) 0.1 $\times$ 10^{-8} F $\times$ 10^{12} = 1000 pF   (b) 0.000025 F $\times$ 10^{12} = 25 $\times$ 10^6 pF
(c) 0.01 μF $\times$ 10^6 = 10,000 pF   (d) 0.005 μF $\times$ 10^6 = 5000 pF

How a Capacitor Stores Energy

A capacitor stores energy in the form of an *electric field* that is established by the opposite charges on the two plates. The electric field is represented by *lines of force*

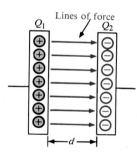

FIGURE 9–3
An electric field exists between the plates of a charged capacitor.

FIGURE 9–4
A force exists between two charged bodies.

between the positive and negative charges and concentrated within the dielectric, as shown in Figure 9–3.

Coulomb's Law

> **A force exists between two charged bodies that is directly proportional to the product of the two charges and inversely proportional to the square of the distance between the bodies.**

This relationship is expressed as

$$F = \frac{kQ_1Q_2}{d^2} \qquad (9\text{--}2)$$

where F is the force in newtons, Q_1 and Q_2 are the charges in coulombs, d is the distance between the charges in meters, and k is a proportionality constant equal to 9×10^9. Figure 9–4 illustrates the line of force between a positive and a negative charge. Many opposite charges on the plates of a capacitor create many lines of force, which form an *electric field* between the plates that stores energy within the dielectric.

The greater the forces between the charges on the plates of a capacitor, the more energy is stored. The amount of energy stored therefore is directly proportional to the capacitance, because, from Coulomb's law, the more charge stored, the greater the force.

Also, from the equation $Q = CV$, the amount of charge stored is directly related to the voltage as well as the capacitance. The amount of energy stored is dependent on the square of the voltage across the plates of the capacitor. The formula for the energy stored by a capacitor is as follows:

$$W = \frac{1}{2} CV^2 \qquad (9\text{--}3)$$

where the energy, W, is in joules when C is in farads and V is in volts.

Voltage Rating

Every capacitor has a limit on the amount of voltage that it can withstand across its plates. The *voltage rating* specifies the maximum dc voltage that can be applied without risk of damage to the device. If this maximum voltage, commonly called the *breakdown voltage* or *working voltage,* is exceeded, permanent damage to the capacitor can result.

Both the capacitance and the voltage rating must be taken into consideration before a capacitor is used in a circuit application. The choice of capacitance value is based on particular circuit requirements (and on factors that are studied later). The voltage rating should always be well above the maximum voltage expected in a particular application.

Dielectric Strength

The breakdown voltage of a capacitor is determined by the dielectric strength of the dielectric material used. The dielectric strength is expressed in volts/mil (1 mil = 0.001 in.). Table 9–2 lists typical values for several materials. Exact values vary depending on the specific composition of the material.

TABLE 9–2
Some common dielectric materials and their dielectric strengths.

Material	Dielectric Strength (volts/mil)
Air	80
Oil	375
Ceramic	1000
Paper (paraffined)	1200
Teflon®	1500
Mica	1500
Glass	2000

The dielectric strength can best be explained by an example. Assume that a certain capacitor has a plate separation of 1 mil and that the dielectric material is ceramic. This particular capacitor can withstand a maximum voltage of 1000 V, because its dielectric strength is 1000 V/mil. If the maximum voltage is exceeded, the dielectric may break down and conduct current, causing permanent damage to the capacitor. If the ceramic capacitor has a plate separation of 2 mils, its breakdown voltage is 2000 V.

Temperature Coefficient

The temperature coefficient indicates the amount and direction of a change in capacitance value with temperature. A *positive* temperature coefficient means that the capacitance increases with an increase in temperature or decreases with a decrease in temperature. A *negative* coefficient means that the capacitance decreases with an increase in temperature or increases with a decrease in temperature.

Temperature coefficients typically are specified in *parts per million per degree Celsius* (ppm/°C). For example, a negative temperature coefficient of 150 ppm/°C for a 1-μF capacitor means that for every degree rise in temperature, the capacitance decreases by 150 pF (there are one million picofarads in one microfarad).

Leakage

No insulating material is perfect. The dielectric of any capacitor will conduct some very small amount of current. Thus, the charge on a capacitor will eventually leak off. Some types of capacitors have higher leakages than others. An equivalent circuit for a nonideal capacitor is shown in Figure 9–5. The parallel resistor represents the extremely high resistance of the dielectric material through which leakage current flows.

Physical Characteristics of a Capacitor

The following parameters are important in establishing the capacitance and the voltage rating of a capacitor: plate area, plate separation, and dielectric constant.

Plate Area

Capacitance is directly proportional to the physical size of the plates as determined by the plate area, A. A larger plate area produces a larger capacitance, and vice versa. Figure 9–6(a) shows that the plate area of a parallel plate capacitor is the area of *one*

FIGURE 9–5
Equivalent circuit for a nonideal capacitor.

FIGURE 9–6

Capacitance is directly proportional to plate area (*A*).

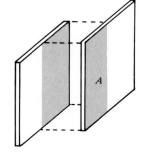

(a) Full plate area:
more capacitance

(b) Reduced plate area:
less capacitance

of the plates. If the plates are moved in relation to each other, as shown in part (b), the *overlapping area* determines the effective plate area. This variation in effective plate area is the basis for a certain type of variable capacitor.

Plate Separation

Capacitance is inversely proportional to the distance between the plates. The plate separation is designated *d,* as shown in Figure 9–7. A greater separation of the plates produces a smaller capacitance, as illustrated in the figure. The breakdown voltage is directly proportional to the plate separation. The further the plates are separated, the greater the breakdown voltage.

FIGURE 9–7

Capacitance is inversely proportional to the distance *d* between the plates.

(a) More capacitance   (b) Less capacitance

Dielectric Constant

As you know, the insulating material between the plates of a capacitor is called the *dielectric*. Every dielectric material has the ability to concentrate the lines of force of the electric field existing between the oppositely charged plates of a capacitor and thus increase the capacity for energy storage. The measure of a material's ability to establish an electric field is called the *dielectric constant* or *relative permittivity*, symbolized by ϵ_r (the Greek letter epsilon).

Capacitance is directly proportional to the dielectric constant. The dielectric constant of a vacuum is defined as 1, and that of air is very close to 1. These values are used as a reference, and all other materials have values of ϵ_r specified with respect to that of a vacuum or air. For example, a material with $\epsilon_r = 8$ can have a capacitance eight times greater than that of air, with all other factors being equal.

TABLE 9–3
Some common dielectric materials and their dielectric constants.

Material	Typical ϵ_r Values
Air (vacuum)	1.0
Teflon®	2.0
Paper (paraffined)	2.5
Oil	4.0
Mica	5.0
Glass	7.5
Ceramic	1200

Table 9–3 lists several common dielectric materials and typical dielectric constants for each. Values can vary because they depend on the specific composition of the material.

The dielectric constant (relative permittivity) is dimensionless, because it is a relative measure and is a ratio of the *absolute permittivity*, ϵ, of a material to the *absolute permittivity*, ϵ_0, of a vacuum, as expressed by the following formula:

$$\epsilon_r = \frac{\epsilon}{\epsilon_0} \tag{9–4}$$

The value of ϵ_0 is 8.85×10^{-12} F/m (farads per meter).

Formula for Capacitance in Terms of Physical Parameters

You have seen how capacitance is directly related to plate area A and the dielectric constant ϵ_r, and inversely related to plate separation d. An exact formula for calculating the capacitance in terms of these three quantities is as follows:

$$C = \frac{A\epsilon_r(8.85 \times 10^{-12} \text{ F/m})}{d} \tag{9–5}$$

where A is in square meters (m^2), d is in meters (m), and C is in farads (F). Recall that 8.85×10^{-12} F/m is the absolute permittivity, ϵ_0, of a vacuum and that $\epsilon_r(8.85 \times 10^{-12})$ F/m is the absolute permittivity of a dielectric, as derived from Equation (9–4).

EXAMPLE 9–4

Determine the capacitance of a parallel plate capacitor having a plate area of 0.01 m^2 and a plate separation of 0.02 m. The dielectric is mica, which has a dielectric constant of 5.0.

Solution
Use Equation (9–5):

$$C = \frac{A\epsilon_r(8.85 \times 10^{-12} \text{ F/m})}{d}$$
$$= \frac{(0.01 \text{ m}^2)(5.0)(8.85 \times 10^{-12} \text{ F/m})}{0.02 \text{ m}} = 22.13 \text{ pF}$$

The calculator sequence is

. 0 1 × 5 × 8 . 8 5 EXP +/− 1 2 ÷ . 0 2 =

1. Define *capacitance*.
2. (a) How many microfarads in a farad?
   (b) How many picofarads in a farad?
   (c) How many picofarads in a microfarad?
3. (a) When the plate area of a capacitor is increased, does the capacitance increase or decrease?
   (b) When the distance between the plates is increased, does the capacitance increase or decrease?

9–2

TYPES OF CAPACITORS

Capacitors normally are classified according to the type of dielectric material. The most common types of dielectric materials are mica, ceramic, paper/plastic, and electrolytic (aluminum oxide and tantalum oxide). In this section, the characteristics and construction of each of these types of capacitors and the variable capacitors are examined.

Mica Capacitors

There are two types of mica capacitors: stacked-foil and silver-mica. The basic construction of the stacked-foil type is shown in Figure 9–8. It consists of alternate layers of metal foil and thin sheets of mica. The metal foil forms the plate, with alternate foil sheets connected together to increase the plate area. More layers are used to increase the plate area, thus increasing the capacitance. The mica/foil stack is encapsulated in an insulating material such as Bakelite®, as shown in part (b) of the figure. The silver-mica capacitor is formed in a similar way by stacking mica sheets with silver electrode material screened on them.

FIGURE 9–8
Construction of a typical radial-lead mica capacitor.

(a) Stacked layer arrangement

(b) Layers are pressed together and encapsulated

Mica capacitors are available with capacitance values ranging from 1 pF to 0.1 μF and voltage ratings from 100 to 2500 V dc. Temperature coefficients from -20 to $+100$ ppm/°C are common. Mica has a typical dielectric constant of 5.

Ceramic Capacitors

Ceramic dielectrics provide very high dielectric constants (1200 is typical). As a result, comparatively high capacitance values can be achieved in a small physical size. Ceramic capacitors are commonly available in a ceramic disk form, as shown in Figure 9–9, in a multilayer radial-lead configuration, as shown in Figure 9–10, or in a leadless ceramic chip form for surface mounting on PC boards.

Ceramic capacitors typically are available in capacitance values ranging from 1 pF to 2.2 μF with voltage ratings up to 6 kV. A typical temperature coefficient for ceramic capacitors is 200,000 ppm/°C.

FIGURE 9–9
A ceramic disk capacitor and its basic construction.

FIGURE 9–10
Ceramic capacitors. (a) Typical capacitors with radial leads. (b) Construction view (courtesy of KEMET Electronics Corporation).

Paper/Plastic Capacitors

There are several types of plastic-film capacitors and the older paper dielectric capacitors. Polycarbonate, parylene, polyester, polystyrene, polypropylene, mylar, and paper are some of the more common dielectric materials used. Some of these types have capacitance values up to 100 μF.

FIGURE 9–11
Basic construction of axial-lead tubular paper/plastic dielectric capacitors.

Figure 9–11 shows a common basic construction used in many plastic-film and paper capacitors. A thin strip of plastic-film dielectric is sandwiched between two thin metal strips which act as plates. One lead is connected to the inner plate and one to the outer plate as indicated. The strips are then rolled in a spiral configuration and encapsulated in a molded case. Thus, a large plate area can be packaged in a relatively small physical size, thereby achieving large capacitance values. Figure 9–12 shows a construction view for one type of plastic-film capacitor.

(a)

(b)

FIGURE 9–12
Film capacitors. (a) Typical example (courtesy of Siemens Corp.); (b) Construction view of an axial-lead plastic film capacitor (courtesy of KEMET Electronics Corporation).

Electrolytic Capacitors

Electrolytic capacitors are *polarized* so that one plate is positive and the other negative. These capacitors are used for high capacitance values up to over 200,000 μF, but they have relatively low breakdown voltages (350 V is a typical maximum) and high amounts of leakage.

Electrolytic capacitors are available in two types: *aluminum* and *tantalum*. The basic construction of an electrolytic capacitor is shown in Figure 9–13(a). The capacitor consists of two strips of either aluminum or tantalum foil separated by a paper or gauze strip saturated with an electrolyte. During manufacturing, an electrochemical reaction is induced which causes an oxide layer (either aluminum oxide or tantalum oxide) to form on the inner surface of the positive plate. This oxide layer acts as the dielectric. Part (b) of Figure 9–13 shows several typical electrolytic capacitors.

Foil (aluminum or tantalum)

Paper or gauze saturated with electrolyte

Oxide dielectric

(a) Expanded view of basic construction

(b)   Typical units

FIGURE 9–13
Electrolytic capacitors.

Since an electrolytic capacitor is polarized, *the positive plate must always be connected to the positive side of a circuit*. The positive end is indicated by plus signs or some other obvious marking such as an arrow. Be very careful to make the correct connection and to install the capacitor only in a dc, not ac, circuit.

Variable Capacitors

Variable capacitors are used in a circuit when there is a need to adjust the capacitance value either manually or automatically, for example, in radio or TV tuners. The major types of variable or adjustable capacitors are now introduced.

Air Capacitor

Variable capacitors with air dielectrics, such as the one shown in Figure 9–14, are sometimes used as tuning capacitors in applications requiring frequency selection. This

FIGURE 9–14
A typical variable air capacitor.

FIGURE 9–15
Schematic symbol for a variable capacitor.

type of capacitor is constructed of several plates that mesh together. One set of plates can be moved relative to the other, thus changing the effective plate area and the capacitance. The movable plates are linked together mechanically so that they all move when a shaft is rotated. The schematic symbol for a variable capacitor is shown in Figure 9–15.

Trimmers and Padders

These adjustable capacitors normally have screwdriver adjustments and are used for very fine adjustments in a circuit. Ceramic or mica is a common dielectric in these types of capacitors, and the capacitance usually is changed by adjusting the plate separation. Figure 9–16 shows some typical devices.

FIGURE 9–16
Trimmer capacitors (courtesy of Murata Erie North America, Inc.).

Varactors

The varactor is a semiconductor device that exhibits a capacitance characteristic that is varied by changing the voltage across its terminals. This device usually is covered in detail in a course on electronic devices.

Capacitor Labeling

Capacitor values are indicated on the body of the capacitor either by *typographical labels* or by *color codes*. Typographical labels consist of letters and numbers that indicate various parameters such as capacitance, voltage rating, tolerance, and others.

Some capacitors carry no unit designation for capacitance. In these cases, the units are implied by the value indicated. For example, a ceramic capacitor marked .001 or .01 has units of microfarads because picofarad values that small are not available. As another example, a ceramic capacitor labeled 50 or 330 has units of picofarads because microfarad units that large normally are not available in this type.

In some instances, the units are labeled as pF or μF; sometimes the microfarad unit is labeled as MF or MFD. Voltage rating appears on some types of capacitors and is omitted on others. When it is omitted, the voltage rating can be determined from information supplied by the manufacturer. The tolerance of the capacitor is usually labeled as a percentage, such as $\pm 10\%$. The temperature coefficient is indicated by a *parts per million* marking. This type of label consists of a *P* or *N* followed by a number. For example, N750 means a negative temperature coefficient of 750 ppm/°C, and P30 means a positive temperature coefficient of 30 ppm/°C.

SECTION REVIEW 9–2

1. How are capacitors commonly classified?
2. What is the difference between a fixed and a variable capacitor?
3. What type of capacitor normally is polarized?
4. What precautions must be taken when a polarized capacitor is installed in a circuit?

9–3

SERIES CAPACITORS

When capacitors are connected in series, *the total capacitance is less than the smallest capacitance value,* because the *effective* plate separation increases. The calculation of total series capacitance is analogous to the calculation of total resistance of parallel resistors.

To start, we will use two capacitors in series to show how the total capacitance is determined. Figure 9–17 shows two capacitors, which initially are uncharged, connected in series with a dc voltage source. When the switch is closed, as shown in part (a), current begins to flow.

Recall that (1) current is the same at all points in a series circuit and (2) current is defined as the rate of flow of charge ($I = Q/t$). In a certain period of time, a certain amount of charge moves through the circuit. Since current is the same everywhere in the circuit of Figure 9–17(a), the same amount of charge is moved from the negative side of the source to plate A of C_1, and from plate B of C_1 to plate A of C_2, and from plate B of C_2 to the positive side of the source. As a result, of course, the same amount of charge is deposited on the plates of both capacitors in a given period of time, and the total charge (Q_T) moved through the circuit in that period of time equals the charge stored by C_1 and also equals the charge stored by C_2:

$$Q_T = Q_1 = Q_2$$

As the capacitors charge, the voltage across each one increases as indicated.

(a) While charging, $I = Q/t$ is the same at all points.

(b) Both capacitors store the same amount of charge ($Q_T = Q_1 = Q_2$).

FIGURE 9-17
Capacitors in series produce a total capacitance that is less than the smallest value.

Figure 9–17(b) shows the capacitors after they have been completely charged and the current has ceased. Both capacitors store an equal amount of charge (Q), and the voltage across each one depends on its capacitance value ($V = Q/C$). By Kirchhoff's voltage law, which applies to capacitive circuits as well as to resistive circuits, the sum of the capacitor voltages equals the source voltage:

$$V_S = V_1 + V_2$$

Using the fact that $V = Q/C$, we can substitute into the formula for Kirchhoff's law and get the following relationship (where $Q = Q_T = Q_1 = Q_2$):

$$\frac{Q}{C_T} = \frac{Q}{C_1} + \frac{Q}{C_2}$$

The Q can be factored out of the right side of the equation and canceled with the Q on the left side as follows:

$$\frac{\cancel{Q}}{C_T} = \cancel{Q}\left(\frac{1}{C_1} + \frac{1}{C_2}\right)$$

Thus we have the following relationship for two capacitors in series:

$$\frac{1}{C_T} = \frac{1}{C_1} + \frac{1}{C_2} \tag{9-6}$$

Taking the reciprocal of both sides of Equation (9–6) gives the formula for the total capacitance:

$$C_T = \frac{1}{(1/C_1) + (1/C_2)} \tag{9-7}$$

EXAMPLE 9–5   Find C_T in Figure 9–18.

FIGURE 9–18

Solution

$$C_T = \frac{1}{(1/C_1) + (1/C_2)}$$

$$= \frac{1}{(1/100 \text{ pF}) + (1/300 \text{ pF})} = 75 \text{ pF}$$

The calculator sequence is

General Formula for Series Capacitance

Equations (9–6) and (9–7) can be extended to any number of capacitors in series, as shown in Figure 9–19. The expanded formulas are as follows, where the subscript n can be any number:

$$\frac{1}{C_T} = \frac{1}{C_1} + \frac{1}{C_2} + \frac{1}{C_3} + \cdots + \frac{1}{C_n} \qquad (9\text{–}8)$$

$$C_T = \frac{1}{(1/C_1) + (1/C_2) + (1/C_3) + \cdots + (1/C_n)} \qquad (9\text{–}9)$$

FIGURE 9–19
General series circuit with n capacitors.

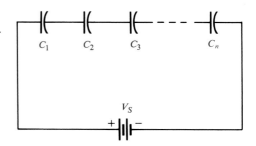

Remember,

The total series capacitance is always less than the smallest capacitance.

EXAMPLE 9–6   Determine the total capacitance in Figure 9–20.

FIGURE 9–20

Solution

$$\frac{1}{C_T} = \frac{1}{C_1} + \frac{1}{C_2} + \frac{1}{C_3} = \frac{1}{10 \ \mu F} + \frac{1}{5 \ \mu F} + \frac{1}{8 \ \mu F}$$

Taking the reciprocal of both sides yields

$$C_T = \frac{1}{(1/10 \ \mu F) \ + \ (1/5 \ \mu F) \ + \ (1/8 \ \mu F)}$$

$$= \frac{1}{0.425} \ \mu F = 2.35 \ \mu F$$

How Series Capacitors Divide the Voltage

The voltage across each capacitor in a series connection depends on its capacitance value according to the formula $V = Q/C$. The largest-value capacitor will have the smallest voltage because of the reciprocal relationship. Likewise, the smallest capacitance value will have the largest voltage. The voltage across any individual capacitor in a series connection can be determined using the following formula:

$$V_x = \left(\frac{C_T}{C_x}\right) V_S \qquad (9–10)$$

where C_x is any capacitor, such as C_1, C_2, C_3, and so on.

EXAMPLE 9–7   Find the voltage across each capacitor in Figure 9–21.

FIGURE 9–21

Solution

$$\frac{1}{C_T} = \frac{1}{C_1} + \frac{1}{C_2} + \frac{1}{C_3}$$

$$= \frac{1}{0.1 \ \mu F} + \frac{1}{0.5 \ \mu F} + \frac{1}{0.2 \ \mu F}$$

$$C_T = \frac{1}{17} \ \mu F = 0.0588 \ \mu F$$

$$V_1 = \left(\frac{C_T}{C_1}\right)V = \left(\frac{0.0588 \ \mu F}{0.1 \ \mu F}\right)25 \ V = 14.7 \ V$$

$$V_2 = \left(\frac{C_T}{C_2}\right)V = \left(\frac{0.0588 \ \mu F}{0.5 \ \mu F}\right)25 \ V = 2.94 \ V$$

$$V_3 = \left(\frac{C_T}{C_3}\right)V = \left(\frac{0.0588 \ \mu F}{0.2 \ \mu F}\right)25 \ V = 7.35 \ V$$

The calculator sequence for V_1 is

$[.]\ [0]\ [5]\ [8]\ [8]\ [\text{EXP}]\ [+/-]\ [6]\ [\div]\ [.]\ [1]\ [\text{EXP}]\ [+/-]\ [6]\ [\times]\ [2]\ [5]\ [=]$

SECTION REVIEW
9–3

1. Is the total capacitance of a series connection less than or greater than the value of the smallest capacitor?
2. The following capacitors are in series: 100 pF, 250 pF, and 500 pF. What is the total capacitance?
3. A 0.01-μF and a 0.015-μF capacitor are in series. Determine the total capacitance.
4. Determine the voltage across the 0.01-μF capacitor in Problem 3 if 10 V are connected across the two series capacitors.

9–4

PARALLEL CAPACITORS

When capacitors are connected in parallel, *the total capacitance is the sum of the individual capacitances,* because the *effective* plate area increases. The calculation of total parallel capacitance is analogous to the calculation of total series resistance.

Figure 9–22 shows two parallel capacitors connected to a dc voltage source. When the switch is closed, as shown in part (a), current begins to flow. A total amount of charge (Q_T) moves through the circuit in a certain period of time. Part of the total charge is stored by C_1 and part by C_2. The portion of the total charge that is stored by each capacitor depends on its capacitance value according to the relationship $Q = CV$.

Figure 9–22(b) shows the capacitors after they have been completely charged and the current has stopped. Since the voltage across both capacitors is the same, the larger capacitor stores more charge. If the capacitors are equal in value, they store an equal amount of charge. The charge stored by both of the capacitors together equals the total charge that was delivered from the source:

$$Q_T = Q_1 + Q_2$$

(a) The amount of charge on each capacitor is directly proportional to its capacitance value.

(b) $Q_T = Q_1 + Q_2$

FIGURE 9–22
Capacitors in parallel produce a total capacitance that is the sum of the individual capacitances.

Using the fact that $Q = CV$, we can substitute into the preceding formula and get the following relationship:

$$C_T V_S = C_1 V_S + C_2 V_S$$

Because all the V_S terms are equal, they can be canceled, leaving

$$C_T = C_1 + C_2 \qquad\qquad (9\text{–}11)$$

EXAMPLE 9–8

What is the total capacitance in Figure 9–23?

FIGURE 9–23

Solution

$$C_T = C_1 + C_2 = 330 \text{ pF} + 220 \text{ pF} = 550 \text{ pF}$$

General Formula for Parallel Capacitance

Equation (9–11) can be extended to any number of capacitors in parallel, as shown in Figure 9–24. The expanded formula is as follows, where the subscript n can be any number:

$$C_T = C_1 + C_2 + C_3 + \cdots + C_n \qquad\qquad (9\text{–}12)$$

FIGURE 9–24
General parallel circuit with n capacitors.

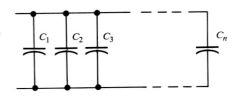

EXAMPLE 9–9

Determine C_T in Figure 9–25.

FIGURE 9–25

Solution

$$
\begin{aligned}
C_T &= C_1 + C_2 + C_3 + C_4 + C_5 + C_6 \\
&= 0.01 \ \mu F + 0.02 \ \mu F + 0.01 \ \mu F + 0.05 \ \mu F + 0.02 \ \mu F + 0.02 \ \mu F \\
&= 0.13 \ \mu F
\end{aligned}
$$

**SECTION REVIEW
9–4**

1. How is total parallel capacitance determined?
2. In a certain application, you need 0.05 μF. The only values available are 0.01 μF, which are available in large quantities. How can you get the total capacitance that you need?
3. The following capacitors are in parallel: 10 pF, 5 pF, 33 pF, and 50 pF. What is C_T?

9–5

CAPACITORS IN DC CIRCUITS

In this section, the response during charging and discharging of a simple capacitive circuit with a dc source is examined. Figure 9–26(a) shows a capacitor connected in series with a resistor and a switch to a dc voltage source. Initially, the switch is open and the capacitor is uncharged with zero volts across its plates. At the instant the switch is closed, the current jumps to its maximum value and the capacitor begins to charge. The current is maximum initially because the capacitor has zero volts across it and, therefore, appears as a *short;* thus, the current is limited only by the resistance. As time passes and the capacitor charges, the current decreases and the voltage V_C across the capacitor increases. The resistor voltage is proportional to the current during this charging period.

Current jumps to maximum
at instant switch is
closed; then it decreases.

Voltage is zero
at instant switch
is closed; then
it increases.

(a) Charging: Capacitor voltage increases
    as the current and resistor voltage decrease.

(b) Fully charged: Capacitor voltage equals
    source voltage. The current is zero.

Current jumps to maximum at instant
switch is closed; then it decreases.

Voltage decreases
as capacitor
discharges.

(c) Discharging: Capacitor voltage, resistor
    voltage, and the current decrease from
    initial maximums. Note that the discharge
    current is opposite to the charge current.

FIGURE 9–26
Charging and discharging of a capacitor.

After a certain period of time, the capacitor reaches full charge. At this point, the current is zero and the capacitor voltage is equal to the dc source voltage, as shown in Figure 9–26(b). If the switch were opened now, the capacitor would retain its full charge (neglecting any leakage).

In Figure 9–26(c), the voltage source has been removed. When the switch is closed, the capacitor begins to discharge. Initially, the current jumps to a maximum but in a direction opposite to its direction during charging. As time passes, the current and capacitor voltage decrease. The resistor voltage is always proportional to the current. When the capacitor has fully discharged, the current and the capacitor voltage are zero.

Remember the following about capacitors in dc circuits:

1. Voltage across a capacitor *cannot* change instantaneously.
2. Current in a capacitive circuit *can* change instantaneously.
3. A fully charged capacitor appears as an *open* to nonchanging current.
4. An uncharged capacitor appears as a *short* to an instantaneous change in current.

Now we will examine in more detail how the voltage and current change with time in a capacitive circuit.

The *RC* Time Constant

As you have seen, when a capacitor charges or discharges through a resistance, a certain time is required for the capacitor to charge fully or discharge fully. *The voltage across a capacitor cannot change instantaneously,* because a finite time is required to move charge from one point to another. The rate at which the capacitor charges or discharges is determined by the *time constant* of the circuit.

> **The time constant of a series *RC* circuit is a time interval that equals the product of the resistance and the capacitance.**

The time constant is symbolized by τ (Greek letter tau), and the formula is as follows:

$$\tau = RC \qquad\qquad (9\text{--}13)$$

Recall that $I = Q/t$. The current is the amount of charge moved in a given time. When the resistance is increased, the charging current is reduced, thus increasing the charging time of the capacitor. When the capacitance is increased, the amount of charge increases; thus, for the same current, more time is required to charge the capacitor.

EXAMPLE 9–10

A series *RC* circuit has a resistance of 1 MΩ and a capacitance of 5 μF. What is the time constant?

Solution

$$\tau = RC = (1 \times 10^6 \ \Omega)(5 \times 10^{-6} \ \text{F}) = 5 \ \text{s}$$

During one time constant interval, the charge on a capacitor changes approximately 63%. Therefore, an uncharged capacitor charges to 63% of its fully charged voltage in one time constant. When a capacitor is discharging, its voltage drops to approximately 37% (100% − 63%) of its initial value in one time constant. This change also corresponds to a 63% change.

The Charging and Discharging Curves

A capacitor charges and discharges following a nonlinear curve, as shown in Figure 9–27. In these graphs, the percentage of full charge is shown at each time constant interval. This type of curve follows a precise mathematical formula and is called an *exponential curve*. The charging curve is an *increasing exponential*, and the discharging curve is a *decreasing exponential*. As you can see, it takes *five time constants* to approximately reach the final value. Five time constants is *accepted* as the time to fully charge or discharge a capacitor and is called the *transient time*.

(a) Charging curve

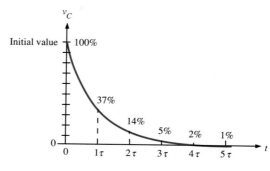

(b) Discharging curve

FIGURE 9–27
Charging and discharging exponential curves for an *RC* circuit. These curves apply to both voltage and current.

General Formula

The general expressions for either increasing or decreasing exponential curves are given in the following equations for both voltage and current:

$$v = V_F + (V_i - V_F)e^{-t/\tau} \qquad (9\text{–}14)$$

$$i = I_F + (I_i - I_F)e^{-t/\tau} \qquad (9\text{–}15)$$

where V_F and I_F are the *final* values and V_i and I_i are the *initial* values. The lowercase letters v and i are the instantaneous values of the capacitor voltage and current at time t, and e is the base of natural logarithms with a value of 2.718. The e^x function on your calculator makes it easy to work with these formulas.

The Charging Curve

The formula for the special case in which an increasing exponential voltage curve begins at zero ($V_i = 0$) is given in Equation (9–16). It is developed as follows, starting with the general formula:

$$v = V_F + (V_i - V_F)e^{-t/\tau}$$
$$= V_F + (0 - V_F)e^{-t/RC}$$
$$v = V_F(1 - e^{-t/RC}) \qquad (9\text{–}16)$$

Using Equation (9–16), we can calculate the value of the charging voltage of a capacitor at any instant of time. The same is true for an increasing current.

EXAMPLE 9–11   In Figure 9–28, determine the capacitor voltage 50 μs after the switch is closed if the capacitor initially is uncharged. Sketch the charging curve.

FIGURE 9–28

Solution
The time constant is $RC = (8\ k\Omega)(0.01\ \mu F) = 80\ \mu s$. The voltage to which the capacitor will fully charge is 50 V (this is V_F). The initial voltage is zero. Notice that 50 μs is less than one time constant; so the capacitor will charge less than 63% of the full voltage in that time.

$$v_C = V_F(1 - e^{-t/RC}) = 50\ V(1 - e^{-50\mu s/80\mu s})$$
$$= 50\ V(1 - e^{-0.625}) = 50\ V(1 - 0.535) = 23.2\ V$$

We determine the value of $e^{-0.625}$ on the calculator by entering -0.625 and then pressing [2nd] [ln x] or [e^x].
The charging curve for the capacitor is shown in Figure 9–29.

FIGURE 9–29

The calculator sequence is

[5] [0] [÷] [8] [0] [=] [+/−] [2nd] [ln x] [+/−] [+] [1] [=] [×] [5] [0] [=]

The Discharging Curve

The formula for the special case in which a decreasing exponential voltage curve ends at zero is derived from the general formula as follows:

$$v = V_F + (V_i - V_F)e^{-t/\tau}$$
$$= 0 + (V_i - 0)e^{-t/RC}$$
$$v = V_i e^{-t/RC} \tag{9–17}$$

where V_i is the voltage at the beginning of the discharge. We can use this formula to calculate the discharging voltage at any instant, as Example 9–12 illustrates.

EXAMPLE 9–12

Determine the capacitor voltage in Figure 9–30 at a point in time 6 ms after the switch is closed. Sketch the discharging curve.

FIGURE 9–30

Solution

The discharge time constant is $RC = (10 \text{ k}\Omega)(2 \text{ }\mu\text{F}) = 20$ ms. The initial capacitor voltage is 10 V. Notice that 6 ms is less than one time constant, so the capacitor will discharge less than 63%. Therefore, it will have a voltage greater than 37% of the initial voltage at 6 ms.

$$v_C = V_i e^{-t/RC} = 10e^{-6 \text{ ms}/20 \text{ ms}}$$
$$= 10e^{-0.3} = 10(0.741) = 7.41 \text{ V}$$

Again, the value of $e^{-0.3}$ can be determined with a calculator.

The discharging curve for the capacitor is shown in Figure 9–31.

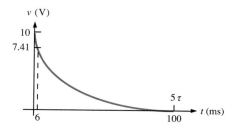

FIGURE 9–31

Universal Exponential Curves

The universal curves in Figure 9–32 provide a graphical solution of the charge and discharge of capacitors. Example 9–13 illustrates this graphical method.

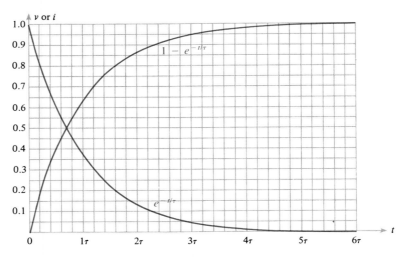

FIGURE 9–32
Universal exponential curves.

EXAMPLE 9–13

How long will it take the capacitor in Figure 9–33 to charge to 75 V? What is the capacitor voltage 2 ms after the switch is closed? Use the universal curves in Figure 9–32 to determine the answers.

FIGURE 9–33

Solution
The full charge voltage is 100 V, which is the 100% level (1.0) on the graph. 75 V is 75% of the maximum, or 0.75 on the graph. You can see that this value occurs at 1.4 time constants. One time constant is 1 ms. Therefore, the capacitor voltage reaches 75 V at 1.4 ms after the switch is closed.

The capacitor is at approximately 87 V in 2 ms. These graphical solutions are shown in Figure 9–34.

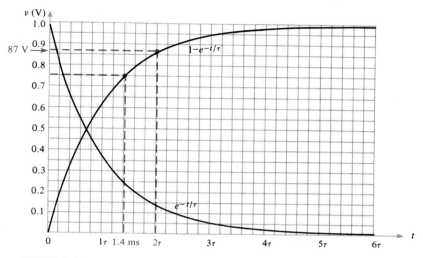

FIGURE 9–34

SECTION REVIEW 9–5

1. Determine the time constant when $R = 1.2$ kΩ and $C = 1000$ pF.
2. If the circuit in Problem 1 is charged with a 5-V source, how long will it take the initially uncharged capacitor to reach full charge? At full charge, what is the capacitor voltage?
3. A certain circuit has a time constant of 1 ms. If it is charged with a 10-V battery, what will the capacitor voltage be at each of the following intervals: 2 ms, 3 ms, 4 ms, and 5 ms? The capacitor is initially uncharged.
4. A capacitor is charged to 100 V. If it is discharged through a resistor, what is the capacitor voltage at one time constant?

9–6

CAPACITORS IN AC CIRCUITS

The Phase Relationship of Current and Voltage

We will illustrate the relationship of current and capacitor voltage in an ac circuit by connecting a capacitor in series with a resistor to a sine wave voltage source and then observing the capacitor voltage and the resistor voltage with an oscilloscope. The connections are shown in Figure 9–35(a) and (b). For each measurement, the positions of R and C are interchanged so that one side of the component across which we are measuring is connected to ground. This is necessary to prevent the ground lead of the scope

(a) Measuring V_C

(b) Measuring V_R

(c) An attempt to measure across R results in one end of the resistor shorted to ground through the scope's ground lead, thus shorting the capacitor.

FIGURE 9–35
Oscilloscope display showing that the current leads the capacitor voltage by 90°. The current is in phase with the resistor voltage. Components are switched so that V_C and V_R can be measured relative to ground. The external trigger (EXT TRIG) ensures that both waveforms are displayed relative to the same point.

probe from shorting a nonground point to ground, as illustrated in part (c) of the figure. Recall that the resistor voltage and the current are always in phase with each other. Therefore, by observing the resistor voltage, we are also indirectly observing the current in terms of its phase relationship to the capacitor voltage. The frequency of the voltage and of the current are always the same in an ac circuit.

Why the Current Leads the Capacitor Voltage by 90°

A sine wave voltage is shown in Figure 9–36. Notice that the rate at which the voltage is changing varies along the sine wave curve, as indicated by the "steepness" of the curve. At the zero crossings, the curve is changing at a faster rate than anywhere else along the curve. At the peaks, the curve has a zero rate of change because it has just reached its maximum and is at the point of changing direction.

FIGURE 9–36
The rates of change of a sine wave.

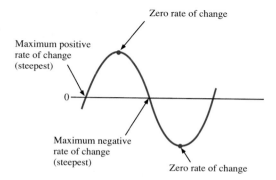

The amount of charge stored by a capacitor determines the voltage across it. Therefore, the rate at which the charge is moved (Q/t = current) from one plate to the other determines the rate at which the voltage changes. When the current is changing at its maximum rate (at the zero crossings), the voltage is at its maximum value (peak). When the current is changing at its minimum rate (zero at the peaks), the voltage is at its minimum value (zero). This relationship is illustrated in Figure 9–37. As you can see, *the current peaks occur a quarter of a cycle before the voltage peaks*. Thus, the current *leads* the voltage by 90°.

FIGURE 9–37
Current is always leading the capacitor voltage by 90°.

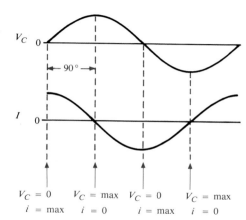

Capacitive Reactance, X_C

Capacitive reactance is the opposition to sinusoidal current, expressed in ohms and symbolized by X_C.

In Figure 9–38, a capacitor is connected to a sine wave voltage source. Note that when the source voltage is held at a constant amplitude value and its frequency is increased, the amplitude of the current increases. Also, when the frequency of the source is decreased, the current amplitude decreases. The reason is as follows: When the frequency of the voltage increases, its rate of change also increases. This relationship is illustrated in Figure 9–39, where the frequency is doubled.

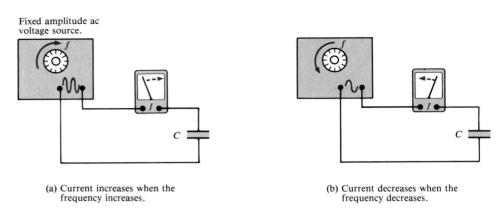

Fixed amplitude ac voltage source.

(a) Current increases when the frequency increases.

(b) Current decreases when the frequency decreases.

FIGURE 9–38
The current in a capacitive circuit varies directly with the frequency of the applied voltage.

FIGURE 9–39
Rate of change increases with frequency.

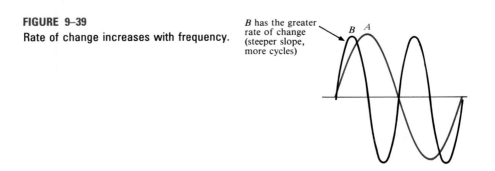

B has the greater rate of change (steeper slope, more cycles)

Now, if the rate at which the voltage is changing increases, the amount of charge moving through the circuit in a given period of time must also increase. More charge in a given period of time means more current. For example, a tenfold increase in frequency means that the capacitor is charging and discharging 10 times as much in a given time interval. Therefore, since the rate of charge movement has increased 10 times, the current must increase by 10 because $I = Q/t$.

An *increase* in the amount of current for a fixed amount of voltage indicates that *opposition* to the current has *decreased*. Therefore, the capacitor offers opposition to current, which varies *inversely* with frequency. The opposition to sinusoidal current is called *capacitive reactance*. The symbol for capacitive reactance is X_C, and its unit is the ohm (Ω).

You have just seen how frequency affects the opposition to current (capacitive reactance) in a capacitor. Now let's see how the capacitance affects the reactance. Figure 9–40(a) shows that when a sine wave voltage with a fixed amplitude and frequency is applied to a 1-μF capacitor, a certain amount of current flows. When the capacitance value is increased to 2 μF, the current increases, as shown in part (b). Thus, when the capacitance increases, the opposition to current (capacitive reactance) decreases. Therefore, not only is the capacitive reactance inversely proportional to frequency, but it is also inversely proportional to capacitance:

$$X_C \text{ is proportional to } \frac{1}{fC}$$

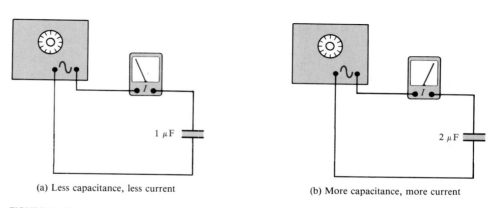

(a) Less capacitance, less current   (b) More capacitance, more current

FIGURE 9–40
For a fixed voltage and frequency, the current varies directly with capacitance.

It can be proven that the constant of proportionality is $1/2\pi$. Therefore, the formula for X_C is

$$X_C = \frac{1}{2\pi fC} \tag{9–18}$$

The 2π term comes from the fact that a sine wave can be described in terms of rotational motion, and one revolution contains 2π radians.

EXAMPLE 9–14

A sinusoidal voltage is applied to a capacitor, as shown in Figure 9–41. The frequency of the sine wave is 1 kHz. Determine the capacitive reactance.

FIGURE 9–41

V_s 0.005 μF

Solution

$$X_C = \frac{1}{2\pi f C} = \frac{1}{2\pi(1 \times 10^3 \text{ Hz})(0.005 \times 10^{-6} \text{ F})}$$

$$= 31.83 \text{ k}\Omega$$

The calculator sequence is

2 $\times$ 2nd π $\times$ 1 EXP 3 $\times$ $\cdot$ 0 0 5 EXP $+/-$ 6 $=$ $1/x$

Ohm's Law in Capacitive Circuits

The reactance of a capacitor is analogous to the resistance of a resistor. In fact, both are expressed in ohms. Since both are forms of opposition to current, Ohm's law applies to capacitive circuits as well as to resistive circuits and is stated as follows for capacitive reactance:

$$V = IX_C \tag{9-19}$$

When applying Ohm's law in ac circuits, you must express both the current and the voltage in the same way, that is, both in rms, both in peak, and so on.

EXAMPLE 9–15   Determine the rms current in Figure 9–42.

FIGURE 9–42

$V_{rms} = 5$ V

$f = 10$ kHz

0.005 μF

Solution

$$X_C = \frac{1}{2\pi f C} = \frac{1}{2\pi(10 \times 10^3 \text{ Hz})(0.005 \times 10^{-6} \text{ F})}$$

$$= 3.18 \text{ k}\Omega$$

Applying Ohm's law:

$$I_{rms} = \frac{V_{rms}}{X_C} = \frac{5 \text{ V}}{3.18 \text{ k}\Omega} = 1.57 \text{ mA}$$

Power in a Capacitor

As discussed earlier in this chapter, a charged capacitor stores energy in the electric field within the dielectric. An ideal capacitor does not dissipate energy; it only stores it. When an ac voltage is applied to a capacitor, energy is stored by the capacitor during a portion of the voltage cycle. Then the stored energy is *returned to the source* during another portion of the cycle. *There is no net energy loss.* Figure 9–43 shows the power curve that results from one cycle of capacitor voltage and current.

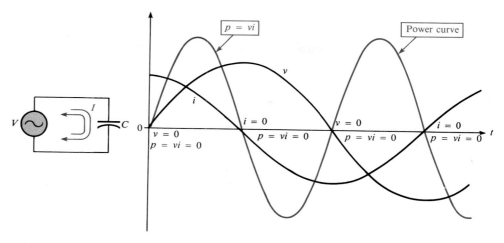

FIGURE 9–43
Power curve.

Instantaneous Power (p)

The product of instantaneous voltage, v, and instantaneous current, i, gives instantaneous power, p. At points where v or i is zero, p is also zero. When both v and i are positive, p is also positive. When either v or i is positive and the other negative, p is negative. When both v and i are negative, p is positive. As you can see, the power follows a sinusoidal-type curve. Positive values of power indicate that energy is stored by the capacitor. Negative values of power indicate that energy is returned from the capacitor to the source. Note that the power fluctuates at a frequency twice that of the voltage or current, as energy is alternately stored and returned to the source.

True Power (P_{true})

Ideally, all of the energy stored by a capacitor during the positive portion of the power cycle is returned to the source during the negative portion. *No net energy is consumed in the capacitor, so the true power is zero.* Actually, because of leakage and foil resistance in a practical capacitor, a small percentage of the total power is dissipated.

Reactive Power (P_r)

The rate at which a capacitor stores or returns energy is called its *reactive power, P_r.* The reactive power is a nonzero quantity, because at any instant in time, the capacitor is actually taking energy from the source or returning energy to it. Reactive power does not represent an energy loss. The following formulas apply:

$$P_r = V_{rms} I_{rms} \qquad (9\text{–}20)$$

$$P_r = \frac{V_{rms}^2}{X_C} \qquad (9\text{–}21)$$

$$P_r = I_{rms}^2 X_C \qquad (9\text{–}22)$$

Notice that these equations are of the same form as those for true power in a resistor. The voltage and current are expressed in rms. The unit of reactive power is *volt-amperes reactive* (VAR).

EXAMPLE 9–16

Determine the true power and the reactive power in Figure 9–44.

FIGURE 9–44

V_{rms}
2 V
$f = 2$ kHz
0.01 μF

Solution

The true power P_{true} is *always zero for a capacitor*. The reactive power is as follows:

$$X_C = \frac{1}{2\pi f C} = \frac{1}{2\pi(2 \times 10^3 \text{ Hz})(0.01 \times 10^{-6} \text{ F})}$$

$$= 7.958 \text{ k}\Omega$$

$$P_r = \frac{V_{rms}^2}{X_C} = \frac{(2 \text{ V})^2}{7.958 \text{ k}\Omega}$$

$$= 0.503 \times 10^{-3} \text{ VAR} = 0.503 \text{ mVAR}$$

SECTION REVIEW 9–6

1. State the phase relationship between current and voltage in a capacitor.
2. Calculate X_C for $f = 5$ kHz and $C = 50$ pF.
3. At what frequency is the reactance of a 0.1-μF capacitor equal to 2 kΩ?
4. Calculate the rms current in Figure 9–45.
5. A 1-μF capacitor is connected to an ac voltage source of 12 V rms. What is the true power?
6. In Problem 5, determine reactive power at a frequency of 500 Hz.

FIGURE 9–45

$V_{rms} = 1$ V
$f = 1$ MHz
0.1 μF

9–7

CAPACITOR APPLICATIONS

Capacitors are very widely used in electrical and electronic applications. A few typical applications are discussed here to illustrate the usefulness of this component.

Power Supply Filter

A device that converts the 60-Hz sine wave voltage from your wall outlet to a pulsating dc voltage is called a *rectifier*. The basic concept of a power supply with a rectifier is

FIGURE 9–46
Basic concept of a power supply which converts ac to dc.

shown in the block diagram of Figure 9–46. In order to be useful in powering most systems such as the radio, TV, or computer, the pulsating dc voltage must be converted to a nearly constant level. This conversion is accomplished with a power supply filter as indicated.

A basic power supply filter is implemented with a capacitor as shown in Figure 9–47(a). As shown in part (b), the basic operation is as follows: The capacitor charges as the rectifier output voltage increases. When the peak is reached and the output voltage starts to decrease, the capacitor begins to discharge through the load resistance. The value of the capacitance is selected so that the time constant is very long compared to the period of the rectifier output voltage. The rectifier circuitry allows current only in a direction to charge the capacitor and prevents discharge current. By the time the output voltage nears its next peak, the capacitor has discharged only a small amount and requires only a small amount of recharging to get it back to the peak. This action results in an almost constant dc voltage, as shown. The small fluctuation is caused by the slight discharging and recharging of the capacitor. This fluctuation is called *ripple voltage*.

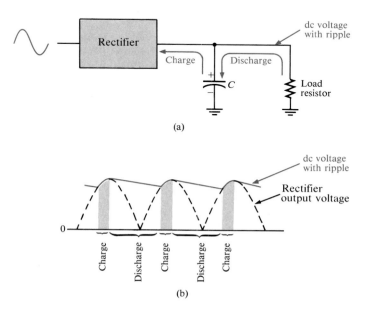

FIGURE 9–47
Basic operation of a capacitor power supply filter.

Coupling and Bypass Capacitors

Many applications, such as transistor amplifiers, require that an ac voltage be superimposed on a dc voltage at a certain point in the circuit, while at other points, the ac voltage must be removed without affecting the dc voltage.

The first situation is illustrated in Figure 9–48(a), where a capacitor is used to *couple* an ac voltage from the source to a point on a voltage divider that has a dc voltage. Since a capacitor blocks dc, the ac source is unaffected by the dc level, but the ac signal is passed through the capacitor and superimposed on the dc level.

The second situation is illustrated in Figure 9–48(b), where a capacitor is used to *bypass* the ac voltage to ground, leaving only the dc voltage. Details of electronic amplifiers with ac coupling and bypass circuits are covered in a later course.

(a) Coupling capacitor (b) Bypass capacitor

FIGURE 9–48
Coupling and bypass capacitors.

Tuned Circuits

Capacitors are used in conjunction with other components such as the inductor (to be covered in the next chapter) to provide *frequency selection* in communications systems. These *tuned circuits* allow a narrow band of frequencies to be selected while all other frequencies are rejected. A tuned circuit is one form of *filter*. The tuners in your TV and radio receivers are based on this principle and permit you to select one channel or station out of the many that are available.

Frequency selectivity is based on the fact that the reactance of a capacitor depends on the frequency. The basic concept of a tuned circuit is shown in Figure 9–49. This topic will be covered in detail in Chapter 14.

FIGURE 9–49
Basic concept of a tuned circuit.

Many frequencies *in*

Tuned circuit containing a variable capacitor

One frequency *out*

Computer Memories

Many computer memories utilize capacitors as the *storage element* for binary data, which consist of arrangements of only two types of digits, 1s and 0s. A charged capacitor can represent a 1, and a discharged capacitor can represent a 0. Patterns of 1s and 0s can be stored in a memory that consists of an array of capacitors with associated circuitry. You will study this topic later in a computer or digital fundamentals course.

SECTION REVIEW 9–7

1. Explain how pulsating dc voltage is smoothed out by a capacitor in a power supply filter.
2. How can you use a capacitor to remove ac voltage from a given point in a circuit?

9–8

TESTING CAPACITORS

Capacitors are generally very reliable devices. Their useful life can be extended significantly by operating well within the voltage rating and at moderate temperatures.

Failures can be categorized into two areas: *catastrophic* and *degradation*. The catastrophic failures are usually a short circuit caused by dielectric breakdown or an open circuit caused by connection failure. Degradation usually results in a gradual decrease in leakage resistance, and hence an increase in leakage current.

Ohmmeter Check

When there is a suspected problem, the capacitor can be removed from the circuit and checked with an ohmmeter. First, to be sure that the capacitor is discharged, short its leads, as indicated in Figure 9–50(a). Connect the meter, set on a high ohms range such as R × 1M, to the capacitor, as shown in Figure 9–50(b), and observe the needle. It should initially indicate near zero ohms. Then it should begin to move toward the high-resistance end of the scale as the capacitor charges from the ohmmeter's battery, as shown in Figure 9–50(c). When the capacitor is fully charged, the meter will indicate an extremely high resistance, as shown in Figure 9–50(d).

(a) Discharging

(b) Initially: The pointer jumps to zero.

(c) Charging: The pointer slowly moves back.

(d) Fully charged

FIGURE 9–50
Checking a capacitor with an analog ohmmeter. This check shows a good capacitor.

As mentioned, the capacitor charges from the internal battery of the ohmmeter, and the meter responds to the charging current. The larger the capacitance value, the more slowly the capacitor will charge, as indicated by the needle movement. For very small pF values, the meter response may be insufficient to indicate the fast charging action.

If the capacitor is internally shorted, the meter will go to zero and stay there. If it is leaky, the final meter reading will be much less than normal. Most capacitors have a resistance of several hundred megohms. The exception is the electrolytic, which may normally have less than one megohm of leakage resistance. If the capacitor is open, no charging action will be observed, and the meter will indicate an infinite resistance. The actual capacitance value can be measured on an instrument such as the *LCR* meter (inductance-capacitance-resistance). An example of this type of instrument is shown in Figure 9–51.

FIGURE 9–51
LCR meter for measuring capacitance (courtesy of Hewlett-Packard).

SECTION REVIEW
9–8

1. How can a capacitor be discharged after removal from the circuit?
2. Describe how the needle of an ohmmeter responds when a good capacitor is checked.

SUMMARY

☐ A capacitor blocks dc and passes ac.
☐ A capacitor is composed of two parallel conductive plates separated by an insulating material called the *dielectric*.
☐ A capacitor stores electrical charge on its plates.
☐ Energy is stored in a capacitor by the electric field created between the charged plates.
☐ Capacitance is measured in units of farads (F).
☐ Capacitance is directly proportional to the plate area and the dielectric constant and inversely proportional to the distance between the plates (the dielectric thickness).
☐ The dielectric constant is an indication of the ability of a material to establish an electric field.
☐ The dielectric strength is one factor that determines the breakdown voltage of a capacitor.
☐ Capacitors are commonly classified according to the dielectric material. Typical materials are mica, ceramic, paper/plastic, and electrolytic (aluminum oxide and tantalum oxide).
☐ The total capacitance of series capacitors is less than the smallest capacitance.
☐ The total capacitance of parallel capacitors is the sum of all the capacitances.
☐ The time constant determines the charging and discharging time of a capacitor with resistance in series.

- In a circuit with capacitance and resistance, the voltage and current during charging and discharging make a 63% change during each time constant interval.
- Five time constants are required for a capacitor to fully charge or discharge. This is called the *transient time*.
- Current in a capacitor leads the voltage by 90°.
- Capacitive reactance (X_C) is inversely proportional to frequency and to capacitance.
- Ideally, there is no energy loss in a capacitor and, thus, the true power (watts) is zero. However, in most capacitors there is some small energy loss due to leakage.

GLOSSARY

Capacitance   The ability of a capacitor to store electrical charge.

Capacitive reactance   The opposition of a capacitor to sinusoidal current. The unit is the ohm.

Capacitor   An electrical device consisting of two conductive plates separated by an insulating material and possessing the property of capacitance.

Coulomb (C)   The unit of electrical charge.

Coulomb's law   A physical law that states a force exists between two charged bodies that is directly proportional to the product of the two charges and inversely proportional to the square of the distance between them.

Dielectric   The insulating material between the plates of a capacitor.

Dielectric constant   A measure of the ability of a dielectric material to establish an electric field.

Dielectric strength   A measure of the ability of a dielectric material to withstand voltage.

Farad (F)   The unit of capacitance.

Phase   The relative displacement of a time-varying quantity with respect to a given reference.

Reactive power   The rate at which energy is stored and alternately returned to the source by a capacitor. The unit is the VAR.

Temperature coefficient   A constant specifying the amount of change in the value of a quantity for a given change in temperature.

Time constant   A fixed interval, set by R and C values, that determines the time response of a circuit.

Trimmer   A small variable capacitor.

VAR   Volt-ampere reactive.

FORMULAS

$$C = \frac{Q}{V} \qquad \text{Capacitance in terms of charge and voltage} \tag{9-1}$$

$$F = \frac{kQ_1Q_2}{d^2} \qquad \text{Coulomb's law (force between two charged bodies)} \tag{9-2}$$

$$W = \frac{1}{2}CV^2 \qquad \text{Energy stored by a capacitor} \tag{9-3}$$

$$\epsilon_r = \frac{\epsilon}{\epsilon_0} \qquad \text{Dielectric constant (relative permittivity)} \tag{9-4}$$

$$C = \frac{A\epsilon_r(8.85 \times 10^{-12} \text{ F/m})}{d} \qquad \text{Capacitance in terms of physical parameters} \tag{9-5}$$

$$\frac{1}{C_T} = \frac{1}{C_1} + \frac{1}{C_2} \qquad \text{Reciprocal of total series capacitance (two capacitors)} \qquad (9\text{--}6)$$

$$C_T = \frac{1}{(1/C_1) + (1/C_2)} \qquad \text{Total series capacitance (two capacitors)} \qquad (9\text{--}7)$$

$$\frac{1}{C_T} = \frac{1}{C_1} + \frac{1}{C_2} + \frac{1}{C_3} + \ldots + \frac{1}{C_n} \qquad \begin{array}{l}\text{Reciprocal of total series capacitance}\\ \text{(general)}\end{array} \qquad (9\text{--}8)$$

$$C_T = \frac{1}{(1/C_1) + (1/C_2) + (1/C_3) + \ldots + (1/C_n)} \qquad \begin{array}{l}\text{Total series capacitance}\\ \text{(general)}\end{array} \qquad (9\text{--}9)$$

$$V_x = \left(\frac{C_T}{C_x}\right)V_S \qquad \text{Voltage across series capacitor} \qquad (9\text{--}10)$$

$$C_T = C_1 + C_2 \qquad \text{Two capacitors in parallel} \qquad (9\text{--}11)$$

$$C_T = C_1 + C_2 + \ldots + C_n \qquad n \text{ capacitors in parallel} \qquad (9\text{--}12)$$

$$\tau = RC \qquad \text{Time constant} \qquad (9\text{--}13)$$

$$v = V_F + (V_i - V_F)e^{-t/\tau} \qquad \text{Exponential voltage (general)} \qquad (9\text{--}14)$$

$$i = I_F + (I_i - I_F)e^{-t/\tau} \qquad \text{Exponential current (general)} \qquad (9\text{--}15)$$

$$v = V_F(1 - e^{-t/RC}) \qquad \text{Increasing exponential voltage beginning at zero} \qquad (9\text{--}16)$$

$$v = V_i e^{-t/RC} \qquad \text{Decreasing exponential voltage ending at zero} \qquad (9\text{--}17)$$

$$X_C = \frac{1}{2\pi f C} \qquad \text{Capacitive reactance} \qquad (9\text{--}18)$$

$$V = IX_C \qquad \text{Ohm's law} \qquad (9\text{--}19)$$

$$P_r = V_{\text{rms}}I_{\text{rms}} \qquad \text{Reactive power} \qquad (9\text{--}20)$$

$$P_r = \frac{V_{\text{rms}}^2}{X_C} \qquad \text{Reactive power} \qquad (9\text{--}21)$$

$$P_r = I_{\text{rms}}^2 X_C \qquad \text{Reactive power} \qquad (9\text{--}22)$$

SELF-TEST

1. The following statement(s) accurately describes a capacitor:
   (a) The plates are conductive.
   (b) The dielectric is an insulator between the plates.
   (c) Constant dc flows through a fully charged capacitor.
   (d) A practical capacitor stores charge indefinitely when disconnected from the source.
   (e) none of the above
   (f) all the above
   (g) only a and b

2. Which one of the following statements is true?
   (a) There is current through the dielectric of a charging capacitor.
   (b) When a capacitor is connected to a dc voltage source, it will charge to the value of the source.
   (c) An ideal capacitor can be discharged by disconnecting it from the voltage source.

3. A capacitance of 0.01 μF is larger than
   (a) 0.00001 F   (b) 100,000 pF   (c) 1000 pF   (d) all of the above

4. A capacitance of 1000 pF is smaller than
   (a) 0.01 μF   (b) 0.001 μF   (c) 0.00000001 F   (d) a and c

5. When the voltage across a capacitor is increased, the stored charge
   (a) increases   (b) decreases   (c) remains constant   (d) fluctuates

6. When the voltage across a capacitor is doubled, the stored charge
   (a) stays the same   (b) is halved   (c) increases by four   (d) doubles

7. The voltage rating of a capacitor is increased by
   (a) increasing the plate separation   (b) decreasing the plate separation
   (c) increasing the plate area   (d) b and c

8. The capacitance value is increased by
   (a) decreasing plate area   (b) increasing plate separation
   (c) decreasing plate separation   (d) increasing plate area
   (e) a and b   (f) c and d

9. A 1-μF, a 2.2-μF, and a 0.05-μF capacitor are connected in series. The total capacitance is less than
   (a) 1 μF   (b) 2.2 μF   (c) 0.05 μF   (d) 0.001 μF

10. Four 0.02-μF capacitors are in parallel. The total capacitance is
    (a) 0.02 μF   (b) 0.08 μF   (c) 0.05 μF   (d) 0.04 μF

11. An uncharged capacitor and a resistor are connected in series with a switch and a 12-V battery. At the instant the switch is closed, the voltage across the capacitor is
    (a) 12 V   (b) 6 V   (c) 24 V   (d) 0 V

12. In Question 11, the voltage across the capacitor when it is fully charged is
    (a) 12 V   (b) 6 V   (c) 24 V   (d) -6 V

13. In Question 11, the capacitor will reach full charge in a time equal to appoximately
    (a) RC (b) $5RC$ (c) $12RC$ (d) cannot be predicted

14. A sine wave voltage is applied across a capacitor. When the frequency of the voltage is increased, the current
    (a) increases   (b) decreases   (c) remains constant   (d) ceases

15. A capacitor and a resistor are connected in series to a sine wave generator. The frequency is set so that the capacitive reactance is equal to the resistance and, thus, an equal amount of voltage appears across each component. If the frequency is decreased,
    (a) $V_R > V_C$ (b) $V_C > V_R$ (c) $V_R = V_C$ (d) $V_C < V_R$

16. An ohmmeter is connected across a discharged capacitor and the needle stabilizes at approximately 50 kΩ. The capacitor is
    (a) good   (b) charged   (c) too large   (d) leaky

PROBLEM SET A

Section 9–1   The Capacitor

1. (a) Find the capacitance when $Q = 50~\mu C$ and $V = 10$ V.
    (b) Find the charge when $C = 0.001~\mu F$ and $V = 1$ kV.
    (c) Find the voltage when $Q = 2$ mC and $C = 200~\mu F$.

2. Convert the following values from microfarads to picofarads:
    (a) $0.1~\mu F$ (b) $0.0025~\mu F$ (c) $5~\mu F$

3. Convert the following values from picofarads to microfarads:
    (a) 1000 pF   (b) 3500 pF   (c) 250 pF

4. Convert the following values from farads to microfarads:
    (a) 0.0000001 F   (b) 0.0022 F   (c) 0.0000000015 F

5. What size capacitor is capable of storing 10 mJ of energy with 100 V across its plates?

6. A mica capacitor has a plate area of $0.04~m^2$ and a dielectric thickness of 0.008 m. What is its capacitance?

7. An air capacitor has plates with an area of $0.1~m^2$. The plates are separated by 0.01 m. Calculate the capacitance.

8. At ambient temperature (25°C), a certain capacitor is specified to be 1000 pF. It has a negative temperature coefficient of 200 ppm/°C. What is its capacitance at 75°C?

9. A 0.001-μF capacitor has a positive temperature coefficient of 500 ppm/°C. How much change in capacitance will a 25°C increase in temperature cause?

Section 9–2   Types of Capacitors

10. In the construction of a stacked-foil mica capacitor, how is the plate area increased?

11. What type of capacitor has the higher dielectric constant, mica or ceramic?

12. Show how to connect an electrolytic capacitor across points A and B in Figure 9–52.

FIGURE 9–52

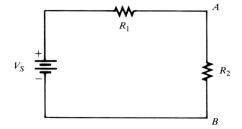

13. Determine the value of the typographically labeled ceramic disk capacitors in Figure 9–53.

FIGURE 9–53

(a)   (b)   (c)   (d)

14. Name two types of electrolytic capacitors. How do electrolytics differ from other capacitors?

15. Identify the parts of the ceramic disk capacitor shown in the cutaway view of Figure 9–54.

FIGURE 9–54

Section 9–3   Series Capacitors

16. Five 1000-pF capacitors are in series. What is the total capacitance?

17. Find the total capacitance for each circuit in Figure 9–55.

18. For each circuit in Figure 9–55, determine the voltage across each capacitor.

FIGURE 9–55

19. The total charge stored by the series capacitors in Figure 9–56 is 10 μC. Determine the voltage across each of the capacitors.

FIGURE 9–56

Section 9–4   Parallel Capacitors

20. Determine C_T for each circuit in Figure 9–57.

FIGURE 9–57

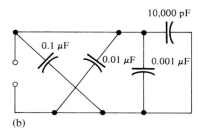

21. Determine C_T for each circuit in Figure 9–58.

22. What is the voltage between points A and B in each circuit in Figure 9–58?

(a)

(b)

(c) $C = 1\ \mu F$ for each capacitor

FIGURE 9–58

Section 9–5   Capacitors in dc Circuits

23. Determine the time constant for each of the following series RC combinations:
  (a) $R = 100\ \Omega$, $C = 1\ \mu F$ **(b)** $R = 10\ M\Omega$, $C = 50$ pF
  (c) $R = 4.7\ k\Omega$, $C = 0.005\ \mu F$ **(d)** $R = 1.5\ M\Omega$, $C = 0.01\ \mu F$

24. Determine how long it takes the capacitor to reach full charge for each of the following combinations:
  (a) $R = 50\ \Omega$, $C = 50\ \mu F$ **(b)** $R = 3300\ \Omega$, $C = 0.015\ \mu F$
  (c) $R = 22\ k\Omega$, $C = 100$ pF   **(d)** $R = 5\ M\Omega$, $C = 10$ pF

25. In the circuit of Figure 9–59, the capacitor initially is uncharged. Determine the capacitor voltage at the following times after the switch is closed:
  (a) 10 μs **(b)** 20 μs **(c)** 30 μs **(d)** 40 μs **(e)** 50 μs

26. In Figure 9–60, the capacitor is charged to 25 V. Find the capacitor voltage after the following times when the switch is closed:
  (a) 1.5 ms   **(b)** 4.5 ms   **(c)** 6 ms   **(d)** 7.5 ms

FIGURE 9–59   **FIGURE 9–60**

27. Repeat Problem 25 for the following time intervals:
  (a) 2 μs **(b)** 5 μs **(c)** 15 μs

28. Repeat Problem 26 for the following times:
  (a) 0.5 ms   **(b)** 1 ms   **(c)** 2 ms

Section 9–6   Capacitors in ac Circuits

29. Determine X_C for a 0.047-μF capacitor at each of the following frequencies:
   (a) 10 Hz   **(b)** 250 Hz   **(c)** 5 kHz   **(d)** 100 kHz

30. What is the value of the total capacitive reactance in each circuit in Figure 9–61?

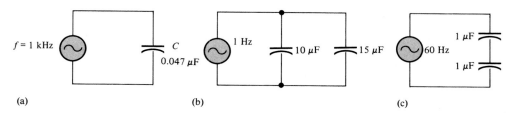

(a)   (b)   (c)

FIGURE 9–61

31. In Figure 9–58, each dc voltage source is replaced by a 10-V rms, 2-kHz ac source. Determine the reactance in each case.

32. In each circuit of Figure 9–61, what frequency is required to produce an X_C of 100 Ω? An X_C of 1 kΩ?

33. A sinusoidal voltage of 20 V rms produces an rms current of 100 mA when connected to a certain capacitor. What is the reactance?

34. A 10-kHz voltage is applied to a 0.0047-μF capacitor, and 1 mA of rms current is measured. What is the value of the voltage?

35. Determine the true power and the reactive power in Problem 34.

Section 9–7   Capacitor Applications

36. If another capacitor is connected in parallel with the existing capacitor in the power supply filter of Figure 9–47, how is the ripple voltage affected?

37. Ideally, what should the reactance of a bypass capacitor be in order to eliminate a 10-kHz ac voltage at a given point in an amplifier circuit?

Section 9–8   Testing Capacitors

38. Assume that you are checking a capacitor with an ohmmeter, and when you connect the leads across the capacitor, the pointer does not move from its left-end scale position. What is the problem?

39. In checking a capacitor with the ohmmeter, you find that the pointer goes all the way to the right end of the scale and stays there. What is the problem?

PROBLEM SET B

40. Two series capacitors (one 1-μF, the other of unknown value) are charged from a 12-V source. The 1-μF capacitor is charged to 8 V, and the other to 4 V. What is the value of the unknown capacitor?

41. How long does it take C to discharge to 3 V in Figure 9–60?

42. How long does it take C to charge to 8 V in Figure 9–59?

43. Determine the time constant for the circuit in Figure 9–62.

44. In Figure 9–63, the capacitor initially is uncharged. At $t = 10$ μs after the switch is closed, the instantaneous capacitor voltage is 7.2 V. Determine the value of R.

FIGURE 9–62

FIGURE 9–63

45. (a) The capacitor in Figure 9–64 is uncharged when the switch is thrown into position 1. The switch remains in position 1 for 10 ms and then is thrown into position 2, where it remains indefinitely. Sketch the complete waveform for the capacitor voltage.

(b) If the switch is thrown back to position 1 after 5 ms in position 2, and then is left in position 1, how will the waveform appear?

FIGURE 9–64

46. Determine the ac voltage across each capacitor and the current in each branch of the circuit in Figure 9–65.

FIGURE 9–65

47. Find the value of C_1 in Figure 9–66.

FIGURE 9–66

48. How much does the voltage across C_5 and C_6 change when the ganged switch is thrown from position 1 to position 2 in Figure 9–67?

49. If C_4 in Figure 9–65 opened, determine the voltages that would be measured across the other capacitors.

FIGURE 9–67

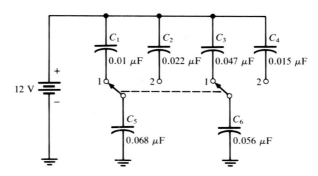

ANSWERS TO SECTION REVIEWS

Section 9–1

1. The ability (capacity) to store electrical charge.
2. (a) 10^6 (b) 10^{12} (c) 10^6 3. (a) increase; (b) decrease.

Section 9–2

1. By the dielectric material.
2. A fixed capacitance cannot be changed; a variable can.   3. Electrolytic.
3. Be sure that the voltage rating is sufficient. Connect the positive end to the positive side of the circuit.

Section 9–3

1. Less.   2. 62.5 pF.   3. 0.006 μF.   4. 6 V.

Section 9–4

1. The individual capacitors are added.
2. By using five 0.01-μF capacitors in parallel.   3. 98 pF.

Section 9–5

1. 1.2 μs.   2. 6 μs, 5 V.   3. 8.6 V, 9.5 V, 9.8 V, 9.9 V.   4. 37 V.

Section 9–6

1. Current leads voltage by 90°.   2. 637 kΩ.   3. 796 Hz.   4. 628 mA.
5. 0.   6. 0.453 VAR.

Section 9–7

1. Once the capacitor charges to the peak voltage, it discharges very little before the next peak.
2. By selecting a capacitor that has a reactance of almost zero at the frequency of the ac voltage and connecting that capacitor from that point to ground.

Section 9–8

1. Short its leads.
2. Initially, the needle jumps to zero; then it moves to the high-resistance end of the scale.

10 Inductors

In this chapter, the basic inductor is introduced and its characteristics are studied. Various types of inductors are covered in terms of their physical construction and their electrical properties. The basic behavior of inductors in both dc and ac circuits is studied, and series and parallel combinations are analyzed. A method of testing inductors is discussed.

After completing this chapter, you will be able to

☐ Describe the basic construction of an inductor.
☐ Define *inductance*.
☐ Explain how an inductor stores energy.
☐ Relate various physical parameters to inductance value.
☐ Explain why practical inductors have both resistance and capacitance.
☐ State Lenz's law and Faraday's law.
☐ Identify various types of inductors.
☐ Determine total series inductance.
☐ Determine total parallel inductance.
☐ Define *time constant* as related to an inductive circuit.
☐ Explain how an inductor causes a phase shift between current and voltage.
☐ Define inductive reactance and determine its value in a circuit.
☐ Explain the significance of reactive power in an inductive circuit.
☐ Describe some common inductor applications.
☐ Check out an inductor with an ohmmeter.

Inductance is the property of a coil of wire that opposes a change in current. The basis for inductance is the electromagnetic field that surrounds any conductor when there is current through it. The electrical component designed to have the property of inductance is called an *inductor, coil,* or *choke.* All of these terms refer to the same type of device.

APPLICATION ASSIGNMENT

A defective power supply is returned for repair. A quick check of the output voltage reveals that there is excessive ripple voltage. Your assignment is to check out both the capacitor and the inductor in the power supply filter to see if either or both are defective. After completing this chapter, you will be able to carry out this assignment.

10–1

THE INDUCTOR

When a length of wire is formed into a coil, as shown in Figure 10–1, it becomes a basic *inductor*. Current through the coil produces an electromagnetic field. The magnetic lines of force form a strong magnetic field within and around the coil, as shown. The net direction of the total magnetic field creates a north and a south pole, as indicated.

FIGURE 10–1
A coil of wire forms an inductor. When current flows through it, a three-dimensional electromagnetic field is created, surrounding the coil in all directions.

Self-Inductance

When there is current through an inductor, an electromagnetic field is established. When the current changes, the electromagnetic field also changes. An increase in current expands the field, and a decrease in current reduces it. Therefore, a changing current produces a changing electromagnetic field around the inductor (coil). In turn, the changing electromagnetic field induces a voltage across the coil in a direction to oppose the change in current. This property is called *self-inductance*, but is usually referred to as simply *inductance*. Inductance is symbolized by *L*.

> **Inductance is a measure of a coil's ability to establish an induced voltage as a result of a change in its current and that induced voltage is in a direction to oppose that change in current.**

$$V = -L \times \frac{\Delta I}{\Delta t}$$

The Unit of Inductance

The *henry*, symbolized by H, is the basic unit of inductance. By definition, the inductance is *one henry* when current through the coil, changing at the rate of *one ampere per second*, induces *one volt* across the coil. In many practical applications, *millihenries* (mH) and microhenries (μH) are the more common units. A common schematic symbol for the inductor is shown in Figure 10–2.

FIGURE 10–2
A symbol for the inductor.

L

Energy Storage

An inductor stores energy in the magnetic field created by the current. The energy stored is expressed as follows:

$$\text{energy} \qquad W = \frac{1}{2}LI^2 \qquad (10\text{–}1)$$

As you can see, the energy stored is proportional to the inductance and the square of the current. When I is in amperes and L is in henries, the energy is in joules.

Physical Characteristics

The following characteristics are important in establishing the inductance of a coil: the core material, the number of turns of wire, the length, and the cross-sectional area.

Core Material

As discussed earlier, an inductor is basically a coil of wire. The material around which the coil is formed is called the *core*. Coils are wound on either nonmagnetic or magnetic materials. Examples of nonmagnetic materials are air, wood, copper, plastic, and glass. The permeabilities of these materials are the same as for a vacuum. Examples of magnetic materials are iron, nickel, steel, cobalt, or alloys. These materials have permeabilities that are hundreds or thousands of times greater than that of a vacuum and are classified as *ferromagnetic*. A ferromagnetic core provides a better path for the magnetic lines of force and thus permits a stronger magnetic field.

As you learned in Chapter 7, the permeability (μ) of the core material determines how easily a magnetic field can be established. *The inductance is directly proportional to the permeability of the core material.*

Parameters

As indicated in Figure 10–3, the number of turns of wire, the length, and the cross-sectional area of the core are factors in setting the value of inductance. The inductance

FIGURE 10–3
Factors that determine the inductance of a coil.

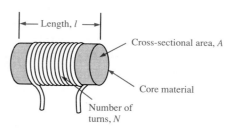

Length, l

Cross-sectional area, A

Core material

Number of turns, N

is inversely proportional to the length of the core and directly proportional to the cross-sectional area. Also, the inductance is directly related to the number of turns squared. This relationship is as follows:

$$L = \frac{N^2 \mu A}{l} \tag{10-2}$$

where L is the inductance in henries, N is the number of turns, μ is the permeability, A is the cross-sectional area in meters squared, and l is the core length in meters.

EXAMPLE 10-1

Determine the inductance of the coil in Figure 10–4. The permeability of the core is 0.25×10^{-3}.

FIGURE 10–4

0.01 m

0.1 m²

$N = 4$

Solution

$$L = \frac{N^2 \mu A}{l} = \frac{(4)^2 (0.25 \times 10^{-3})(0.1)}{0.01} = 40 \text{ mH}$$

The calculator sequence is

Winding Resistance

When a coil is made of a certain material, for example, insulated copper wire, that wire has a certain *resistance* per unit of length. When many turns of wire are used to construct a coil, the total resistance may be significant. This inherent resistance is called the *dc resistance* or the *winding resistance* (R_W). Although this resistance is distributed along the length of the wire, it effectively appears in series with the inductance of the coil, as shown in Figure 10–5. In many applications, the winding resistance can be ignored and the coil considered as an ideal inductor. In other cases, the resistance must be considered.

R_W L

(a) The wire has resistance.   (b) Equivalent circuit

FIGURE 10–5
Winding resistance of a coil.

Winding Capacitance

When two conductors are placed side by side, there is always some capacitance between them. Thus, when many turns of wire are placed close together in a coil, a certain amount of *stray* capacitance is a natural side effect. In many applications, this stray capacitance is very small and has no significant effect. In other cases, particularly at high frequencies, it may become quite important.

The equivalent circuit for an inductor with both its winding resistance (R_W) and its winding capacitance (C_W) is shown in Figure 10–6. The capacitance effectively acts in parallel.

(a) Stray capacitance between each loop appears as a total parallel capacitance.   (b) Equivalent circuit

FIGURE 10–6
Winding capacitance of a coil.

Review of Faraday's law

The principle of Faraday's law was introduced in Chapter 7 and is reviewed here because of its importance in the study of inductors. Faraday found that by moving a magnet through a coil of wire, a voltage was induced across the coil, and that when a complete path was provided, the induced voltage caused an induced current.

> **The amount of induced voltage is directly proportional to the rate of change of the magnetic field with respect to the coil.**

This principle is illustrated in Figure 10–7, where a bar magnet is moved through a coil of wire. An induced voltage is indicated by the voltmeter connected across the coil. The *faster* the magnet is moved, the *greater* is the induced voltage.

When a wire is formed into a certain number of loops or turns and is exposed to a *changing magnetic field*, a voltage is induced across the coil. The induced voltage is proportional to the *number of turns* of wire in the coil, N, and to the *rate* at which the magnetic field changes.

FIGURE 10–7
Induced voltage is created by a changing magnetic field.

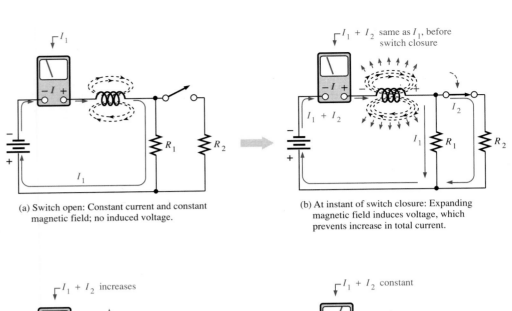

(a) Switch open: Constant current and constant magnetic field; no induced voltage.

(b) At instant of switch closure: Expanding magnetic field induces voltage, which prevents increase in total current.

(c) Right after switch closure: The rate of expansion of the magnetic field decreases, allowing the current to increase as induced voltage decreases.

(d) Switch remains closed: Current and magnetic field reach constant value.

(e) At instant of switch opening: Magnetic field begins to collapse, creating an induced voltage, which prevents decrease in current.

(f) After switch opening: Rate of collapse of magnetic field decreases, allowing current to decrease back to original value.

FIGURE 10–8

Demonstration of Lenz's law: When the current tries to change suddenly, the electromagnetic field changes and induces a voltage in a direction that opposes that change in current.

Lenz's Law

Lenz's law adds to Faraday's law by defining the direction of induced voltage as follows:

> **When the current through a coil changes and an induced voltage is created as a result of the changing magnetic field, the direction of the induced voltage is such that it always opposes the change in current.**

In Figure 10–8(a), the current is constant and is limited by R_1. There is no induced voltage because the magnetic field is unchanging. In part (b), the switch suddenly is closed, placing R_2 in parallel with R_1 and thus reducing the resistance. Naturally, the current tries to increase and the magnetic field begins to expand, but the induced voltage opposes this attempted increase in current for an instant.

In part (c), the induced voltage gradually decreases, allowing the current to increase. In part (d), the current has reached a constant value as determined by the parallel resistors, and the induced voltage is zero. In part (e), the switch has been suddenly opened, and, for an instant, the induced voltage prevents any decrease in current. In part (f), the induced voltage gradually decreases, allowing the current to decrease back to a value determined by R_1. Notice that the induced voltage has a polarity that opposes any current change. The polarity of the induced voltage is opposite that of the battery voltage for an increase in current and aids the battery voltage for a decrease in current.

**SECTION REVIEW
10–1**

1. Describe what happens to L when
   (a) N is increased
   (b) the core length is increased
   (c) the cross-sectional area of the core is decreased
   (d) a ferromagnetic core is replaced by an air core
2. Explain why inductors inevitably have some winding resistance.

10–2 TYPES OF INDUCTORS

Inductors are made in a variety of shapes and sizes. Basically, they fall into two general categories: *fixed* and *variable*. The standard schematic symbols are shown in Figure 10–9.

FIGURE 10–9
Symbols for (a) fixed and (b) variable inductors.

(a) Fixed   (b) Variable

Both fixed and variable inductors can be classified according to the type of core material. Three common types are the air core, the iron core, and the ferrite core. Each has a unique symbol, as shown in Figure 10–10.

FIGURE 10–10
Inductor symbols.

(a) Air core   (b) Iron core   (c) Ferrite core

Adjustable (variable) inductors usually have a screw-type adjustment that moves a sliding core in and out, thus changing the inductance. A wide variety of inductors exists, some of which are shown in Figure 10–11.

(a)   (b)   (c)

FIGURE 10–11
Typical inductors. (a) Fixed molded inductors; (b) variable coils; (c) toroid inductor [(a) and (b) courtesy of Delevan/American Precision. (c) courtesy of Dale Electronics].

SECTION REVIEW
10–2

1. Name two general categories of inductors.
2. Identify the inductor symbols in Figure 10–12.

FIGURE 10–12

(a)   (b)   (c)

10–3

SERIES INDUCTORS

When inductors are connected in series, as in Figure 10–13, *the total inductance, L_T, is the sum of the individual inductances*. The formula for L_T is expressed in the following equation for the general case of *n* inductors in series:

$$L_T = L_1 + L_2 + L_3 + \cdots + L_n \tag{10–3}$$

Notice that the formula for inductance in series is similar to the formula for resistance in series.

L_1 L_2 L_3 L_n

FIGURE 10–13
Inductors in series.

(handwritten annotations:)

$L_T = L_1 + L_2 \pm 2M$

$L_1 = 3H$
$L_2 = 5H$
$M = 2H$

aiding = $L_T = 3H + 5H + 2(2H)$
$= 12 H$

opposing = $L_T = L_1 + L_2 - 2(2H)$
$= 4 H$

= mutual Inductance M

EXAMPLE 10–2   Determine the total inductance for each of the series connections in Figure 10–14.

1 H   2 H   1.5 H   5 H   5 mH   2 mH   10 mH   1000 μH

(a)   (b)

FIGURE 10–14

Solution
(a) $L_T = 1\,H + 2\,H + 1.5\,H + 5\,H = 9.5\,H$
(b) $L_T = 5\,mH + 2\,mH + 10\,mH + 1\,mH = 18\,mH$
Note: 1000 μH = 1 mH

SECTION REVIEW   1. State the rule for combining inductors in series.
10–3   2. What is L_T for a series connection of 100 μH, 500 μH, and 2 mH?

10–4

PARALLEL INDUCTORS

When inductors are connected in parallel, as in Figure 10–15, *the total inductance is less than the smallest inductance.* The formula for total inductance in parallel is similar to that for total parallel resistance or total series capacitance:

$$\frac{1}{L_T} = \frac{1}{L_1} + \frac{1}{L_2} + \frac{1}{L_3} + \cdots + \frac{1}{L_n} \qquad (10–4)$$

FIGURE 10–15
Inductors in parallel.

This general formula states that the reciprocal of the total inductance is equal to the sum of the reciprocals of the individual inductances. L_T can be found by taking the reciprocals of both sides of Equation (10–4):

$$L_T = \frac{1}{(1/L_1) + (1/L_2) + (1/L_3) + \cdots + (1/L_n)} \qquad (10–5)$$

EXAMPLE 10–3

Determine L_T in Figure 10–16.

FIGURE 10–16

Solution

$$L_T = \frac{1}{(1/10 \text{ mH}) + (1/5 \text{ mH}) + (1/2 \text{ mH})}$$

$$= \frac{1}{0.8 \text{ mH}} = 1.25 \text{ mH}$$

The calculator sequence is

[1] [0] [EXP] [+/−] [3] [1/x] [+] [5] [EXP] [+/−] [3] [1/x] [+]
[2] [EXP] [+/−] [3] [1/x] [=] [1/x]

**SECTION REVIEW
10–4**

1. Compare the total inductance in parallel with the smallest-value individual inductor.
2. The calculation of total parallel inductance is similar to that for parallel resistance (T or F).
3. Determine L_T for each parallel combination:
   (a) 100 mH, 50 mH, and 10 mH
   (b) 40 μH and 60 μH
   (c) Ten 1-H coils

10–5

INDUCTORS IN DC CIRCUITS

When there is constant direct current in an inductor, there is no induced voltage. There is, however, a voltage drop due to the winding resistance of the coil. The inductance itself appears as a *short* to dc. Energy is stored in the magnetic field according to the

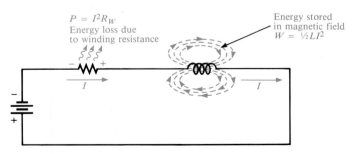

FIGURE 10–17
Energy storage and loss in an inductor. The only dc voltage drop across the coil is due to the winding resistance.

formula $W = \frac{1}{2}LI^2$. The only energy loss occurs in the winding resistance ($P = I^2R_W$). This condition is illustrated in Figure 10–17.

Time Constant

Because the inductor's basic action is to *oppose a change in its current*, it follows that *current cannot change instantaneously in an inductor*. A certain time is required for the current to make a change from one value to another. The rate at which the current changes is determined by the *time constant*. The time constant for a series *RL* circuit is

$$\tau = \frac{L}{R}$$ (10–6)

where τ is in seconds when L is in henries and R is in ohms.

EXAMPLE 10–4

A series *RL* circuit has a resistance of 1 kΩ and an inductance of 1 mH. What is the time constant?

Solution

$$\tau = \frac{L}{R} = \frac{1 \text{ mH}}{1 \text{ k}\Omega} = \frac{1 \times 10^{-3} \text{ H}}{1 \times 10^3 \ \Omega}$$
$$= 1 \times 10^{-6} \text{ s} = 1 \ \mu s$$

Energizing Current in an Inductor

In a series *RL* circuit, the current will increase to 63% of its full value in one time constant interval after the switch is closed. This buildup of current is analogous to the buildup of capacitor voltage during the charging in an *RC* circuit; they both follow an exponential curve and reach the approximate percentages of final value as indicated in Figure 10–18.

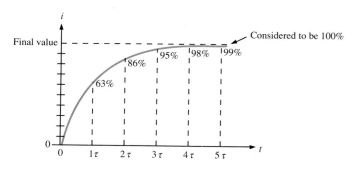

FIGURE 10–18
Energizing current in an inductor.

The change in current over five time constant intervals is illustrated in Figure 10–19. When the current reaches its final value at approximately 5τ, it ceases to change.

At this time, *the inductor acts as a short (except for winding resistance) to the constant current.* The final value of the current is $V_S/R_W = 10 \text{ V}/10 \text{ }\Omega = 1$ A.

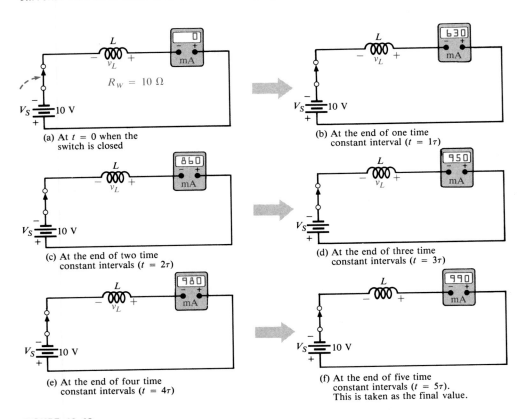

(a) At $t = 0$ when the switch is closed

(b) At the end of one time constant interval ($t = 1\tau$)

(c) At the end of two time constant intervals ($t = 2\tau$)

(d) At the end of three time constant intervals ($t = 3\tau$)

(e) At the end of four time constant intervals ($t = 4\tau$)

(f) At the end of five time constant intervals ($t = 5\tau$). This is taken as the final value.

FIGURE 10–19
Illustration of the exponential buildup of current in an inductor. The current increases another 63% during each time constant interval. A winding resistance of 10 Ω is assumed. A voltage (v_L) is induced in the coil that tends to oppose the increase in current.

EXAMPLE 10–5

Calculate the time constant for Figure 10–20. Then determine the current and the time at each time constant interval, measured from the instant the switch is closed.

FIGURE 10–20

Solution

$$I_{\text{final}} = \frac{V_S}{R} = \frac{20\ \text{V}}{100\ \Omega} = 0.2\ \text{A}$$

$$\tau = \frac{L}{R} = \frac{50\ \text{mH}}{100\ \Omega} = 0.5\ \text{ms}$$

At $1\tau = 0.5$ ms: $i = 0.63(0.2\ \text{A}) = 0.126\ \text{A}$
At $2\tau = 1$ ms: $i = 0.86(0.2\ \text{A}) = 0.172\ \text{A}$
At $3\tau = 1.5$ ms: $i = 0.95(0.2\ \text{A}) = 0.190\ \text{A}$
At $4\tau = 2$ ms: $i = 0.98(0.2\ \text{A}) = 0.196\ \text{A}$
At $5\tau = 2.5$ ms: $i = 0.99(0.2\ \text{A}) = 0.198\ \text{A}$
$$\cong 0.2\ \text{A}$$

Deenergizing Current in an Inductor

Current in an inductor decreases exponentially according to the approximate percentage values shown on the curve in Figure 10–21.

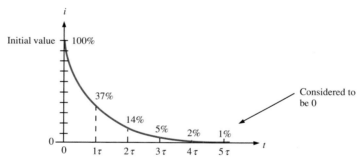

FIGURE 10–21
Deenergizing current in an inductor.

Figure 10–22(a) shows a constant current of 1 A (1000 mA) through an inductor. Switch 1 (SW1) is opened and switch 2 (SW2) is closed simultaneously, and for an instant the induced voltage keeps the 1 A flowing. During the first time constant interval, the current decreases by 63% down to 370 mA (37% of its initial value), as indicated in part (b). During the second time constant interval, the current decreases by another 63% to 140 mA (14% of its initial value), as shown in part (c). The continued decrease in the current is illustrated by the remaining parts of Figure 10–22. Part (f) shows that only 1% of the initial current is left at the end of five time constants. Traditionally, this value is accepted as the final value and is approximated as zero current. Notice that until after the five time constants have elapsed, there is an induced voltage across the coil which is trying to keep the current flowing. This voltage follows a decreasing exponential curve, as will be shown later.

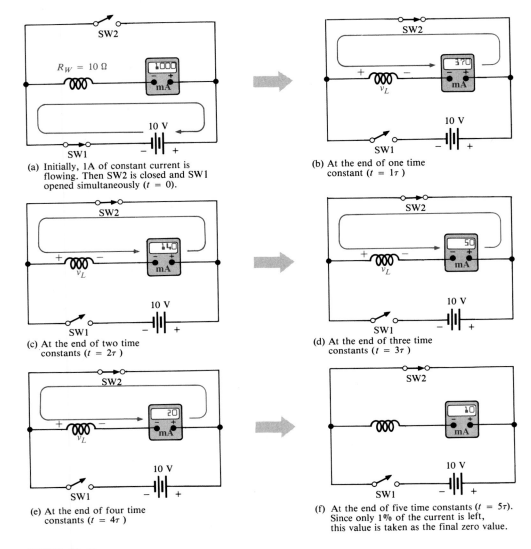

(a) Initially, 1A of constant current is flowing. Then SW2 is closed and SW1 opened simultaneously ($t = 0$).

(b) At the end of one time constant ($t = 1\tau$)

(c) At the end of two time constants ($t = 2\tau$)

(d) At the end of three time constants ($t = 3\tau$)

(e) At the end of four time constants ($t = 4\tau$)

(f) At the end of five time constants ($t = 5\tau$). Since only 1% of the current is left, this value is taken as the final zero value.

FIGURE 10–22

Illustration of the exponential decrease of current in an inductor. The current decreases another 63% during each time constant interval.

EXAMPLE 10–6

In Figure 10–23, switch 1 (SW1) is opened at the instant that switch 2 (SW2) is closed.
(a) What is the time constant?
(b) What is the initial coil current at the instant of switching?
(c) What is the coil current at 1τ?
Assume steady state current through the coil prior to switch change.

Solution
(a) $\tau = L/R = 200\ \mu H/10\ \Omega = 20\ \mu s$.
(b) Current cannot change instantaneously in an inductor. Therefore, the current at the instant of the switch change is the same as the steady state current:

FIGURE 10–23

$$I = \frac{5\ V}{10\ \Omega} = 0.5\ A$$

(c) At 1τ, the current has decreased to 37% of its initial value:

$$i = 0.37(0.5\ A) = 0.185\ A$$

Induced Voltage in the Series *RL* Circuit

As you know, when current changes in an inductor, a voltage is induced. We now examine what happens to the voltages across the resistor and the coil in a series circuit when a change in current occurs.

Look at the circuit in Figure 10–24(a). When the switch is open, there is no current, and the resistor voltage and the coil voltage are both zero. At the instant the switch is closed, as indicated in part (b), v_R is zero and v_L is 10 V. The reason for this change is that the induced voltage across the coil is equal and opposite to the applied voltage, preventing the current from changing instantaneously. Therefore, at the instant of switch closure, *L effectively acts as an open with all the applied voltage across it.*

(a) Before switch is closed.

(b) At instant switch is closed, v_L is equal and opposite to V_S.

FIGURE 10–24(a) and (b)
Illustration of how the coil voltage and the resistor voltage change in response to an increase in current.

(c) During the first five time constants, v_R increases exponentially with current, and v_L decreases exponentially.

(d) After the first five time constants, $v_R = 10$ V and $v_L \approx 0$ V. The current is at a constant maximum value.

FIGURE 10–24(c) and (d)

Illustration of how the coil voltage and the resistor voltage change in response to an increase in current.

(a) Initially, a constant 1 A flows and the voltage is dropped across R.

(b) At the instant that SW1 is opened and SW2 closed, 10 V is induced across L.

(c) During the five time constant interval, v_R and v_L decrease exponentially with the current.

(d) After five time constants, V_R, V_L, and i are all zero.

FIGURE 10–25

Illustration of how the coil voltage and the resistor voltage change in response to a decrease in current.

During the first five time constants, the current is building up exponentially, and the induced coil voltage is decreasing. The resistor voltage increases with the current, as Figure 10–24(c) illustrates.

After five time constants have elapsed, the current has reached its final value, V_S/R. At this time, all of the applied voltage is dropped across the resistor and none across the coil. Thus, L effectively acts as a *short* to nonchanging current, as Figure 10–24(d) illustrates. Keep in mind that the inductor always reacts to a change in current by creating an induced voltage in order to counteract that change.

Now let us examine the case illustrated in Figure 10–25, where the current is switched out, and the inductor discharges through another path. Part (a) shows the steady state condition, and part (b) illustrates the instant at which the source is removed by opening SW1 and the discharge path is connected by the closure of SW2. There was 1 A through L prior to this. Notice that 10 V are induced in L in the direction to aid the 1 A in an effort to keep it from changing. Then, as shown in part (c), the current decays exponentially, and so do v_R and v_L. After 5τ, as shown in part (d), all the energy stored in the magnetic field of L is dissipated, and all values are zero.

EXAMPLE 10–7

(a) In Figure 10–26(a), what is v_L at the instant the switch is closed? What is v_L after 5τ?

(b) In Figure 10–26(b), what is v_L at the instant SW1 opens and SW2 closes? What is v_L after 5τ?

FIGURE 10–26

Solution

(a) At the instant the switch is closed, all the source voltage is across L. Thus, $v_L = $ 25 V, with the polarity as shown. After 5τ, L acts as a short, so $v_L = 0$ V.

(b) With S_1 closed and S_2 open, the steady-state current is

$$I = \frac{25 \text{ V}}{12.5 \text{ } \Omega} = 2 \text{ A}$$

When the switches are thrown, an induced voltage is created across L sufficient to keep this 2-A current for an instant. In this case, it takes an induced voltage

$$v_L = IR_2 = (2 \text{ A})(100 \text{ } \Omega) = 200 \text{ V}$$

After 5τ, the inductor voltage is zero.

The Exponential Formulas

The formulas for the exponential current and voltage in an *RL* circuit are similar to those used in the last chapter for the *RC* circuit, and the universal exponential curves in Figure 9–32 apply to inductors as well as capacitors. The general formulas for *RL* circuits are stated as follows:

$$v = V_F + (V_i - V_F)e^{-Rt/L} \qquad (10\text{–}7)$$

$$i = I_F + (I_i - I_F)e^{-Rt/L} \qquad (10\text{–}8)$$

where V_F and I_F are the final values, V_i and I_i are the initial values, and v and i are the instantaneous values of the inductor voltage or current at time t.

Increasing Current

The formula for the special case in which an increasing exponential current curve begins at zero ($I_i = 0$) is

$$i = I_F(1 - e^{-Rt/L}) \qquad (10\text{–}9)$$

Using Equation (10–9), we can calculate the value of the increasing inductor current at any instant of time. The same is true for voltage.

EXAMPLE 10–8   In Figure 10–27, determine the inductor current 30 μs after the switch is closed.

FIGURE 10–27

Solution
The time constant is $L/R = 100$ mH/2 kΩ = 50 μs. The final current is $V_S/R = $ 12 V/2 kΩ = 6 mA. The initial current is zero. Notice that 30 μs is less than one time constant, so the current will reach less than 63% of its final value in that time:

$$i_L = I_F(1 - e^{-Rt/L}) = 6 \text{ mA}(1 - e^{-0.6})$$
$$= 6 \text{ mA}(1 - 0.549) = 2.71 \text{ mA}$$

The calculator sequence is

[30] [EXP] [+/−] [6] [÷] [50] [EXP] [+/−] [6] [=] [+/−] [2nd] [ln x] [+/−]
[+] [1] [=] [×] [6] [EXP] [+/−] [3] [=]

Decreasing Current

The formula for the special case in which a decreasing exponential current has a final value of zero is as follows:

$$i = I_i e^{-Rt/L}$$

(10–10)

This formula can be used to calculate the deenergizing current at any instant, as the following example shows.

EXAMPLE 10–9

Determine the inductor current in Figure 10–28 at a point in time 2 ms after the switches are thrown (SW1 opened and SW2 closed).

FIGURE 10–28

Solution

The deenergizing time constant is $L/R = 200$ mH/50 $\Omega = 4$ ms. The initial current in the inductor is 100 mA. Notice that 2 ms is less than one time constant, so the current will show a decrease less than 63%. Therefore, the current will be greater than 37% of its initial value at 2 ms after the switches are thrown:

$$i = I_i e^{-Rt/L} = (100 \text{ mA})e^{-0.5} = 60.65 \text{ mA}$$

The calculator sequence is

[2] [EXP] [+/−] [3] [÷] [4] [EXP] [+/−] [3] [=] [+/−] [2nd] [ln x]
[×] [1] [0] [0] [EXP] [+/−] [3] [=]

SECTION REVIEW 10–5

1. A 15-mH inductor with a winding resistance of 10 Ω has a constant direct current of 10 mA through it. What is the voltage drop across the inductor?

2. A 20-V dc source is connected to a series RL circuit with a switch. At the instant of switch closure, what are the values of v_R and v_L?

3. In the same circuit, after a time interval equal to 5τ from switch closure, what are v_R and v_L?

4. In a series RL circuit where $R = 1$ kΩ and $L = 500$ μH, what is the time constant? Determine the current 0.25 μs after a switch connects 10 V across the circuit.

10–6 INDUCTORS IN AC CIRCUITS

We will illustrate the relationship of current and inductor voltage in an ac circuit by connecting an inductor in series with a resistor to a sine wave voltage source and then observing the inductor voltage and the resistor voltage with an oscilloscope, as shown in Figure 10–29. Recall that the resistor voltage and the current are always in phase with each other, so if you are looking at the resistor voltage, you are also effectively looking at the current in terms of its phase relationship to the inductor voltage. In this

(a) Measuring V_L

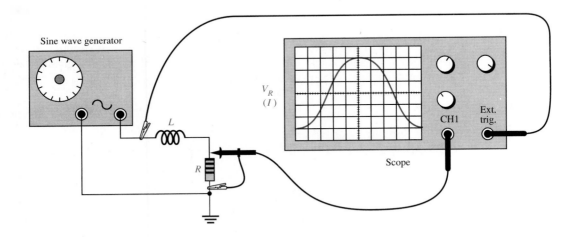

(b) Measuring V_R

FIGURE 10–29
Oscilloscope display showing that the current lags the inductor voltage by 90°. The current is in phase with the resistor voltage. Components are switched so that V_R and V_L can be measured relative to ground. The external trigger ensures that both waveforms are displayed relative to the same point.

case, it is assumed that the winding resistance is negligible. Notice that the frequency of the voltage and current are always the same in an ac circuit.

Why the Current Lags the Inductor Voltage by 90°

As you know, a sine wave voltage has a maximum rate of change at its zero crossings and a zero rate of change at the peaks. From Faraday's law (Chapter 7) you know that the amount of voltage induced across a coil is directly proportional to the rate at which the current is changing. Therefore, the coil voltage is maximum at the zero crossings of the current where the rate of change of the current is the greatest. Also, the amount of voltage is zero at the peaks of the current where its rate of change is zero. This relationship is illustrated in Figure 10–30. As you can see, the current peaks occur a quarter cycle after the voltage peaks. Thus, the current *lags* the voltage by 90°.

FIGURE 10–30
Current is always lagging the inductor voltage by 90°.

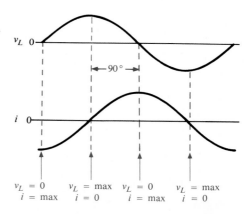

Inductive Reactance, X_L

In Figure 10–31, an inductor is connected to a sine wave source. Note that when the source voltage is held at a constant amplitude value and its frequency is increased, the

Fixed-amplitude ac voltage source

(a) Current decreases when the frequency increases.

(b) Current increases when the frequency decreases.

FIGURE 10–31
The current in an inductive circuit varies inversely with the frequency of the applied voltage.

amplitude of the current decreases. Also, when the frequency of the source is decreased, the current amplitude increases. The reason is as follows: When the frequency of the applied voltage increases, its rate of change also increases, as you already know. Now, if the frequency of the applied voltage is increased, the frequency of the current also increases. According to Faraday's and Lenz's laws, this increase in frequency induces more voltage across the inductor in a direction to oppose the current and cause it to decrease in amplitude. Similarly, a decrease in frequency will cause an increase in current.

A *decrease* in the amount of current for a fixed amount of voltage indicates that *opposition* to the current has *increased*. Thus, the inductor offers opposition to current which varies *directly* with frequency.

The opposition to sinusoidal current in an inductor is called inductive reactance.

The symbol for inductive reactance is X_L, and its unit is the ohm (Ω).

You have just seen how frequency affects the opposition to current (inductive reactance) in an inductor. Now let's see how the *inductance, L,* affects the reactance. Figure 10–32(a) shows that when a sine wave voltage with a fixed amplitude and frequency is applied to a 1-mH inductor, a certain amount of current flows. When the inductance value is increased to 2 mH, the current decreases, as shown in part (b). Thus, when the inductance increases, the opposition to current (inductive reactance) increases. So not only is the inductive reactance directly proportional to frequency, but it is also directly proportional to inductance:

$$X_L \text{ is proportional to } fL$$

(a) Less inductance, more current   (b) More inductance, less current

FIGURE 10–32
For a fixed voltage and frequency, the current varies inversely with inductance.

It can be proven that the constant of proportionality is 2π, so the formula for X_L is

$$X_L = 2\pi fL \tag{10–11}$$

As with capacitive reactance, the 2π term comes from the relationship of the sine wave to rotational motion.

EXAMPLE 10–10

A sinusoidal voltage is applied to the circuit in Figure 10–33. The frequency is 1 kHz. Determine the inductive reactance.

FIGURE 10–33

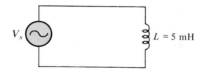

Solution

$$f = 1 \text{ kHz} = 1 \times 10^3 \text{ Hz}$$
$$L = 5 \text{ mH} = 5 \times 10^{-3} \text{ H}$$
$$X_L = 2\pi f L = 2\pi(1 \times 10^3 \text{ Hz})(5 \times 10^{-3} \text{ H}) = 31.4 \ \Omega$$

Ohm's Law in Inductive Circuits

The reactance of an inductor is analogous to the resistance of a resistor. In fact, X_L, just like X_C and R, is expressed in ohms. Since inductive reactance is a form of opposition to current, Ohm's law applies to inductive circuits as well as to resistive circuits and capacitive circuits, and it is stated as follows:

$$V = IX_L \qquad \qquad \textbf{(10–12)}$$

When applying Ohm's law in ac circuits, you must express both the current and the voltage in the same way, that is, both in rms, both in peak, and so on.

EXAMPLE 10–11

Determine the rms current in Figure 10–34.

FIGURE 10–34

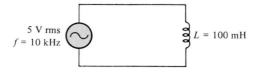

Solution

$$f = 10 \text{ kHz} = 10 \times 10^3 \text{ Hz}$$
$$L = 100 \text{ mH} = 100 \times 10^{-3} \text{ H}$$

First calculate X_L:

$$X_L = 2\pi f L = 2\pi(10 \times 10^3 \text{ Hz})(100 \times 10^{-3} \text{ H}) = 6283 \ \Omega$$

Using Ohm's law, we obtain

$$I_{rms} = \frac{V_{rms}}{X_L} = \frac{5 \text{ V}}{6283 \ \Omega} = 795.8 \ \mu A$$

Power in an Inductor

As discussed earlier, an inductor stores energy in its magnetic field when there is current through it. An ideal inductor (assuming no winding resistance) does not dissipate energy; it only stores it. When an ac voltage is applied to an inductor, energy is stored by the inductor during a portion of the cycle; then the stored energy is returned to the source during another portion of the cycle. *There is no net energy loss.* Figure 10–35 shows the power curve that results from one cycle of inductor current and voltage.

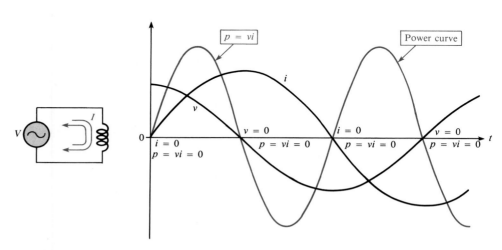

FIGURE 10–35
Power curve.

Instantaneous Power (*p*)

The product of instantaneous voltage, v, and instantaneous current, i, gives instantaneous power, p. At points where v or i is zero, p is also zero. When both v and i are positive, p is also positive. When either v or i is positive and the other negative, p is negative. When both v and i are negative, p is positive. As you can see in Figure 10–35, the power follows a sinusoidal-type curve. Positive values of power indicate that energy is stored by the inductor. Negative values of power indicate that energy is returned from the inductor to the source. Note that the power fluctuates at a frequency twice that of the voltage or current, as energy is alternately stored and returned to the source.

True Power (*P*_{true})

Ideally, all of the energy stored by an inductor during the positive portion of the power cycle is returned to the source during the negative portion. *No net energy is consumed in the inductance,* so the power is zero. Actually, because of winding resistance in a practical inductor, some power is always dissipated:

$$P_{\text{true}} = I_{\text{rms}}^2 R_W \tag{10–13}$$

Reactive Power (P_r)

The rate at which an inductor stores or returns energy is called its *reactive power*, P_r. The reactive power is a nonzero quantity, because at any instant in time, the inductor is actually taking energy from the source or returning energy to it. Reactive power does not represent an energy loss. The following formulas apply:

$$P_r = V_{rms}I_{rms} \qquad (10\text{--}14)$$

$$P_r = \frac{V_{rms}^2}{X_L} \qquad (10\text{--}15)$$

$$P_r = I_{rms}^2 X_L \qquad (10\text{--}16)$$

EXAMPLE 10–12

A 10-V rms signal with a frequency of 1 kHz is applied to a 10-mH coil with a winding resistance of 5 Ω. Determine the reactive power (P_r) and the true power (P_{true}).

Solution

$$X_L = 2\pi fL = 2\pi(1 \text{ kHz})(10 \text{ mH}) = 62.8 \ \Omega$$

$$I = \frac{V_S}{X_L} = \frac{10 \text{ V}}{63 \ \Omega} = 0.159 \text{ A}$$

$$P_r = I^2 X_L = (0.159 \text{ A})^2(62.8 \ \Omega) = 1.59 \text{ VAR}$$

$$P_{true} = I^2 R_W = (0.159 \text{ A})^2 \ 5 \ \Omega = 0.126 \text{ W}$$

SECTION REVIEW 10–6

1. State the phase relationship between current and voltage in an inductor.
2. Calculate X_L for $f = 5$ kHz and $L = 100$ mH.
3. At what frequency is the reactance of a 50-μH inductor equal to 800 Ω?
4. Calculate the rms current in Figure 10–36.
5. A 50-mH inductor is connected to a 12-V rms source. What is the true power? What is the reactive power at a frequency of 1 kHz?

FIGURE 10–36

10–7

INDUCTOR APPLICATIONS

Inductors are not as versatile as capacitors and tend to be more limited in their applications. However, there are many practical uses for inductors (coils), such as those discussed in Chapter 7. Recall that relay and solenoid coils, recording and pick-up heads, and sensing elements were introduced as electromagnetic applications of coils. In this section, some additional uses of inductors are presented.

Power Supply Filter

In Chapter 9, you saw that a capacitor is used to filter the pulsating dc on the output of the rectifier in a power supply. The final output voltage was a dc voltage with a small amount of *ripple*. In many cases, an inductor is used in the filter, as shown in Figure 10–37(a), to smooth out the ripple voltage. The inductor, placed in series with the load as shown, tends to oppose the current fluctuations caused by the ripple voltage, and thus the voltage developed across the load is more constant, as shown in Figure 10–37(b).

FIGURE 10–37
Basic capacitor power supply filter with a series inductor.

rf Choke

Certain types of inductors are used in applications where radio frequencies (rf) must be prevented from getting into parts of a system, such as the power supply or the audio section of a receiver. In these situations, an inductor is used as a series filter and "chokes" off any unwanted rf signals that may be picked up on a line. This filtering action is based on the fact that the reactance of a coil increases with frequency. When the frequency of the current is sufficiently high, the reactance of the coil becomes extremely large and essentially blocks the current. A basic illustration of an inductor used as an rf choke is shown in Figure 10–38.

FIGURE 10–38
An inductor used as an rf choke to minimize interfering signals on the power supply line.

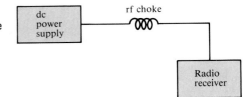

Tuned Circuits

As mentioned in Chapter 9, inductors are used in conjunction with capacitors to provide *frequency selection* in communications systems. These *tuned circuits* allow a narrow band of frequencies to be selected while all other frequencies are rejected. The tuners in your TV and radio receivers are based on this principle and permit you to select one channel or station out of the many that are available.

Frequency selectivity is based on the fact that the reactances of both capacitors and inductors depend on the frequency and on the interaction of these two components when connected in series or parallel. Since the capacitor and the inductor produce opposite phase shifts, their combined opposition to current can be used to obtain a desired response at a selected frequency. Tuned *LC* circuits are covered in detail in a later chapter.

SECTION REVIEW
10–7

1. Explain how the ripple voltage from a power supply filter can be reduced through use of an inductor.

2. How does an inductor connected in series act as an rf choke?

10–8

TESTING INDUCTORS

The most common failure in an inductor is an open coil. To check for an open, remove the coil from the circuit. If there is an open, an ohmmeter check will indicate infinite resistance, as shown in Figure 10–39(a). If the coil is good, the ohmmeter will show the winding resistance. The value of the winding resistance depends on the wire size and length of the coil. It can be anywhere from one ohm to several hundred ohms. Part (b) shows a good reading.

Occasionally, when an inductor is overheated with excessive current, the wire insulation will melt, and some coils will short together. This produces a reduction in the inductance by reducing the effective number of turns and a corresponding reduction in winding resistance, as shown in part (c).

(a) Open, reads ∞.

(b) Good, reads R_W.

(c) Shorted windings, reads low R_W or near zero.

FIGURE 10–39
Checking a coil by measuring the resistance.

SECTION REVIEW
10–8

1. When a coil is checked, a reading of infinity on the ohmmeter indicates a partial short (T or F).

2. An ohmmeter check of a good coil will indicate the value of the inductance (T or F).

SUMMARY

☐ An inductor passes dc and opposes ac.

☐ An inductor is also called a *coil* and, in some applications, a *choke*.

☐ An inductor consists of a coil of wire.

☐ Energy is stored in an inductor by the electromagnetic field created when current flows through it.

☐ Inductance is measured in units of henries (H).

☐ The inductance of a coil is directly proportional to the number of turns squared and to the permeability and cross-sectional area of the core, and it is inversely proportional to the length of the core.

☐ The total inductance of series inductors is the sum of all the inductances.

☐ The total inductance of parallel inductors is less than the smallest inductance.

☐ In an *RL* circuit, the voltage and current in an energizing or deenergizing inductor make a 63% change during each time constant interval.

☐ Current in an inductor lags the voltage by 90°.

☐ Inductive reactance X_L is directly proportional to frequency and to inductance.

☐ Ideally, there is no energy loss in an inductor, and thus the true power is zero. However, in all inductors there is some energy loss due to the winding resistance.

GLOSSARY

Choke In inductor. The term is more commonly used in connection with inductors used to block or choke off high frequencies.

Coil A common term for an inductor.

Core The structure around which the winding of an inductor is formed. The core material influences the electromagnetic characteristics of the inductor.

Henry (H) The unit of inductance.

Induced voltage Voltage produced as a result of a changing magnetic field.

Inductance The property of an inductor that produces an opposition to any change in current.

Inductor An electrical device formed by a wire wound around a core having the property of inductance. Also known as a coil or a choke.

Inductive reactance The opposition of an inductor to sinusoidal current. The unit is the ohm.

Lenz's law A physical law that states when the current through a coil changes, an induced voltage is created in a direction to oppose the change in current. The current cannot change instantaneously.

Time constant A fixed time interval, set by the *L* and *R* values, that determines the time response of a circuit.

Winding The loops of wire in an inductor.

Reactive power The rate at which energy is alternately stored and returned to the source by an inductor.

FORMULAS

$$W = \frac{1}{2}LI^2$$ Energy stored by an inductor   (10–1)

$$L = \frac{N^2 \mu A}{l}$$ Inductance in terms of physical parameters   (10–2)

$$L_T = L_1 + L_2 + L_3 + \cdots + L_n \qquad \text{Series inductance} \tag{10-3}$$

$$\frac{1}{L_T} = \frac{1}{L_1} + \frac{1}{L_2} + \frac{1}{L_3} + \cdots + \frac{1}{L_n} \qquad \text{Reciprocal of total parallel inductance} \tag{10-4}$$

$$L_T = \frac{1}{(1/L_1) + (1/L_2) + (1/L_3) + \cdots + (1/L_n)} \qquad \text{Total parallel inductance} \tag{10-5}$$

$$\tau = \frac{L}{R} \qquad \text{Time constant} \tag{10-6}$$

$$v = V_F + (V_i - V_F)e^{-Rt/L} \qquad \text{Exponential voltage (general)} \qquad \times \tag{10-7}$$

$$i = I_F + (I_i - I_F)e^{-Rt/L} \qquad \text{Exponential current (general)} \qquad \times \tag{10-8}$$

$$i = I_F(1 - e^{-Rt/L}) \qquad \text{Increasing exponential current beginning at zero} \qquad \times \tag{10-9}$$

$$i = I_i e^{-Rt/L} \qquad \text{Decreasing exponential current ending at zero} \qquad \lor \tag{10-10}$$

$$X_L = 2\pi f L \qquad \text{Inductive reactance} \tag{10-11}$$

$$V = I X_L \qquad \text{Ohm's law} \tag{10-12}$$

$$P_{\text{true}} = I_{\text{rms}}^2 R_W \qquad \text{True power} \tag{10-13}$$

$$P_r = V_{\text{rms}} I_{\text{rms}} \qquad \text{Reactive power} \tag{10-14}$$

$$P_r = \frac{V_{\text{rms}}^2}{X_L} \qquad \text{Reactive power} \tag{10-15}$$

$$P_r = I_{\text{rms}}^2 X_L \qquad \text{Reactive power} \tag{10-16}$$

SELF-TEST

1. An inductance of 0.05 μH is larger than
   (a) 0.0000005 H   (b) 0.000005 H   (c) 0.000000008 H
   (d) 0.00005 mH

2. An inductance of 0.33 mH is smaller than
   (a) 33 μH   (b) 330 μH   (c) 0.05 mH   (d) 0.0005 H

3. When the current through an inductor increases, the amount of energy stored in the electromagnetic field
   (a) decreases   (b) remains constant   (c) increases   (d) doubles

4. When the current through an inductor doubles, the stored energy
   (a) doubles   (b) quadruples   (c) is halved   (d) does not change

5. The winding resistance of a coil can be decreased by
   (a) reducing the number of turns
   (b) using a larger wire
   (c) changing the core material
   (d) a or b

6. The inductance of an iron-core coil increases if
    (a) the number of turns is increased
    (b) the iron core is removed
    (c) the length of the core is increased
    (d) larger wire is used

7. Four 10-mH inductors are in series. The total inductance is
    (a) 40 mH   (b) 2.5 mH   (c) 40,000 μH   (d) a and c

8. A 1-mH, a 3.3-mH, and a 0.1-mH inductor are connected in parallel. The total inductance is
    (a) 4.4 mH   (b) greater than 3.3 mH   (c) less than 0.1 mH   (d) a and b

9. An inductor, a resistor, and a switch are connected in series to a 12-V battery. At the instant the switch is closed, the inductor voltage is
    (a) 0 V   (b) 12 V   (c) 6 V   (d) 4 V

10. A sine wave voltage is applied across an inductor. When the frequency of the voltage is increased, the current
    (a) decreases   (b) increases   (c) does not change
    (d) momentarily goes to zero

11. An inductor and a resistor are in series with a sine wave voltage source. The frequency is set so that the inductive reactance is equal to the resistance. If the frequency is increased, then
    (a) $V_R > V_L$ (b) $V_L < V_R$ (c) $V_L = V_R$ (d) $V_L > V_R$

12. An ohmmeter is connected across an inductor and the pointer indicates an infinite value. The inductor is
    (a) good   (b) open   (c) shorted   (d) resistive

PROBLEM SET A   Section 10–1   **The Inductor**

1. Convert the following to millihenries:
    (a) 1 H   (b) 250 μH   (c) 10 μH   (d) 0.0005 H

2. Convert the following to microhenries:
    (a) 300 mH   (b) 0.08 H   (c) 5 mH   (d) 0.00045 mH

3. How many turns are required to produce 30 mH with a coil wound on a cylindrical core having a cross-sectional area of 10×10^{-5} m^2 and a length of 0.05 m? The core has a permeability of 1.2×10^{-6}.

4. A 12-V battery is connected across a coil with a winding resistance of 12 Ω. How much current is there in the coil?

5. How much energy is stored by a 100-mH inductor with a current of 1 A?

6. The current through a 100-mH coil is changing at a rate of 200 mA/s. How much voltage is induced across the coil?

Section 10–3   **Series Inductors**

7. Five inductors are connected in series. The lowest value is 5 μH. If the value of each inductor is twice that of the preceding one, and if the inductors are connected in order of ascending values, what is the total inductance?

8. Suppose that you require a total inductance of 50 mH. You have available a 10-mH coil and a 22-mH coil. How much additional inductance do you need?

Section 10–4   Parallel Inductors

9. Determine the total parallel inductance for the following coils in parallel: 75 μH, 50 μH, 25 μH, and 15 μH.

10. You have a 12-mH inductor, and it is your smallest value. You need an inductance of 8 mH. What value can you use in parallel with the 12-mH to obtain 8 mH?

11. Determine the total inductance of each circuit in Figure 10–40.

12. Determine the total inductance of each circuit in Figure 10–41.

(a)   (b)   (c)

FIGURE 10–40

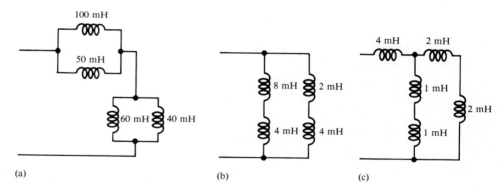

(a)   (b)   (c)

FIGURE 10–41

Section 10–5   Inductors in dc Circuits

13. Determine the time constant for each of the following series RL combinations:
    (a) $R = 100\ \Omega$, $L = 100\ \mu$H   (b) $R = 4.7\ k\Omega$, $L = 10$ mH
    (c) $R = 1.5\ M\Omega$, $L = 3$ H

14. In a series RL circuit, determine how long it takes the current to build up to its full value for each of the following:
    (a) $R = 50\ \Omega$, $L = 50\ \mu$H   (b) $R = 3300\ \Omega$, $L = 15$ mH
    (c) $R = 22\ k\Omega$, $L = 100$ mH

15. In the circuit of Figure 10–42, there is initially no current. Determine the inductor voltage at the following times after the switch is closed:
    (a) 10 μs   (b) 20 μs   (c) 30 μs   (d) 40 μs   (e) 50 μs

FIGURE 10–42

16. In Figure 10–43, there are 100 mA through the coil. When SW1 is opened and SW2 simultaneously closed, find the inductor voltage at the following times:
   (a) initially   (b) 1.5 ms   (c) 4.5 ms   (d) 6 ms

FIGURE 10–43

Section 10–6   Inductors in ac Circuits

17. Find the total reactance for each circuit in Figure 10–40 when a voltage with a frequency of 5 kHz is applied across the terminals.

18. Find the total reactance for each circuit in Figure 10–41 when a 400-Hz voltage is applied.

19. Determine the total rms current in Figure 10–44. What are the currents through L_2 and L_3?

FIGURE 10–44

20. What frequency will produce 500-mA total rms current in each circuit of Figure 10–41 with an rms input voltage of 10 V?

21. Determine the reactive power in Figure 10–44 neglecting the winding resistance.

Section 10–8   Testing Inductors

22. A certain coil that is supposed to have a 5-Ω winding resistance is measured with an ohmmeter. The meter indicates 2.8 Ω. What is the problem with the coil?

23. Determine the meter indication corresponding to each of the following failures in a coil:
   (a) open   (b) shorted   (c) partially shorted

PROBLEM SET B

24. Determine the time constant for the circuit in Figure 10–45.

25. Find the inductor current 10 μs after the switch is thrown from position 1 to position 2 in Figure 10–46. For simplicity, assume that the switch makes contact with position 2 at the same instant it breaks contact with position 1.

26. In Figure 10–47, SW1 is opened and SW2 is closed at the same instant (t_0). What is the instantaneous voltage across R_2 at t_0?

FIGURE 10–45

FIGURE 10–46

FIGURE 10–47

27. Determine the value of I_{L2} in Figure 10–48.

28. What is the total inductance between points A and B for each switch position in Figure 10–49?

FIGURE 10–48

FIGURE 10–49

ANSWERS TO
SECTION REVIEWS

Section 10–1
1. (a) L increases; (b) L decreases; (c) L decreases; (d) L decreases.
2. All wire has some resistance, and since inductors are made from turns of wire, there is always resistance.

Section 10–2
1. Fixed and variable. 2. (a) Air core; (b) iron core; (c) variable.

Section 10–3
1. Inductances are added in series. 2. 2600 μH.

Section 10–4
1. The total parallel inductance is smaller than that of the smallest individual inductor in parallel.
2. T.
3. (a) 7.69 mH; (b) 24 μH; (c) 0.1 H.

Section 10–5
1. 0.1 V. 2. $v_R = 0$ V, $v_L = 20$ V. 3. $v_R = 20$ V, $v_L = 0$ V.
4. 0.5 μs, 3.93 mA.

Section 10–6
1. Voltage leads current by 90°. 2. 3.14 kΩ. 3. 2.55 MHz. 4. 15.9 mA.
5. 0, 458.4 mVAR.

Section 10–7
1. The inductor tends to level out the ripple because of its opposition to changes in current.
2. The inductive reactance is very high at radio frequencies and blocks these frequencies.

Section 10–8
1. F. 2. F.

AC Color Insert: Actual Instruments and Circuits for Selected Problems.

The crucial control settings are indicated where it is difficult to take the reading directly from the instrument itself.

FIGURE 8-84
For problems 42, 43, and 44.

Oscilloscope Setting:
VOLTS/DIV = 2 V
TIME/DIV = 20 μs

FIGURE 12-83
For Problem 46.

Oscilloscope Setting:
VOLTS/DIV = 1 V
TIME/DIV = 1 μs

Frequency = 100 kHz

Generator Setting:
SINE WAVE

0.1μF

Closeup

Closeup

FIGURE 13-62
For Problem 41.

Oscilloscope Setting:
VOLTS/DIV = 0.1 V
TIME/DIV = 20 µs

Frequency = 10 kHz

Generator Setting:
SINE WAVE

FIGURE 15-67
For Problem 26.

Oscilloscope Setting:
VOLTS/DIV = 2 V
TIME/DIV = 2 μs

Frequency = 100 kHz

Generator Setting:
SQUARE WAVE

Closeup

11

Transformers

In the last chapter, self-inductance was studied. In this chapter, another inductive effect, called *mutual inductance,* is introduced. Mutual inductance occurs when two or more coils are very close together. It is based on the principle of magnetic coupling. A basic transformer is formed by two coils that are magnetically coupled but have no electrical connection between them. The transfer of energy from one coil to the other occurs entirely through a changing magnetic field. There are many applications for transformers, and a basic understanding of their operation is essential.

After completing this chapter, you will be able to

Explain *mutual inductance* and its importance in transformer operation.
Discuss the significance of the *turns ratio.*
Explain how a basic transformer works.
Relate physical characteristics such as core construction and turns ratio to the performance of a transformer.
Illustrate how the direction of windings determine the phase relationships of the transformer voltages.
☐ Show how transformers can be used to increase or decrease ac voltage.
☐ Define the terms *primary* and *secondary.*
☐ Describe the reflected load concept in transformers.
☐ State the maximum power transfer theorem.
☐ Use a transformer to match a source to a load.
☐ Use a transformer as an isolation device.
☐ Explain how an actual transformer differs from the ideal model.
☐ Describe tapped transformers, multiple-winding transformers, and autotransformers.
☐ Discuss several transformer applications.
☐ Troubleshoot a transformer for common failures.

APPLICATION ASSIGNMENT

You are in charge of modifying the sound system for a sports arena. The audio amplifier has a 50-Ω internal resistance (output resistance) and must drive four 8-Ω speakers (in parallel). You have been requested to get the maximum power possible delivered to the speakers.

After completing this chapter, you will be able to carry out this assignment.

11–1

THE BASIC TRANSFORMER

When two coils are placed close together and a sinusoidal current is applied to one coil, the changing magnetic field produced by the current causes an induced voltage in the other coil, in accordance with Faraday's law (Chapter 7). This situation basically describes how a *transformer* operates. Thus, two coils that are *magnetically linked* to each other make up a transformer. There is no *electrical* connection between the coils; therefore, they are *electrically isolated*.

Mutual Inductance

Recall that the electromagnetic field surrounding a coil expands, collapses, and reverses as the current increases, decreases, and reverses.

When a second coil is placed very close to the first coil so that the changing magnetic flux lines cut through the second coil, a voltage is induced, as indicated in Figure 11–1. If the current in the first coil is a sine wave, the voltage induced in the second coil is also a sine wave. The amount of voltage induced in the second coil as a result of the current in the first coil is dependent on the *mutual inductance, L_M,* between the two coils. *The mutual inductance is established by the inductance of each coil (L_1 and L_2) and by the amount of coupling (k) between the two coils.*

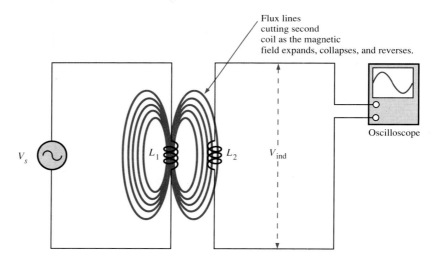

FIGURE 11–1
A voltage is induced in the second coil as a result of the changing current, which produces a changing magnetic field in the first coil.

Coefficient of Coupling

The coefficient of coupling k between two coils is the ratio of the flux lines produced by coil 1 that link coil 2 (ϕ_{12}) to the total flux produced by coil 1 (ϕ_1):

$$k = \frac{\phi_{12}}{\phi_1}$$

(11–1)

For example, if half of the total flux produced by coil 1 links coil 2, then $k = 0.5$. A greater value of k means that more voltage is induced in coil 2 for a certain rate of change of current in coil 1. Note that k *has no units. Recall that the unit of magnetic lines of force (flux) is the weber, abbreviated Wb.*

The coefficient k depends on the physical closeness of the coils and the type of core material on which they are wound. Also, the contruction and shape of the cores are factors.

Formula for Mutual Inductance

The three factors influencing L_M (k, L_1, and L_2) are shown in Figure 11–2. The formula for L_M is

$$L_M = k\sqrt{L_1 L_2} \qquad (11\text{--}2)$$

FIGURE 11–2
The mutual inductance of two coils.

EXAMPLE 11–1

One coil produces a total magnetic flux of 50 μWb, and 20 μWb link coil 2. What is k?

Solution

$$k = \frac{20\ \mu\text{Wb}}{50\ \mu\text{Wb}} = 0.4$$

EXAMPLE 11–2

Two coils are wound on a single core, and the coefficient of coupling is 0.3. The inductance of coil 1 is 10 μH, and the inductance of coil 2 is 15 μH. What is L_M?

Solution

$$L_M = k\sqrt{L_1 L_2} = 0.3\sqrt{(10\ \mu\text{H})(15\ \mu\text{H})}$$
$$= 3.67\ \mu\text{H}$$

Basic Transformer Construction

A schematic diagram of a transformer is shown in Figure 11–3(a). One coil is called the *primary* winding and one the *secondary* winding as indicated. In standard operation, the source voltage is applied to the primary, and a load is connected to the secondary, as shown in part (b). The windings of a transformer are formed around the *core*. The purpose of the core is to provide both a physical structure for placement of the windings and a magnetic path so that the magnetic flux lines are concentrated close to the coils.

FIGURE 11-3
The basic transformer.

(a) Schematic symbol   (b) Source/load connections

There are three general categories of core material: air, ferrite, and iron. Schematic symbols for each type are shown in Figure 11–4.

(a) Air-core   (b) Ferrite-core   (c) Iron-core

FIGURE 11-4
Schematic symbols based on type of core.

Air-core and ferrite-core transformers generally are used for high-frequency applications and consist of windings on an insulating shell which is hollow (air) or constructed of ferrite, such as depicted in Figure 11–5. The wire is typically covered by a varnish-type coating to prevent the windings from shorting together. The amount of magnetic coupling between the primary and the secondary is set by the type of core material and by the relative positions of the windings. In part (a) of the figure, the windings are loosely coupled because they are separated, and in part (b) they are tightly coupled because they are overlapping. The tighter the coupling, the greater the induced voltage in the secondary for a given current in the primary.

Air or ferrite core

(a) Loosely coupled windings   (b) Tightly coupled windings. Cutaway view shows both windings.

FIGURE 11-5
Transformers with cylindrical-shaped cores.

Iron-core transformers generally are used for audio frequency (af) and power applications. These transformers consist of windings on a core constructed from laminated sheets of ferromagnetic material insulated from each other, as shown in Figure 11–6(a). This construction provides an easy path for the magnetic flux and increases the amount of coupling between the windings. Figures 11–6(b) and 11–6(c) show the basic construction of two major configurations of iron-core transformer. In the core-type construction, shown in part (b), the windings are on separate legs of the laminated core. In the shell-type construction, shown in part (c), both windings are on the same leg.

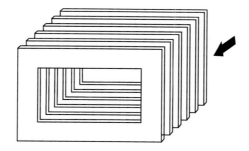

(a) In laminated iron-core construction, the layers are insulated from each other.

(b) Core-type has each winding on a separate leg.

(c) Shell-type has both windings on the same leg.

FIGURE 11–6
Iron-core transformer construction with multilayer windings.

FIGURE 11–7
Some common types of transformers (courtesy of Litton Triad-Utrad).

A variety of transformers are shown in Figure 11–7.

Turns Ratio

An important parameter of a transformer is its *turns ratio*. In this book, *the turns ratio (n) is defined as the ratio of the number of turns in the secondary winding (N_s) to the number of turns in the primary winding (N_p):*

$$n = \frac{N_s}{N_p} \tag{11–3}$$

In the following sections, you will see how the turns ratio affects the voltages and currents in a transformer. You may find that some books define the turns ratio as N_P/N_s.

EXAMPLE 11–3

A certain transformer used in a terrain-following radar system has a primary winding with 100 turns and a secondary winding with 400 turns. What is the turns ratio?

Solution

$$N_s = 400 \quad \text{and} \quad N_p = 100$$
$$n = \frac{N_s}{N_p} = \frac{400}{100} = \frac{4}{1} = 4$$

A turns ratio of 4 is usually expressed as $1:4$ on a schematic.

Direction of Windings

Another important transformer parameter is the direction in which the windings are placed around the core. As illustrated in Figure 11–8, the direction of the windings determines the polarity of the secondary voltage with respect to the primary voltage. *Phase dots* are used on the schematic symbols to indicate polarities, as shown in Figure 11–9. If the primary is positive at the dot, the secondary is positive at the dot. If the primary is negative at the dot, the secondary is negative at the dot.

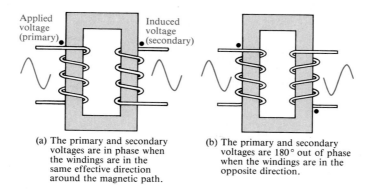

Applied voltage (primary)   Induced voltage (secondary)

(a) The primary and secondary voltages are in phase when the windings are in the same effective direction around the magnetic path.

(b) The primary and secondary voltages are 180° out of phase when the windings are in the opposite direction.

FIGURE 11–8
Relative polarities of the voltages are determined by the direction of the windings.

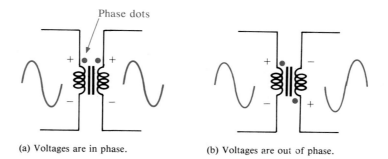

Phase dots

(a) Voltages are in phase.

(b) Voltages are out of phase.

FIGURE 11–9
Phase dots indicate corresponding polarities of primary and secondary voltages.

**SECTION REVIEW
11–1**

1. Define *mutual inductance*.
2. Two 50-mH coils have $k = 0.9$. What is L_M?
3. If k is increased, what happens to the voltage induced in one coil as a result of a current change in the other coil?
4. Define *turns ratio*.
5. Why are the directions of the windings of a transformer important?

11–2   STEP-UP TRANSFORMERS

A transformer in which the secondary voltage is greater than the primary voltage is called a step-up transformer. The amount that the voltage is stepped up depends on the turns ratio.

For any transformer, the ratio of secondary voltage to primary voltage is equal to the ratio of the number of secondary turns to the number of primary turns:

$$\frac{V_s}{V_p} = \frac{N_s}{N_p}$$

where V_s is the secondary voltage and V_p is the primary voltage.

From this relationship, we get

$$V_s = \left(\frac{N_s}{N_p}\right)V_p \qquad \text{(11–4)}$$

This equation shows that the secondary voltage is equal to the turns ratio times the primary voltage. This condition assumes that *all* the magnetic flux lines produced in the primary pass through the secondary.

The turns ratio for a step-up transformer is always *greater* than 1.

EXAMPLE 11–4

The transformer in Figure 11–10 has a 200-turn primary winding and a 600-turn secondary winding. What is the voltage across the secondary?

FIGURE 11–10

Solution

$$N_p = 200 \quad \text{and} \quad N_s = 600$$

$$V_s = \left(\frac{N_s}{N_p}\right)V_p = \left(\frac{600}{200}\right)V_p = 3V_p = 3(120 \text{ V}) = 360 \text{ V}$$

Note that the turns ratio of 3 is indicated on the schematic as 1:3, meaning that there are 3 secondary turns for each primary turn.

The calculator sequence is

⑥ ⓪ ⓪ ÷ ② ⓪ ⓪ × ① ② ⓪ =

SECTION REVIEW 11–2

1. What does a step-up transformer do?
2. If the turns ratio is 5, how much greater is the secondary voltage than the primary voltage?
3. When 240 V ac are applied to a transformer with a turns ratio of 10, what is the secondary voltage?

11-3

STEP-DOWN TRANSFORMERS

A transformer in which the secondary voltage is less than the primary voltage is called a step-down transformer. The amount by which the voltage is stepped down depends on the turns ratio. Equation (11–4) also applies to a step-down transformer; the turns ratio, however, is always *less* than 1.

EXAMPLE 11-5

The transformer in Figure 11–11 is part of a laboratory power supply and has 50 turns in the primary and 10 turns in the secondary. What is the secondary voltage?

FIGURE 11–11

120 V rms   5:1

Solution

$$N_p = 50 \quad \text{and} \quad N_s = 10$$

$$\text{Turns ratio} = \frac{N_s}{N_p} = \frac{10}{50} = 0.2$$

$$V_s = \left(\frac{N_s}{N_p}\right) V_p$$

$$V_s = 0.2 \, V_p = 0.2(120 \text{ V}) = 24 \text{ V}$$

**SECTION REVIEW
11-3**

1. What does a step-down transformer do?
2. One hundred twenty volts are applied to the primary of a transformer with a turns ratio of 0.5. What is the secondary voltage?
3. A primary voltage of 120 V is reduced to 12 V. What is the turns ratio?

11-4

LOADING THE SECONDARY

When a load resistor (R_L) is connected to the secondary, as shown in Figure 11–12, secondary current will flow because of the voltage induced in the secondary coil. It can be shown that the ratio of the primary current I_p to the secondary current I_s is equal to the turns ratio, as expressed in the following equation:

$$\frac{I_p}{I_s} = \frac{N_s}{N_p}$$

Inverting both sides of the equation gives

$$\frac{I_s}{I_p} = \frac{N_p}{N_s}$$

AC CIRCUITS

(a) Step-up transformer: $V_s > V_p$ and $I_s < I_p$

(b) Step-down transformer: $V_s < V_p$ and $I_s > I_p$

FIGURE 11–12
Illustration of voltages and currents in transformers with loaded secondaries.

Solving for I_s, we have

$$I_s = \left(\frac{N_p}{N_s}\right)I_p \qquad (11\text{--}5)$$

Note that N_p/N_s is the *reciprocal* of the turns ratio. Thus, as shown in Figure 11–12(a), for a step-up transformer, in which N_s/N_p is greater than 1, the secondary current is less than the primary current. For a step-down transformer, shown in part (b), N_s/N_p is less than 1, and I_s is greater than I_p.

EXAMPLE 11–6

The transformers in Figures 11–13(a) and 11–13(b) have loaded secondaries. If the primary current is 100 mA in each case, how much current flows through the load?

FIGURE 11–13

Solution

(a) $I_s = \left(\frac{N_p}{N_s}\right)I_p = 0.1(100 \text{ mA}) = 10 \text{ mA}$

(b) $I_s = \left(\frac{N_p}{N_s}\right)I_p = 2(100 \text{ mA}) = 200 \text{ mA}$

Primary Power Equals Secondary Power

When a load is connected to the secondary of a transformer, the power transferred to the load can never be greater than the power that is put into the primary. For an ideal transformer, the power in the secondary equals the power in the primary. When losses are considered, the secondary power is always less.

Power is dependent on voltage and current, and there can be no increase in power. Therefore, if the voltage is stepped up, the current is stepped down and vice versa. In an ideal transformer, the secondary power is the same as the primary power regardless of the turns ratio, as shown in the following steps:

$$P_p = V_p I_p \quad \text{and} \quad P_s = V_s I_s$$

$$I_s = \left(\frac{N_p}{N_s}\right) I_p \quad \text{and} \quad V_s = \left(\frac{N_s}{N_p}\right) V_p$$

By substitution, we obtain

$$P_s = \left(\frac{\cancel{N_p}}{\cancel{N_s}}\right)\left(\frac{\cancel{N_s}}{\cancel{N_p}}\right) V_p I_p$$

Canceling yields

$$P_s = V_p I_p = P_p$$

This result is closely approached in practice because of the very high efficiencies of transformers.

**SECTION REVIEW
11–4**

1. If the turns ratio of a transformer is 2, is the secondary current greater than or less than the primary current? By how much?
2. A transformer has 100 primary turns and 25 secondary turns, and I_p is 0.5 A. What is the value of I_s?

11–5

REFLECTED LOAD IN A TRANSFORMER

From the viewpoint of the primary, a resistive load connected across the secondary of a transformer appears to have a value that is not necessarily equal to the actual resistance value of the load, but rather depends on the turns ratio. This load that is *reflected* into the primary is what the source effectively sees, and it determines the amount of primary current. The concept of the *reflected load* is illustrated in Figure 11–14.

From the relationships of the currents, voltages, and the turns ratio in a transformer, the equation for the reflected load resistance can be derived. The result is given in the following formula:

$$R_p = \left(\frac{1}{n}\right)^2 R_L \tag{11–6}$$

where $1/n$ is the reciprocal of the turns ratio (N_p/N_s).

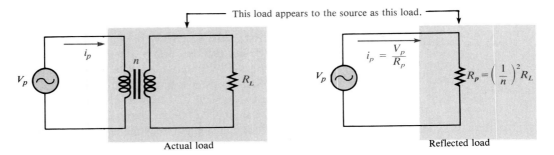

FIGURE 11–14
The load in the secondary of a transformer is reflected into the primary by transformer action. It appears to the source to be a resistance (R_p) with a value determined by the reciprocal of the turns ratio squared times the value of the load resistance.

EXAMPLE 11–7

Figure 11–15 shows a source that is transformer-coupled to a load resistor of 100 Ω. The transformer has a turns ratio of 4. What is the reflected resistance seen by the source?

FIGURE 11–15

Solution
The reflected resistance is determined as follows:

$$R_p = \left(\frac{1}{4}\right)^2 R_L = \left(\frac{1}{16}\right)(100 \ \Omega) = 6.25 \ \Omega$$

The source sees a resistance of 6.25 Ω just as if it were connected directly, as shown in the equivalent circuit of Figure 11–16.

FIGURE 11–16

Resistance "reflected"
from secondary

$R_p = 6.25 \ \Omega$

The calculator sequence is

$\boxed{4}\ \boxed{1/x}\ \boxed{x^2}\ \boxed{\times}\ \boxed{1}\ \boxed{0}\ \boxed{0}\ \boxed{=}$

EXAMPLE 11–8

In Figure 11–15, if a transformer is used having 40 primary turns and 10 secondary turns, what is the reflected resistance?

Solution

$$\text{Turns ratio} = 0.25$$

$$R_p = \left(\frac{1}{0.25}\right)^2 (100\ \Omega) = (4)^2(100\ \Omega) = 1600\ \Omega$$

Examples 11–7 and 11–8 illustrate that in a step-up transformer ($n > 1$), the reflected resistance is less than the actual load resistance; and in a step-down transformer ($n < 1$), the reflected resistance is greater than the load resistance.

**SECTION REVIEW
11–5**

1. Define *reflected resistance*.
2. What transformer characteristic determines the reflected resistance?
3. A given transformer has a turns ratio of 10, and the load is 50 Ω. How much resistance is reflected into the primary?
4. What is the terms ratio required to reflect a 4-Ω load resistance into the primary as 400 Ω?

11–6

MATCHING THE LOAD RESISTANCE TO THE SOURCE RESISTANCE

One application of transformers is in *impedance matching* a load to a source in order to achieve maximum transfer of power from the source to the load. The term *impedance* will become very familiar to you in the next chapter. Basically, impedance is a general term for the opposition to current, including the effects of both resistance and reactance combined. However, for the time being, we will confine our usage to resistance only.

The concept of impedance matching is illustrated in the basic circuit of Figure 11–17. Part (a) shows an ac voltage source with a series resistance representing its internal resistance. Some internal resistance is inherent in all sources due to their internal circuitry or physical makeup. When the source is connected directly to a load, as shown in part (b), often the objective is to transfer as much of the power produced by the source to the load as possible. However, a certain amount of the power produced by the source is lost in its internal resistance, and the remaining power goes to the load.

(a) Voltage source with internal resistance, R_{int}

(b) A portion of the total power is lost in R_{int}.

FIGURE 11–17
Power transfer from a nonideal voltage source to a load.

Maximum Power Transfer Theorem

When a source is connected to a load, maximum power is delivered to the load when the load resistance is equal to the source resistance.

We will demonstrate this theorem by finding the load power for various values of load resistance and a specific source resistance in Example 11–9.

EXAMPLE 11–9

The source in Figure 11–18 has an internal resistance of 75 Ω. Determine the power in each of the following values of load resistor:

(a) 25 Ω
(b) 50 Ω
(c) 75 Ω
(d) 100 Ω
(e) 125 Ω

Draw a graph showing the load power versus the load resistance.

FIGURE 11–18

Solution

We will use Ohm's law ($I = V/R$) and the power formula ($P = I^2R$) to find the load power for each value of load resistance.

(a) $R_L = 25\ \Omega$:

$$I = \frac{V_s}{R_{int} + R_L} = \frac{10\ V}{100\ \Omega} = 0.1\ A$$

$$P_L = I^2R_L = (0.1\ A)^2(25\ \Omega) = 0.25\ W$$

(b) $R_L = 50\ \Omega$:

$$I = \frac{V_s}{R_{int} + R_L} = \frac{10\ V}{125\ \Omega} = 0.08\ A$$

$$P_L = I^2R_L = (0.08\ A)^2(50\ \Omega) = 0.32\ W$$

(c) $R_L = 75\ \Omega$:

$$I = \frac{V_s}{R_{int} + R_L} = \frac{10\ V}{150\ \Omega} = 0.067\ A$$

$$P_L = I^2R_L = (0.067)^2(75\ \Omega) = 0.337\ W$$

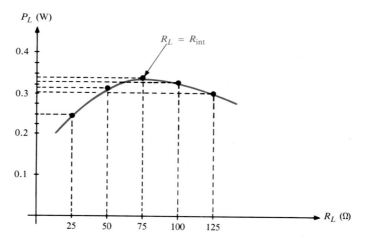

FIGURE 11–19
Curve showing that the load power is maximum when $R_L = R_{int}$.

(d) $R_L = 100 \ \Omega$:

$$I = \frac{V_s}{R_{int} + R_L} = \frac{10 \text{ V}}{175 \ \Omega} = 0.057 \text{ A}$$

$$P_L = I^2 R_L = (0.057 \text{ A})^2 (100 \ \Omega) = 0.33 \text{ W}$$

(e) $R_L = 125 \ \Omega$:

$$I = \frac{V_s}{R_{int} + R_L} = \frac{10 \text{ V}}{200 \ \Omega} = 0.05 \text{ A}$$

$$P_L = I^2 R_L = (0.05 \text{ A})^2 (125 \ \Omega) = 0.31 \text{ W}$$

Notice that the load power is greatest when $R_L = 75 \ \Omega$, which is the same as the internal source resistance. When the load resistance is less than or greater than this value, the power drops off, as the curve in Figure 11–19 graphically illustrates.

The calculator sequence for (a) is

[1] [0] [÷] [(] [7] [5] [+] [2] [5] [)] [=] [x^2] [×] [2] [5] [=]

A Practical Application

In most practical situations, the internal source resistance of various types of sources is fixed. Also, in many cases, the resistance of a device that acts as a load is fixed and cannot be altered. If you need to connect a given source to a given load, remember that only by chance will their resistances match. In this situation a transformer comes in handy. You can use the reflected-resistance characteristic of a transformer to make the load resistance appear to have the same value as the source resistance, thereby "fooling" the source into "thinking" that there is a match.

Let's take a practical, everyday situation to illustrate. The typical resistance of the input to a TV receiver is 300 Ω. An antenna must be connected to this input by a lead-

FIGURE 11–20

An antenna directly coupled
to a TV receiver.

(a) The antenna/lead-in is the source; the TV input is the load.

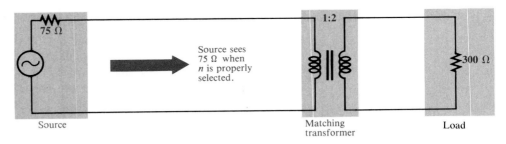

(b) Circuit equivalent of antenna and TV receiver system

in cable in order to receive TV signals. In this situation, the antenna and the lead-in act as the source, and the input resistance of the TV receiver is the load, as illustrated in Figure 11–20.

It is common for an antenna system to have a *characteristic resistance* of 75 Ω. Thus, if the 75-Ω source (antenna and lead-in) is connected directly to the 300-Ω TV input, maximum power will not be delivered to the input to the TV, and you will have poor signal reception. The solution is to use a *matching transformer* connected as indicated in Figure 11–21, in order to match the 300-Ω load resistance to the 75-Ω source resistance.

FIGURE 11–21

Example of a load matched to a source by transformer coupling for maximum power transfer.

To match the resistances, we must select a proper value of turns ratio (n). We want the 300-Ω load to look like 75 Ω to the source. We solve equation (11–6) for n, using 300 Ω for R_L and 75 Ω for R_p:

$$R_p = \left(\frac{1}{n}\right)^2 R_L$$

Transposing terms, we get

$$\left(\frac{1}{n}\right)^2 = \frac{R_p}{R_L}$$

Taking the square root of both sides, we have

$$\frac{1}{n} = \sqrt{\frac{R_p}{R_L}}$$

Inverting both sides and solving for n gives

$$n = \sqrt{\frac{R_L}{R_p}} \tag{11-7}$$

$$n = \sqrt{\frac{300\ \Omega}{75\ \Omega}} = \sqrt{4} = 2$$

Therefore, a matching transformer with a turns ratio of 2 must be used in this application.

EXAMPLE 11–10

An amplifier has an 800-Ω internal resistance. In order to provide maximum power to an 8-Ω speaker, what turns ratio must be used in the coupling transformer?

Solution
The reflected resistance must equal 800 Ω. Thus,

$$n = \sqrt{\frac{R_L}{R_p}} = \sqrt{\frac{8\ \Omega}{800\ \Omega}} = \sqrt{0.01} = 0.1$$

There must be ten primary turns for each secondary turn. The diagram and its equivalent reflected circuit are shown in Figure 11–22.

FIGURE 11–22

The calculator sequence is

8 ÷ 8 0 0 = 2nd √x̄

1. What does *impedance matching* mean?
2. What is the advantage of matching the load impedance to the impedance of a source?
3. A transformer has 100 primary turns and 50 secondary turns. What is the reflected impedance with 100 Ω across the secondary?

√11–7   THE TRANSFORMER AS AN ISOLATION DEVICE

Transformers are useful in providing *electrical isolation* between the primary and the secondary, since there is no electrical connection between the two windings. As you know, energy is transferred entirely by magnetic coupling.

dc Isolation

If a nonchanging direct current flows through the primary of a transformer, nothing happens in the secondary, as indicated in Figure 11–23(a). The reason is that a changing primary current is necessary to create a changing magnetic field, which will cause voltage to be induced in the secondary, as indicated in part (b) of the figure. Therefore, the transformer *isolates* the secondary from any dc in the primary.

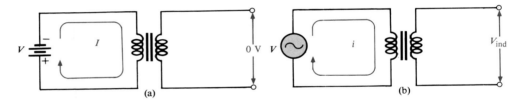

(a)   (b)

FIGURE 11–23
dc isolation and ac coupling.

For example, in some amplifiers, transformers are used to prevent dc voltages at certain points in the circuit, while allowing the signal (ac) voltage to get through. Figure 11–24 is a simplified illustration of transformer coupling in a two-stage audio amplifier.

FIGURE 11–24
Audio amplifier stages with transformer coupling for dc isolation.

Power Line Isolation

Transformers are often used to electrically isolate electronic equipment from the 60-Hz, 120-V ac power line. The reason for using an isolation transformer to couple the 60-Hz ac to an instrument is to prevent a possible shock hazard if the 120-V line is connected to the metal chassis of the equipment. This condition is possible if the line cord plug can be inserted into an outlet either way. Incidentally, to prevent this situation, many plugs have keyed prongs so that they can be plugged in only one way.

Figure 11–25 illustrates how a transformer can prevent the metal chassis from being connected to the 120-V line rather than to neutral (ground), no matter how the cord is plugged into the outlet. When an isolation transformer is used, the secondary is said to be *floating* because it is not referenced to the power line ground. Should a person come in contact with the secondary voltage, there is no complete current path back to ground, and therefore there is no shock hazard. Current must flow through your body in order for you to receive an electrical shock.

Some TV sets, for example, do not have isolation transformers for reasons of economy. When working on such a unit, be careful to use an external isolation transformer or be sure that the outlet connection does not put the chassis at 120 V.

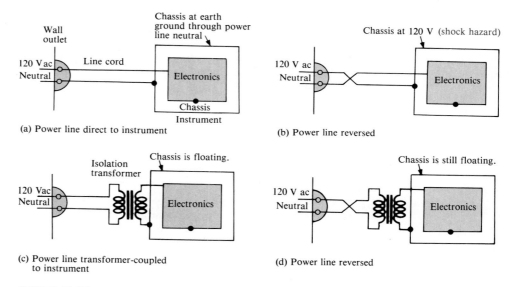

(a) Power line direct to instrument

(b) Power line reversed

(c) Power line transformer-coupled to instrument

(d) Power line reversed

FIGURE 11–25
The use of an isolation transformer to prevent a shock hazard.

**SECTION REVIEW
11–7**

1. What does the term *electrical isolation* mean?
2. Can a dc voltage be coupled by a transformer?

11–8

NONIDEAL TRANSFORMER CHARACTERISTICS

Up to this point, transformer operation has been presented from an ideal point of view, and this approach is valid when you are learning new concepts. However, you should be aware of the nonideal characteristics of practical transformers and how they affect performance.

Winding Resistance

Both the primary and the secondary windings of a transformer have winding resistance. (You learned about winding resistance in Chapter 10 on inductors.) The winding resistances of a transformer are represented as resistors in series with the windings as shown in Figure 11–26.

FIGURE 11–26
Winding resistance in a nonideal transformer.

Winding resistance in a transformer results in less voltage across a secondary load. Voltage drops due to the winding resistance effectively subtract from the primary and secondary voltages and result in a load voltage that is less than that predicted by the relationship $V_s = (N_2/N_1)V_p$. In most cases the effect is relatively small and is not considered.

Losses in the Core

There is always some energy loss in the core material of a nonideal transformer. This loss is seen as a heating of ferrite and iron cores, but it does not occur in air cores. Part of this energy is consumed in the continuous reversal of the magnetic field due to the changing direction of the primary current; this energy loss is called *hysteresis loss*. The rest of the energy loss is caused by *eddy currents* induced in the core material by the changing magnetic flux. The eddy-currents flow in circular patterns in the core resistance, thus causing the energy loss. This loss is greatly reduced by the use of laminated construction of iron cores. The thin layers of ferromagnetic material are insulated from each other to minimize the buildup of eddy currents by confining them to a small area and to keep core losses to a minimum.

Magnetic Flux Leakage

In an ideal transformer, all the magnetic flux produced by the primary current is assumed to pass through the core to the secondary winding and vice versa. In an actual transformer, some of the magnetic flux lines break out of the core and pass through the surrounding air back to the other end of the winding, as illustrated in Figure 11–27 for the magnetic field produced by the primary current. Magnetic flux leakage results in a reduced secondary voltage.

The percentage of magnetic flux that actually reaches the secondary determines the *coefficient of coupling* of the transformer. For example, if nine out of ten flux lines remain inside the core, the coefficient of coupling is 0.90 or 90%. Most iron-core transformers have very high coefficients of coupling (greater than 0.99), whereas ferrite-core and air-core devices have lower values.

FIGURE 11–27
Flux leakage in a nonideal transformer.

Winding Capacitance

As you learned in Chapter 10, there is always some stray capacitance between adjacent turns of a winding. These stray capacitances result in an effective capacitance in parallel with each winding of a transformer, as indicated in Figure 11–28.

FIGURE 11–28
Winding capacitance in a nonideal transformer.

These stray capacitances have very little effect on the transformer's operation at low frequencies, because the reactances (X_C) are very high. However, at higher frequencies, the reactances decrease and begin to produce a bypassing effect across the primary winding and across the secondary load. As a result, less of the total primary current flows through the primary winding, and less of the total secondary current flows through the load. This effect reduces the load voltage as the frequency goes up.

Transformer Power Rating

A transformer is typically rated in volt-amperes (VA), primary/secondary voltage, and operating frequency. For example, a given transformer rating may be specified as 2 kVA, 500/50, 60 Hz. The 2-kVA value is the *apparent power rating*. The 500 and the 50 can be either secondary or primary voltages.

Let's assume, for example, that 50 V is the secondary voltage. In this case the load current is $I_L = P_s/V_s = 2$ kVA/50 V $= 40$ A. On the other hand, if 500 V is the secondary voltage, then $I_L = P_s/V_s = 2$ kVA/500 V $= 4$ A. These are the *maximum* currents that the secondary can handle in either case.

The reason that the power rating is in volt-amperes (VA) rather than in watts (true power) is as follows: If the transformer load is purely capacitive or purely inductive, the true power (watts) delivered to the load is ideally zero. However, the current for $V_s = 500$ V and $X_C = 100$ Ω at 60 Hz, for example, is 5 A. This current exceeds the maximum of 4 A that the 2-kVA secondary can handle, and even though the true power is zero, the transformer may be damaged. So it is meaningless to specify power in watts.

Transformer Efficiency

Recall that the secondary power is equal to the primary power in an ideal transformer. Because the nonideal characteristics just discussed result in a power loss in the transformer, the secondary (output) power is always less than the primary (input) power. The *efficiency* (η) of a transformer is a measure of the percentage of the input power that is delivered to the output:

$$\eta = \left(\frac{P_{out}}{P_{in}}\right)100\% \qquad (11\text{–}8)$$

Most power transformers have efficiencies in excess of 95%.

EXAMPLE 11–11

A certain type of transformer has a primary current of 5 A and a primary voltage of 4800 V. The secondary current is 90 A and the secondary voltage is 240 V. Determine the efficiency of this transformer.

Solution
The input power is

$$P_{in} = V_p I_p = (4800 \text{ V})(5 \text{ A}) = 24 \text{ kVA}$$

The output power is

$$P_{out} = V_s I_s = (240 \text{ V})(90 \text{ A}) = 21.6 \text{ kVA}$$

The efficiency is

$$\eta = \left(\frac{P_{out}}{P_{in}}\right)100\% = \left(\frac{21.6 \text{ kVA}}{24 \text{ kVA}}\right)100\% = 90\%$$

SECTION REVIEW 11–8

1. Explain how an actual transformer differs from the ideal model.
2. The coefficient of coupling of a certain transformer is 0.85. What does this mean?
3. A certain transformer has a rating of 10 kVA. If the secondary voltage is 250 V, how much load current can the transformer handle?

11–9   OTHER TYPES OF TRANSFORMERS

There are several important variations of the basic transformer that you have studied so far. They include

  □ tapped transformers,
  □ multiple-winding transformers, and
  □ autotransformers.

Tapped Transformers

A schematic diagram of a transformer with a *center-tapped* secondary winding is shown in Figure 11–29(a). The center tap (CT) is equivalent to two secondary windings with half the total voltage across each.

(a) Center-tapped transformer

(b) Output voltages with respect to the center tap are 180° out of phase with each other and are one-half the magnitude of the secondary voltage.

FIGURE 11–29
Operation of a center-tapped transformer.

The voltages between either end of the secondary and the center tap are, at any instant, equal in magnitude but opposite in polarity, as illustrated in Figure 11–29(b). Here, for example, at some instant on the sine wave voltage, the polarity across the entire secondary is as shown (top end +, bottom −). At the center tap, the voltage is less positive than the top end but more positive than the bottom end of the secondary. Therefore, measured *with respect to the center tap,* the top end of the secondary is positive, and the bottom end is negative. This center-tapped feature is used in some power supply rectifiers in which the ac voltage is converted to dc, as illustrated in Figure 11–30.

FIGURE 11–30
Application of a center-tapped transformer in ac-to-dc conversion.

Some tapped transformers have taps on the secondary winding at points other than the electrical center. Also, multiple primary and secondary taps are sometimes used in certain applications. Examples of these types of transformers are shown in Figure 11–31.

(a)   (b)   (c)

FIGURE 11–31
Tapped transformers.

One example of a transformer with a multiple-tap primary and a center-tapped secondary is the utility-pole transformer used by power companies to step down the high voltage from the power line to 120 V/240 V service for residential and commercial customers, as shown in Figure 11–32. The multiple taps on the primary are used for minor adjustments in the turns ratio in order to overcome line voltages that are slightly too high or too low.

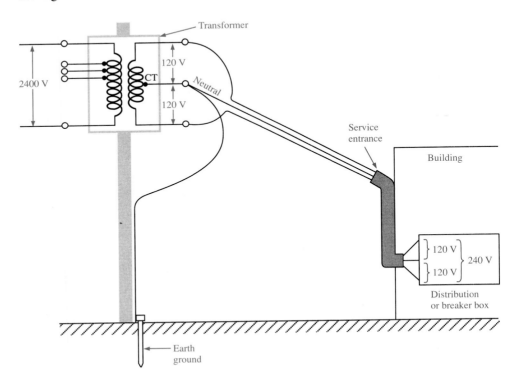

FIGURE 11–32
Utility-pole transformer in a typical power distribution system.

Multiple-Winding Transformers

Some transformers are designed to operate from either 120-V ac or 240-V ac lines. These transformers usually have two primary windings, each of which is designed for 120 V ac. When the two are connected in series, the transformer can be used for 240-V ac operation, as illustrated in Figure 11–33.

(a) Two primaries

(b) Primaries in parallel for 120-V ac operation

(c) Primaries in series for 240-V ac operation

FIGURE 11–33
Multiple-primary transformer.

FIGURE 11–34
Multiple-secondary transformer.

More than one secondary can be wound on a common core. Transformers with several secondaries are often used to achieve several voltages by either stepping up or stepping down the primary voltage. These types are commonly used in power supply applications in which several voltage levels are required for the operation of an electronic instrument.

A typical schematic of a multiple-secondary transformer is shown in Figure 11–34; this transformer has three secondaries. Sometimes you will find combinations of multiple-primary, multiple-secondary, and tapped transformers all in one unit.

EXAMPLE 11–12

The transformer shown in Figure 11–35 has the numbers of turns (denoted as T) indicated. One of the secondaries is also center-tapped. If 120 V ac are connected to the primary, determine each secondary voltage and the voltages with respect to CT on the middle secondary.

FIGURE 11–35

Solution

$$V_{AB} = \left(\frac{5}{100}\right) 120 \text{ V} = 6 \text{ V}$$

$$V_{CD} = \left(\frac{200}{100}\right) 120 \text{ V} = 240 \text{ V}$$

$$V_{CTC} = V_{CTD} = \frac{240 \text{ V}}{2} = 120 \text{ V}$$

$$V_{EF} = \left(\frac{10}{100}\right) 120 \text{ V} = 12 \text{ V}$$

Autotransformers

In an autotransformer, *one winding* serves as both the primary and the secondary. The winding is tapped at the proper points to achieve the desired turns ratio for stepping up or stepping down the voltage.

Autotransformers differ from conventional transformers in that there is no electrical isolation between the primary and the secondary because both are on one winding. Autotransformers normally are smaller and lighter than equivalent conventional transformers because they require a much lower kVA rating for a given load. Many autotransformers provide an adjustable tap using a sliding contact mechanism so that the output voltage can be varied (these are often called *variacs*). Figure 11–36 shows a typical autotransformer and several schematic symbols.

The following example illustrates why an autotransformer has a kVA requirement that is less than the input or output kVA.

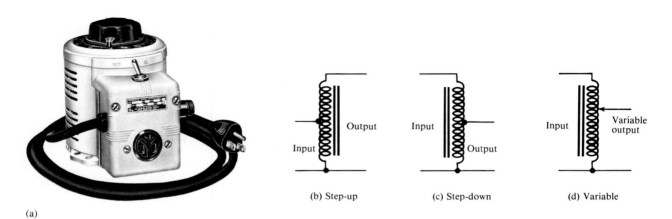

(a)

(b) Step-up   (c) Step-down   (d) Variable

FIGURE 11–36
The autotransformer (courtesy of Superior Electric Co.).

EXAMPLE 11–13

A certain autotransformer is used to change a source voltage of 240 V to a load voltage of 160 V across an 8-Ω load resistance. Determine the input and output power in kVA, and show that the actual kVA requirement is less than this value. Assume that this transformer is ideal.

Solution

The circuit is shown in Figure 11–37 with the voltages and currents indicated. The current directions have been assigned arbitrarily for convenience.

FIGURE 11–37

The load current I_3 is determined as follows:

$$I_3 = \frac{V_3}{R_L} = \frac{160 \text{ V}}{8 \text{ }\Omega} = 20 \text{ A}$$

The input power is the total source voltage (V_1) times the total current from the source (I_1):

$$P_{\text{in}} = V_1 I_1$$

The output power is the load voltage V_3 times the load current I_3:

$$P_{\text{out}} = V_3 I_3$$

For an ideal transformer, $P_{\text{in}} = P_{\text{out}}$; thus,

$$V_1 I_1 = V_3 I_3$$

Solving for I_1 yields

$$I_1 = \frac{V_3 I_3}{V_1} = \frac{(160 \text{ V})(20 \text{ A})}{240 \text{ V}} = 13.33 \text{ A}$$

Applying Kirchhoff's current law at the tap junction, we get

$$I_1 = I_2 + I_3$$

Solving for I_2 yields

$$I_2 = I_1 - I_3 = 13.33 \text{ A} - 20 \text{ A} = -6.67 \text{ A}$$

The minus sign can be dropped because the current directions are arbitrary. The input and output power are

$$P_{in} = P_{out} = V_3I_3 = (160 \text{ V})(20 \text{ A}) = 3.2 \text{ kVA}$$

The power in winding A is

$$V_2I_1 = (80 \text{ V})(13.33 \text{ A}) = 1.07 \text{ kVA}$$

The power in winding B is

$$V_3I_2 = (160 \text{ V})(6.67 \text{ A}) = 1.07 \text{ kVA}$$

Thus, the power rating required for each winding is less than the power that is delivered to the load.

SECTION REVIEW 11–9

1. A certain transformer has two secondaries. The turns ratio from the primary to the first secondary is 10. The turns ratio from the primary to the other secondary is 0.2. If 240 V ac are applied to the primary, what are the secondary voltages?
2. Name one advantage and one disadvantage of an autotransformer over a conventional transformer.

11–10

TROUBLESHOOTING TRANSFORMERS

The common failures in transformers are opens, shorts, or partial shorts in either the primary or the secondary windings. One cause of such failures is the operation of the device under conditions that exceed its ratings. A few transformer failures and the typical symptoms are discussed in this section.

Open Primary Winding

When there is an open primary winding, there is no primary current and, therefore, no induced voltage or current in the secondary. This condition is illustrated in Figure 11–38(a), and the method of checking with an ohmmeter is shown in part (b).

Open Secondary Winding

When there is an open secondary winding, there is no current in the secondary and, as a result, no voltage across the load. Also, an open secondary causes the primary current to be very small (only a small magnetizing current flows). In fact, the primary current may be practically zero. This condition is illustrated in Figure 11–39(a), and the ohmmeter check is shown in part (b).

(a) Conditions when the primary is open

(b) Checking the primary with an ohmmeter

FIGURE 11–38
Open primary winding.

(a) Conditions when the secondary is open

(b) Checking the secondary with an ohmmeter

FIGURE 11–39
Open secondary winding.

Shorted or Partially Shorted Secondary

In this case, there is an excessive primary current because of the low reflected impedance due to the short. Often, this excessive current will burn out the primary and result
in an open. The short-circuit current in the secondary causes the load current to be zero
(full short) or smaller than normal (partial short), as demonstrated in Figures 11–40(a)
and 11–40(b). The ohmmeter check for this condition is shown in part (c).

Normally, when a transformer fails, it is very difficult to repair, and therefore the
simplest procedure is to replace it.

(a) Secondary completely shorted

(b) Secondary partially shorted

(c) Checking the secondary with an ohmmeter

FIGURE 11–40
Shorted secondary winding.

Shorted or Partially Shorted Primary

A completely shorted primary will draw excessive current from the source and, unless
there is a breaker or a fuse in the circuit, either the source or the transformer or both
will burn out. A partial short in the primary can cause higher than normal or even
excessive primary current.

**SECTION REVIEW
11–10**

1. Name two possible failures in a transformer.
2. What is often the cause of transformer failure?

SUMMARY

- ☐ A transformer consists of two or more coils that are magnetically coupled on a common core.
- ☐ There is mutual inductance between two magnetically coupled coils.
- ☐ When current in one coil changes, voltage is induced in the other coil.
- ☐ The primary is the winding connected to the source, and the secondary is the winding connected to the load.
- ☐ The number of turns in the primary and the number of turns in the secondary determine the *turns ratio*.
- ☐ The relative polarities of the primary and secondary voltages are determined by the direction of the windings around the core.
- ☐ A step-up transformer has a turns ratio greater than 1.
- ☐ A step-down transformer has a turns ratio less than 1.
- ☐ A transformer cannot increase power.
- ☐ In an ideal transformer, the power from the source (input power) is equal to the power delivered to the load (output power).
- ☐ If the voltage is stepped up, the current is stepped down, and vice versa.
- ☐ A load in the secondary of a transformer appears to the source as a *reflected* load having a value dependent on the reciprocal of the turns ratio squared.
- ☐ A transformer can match a load resistance to a source resistance to achieve maximum power transfer to the load by selection of the proper turns ratio.
- ☐ A transformer does not respond to dc.
- ☐ Energy losses in an actual transformer result from winding resistances, hysteresis loss in the core, eddy currents in the core, and flux leakage.

GLOSSARY

Apparent power   The product of the voltage times the current, expressed in volt-amperes (VA). In a purely resistive circuit it is the same as the true power.

Autotransformer   A transformer in which the primary and secondary are in a single winding.

Center tap   A connection at the midpoint of the secondary of a transformer.

Coefficient of coupling   A constant associated with transformers that is the ratio of secondary flux to primary flux. The ideal value of 1 indicates that all the flux in the primary is coupled into the secondary.

Core   The physical structure around which the windings are placed.

Mutual inductance   The inductance between two separate coils, such as a transformer.

Maximum power transfer theorem   A theorem that states the maximum power is transferred from a source to a load when the load resistance equals the source resistance.

Primary   The input winding of a transformer.

Reflected load   The load as it appears to the source in the primary.

Secondary   The output winding of a transformer.

Step-down transformer   A transformer in which the secondary voltage is less than the primary voltage.

Step-up transformer   A transformer in which the secondary voltage is greater than the primary voltage.

Transformer   A device formed by two or more windings that are magnetically coupled to each other and providing a transfer of power electromagnetically from one winding to another.

Turns ratio   The ratio of secondary turns to primary turns.

FORMULAS

$$k = \frac{\phi_{12}}{\phi_1} \qquad \text{Coefficient of coupling} \qquad (11\text{–}1)$$

$$L_M = k\sqrt{L_1 L_2} \qquad \text{Mutual inductance} \qquad (11\text{–}2)$$

$$n = \frac{N_s}{N_p} \qquad \text{Turns ratio} \qquad (11\text{–}3)$$

$$V_s = \left(\frac{N_s}{N_p}\right) V_p \qquad \text{Secondary voltage} \qquad (11\text{–}4)$$

$$I_s = \left(\frac{N_p}{N_s}\right) I_p \qquad \text{Secondary current} \qquad (11\text{–}5)$$

$$R_p = \left(\frac{1}{n}\right)^2 R_L \qquad \text{Reflected load resistance} \qquad (11\text{–}6)$$

$$n = \sqrt{\frac{R_L}{R_p}} \qquad \text{Turns ratio for matching} \qquad (11\text{–}7)$$

$$\eta = \left(\frac{P_{\text{out}}}{P_{\text{in}}}\right) 100\% \qquad \text{Transformer efficiency} \qquad (11\text{–}8)$$

SELF-TEST

1. A transformer is used for
   (a) dc voltages   (b) ac voltages   (c) both dc and ac

2. The turns ratio of a transformer affects the following:
   (a) primary voltage   (b) dc voltage
   (c) secondary voltage   (d) none of these

3. If the windings of a certain transformer with a turns ratio of 1 are in opposite directions around the core, the secondary voltage is
   (a) in phase with the primary voltage
   (b) less than the primary voltage
   (c) greater than the primary voltage
   (d) out of phase with the primary voltage

4. When the turns ratio of a transformer is 10 and the primary ac voltage is 6 V, the secondary voltage is
   (a) 60 V   (b) 0.6 V   (c) 6 V   (d) 36 V

5. When the turns ratio of a transformer is 0.5 and the primary ac voltage is 100 V, the secondary voltage is
   (a) 200 V   (b) 50 V   (c) 10 V   (d) 100 V

6. A certain transformer has 500 turns in the primary winding and 2500 turns in the secondary winding. The turns ratio is
   (a) 0.2   (b) 2.5   (c) 5   (d) 0.5

7. If 10 W of power are applied to the primary of an ideal transformer with a turns ratio of 5, the power delivered to the secondary load is
   (a) 50 W   (b) 0.5 W   (c) 0 W   (d) 10 W

8. In a certain loaded transformer, the secondary voltage is one-third the primary voltage. The secondary current is
   (a) one-third the primary current
   (b) three times the primary current
   (c) equal to the primary current
   (d) b and d

9. When a 1-kΩ load resistor is connected across the secondary winding of a transformer with a turns ratio of 2, the source "sees" a reflected load of
   (a) 250 Ω   (b) 2 kΩ   (c) 4 kΩ   (d) 1 kΩ $R_p = (\frac{1}{2})^2 1000 \Omega$

10. In Problem 9, if the turns ratio is 0.5, the source "sees" a reflected load of
    (a) 1 kΩ   (b) 2 kΩ   (c) 4 kΩ   (d) 500 Ω

11. The turns ratio required to match a 50-Ω source to a 200-Ω load is
    (a) 0.25   (b) 0.5   (c) 4   (d) 2

12. Maximum power is transferred from a source to a load when
    (a) $R_L > R_s$ (b) $R_L < R_s$ (c) $R_L = R_s$ (d) $R_L = nR_s$

13. When a 12-V battery is connected across the primary of a transformer with a turns ratio of 4, the secondary voltage is
    (a) 0 V   (b) 12 V   (c) 48 V   (d) 3 V

14. A certain transformer has a turns ratio of 1 and a 0.95 coefficient of coupling. When 1 V ac is applied to the primary, the secondary voltage is
    (a) 1 V   (b) 1.95 V   (c) 0.95 V

PROBLEM SET A   Section 11–1   The Basic Transformer

1. What is the turns ratio of a transformer having 12 primary turns and 36 secondary turns?

2. What is the turns ratio of a transformer having 250 primary turns and 1000 secondary turns? What is the turns ratio when the primary has 400 turns and the secondary has 100 turns?

3. Determine the phase of the secondary voltage with respect to the primary voltage for each transformer in Figure 11–41.

(a)

(b)

(c)

(d)

FIGURE 11–41

Section 11–2   Step-Up Transformers

4. If 120 V ac are connected to the primary of a transformer with 10 primary turns and 15 secondary turns, what is the secondary voltage?

5. A certain transformer has 25 turns in its primary. In order to double the voltage, how many turns must be in the secondary?

6. How many primary volts must be applied to a transformer with a turns ratio of 10 to obtain a secondary voltage of 60 V ac?

7. For each transformer in Figure 11–42, sketch the secondary voltage showing its relationship to the primary voltage. Also indicate the amplitude.

FIGURE 11–42

(a)   (b)

Section 11–3   Step-Down Transformers

8. To step 120 V down to 30 V, what must the turns ratio be?

9. The primary of a transformer has 1200 V across it. What is the secondary voltage if the turns ratio is 0.2?

10. How many primary volts must be applied to a transformer with a turns ratio of 0.1 to obtain a secondary voltage of 6 V ac?

Section 11–4   Loading the Secondary

11. Determine I_s in Figure 11–43.

12. Determine the following quantities in Figure 11–44:
  (a) primary current   **(b)** secondary current
  (c) secondary voltage   **(d)** power in the load

FIGURE 11–43

FIGURE 11–44

Section 11–5   Reflected Load in a Transformer

13. What is the load resistance as seen by the source in Figure 11–45?

14. What is the resistance reflected into the primary of the circuit in Figure 11–46?

FIGURE 11–45

FIGURE 11–46

15. What is the primary current (rms) in Figure 11–46 if the rms source voltage is 115 V?

16. What must the turns ratio be in Figure 11–47 in order to reflect 300 Ω into the primary?

FIGURE 11–47

Section 11–6   Matching the Load Resistance to the Source Resistance

17. For the circuit in Figure 11–48, find the turns ratio required to deliver maximum power to the 4-Ω speaker.

FIGURE 11–48

18. Determine the value to which R_L must be adjusted in Figure 11–49 for maximum power transfer. The internal resistance of the source is 50 Ω.

19. In Figure 11–48, what is the maximum power in watts delivered to the speaker?

FIGURE 11–49

Section 11–7   The Transformer as an Isolation Device

20. What is the voltage across the load in each circuit of Figure 11–50?

21. If the bottom of each secondary in Figure 11–50 were grounded, would the values of the load voltages be changed?

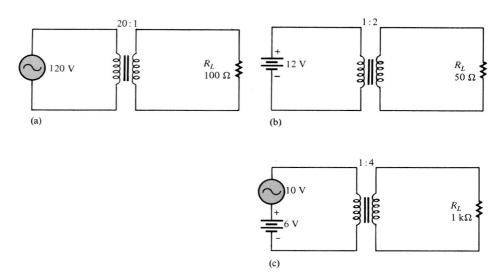

FIGURE 11–50

22. Determine the unspecified meter readings in Figure 11–51.

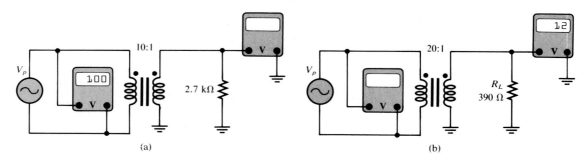

FIGURE 11–51

Section 11–8   Nonideal Transformer Characteristics

23. In a certain transformer, the input power to the primary is 100 W. If 5.5 W are lost in the winding resistances, what is the output power to the load, neglecting any other losses?

24. What is the efficiency of the transformer in Problem 23?

25. Determine the coefficient of coupling for a transformer in which 2% of the total flux generated in the primary does not pass through the secondary.

26. A certain transformer is rated at 1 kVA. It operates on 60 Hz, 120 V ac. The secondary voltage is 600 V.
(a) What is the maximum load current?
(b) What is the smallest R_L that you can drive?
(c) What is the largest capacitor that can be connected as a load?

27. What kVA rating is required for a transformer that must handle a maximum load current of 10 A with a secondary voltage of 2.5 kV?

Section 11–9   Other Types of Transformers

28. Determine each unknown voltage indicated in Figure 11–52.

29. Using the indicated secondary voltages in Figure 11–53, determine the turns ratio of the primary to each tapped section.

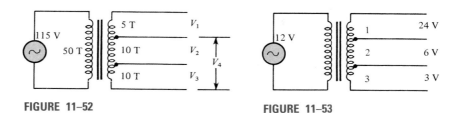

FIGURE 11–52   **FIGURE 11–53**

30. In Figure 11–54, each primary can accommodate 120 V ac. How should the primaries be connected for 240-V ac operation? Determine each secondary voltage.

FIGURE 11–54

31. Find the secondary voltage for each autotransformer in Figure 11–55.

(a)   (b)

FIGURE 11–55

Section 11–10   Troubleshooting Transformers

32. When you apply 120 V ac across the primary of a transformer and check the voltage across the secondary, you get 0 V. Further investigation shows no primary or secondary current. List the possible faults. What is your next step in isolating the problem?

33. What is likely to happen if the primary of a transformer shorts?

34. In checking out a transformer circuit you find that the secondary voltage is less than it should be, although it is not zero. What is the most likely fault?

PROBLEM SET B

35. For the loaded, tapped-secondary transformer in Figure 11–56, determine the following:
  (a) all load voltages and currents
  (b) the impedance looking into the primary

FIGURE 11–56

36. A certain transformer is rated at 5 kVA, 2400/120 V, at 60 Hz.
  (a) What is the turns ratio if the 120 V is the secondary voltage?
  (b) What is the current rating of the secondary if 2400 V is the primary voltage?
  (c) What is the current rating of the primary if 2400 V is the primary voltage?

37. Determine the voltage measured by each voltmeter in Figure 11–57. The bench-type meters have one terminal that connects to ground as indicated.

FIGURE 11–57

38. Find the appropriate turns ratio for each switch position in Figure 11–58 in order to transfer the maximum power to each load when the source resistance is 10 Ω. Specify the number of turns for the secondary if the primary has 1000 turns.

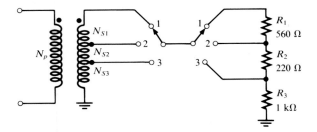

FIGURE 11–58

39. What must the turns ratio be in Figure 11–46 to limit the primary current to 3 mA with a source voltage of 115 V? Assume an ideal transformer and source.

ANSWERS TO
SECTION REVIEWS

Section 11–1

1. The inductance between two coils, established by the amount of coupling between the coils.
2. 45 mH.
3. It increases.
4. The ratio of secondary turns to primary turns.
5. The directions determine the relative polarities of the voltages.

Section 11–2

1. Increases voltage.   2. Five times greater.   3. 2400 V.

Section 11–3

1. Decreases voltage.   2. 60 V.   3. 0.1.

Section 11–4

1. Less, half.   2. 2 A.

Section 11–5

1. The resistance in the secondary as seen from the primary as a function of the turns ratio.
2. The turns ratio.   3. 0.5 Ω.
4. 0.1.

Section 11–6

1. Making the load impedance equal the source impedance.
2. Maximum power is delivered to the load.   3. 400 Ω.

Section 11–7

1. There is no electrical connection between primary and secondary.   2. No.

Section 11–8

1. In an actual transformer, energy loss reduces the efficiency. An ideal model has an efficiency of 100%.
2. 85% of the magnetic flux generated in the primary passes through the secondary.
3. 40 A.

Section 11–9

1. 2400 V, 48 V.   2. Smaller and lighter for same rating, no electrical isolation.

Section 11–10

1. Open primary, open secondary.   2. Operating above rated values.

12 Frequency Response of *RC* Circuits

An *RC* circuit contains both resistance and capacitance. It is one of the basic types of *reactive* circuits that we will study. In this chapter, basic series and parallel *RC* circuits and their responses to sine wave voltages are covered. Series-parallel combinations are also examined. Power considerations in *RC* circuits are introduced, and practical aspects of power ratings are discussed. Several basic areas of application are presented to give you an idea of how simple combinations of resistors and capacitors can be applied.

After completing this chapter, you will be able to

□ Analyze the response of an *RC* circuit with a sine wave input.
□ Define *impedance*.
□ Determine the impedance of both series and parallel *RC* circuits.
□ Determine the phase difference between current and voltage in an *RC* circuit.
□ Convert a parallel circuit to an equivalent series circuit.
□ Determine the effects of frequency on an *RC* circuit in terms of changes in impedance and phase angle.
□ Define *conductance, capacitive susceptance,* and *admittance.*
□ Apply Ohm's law to *RC* circuits in order to find current and voltage values.
□ Determine the true power, reactive power, and apparent power.
□ Define *power factor* and explain its significance.
□ Explain how *RC* circuits are used as phase-shifting circuits (lead and lag networks).
□ Explain how *RC* circuits are used as frequency selective circuits (filters).
□ Discuss how *RC* circuits are used in amplifiers for signal coupling.
□ Troubleshoot *RC* circuits for basic component failures.

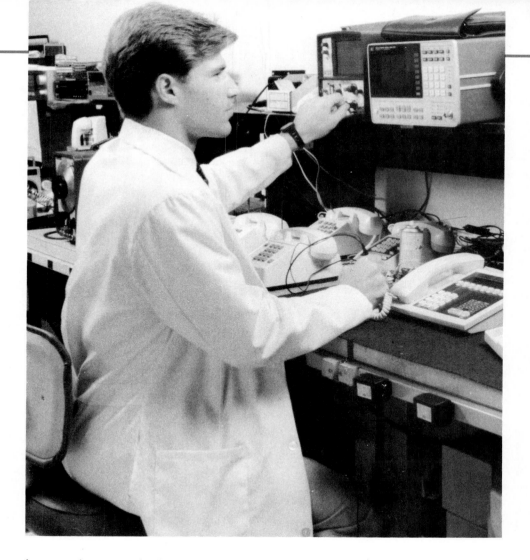

In a certain communications system, a section of audio amplifiers is being upgraded to accommodate a wider bandwidth so that a lower range of audio frequencies can be amplified without significant loss due to coupling capacitors. As lead technician, you have been asked to examine the coupling networks between each amplifier stage to ensure that there is no significant signal loss at the lower frequencies. If there is, you are to make the appropriate modifications. After studying this chapter, you will be able to accomplish this assigned task.

12–1   SINUSOIDAL RESPONSE OF *RC* CIRCUITS

When a sinusoidal voltage is applied to an *RC* circuit, *each resulting voltage drop and the current in the circuit is also sinusoidal with the same frequency as that of the applied voltage.* As shown in Figure 12–1, the resistor voltage, the capacitor voltage, and the current are all sine waves with the frequency of the source.

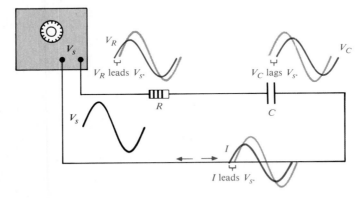

FIGURE 12–1
Illustration of sinusoidal response with general phase relationships of V_R, V_C, and I relative to the source voltage. V_R and I are in phase; V_R leads V_s; V_C lags V_s; and V_R and V_C are 90° out of phase.

Phase shifts are introduced because of the capacitance. As you will learn, the resistor voltage and current are in phase with each other and lead the source voltage in phase. The capacitor voltage lags the source voltage. The phase angle between the current and the capacitor voltage is always 90°. These generalized phase relationships are indicated in the figure.

The amplitudes and the phase relationships of the voltages and current depend on the values of the resistance and the capacitive reactance (defined in Section 9–6). When a circuit is purely resistive, the phase angle between the applied (source) voltage and the total current is zero. When a circuit is purely capacitive, the phase angle between the applied voltage and the total current is 90°, with the current *leading* the voltage. When there is a combination of both resistance and capacitive reactance in a circuit, the phase angle between the applied voltage and the total current is somewhere between zero and 90°, depending on the relative values of the resistance and the reactance.

Signal Generators

When a circuit is hooked up for a laboratory experiment or for troubleshooting, a signal generator similar to those shown in Figure 12–2 is used to provide the source voltage. These instruments, depending on their capability, are classified as *sine wave generators,* which produce only sine waves; *sine/square generators,* which produce both sine waves and square waves; or *function generators,* which produce sine waves, pulse waveforms, and triangular (ramp) waveforms.

(a)

(b)

(c)

FIGURE 12–2
Typical signal (function) generators used in circuit testing and troubleshooting [(a) courtesy of Hewlett-Packard Company. (b) courtesy of Wavetek. (c) courtesy of B&K Precision.].

SECTION REVIEW 12–1

1. A 60-Hz sinusoidal voltage is applied to an *RC* circuit. What is the frequency of the capacitor voltage? The current?

2. When the resistance in an *RC* circuit is greater than the capacitive reactance, is the phase angle between the applied voltage and the total current closer to 0° or to 90°?

12–2

IMPEDANCE AND PHASE ANGLE OF A SERIES *RC* CIRCUIT

Impedance is the total opposition to sinusoidal current and is expressed in ohms.

In a purely resistive circuit, the impedance is simply equal to the total resistance. In a purely capacitive circuit, the impedance is the total capacitive reactance. The impedance of a series *RC* circuit is determined by *both the resistance and the capacitive reactance*. These cases are illustrated in Figure 12–3. The magnitude of the impedance is symbolized by *Z*.

AC CIRCUITS

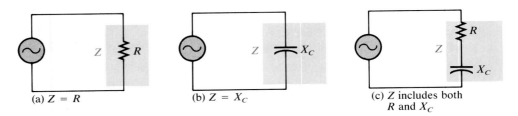

(a) $Z = R$ (b) $Z = X_C$ (c) Z includes both
                                                                    R and X_C

FIGURE 12–3
Three cases of impedance.

The Impedance Triangle

In ac analysis, both R and X_C are treated as *phasor quantities*, as shown in the phasor diagram of Figure 12–4(a), with X_C appearing at a $-90°$ angle with respect to R. This relationship comes from the fact that the capacitor voltage in a series *RC* circuit *lags* the current, and thus the resistor voltage, by 90°.

Since Z is the *phasor sum* of R and X_C, its phasor representation is as shown in Figure 12–4(b). A repositioning of the phasors, as shown in part (c), forms a right triangle, called the *impedance triangle*. The length of each phasor represents the *magnitude* in ohms, and the angle θ (the Greek letter theta) is the phase angle of the *RC* circuit and represents the phase difference between the applied (source) voltage and the current.

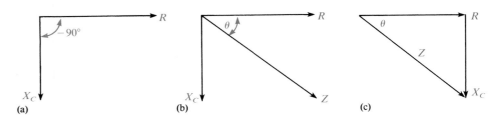

(a)   (b)   (c)

FIGURE 12–4
Development of the impedance triangle for a series *RC* circuit.

From right angle trigonometry (Pythagorean theorem), the magnitude of the impedance can be expressed in terms of the resistance and reactance as

$$Z = \sqrt{R^2 + X_C^2} \tag{12–1}$$

This is the *magnitude* of Z, as shown in Figure 12–5, and is expressed in ohms.
The phase angle, θ, is expressed as

$$\theta = \tan^{-1}\left(\frac{X_C}{R}\right) \tag{12–2}$$

The symbol $\tan^{-1}$ stands for *inverse tangent* and can be found on some calculators by pressing (2nd) (tan) and on others by pressing (INV) (tan). We will use the (2nd) (tan) sequence in this book. Another term for inverse tangent is *arctangent* (arctan).

FIGURE 12–5
Impedance of a series *RC* circuit.

EXAMPLE 12–1   Determine the impedance and the phase angle of the circuit in Figure 12–6.

FIGURE 12–6

Solution
The impedance is

$$Z = \sqrt{R^2 + X_C^2} = \sqrt{(50 \ \Omega)^2 + (100 \ \Omega)^2} = 111.8 \ \Omega$$

The phase angle is

$$\theta = \tan^{-1}\left(\frac{X_C}{R}\right) = \tan^{-1}\left(\frac{100}{50}\right) = \tan^{-1}(2) = 63.4°$$

The applied voltage *lags* the current by 63.4°.
     The calculator sequences are as follows:
Z: ⑤ ⓪ x^2 + ① ⓪ ⓪ x^2 = 2nd $\sqrt{x}$
θ: ① ⓪ ⓪ ÷ ⑤ ⓪ = 2nd tan

SECTION REVIEW   1. Define *impedance*.
12–2   2. Does the source voltage lead or lag the current in a series *RC* circuit?
                     3. What causes the phase angle in the *RC* circuit?
                     4. A series *RC* circuit has a resistance of 33 kΩ and a capacitive reactance of 50 kΩ. What is
                         the value of the impedance? What is the phase angle?

12–3

ANALYSIS OF SERIES *RC* CIRCUITS

In the previous section, you learned how to express the impedance and the phase angle
of a series *RC* circuit. In this section, Ohm's law and Kirchhoff's voltage law are
utilized in the analysis of *RC* circuits.

Ohm's Law

The application of Ohm's law to series RC circuits involves the use of the quantities of Z, V, and I. The three equivalent forms of Ohm's law are as follows:

$$V = IZ \qquad (12\text{--}3)$$

$$I = \frac{V}{Z} \qquad (12\text{--}4)$$

$$Z = \frac{V}{I} \qquad (12\text{--}5)$$

The following two examples illustrate the use of Ohm's law.

EXAMPLE 12–2

If the current in Figure 12–7 is 0.2 mA, determine the source voltage and the phase angle.

FIGURE 12–7

Solution
The capacitive reactance is

$$X_C = \frac{1}{2\pi f C} = \frac{1}{2\pi(1000 \text{ Hz})(0.01 \text{ }\mu\text{F})} = 15.9 \text{ k}\Omega$$

The impedance is

$$Z = \sqrt{R^2 + X_C^2} = \sqrt{(10 \text{ k}\Omega)^2 + (15.9 \text{ k}\Omega)^2} = 18.78 \text{ k}\Omega$$

Applying Ohm's law, we obtain

$$V_s = IZ = (0.2 \text{ mA})(18.78 \text{ k}\Omega) = 3.76 \text{ V}$$

The phase angle is

$$\theta = \tan^{-1}\left(\frac{X_C}{R}\right) = \tan^{-1}\left(\frac{15.9 \text{ k}\Omega}{10 \text{ k}\Omega}\right) = 57.83°$$

The source voltage has a magnitude of 3.76 V and lags the current by 57.83°.
The calculator sequence for V_s is

$$\boxed{2}\ \boxed{\times}\ \boxed{\text{2nd}}\ \boxed{\pi}\ \boxed{\times}\ \boxed{1}\ \boxed{0}\ \boxed{0}\ \boxed{0}\ \boxed{\times}\ \boxed{\cdot}\ \boxed{0}\ \boxed{1}\ \boxed{\text{EXP}}\ \boxed{+/-}\ \boxed{6}\ \boxed{=}\ \boxed{1/x}$$

$$\boxed{x^2}\ \boxed{+}\ \boxed{1}\ \boxed{0}\ \boxed{\text{EXP}}\ \boxed{3}\ \boxed{x^2}\ \boxed{=}\ \boxed{\text{2nd}}\ \boxed{\sqrt{x}}\ \boxed{\times}\ \boxed{\cdot}\ \boxed{2}\ \boxed{\text{EXP}}\ \boxed{+/-}\ \boxed{3}\ \boxed{=}$$

EXAMPLE 12–3   Determine the current in the circuit of Figure 12–8.

FIGURE 12–8

Solution

$$X_C = \frac{1}{2\pi f C} = \frac{1}{2\pi (1.5 \text{ kHz})(0.02 \text{ } \mu F)} = 5.3 \text{ k}\Omega$$

The impedance is

$$Z = \sqrt{R^2 + X_C^2} = \sqrt{(2.2 \text{ k}\Omega)^2 + (5.3 \text{ k}\Omega)^2}$$
$$= 5.74 \text{ k}\Omega$$

Applying Ohm's law, we obtain

$$I = \frac{V}{Z} = \frac{10 \text{ V}}{5.74 \text{ k}\Omega} = 1.74 \text{ mA}$$

Relationships of the Current and Voltages in a Series *RC* Circuit

In a series circuit, the current is the same through both the resistor and the capacitor. Thus, *the resistor voltage is in phase with the current, and the capacitor voltage lags the current by 90°*. Therefore, there is a phase difference of 90° between the resistor voltage, V_R, and the capacitor voltage, V_C, as shown in the waveform diagram of Figure 12–9.

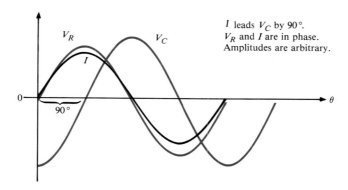

FIGURE 12–9
Phase relation of voltages and current in a series *RC* circuit.

We know from Kirchhoff's voltage law that the sum of the voltage drops must equal the applied voltage. However, since V_R and V_C are not in phase with each other, they must be added as phasor quantities, with V_C lagging V_R by 90°, as shown in Figure 12–10(a). As shown in part (b), V_s is the phasor sum of V_R and V_C, as expressed in the following equation:

$$V_s = \sqrt{V_R^2 + V_C^2} \qquad (12\text{–}6)$$

FIGURE 12–10
Voltage phasor diagram for a series *RC* circuit.

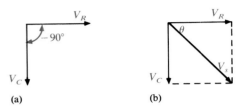

(a)   (b)

The phase angle between the resistor voltage and the source voltage is

$$\theta = \tan^{-1}\left(\frac{V_C}{V_R}\right) \qquad (12\text{–}7)$$

Since the resistor voltage and the current are in phase, θ also represents the phase angle between the source voltage and the current. Figure 12–11 shows a complete voltage and current phasor diagram representing the waveform diagram of Figure 12–9.

FIGURE 12–11
Voltage and current phasor diagram for the waveforms in Figure 12–9.

EXAMPLE 12–4   Determine the source voltage and the phase angle in Figure 12–12.

FIGURE 12–12

Solution

Since V_R and V_C are 90° out of phase, they cannot be added directly. The source voltage is the *phasor sum* of V_R and V_C:

$$V_s = \sqrt{V_R^2 + V_C^2} = \sqrt{(10 \text{ V})^2 + (15 \text{ V})^2} = 18 \text{ V}$$

The angle between the current and source voltage is

$$\theta = \tan^{-1}\left(\frac{V_C}{V_R}\right) = \tan^{-1}\left(\frac{15 \text{ V}}{10 \text{ V}}\right) = 56.3°$$

The calculator sequences are as follows:

V_s: $\boxed{1}\ \boxed{0}\ \boxed{x^2}\ \boxed{+}\ \boxed{1}\ \boxed{5}\ \boxed{x^2}\ \boxed{=}\ \boxed{\text{2nd}}\ \boxed{\sqrt{x}}$

θ: $\boxed{1}\ \boxed{5}\ \boxed{\div}\ \boxed{1}\ \boxed{0}\ \boxed{=}\ \boxed{\text{2nd}}\ \boxed{\text{tan}}$

Variation of Voltage and Current Magnitudes with Frequency

As you know, capacitive reactance varies *inversely* with frequency. Since $Z = \sqrt{R^2 + X_C^2}$, you can see that when X_C increases, the entire term under the square root sign increases and thus the total impedance also increases; and when X_C decreases, the total impedance also decreases. Therefore, *in an RC circuit, Z is inversely dependent on frequency*.

Figure 12–13 illustrates how the voltages and current in a series *RC* circuit vary as the frequency increases or decreases, with the source voltage held at a constant value. In part (a), as the frequency is increased, X_C decreases, so less voltage is dropped across the capacitor. Also, Z decreases as X_C decreases, causing the current to increase. An increase in the current causes more voltage across R.

(a) As frequency is increased   (b) As frequency is decreased

FIGURE 12–13

An illustration of how the voltages and current change as the source frequency is varied. The source voltage is held at a constant amplitude.

In part (b) of the figure, as the frequency is decreased X_C increases, so more voltage is dropped across the capacitor. Also, Z increases as X_C increases, causing the current to decrease. A decrease in the current causes less voltage across R.

Changes in Z and X_C can be observed as shown in Figure 12–14. As the frequency increases, the voltage across Z remains constant because V_s is constant. Also, the voltage across C decreases. The increasing current indicates that Z is decreasing. It does so because of the inverse relationship stated in Ohm's law ($Z = V_Z/I$). The increasing current also indicates that X_C is decreasing ($X_C = V_C/I$). The decrease in V_C corresponds to the decrease in X_C.

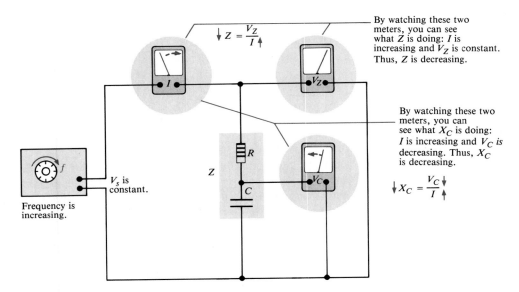

FIGURE 12–14
An illustration of how Z and X_c change with frequency.

Variation of the Phase Angle with Frequency

Since X_C is the factor that introduces the phase angle in a series RC circuit, a change in X_C produces a change in the phase angle. As the frequency is increased, X_C becomes smaller, and thus the phase angle decreases. As the frequency is decreased, X_C becomes larger, and thus the phase angle increases. This variation is illustrated in Figure 12–15 where a phase meter is connected across the capacitor to measure the angle between V_s and V_R. This angle is the phase angle of the circuit because I is in phase with V_R. By measuring the phase of V_R, we are effectively measuring the phase of I. We can use the phase-meter portion of a vector voltmeter such as the one shown in part (c) of the figure to measure the phase of one voltage relative to another. You can also see the variation of phase angle with frequency by using the impedance triangle as shown in Figure 12–16.

(a) As frequency is increased, the phase angle decreases.

(b) As frequency is decreased, the phase angle increases.

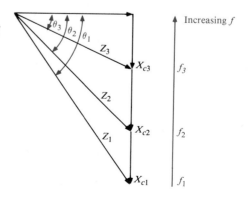

(c)

FIGURE 12–15

The phase meter on a vector voltmeter shows the effect of frequency on the phase angle of a circuit. The phase meter indicates the phase angle between two voltages, V_s and V_R. Since V_R and I are in phase, this angle is the same as the angle between V_s and I. The vector voltmeter also indicates voltage magnitudes (photo courtesy of Hewlett-Packard Co.).

FIGURE 12–16

As the frequency increases, X_C decreases, Z decreases, and θ decreases. Each value of frequency can be visualized as forming a different impedance triangle.

EXAMPLE 12–5

For the series RC circuit in Figure 12–17, determine the impedance and phase angle for each of the following values of frequency:

(a) 10 kHz   (b) 20 kHz   (c) 30 kHz

FIGURE 12–17

Solution

(a) For $f = 10$ kHz:

$$X_C = \frac{1}{2\pi fC} = \frac{1}{2\pi(10 \text{ kHz})(0.01 \text{ } \mu\text{F})} = 1592 \text{ } \Omega$$

$$Z = \sqrt{R^2 + X_C^2} = \sqrt{(1000 \text{ } \Omega)^2 + (1592 \text{ } \Omega)^2} = 1880 \text{ } \Omega$$

$$\theta = \tan^{-1}\left(\frac{X_C}{R}\right) = \tan^{-1}\left(\frac{1592 \text{ } \Omega}{1000 \text{ } \Omega}\right) = 57.87°$$

(b) For $f = 20$ kHz:

$$X_C = \frac{1}{2\pi(20 \text{ kHz})(0.01 \text{ } \mu\text{F})} = 796 \text{ } \Omega$$

$$Z = \sqrt{(1000 \text{ } \Omega)^2 + (796 \text{ } \Omega)^2} = 1278 \text{ } \Omega$$

$$\theta = \tan^{-1}\left(\frac{796 \text{ } \Omega}{1000 \text{ } \Omega}\right) = 38.52°$$

(c) For $f = 30$ kHz:

$$X_C = \frac{1}{2\pi(30 \text{ kHz})(0.01 \text{ } \mu\text{F})} = 531 \text{ } \Omega$$

$$Z = \sqrt{(1000 \text{ } \Omega)^2 + (531 \text{ } \Omega)^2} = 1132 \text{ } \Omega$$

$$\theta = \tan^{-1}\left(\frac{531 \text{ } \Omega}{1000 \text{ } \Omega}\right) = 27.97°$$

Notice that as the frequency increases, X_C, Z, and θ decrease.

SECTION REVIEW
12–3

1. In a certain series RC circuit, $V_R = 4$ V and $V_C = 6$ V. What is the magnitude of the total voltage?
2. In Problem 1, what is the phase angle?
3. What is the phase difference between the capacitor voltage and the resistor voltage in a series RC circuit?

4. When the frequency of the applied voltage in a series *RC* circuit is increased, what happens to each of the following?
   (a) the capacitive reactance   (b) the impedance   (c) the phase angle

---

12-4

IMPEDANCE AND PHASE ANGLE OF A PARALLEL *RC* CIRCUIT

A basic parallel *RC* circuit is shown in Figure 12–18. The expression for the impedance is

$$Z = \frac{RX_C}{\sqrt{R^2 + X_C^2}} \qquad (12\text{–}8)$$

FIGURE 12–18
Parallel *RC* circuit.

The phase angle between the applied voltage and the total current is

$$\theta = \tan^{-1}\left(\frac{R}{X_C}\right) \qquad (12\text{–}9)$$

EXAMPLE 12–6   For each circuit in Figure 12–19, determine the impedance and the phase angle.

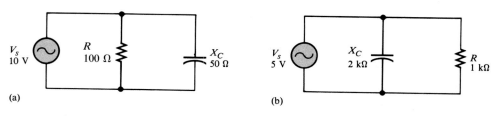

(a)

(b)

FIGURE 12–19

Solution

(a) $Z = \dfrac{RX_C}{\sqrt{R^2 + X_C^2}} = \dfrac{(100\ \Omega)(50\ \Omega)}{\sqrt{(100\ \Omega)^2 + (50\ \Omega)^2}} = 44.72\ \Omega$

$\theta = \tan^{-1}\left(\dfrac{R}{X_C}\right) = \tan^{-1}\left(\dfrac{100\ \Omega}{50\ \Omega}\right) = 63.43°$

(b) $Z = \dfrac{(1\ k\Omega)(2\ k\Omega)}{\sqrt{(1\ k\Omega)^2 + (2\ k\Omega)^2}} = 894.4\ \Omega$

$\theta = \tan^{-1}\left(\dfrac{1\ k\Omega}{2\ k\Omega}\right) = 26.57°$

The calculator sequences for part (a) are as follows:

Z: $\boxed{1}\,\boxed{0}\,\boxed{0}\,\boxed{\times}\,\boxed{5}\,\boxed{0}\,\boxed{\div}\,\boxed{(}\,\boxed{(}\,\boxed{1}\,\boxed{0}\,\boxed{0}\,\boxed{x^2}$
   $\boxed{+}\,\boxed{5}\,\boxed{0}\,\boxed{x^2}\,\boxed{)}\,\boxed{\text{2nd}}\,\boxed{\sqrt{x}}\,\boxed{)}\,\boxed{=}$

θ: $\boxed{1}\,\boxed{0}\,\boxed{0}\,\boxed{\div}\,\boxed{5}\,\boxed{0}\,\boxed{=}\,\boxed{\text{2nd}}\,\boxed{\text{tan}}$

Conductance, Susceptance, and Admittance

Conductance (G) is the reciprocal of resistance and is expressed as

$$G = \frac{1}{R} \tag{12–10}$$

Two other terms are convenient for use in parallel *RC* circuits. *Capacitive susceptance (B_C)* is the reciprocal of capacitive reactance and is expressed as

$$B_C = \frac{1}{X_C} \tag{12–11}$$

Admittance (Y) is the reciprocal of impedance and is expressed as

$$Y = \frac{1}{Z} \tag{12–12}$$

The unit of each of these quantities is the *siemen* (S), which is the reciprocal of the ohm.

In working with parallel circuits, you will often find it easier to use G, B_C, and Y rather than R, X_C, and Z, as we now discuss. In a parallel *RC* circuit, as shown in Figure 12–20, the total admittance is simply the phasor sum of the conductance and the susceptance:

$$Y = \sqrt{G^2 + B_C^2} \tag{12–13}$$

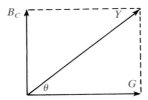

FIGURE 12–20
Admittance is the reciprocal of impedance in a parallel *RC* circuit.

EXAMPLE 12–7   Determine the admittance in Figure 12–21, and then convert it to impedance.

FIGURE 12–21

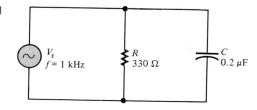

Solution
Since $R = 330 \ \Omega$, then $G = 1/R = 1/330 \ \Omega = 0.003 \ \text{S}$.

$$X_C = \frac{1}{2\pi(1000 \ \text{Hz})(0.2 \ \mu\text{F})} = 796 \ \Omega$$

$$B_C = \frac{1}{X_C} = \frac{1}{796 \ \Omega} = 0.00126 \ \text{S}$$

$$Y = \sqrt{G^2 + B_C^2} = \sqrt{(0.003 \ \text{S})^2 + (0.00126 \ \text{S})^2} = 0.00325 \ \text{S}$$

This can be converted to impedance as follows:

$$Z = \frac{1}{Y} = \frac{1}{0.00325 \ \text{S}} = 307.69 \ \Omega$$

The calculator sequence for Y is

⎡·⎤⎡0⎤⎡0⎤⎡3⎤⎡x^2⎤⎡+⎤⎡·⎤⎡0⎤⎡0⎤⎡1⎤⎡2⎤⎡6⎤⎡x^2⎤⎡=⎤⎡2nd⎤⎡$\sqrt{x}$⎤

SECTION REVIEW
12–4

1. Define *conductance, capacitive susceptance,* and *admittance*.
2. If $Z = 100 \ \Omega$, what is the value of Y?
3. In a certain parallel *RC* circuit, $R = 50 \ \Omega$ and $X_C = 75 \ \Omega$. Determine Y.

12–5

ANALYSIS OF PARALLEL *RC* CIRCUITS

For convenience in the analysis of parallel circuits, the Ohm's law formulas using impedance, previously stated, can be rewritten for admittance using the relation $Y = 1/Z$:

$$V = \frac{I}{Y} \tag{12–14}$$

$$I = VY \tag{12–15}$$

$$Y = \frac{I}{V} \tag{12–16}$$

EXAMPLE 12–8   Determine the total current and the phase angle in Figure 12–22.

FIGURE 12–22

Solution

$$X_C = \frac{1}{2\pi(1.5 \text{ kHz})(0.02 \text{ } \mu\text{F})} = 5.3 \text{ k}\Omega$$

The susceptance is

$$B_C = \frac{1}{X_C} = \frac{1}{5.3 \text{ k}\Omega} = 0.189 \text{ mS}$$

The conductance is

$$G = \frac{1}{R} = \frac{1}{2.2 \text{ k}\Omega} = 0.455 \text{ mS}$$

The total admittance is

$$Y = \sqrt{G^2 + B_C^2} = \sqrt{(0.455 \text{ mS})^2 + (0.189 \text{ mS})^2}$$
$$= 0.493 \text{ mS}$$

Applying Ohm's law, we obtain

$$I_T = VY = (10 \text{ V})(0.493 \text{ mS}) = 4.93 \text{ mA}$$

The phase angle is

$$\theta = \tan^{-1}\left(\frac{R}{X_C}\right) = \tan^{-1}\left(\frac{2.2 \text{ k}\Omega}{5.3 \text{ k}\Omega}\right) = 22.56°$$

The total current is 4.93 mA, and it *leads* the applied voltage by 22.56°.

Relationships of the Currents and Voltages in a Parallel *RC* Circuit

Figure 12–23(a) shows all the currents and voltages in a basic parallel *RC* circuit. As you can see, the applied voltage, V_s, appears across both the resistive and the capacitive branches, so V_s, V_R, and V_C are all in phase and of the same magnitude. The total current, I_T, divides at the junction into the two branch currents, I_R and I_C.

The current through the resistor is in phase with the voltage. The current through the capacitor leads the voltage, and thus the resistive current, by 90°. By Kirchhoff's

(a)

(b)

FIGURE 12–23
Currents and voltages in a parallel RC circuit. The current directions shown in (a) are instanta-neous.

current law, the total current is the phasor sum of the two branch currents, as shown by the phasor diagram in Figure 12–23(b). The total current is expressed as

$$I_T = \sqrt{I_R^2 + I_C^2} \qquad (12\text{--}17)$$

The phase angle can be found from the current values as follows:

$$\theta = \tan^{-1}\left(\frac{I_C}{I_R}\right) \qquad (12\text{--}18)$$

which is equivalent to the previous expression, $\theta = \tan^{-1}(R/X_C)$.

Figure 12–24 shows a complete current and voltage phasor diagram. Notice that I_C leads I_R by 90° and that I_R is in phase with the voltage ($V_s = V_R = V_C$).

FIGURE 12–24
Current and voltage phasor diagram for a parallel RC circuit (amplitudes are arbitrary).

EXAMPLE 12–9

Determine the value of each current in Figure 12–25, and describe the phase relation-ship of each with the applied voltage.

FIGURE 12–25

Solution

$$I_R = \frac{V_s}{R} = \frac{12 \text{ V}}{220 \text{ }\Omega} = 54.55 \text{ mA}$$

$$I_C = \frac{V_s}{X_C} = \frac{12 \text{ V}}{150 \text{ }\Omega} = 80 \text{ mA}$$

$$I_T = \sqrt{I_R^2 + I_C^2} = \sqrt{(54.55 \text{ mA})^2 + (80 \text{ mA})^2} = 96.83 \text{ mA}$$

$$\theta = \tan^{-1}\left(\frac{I_C}{I_R}\right) = \tan^{-1}\left(\frac{80 \text{ mA}}{54.55 \text{ mA}}\right) = 55.7°$$

I_R is in phase with the voltage, I_C leads the voltage by 90°, and I_T leads the voltage by 55.7°.

Conversion from Parallel to Series Form

For every parallel RC circuit, there is an equivalent series RC circuit for any given frequency. Two circuits are equivalent when they both present an equal impedance and phase angle. To obtain the equivalent series circuit for a given parallel *RC* circuit, first find the impedance and phase angle of the parallel circuit. Then use the values of Z and θ to construct an impedance triangle shown in Figure 12–26. The unknown sides of the triangle (black) represent the equivalent series resistance and capacitive reactance as indicated. These values can be found using the following trigonometric relationships:

$$R_{eq} = Z \cos \theta \tag{12–19}$$

$$X_{C(eq)} = Z \sin \theta \tag{12–20}$$

The cosine (cos) and the sine (sin) functions are available on your calculator.

FIGURE 12–26
Impedance triangle for the series equivalent of a parallel *RC* circuit. Z and θ are the known values for the parallel circuit. R_{eq} and $X_{C(eq)}$ are the series equivalent values.

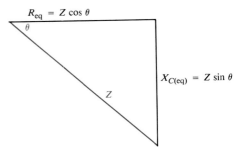

EXAMPLE 12–10   Convert the parallel circuit in Figure 12–27 to an equivalent series form.

FIGURE 12–27

Solution

First, find the impedance of the parallel circuit:

$$G = \frac{1}{R} = \frac{1}{18 \ \Omega} = 0.055 \text{ S}$$

$$B_C = \frac{1}{X_C} = \frac{1}{27 \ \Omega} = 0.037 \text{ S}$$

$$Y = \sqrt{G^2 + B_C^2} = \sqrt{(0.055 \text{ S})^2 + (0.037 \text{ S})^2} = 0.066 \text{ S}$$

The total impedance is

$$Z = \frac{1}{Y} = \frac{1}{0.066 \text{ S}} = 15.15 \ \Omega$$

The phase angle is

$$\theta = \tan^{-1}\left(\frac{R}{X_C}\right) = \tan^{-1}\left(\frac{18 \ \Omega}{27 \ \Omega}\right) = 33.69°$$

The equivalent series values are

$$R_{eq} = Z \cos \theta = 15.15 \cos(33.69°) = 12.6 \ \Omega$$

$$X_{C(eq)} = Z \sin \theta = 15.15 \sin(33.69°) = 8.4 \ \Omega$$

The series equivalent circuit is shown in Figure 12–28. Note that the value of C can be determined if the frequency is known.

FIGURE 12–28

The calculator sequence for R_{eq} is

[3] [3] [·] [6] [9] [cos] [×] [1] [5] [·] [1] [5] [=]

It is interesting to observe that a parallel *RC* circuit becomes *less* reactive when X_C is increased. That is, the circuit phase angle becomes smaller. The reason for this effect is that when X_C is increased relative to R, less current flows through the capacitive branch, and although the in-phase or resistive current does not increase, it becomes a greater percentage of the total current.

**SECTION REVIEW
12–5**

1. The admittance of an *RC* circuit is 0.0035 S, and the applied voltage is 6 V. What is the total current?

2. In a certain parallel RC circuit, the resistor current is 10 mA, and the capacitor current is 15 mA. Determine the phase angle and the total current.

3. What is the phase angle between the capacitor current and the applied voltage in a parallel RC circuit?

12-6

SERIES-PARALLEL ANALYSIS

In this section, we use the concepts studied in the previous sections to analyze circuits with combinations of both series and parallel R and C elements. The following example demonstrates the procedures.

EXAMPLE 12-11

In the circuit of Figure 12–29, determine the following:
(a) total impedance   (b) total current   (c) phase angle

FIGURE 12-29

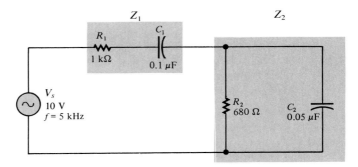

Solution

(a) First, calculate the magnitudes of capacitive reactances:

$$X_{C1} = \frac{1}{2\pi(5 \text{ kHz})(0.1 \ \mu F)} = 318.3 \ \Omega$$

$$X_{C2} = \frac{1}{2\pi(5 \text{ kHz})(0.05 \ \mu F)} = 636.6 \ \Omega$$

One approach is to find the series equivalent resistance and capacitive reactance for the parallel portion of the circuit, and then add the resistances to get total resistance and add the reactances to get total reactance. From this, the total impedance can be determined.

We find the impedance of the parallel portion (Z_2) by first finding the admittance:

$$G_2 = \frac{1}{R_2} = \frac{1}{680 \ \Omega} = 0.00147 \text{ S}$$

$$B_{C2} = \frac{1}{X_{C2}} = \frac{1}{636.6 \ \Omega} = 0.00157 \text{ S}$$

$$Y_2 = \sqrt{G_2^2 + B_{C2}^2} = \sqrt{(0.00147 \text{ S})^2 + (0.00157 \text{ S})^2} = 2.15 \text{ mS}$$

$$Z_2 = \frac{1}{Y_2} = \frac{1}{0.00215 \text{ S}} = 465.12 \ \Omega$$

The phase angle associated with the parallel portion of the circuit is

$$\theta_p = \tan^{-1}\left(\frac{R_2}{X_{C2}}\right) = \tan^{-1}\left(\frac{680\ \Omega}{636.6\ \Omega}\right) = 46.89°$$

The series equivalent values for the parallel portion are

$$R_{eq} = Z_2\cos\theta_p = 465.12\cos(46.89°) = 317.9\ \Omega$$
$$X_{C(eq)} = Z_2\sin\theta_p = 465.12\sin(46.89°) = 339.6\ \Omega$$

The total resistance is

$$R_T = R_1 + R_{eq} = 1000\ \Omega + 317.9\ \Omega = 1317.9\ \Omega$$

The total reactance is

$$X_{CT} = X_{C1} + X_{C(eq)} = 318.3\ \Omega + 339.6\ \Omega = 657.9\ \Omega$$

The total impedance is

$$Z_T = \sqrt{R_T^2 + X_{CT}^2} = \sqrt{(1317.9\ \Omega)^2 + (657.9\ \Omega)^2} = 1473\ \Omega$$

(b) We find the total current by using Ohm's law:

$$I_T = \frac{V_s}{Z_T} = \frac{10\ \text{V}}{1473\ \Omega} = 6.79\ \text{mA}$$

(c) To get the phase angle, view the circuit as a series combination of R_T and X_{CT}:

$$\theta = \tan^{-1}\left(\frac{X_{CT}}{R_T}\right) = \tan^{-1}\left(\frac{657.9\ \Omega}{1317.9\ \Omega}\right) = 26.53°$$

The total current, of course, *leads* the source voltage by 26.53°.

Measurement of Z_T and θ

Now, let's see how the values of Z_T and θ for the circuit in Example 12–11 can be determined by measurement. First, the total impedance is measured as outlined in the following steps and as illustrated in Figure 12–30 (other ways are also possible):

1. Using a sine wave generator, set the source voltage to a known value (10 V) and the frequency to 5 kHz. It is advisable to check the voltage with an ac voltmeter and the frequency with a frequency counter rather than relying on the marked values on the generator controls.
2. Connect an ac ammeter as shown in the figure, and measure the total current.
3. Calculate the total impedance by using Ohm's law.

Although we could use a phase meter to measure the phase angle, we will use an oscilloscope in this illustration because it is more commonly available. To measure the phase angle, we must have the source voltage and the total current displayed on the screen in the proper time relationship. Two basic types of scope probes are available to measure the quantities with an oscilloscope: the voltage probe and the current probe.

FIGURE 12–30
Determining Z_T by measurement of V_s and I_T.

$$Z_T = \frac{V_s}{I_T} = \frac{10 \text{ V}}{6.79 \text{ mA}} = 1473 \ \Omega$$

(a) Circuit with ground reference

(b) Ground lead on scope probe grounds point B

(c) The effect of grounding point B is to short out the rest of the circuit.

FIGURE 12–31
Effects of measuring directly across a component when the instrument and the circuit are grounded.

Although the current probe is a convenient device, it is often not as readily available as a voltage probe. For this reason, we will confine our phase measurement technique to the use of voltage probes in conjunction with the oscilloscope. A typical oscilloscope voltage probe has two points that are connected to the circuit: the probe tip and the ground lead. Thus, all voltage measurements must be referenced to ground.

Since only voltage probes are to be used, the total current cannot be measured directly. However, for phase measurement, the voltage across R_1 is in phase with the total current and can be used to establish the phase angle of the current. In setting up this circuit, we take the lower side of the source as circuit ground, as shown in Figure 12–31(a).

Before proceeding with the actual phase measurement, there is a problem with displaying V_{R1}. If the scope probe is connected across the resistor, as indicated in Figure 12–31(b), the ground lead of the scope will short point B to ground, thus bypassing the rest of the components and effectively removing them from the circuit electrically, as illustrated in part (c) (assuming that the scope is not isolated from power line ground).

To avoid this problem, we can reposition R_1 in the circuit (when possible) so that one end of it is connected to ground, as shown in Figure 12–32(a). This connection

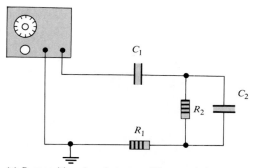

(a) R_1 repositioned so that one end is grounded

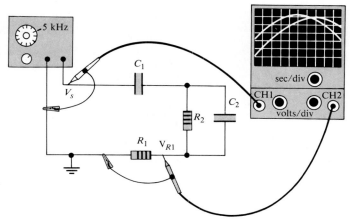

(b) The scope displays a half-cycle of V_{R1} and V_s.
V_{R1} represents the phase of the total current.

FIGURE 12–32

Repositioning R_1 so that a direct voltage measurement can be made with respect to ground without shorting out part of the circuit.

does not alter the circuit electrically, because R_1 still has the same series relationship with the rest of the circuit. Now the scope can be connected across it to display V_{R1}, as indicated in part (b) of the figure. The other probe is connected across the voltage source to display V_s as indicated. Now channel 1 of the scope has V_s as an input, and channel 2 has V_{R1}. The *trigger source* switch on the scope should be on internal so that each trace on the screen will be triggered by one of the inputs and the other will then be shown in the proper time relationship to it. Since amplitudes are not important, the *volts/div* settings are arbitrary. The *time/div* settings should be adjusted so that one half-cycle of the waveforms appears on the screen.

Before connecting the probes to the circuit, we must align the two horizontal lines (traces) so that they appear as a single line across the center of the screen. To do so, ground the probe tips and adjust the *vertical position* knobs to move the traces toward the center line of the screen until they are superimposed. The reason for this procedure is to ensure that both waveforms have the same zero crossing so that an accurate phase measurement can be made.

The resulting oscilloscope display is shown in Figure 12–33. Since there is 180° in one half-cycle, each of the ten horizontal divisions across the screen represents 18°. Thus, the horizontal distance between the corresponding points of the two waveforms is the phase angle in degrees as indicated.

FIGURE 12–33
Measurement of the phase angle on the oscilloscope.

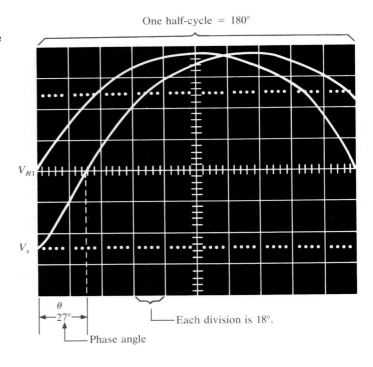

SECTION REVIEW 12–6

1. Determine the series equivalent circuit for the series-parallel circuit in Figure 12–29.
2. What is the voltage across R_1 in Figure 12–29?

12–7

POWER IN *RC* CIRCUITS

In a purely resistive ac circuit, all the energy delivered by the source is dissipated in the form of heat by the resistance. In a purely capacitive ac circuit, all the energy delivered by the source is stored by the capacitor during a portion of the voltage cycle and then is returned to the source during another portion of the cycle so that there is no net energy transferred. What happens when both resistance and capacitance exist in a circuit? *Some of the energy is alternately stored and returned by the capacitance, and some is dissipated by the resistance.* The amount of energy dissipated is determined by the relative values of the resistance and the capacitive reactance.

It is reasonable to assume that when the resistance is greater than the reactance, more of the total energy delivered by the source is dissipated by the resistance than is stored by the capacitance. Likewise, when the reactance is greater than the resistance, more of the total energy is stored and returned than is dissipated.

The formulas for power in a resistor, sometimes called *true power* (P_{true}), and the power in a capacitor, called *reactive power* (P_r), are restated here. The unit of true power is the *watt* (W), and the unit of reactive power is the *volt-ampere reactive* (VAR):

$$P_{true} = I_T^2 R \qquad (12-21)$$

$$P_r = I_T^2 X_C \qquad (12-22)$$

The Power Triangle

The generalized impedance phasor diagram is shown in Figure 12–34(a). A phasor relationship for power can also be represented by a similar diagram, because the respective magnitudes of the powers, P_{true} and P_r, differ from R and X_C by a factor of I_T^2, as shown in Figure 12–34(b).

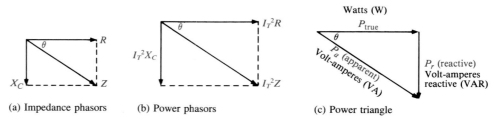

(a) Impedance phasors   (b) Power phasors   (c) Power triangle

FIGURE 12–34
Development of the power triangle for an *RC* circuit.

The resultant power phasor, $I_T^2 Z$, represents the *apparent power*, P_a. At any instant in time, P_a is the total power that *appears* to be transferred between the source and the *RC* circuit. Part of the apparent power is true power, and part of it is reactive power. The unit of apparent power is the *volt-ampere* (VA). The expression for apparent power is

$$P_a = I_T^2 Z \qquad (12-23)$$

The power phasor diagram in Figure 12–34(b) can be rearranged in the form of a right triangle as shown in part (c), which is called the *power triangle*. Using the rules of trigonometry, P_{true} can be expressed as

$$P_{\text{true}} = P_a\cos\theta$$

Since P_a equals $I_T^2 Z$ or $V_s I_T$, the equation for the true power can be written as

$$P_{\text{true}} = V_s I_T \cos\theta \qquad (12\text{–}24)$$

where V_s is the applied voltage and I_T is the total current.

For the case of a purely resistive circuit, $\theta = 0°$ and $\cos 0° = 1$, so P_{true} equals $V_s I_T$. For the case of a purely capacitive circuit, $\theta = 90°$ and $\cos 90° = 0$, so P_{true} is zero. As you already know, there is no power dissipated in an ideal capacitor.

Power Factor

The term $\cos\theta$ is called the *power factor* and is stated as follows:

$$PF = \cos\theta \qquad (12\text{–}25)$$

As the phase angle between applied voltage and total current increases, the power factor decreases, indicating an increasingly reactive circuit. The smaller the power factor, the smaller the power dissipation.

The power factor can vary from 0 for a purely reactive circuit to 1 for a purely resistive circuit. In an *RC* circuit, the power factor is referred to as a *leading* power factor because the current leads the voltage.

EXAMPLE 12–12   Determine the power factor and the true power in the circuit of Figure 12–35.

FIGURE 12–35

$$V_s = 15\text{ V}$$
$$f = 10\text{ kHz}$$

Solution

$$X_C = \frac{1}{2\pi f C} = \frac{1}{2\pi(10\text{ kHz})(0.005\ \mu\text{F})} = 3183\ \Omega$$

$$\theta = \tan^{-1}\left(\frac{X_C}{R}\right) = \tan^{-1}\left(\frac{3183\ \Omega}{1000\ \Omega}\right) = 72.56°$$

$$PF = \cos\theta = \cos(72.56°) = 0.2997$$

$$Z = \sqrt{R^2 + X_C^2} = \sqrt{(1000\ \Omega)^2 + (3183\ \Omega)^2} = 3336.4\ \Omega$$

$$I = \frac{V_s}{Z} = \frac{15\text{ V}}{3336.4\ \Omega} = 4.496\text{ mA}$$

The true power is

$$P_{\text{true}} = V_s I \cos \theta = (15 \text{ V})(4.496 \text{ mA})(0.2997) = 20.21 \text{ mW}$$

EXAMPLE 12–13

For the circuit in Figure 12–36, find the true power, the reactive power, and the apparent power. X_C has been determined to be 2 kΩ.

FIGURE 12–36

Solution

We first find the total impedance so that the current can be calculated.

$$Z = \sqrt{R^2 + X_C^2} = \sqrt{(1000 \text{ } \Omega)^2 + (2000 \text{ } \Omega)^2} = 2236 \text{ } \Omega$$

$$I = \frac{V_s}{Z} = \frac{10 \text{ V}}{2236 \text{ } \Omega} = 4.47 \text{ mA}$$

The phase angle, θ, is

$$\theta = \tan^{-1}\left(\frac{2 \text{ k}\Omega}{1 \text{ k}\Omega}\right) = 63.44°$$

The true power is

$$P_{\text{true}} = V_s I \cos \theta = (10 \text{ V})(4.47 \text{ mA}) \cos(63.44°) = 19.99 \text{ mW}$$

(The same result is realized using the formula $P_{\text{true}} = I^2 R$.)
    The reactive power is

$$P_r = I^2 X_C = (4.47 \text{ mA})^2 (2 \text{ k}\Omega) = 39.96 \text{ mVAR}$$

The apparent power is

$$P_a = I^2 Z = (4.47 \text{ mA})^2 (2236 \text{ } \Omega) = 44.68 \text{ mVA}$$

The apparent power is also the phasor sum of P_{true} and P_r:

$$P_a = \sqrt{P_{\text{true}}^2 + P_r^2} = 44.68 \text{ mVA}$$

The Significance of Apparent Power

As mentioned, apparent power is the power that *appears* to be transferred between the source and the load, and it consists of two components: a true power component and a reactive power component.

In all electrical and electronic systems, it is the true power that does the work. The reactive power is simply shuttled back and forth between the source and load. Ideally, in terms of performing useful work, all of the power transferred to the load should be true power and none of it reactive power. However, in most practical situations the load must have some reactance associated with it, and therefore we must deal with both power components.

In the last chapter, the use of apparent power was discussed in relation to transformers. For any reactive load, there are two components of the total current: the resistive component and the reactive component. If we consider only the true power (watts) in a load, we are dealing with only a portion of the total current that the load demands from a source. In order to have a realistic picture of the actual current that a load will draw, we must consider apparent power (in VA).

A source such as an ac generator can provide current to a load up to some maximum value. *If the load draws more than this maximum value, the source can be damaged.* Figure 12–37(a) shows a 120-V generator that can deliver a maximum current of 5 A to a load. Assume that the generator is rated at 600 W and is connected to a purely resistive load of 24 Ω (power factor of 1). The ammeter shows that the current is 5 A, and the wattmeter indicates that the power is 600 W. The generator has no problem under these conditions, although it is operating at maximum current and power.

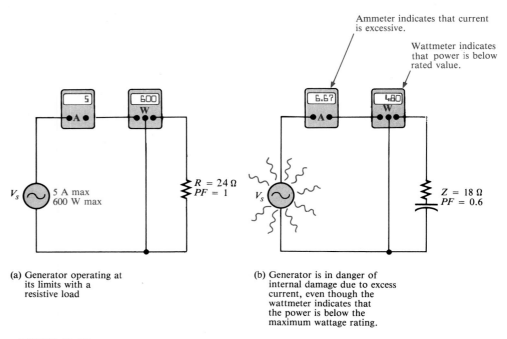

(a) Generator operating at its limits with a resistive load

(b) Generator is in danger of internal damage due to excess current, even though the wattmeter indicates that the power is below the maximum wattage rating.

FIGURE 12–37

The wattage rating of a source is inappropriate when the load is reactive. The rating should be in VA rather than in watts.

Now, consider what happens if the load is changed to a reactive one with an impedance of 18 Ω and a power factor of 0.6, as indicated in Figure 12–37(b). The current is 120 V/18 Ω = 6.67 A, which *exceeds* the maximum. Even though the wattmeter reads 480 W, which is less than the power rating of the generator, the excessive current probably will cause damage. This example shows that a true power rating can be deceiving and is inappropriate for ac sources. The ac generator should be rated at 600 VA, a rating that manufacturers generally use, rather than 600 W.

SECTION REVIEW
12–7

1. To which component in an *RC* circuit is the power dissipation due?
2. If the phase angle, θ, is 45°, what is the power factor?
3. A certain series *RC* circuit has the following parameter values: $R = 300\ \Omega$, $X_C = 460\ \Omega$, and $I = 2$ A. Determine the true power, the reactive power, and the apparent power.

12–8

BASIC APPLICATIONS OF *RC* CIRCUITS

In this section, several basic types of *RC* circuits that have applications in a broad range of areas are introduced. The basic circuits to be discussed are *phase-shift networks*, *filters*, and *signal-coupling networks*.

The *RC* Lag Network

The *RC* lag network is a phase-shift circuit in which the output voltage *lags* the input voltage by a specified angle. Phase-shift circuits are commonly used in electronic communication systems and in other application areas.

A basic *RC* lag network is shown in Figure 12–38(a). Notice that *the output voltage is taken across the capacitor,* and the input voltage is the total voltage applied across the circuit. The relationship of the voltages is shown in the phasor diagram in Figure 12–38(b). Keep in mind that the circuit phase angle is the angle between V_{in} and

(a) A basic *RC* lag network

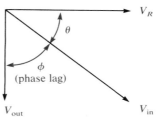

(b) Phasor voltage diagram showing the phase lag between V_{in} and V_{out}

FIGURE 12–38
The *RC* lag network ($V_{out} = V_C$).

the current. Notice that V_C, which is the output voltage, *lags* V_{in} by an angle that is the difference between 90° and the circuit phase angle θ.

Since $\theta = \tan^{-1}(X_C/R)$, the phase lag, ϕ, can be expressed as follows:

$$\phi = 90° - \tan^{-1}\left(\frac{X_C}{R}\right) \tag{12-26}$$

When the input and output waveforms are displayed on an oscilloscope, a relationship similar to that in Figure 12–39 is observed. The exact amount of phase lag between the input and the output depends on the values of the resistance and the capacitive reactance. The magnitude of the output voltage depends on these values also.

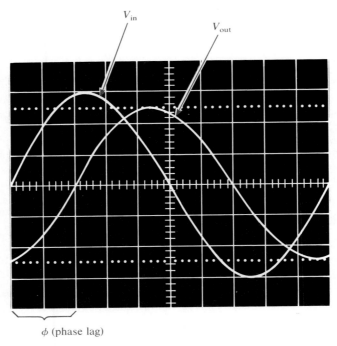

ϕ (phase lag)

FIGURE 12–39
General oscilloscope display of the input and output waveforms of a lag network (V_{out} lags V_{in}).

EXAMPLE 12–14   Determine the amount of phase lag from input to output in the lag network in Figure 12–40.

FIGURE 12–40

Solution

$$X_C = \frac{1}{2\pi f C} = \frac{1}{2\pi(1\ \text{kHz})(0.1\ \mu\text{F})} = 1592\ \Omega$$

$$\phi = 90° - \tan^{-1}\left(\frac{X_C}{R}\right) = 90° - \tan^{-1}\left(\frac{1592\ \Omega}{680\ \Omega}\right) = 23.13°$$

The output lags the input by 23.13°.

The phase-lag network can be viewed as a voltage divider with a portion of the input voltage dropped across R and a portion across C. The output voltage can be determined with the following formula:

$$V_{\text{out}} = \left(\frac{X_C}{\sqrt{R^2 + X_C^2}}\right)V_{\text{in}} \qquad (12\text{--}27)$$

EXAMPLE 12–15

For the lag network in Figure 12–40 of Example 12–14, determine the output voltage in phasor form when the input voltage has an rms value of 10 V. Sketch the input and output waveforms showing the proper relationships. ϕ was found in Example 12–14 to be 23.13°.

Solution

$$V_{\text{out}} = \left(\frac{X_C}{\sqrt{R^2 + X_C^2}}\right)V_{\text{in}}$$

$$= \left(\frac{1592\ \Omega}{\sqrt{(680\ \Omega)^2 + (1592\ \Omega)^2}}\right)10\ \text{V} = 9.2\ \text{V}$$

The waveforms are shown in Figure 12–41.

FIGURE 12–41

Effects of Frequency on the Phase Shift and Output Voltage of the Lag Network

Since the circuit phase angle, θ, decreases as frequency increases, the phase lag between the input and the output voltages *increases*. You can see this relationship by examining Equation (12–26). Also, the magnitude of V_{out} decreases as the frequency increases, because X_C becomes smaller and less of the total input voltage is dropped across the capacitor. Figure 12–42 illustrates the changes in phase lag and output voltage magnitude with frequency.

(a) When f increases, the phase lag ϕ increases and V_{out} decreases.

(b) When f decreases, ϕ decreases and V_{out} increases.

FIGURE 12–42
Illustration of how the frequency affects the phase lag and the output voltage in an *RC* lag network, with the amplitude of V_{in} held constant.

The *RC* Lead Network

The *RC* lead network is a phase-shift circuit in which the output voltage *leads* the input voltage by a specified angle. A basic *RC* lead network is shown in Figure 12–43(a). Notice how it differs from the lag network. Here *the output voltage is taken across the resistor*. The relationship of the voltages is given in the phasor diagram in Figure 12–43(b). The output voltage, V_R, *leads* V_{in} by an angle that is the same as the circuit phase angle, because V_R and I are in phase with each other.

(a) A basic *RC* lead network

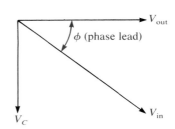

(b) Phasor voltage diagram showing the phase lead between V_{in} and V_{out}

FIGURE 12–43
The *RC* lead network ($V_{out} = V_R$).

When the input and output waveforms are displayed on an oscilloscope, a relationship similar to that in Figure 12–44 is observed. Of course, the exact amount of phase lead and the output voltage magnitude depend on the values of R and X_C. The phase lead, ϕ, is expressed as follows:

$$\phi = \tan^{-1}\left(\frac{X_C}{R}\right) \tag{12–28}$$

The output voltage is expressed as

$$V_{out} = \left(\frac{R}{\sqrt{R^2 + X_C^2}}\right)V_{in} \tag{12–29}$$

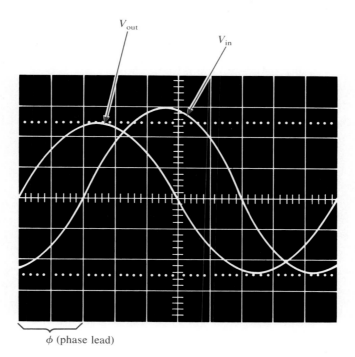

FIGURE 12–44
General oscilloscope display of the input and output waveforms of a lead network (V_{out} leads V_{in}).

EXAMPLE 12–16

Calculate the phase lead and the output voltage for the circuit in Figure 12–45.

FIGURE 12–45

Solution

$$\phi = \tan^{-1}\left(\frac{X_C}{R}\right) = \tan^{-1}\left(\frac{150\ \Omega}{220\ \Omega}\right) = 34.29°$$

The output leads the input by 34.29°.

$$V_{out} = \left(\frac{R}{\sqrt{R^2 + X_C^2}}\right)V_{in} = \left(\frac{220\ \Omega}{\sqrt{(220\ \Omega)^2 + (150\ \Omega)^2}}\right)10\ \text{V}$$

$$= 8.26\ \text{V}$$

The calculator sequence for V_{out} is

Effects of Frequency on the Phase Shift and Output Voltage of the Lead Network

Since the phase lead is the same as the circuit phase angle θ, it *decreases* as frequency increases. The output voltage increases with frequency because as X_C becomes smaller,

Phase meter measures angle between V_{in} and V_{out} (phase lead).

Voltmeter measures V_{out} with respect to ground.

(a) When f increases, the phase lead ϕ decreases and V_{out} increases.

(b) When f decreases, ϕ increases and V_{out} decreases.

FIGURE 12–46
Illustration of how the frequency affects the phase lead and output voltage in an *RC* lead network with the amplitude of V_{in} held constant.

more of the input voltage is dropped across the resistor. Figure 12–46 illustrates this relationship.

The *RC* Circuit as a Filter

Filters are frequency-selective circuits that permit signals of certain frequencies to pass from the input to the output while blocking all others. That is, all frequencies but the selected ones are *filtered* out.

Series *RC* circuits exhibit a frequency-selective characteristic and therefore act as simple passive filters. There are two types. The first one that we examine, called a *low-pass filter,* is realized by taking the output across the capacitor, just as in a lag network. The second type, called a *high-pass filter,* is implemented by taking the output across the resistor, as in a lead network.

The *RC* Low-Pass Filter

You have seen what happens to the phase and the output voltage in the lag network. In terms of its filtering action, we are interested primarily in how the magnitude of the output voltage varies with frequency.

To illustrate low-pass filter action, Figure 12–47 shows a specific series of measurements in which the frequency starts at zero (dc) and is increased in increments up to 20 kHz. At each value of frequency, the output voltage is measured. As you can see, the capacitive reactance decreases as frequency goes up, thus dropping less voltage across the capacitor while the input voltage is held at a constant 10 V throughout each step. Table 12–1 summarizes the variation of the circuit parameters with frequency.

(a) $f = 0$ Hz, $X_C = \infty$, $V_{out} = 10$ V

(b) $f = 1$ kHz, $X_C = 159$ Ω, $V_{out} = 8.5$ V

(c) $f = 10$ kHz, $X_C = 15.9$ Ω, $V_{out} = 1.57$ V

(d) $f = 20$ kHz, $X_C = 7.96$ Ω, $V_{out} = 0.79$ V

FIGURE 12–47
Example of low-pass filter action. As frequency increase, V_{out} decreases.

TABLE 12–1

f	X_C (Ω)	Z_T (Ω)	I (mA)	V_{out} (V)
0	∞	∞	0	10
1 kHz	159	187.8	53.25	8.5
10 kHz	15.9	101.3	98.7	1.57
20 kHz	7.96	100.3	99.7	0.79

In Figure 12–48, the measured values are plotted on a graph of V_{out} versus frequency, and a smooth curve is drawn connecting the points. This graph, called a *response curve*, shows that the output voltage is greater at the lower frequencies and decreases as the frequency increases.

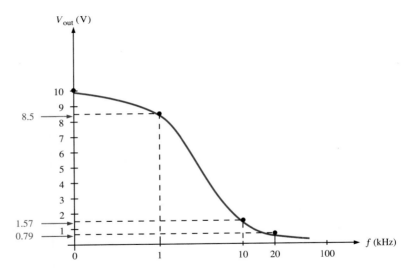

FIGURE 12–48
Frequency response curve for the low-pass filter in Figure 12–47.

The *RC* High-Pass Filter

To illustrate high-pass filter action, Figure 12–49 shows a series of specific measurements. Again, the frequency starts at zero (dc) and is increased in increments up to 10 kHz. As you can see, the capacitive reactance decreases as the frequency goes up, thus causing more of the total input voltage to be dropped across the resistor. Table 12–2 summarizes the variation of circuit parameters with frequency.

TABLE 12–2

f	X_C (Ω)	Z_T (Ω)	I (mA)	V_{out} (V)
0	∞	∞	0	0
100 Hz	1590	1593	6.3	0.63
1 kHz	159	187.8	53.2	5.32
10 kHz	15.9	101.3	98.7	9.88

(a) $f = 0$ Hz, $X_C = \infty$, $V_{out} = 0$ V

(b) $f = 100$ Hz, $X_C = 1.59$ kΩ, $V_{out} = 0.63$ V

(c) $f = 1$ kHz, $X_C = 159$ Ω, $V_{out} = 5.32$ V

(d) $f = 10$ kHz, $X_C = 15.9$ Ω, $V_{out} = 9.87$ V

FIGURE 12–49
Example of high-pass filter action. As frequency increases, V_{out} increases.

In Figure 12–50, the measured values have been plotted to produce a response curve for this circuit. As you can see, the output voltage is greater at the higher frequencies and decreases as the frequency is reduced.

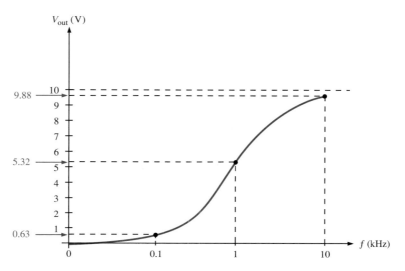

FIGURE 12–50
Frequency response curve for the high-pass filter in Figure 12–49.

The Cutoff Frequency and the Bandwidth of a Filter

The frequency at which the capacitive reactance equals the resistance in a low-pass or high-pass *RC* filter is called the *cutoff frequency* and is designated f_c. This condition is expressed as $1/(2\pi f_c C) = R$. Solving for f_c, we get the following formula:

$$f_c = \frac{1}{2\pi RC} \tag{12–30}$$

At f_c, the output voltage of the filter is 70.7% of its maximum value. It is standard practice to consider the cutoff frequency as the limit of a filter's performance in terms of passing or rejecting frequencies. For example, in a high-pass filter, all frequencies above f_c are considered to be passed by the filter, and all those below f_c are considered to be rejected. The reverse is true for a low-pass filter.

The range of frequencies that are considered to be passed by a filter is called the *bandwidth*. Figure 12–51 illustrates the bandwidth and the cutoff frequency for a low-pass filter.

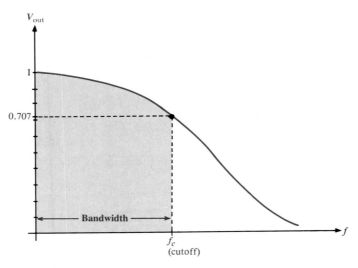

FIGURE 12–51
Normalized general response curve of a low-pass filter showing the cutoff frequency and the bandwidth.

Coupling an ac Signal into a dc Bias Network

Figure 12–52 shows an *RC* network that is used to create a dc voltage level with an ac voltage superimposed on it. This type of circuit is commonly found in amplifiers in which the dc voltage is required to *bias* the amplifier to the proper operating point and the signal voltage to be amplified is coupled through a capacitor and superimposed on the dc level. The capacitor prevents the low internal resistance of the signal source from affecting the dc bias voltage.

In this type of application, a relatively high value of capacitance is selected so that for the frequencies to be amplified, the reactance is very small compared to the

FIGURE 12–52
Amplifier bias and signal-coupling circuit.

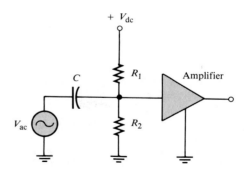

resistance of the bias network. When the reactance is very small (ideally zero), there is practically no phase shift or signal voltage dropped across the capacitor. Therefore, all of the signal voltage passes from the source to the input to the amplifier.

Figure 12–53 illustrates the application of the superposition principle to the circuit in Figure 12–52. In part (a), we have effectively removed the ac source from the circuit by replacing it with a short to represent its ideal internal resistance. Since C is open to dc, the voltage at point A is determined by the voltage divider action of R_1 and R_2 and the dc voltage source.

In part (b), we have effectively removed the dc source from the circuit by replacing it with a short to represent its ideal internal resistance. Since C appears as a short at the frequency of the ac, the signal voltage is coupled directly to point A and appears across the parallel combination of R_1 and R_2. Part (c) illustrates that the combined effect of the superposition of the dc and the ac voltages results in the signal voltage "riding" on the dc level, as was discussed in Chapter 8.

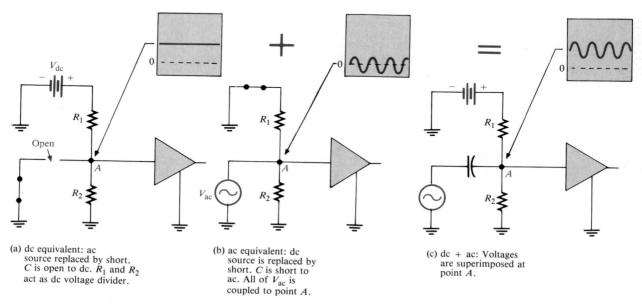

(a) dc equivalent: ac source replaced by short. C is open to dc. R_1 and R_2 act as dc voltage divider.

(b) ac equivalent: dc source is replaced by short. C is short to ac. All of V_{ac} is coupled to point A.

(c) dc + ac: Voltages are superimposed at point A.

FIGURE 12–53
The superposition of dc and ac voltages in an *RC* bias and coupling circuit.

1. A certain *RC* lag network consists of a 4.7-kΩ resistor and a 0.022-μF capacitor. Determine the phase lag between input and output at a frequency of 3 kHz.

2. An *RC* lead network has the same component values as the lag network in Problem 1. What is the magnitude of the output voltage at 3 kHz when the input is 10 V rms?

3. When an *RC* circuit is used as a low-pass filter, across which component is the output taken?

12–9

TROUBLESHOOTING *RC* CIRCUITS

In this section, we consider the effects that typical component failures or degradation have on the response of basic *RC* circuits. The cases to be considered are open resistor, open capacitor, shorted capacitor, and excessive leakage in a capacitor.

Effects of an Open Resistor

It is very easy to see how an open resistor affects the operation of a basic series *RC* circuit, as shown in Figure 12–54. Obviously, there is no path for current, so the capacitor voltage remains at zero; thus, the total voltage, V_s, appears across the open resistor.

FIGURE 12–54
Effect of an open resistor.

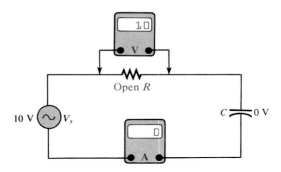

Effects of an Open Capacitor

When the capacitor is open, there is no current; thus, the resistor voltage remains at zero. The total source voltage is across the open capacitor, as shown in Figure 12–55.

FIGURE 12–55
Effect of an open capacitor.

Effects of a Shorted Capacitor

When a capacitor shorts out, the voltage across it is zero, the current equals V_s/R, and the total voltage appears across the resistor, as shown in Figure 12–56.

FIGURE 12–56
Effect of a shorted capacitor.

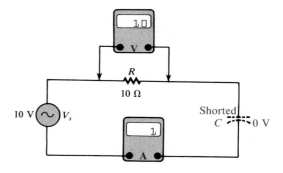

Effects of a Leaky Capacitor

When a capacitor exhibits a high leakage current, the leakage resistance effectively appears in parallel with the capacitor, as shown in Figure 12–57(a). When the leakage

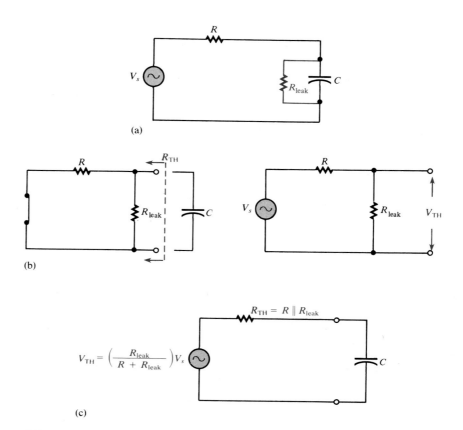

FIGURE 12–57
Effects of a leaky capacitor.

AC CIRCUITS

resistance is comparable in value to the circuit resistance, R, the circuit response is drastically affected. The circuit, looking from the capacitor toward the source, can be Thevenized, as shown in Figure 12–57(b). The Thevenin equivalent resistance is R in parallel with R_{leak} (the source appears as a short), and the Thevenin equivalent voltage is determined by the voltage divider action of R and R_{leak}. The Thevenin equivalent circuit is shown in Figure 12–57(c).

$$R_{TH} = R \parallel R_{leak}$$

$$R_{TH} = \frac{RR_{leak}}{R + R_{leak}} \tag{12–31}$$

$$V_{TH} = \frac{R_{leak}V_{in}}{R + R_{leak}} \tag{12–32}$$

As you can see, the voltage to which the capacitor will charge is reduced, since $V_{TH} < V_s$. Also, the circuit time constant is reduced, and the current is increased.

EXAMPLE 12–17

Assume that the capacitor in Figure 12–58 is degraded to a point where its leakage resistance is 10 kΩ. Determine the phase shift from input to output and the output voltage under the degraded condition.

FIGURE 12–58

Solution

The effective circuit resistance is

$$R_{TH} = \frac{RR_{leak}}{R + R_{leak}} = \frac{(4.7 \text{ k}\Omega)(10 \text{ k}\Omega)}{14.7 \text{ k}\Omega} = 3.2 \text{ k}\Omega$$

The phase shift is

$$\phi = 90° - \tan^{-1}\left(\frac{X_C}{R_{TH}}\right) = 90° - \tan^{-1}\left(\frac{5 \text{ k}\Omega}{3.2 \text{ k}\Omega}\right) = 32.62°$$

$$V_{TH} = \left(\frac{R_{leak}}{R + R_{leak}}\right)V_{in} = \left(\frac{10 \text{ k}\Omega}{14.7 \text{ k}\Omega}\right)10 \text{ V} = 6.8 \text{ V}$$

$$V_{out} = \left(\frac{X_C}{\sqrt{R_{TH}^2 + X_C^2}}\right)V_{TH}$$

$$= \left(\frac{5 \text{ k}\Omega}{\sqrt{(3.2 \text{ k}\Omega)^2 + (5 \text{ k}\Omega)^2}}\right)(6.8 \text{ V}) = 5.73 \text{ V}$$

<table>
<tr><td>

SECTION REVIEW
12–9

</td><td>

1. Describe the effect of a leaky capacitor on the response of an *RC* circuit.
2. In a series *RC* circuit, all the applied voltage appears across an open capacitor (T or F).

</td></tr>
</table>

SUMMARY

☐ When a sine wave voltage is applied to an *RC* circuit, the current and all the voltage drops are also sine waves.

☐ Total current in an *RC* circuit always leads the source voltage.

☐ The resistor voltage is always in phase with the current.

☐ The capacitor voltage always lags the current by 90°.

☐ In an *RC* circuit, the impedance is determined by both the resistance and the capacitive reactance combined.

☐ Impedance is expressed in units of ohms.

☐ The circuit phase angle is the angle between the total current and the applied (source) voltage.

☐ The impedance of a series *RC* circuit varies inversely with frequency.

☐ The phase angle *(θ)* of a series *RC* circuit varies inversely with frequency.

☐ For each parallel *RC* circuit, there is an equivalent series circuit for any given frequency.

☐ The impedance of a circuit can be determined by measuring the applied voltage and the total current and then applying Ohm's law.

☐ In an *RC* circuit, part of the power is resistive and part reactive.

☐ The phasor combination of resistive power (true power) and reactive power is called *apparent power*.

☐ Apparent power is expressed in volt-amperes (VA).

☐ The power factor *(PF)* indicates how much of the apparent power is true power.

☐ A power factor of 1 indicates a purely resistive circuit, and a power factor of 0 indicates a purely reactive circuit.

☐ In a lag network, the output voltage lags the input voltage in phase.

☐ In a lead network, the output voltage leads the input voltage.

☐ A filter passes certain frequencies and rejects others.

GLOSSARY

Admittance   A measure of the ability of a reactive circuit to permit current. The reciprocal of impedance with the unit of siemens (S).

Apparent power   The product of volts and amperes with the unit of volt-amperes (VA).

Bandwidth   The range of frequencies that are considered to be passed by a filter.

Bias   The application of a dc voltage to an electronic device to produce a desired mode of operation.

Capacitive susceptance   The ability of a capacitor to permit current.

Cutoff frequency   The frequency at which the output voltage of a filter is 70.7% of the maximum output voltage.

Filter   A type of circuit that passes certain frequencies and rejects all others.

High-pass   The characteristic of a certain type of filter whereby higher frequencies are passed and lower frequencies are rejected.

Impedance   The total opposition to sinusoidal current expressed in ohms.

Lag   A condition of the phase or time relationship of waveforms in which one waveform is behind the other in phase or time.

Lead   A condition of the phase or time relationship of waveforms in which one waveform is ahead of the other in phase or time.

Low-pass   The characteristic of a certain type of filter whereby lower frequencies are passed and higher frequencies are rejected.

Power factor   The relationship between volt-amperes and true power or watts. Volt-amperes multiplied by the power factor equals true power.

Response   In electrical circuits, the reaction of a circuit to a given input.

FORMULAS

$$Z = \sqrt{R^2 + X_C^2} \qquad \text{Series } RC \text{ impedance} \tag{12–1}$$

$$\theta = \tan^{-1}\left(\frac{X_C}{R}\right) \qquad \text{Series } RC \text{ phase angle} \tag{12–2}$$

$$V = IZ \qquad \text{Ohm's law} \tag{12–3}$$

$$I = \frac{V}{Z} \qquad \text{Ohm's law} \tag{12–4}$$

$$Z = \frac{V}{I} \qquad \text{Ohm's law} \tag{12–5}$$

$$V_s = \sqrt{V_R^2 + V_C^2} \qquad \text{Total voltage in series } RC \tag{12–6}$$

$$\theta = \tan^{-1}\left(\frac{V_C}{V_R}\right) \qquad \text{Series } RC \text{ phase angle} \tag{12–7}$$

$$Z = \frac{RX_C}{\sqrt{R^2 + X_C^2}} \qquad \text{Parallel } RC \text{ impedance} \tag{12–8}$$

$$\theta = \tan^{-1}\left(\frac{R}{X_C}\right) \qquad \text{Parallel } RC \text{ phase angle} \tag{12–9}$$

$$G = \frac{1}{R} \qquad \text{Conductance} \tag{12–10}$$

$$B_C = \frac{1}{X_C} \qquad \text{Capacitive susceptance} \tag{12–11}$$

$$Y = \frac{1}{Z} \qquad \text{Admittance} \tag{12–12}$$

$$Y = \sqrt{G^2 + B_C^2} \qquad \text{Admittance} \tag{12–13}$$

$$V = \frac{I}{Y} \qquad \text{Ohm's law} \tag{12–14}$$

$$I = VY \qquad \text{Ohm's law} \tag{12–15}$$

$$Y = \frac{I}{V} \quad \text{Ohm's law} \tag{12-16}$$

$$I_T = \sqrt{I_R^2 + I_C^2} \quad \text{Total current in parallel } RC \text{ circuits} \tag{12-17}$$

$$\theta = \tan^{-1}\left(\frac{I_C}{I_R}\right) \quad \text{Parallel } RC \text{ phase angle} \tag{12-18}$$

$$R_{eq} = Z \cos \theta \quad \text{Equivalent series resistance} \tag{12-19}$$

$$X_{C(eq)} = Z \sin \theta \quad \text{Equivalent series reactance} \tag{12-20}$$

$$P_{true} = I_T^2 R \quad \text{True power (W)} \tag{12-21}$$

$$P_r = I_T^2 X_C \quad \text{Reactive power (VAR)} \tag{12-22}$$

$$P_a = I_T^2 Z \quad \text{Apparent power (VA)} \tag{12-23}$$

$$P_{true} = V_s I_T \cos \theta \quad \text{True power} \tag{12-24}$$

$$PF = \cos \theta \quad \text{Power factor} \tag{12-25}$$

$$\phi = 90° - \tan^{-1}\left(\frac{X_C}{R}\right) \quad \text{Phase lag} \tag{12-26}$$

$$V_{out} = \left(\frac{X_C}{\sqrt{R^2 + X_C^2}}\right) V_{in} \quad \text{Output of lag network} \tag{12-27}$$

$$\phi = \tan^{-1}\left(\frac{X_C}{R}\right) \quad \text{Phase lead} \tag{12-28}$$

$$V_{out} = \left(\frac{R}{\sqrt{R^2 + X_C^2}}\right) V_{in} \quad \text{Output of lead network} \tag{12-29}$$

$$f_c = \frac{1}{2\pi RC} \quad \text{Cutoff frequency of an } RC \text{ filter} \tag{12-30}$$

$$R_{TH} = \frac{R R_{leak}}{R + R_{leak}} \quad \text{Thevenin equivalent resistance} \tag{12-31}$$

$$V_{TH} = \frac{R_{leak} V_{in}}{R + R_{leak}} \quad \text{Thevenin equivalent voltage} \tag{12-32}$$

SELF-TEST

1. In a series *RC* circuit, the voltage across the resistance is
   (a) in phase with the source voltage
   (b) lagging the source voltage by 90°
   (c) in phase with the current
   (d) lagging the current by 90°

2. In a series *RC* circuit, the voltage across the capacitor is
   (a) in phase with the source voltage
   (b) lagging the resistor voltage by 90°
   (c) in phase with the current
   (d) lagging the source voltage by 90°

3. When the frequency of the voltage applied to a series *RC* circuit is increased, the impedance
   (a) increases (b) decreases (c) remains the same (d) doubles

4. When the frequency of the voltage applied to a series *RC* circuit is decreased, the phase angle
   (a) increases (b) decreases (c) remains the same (d) becomes erratic

5. In a series *RC* circuit when the frequency and the resistance are doubled, the impedance
   (a) doubles (b) is halved (c) is quadrupled
   (d) cannot be determined without values

6. In a series *RC* circuit, 10 V rms is measured across the resistor and 10 V rms is also measured across the capacitor. The rms source voltage is
   (a) 20 V (b) 14.14 V (c) 28.28 V (d) 10 V

7. The voltages in Problem 6 are measured at a certain frequency. To make the resistor voltage greater than the capacitor voltage, the frequency
   (a) must be increased (b) must be decreased
   (c) is held constant (d) has no effect

8. When $R = X_C$, the phase angle is
   (a) 0° (b) +90° (c) −90° (d) 45°

9. To decrease the phase angle below 45°, the following condition must exist:
   (a) $R = X_C$ (b) $R < X_C$ (c) $R > X_C$ (d) $R = 10X_C$

10. When the frequency of the source voltage is increased, the impedance of a parallel *RC* circuit
    (a) increases (b) decreases (c) does not change

11. In a parallel *RC* circuit, there is 1 A rms through the resistive branch and 1 A rms through the capacitive branch. The total rms current is
    (a) 1 A (b) 2 A (c) 2.28 A (d) 1.414 A

12. A power factor of 1 indicates that the circuit phase angle is
    (a) 90° (b) 45° (c) 180° (d) 0°

13. For a certain load, the true power is 100 W and the reactive power is 100 VAR. The apparent power is
    (a) 200 VA (b) 100 VA (c) 141.4 VA (d) 141.4 W

14. Energy sources are normally rated in
    (a) watts (b) volt-amperes (c) volt-amperes reactive
    (d) none of these

15. If the bandwidth of a certain low-pass filter is 1 kHz, the cutoff frequency is
    (a) 0 Hz (b) 500 Hz (c) 2 kHz (d) 1000 Hz

PROBLEM SET A

Section 12–1 Sinusoidal Response of *RC* Circuits

1. A 8-kHz sinusoidal voltage is applied to a series *RC* circuit. What is the frequency of the voltage across the resistor? The capacitor?

2. What is the waveshape of the current in the circuit of Problem 1?

Section 12–2 Impedance and Phase Angle of a Series *RC* Circuit

3. Find the total impedance of each circuit in Figure 12–59.

4. Determine the impedance and the phase angle in each circuit in Figure 12–60.

FIGURE 12-59

(a)

(b)

(a)

(b)

FIGURE 12-60

Section 12-3 Analysis of Series *RC* Circuits

5. Calculate the total current in each circuit of Figure 12–59.

6. Repeat Problem 5 for the circuits in Figure 12–60.

7. For the circuit in Figure 12–61, draw the phasor diagram showing all voltages and the total current. Indicate the phase angles.

8. For the circuit in Figure 12–62, determine the following:
   (a) Z **(b)** I **(c)** V_R **(d)** V_C

FIGURE 12-61

FIGURE 12-62

9. To what value must the rheostat be set in Figure 12–63 to make the total current 10 mA? What is the resulting phase angle?

FIGURE 12-63

Section 12–4   Impedance and Phase Angle of a Parallel *RC* Circuit

10. Determine the impedance for the circuit in Figure 12–64.

11. Determine the impedance and the phase angle in Figure 12–65.

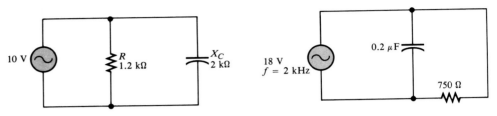

FIGURE 12–64   FIGURE 12–65

12. Repeat Problem 11 for the following frequencies:
    (a) 1.5 kHz   (b) 3 kHz   (c) 5 kHz   (d) 10 kHz

13. Determine the impedance and phase angle in Figure 12–66.

FIGURE 12–66

Section 12–5   Analysis of Parallel *RC* Circuits

14. For the circuit in Figure 12–67, find all the currents and voltages.

15. For the parallel circuit in Figure 12–68, find each branch current and the total current. What is the phase angle between the applied voltage and the total current?

FIGURE 12–67

FIGURE 12–68

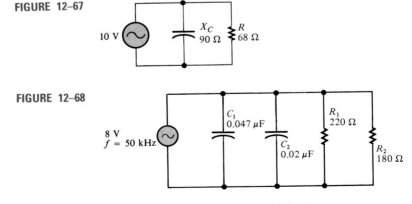

16. For the circuit in Figure 12–69, determine the following:
    (a) Z (b) I_R (c) I_C (d) I_T (e) θ

17. Repeat Problem 16 for $R = 5$ kΩ, $C = 0.05$ μF, and $f = 500$ Hz.

18. Convert the circuit in Figure 12–70 to an equivalent series form.

FIGURE 12–69   FIGURE 12–70

Section 12–6   Series-Parallel Analysis

19. Determine the voltages across each element in Figure 12–71. Sketch the voltage phasor diagram.

20. Is the circuit in Figure 12–71 predominantly resistive or predominantly capacitive?

21. Find the current through each branch and the total current in Figure 12–71. Sketch the current phasor diagram.

FIGURE 12–71

22. For the circuit in Figure 12–72, determine the following:

    (a) I_T **(b)** θ **(c)** V_{R1} **(d)** V_{R2} **(e)** V_{R3} **(f)** V_C

FIGURE 12–72

Section 12–7   Power in *RC* Circuits

23. In a certain series *RC* circuit, the true power is 2 W, and the reactive power is 3.5 VAR. Determine the apparent power.

24. In Figure 12–62, what is the true power and the reactive power?

25. What is the power factor for the circuit of Figure 12–70?

26. Determine P_{true}, P_r, P_a, and *PF* for the circuit in Figure 12–72. Sketch the power triangle.

Section 12–8   Basic Applications of *RC* Circuits

27. For the lag network in Figure 12–73, determine the phase lag between the input voltage and the output voltage for each of the following frequencies:

    (a) 1 Hz   **(b)** 100 Hz   **(c)** 1 kHz   **(d)** 10 kHz

28. The lag network in Figure 12–73 also acts as a low-pass filter. Draw a response curve for this circuit by plotting the output voltage versus frequency for 0 Hz to 10 kHz in 1-kHz increments.

FIGURE 12–73

29. Repeat Problem 27 for the lead network in Figure 12–74.

30. Plot the frequency response curve for the lead network in Figure 12–74 for a frequency range of 0 Hz to 10 kHz in 1-kHz increments.

FIGURE 12–74

31. Draw the voltage phasor diagram for each circuit in Figures 12–73 and 12–74 for a frequency of 5 kHz with V_{in} = 1 V rms.

32. The rms value of the signal voltage out of amplifier A in Figure 12–75 is 50 mV. If the input resistance to amplifier B is 10 kΩ, how much of the signal is lost due to the coupling capacitor when the frequency is 3 kHz?

FIGURE 12–75

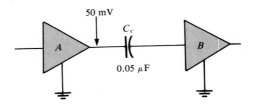

33. Determine the cutoff frequency for each circuit in Figures 12–73 and 12–74.

34. Determine the bandwidth of the circuit in Figure 12–73.

Section 12–9   Troubleshooting *RC* Circuits

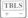

35. Assume that the capacitor in Figure 12–76 is excessively leaky. Show how this degradation affects the output voltage and phase angle, assuming that the leakage resistance is 5 kΩ and the frequency is 10 Hz.

FIGURE 12–76

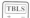

36. Each of the capacitors in Figure 12–77 has developed a leakage resistance of 2 kΩ. Determine the output voltages under this condition for each circuit.

37. Determine the output voltage for the circuit in Figure 12–77(a) for each of the following failure modes, and compare it to the correct output:
(a) R_1 open (b) R_2 open (c) C open (d) C shorted

38. Determine the output voltage for the circuit in Figure 12–77(b) for each of the following failure modes, and compare it to the correct output.
(a) C open (b) C shorted (c) R_1 open
(d) R_2 open (e) R_3 open

FIGURE 12–77

PROBLEM SET B

39. A single 220-V, 60-Hz source drives two loads. Load A has an impedance of 50 Ω and a power factor of 0.85. Load B has an impedance of 72 Ω and a power factor of 0.95.
(a) How much current does each load draw?
(b) What is the reactive power in each load?
(c) What is the true power in each load?
(d) What is the apparent power in each load?
(e) Which load has more voltage drop along the lines connecting it to the source?

40. What value of coupling capacitor is required in Figure 12–78 so that the signal voltage at the input of amplifier 2 is at least 70.7% of the signal voltage at the output of amplifier 1 when the frequency is 20 Hz? Disregard the input resistance of the amplifier.

FIGURE 12–78

41. Determine the value to which R_1 must be adjusted to get a phase angle of 30° between the source voltage and the total current in Figure 12–79.

FIGURE 12–79

42. Sketch the voltage and current phasor diagram for Figure 12–80.

43. A certain load dissipates 1.5 kW of power with an impedance of 12 Ω and a power factor of 0.75. What is its reactive power? What is its apparent power?

FIGURE 12-80

44. Determine the series element or elements that are in the block of Figure 12-81 to meet the following overall circuit requirements:
    (a) $P_{true} = 400$ W   (b) leading power factor (I_T leads V_s)
45. Determine the value of C_2 in Figure 12-82 when $V_A = V_B$.

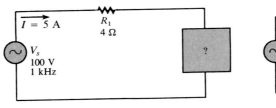

FIGURE 12-81

FIGURE 12-82

46. Draw the schematic for the circuit in Figure 12-83 and determine if the waveform on the scope is correct (a full-color photo appears following page 416). If there is a fault in the circuit, identify it.

ANSWERS TO SECTION REVIEWS

Section 12-1
1. 60 Hz, 60 Hz.   2. Closer to 0°.

Section 12-2
1. The opposition to sinusoidal current.   2. It lags.
3. The capacitive reactance.   4. 59.91 Ω, 56.6°.

Section 12-3
1. 7.2 V.   2. 56.3°.   3. 90°.
4. (a) X_C decreases;   (b) Z decreases;   (c) θ decreases.

Section 12-4
1. *Conductance* is the reciprocal of resistance; *capacitive susceptance* is the reciprocal of capacitive reactance; and *admittance* is the reciprocal of impedance.
2. $Y = 0.01$ S.   3. $Y = 0.024$ S.

Section 12-5
1. 21 mA.   2. 56.3°, 18 mA.   3. 90°.

Section 12–6

1. See Figure 12–84.   **2.** 6.79 V.

FIGURE 12–84

Section 12–7

1. Resistance.   **2.** 0.707.

3. $P_{\text{true}} = 1200$ W, $P_r = 1840$ VAR, $P_a = 2196.73$ VA.

Section 12–8

1. 62.84°.   **2.** 8.9 V rms.   **3.** Capacitor.

Section 12–9

1. The leakage resistance acts in parallel with C, which alters the circuit time constant.
2. T.

13

Frequency Response of *RL* Circuits

An *RL* circuit contains both resistance and inductance. In this chapter, basic series and parallel *RL* circuits and their responses to sine wave voltages are covered. In addition, series-parallel combinations are examined. Power considerations in *RL* circuits are introduced, and practical aspects of the power factor are discussed. A method of improving the power factor is presented, and basic areas of application are covered. As you study this chapter, note the difference in response of *RL* circuits compared to *RC* circuits.

After completing this chapter, you will be able to

☐ Discuss how an *RL* circuit responds to a sine wave input.
☐ Determine the impedance of both series and parallel *RL* circuits.
☐ Determine the phase angle between current and voltage in an *RL* circuit.
☐ Compare the response of an *RL* circuit to that of an *RC* circuit.
☐ Determine how frequency affects impedance and phase angle in an *RL* circuit.
☐ Define *inductive susceptance*.
☐ Apply Ohm's law to *RL* circuits to find currents and voltages.
☐ Determine true power, reactive power, and apparent power.
☐ Discuss further the significance of the power factor.
☐ Adjust the power factor of an inductive load.
☐ Apply *RL* circuits to phase-shift applications.
☐ Apply *RL* circuits to filter applications.
☐ Troubleshoot *RL* circuits for common component failures.

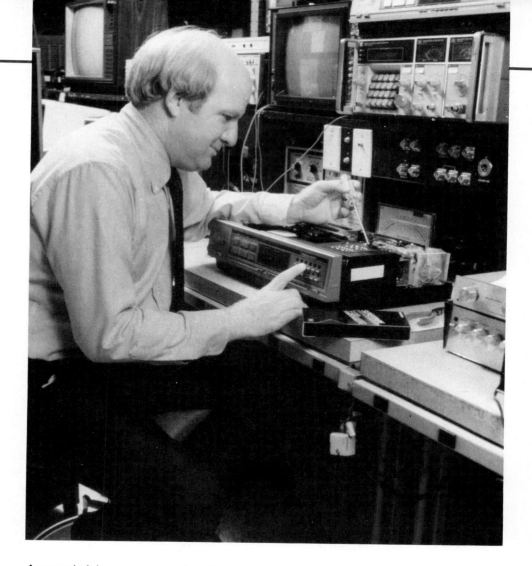

APPLICATION ASSIGNMENT

As a technician, you are assigned to set up a system of servomotors that drive an array of antennas. The purpose of the system is to gather information for a wildlife management study by tracking the movements of wild animals to which transmitters have been attached.

Each of the motors represents an inductive load that is effectively an RL circuit with defined values of R and L. Since the system must be portable, the size of the source that drives each motor must be kept to a minimum. Your project director has asked you to reduce the amount of current that each motor draws from the source and to keep the size of the interconnecting cables to a minimum. The director suggests that reducing the power factor of each motor is a way to accomplish this objective. After completing this chapter, you will be able to carry out this assignment.

13–1

SINUSOIDAL RESPONSE OF *RL* CIRCUITS

As with the *RC* circuit, all currents and voltages in an *RL* circuit are sinusoidal when the input is sinusoidal. Phase shifts are introduced because of the inductance. As you will learn, the resistor voltage and current are in phase with each other but lag the source voltage, and the inductor voltage leads the source voltage. The phase angle between the current and the inductor voltage is always 90°. These generalized phase relationships are indicated in Figure 13–1. Notice that they are opposite from those of the *RC* circuit, as discussed in Chapter 12.

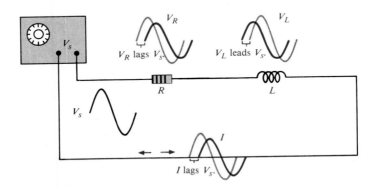

FIGURE 13–1
Illustration of sinusoidal response with general phase relationships of V_R, V_L, and I relative to V_s. V_R and I are in phase, V_R lags V_s, and V_L leads V_s. V_R and V_L are 90° out of phase with each other.

The amplitudes and the phase relationships of the voltages and current depend on the values of the resistance and the inductive reactance. When a circuit is purely inductive, the phase angle between the applied voltage and the total current is 90°, with the current *lagging* the voltage. When there is a combination of both resistance and inductive reactance in a circuit, the phase angle is somewhere between zero and 90°, depending on the relative values of R and X_L.

SECTION REVIEW 13–1

1. A 1-kHz sinusoidal voltage is applied to an *RL* circuit. What is the frequency of the resulting current?
2. When the resistance in an *RL* circuit is greater than the inductive reactance, is the phase angle between the applied voltage and the total current closer to 0° or to 90°?

13–2

IMPEDANCE AND PHASE ANGLE OF SERIES *RL* CIRCUITS

As you know, *impedance is the total opposition to sinusoidal current* in a circuit and is expressed in ohms. The impedance of a series *RL* circuit is determined by the resistance and the inductive reactance, as indicated in Figure 13–2.

FIGURE 13–2
Impedance of a series *RL* circuit.

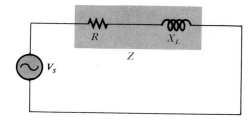

The Impedance Triangle

In ac analysis, both R and X_L are treated as *phasor quantities*, as shown in the phasor diagram of Figure 13–3(a), with X_L appearing at a $+90°$ angle with respect to R. This relationship comes from the fact that the inductor voltage leads the current, and thus the resistor voltage, by 90°. Since Z is the phasor sum of R and X_L, its phasor representation is as shown in Figure 13–3(b). A repositioning of the phasors, as shown in part (c), forms a right triangle. This formation, as you have learned, is called the *impedance triangle*. The length of each phasor represents the magnitude of the quantity, and θ is the phase angle between the applied voltage and the current in the *RL* circuit.

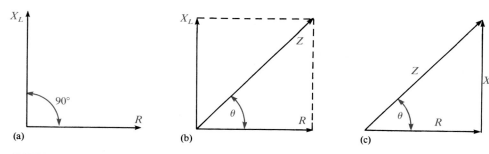

FIGURE 13–3
Development of the impedance triangle for a series *RL* circuit.

The impedance Z of the series *RL* circuit can be expressed in terms of the resistance and reactance as

$$Z = \sqrt{R^2 + X_L^2} \tag{13–1}$$

where Z is expressed in ohms. The phase angle θ is expressed as

$$\theta = \tan^{-1}\left(\frac{X_L}{R}\right) \tag{13–2}$$

EXAMPLE 13–1

Determine the impedance and phase angle of the circuit in Figure 13–4.

Solution
The impedance is

$$Z = \sqrt{R^2 + X_L^2} = \sqrt{(50 \ \Omega)^2 + (100 \ \Omega)^2} = 111.8 \ \Omega$$

FIGURE 13–4

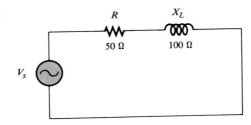

The phase angle is

$$\theta = \tan^{-1}\left(\frac{X_L}{R}\right) = \tan^{-1}\left(\frac{100 \ \Omega}{50 \ \Omega}\right) = \tan^{-1}(2) = 63.4°$$

The applied voltage *leads* the current by 63.4°.
   The calculator sequences are as follows:
Z: ⑤ ⓪ x^2 + ① ⓪ ⓪ x^2 = [2nd] [√x]
θ: ① ⓪ ⓪ ÷ ⑤ ⓪ = [2nd] [tan]

**SECTION REVIEW
13–2**

1. Does the source voltage lead or lag the current in a series *RL* circuit?
2. What causes the phase angle in the *RL* circuit?
3. How does the response of an *RL* circuit differ from that of an *RC* circuit?
4. A series *RL* circuit has a resistance of 33 kΩ and an inductive reactance of 50 kΩ. Determine Z and θ.

13–3

ANALYSIS OF SERIES *RL* CIRCUITS

Ohm's Law

The application of Ohm's law to series *RL* circuits involves the use of the quantities of Z, V, and I. The three equivalent forms of Ohm's law were stated in Chapter 12 for *RC* circuits. They apply also to *RL* circuits and are restated here for convenience:

$$V = IZ, \qquad I = \frac{V}{Z}, \qquad Z = \frac{V}{I}$$

The following example illustrates the use of Ohm's law.

EXAMPLE 13–2

The current in Figure 13–5 is 0.2 mA. Determine the source voltage.

FIGURE 13–5

Solution

From Equation (10–11), the inductive reactance is

$$X_L = 2\pi fL = 2\pi(10 \text{ kHz})(100 \text{ mH}) = 6.28 \text{ k}\Omega$$

The impedance is

$$Z = \sqrt{R^2 + X_L^2} = \sqrt{(10 \text{ k}\Omega)^2 + (6.28 \text{ k}\Omega)^2} = 11.81 \text{ k}\Omega$$

Applying Ohm's law, we have

$$V_s = IZ = (0.2 \text{ mA})(11.81 \text{ k}\Omega) = 2.36 \text{ V}$$

Relationships of the Current and Voltages in a Series *RL* Circuit

In a series *RL* circuit, the current is the same through both the resistor and the inductor. Thus, the resistor voltage is in phase with the current, and *the inductor voltage leads the current by 90°*. Therefore, there is a phase difference of 90° between the resistor voltage, V_R, and the inductor voltage, V_L, as shown in the waveform diagram of Figure 13–6.

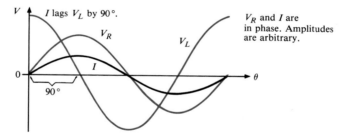

FIGURE 13–6
Phase relation of current and voltages in a series *RL* circuit.

From Kirchhoff's voltage law, the sum of the voltage drops must equal the applied voltage. However, since V_R and V_L are not in phase with each other, they must be added as phasor quantities with V_L leading V_R by 90°, as shown in Figure 13–7(a). As shown in part (b), V_s is the phasor sum of V_R and V_L. This equation can be expressed as

$$V_s = \sqrt{V_R^2 + V_L^2} \tag{13–3}$$

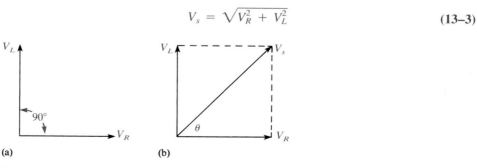

(a)   (b)

FIGURE 13–7
Voltage phasor diagram for a series *RL* circuit.

The phase angle between the resistor voltage and the source voltage is

$$\theta = \tan^{-1}\left(\frac{V_L}{V_R}\right)$$ (13–4)

θ is also the phase angle between the source voltage and the current. Figure 13–8 shows a voltage and current phasor diagram that represents the waveform diagram of Figure 13–6.

FIGURE 13–8
Voltage and current phasor diagram for the waveforms in Figure 13–6.

EXAMPLE 13–3

Determine the source voltage and the phase angle in Figure 13–9.

FIGURE 13–9

Solution
Since V_R and V_L are 90° out of phase, they cannot be added directly and must be added as phasor quantities:

$$V_s = \sqrt{V_R^2 + V_L^2} = \sqrt{(50\text{ V})^2 + (35\text{ V})^2} = 61\text{ V}$$

The angle between the current and the source voltage is

$$\theta = \tan^{-1}\left(\frac{V_L}{V_R}\right) = \tan^{-1}\left(\frac{35\text{ V}}{50\text{ V}}\right) = 34.99°$$

Variation of Impedance and Phase Angle with Frequency

As you know, inductive reactance varies *directly* with frequency. When X_L increases, the total impedance also increases; and when X_L decreases, the total impedance decreases. Thus, *Z is directly dependent on frequency.*

Figure 13–10 illustrates how the voltages and current in a series *RL* circuit vary as the frequency increases or decreases, with the source voltage held at a constant value. In part (a), as the frequency is increased, X_L increases, so more of the total voltage is dropped across the inductor. Also, *Z* increases as X_L increases, causing the current to decrease. A decrease in current causes less voltage across *R*.

In part (b) of the figure, as the frequency is decreased, X_L decreases, so less voltage is dropped across the inductor. Also, *Z* decreases as X_L decreases, causing the current to increase. An increase in current causes more voltage across *R*.

(a) As frequency is increased

(b) As frequency is decreased

FIGURE 13–10
An illustration of how the voltages and current change as the source frequency is varied. The source voltage is held at a constant amplitude.

Changes in *Z* and X_L can be observed as shown in Figure 13–11. As the frequency increases, the voltage across *Z* remains constant because V_s is constant, but the voltage across *L* increases. The decreasing current indicates that *Z* is increasing. It does so because of the inverse relationship stated in Ohm's law ($Z = V_Z/I$). The decreasing current also indicates that X_L is increasing. The increase in V_L corresponds to the increase in X_L.

Variation of the Phase Angle with Frequency

Since X_L is the factor that introduces the phase angle in a series *RL* circuit, a change in X_L produces a change in the phase angle. As the frequency is increased, X_L becomes greater, and thus the phase angle increases. As the frequency is decreased, X_L becomes smaller, and thus the phase angle decreases. This relationship is illustrated in Figure 13–12 where a phase meter is connected to V_s and to V_R. As you have seen before, this connection gives the angle between V_s and V_R, which is the phase angle of the circuit

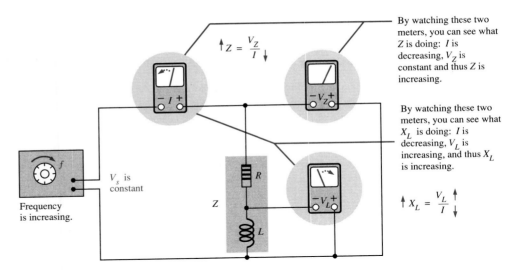

By watching these two meters, you can see what Z is doing: I is decreasing, V_Z is constant and thus Z is increasing.

By watching these two meters, you can see what X_L is doing: I is decreasing, V_L is increasing, and thus X_L is increasing.

$$Z = \frac{V_Z}{I}$$

$$X_L = \frac{V_L}{I}$$

V_s is constant

Frequency is increasing.

FIGURE 13–11
Observing changes in Z and X_L with frequency by watching the meters and recalling Ohm's law.

(a) As frequency is increased, the phase angle increases.

(b) As frequency is decreased, the phase angle decreases.

FIGURE 13–12
A phase meter indicates that the phase angle between V_s and V_R varies directly with frequency. V_R is the same phase as I.

FIGURE 13–13
As the frequency increases, the phase angle θ increases.

because I is in phase with V_R. By measuring the phase of V_R, we are effectively measuring the phase of I.

The variations of phase angle with frequency are illustrated with the impedance triangle as shown in Figure 13–13.

EXAMPLE 13–4

For the series *RL* circuit in Figure 13–14, determine the impedance and the phase angle for each of the following frequencies:
(a) 10 kHz   **(b)** 20 kHz   **(c)** 30 kHz

FIGURE 13–14

Solution
(a) For $f = 10$ kHz:

$$X_L = 2\pi fL = 2\pi(10 \text{ kHz})(20 \text{ mH}) = 1.26 \text{ k}\Omega$$
$$Z = \sqrt{R^2 + X_L^2} = \sqrt{(1 \text{ k}\Omega)^2 + (1.26 \text{ k}\Omega)^2} = 1.61 \text{ k}\Omega$$
$$\theta = \tan^{-1}\left(\frac{X_L}{R}\right) = \tan^{-1}\left(\frac{1.26 \text{ k}\Omega}{1 \text{ k}\Omega}\right) = 51.56°$$

(b) For $f = 20$ kHz:

$$X_L = 2\pi(20 \text{ kHz})(20 \text{ mH}) = 2.52 \text{ k}\Omega$$
$$Z = \sqrt{(1 \text{ k}\Omega)^2 + (2.52 \text{ k}\Omega)^2} = 2.71 \text{ k}\Omega$$
$$\theta = \tan^{-1}\left(\frac{2.52 \text{ k}\Omega}{1 \text{ k}\Omega}\right) = 68.36°$$

(c) For $f = 30$ kHz:

$$X_L = 2\pi(30 \text{ kHz})(20 \text{ mH}) = 3.77 \text{ k}\Omega$$
$$Z = \sqrt{(1 \text{ k}\Omega)^2 + (3.77 \text{ k}\Omega)^2} = 3.9 \text{ k}\Omega$$
$$\theta = \tan^{-1}\left(\frac{3.77 \text{ k}\Omega}{1 \text{ k}\Omega}\right) = 75.14°$$

Notice that as the frequency increases, X_L, Z, and θ also increase.

**SECTION REVIEW
13–3**

1. In a certain series *RL* circuit, $V_R = 2$ V and $V_L = 3$ V. What is the magnitude of the total voltage?
2. In Problem 1, what is the phase angle?
3. When the frequency of the applied voltage in a series *RL* circuit is increased, what happens to the inductive reactance? To the impedance? To the phase angle?

13-4

IMPEDANCE AND PHASE ANGLE OF PARALLEL *RL* CIRCUITS

A basic parallel *RL* circuit is shown in Figure 13–15. The expression for the impedance is

$$Z = \frac{RX_L}{\sqrt{R^2 + X_L^2}} \tag{13-5}$$

FIGURE 13–15
Parallel *RL* circuit.

The phase angle between the applied voltage and the total current is

$$\theta = \tan^{-1}\left(\frac{R}{X_L}\right) \tag{13-6}$$

EXAMPLE 13–5

For each circuit in Figure 13–16, determine the impedance and the phase angle.

FIGURE 13–16

(a)

(b)

Solution

(a) $Z = \dfrac{RX_L}{\sqrt{R^2 + X_L^2}} = \dfrac{(100\ \Omega)(50\ \Omega)}{\sqrt{(100\ \Omega)^2 + (50\ \Omega)^2}} = 44.72\ \Omega$

$\theta = \tan^{-1}\left(\dfrac{R}{X_L}\right) = \tan^{-1}\left(\dfrac{100\ \Omega}{50\ \Omega}\right) = 63.43°$

(b) $Z = \dfrac{(1\ k\Omega)(2\ k\Omega)}{\sqrt{(1\ k\Omega)^2 + (2\ k\Omega)^2}} = 894.4\ \Omega$

$\theta = \tan^{-1}\left(\dfrac{1\ k\Omega}{2\ k\Omega}\right) = 26.57°$

The voltage leads the current, as opposed to the *RC* case where the voltage lags the current.

The calculator sequences for (a) are as follows:

Z: ① ⓪ ⓪ ✕ ⑤ ⓪ ÷ (((① ⓪ ⓪
x^2 + ⑤ ⓪ x^2) 2nd $\sqrt{x}$) =

θ: ① ⓪ ⓪ ÷ ⑤ ⓪ = 2nd tan

Susceptance and Admittance

As you know from Section 12–4, conductance is the reciprocal of resistance, susceptance is the reciprocal of reactance, and admittance is the reciprocal of impedance. For parallel *RL* circuits, inductive susceptance is expressed as

$$B_L = \frac{1}{X_L} \tag{13–7}$$

and admittance as

$$Y = \frac{1}{Z} \tag{13–8}$$

As with the *RC* circuit, the unit for G, B_L, and Y is the siemen (S). In the basic parallel *RL* circuit shown in Figure 13–17, the total admittance is the phasor sum of the conductance and the susceptance and is expressed as follows:

$$Y = \sqrt{G^2 + B_L^2} \tag{13–9}$$

FIGURE 13–17
Admittance is the reciprocal of impedance in a parallel *RL* circuit.

EXAMPLE 13–6

Determine the admittance in Figure 13–18; then convert it to impedance.

FIGURE 13–18

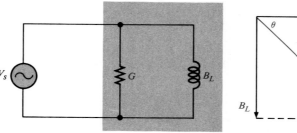

Solution
Since $R = 330\ \Omega$, then

$$G = \frac{1}{R} = \frac{1}{330\ \Omega} = 0.003\ \text{S}$$

$$X_L = 2\pi(1000\ \text{Hz})(100\ \text{mH}) = 628.3\ \Omega$$

$$B_L = \frac{1}{X_L} = 0.00159\ \text{S}$$

$$Y = \sqrt{G^2 + B_L^2} = \sqrt{(0.003\ \text{S})^2 + (0.00159\ \text{S})^2} = 0.0034\ \text{S}$$

Converting to impedance, we get

$$Z = \frac{1}{Y} = \frac{1}{0.0034 \text{ S}} = 294.12 \text{ }\Omega$$

SECTION REVIEW
13–4

1. If $Y = 0.05$ S, what is the value of Z?
2. In a certain parallel RL circuit, $R = 50$ Ω and $X_L = 75$ Ω. Determine the admittance.
3. In the circuit of Problem 2, does the total current lead or lag the applied voltage and by what phase angle?

13–5

ANALYSIS OF PARALLEL *RL* CIRCUITS

The following example applies Ohm's law to the analysis of a parallel *RL* circuit.

EXAMPLE 13–7

Determine the total current and the phase angle in the circuit of Figure 13–19.

FIGURE 13–19

Solution

$$X_L = 2\pi(1.5 \text{ kHz})(150 \text{ mH}) = 1.41 \text{ k}\Omega$$

$$B_L = \frac{1}{X_L} = \frac{1}{1.41 \text{ k}\Omega} = 0.709 \text{ mS}$$

$$G = \frac{1}{R} = \frac{1}{2.2 \text{ k}\Omega} = 0.455 \text{ mS}$$

$$Y = \sqrt{G^2 + B_L^2} = \sqrt{(0.455 \text{ mS})^2 + (0.709 \text{ mS})^2} = 0.842 \text{ mS}$$

Apply Ohm's law to get the total current:

$$I_T = VY = (10 \text{ V})(0.842 \text{ mS}) = 8.42 \text{ mA}$$

$$\theta = \tan^{-1}\left(\frac{R}{X_L}\right) = \tan^{-1}\left(\frac{2.2 \text{ k}\Omega}{1.41 \text{ k}\Omega}\right) = 57.34°$$

The total current lags the source voltage by 57.34°.

Relationships of the Currents and Voltages in a Parallel *RL* Circuit

Figure 13–20(a) shows all the currents and voltages in a basic parallel *RL* circuit. As you can see, the applied voltage, V_s, appears across both the resistive and the inductive branches, so V_s, V_R, and V_L are all in phase and of the same magnitude. The total current, I_T, divides at the junction into the two branch currents, I_R and I_L.

(a)   (b)

FIGURE 13–20

Currents and voltages in a parallel *RL* circuit. The current directions shown are instantaneous.

The current through the resistor is in phase with the voltage. The current through the inductor lags the voltage and the resistor current by 90°. By Kirchhoff's current law, the total current is the phasor sum of the two branch currents, as shown by the phasor diagram in Figure 13–20(b). The total current is expressed as

$$I_T = \sqrt{I_R^2 + I_L^2} \tag{13–10}$$

and the phase angle between the resistor current and the total current is

$$\theta = \tan^{-1}\left(\frac{I_L}{I_R}\right) \tag{13–11}$$

Figure 13–21 shows a complete current and voltage phasor diagram.

FIGURE 13–21

Current and voltage phasor diagram for a parallel *RL* circuit (amplitudes are arbitrary).

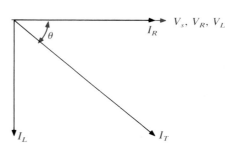

EXAMPLE 13–8

Determine the value of each current in Figure 13–22, and describe the phase relationship of each with the applied voltage.

FIGURE 13–22

Solution

$$I_R = \frac{V_s}{R} = \frac{12 \text{ V}}{220 \text{ }\Omega} = 54.55 \text{ mA}$$

$$I_L = \frac{V_s}{X_L} = \frac{12 \text{ V}}{150 \text{ }\Omega} = 80 \text{ mA}$$

$$I_T = \sqrt{I_R^2 + I_L^2} = \sqrt{(54.55 \text{ mA})^2 + (80 \text{ mA})^2} = 96.83 \text{ mA}$$

$$\theta = \tan^{-1}\left(\frac{R}{X_L}\right) = \tan^{-1}\left(\frac{220 \text{ }\Omega}{150 \text{ }\Omega}\right) = 55.71°$$

The resistor current is 54.55 mA and is in phase with the applied voltage. The inductor current is 80 mA and lags the applied voltage by 90°. The total current is 96.83 mA and lags the voltage by 55.71°.

**SECTION REVIEW
13–5**

1. The admittance of a parallel *RL* circuit is 0.004 S, and the applied voltage is 8 V. What is the total current?

2. In a certain parallel *RL* circuit, the resistor current is 12 mA, and the inductor current is 20 mA. Determine the phase angle and the total current.

3. What is the phase angle between the inductor current and the applied voltage in a parallel *RL* circuit?

13–6

SERIES-PARALLEL ANALYSIS

In this section, we analyze *RL* circuits with combinations of both series and parallel *R* and *L* elements. Examples illustrate the procedures.

EXAMPLE 13–9

In the circuit of Figure 13–23, determine the values of the following:
(a) Z_T (b) I_T (c) θ

FIGURE 13–23

Solution
(a) First, the inductive reactances are calculated:

$$X_{L1} = 2\pi f L_1 = 2\pi(5 \text{ kHz})(250 \text{ mH}) = 7.85 \text{ k}\Omega$$
$$X_{L2} = 2\pi f L_2 = 2\pi(5 \text{ kHz})(100 \text{ mH}) = 3.14 \text{ k}\Omega$$

The approach we will use is to find the series equivalent resistance and inductive reactance for the parallel portion of the circuit. Then we will add the resistances to get the total resistance, add the reactances to get the total reactance, and then determine the total impedance.

The impedance of the parallel portion (Z_2) is determined as follows:

$$G_2 = \frac{1}{R_2} = \frac{1}{3.3 \text{ k}\Omega} = 303 \ \mu S$$

$$B_{L2} = \frac{1}{X_{L2}} = \frac{1}{3.14 \text{ k}\Omega} = 318.5 \ \mu S$$

$$Y_2 = \sqrt{G_2^2 + B_L^2} = \sqrt{(303 \ \mu S)^2 + (318.5 \ \mu S)^2} = 439.6 \ \mu S$$

Then

$$Z_2 = \frac{1}{Y_2} = \frac{1}{439.6 \ \mu S} = 2.28 \text{ k}\Omega$$

The phase angle associated with the parallel portion of the circuit is

$$\theta_p = \tan^{-1}\left(\frac{R_2}{X_{L2}}\right) = \tan^{-1}\left(\frac{3.3 \text{ k}\Omega}{3.14 \text{ k}\Omega}\right) = 46.42°$$

The series equivalent values for the parallel portion are

$$R_{eq} = Z_2\cos\theta_p = (2.28 \text{ k}\Omega) \cos(46.42°) = 1.57 \text{ k}\Omega$$

$$X_{L(eq)} = Z_2\sin\theta_p = (2.28 \text{ k}\Omega) \sin(46.42°) = 1.65 \text{ k}\Omega$$

The total circuit resistance is

$$R_T = R_1 + R_{eq} = 4.7 \text{ k}\Omega + 1.57 \text{ k}\Omega = 6.27 \text{ k}\Omega$$

The total circuit reactance is

$$X_{LT} = X_{L1} + X_{L(eq)} = 7.85 \text{ k}\Omega + 1.65 \text{ k}\Omega = 9.5 \text{ k}\Omega$$

The total circuit impedance is

$$Z_T = \sqrt{R_T^2 + X_{LT}^2} = \sqrt{(6.27 \text{ k}\Omega)^2 + (9.5 \text{ k}\Omega)^2} = 11.38 \text{ k}\Omega$$

(b) Use Ohm's law to find the total current:

$$I_T = \frac{V_s}{Z_T} = \frac{10 \text{ V}}{11.38 \text{ k}\Omega} = 0.879 \text{ mA}$$

(c) To find the phase angle, view the circuit as a series combination of R_T and X_{LT}:

$$\theta = \tan^{-1}\left(\frac{X_{LT}}{R_T}\right) = \tan^{-1}\left(\frac{9.5 \text{ k}\Omega}{6.27 \text{ k}\Omega}\right) = 56.58°$$

The total current *lags* the source voltage by 56.58°.

EXAMPLE 13–10

Determine the voltage across each element in Figure 13–24. Sketch a voltage phasor diagram.

FIGURE 13–24

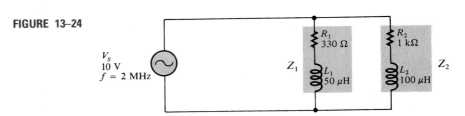

Solution

First calculate X_{L1} and X_{L2}:

$$X_{L1} = 2\pi fL_1 = 2\pi(2 \text{ MHz})(50 \text{ } \mu\text{H}) = 628.3 \text{ } \Omega$$
$$X_{L2} = 2\pi fL_2 = 2\pi(2 \text{ MHz})(100 \text{ } \mu\text{H}) = 1.257 \text{ k}\Omega$$

Now, determine the impedance of each branch:

$$Z_1 = \sqrt{R_1^2 + X_{L1}^2} = \sqrt{(330 \text{ } \Omega)^2 + (628.3 \text{ } \Omega)^2} = 709.69 \text{ } \Omega$$
$$Z_2 = \sqrt{R_2^2 + X_{L2}^2} = \sqrt{(1 \text{ k}\Omega)^2 + (1.257 \text{ k}\Omega)^2} = 1.61 \text{ k}\Omega$$

Calculate each branch current:

$$I_1 = \frac{V_s}{Z_1} = \frac{10 \text{ V}}{709.69 \text{ } \Omega} = 14.09 \text{ mA}$$

$$I_2 = \frac{V_s}{Z_2} = \frac{10 \text{ V}}{1.61 \text{ k}\Omega} = 6.21 \text{ mA}$$

Now, use Ohm's law to find the voltage across each element:

$$V_{R1} = I_1R_1 = (14.09 \text{ mA})(330 \text{ } \Omega) = 4.65 \text{ V}$$
$$V_{L1} = I_1X_{L1} = (14.09 \text{ mA})(628.3 \text{ } \Omega) = 8.85 \text{ V}$$
$$V_{R2} = I_2R_2 = (6.21 \text{ mA})(1 \text{ k}\Omega) = 6.21 \text{ V}$$
$$V_{L2} = I_2X_{L2} = (6.21 \text{ mA})(1.257 \text{ k}\Omega) = 7.81 \text{ V}$$

The angles associated with each parallel branch are now determined:

$$\theta_1 = \tan^{-1}\left(\frac{X_{L1}}{R_1}\right) = \tan^{-1}\left(\frac{628.3 \text{ } \Omega}{330 \text{ } \Omega}\right) = 62.29°$$

$$\theta_2 = \tan^{-1}\left(\frac{X_{L2}}{R_2}\right) = \tan^{-1}\left(\frac{1.257 \text{ k}\Omega}{1 \text{ k}\Omega}\right) = 51.5°$$

Thus, I_1 lags V_s by 62.29°, and I_2 lags V_s by 51.5°, as indicated in Figure 13–25(a), where the negative signs indicate lagging angles.

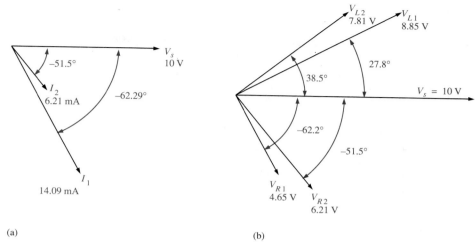

(a)

(b)

FIGURE 13–25

The phase relationships of the voltages are determined as follows:

☐ V_{R1} is in phase with I_1 and therefore lags V_s by 62.2°.
☐ V_{L1} leads I_1 by 90°, so its angle is 90° − 62.2° = 27.8°.
☐ V_{R2} is in phase with I_2 and therefore lags V_s by 51.5°.
☐ V_{L2} leads I_2 by 90°, so its angle is 90° − 51.5° = 38.5°.

These phase relationships are shown in Figure 13–25(b).

The two previous examples should give you a feel for how complex *RL* networks respond to sine wave inputs.

**SECTION REVIEW
13–6**

1. What is the total impedance of the circuit in Figure 13–24?
2. Determine the total current for the circuit in Figure 13–24.

13–7

POWER IN *RL* CIRCUITS

As you know, in a purely resistive ac circuit, all of the energy that is delivered by the source is dissipated in the form of heat by the resistance. In a purely inductive ac circuit, all of the energy delivered by the source is stored by the inductor in its magnetic field during a portion of the voltage cycle and then is returned to the source during another portion of the cycle so that no net energy is transferred.

When there is both resistance and inductance in a circuit, some of the energy is alternately stored and returned by the inductance, and some is dissipated by the resistance. The amount of energy dissipated is determined by the relative values of resistance and inductive reactance.

When the resistance is greater than the inductive reactance, more of the total energy delivered by the source is dissipated by the resistance than is stored by the inductor, and when the reactance is greater than the resistance, more of the total energy is stored and returned than is dissipated.

As you know, the power in a resistor is called the *true power*. The power in an inductor is *reactive power* and is expressed as

$$P_r = I^2 X_L \qquad \qquad (13\text{--}12)$$

The Power Triangle

The generalized power triangle for the *RL* circuit is shown in Figure 13–26. The apparent power, P_a, is the resultant of the true power, P_{true}, and the reactive power, P_r.

FIGURE 13–26
Generalized power triangle for an *RL* circuit.

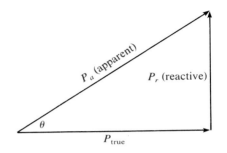

Recall that the power factor equals the cosine of θ ($PF = \cos \theta$). As the phase angle between the applied voltage and the total current increases, the power factor decreases, indicating an increasingly reactive circuit. The smaller the power factor, the smaller the true power is compared to the reactive power. The power factor of inductive loads is called a *lagging power factor* because the current lags the source voltage.

EXAMPLE 13–11

Determine the power factor, the true power, the reactive power, and the apparent power in Figure 13–27.

FIGURE 13–27

Solution
The impedance of the circuit is

$$Z = \sqrt{R^2 + X_L^2} = \sqrt{(1 \text{ k}\Omega)^2 + (2 \text{ k}\Omega)^2} = 2236 \ \Omega$$

The current is

$$I = \frac{V_s}{Z} = \frac{10 \text{ V}}{2236 \text{ } \Omega} = 4.47 \text{ mA}$$

The phase angle is

$$\theta = \tan^{-1}\left(\frac{X_L}{R}\right) = \tan^{-1}\left(\frac{2 \text{ k}\Omega}{1 \text{ k}\Omega}\right) = 63.4°$$

Therefore, the power factor is

$$PF = \cos\theta = \cos(63.4°) = 0.448$$

The true power is

$$P_{\text{true}} = V_s I \cos\theta = (10 \text{ V})(4.47 \text{ mA})(0.448) = 20 \text{ mW}$$

The reactive power is

$$P_r = I^2 X_L = (4.47 \text{ mA})^2(2 \text{ k}\Omega) = 39.96 \text{ mVAR}$$

The apparent power is

$$P_a = I^2 Z = (4.47 \text{ mA})^2(2236 \text{ } \Omega) = 44.68 \text{ mVA}$$

Significance of the Power Factor

As you learned in Chapter 12, the power factor *(PF)* is very important in determining how much useful power (true power) is transferred to a load. The highest power factor is 1, which indicates that all of the current to a load is in-phase with the voltage (resistive). When the power factor is 0, all of the current to a load is 90° out of phase with the voltage (reactive).

Generally, a power factor as close to 1 as possible is desirable, because then most of the power transferred from the source to the load is useful or true power. True power goes only one way—from source to load—and performs work on the load in terms of energy utilization. Reactive power simply goes back and forth between the source and the load with no net work being done. Energy must be used in order for work to be done.

Many practical loads have inductance as a result of their particular function, and it is essential for their proper operation. Examples are transformers, electric motors, and speakers, to name a few. Therefore, inductive (and capacitive) loads are a fact of life and we must live with them.

To see the effect of the power factor on system requirements, refer to Figure 13–28 which shows a representation of a typical inductive load consisting effectively of inductance and resistance in parallel. Part (a) shows a load with a relatively low power factor (0.75), and part (b) shows a load with a relatively high power factor (0.95). Both loads dissipate equal amounts of power as indicated by the wattmeters. Thus, an equal amount of work is done on both loads.

Although both loads are equivalent in terms of the amount of work done (true power), the low–power factor load in part (a) draws more current from the source than does the high–power factor load in part (b), as indicated by the ammeters. Therefore, the source in part (a) must have a higher VA rating than the one in part (b). Also, the lines connecting the source to the load must be a larger wire gage than those in part (b), a condition that becomes significant when very long transmission lines are required, such as in power distribution.

Figure 13–28 has demonstrated that a higher power factor is an advantage in delivering power more efficiently to a load.

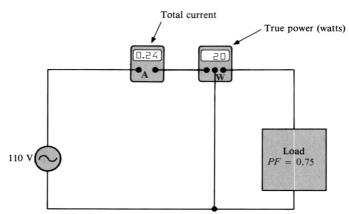

(a) A lower power factor means more total current for a given power dissipation (watts). A larger source is required to deliver the watts.

(b) A higher power factor means less total current for a given power dissipation. A smaller source can deliver the same true power (watts).

FIGURE 13–28

Illustration of the effect of the power factor on system requirements such as source rating (VA) and conductor size.

Power Factor Correction

The power factor of an inductive load can be increased by the addition of a capacitor in parallel, as shown in Figure 13–29. The capacitor compensates for the phase lag of the total current by creating a capacitive component of current that is 180° out of phase with the inductive component. This has a canceling effect and reduces the phase angle, thus increasing the power factor as well as the total current, as illustrated in the figure.

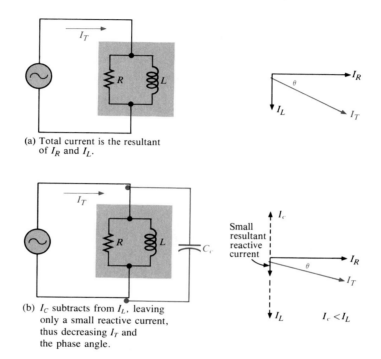

(a) Total current is the resultant of I_R and I_L.

(b) I_C subtracts from I_L, leaving only a small reactive current, thus decreasing I_T and the phase angle.

FIGURE 13–29
Example of how the power factor can be increased by the addition of a compensating capacitor.

SECTION REVIEW
13–7

1. To which component in an *RL* circuit is the power dissipation due?
2. Calculate the power factor when $\theta = 50°$.
3. A certain series *RL* circuit consists of a 470-Ω resistor and an inductive reactance of 620 Ω at the operating frequency. Determine P_{true}, P_r, and P_a when $I = 100$ mA.

13–8

BASIC APPLICATIONS OF *RL* CIRCUITS

Like *RC* circuits, *RL* circuits can also be applied in phase-shifting and filter applications.

The *RL* Lag Network

The *RL* lag network is a phase-shift circuit in which the output voltage *lags* the input voltage by a specified angle, ϕ. A basic *RL* lag network is shown in Figure 13–30(a).

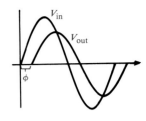

(a) A basic *RL* lead network

(b) Phasor voltage diagram showing phase lead between V_{in} and V_{out}

(c) Input and output waveforms

FIGURE 13–30
The *RL* lag network ($V_{out} = V_R$).

Notice that the output voltage is taken across the *resistor* and the input voltage is the total voltage applied across the circuit. The relationship of the voltages is shown in the phasor diagram in Figure 13–30(b), and a waveform diagram is shown in part (c). Notice that V_R, which is the output voltage, *lags* V_{in} by an angle that is the same as the circuit phase angle. The angles are equal, of course, because V_R and I are in phase with each other.

The phase lag, ϕ, can be expressed as

$$\phi = \tan^{-1}\left(\frac{X_L}{R}\right) \tag{13–13}$$

EXAMPLE 13–12   Calculate the phase lag for each circuit in Figure 13–31.

(a)

(b)

FIGURE 13–31

Solution

(a) $\phi = \tan^{-1}\left(\frac{X_L}{R}\right) = \tan^{-1}\left(\frac{5\ k\Omega}{15\ k\Omega}\right) = 18.43°$

The output lags the input by 18.43°.

(b) $X_L = 2\pi f L = 2\pi(1\ kHz)(1\ mH) = 6.28\ \Omega$

$\phi = \tan^{-1}\left(\frac{X_L}{R}\right) = \tan^{-1}\left(\frac{6.28\ \Omega}{10\ \Omega}\right) = 32.13°$

The output lags the input by 32.13°.

The phase-lag network can be considered as a voltage divider with a portion of the input voltage dropped across L and a portion across R. The output voltage can be determined with the following formula:

$$V_{out} = \left(\frac{R}{\sqrt{R^2 + X_L^2}}\right) V_{in} \qquad (13\text{--}14)$$

EXAMPLE 13–13

The input voltage in Figure 13–31(b) of Example 13–12 has an rms value of 10 V. Determine the output voltage. Sketch the waveform relationships for the input and output voltages.

Solution
The phase lag was found to be 32.13° in Example 13–12.

$$V_{out} = \left(\frac{R}{\sqrt{R^2 + X_L^2}}\right) V_{in} = \left(\frac{10 \ \Omega}{11.81 \ \Omega}\right) 10 \ V = 8.47 \ V \ \text{rms}$$

The waveforms are shown in Figure 13–32.

FIGURE 13–32

The calculator sequence is

Effects of Frequency on the Phase Shift and Output Voltage of the Lag Network

Since the circuit phase angle and the phase lag are the same, an increase in frequency causes an increase in phase lag. Also, an increase in frequency causes a decrease in the magnitude of the output voltage, because X_L becomes greater and more of the total voltage is dropped across the inductor and less across the resistor. Figure 13–33 illustrates the changes in phase lag and output voltage magnitude with frequency.

The *RL* Lead Network

The *RL* lead network is a phase-shift circuit in which the output voltage *leads* the input voltage by a specified angle, ϕ. A basic *RL* lead network is shown in Figure 13–34(a). Notice how this network differs from the lag network. Here the output voltage is taken across the inductor rather than across the resistor. The relationship of the voltages is shown in the phasor diagram of Figure 13–34(b) and in the waveform plot of part (c).

Phase meter measures angle between V_{in} and V_{out}.

Voltmeter measures V_{out} with respect to ground.

(a) When f increases, the phase lag ϕ increases and V_{out} decreases.

(b) When f decreases, ϕ decreases and V_{out} increases.

FIGURE 13–33
Illustration of how the frequency affects the phase lag and the output voltage in an *RL* lag network with the amplitude of V_{in} held constant.

(a) A basic *RL* lag network

(b) Phasor voltage diagram showing V_{out} leading V_{in}

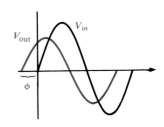

(c) Input and output waveforms

FIGURE 13–34
The *RL* lead network ($V_{out} = V_L$).

The output voltage V_L leads V_{in} by an angle (phase lead) that is the difference between 90° and the circuit phase angle θ, as you can see in the figure [$\theta = \tan^{-1}(X_L/R)$]. The amount of phase lead is expressed as follows:

$$\phi = 90° - \tan^{-1}\left(\frac{X_L}{R}\right)$$ (13–15)

Again, the phase-lead network can be considered as a voltage divider with the voltage across L being the output. The expression for the output voltage is

$$V_{out} = \left(\frac{X_L}{\sqrt{R^2 + X_L^2}}\right)V_{in}$$ (13–16)

EXAMPLE 13–14

Determine the amount of phase lead and output voltage in the lead network in Figure 13–35.

FIGURE 13–35

Solution

$$X_L = 2\pi f L = 2\pi(1 \text{ kHz})(50 \text{ mH}) = 314.2 \ \Omega$$

$$\phi = 90° - \tan^{-1}\left(\frac{X_L}{R}\right) = 90° - \tan^{-1}\left(\frac{314.2 \ \Omega}{680 \ \Omega}\right) = 65.2°$$

The output leads the input by 65.2°.

$$V_{out} = \left(\frac{X_L}{\sqrt{R^2 + X_L^2}}\right)V_{in} = \left(\frac{314.2 \ \Omega}{\sqrt{(680 \ \Omega)^2 + (314.2 \ \Omega)^2}}\right)5 \text{ V}$$
$$= 2.1 \text{ V}$$

Effects of Frequency on the Phase Shift and Output Voltage of the Lead Network

Since the circuit phase angle, θ, increases as frequency increases, the phase lead between the input and the output voltages *decreases*. You can see this relationship by examining Equation (13–15). Also, the amplitude of the output voltage increases as the frequency increases because X_L becomes greater and more of the total input voltage is dropped across the inductor. Figure 13–36 illustrates the changes in the phase lead and output voltage with frequency.

(a) When f increases, the phase lead ϕ decreases and V_{out} increases.

(b) When f decreases, ϕ increases and V_{out} decreases.

FIGURE 13–36
Illustration of how the frequency affects the phase lead and output voltage in an *RL* lead network with the amplitude of V_{in} held constant.

The *RL* Circuit as a Filter

Like *RC* circuits, series *RL* circuits also exhibit a frequency-selective characteristic and therefore act as basic filters.

The *RL* Low-Pass Filter

You have seen what happens to the phase and the output voltage in the lag network. In terms of its filtering action, the change in the amplitude of the output voltage with frequency is the primary consideration.

    To illustrate low-pass filter action, Figure 13–37 shows a specific series of measurements in which the frequency starts at zero (dc) and is increased in increments up to 20 kHz. At each value of frequency the output voltage is measured. As you can see, the inductive reactance increases as frequency goes up, thus causing less voltage to be dropped across the resistor while the input voltage is held at a constant 10 V throughout each step. The response curve for these particular values would appear the same as the response curve in Figure 12–48 for the *RC* low-pass filter.

The *RL* High-Pass Filter

To illustrate *RL* high-pass filter action, Figure 13–38 shows a series of specific measurements. Again, the frequency starts at zero (dc) and is increased in increments up to

FIGURE 13–37
Example of low-pass filter action. Winding resistance has been neglected.

FIGURE 13–38
Example of high-pass filter action. Winding resistance has been neglected.

10 kHz. As you can see, the inductive reactance increases as the frequency goes up, thus causing more voltage to be dropped across the inductor. Again, when the values are plotted, the response curve is the same as the one for the *RC* high-pass filter as shown in Figure 12–50.

The Cutoff Frequency of an *RL* Filter

The frequency at which the inductive reactance equals the resistance in a low-pass or high-pass *RL* filter is called the *cutoff frequency* and is designated f_c. This condition is expressed as $2\pi f_c L = R$. Solving for f_c, we get the following formula:

$$f_c = \frac{R}{2\pi L} \qquad (13\text{--}17)$$

As with the *RC* filter, the output voltage is 70.7% of its maximum value at f_c. In a high-pass filter, all frequencies above f_c are considered to be passed by the filter, and all those below f_c are considered to be rejected. The reverse, of course, is true for a low-pass filter. *Filter bandwidth,* defined in Chapter 12, applies to both *RC* and *RL* filter circuits.

**SECTION REVIEW
13–8**

1. A certain *RL* lead network consists of a 3.3-kΩ resistor and a 15-mH inductor. Determine the phase lead between input and output at a frequency of 5 kHz.
2. An *RL* lag network has the same component value as the lead network in Problem 1. What is the magnitude of the output voltage at 5 kHz when the input is 10 V rms?
3. When an *RL* circuit is used as a low-pass filter, across which component is the output taken?

13–9

TROUBLESHOOTING *RL* CIRCUITS

Effects of an Open Inductor

The most common failure mode for inductors occurs when the winding opens as a result of excessive current or a mechanical contact failure. It is easy to see how an open coil affects the operation of a basic series *RL* circuit, as shown in Figure 13–39. Obviously,

FIGURE 13–39

Effect of an open coil.

there is no current path—so the resistor voltage is zero, and the total applied voltage appears across the inductor.

Effects of an Open Resistor

When the resistor is open, there is no current, and the inductor voltage is zero. The total input voltage is across the open resistor, as shown in Figure 13–40.

FIGURE 13–40
Effect of an open resistor.

Open Components in Parallel Circuits

In a parallel *RL* circuit, an open resistor or inductor will cause the total current to decrease, because the total impedance will increase. Obviously, the branch with the open component will have zero current. Figure 13–41 illustrates these conditions.

(a) Before failure   (b) After *L* opens

FIGURE 13–41
Effect of an open component in a parallel circuit with V_s constant.

Effects of an Inductor with Shorted Windings

It is possible for some of the windings of coils to short together as a result of damaged insulation. This failure mode is much less likely than the open coil. Shorted windings result in a reduction in inductance, since the inductance of a coil is proportional to the square of the number of turns.

SECTION REVIEW
13–9

1. Describe the effect of an inductor with shorted windings on the response of a series *RL* circuit.

2. In the circuit of Figure 13–42, indicate whether I_T, V_{R1}, and V_{R2} increase or decrease as a result of *L* opening.

FIGURE 13–42

SUMMARY

☐ When a sine wave voltage is applied to an *RL* circuit, the current and all the voltage drops are also sine waves.

☐ Total current in an *RL* circuit always lags the source voltage.

☐ The resistor voltage is always in phase with the current.

☐ The inductor voltage always leads the current by 90°.

☐ In an *RL* circuit, the impedance is determined by both the resistance and the inductive reactance combined.

☐ Impedance is expressed in units of ohms.

☐ The impedance of an *RL* circuit varies directly with frequency.

☐ The phase angle *(θ)* of a series *RL* circuit varies directly with frequency.

☐ You can determine the impedance of a circuit by measuring the applied voltage and the total current and then applying Ohm's law.

☐ In an *RL* circuit, part of the power is resistive and part reactive.

☐ The phasor combination of resistive power (true power) and reactive power is called *apparent power*.

☐ The power factor indicates how much of the apparent power is true power.

☐ A power factor of 1 indicates a purely resistive circuit, and a power factor of 0 indicates a purely reactive circuit.

☐ In a lag network, the output voltage lags the input voltage in phase.

☐ In a lead network, the output voltage leads the input voltage in phase.

☐ A filter passes certain frequencies and rejects others.

GLOSSARY

Admittance   A measure of the ability of a reactive circuit to permit current. The reciprocal of impedance. The unit is siemens (S).

Apparent power   The product of volts and amperes, with the unit of volt-amperes (VA).

Choke   An inductor.

Coil   An inductor.

Core   The material within the windings of an inductor.

Inductance   The property of an inductor whereby a change in current causes the inductor to produce a voltage that opposes the change in current.

Inductive reactance   The opposition of an inductor to sinusoidal current. The unit is ohms.

Inductive susceptance   The reciprocal of inductive reactance. The unit is siemens.

Inductor   An electrical device formed by a coil of wire around a core material having the property of inductance.

FORMULAS

$$Z = \sqrt{R^2 + X_L^2} \qquad \text{Series } RL \text{ impedance} \tag{13-1}$$

$$\theta = \tan^{-1}\left(\frac{X_L}{R}\right) \qquad \text{Series } RL \text{ phase angle} \tag{13-2}$$

$$V_s = \sqrt{V_R^2 + V_L^2} \qquad \text{Total voltage in a series } RL \text{ circuit} \tag{13-3}$$

$$\theta = \tan^{-1}\left(\frac{V_L}{V_R}\right) \qquad \text{Series } RL \text{ phase angle} \tag{13-4}$$

$$Z = \frac{RX_L}{\sqrt{R^2 + X_L^2}} \qquad \text{Parallel } RL \text{ impedance} \tag{13-5}$$

$$\theta = \tan^{-1}\left(\frac{R}{X_L}\right) \qquad \text{Parallel } RL \text{ phase angle} \tag{13-6}$$

$$B_L = \frac{1}{X_L} \qquad \text{Inductive susceptance} \tag{13-7}$$

$$Y = \frac{1}{Z} \qquad \text{Admittance} \tag{13-8}$$

$$Y = \sqrt{G^2 + B_L^2} \qquad \text{Admittance} \tag{13-9}$$

$$I_T = \sqrt{I_R^2 + I_L^2} \qquad \text{Total current in parallel } RL \text{ circuit} \tag{13-10}$$

$$\theta = \tan^{-1}\left(\frac{I_L}{I_R}\right) \qquad \text{Parallel } RL \text{ phase angle} \tag{13-11}$$

$$P_r = I^2 X_L \qquad \text{Reactive power} \tag{13-12}$$

$$\phi = \tan^{-1}\left(\frac{X_L}{R}\right) \qquad \text{Phase lag} \tag{13-13}$$

$$V_{\text{out}} = \left(\frac{R}{\sqrt{R^2 + X_L^2}}\right)V_{\text{in}} \qquad \text{Output of lag network} \tag{13-14}$$

$$\phi = 90° - \tan^{-1}\left(\frac{X_L}{R}\right) \qquad \text{Phase lead} \tag{13-15}$$

$$V_{\text{out}} = \left(\frac{X_L}{\sqrt{R^2 + X_L^2}}\right)V_{\text{in}} \qquad \text{Output of lead network} \tag{13-16}$$

$$f_c = \frac{R}{2\pi L} \qquad \text{Cutoff frequency of } RL \text{ filter} \tag{13-17}$$

SELF-TEST

1. In a series *RL* circuit, the resistor voltage
   (a) leads the applied voltage
   (b) lags the applied voltage
   (c) is in phase with the applied voltage
   (d) is in phase with the current
   (e) a and d
   (f) b and d

2. When the frequency of the voltage applied to a series *RL* circuit is increased, the impedance
   (a) decreases   (b) increases   (c) does not change

3. When the frequency of the voltage applied to a series *RL* circuit is decreased, the phase angle
   (a) decreases   (b) increases   (c) does not change

4. If the frequency is doubled and the resistance is halved, the impedance of a series *RL* circuit
   (a) doubles (b) halves (c) remains constant (d) cannot be determined without values

5. To reduce the current in a series *RL* circuit, the frequency should be
   (a) increased   (b) decreased   (c) constant

6. In a series *RL* circuit, 10 V rms is measured across the resistor, and 10 V rms is measured across the inductor. The peak value of the source voltage is
   (a) 14.14 V   (b) 28.28 V   (c) 10 V   (d) 20 V

7. The voltages in Problem 6 are measured at a certain frequency. To make the resistor voltage greater than the inductor voltage, the frequency is
   (a) increased   (b) decreased   (c) doubled   (d) not a factor

8. When the resistor voltage in a series *RL* circuit becomes greater than the inductor voltage, the phase angle
   (a) increases   (b) decreases   (c) is not affected

9. When the frequency is increased, the impedance of a parallel *RL* circuit
   (a) increases   (b) decreases   (c) remains constant

10. In a parallel *RL* circuit, there are 2 A rms in the resistive branch and 2 A rms in the inductive branch. The total rms current is
    (a) 4 A   (b) 5.656 A   (c) 2 A   (d) 2.828 A

11. You are observing two voltage waveforms on an oscilloscope. The time base (time/division) of the scope is adjusted so that one half-cycle of the waveforms covers the ten horizontal divisions. The positive-going zero crossing of one waveform is at the leftmost division, and the positive-going zero crossing of the other is three divisions to the right. The phase angle between these two waveforms is
    (a) 18°   (b) 36°   (c) 54°   (d) 180°

12. Which of the following power factors results in less energy loss in an *RL* circuit?
    (a) 1   (b) 0.9   (c) 0.5   (d) 0.1

13. If a load is purely inductive and the reactive power is 10 VAR, the apparent power is
    (a) 0 VA   (b) 10 VA   (c) 14.14 VA   (d) 3.16 VA

14. For a certain load, the true power is 10 W and the reactive power is 10 VAR. The apparent power is
    (a) 5 VA   (b) 20 VA   (c) 14.14 VA   (d) 100 VA

15. The cutoff frequency of a certain low-pass *RL* filter is 20 kHz. The filter's bandwidth is
    (a) 20 kHz   (b) 40 kHz   (c) 0 kHz   (d) unknown

PROBLEM SET A   **Section 13–1   Sinusoidal Response of *RL* Circuits**

1. A 15-kHz sinusoidal voltage is applied to a series *RL* circuit. Determine the frequency of *I*, V_R, and V_L.

2. What are the waveshapes of *I*, V_R, and V_L in Problem 1?

Section 13-2   Impedance and Phase Angle of Series *RL* Circuits

3. Find the impedance of each circuit in Figure 13–43.
4. Determine the impedance and phase angle in each circuit in Figure 13–44.

FIGURE 13–43

(a)   (b)

(a)

(b)

FIGURE 13–44

5. In Figure 13–45, determine the impedance at each of the following frequencies:
   (a)   100 Hz   (b)   500 Hz   (c)   1 kHz   (d)   2 kHz

FIGURE 13–45

6. Determine the values of R and X_L in a series *RL* circuit for the following values of impedance and phase angle:
   (a) $Z = 20\ \Omega,\ \theta = 45°$ (b) $Z = 500\ \Omega,\ \theta = 35°$
   (c) $Z = 2.5\ k\Omega,\ \theta = 72.5°$ (d) $Z = 998\ \Omega,\ \theta = 45°$

Section 13-3   Analysis of Series *RL* Circuits

7. Find the current for each circuit of Figure 13–43.
8. Calculate the total current in each circuit of Figure 13–44.
9. Determine θ for the circuit in Figure 13–46.
10. If the inductance in Figure 13–46 is doubled, does θ increase or decrease, and by how
    many degrees?
11. Sketch the waveforms for V_s, V_R, and V_L in Figure 13–46. Show the proper phase relation-
    ships.

FIGURE 13–46

12. For the circuit in Figure 13–47, find V_R and V_L for each of the following frequencies:
    (a) 60 Hz   **(b)** 200 Hz   **(c)** 500 Hz   **(d)** 1 kHz

FIGURE 13–47

Section 13–4   Impedance and Phase Angle of Parallel *RL* Circuits

13. What is the impedance for the circuit in Figure 13–48?
14. Repeat Problem 13 for the following frequencies:
    (a) 1.5 kHz   **(b)** 3 kHz   **(c)** 5 kHz   **(d)** 10 kHz
15. At what frequency does X_L equal R in Figure 13–48?

FIGURE 13–48

Section 13–5   Analysis of Parallel *RL* Circuits

16. Find the total current and each branch current in Figure 13–49.
17. Determine the following quantities in Figure 13–50:
    (a) Z **(b)** I_R **(c)** I_L **(d)** I_T **(e)** θ

FIGURE 13–49   **FIGURE 13–50**

18. Convert the circuit in Figure 13–51 to an equivalent series form.

FIGURE 13–51

Section 13–6   Series-Parallel Analysis

19. Determine the voltage across each element in Figure 13–52.
20. Is the circuit in Figure 13–52 predominantly resistive or predominantly inductive?
21. Find the current in each branch and the total current in Figure 13–52.

FIGURE 13–52

Section 13–7 Power in *RL* Circuits

22. In a certain *RL* circuit, the true power is 100 mW, and the reactive power is 340 mVAR. What is the apparent power?

23. Determine the true power and the reactive power in Figure 13–46.

24. What is the power factor in Figure 13–49?

25. Determine P_{true}, P_r, P_a, and *PF* for the circuit in Figure 13–52. Sketch the power triangle.

Section 13–8 Basic Applications of *RL* Circuits

26. For the lag network in Figure 13–53, determine the phase lag of the output voltage with respect to the input for the following frequencies:
    (a) 1 Hz   (b) 100 Hz   (c) 1 kHz   (d) 10 kHz

27. Plot the response curve for the circuit in Figure 13–53. Show the output voltage versus frequency in 1-kHz increments from 0 Hz to 5 kHz.

28. Repeat Problem 26 for the lead network in Figure 13–54.

29. Using the same procedure as in Problem 27, plot the response curve for Figure 13–54.

30. Sketch the voltage phasor diagram for each circuit in Figures 13–53 and 13–54 for a frequency of 8 kHz.

FIGURE 13–53

FIGURE 13–54

Section 13–9 Troubleshooting *RL* Circuits

31. Determine the voltage across each element in Figure 13–55 if L_1 were open.

32. Determine the output voltage in Figure 13–56 for each of the following failure modes:
    (a) L_1 open   (b) L_2 open   (c) R_1 open   (d) a short across R_2

FIGURE 13–55

FIGURE 13–56

PROBLEM SET B   **33.** Determine the voltage across the inductors in Figure 13–57.

34. Is the circuit in Figure 13–57 predominantly resistive or predominantly inductive?

35. Find the total current in Figure 13–57.

FIGURE 13–57

36. For the circuit in Figure 13–58, determine the following:
(a) Z_T (b) I_T (c) θ (d) V_L
(e) V_{R3}

FIGURE 13–58

37. For the circuit in Figure 13–59, determine the following:
(a) I_{R1} (b) I_{L1} (c) I_{L2} (d) I_{R2}

FIGURE 13–59

38. Determine the phase shift and attenuation (ratio of V_{out} to V_{in}) from the input to the output for the network in Figure 13–60.

FIGURE 13–60

39. Determine the phase shift and attenuation from the input to the output for the ladder network in Figure 13–61.

FIGURE 13–61

40. Design an ideal inductive switching circuit that will provide a momentary voltage of 2.5 kV from a 12-V dc source when a switch is thrown instantaneously from one position to another. The drain on the source must not exceed 1 A.

COLOR INSERT TBLS + −

41. Draw the schematic for the circuit in Figure 13–62 and determine if the waveforms on the scope are correct (a full-color photo appears following page 416). If there is a fault in the circuit, identify it.

ANSWERS TO SECTION REVIEWS

Section 13–1
1. 1 kHz.   **2.** Closer to 0°.

Section 13–2
1. Leads.   **2.** The inductive reactance.   **3.** Opposite phase angles.
4. 59.9 kΩ, 56.58°.

Section 13–3
1. 3.61 V.   **2.** 56.31°.   **3.** X_L increases, Z increases, and θ increases.

Section 13–4
1. 20 Ω.   **2.** 0.024 S.   **3.** Lags by 33.69°.

Section 13–5
1. 32 mA.   **2.** 59.04°, 23.32 mA.   **3.** 90°.

Section 13–6
1. $Z = 494.6$ Ω.   **2.** $I_T = 20.22$ mA.

Section 13–7
1. Resistor.   **2.** 0.643.   **3.** $P_{true} = 4.7$ W, $P_r = 6.2$ VAR, $P_a = 7.78$ VA.

Section 13–8
1. 81.87°.   **2.** 1.41 V.   **3.** Resistor.

Section 13–9
1. Shorted windings reduce L and thereby reduce X_L at any given frequency.
2. I_T decreases, V_{R1} decreases, V_{R2} increases.

14 *RLC* Circuits and Resonance

In this chapter, we introduce frequency response of *RLC* circuits—that is, circuits with combinations of resistance, capacitance, and inductance. We will discuss both series and parallel *RLC* circuits, including the concepts of series and parallel resonance.

Because it is the basis for frequency selectivity, resonance in electrical circuits is very important to the operation of many types of electronic systems, particularly in the area of communications. For example, the ability of a radio or television receiver to select a certain frequency transmitted by a certain station and to eliminate frequencies from other stations is based on the principle of resonance.

The operation of both band-pass and band-stop filters is based on resonance of circuits containing inductance and capacitance, and these filters are discussed in this chapter. Also, certain system applications are presented.

After completing this chapter, you will be able to

- ☐ Determine the impedance and phase angle in series and parallel *RLC* circuits.
- ☐ Analyze *RLC* circuits for voltage and current values.
- ☐ Define the term *resonance*.
- ☐ Discuss the characteristics of the series and parallel resonant conditions.
- ☐ Determine the frequency at which resonance occurs in a given circuit.
- ☐ Explain the difference between series and parallel resonance.
- ☐ Analyze a nonideal parallel resonant circuit that has resistance in its inductive branch.
- ☐ Describe the effects of parallel resistance on a parallel resonant circuit.
- ☐ Define the term Q (quality factor).
- ☐ Determine Q of both series and parallel resonant circuits.
- ☐ Recognize the response characteristics of band-pass and band-stop filters
- ☐ Explain how the Q affects the bandwidth (*BW*), resonant frequency (f_r), and impedance of a resonant filter.
- ☐ Explain what a *decibel* (dB) is.
- ☐ Discuss how resonant circuits are applied in several areas.

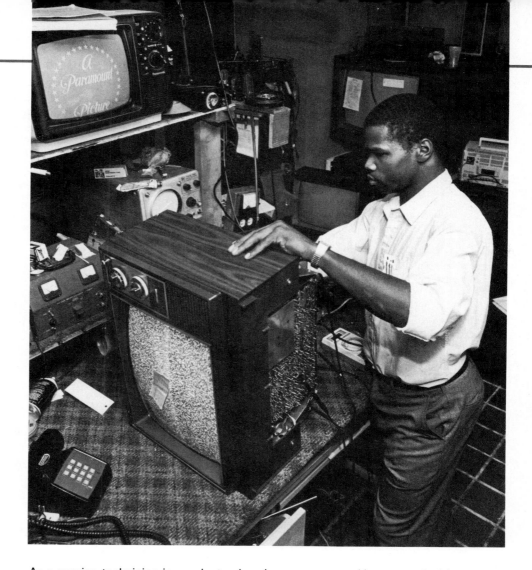

APPLICATION ASSIGNMENT

As a service technician in an electronics shop, you are working on a television receiver that has dark bars across the picture screen. Being an inexperienced technician, you are not sure what is causing this problem. Your supervisor tells you that the dark bars are *sound bars* caused by the audio signal getting through to the picture tube and interfering with the video signal. After studying this chapter, you will know the most likely cause of the problem.

14–1   IMPEDANCE AND PHASE ANGLE OF SERIES *RLC* CIRCUITS

A series *RLC* circuit is shown in Figure 14–1. It contains resistance, inductance, and capacitance. As you know, X_L causes the total current to lag the applied voltage. X_C has the opposite effect: It causes the current to lead the voltage. Thus, X_L and X_C tend to offset each other. When they are equal, they cancel, and the total reactance is zero. In any case, the total reactance in the series circuit is

$$X_T = |X_L - X_C| \tag{14–1}$$

FIGURE 14–1
Series *RLC* circuit.

The term $|X_L - X_C|$ means the *absolute value* of the difference of the two reactances. That is, the sign of the result is considered positive no matter which reactance is greater. For example, $3 - 7 = -4$, but the absolute value is

$$|3 - 7| = 4$$

When $X_L > X_C$, the circuit is predominantly inductive, and when $X_C > X_L$, the circuit is predominantly capacitive.

The total impedance for the series *RLC* circuit is given by

$$Z = \sqrt{R^2 + (X_L - X_C)^2} \tag{14–2}$$

and the phase angle is

$$\theta = \tan^{-1}\left(\frac{X_T}{R}\right) \tag{14–3}$$

EXAMPLE 14–1   Determine the total impedance and the phase angle in Figure 14–2.

FIGURE 14–2

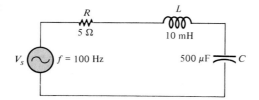

Solution
First find X_C and X_L:

$$X_C = \frac{1}{2\pi f C} = \frac{1}{2\pi(100 \text{ Hz})(500 \ \mu\text{F})} = 3.18 \ \Omega$$

$$X_L = 2\pi f L = 2\pi(100 \text{ Hz})(10 \text{ mH}) = 6.28 \ \Omega$$

In this case, X_L is greater than X_C, and thus the circuit is more inductive than capacitive. The magnitude of the total reactance is

$$X_T = |X_L - X_C| = |6.28 \ \Omega - 3.18 \ \Omega| = 3.1 \ \Omega \qquad \text{(inductive)}$$

The total circuit impedance is

$$Z = \sqrt{R^2 + X_T^2} = \sqrt{(5 \ \Omega)^2 + (3.1 \ \Omega)^2} = 5.88 \ \Omega$$

The phase angle (between I and V_s) is

$$\theta = \tan^{-1}\left(\frac{X_T}{R}\right) = \tan^{-1}\left(\frac{3.1 \ \Omega}{5 \ \Omega}\right)$$
$$= 31.8° \qquad \text{(current lagging } V_s\text{)}$$

The calculator sequences are as follows:

Z: [2] [×] [2nd] [π] [×] [1] [0] [0] [×] [1] [0]
[EXP] [+/−] [3] [=] [STO] [2] [×] [2nd] [π] [×]
[1] [0] [0] [×] [5] [0] [0] [EXP] [+/−] [6]
[=] [1/x] [+/−] [SUM] [RCL] [x²] [+] [5] [x²] [=]
[2nd] [√x]

θ: [RCL] [÷] [5] [=] [2nd] [tan]

As you have seen, when the inductive reactance is greater than the capacitive reactance, the circuit appears inductive, and the current lags the applied voltage. When the capacitive reactance is greater, the circuit appears capacitive, and the current leads the applied voltage.

SECTION REVIEW 14–1

1. In a given series *RLC* circuit, X_C is 150 Ω and X_L is 80 Ω. What is the total reactance in ohms? Is it inductive or capacitive?
2. Determine the impedance for the circuit in Problem 1 when $R = 45 \ \Omega$. What is the phase angle? Is the current leading or lagging the applied voltage?

14–2

ANALYSIS OF SERIES *RLC* CIRCUITS

Recall that capacitive reactance varies inversely with frequency and inductive reactance varies directly. As shown in Figure 14–3, for a typical series *RLC* circuit the total reactance behaves as follows: Starting at a very low frequency, X_C is high, X_L is low, and the circuit is predominantly capacitive. As the frequency is increased, X_C decreases and X_L increases until a value is reached where $X_C = X_L$ and the two reactances cancel, making the circuit purely resistive. This condition is *series resonance* and will be studied in the next section. As the frequency is increased further, X_L becomes greater than X_C, and the circuit is predominantly inductive. Example 14–2 illustrates how the impedance and phase angle change as the source frequency is varied.

The reason that the graph of X_L is a straight line and the graph of X_C is curved in Figure 14–3 is as follows: The general equation for a straight line is $y = mx + b$,

FIGURE 14–3
How X_C and X_L vary with frequency.

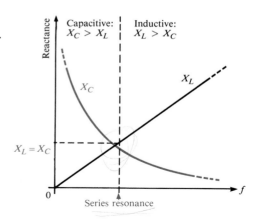

where m is the slope of the line and b is the y-axis intercept point. The formula $X_L = 2\pi fL$ fits this general straight-line formula, where $y = X_L$ (a variable), $m = 2\pi L$ (a constant), $x = f$ (a variable), and $b = 0$ as follows: $X_L = 2\pi Lf + 0$.

The X_C curve is called a *hyperbola,* and the general equation of a hyperbola is $xy = k$. The equation for capacitive reactance, $X_C = 1/2\pi fC$, can be rearranged as $X_Cf = 1/2\pi C$ where $x = X_C$ (a variable), $y = f$ (a variable), and $k = 1/2\pi C$ (a constant).

EXAMPLE 14–2

For each of the following frequencies of the source voltage, find the impedance and the phase angle for the circuit in Figure 14–4:

(a) $f = 1$ kHz   (b) $f = 3.5$ kHz   (c) $f = 5$ kHz

Note how the impedance and the phase angle change.

FIGURE 14–4

Solution

(a) At $f = 1$ kHz:

$$X_C = \frac{1}{2\pi(1\ \text{kHz})(0.02\ \mu\text{F})} = 7.96\ \text{k}\Omega$$

$$X_L = 2\pi(1\ \text{kHz})(100\ \text{mH}) = 628.3\ \Omega$$

The circuit is highly capacitive because X_C is much larger than X_L. The impedance is

$$Z = \sqrt{R^2 + (X_L - X_C)^2} = \sqrt{(3.3\ \text{k}\Omega)^2 + (628.3\ \Omega - 7.96\ \text{k}\Omega)^2}$$
$$= 8.04\ \text{k}\Omega$$

The phase angle is

$$\theta = \tan^{-1}\left(\frac{X_T}{R}\right) = \tan^{-1}\left(\frac{7.33 \text{ k}\Omega}{3.3 \text{ k}\Omega}\right) = 65.76°$$

I leads V_s by 65.76°.
(b) At $f = 3.5$ kHz:

$$X_C = \frac{1}{2\pi(3.5 \text{ kHz})(0.02 \text{ }\mu\text{F})} = 2.27 \text{ k}\Omega$$

$$X_L = 2\pi(3.5 \text{ kHz})(100 \text{ mH}) = 2.2 \text{ k}\Omega$$

The circuit is very close to being purely resistive but is still slightly capacitive because X_C is slightly larger than X_L.

$$Z = \sqrt{(3.3 \text{ k}\Omega)^2 + (2.2 \text{ k}\Omega - 2.27 \text{ k}\Omega)^2} = 3.3 \text{ k}\Omega$$

$$\theta = \tan^{-1}\left(\frac{X_T}{R}\right) = \tan^{-1}\left(\frac{0.07 \text{ k}\Omega}{3.3 \text{ k}\Omega}\right) = 1.2°$$

I leads V_s by 1.2°.
(c) At $f = 5$ kHz:

$$X_C = \frac{1}{2\pi(5 \text{ kHz})(0.02 \text{ }\mu\text{F})} = 1.59 \text{ k}\Omega$$

$$X_L = 2\pi(5 \text{ kHz})(100 \text{ mH}) = 3.14 \text{ k}\Omega$$

The circuit is now predominantly inductive because $X_L > X_C$.

$$Z = \sqrt{(3.3 \text{ k}\Omega)^2 + (3.14 \text{ k}\Omega - 1.59 \text{ k}\Omega)^2} = 3.64 \text{ k}\Omega$$

$$\theta = \tan^{-1}\left(\frac{X_T}{R}\right) = \tan^{-1}\left(\frac{1.55 \text{ k}\Omega}{3.3 \text{ k}\Omega}\right) = 25.16°$$

I lags V_s by 25.16°.

Notice how the circuit changed from capacitive to inductive as the frequency increased. The phase condition changed from the leading current to the lagging current. It is interesting to note that both the impedance and the phase angle decreased to a minimum and then began increasing again as the frequency went up.

In a series *RLC* circuit, the capacitor voltage and the inductor voltage are *always* 180° out of phase with each other. For this reason, V_C and V_L subtract from each other, and thus the voltage across L and C combined is always less than the larger individual voltage across either element, as illustrated in Figure 14–5 and in the waveform diagram of Figure 14–6.

In the next example, we use Ohm's law to find the voltages in a series *RLC* circuit.

FIGURE 14–5
The voltage across the series combination of C and L is always less than the *larger* individual voltage.

FIGURE 14–6
V_L and V_C effectively subtract.

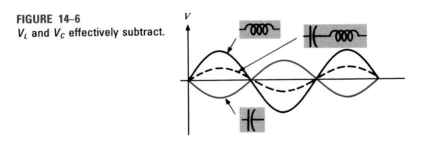

EXAMPLE 14–3

Find the voltages across each element in Figure 14–7, and draw a complete voltage phasor diagram.

FIGURE 14–7

Solution

First find the total impedance:

$$Z = \sqrt{R^2 + (X_L - X_C)^2} = \sqrt{(75\ \Omega)^2 + (25\ \Omega - 60\ \Omega)^2}$$
$$= \sqrt{(75\ \Omega)^2 + (35\ \Omega)^2} = 82.76\ \Omega$$

Apply Ohm's law to find the current:

$$I = \frac{V_s}{Z} = \frac{10\ V}{82.76\ \Omega} = 0.121\ A$$

Now apply Ohm's law to find the voltages across R, L, and C:

$$V_R = IR = (0.121\ A)(75\ \Omega) = 9.08\ V$$
$$V_L = IX_L = (0.121\ A)(25\ \Omega) = 3.03\ V$$
$$V_C = IX_C = (0.121\ A)(60\ \Omega) = 7.26\ V$$

The voltage across L and C combined is

$$V_{LC} = V_C - V_L = 7.26\ V - 3.03\ V = 4.23\ V$$

The circuit phase angle is

$$\theta = \tan^{-1}\left(\frac{X_T}{R}\right) = \tan^{-1}\left(\frac{35\ \Omega}{75\ \Omega}\right) = 25°$$

Since the circuit is capacitive ($X_C > X_L$), the current leads the source voltage by 25°.

The phasor diagram is shown in Figure 14–8. Notice that V_L is leading V_R by 90°, and V_C is lagging V_R by 90°. Also, there is a 180° phase difference between V_L and V_C. If the current phasor were shown, it would be at the same angle as V_R.

FIGURE 14–8

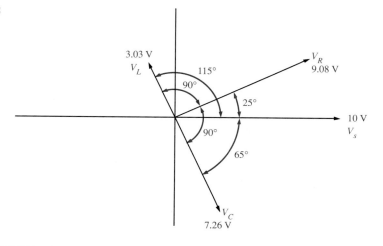

**SECTION REVIEW
14–2**

1. The following voltages occur in a certain series *RLC* circuit: $V_R = 24$ V, $V_L = 15$ V, $V_C = 45$ V. Determine the source voltage.

2. When $R = 10\ \Omega$, $X_C = 18\ \Omega$, and $X_L = 12\ \Omega$, does the current lead or lag the applied voltage?

3. Determine the total reactance in Problem 2.

14–3

SERIES RESONANCE

In a series *RLC* circuit, *series resonance* occurs when $X_L = X_C$. The frequency at which resonance occurs is called the *resonant frequency*, f_r. Figure 14–9 illustrates the series resonant condition. Since $X_L = X_C$, the reactances effectively cancel, and the *impedance is purely resistive*. These resonant conditions are stated in the following equations:

$$X_L = X_C \tag{14–4}$$

$$Z_r = R \tag{14–5}$$

FIGURE 14–9
At the resonant frequency f_r, the reactances effectively cancel, leaving $Z = R$.

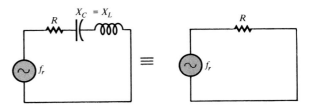

EXAMPLE 14–4

For the series *RLC* circuit in Figure 14–10, determine X_C and Z at resonance.

FIGURE 14–10

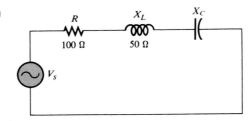

Solution

$$X_C = X_L = 50\ \Omega \qquad \text{at the resonant frequency}$$

$$Z_r = R = 100\ \Omega$$

Why X_L and X_C Effectively Cancel at Resonance

At the series resonant frequency, the voltages across C and L are equal in magnitude because the reactances are equal and because the same current flows through both since they are in series ($IX_C = IX_L$). Also, V_L and V_C are always 180° out of phase with each other.

During any given cycle, the polarities of the voltages across C and L are opposite, as shown in Figures 14–11(a) and 14–11(b). The equal and opposite voltages across C and L cancel, leaving zero volts from point A to point B as shown in the figure. Since there is no voltage drop from A to B but there is still current, the total reactance must be zero, as indicated in part (c) of the figure. Also, the voltage phasor diagram in part (d) shows that V_C and V_L are equal in magnitude and 180° out of phase with each other.

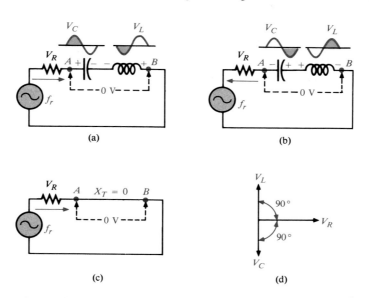

FIGURE 14–11
At the resonant frequency, f_r, the voltages across C and L are equal in magnitude. Since they are 180° out of phase with each other, they cancel, leaving 0 V across the *LC* combination (point A to point B). The section of the circuit from A to B effectively looks like a short at resonance (neglecting winding resistance).

The Series Resonant Frequency

For a given series *RLC* circuit, resonance occurs at only one specific frequency. A formula for this resonant frequency is developed as follows:

$$X_L = X_C$$

Substituting the reactance formulas, we have

$$2\pi f_r L = \frac{1}{2\pi f_r C}$$

Solving for the resonant frequency f_r yields

$$(2\pi f_r L)(2\pi f_r C) = 1$$
$$4\pi^2 f_r^2 LC = 1$$

$$f_r^2 = \frac{1}{4\pi^2 LC}$$

Taking the square root of both sides, we obtain

$$f_r = \frac{1}{2\pi\sqrt{LC}}$$ (14–6)

EXAMPLE 14–5

Find the series resonant frequency for the circuit in Figure 14–12.

FIGURE 14–12

Solution

$$f_r = \frac{1}{2\pi\sqrt{LC}} = \frac{1}{2\pi\sqrt{(5\text{ mH})(50\text{ pF})}} = 318\text{ kHz}$$

The calculator sequence is

[5] [EXP] [+/−] [3] [×] [5] [0] [EXP] [+/−] [1] [2] [=] [2nd] [√x] [×]
[2] [×] [2nd] [π] [=] [1/x]

Voltages and Current in a Series *RLC* Circuit

It is interesting to see how the current and the voltages in a series *RLC* circuit vary as the frequency is increased from below the resonant frequency, through resonance, and then above resonance. Figure 14–13 illustrates the general response of a circuit in terms of changes in the current and in the voltage drops.

Below the Resonant Frequency

Figure 14–13(a) indicates the response as the source frequency is increased from zero toward f_r. At $f = 0$ Hz (dc), the capacitor appears open and blocks the current. Thus, there is no voltage across R or L, and the entire source voltage appears across C. The impedance of the circuit is infinitely large at 0 Hz because X_C is infinite (C is open). As the frequency begins to increase, X_C decreases and X_L increases, causing the total reactance, $X_C - X_L$, to decrease. As a result, the impedance decreases and the current increases. As the current increases, V_R also increases, and both V_C and V_L increase. The voltage across C and L combined decreases from its maximum value of V_s because as V_C and V_L approach the same value, their *difference* becomes less.

At the Resonant Frequency

When the frequency reaches its resonant value, f_r, as shown in Figure 14–13(b), V_C and V_L can each be much larger than the source voltage. V_C and V_L *cancel* leaving 0 V

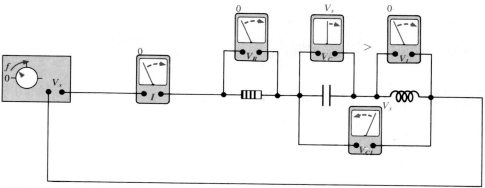

(a) Below resonance ($X_C > X_L$)

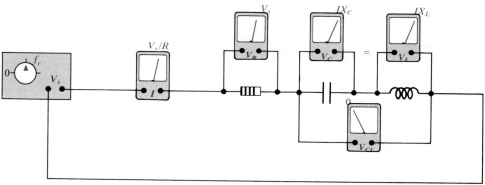

(b) At resonance ($X_C = X_L$)

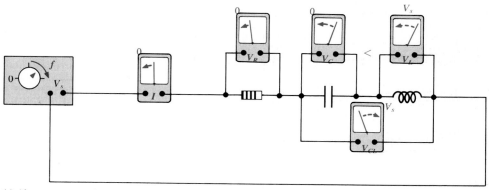

(c) Above resonance ($X_L > X_C$)

FIGURE 14–13

An illustration of how the voltages and the current respond in a series *RLC* circuit as the frequency is increased from below to above its resonant value. The source voltage is held at a constant amplitude.

561

across the combination, because they are equal in magnitude but opposite in phase. At this point the total impedance is equal to R and is at its minimum value because the total reactance is zero. Thus, the current is at its maximum value, V_s/R, and V_R is at its maximum value which is equal to the source voltage.

Above the Resonant Frequency

As the frequency is increased above resonance, as indicated in Figure 14–13(c), X_L continues to increase and X_C continues to decrease, causing the total reactance, $X_L - X_C$, to increase. As a result, there is an increase in impedance and a decrease in current. As the current decreases, V_R also decreases and V_C and V_L decrease. As V_C and V_L decrease, their difference becomes greater, so V_{CL} increases. As the frequency becomes very high, the current approaches zero, both V_R and V_C approach zero, and V_L approaches V_s.

The responses of voltage and current to increasing frequency are summarized as follows: As frequency is increased, the current increases below resonance, reaches a peak at the resonant frequency, and then decreases above resonance. The resistor voltage responds in the same way as the current. The voltage across the C and L combination decreases as the frequency increases below resonance, reaching a minimum of zero at the resonant frequency; then it increases above resonance. The general response curves in Figure 14–14 illustrate this action graphically.

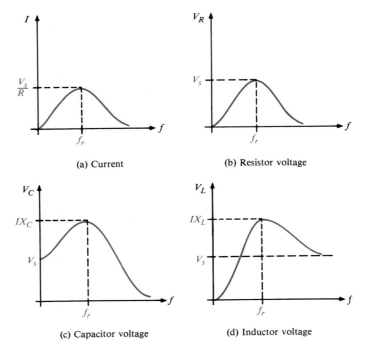

(a) Current

(b) Resistor voltage

(c) Capacitor voltage

(d) Inductor voltage

FIGURE 14–14
Current and voltages as functions of frequency in a series RLC circuit.

EXAMPLE 14–6   Find I, V_R, V_L, and V_C at resonance in Figure 14–15.

FIGURE 14–15

Solution
At resonance, I is maximum and equal to V_s/R:

$$I = \frac{V_s}{R} = \frac{50 \text{ V}}{25 \text{ }\Omega} = 2 \text{ A}$$

Applying Ohm's law, we obtain the following voltages:

$$V_R = IR = (2 \text{ A})(25 \text{ }\Omega) = 50 \text{ V}$$
$$V_L = IX_L = (2 \text{ A})(100 \text{ }\Omega) = 200 \text{ V}$$
$$V_C = IX_C = (2 \text{ A})(100 \text{ }\Omega) = 200 \text{ V}$$

The voltages are maximum at resonance but drop off above and below f_r. The voltages across L and C at resonance are exactly *equal in magnitude but 180° out of phase, so they cancel.* Thus, the total voltage across both L and C is zero, and $V_R = V_s$ at resonance. Individually, V_L and V_C can be much greater than the source voltage, as you see. Keep in mind that V_L and V_C are always opposite in polarity regardless of the frequency, but only at resonance are their magnitudes equal.

The Impedance of a Series *RLC* Circuit

Figure 14–16 shows a general graph of impedance versus frequency superimposed on the curves for X_C and X_L. At zero frequency, both X_C and Z are infinitely large and X_L is zero, because the capacitor looks like an open at 0 Hz and the inductor looks like a

FIGURE 14–16
Series *RLC* impedance as a function of frequency.

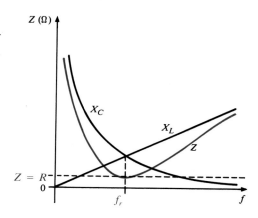

short. As the frequency increases, X_C decreases and X_L increases. Since X_C is larger than X_L at frequencies below f_r, Z decreases along with X_C. At f_r, $X_C = X_L$ and $Z = R$. At frequencies above f_r, X_L becomes increasingly larger than X_C, causing Z to increase.

EXAMPLE 14–7

For the circuit in Figure 14–17, determine the impedance at the following frequencies:
(a) f_r (b) 1000 Hz below f_r (c) 1000 Hz above f_r

FIGURE 14–17

V_s — 10 Ω — 100 mH — 0.01 μF

Solution

$$f_r = \frac{1}{2\pi\sqrt{LC}} = \frac{1}{2\pi\sqrt{(100 \text{ mH})(0.01 \text{ μF})}} = 5.03 \text{ kHz}$$

At 1000 Hz below f_r,

$$f_r - 1 \text{ kHz} = 4.03 \text{ kHz}$$

At 1000 Hz above f_r,

$$f_r + 1 \text{ kHz} = 6.03 \text{ kHz}$$

(a) The impedance at resonance is equal to R:

$$Z = 10 \text{ Ω}$$

(b) The impedance at $f_r - 1$ kHz is

$$Z = \sqrt{R^2 + (X_L - X_C)^2} = \sqrt{(10 \text{ Ω})^2 + (2.53 \text{ kΩ} - 3.95 \text{ kΩ})^2}$$
$$= 1.42 \text{ kΩ}$$

Notice that X_C is greater than X_L, so Z is capacitive.

(c) The impedance at $f_r + 1$ kHz is

$$Z = \sqrt{R^2 + (X_L - X_C)^2} = \sqrt{(10 \text{ Ω})^2 + (3.79 \text{ kΩ} - 2.64 \text{ kΩ})^2}$$
$$= 1.15 \text{ kΩ}$$

X_L is greater than X_C, so Z is inductive.

The Phase Angle of a Series *RLC* Circuit

At frequencies below resonance, $X_C > X_L$, and the current *leads* the source voltage, as indicated in Figure 14–18(a). The phase angle decreases as the frequency approaches the resonant value and is 0° at resonance, as indicated in part (b). At frequencies above resonance, $X_L > X_C$, and the current *lags* the source voltage, as indicated in part (c).

As the frequency goes higher, the phase angle approaches 90°. A plot of phase angle versus frequency is shown in part (d) of the figure.

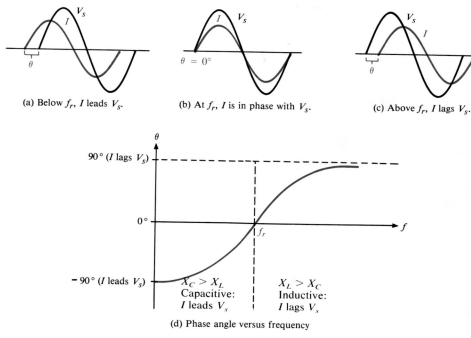

(a) Below f_r, I leads V_s.   (b) At f_r, I is in phase with V_s.   (c) Above f_r, I lags V_s.

(d) Phase angle versus frequency

FIGURE 14–18
The phase angle as a function of frequency in a series *RLC* circuit.

**SECTION REVIEW
14–3**

1. What is the condition for series resonance?
2. Why is the current maximum at the resonant frequency?
3. Calculate the resonant frequency for $C = 1000$ pF and $L = 1000$ μH.
4. In Problem 3, is the circuit inductive or capacitive at 50 kHz?

14–4

SERIES RESONANT FILTERS

A common use of series *RLC* circuits is in filter applications. In this section, you will learn the basic configurations for *band-pass* and *band-stop* filters and several important filter characteristics.

The Band-Pass Filter

A basic series resonant band-pass filter is shown in Figure 14–19. Notice that the series *LC* portion is placed between the input and the output and that the output is taken across the resistor.

This filter allows signals at the resonant frequency and at frequencies within a certain band (or range) extending below and above the resonant value to pass from input

FIGURE 14–19
A series resonant band-pass filter.

to output without a significant reduction in amplitude. Signals at frequencies lying outside this specified band (called the *pass band*) are reduced in amplitude to below a certain level and are considered to be rejected by the filter.

The filtering action is the result of the impedance characteristic of the filter. As you learned in the previous section, the impedance is minimum at resonance and has increasingly higher values below and above the resonant frequency. At very low frequencies, the impedance is very high and tends to block the current. As the frequency increases, the impedance drops, allowing more current and thus more voltage across the

(a) As the frequency increases to f_1, V_{out} increases to 7.07 V.

(b) As the frequency increases from f_1 to f_r, V_{out} increases from 7.07 V to 10 V.

(c) As the frequency increases from f_r to f_2, V_{out} decreases from 10 V to 7.07 V.

(d) As the frequency increases above f_2, V_{out} decreases below 7.07 V.

FIGURE 14–20
Example of the frequency response of a series resonant band-pass filter with the input voltage at a constant 10 V rms. The winding resistance of the coil is neglected.

output resistor. At the resonant frequency, the impedance is very low and equal to the winding resistance of the coil. At this point a maximum current is flowing and the resulting output voltage is maximum. As the frequency goes above resonance, the impedance again increases, causing the current and the resulting output voltage to drop.

Bandwidth of the Pass Band

The bandwidth of a band-pass filter is the range of frequencies for which the current (or output voltage) is equal to or greater than 70.7% of its value at the resonant frequency. Bandwidth is often abbreviated as *BW*.

Figure 14–20 illustrates the general response of a series resonant band-pass filter, and Figure 14–21 shows the bandwidth on the response curve for the filter. Notice that frequency f_1, which is below f_r, is the frequency at which I (or V_{out}) is 70.7% of the resonant value (I_{max}); f_1 is commonly called the *lower cutoff frequency*. At frequency f_2, above f_r, the current (or V_{out}) is again 70.7% of its maximum; f_2 is called the *upper cutoff frequency*. Other names for f_1 and f_2 are -3 *dB frequencies*, *band frequencies*, and *half-power frequencies*. (The term *decibel*, abbreviated dB, is defined on page 568).

The formula for calculating the bandwidth is as follows:

$$BW = f_2 - f_1 \qquad (14\text{–}7)$$

The unit of bandwidth is the hertz (Hz), the same as for frequency.

FIGURE 14–21
Generalized response curve of a series resonant band-pass filter.

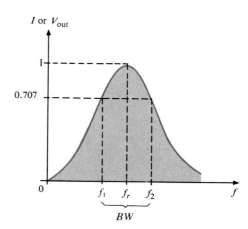

EXAMPLE 14–8

A certain series resonant band-pass filter has a maximum current of 100 mA at the resonant frequency. What is the value of the current at the cutoff frequencies?

Solution
Current at the cutoff frequencies is 70.7% of maximum:

$$I_{f1} = I_{f2} = 0.707 I_{max} = 0.707(100 \text{ mA}) = 70.7 \text{ mA}$$

EXAMPLE 14–9

A resonant circuit has a lower cutoff frequency of 8 kHz and an upper cutoff frequency of 12 kHz. Determine the bandwidth.

Solution

$$BW = f_2 - f_1 = 12 \text{ kHz} - 8 \text{ kHz} = 4 \text{ kHz}$$

Half-Power Points of the Filter Response

As previously mentioned, the upper and lower cutoff frequencies are sometimes called the *half-power frequencies*. This term is derived from the fact that the true power from the source at these frequencies is one-half the power delivered at the resonant frequency. The following steps show that this relationship is true for a series resonant circuit.

At resonance,

$$P_{max} = I_{max}^2 R$$

The power at f_1 or f_2 is

$$P_{f1} = I_{f1}^2 R = (0.707I_{max})^2 R = (0.707)^2 I_{max}^2 R$$
$$= 0.5I_{max}^2 R = 0.5P_{max}$$

Decibel (dB) Measurement

As previously mentioned, another common term for the upper and lower cutoff frequencies is *−3 dB frequencies*. The decibel (dB) is a logarithmic measurement of voltage or power ratios which can be used to express the input-to-output relationship of a filter. The following equation expresses a *voltage* ratio in decibels:

$$dB = 20 \log\left(\frac{V_{out}}{V_{in}}\right) \tag{14–8}$$

The following equation is the decibel formula for a *power* ratio:

$$dB = 10 \log\left(\frac{P_{out}}{P_{in}}\right) \tag{14–9}$$

EXAMPLE 14–10

The output voltage of a filter is 5 V and the input is 10 V. Express the voltage ratio in decibels.

Solution

$$20 \log\left(\frac{V_{out}}{V_{in}}\right) = 20 \log\left(\frac{5 \text{ V}}{10 \text{ V}}\right) = 20 \log(0.5) = -6.02 \text{ dB}$$

The calculator sequence is

$\boxed{5}\ \boxed{\div}\ \boxed{1}\ \boxed{0}\ \boxed{=}\ \boxed{\log}\ \boxed{\times}\ \boxed{2}\ \boxed{0}\ \boxed{=}$

The −3-dB Frequencies

The output of a filter is said to be *down 3 dB* at the cutoff frequencies. Actually, this frequency is the point at which the output voltage is 70.7% of the maximum voltage at resonance. We can show that the 70.7% point is the same as 3 dB below maximum (or −3 dB) as follows. The maximum voltage is the 0-dB reference:

$$20 \log\left(\frac{0.707 V_{max}}{V_{max}}\right) = 20 \log(0.707) = -3 \text{ dB}$$

Selectivity of a Band-Pass Filter

The response curve in Figure 14–21 is also called a *selectivity curve*. Selectivity defines how well a resonant circuit responds to a certain frequency and discriminates against all others. *The narrower the bandwidth, the greater the selectivity.*

We normally assume that a resonant circuit accepts frequencies within its bandwidth and completely eliminates frequencies outside the bandwidth. Such is not actually the case, however, because signals with frequencies outside the bandwidth are not completely eliminated. Their magnitudes, however, are greatly reduced. The further the frequencies are from the cutoff frequencies, the greater is the reduction, as illustrated in Figure 14–22(a). An ideal selectivity curve is shown in Figure 14–22(b).

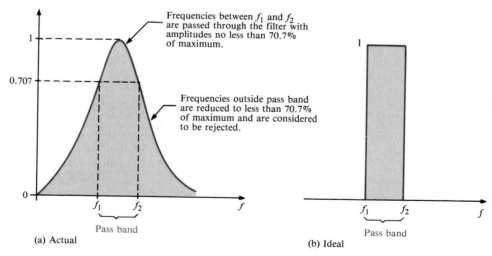

FIGURE 14–22
Generalized selectivity curve of a band-pass filter.

As you can see in Figure 14–23, another factor that influences selectivity is the steepness of the slopes of the curve. The faster the curve drops off at the cutoff frequencies, the more selective the circuit is, because the frequencies outside the pass band are more quickly reduced (attenuated).

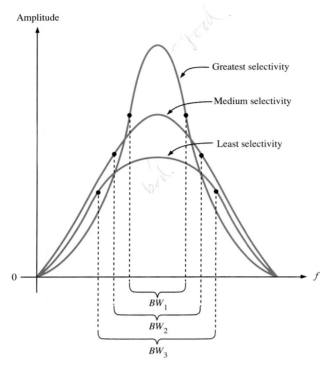

FIGURE 14–23
Comparative selectivity curves.

The Quality Factor (Q) of a Resonant Circuit and Its Effect on Bandwidth

The quality factor (Q) is the ratio of the reactive power in the inductor to the true power in the winding resistance of the coil or the resistance in series with the coil. It is a ratio of the power in L to the power in R. The quality factor is very important in resonant circuits. A formula for Q is developed as follows:

$$Q = \frac{\text{energy stored}}{\text{energy dissipated}}$$

$$= \frac{\text{reactive power}}{\text{true power}}$$

$$= \frac{I^2 X_L}{I^2 R}$$

In a series circuit, I is the same in L and R; thus, the I^2 terms cancel, leaving

$$Q = \frac{X_L}{R} \qquad (14\text{–}10)$$

When the resistance is just the winding resistance of the coil, the circuit Q and the coil Q are the same. Since Q varies with frequency because X_L varies, we are interested mainly in Q at *resonance*. Note that Q is a ratio of like units (ohms) and, therefore, has no unit itself.

EXAMPLE 14–11

Determine Q at resonance for the circuit in Figure 14–24.

FIGURE 14–24

Solution

$$X_L = 2\pi f_r L = 31.6 \ \Omega$$

$$Q = \frac{X_L}{R} = \frac{31.6 \ \Omega}{10 \ \Omega} = 3.16$$

How Q Affects Bandwidth

A higher value of circuit Q results in a smaller bandwidth. A lower value of Q causes a larger bandwidth. A formula for the bandwidth of a resonant circuit in terms of Q is stated in the following equation:

$$BW \cong \frac{f_r}{Q} \tag{14–11}$$

EXAMPLE 14–12

What is the bandwidth of the filter in Figure 14–25?

FIGURE 14–25

Solution
The total resistance is

$$R_T = R + R_W = 90 \ \Omega + 10 \ \Omega = 100 \ \Omega$$

The bandwidth is found as follows:

$$f_r = \frac{1}{2\pi\sqrt{LC}} = 22.5 \text{ kHz}$$

$$Q = \frac{X_L}{R} = \frac{1.41 \text{ k}\Omega}{100 \text{ }\Omega} = 14.1$$

$$BW = \frac{f_r}{Q} = \frac{22.5 \text{ kHz}}{14.1} = 1.6 \text{ kHz}$$

The Band-Stop Filter

A basic series resonant band-stop filter is shown in Figure 14–26. Notice that the output voltage is taken across the *LC* portion of the circuit. This filter is still a series *RLC* circuit, just as the band-pass filter is. The difference is that in this case, the output voltage is taken across the combination of *L* and *C* rather than across *R*.

FIGURE 14–26
A series resonant band-stop filter.

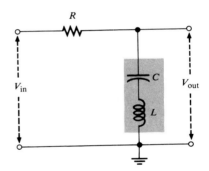

This filter rejects signals with frequencies between the upper and lower cutoff frequencies and passes those signals with frequencies below and above the cutoff values, as shown in the response curve of Figure 14–27. The range of frequencies between the lower and upper cutoff points is called the *stop band*. This type of filter is also referred to as a *band-elimination filter*, *band-reject filter*, or a *notch filter*.

FIGURE 14–27
Generalized response curve for a band-stop filter.

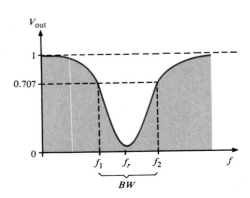

At very low frequencies, the *LC* combination appears as a near open due to the high X_C, thus allowing most of the input voltage to pass through to the output. As the frequency increases, the impedance of the *LC* combination decreases until, at resonance, it is zero (ideally). Thus, the input signal is shorted to ground, and there is very little output voltage. As the frequency goes above its resonant value, the *LC* impedance increases, allowing an increasing amount of voltage to be dropped across it. The general response of a series resonant band-stop filter is illustrated in Figure 14–28.

(a) As frequency increases to f_1, V_{out} decreases from 10 V to 7.07 V.

(b) As frequency increases from f_1 to f_r, V_{out} decreases from 7.07 V to 0 V.

(c) As frequency increases from f_r to f_2, V_{out} increases from 0 V to 7.07 V.

(d) As frequency increases above f_2, V_{out} increases toward 10 V.

FIGURE 14–28
Example of the response of a series resonant band-stop filter with V_{in} at a constant 10 V rms. The winding resistance is neglected.

Characteristics of the Band-Stop Filter

All the characteristics that have been discussed in relation to the band-pass filter (current response, impedance characteristic, bandwidth, selectivity, and Q) apply equally to the band-stop filter, with the exception that the response curve of the output voltage is opposite. For the band-pass filter, V_{out} is maximum at resonance. For the band-stop filter, V_{out} is minimum at resonance.

EXAMPLE 14–13

Find the output voltage at f_r and the bandwidth in Figure 14–29.

FIGURE 14–29

Solution
Since $X_L = X_C$ at resonance, then

$$V_{out} = \left(\frac{R_W}{R + R_W}\right)V_{in} = \left(\frac{2\ \Omega}{52\ \Omega}\right)100\ \text{mV} = 3.85\ \text{mV}$$

$$f_r = \frac{1}{2\pi\sqrt{LC}} = \frac{1}{2\pi\sqrt{(100\ \text{mH})(0.01\ \mu\text{F})}} = 5.03\ \text{kHz}$$

$$Q = \frac{X_L}{R} = \frac{3160\ \Omega}{52\ \Omega} = 60.8$$

$$BW \cong \frac{f_r}{Q} = \frac{5.03\ \text{kHz}}{60.8} = 82.7\ \text{Hz}$$

**SECTION REVIEW
14–4**

1. The output voltage of a certain band-pass filter is 15 V at the resonant frequency. What is its value at the cutoff frequencies?
2. For a certain band-pass filter, $f_r = 120$ kHz and $Q = 12$. What is the bandwidth of the filter?
3. In a band-stop filter, is the current minimum or maximum at resonance? Is the output voltage minimum or maximum at resonance?

14–5

PARALLEL *RLC* CIRCUITS

In this section, a basic parallel *RLC* circuit, shown in Figure 14–30, is examined in terms of its impedance, phase angle, and current relationships.

FIGURE 14–30
Parallel *RLC* circuit.

Impedance and Phase Angle

The circuit in Figure 14–30 consists of the parallel combination of R, L, and C. To find the admittance, add the conductance, G, and the total susceptance, B_T, as phasor quantities. B_T is the difference of the capacitive susceptance and the inductive susceptance. Thus,

$$Y = \sqrt{G^2 + (B_C - B_L)^2} \qquad (14\text{--}12)$$

The total impedance is the reciprocal of the admittance:

$$Z = \frac{1}{Y}$$

The phase angle of the circuit is given by the following formula:

$$\theta = \tan^{-1}\left(\frac{B_T}{G}\right) \qquad (14\text{--}13)$$

When the frequency is above its resonant value ($X_C < X_L$), the impedance of the circuit in Figure 14–30 is predominantly capacitive because the capacitive current is greater, and the total current *leads* the source voltage. When the frequency is below its resonant value ($X_L < X_C$), the impedance of the circuit is predominantly inductive, and the total current *lags* the source voltage.

EXAMPLE 14–14

Determine the total impedance and the phase angle in Figure 14–31.

FIGURE 14–31

Solution

$$G = \frac{1}{R} = \frac{1}{10\ \Omega} = 0.1\ \text{S}$$

$$B_C = \frac{1}{X_C} = \frac{1}{10\ \Omega} = 0.1\ \text{S}$$

$$B_L = \frac{1}{X_L} = \frac{1}{5\ \Omega} = 0.2\ \text{S}$$

$$Y = \sqrt{G^2 + (B_C - B_L)^2} = \sqrt{(0.1\ \text{S})^2 + (0.1\ \text{S} - 0.2\ \text{S})^2}$$
$$= \sqrt{(0.1\ \text{S})^2 + (-0.1\ \text{S})^2} = 0.1414\ \text{S}$$

From Y, we can get Z:

$$Z = \frac{1}{Y} = \frac{1}{0.1414\ \text{S}} = 7.07\ \Omega$$

The phase angle is

$$\theta = \tan^{-1}\left(\frac{B_T}{G}\right) = \tan^{-1}\left(\frac{0.1 \text{ S}}{0.1 \text{ S}}\right) = 45°$$

The total current *lags* V_s by 45° because the circuit is predominantly inductive ($X_L < X_C$).

Current Relationships

In a parallel *RLC* circuit, the current in the capacitive branch and the current in the inductive branch are *always* 180° out of phase with each other (neglecting any coil resistance). For this reason, I_C and I_L subtract from each other, and thus the total current into the parallel branches of L and C is always less than the largest individual branch current, as illustrated in Figure 14–32 and in the waveform diagram of Figure 14–33. Of course, the current in the resistive branch is always 90° out of phase with both reactive currents, as shown in the current phasor diagram of Figure 14–34.

The total current can be expressed as

$$I_T = \sqrt{I_R^2 + (I_C - I_L)^2} \tag{14–14}$$

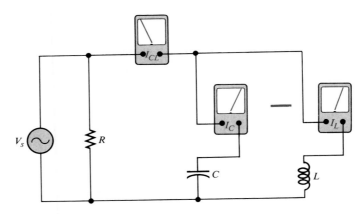

FIGURE 14–32

The total current into the parallel combination of *C* and *L* is the difference of the two branch currents. $I_{CL} = I_C - I_L$

FIGURE 14–33

I_C and I_L effectively subtract.

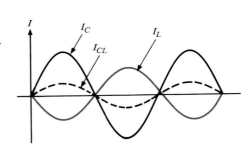

FIGURE 14–34
Current phasor diagram for a parallel *RLC* circuit.

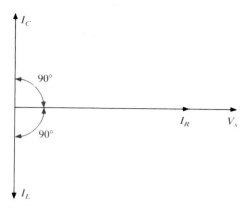

The phase angle can also be expressed in terms of the branch currents as

$$\theta = \tan^{-1}\left(\frac{I_{CL}}{I_R}\right)$$ (14–15)

where I_{CL} is $|I_C - I_L|$.

EXAMPLE 14–15

Find each branch current and the total current in Figure 14–35, and show their relationship.

FIGURE 14–35

Solution
Each branch current can be found using Ohm's law:

$$I_R = \frac{V_s}{R} = \frac{5\ V}{2\ \Omega} = 2.5\ A$$

$$I_C = \frac{V_s}{X_C} = \frac{5\ V}{5\ \Omega} = 1\ A$$

$$I_L = \frac{V_s}{X_L} = \frac{5\ V}{10\ \Omega} = 0.5\ A$$

The total current is the phasor sum of the branch currents:

$$I_T = \sqrt{I_R^2 + (I_C - I_L)^2} = \sqrt{(2.5\ A)^2 + (1\ A - 0.5\ A)^2}$$
$$= \sqrt{(2.5\ A)^2 + (0.5\ A)^2} = 2.55\ A$$

The phase angle is

$$\theta = \tan^{-1}\left(\frac{I_{CL}}{I_R}\right) = \tan^{-1}\left(\frac{0.5\ A}{2.5\ A}\right) = 11.31°$$

The total current is 2.55 A, *leading* V_s by 11.31°. Figure 14–36 is the current phasor diagram for the circuit.

FIGURE 14–36

Conversion of Series-Parallel to Parallel

The series-parallel configuration shown in Figure 14–37 is important because it represents a circuit having parallel L and C branches, with the winding resistance of the coil taken into account as a series resistance in the L branch.

It is helpful to view the series-parallel circuit in Figure 14–37 in an *equivalent* purely parallel form, as indicated in Figure 14–38. This form will simplify the analysis of parallel resonant characteristics in the next section.

FIGURE 14–37
An important series-parallel *RLC* circuit ($Q = X_L/R_W$).

FIGURE 14–38
Parallel equivalent form of the circuit in Figure 14–37.

The equivalent inductance L_{eq} and the equivalent parallel resistance $R_{p(eq)}$ are given by the following formulas:

$$L_{eq} = L\left(\frac{Q^2 + 1}{Q^2}\right) \tag{14–16}$$

$$R_{p(eq)} = R_W(Q^2 + 1) \tag{14–17}$$

where Q is the quality factor of the coil, X_L/R_W. Derivations of these formulas are quite involved and thus are not given here. Notice in the equations that for a $Q \geq 10$, the value of L_{eq} is approximately the same as the original value of L. For example, if $L = 10$ mH, then

$$L_{eq} = 10 \text{ mH}\left(\frac{10^2 + 1}{10^2}\right) = 10 \text{ mH}(1.01) = 10.1 \text{ mH}$$

The equivalency of the two circuits means that at a given frequency, when the same value of voltage is applied to both circuits, the same total current flows in both circuits and the phase angles are the same. Basically, an equivalent circuit simply makes circuit analysis more convenient.

EXAMPLE 14–16

Convert the series-parallel circuit in Figure 14–39 to an equivalent parallel form at the given frequency.

FIGURE 14–39

Solution

$$X_L = 2\pi fL = 2\pi(15.9 \text{ kHz})(5 \text{ mH}) = 500 \ \Omega$$

The Q of the coil is

$$Q = \frac{X_L}{R_W} = \frac{500 \ \Omega}{25 \ \Omega} = 20$$

Since $Q > 10$, then $L_{eq} \cong L = 5$ mH.
The equivalent parallel resistance is

$$R_{p(eq)} = R_W(Q^2 + 1) = (25 \ \Omega)(20^2 + 1) = 10.025 \text{ k}\Omega$$

This equivalent resistance appears in parallel with R as shown in Figure 14–40(a). When combined, they give a total parallel resistance (R_{pT}) of 3.38 kΩ, as indicated in Figure 14–40(b).

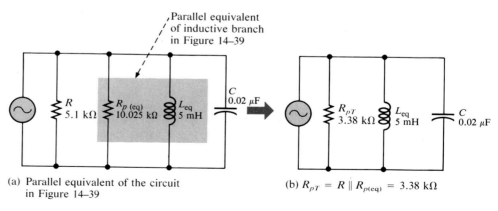

(a) Parallel equivalent of the circuit in Figure 14–39

(b) $R_{pT} = R \parallel R_{p(eq)} = 3.38$ kΩ

FIGURE 14–40

1. In a three-branch parallel circuit, $R = 150\ \Omega$, $X_C = 100\ \Omega$, and $X_L = 50\ \Omega$. Determine the current in each branch when $V_s = 12$ V.

2. Is the circuit in Problem 1 capacitive or inductive? Why?

3. Find the equivalent parallel inductance and resistance for a 20-mH coil with a winding resistance of $10\ \Omega$ at a frequency of 1 kHz.

14–6

PARALLEL RESONANCE

First we will look at the resonant condition in an *ideal* parallel *LC* circuit. Then we will progress to the more realistic case where the resistance of the coil is taken into account.

Ideally, parallel resonance occurs when $X_C = X_L$. The frequency at which resonance occurs is called the *resonant frequency*, just as in the series case. When $X_C = X_L$, the two branch currents, I_C and I_L, are equal in magnitude, and, of course, they are always 180° out of phase with each other. Thus, the two currents cancel and the total current is zero, as presented in Figure 14–41.

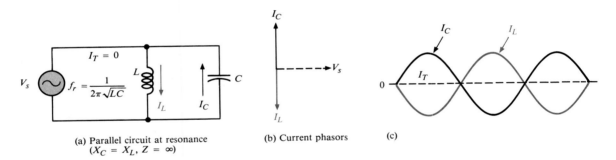

(a) Parallel circuit at resonance
$(X_C = X_L, Z = \infty)$

(b) Current phasors

(c)

FIGURE 14–41
An ideal parallel *LC* circuit at resonance.

Since the total current is zero, the impedance of the parallel *LC* circuit is infinitely large (∞). These ideal resonant conditions are stated in the following equations:

$$X_L = X_C$$
$$Z_r = \infty$$

The Parallel Resonant Frequency

For an ideal parallel resonant circuit, the frequency at which resonance occurs is determined by the same formula as in series resonant circuits:

$$f_r = \frac{1}{2\pi\sqrt{LC}}$$

Currents in a Parallel Resonant Circuit

It is interesting to see how the currents in a parallel *LC* circuit vary as the frequency is increased from below the resonant value, through resonance, and then above the resonant value. Figure 14–42 illustrates the general response of an ideal circuit in terms of changes in the currents. Part (a) indicates the response as the source frequency is increased from zero toward f_r. At very low frequencies, X_C is very high and X_L is very low, so most of the current is through L. As the frequency increases, the current through L decreases and the current through C increases, causing the total current to decrease. At all times, I_L and I_C are 180° out of phase with each other, and thus the total current is the difference of the two branch currents. During this time, the impedance is increasing, as indicated by the decrease in total current.

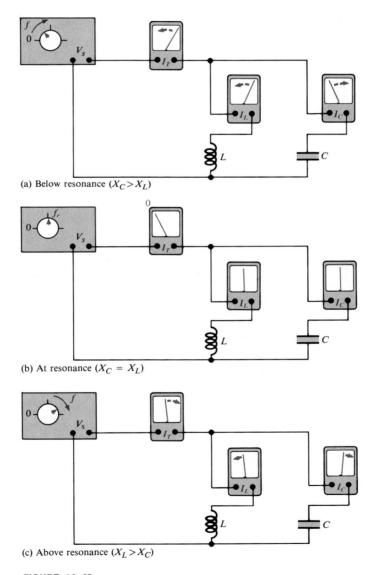

(a) Below resonance $(X_C > X_L)$

(b) At resonance $(X_C = X_L)$

(c) Above resonance $(X_L > X_C)$

FIGURE 14–42

An illustration of how the currents respond in a parallel *LC* circuit as the frequency is varied from below resonance to above resonance. The source voltage amplitude is constant.

When the frequency reaches its resonant value, f_r, as shown in Figure 14–42(b), X_C and X_L are equal, so I_C and I_L cancel because they are equal in magnitude but opposite in phase. At this point the total current is at its minimum value of zero. Since I_T is zero, Z is infinite. Thus, the ideal parallel LC circuit appears as an open at the resonant frequency.

As the frequency is increased above resonance, as indicated in Figure 14–42(c), X_C continues to decrease and X_L continues to increase, causing the branch currents again to be unequal, with I_C being larger. As a result, total current increases and impedance decreases. As the frequency becomes very high, the impedance becomes very small due to the dominance of X_C.

In summary, the current dips to a minimum as the impedance peaks to a maximum at parallel resonance. The expression for the total current into the L and C branches is

$$I_T = |I_C - I_L| \qquad\qquad (14\text{–}18)$$

EXAMPLE 14–17

Find the resonant frequency and the branch currents in the ideal parallel LC circuit of Figure 14–43.

FIGURE 14–43

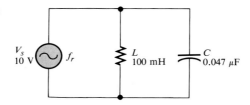

Solution

$$f_r = \frac{1}{2\pi\sqrt{LC}} = \frac{1}{2\pi\sqrt{(100 \text{ mH})(0.047 \ \mu\text{F})}} = 2322 \text{ Hz}$$

$$X_L = 2\pi f_r L = 2\pi(2322 \text{ Hz})(100 \text{ mH}) = 1459 \ \Omega$$

$$X_C = X_L = 1459 \ \Omega$$

$$I_L = \frac{V_s}{X_L} = \frac{10 \text{ V}}{1459 \ \Omega} = 6.85 \text{ mA}$$

$$I_C = I_L = 6.85 \text{ mA}$$

$$I_{LC} = 0$$

Why the Parallel Resonant LC Circuit Is Often Called a Tank Circuit

The term *tank circuit* refers to the fact that the parallel resonant circuit stores energy in the magnetic field of the coil and in the electric field of the capacitor. The stored energy is transferred back and forth between the capacitor and the coil on alternate half-cycles as the current goes first one way and then the other when the inductor deenergizes and the capacitor charges, and vice versa, as illustrated in Figure 14–44.

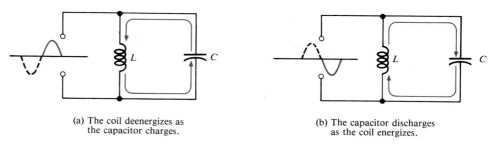

(a) The coil deenergizes as
the capacitor charges.

(b) The capacitor discharges
as the coil energizes.

FIGURE 14-44
Energy storage in an ideal parallel resonant tank circuit.

Parallel Resonant Conditions in a Nonideal Circuit

So far, the resonance of an ideal parallel LC circuit has been examined. Now we will consider resonance in a tank circuit with the resistance of the coil taken into account. Figure 14–45 shows a nonideal tank circuit and its parallel RLC equivalent.

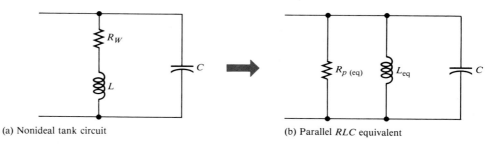

(a) Nonideal tank circuit

(b) Parallel RLC equivalent

FIGURE 14-45
A practical treatment of parallel resonant circuits must include the coil resistance.

The Q of the circuit at resonance is simply the Q of the coil:

$$Q = \frac{X_L}{R_W}$$

The expressions for the equivalent parallel resistance and the equivalent inductance were given in the previous section as

$$R_{p(eq)} = R_W(Q^2 + 1)$$

$$L_{eq} = L\left(\frac{Q^2 + 1}{Q^2}\right)$$

Recall that for $Q \geq 10$, $L_{eq} \cong L$.

At parallel resonance,

$$X_{L(eq)} = X_C$$

In the parallel equivalent circuit, we have $R_{p(eq)}$ in parallel with an ideal coil and a capacitor, so the L and C branches act as an ideal tank circuit which has an infinite impedance at resonance as shown in Figure 14–46. Therefore, the total impedance of the nonideal tank circuit at resonance can be expressed as simply the equivalent parallel resistance:

$$Z_r = R_W(Q^2 + 1) \tag{14–19}$$

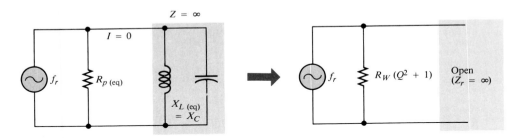

FIGURE 14–46
At resonance, the parallel LC portion appears open and the source sees only $R_{p(eq)} = R_W(Q^2 + 1)$.

EXAMPLE 14–18   Determine the impedance of the circuit in Figure 14–47 at the resonant frequency ($f_r \cong$ 17,794 Hz).

FIGURE 14–47

Solution
$$X_L = 2\pi f_r L = 2\pi(17,794 \text{ Hz})(8 \text{ mH}) = 894 \ \Omega$$

$$Q = \frac{X_L}{R_W} = \frac{894 \ \Omega}{50 \ \Omega} = 17.9$$

$$Z_r = R_W(Q^2 + 1) = 50 \ \Omega(17.9^2 + 1) = 16.1 \text{ k}\Omega$$

The calculator sequence is

Variation of the Impedance with Frequency

The impedance of a parallel resonant circuit is maximum at the resonant frequency and decreases at lower and higher frequencies, as indicated by the curve in Figure 14–48.

At very low frequencies, X_L is very small and X_C is very high, so the total impedance is essentially equal to that of the inductive branch. As the frequency goes up, the impedance also increases, and the inductive reactance dominates (because it is less than X_C) until the resonant frequency is reached. At this point, of course, $X_L \cong X_C$ (for

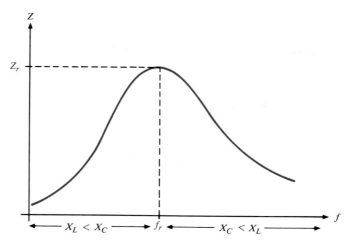

FIGURE 14–48
Generalized impedance curve for a parallel resonant circuit. The circuit is inductive below f_r, resistive at f_r, and capacitive above f_r.

$Q > 10$) and the impedance is at its maximum. As the frequency goes above resonance, the capacitive reactance dominates (because it is less than X_L) and the impedance decreases.

Current and Phase Angle at Resonance

In the ideal tank circuit, the total current from the source at resonance is zero because the impedance is infinite. In the nonideal case, there is some total current at the resonant frequency, and it is determined by the impedance at resonance:

$$I_T = \frac{V_s}{Z_r} \qquad (14\text{–}20)$$

The phase angle of the parallel resonant circuit is $0°$ because the impedance is purely resistive at the resonant frequency.

Parallel Resonant Frequency in a Nonideal Circuit

As you know, when the coil resistance is considered, the resonant condition is

$$X_{L(eq)} = X_C$$

which can be expressed as

$$2\pi f_r L \left(\frac{Q^2 + 1}{Q^2} \right) = \frac{1}{2\pi f_r C}$$

Solving for f_r, we get

$$f_r = \frac{1}{2\pi \sqrt{LC}} \sqrt{\frac{Q^2}{Q^2 + 1}} \qquad (14\text{–}21)$$

When $Q \geq 10$, the term with the Q factors is approximately 1:

$$\sqrt{\frac{Q^2}{Q^2 + 1}} = \sqrt{\frac{100}{101}} = 0.995 \cong 1$$

Therefore, the parallel resonant frequency is approximately the same as the series resonant frequency as long as Q is equal to or greater than 10:

$$f_r \cong \frac{1}{2\pi\sqrt{LC}} \qquad \text{for } Q \geq 10$$

Equation (14–21) was given only to show the effect of Q on the resonant frequency. You cannot use it to calculate the f_r of a given circuit, because we must first know the value of Q. Since $Q = X_L/R_W$, we must know X_L at the resonant frequency. In order to get X_L, we must know f_r. Since f_r is what we are looking for in the first place, there is no way to proceed with only Equation (14–21) at our disposal.

A more precise expression for f_r in terms of the circuit component values is as follows:

$$f_r = \frac{\sqrt{1 - (R_W^2 C/L)}}{2\pi\sqrt{LC}} \tag{14–22}$$

Now, f_r can be found from the component values alone.

EXAMPLE 14–19   Find the frequency, impedance, and total current at resonance for the circuit in Figure 14–49.

FIGURE 14–49

Solution

$$f_r = \frac{\sqrt{1 - (R_W^2 C/L)}}{2\pi\sqrt{LC}} = \frac{\sqrt{1 - [(100\ \Omega)^2(0.05\ \mu F)/0.1\ H]}}{2\pi\sqrt{(0.05\ \mu F)(0.1\ H)}} = 2.25\ \text{kHz}$$

$$X_L = 2\pi f_r L = 2\pi(2.25\ \text{kHz})(0.1\ H) = 1.4\ \text{k}\Omega$$

$$Q = \frac{X_L}{R_W} = \frac{1.4\ \text{k}\Omega}{100\ \Omega} = 14$$

$$Z_r = R_W(Q^2 + 1) = 100\ \Omega(14^2 + 1) = 19.7\ \text{k}\Omega$$

$$I_T = \frac{V_s}{Z_r} = \frac{10\ V}{19.7\ \text{k}\Omega} = 0.51\ \text{mA}$$

Note that since $Q > 10$, the approximate formula $f_r \cong 1/2\pi\sqrt{LC}$ could be used.

The calculator sequence for f_r is

How an External Parallel Load Resistance Affects a Tank Circuit

There are many practical situations in which an external load resistance appears in parallel with a tank circuit as shown in Figure 14–50(a). Obviously, the external resistor (R_L) will dissipate more of the energy delivered by the source and thus will lower the *overall Q* of the circuit. The external resistor effectively appears in parallel with the equivalent parallel resistance of the coil, $R_{p(eq)}$, and both are combined to determine a total parallel resistance, R_{pT}, as indicated in Figure 14–50(b):

$$R_{pT} = R_L \parallel R_{p(eq)}$$

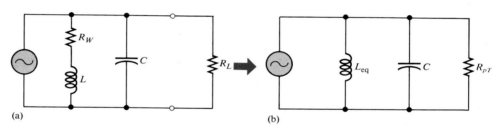

(a)   (b)

FIGURE 14–50
Tank circuit with a parallel load resistor and its equivalent circuit.

The overall parallel *RLC* circuit Q is expressed differently from the Q of a series circuit:

$$Q_O = \frac{R_{pT}}{X_{L(eq)}} \tag{14–23}$$

As you can see, the effect of loading the tank circuit is to reduce its overall Q (which is equal to the coil Q when unloaded).

**SECTION REVIEW
14–6**

1. Is the impedance minimum or maximum at parallel resonance?
2. Is the current minimum or maximum at parallel resonance?
3. If $Q = 25$, $L = 50$ mH, and $C = 1000$ pF, what is f_r?
4. In Problem 3, if $Q = 2.5$, what is f_r?
5. In a certain tank circuit, the coil resistance is 20 Ω. What is the total impedance at resonance if $Q = 20$?

14–7

PARALLEL RESONANT FILTERS

Parallel resonant circuits are commonly applied in band-pass and band-stop filters. In this section, we will examine these applications.

The Band-Pass Filter

A basic parallel resonant band-pass filter is shown in Figure 14–51. Notice that the output is taken across the tank circuit in this application.

The filtering action is as follows: At very low frequencies, the impedance of the tank circuit is very low, and therefore only a small amount of voltage is dropped across it. As the frequency increases, the impedance of the tank circuit increases, and, as a result, the output voltage increases. When the frequency reaches its resonant value, the impedance is at its maximum and so is the output voltage. As the frequency goes above resonance, the impedance begins to decrease, causing the output voltage to decrease. The bandwidth and cutoff frequencies for a parallel resonant band-pass filter are defined in the same way as for the series resonant circuit, and the formulas given in Section 14–4 still apply. General band-pass response curves showing both V_{out} and I_T versus frequency are given in Figure 14–52(a) and 14–52(b), respectively.

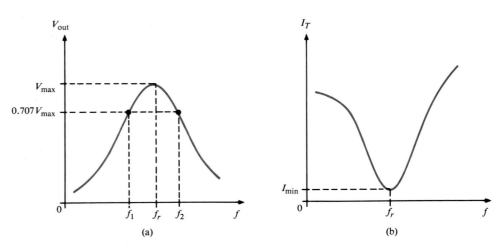

FIGURE 14–52
Generalized response curves for a parallel resonant band-pass filter.

The general response of a parallel resonant band-pass filter is illustrated in Figure 14–53. This illustration pictorially shows how the current and the output voltage change with frequency.

(a) As the frequency increases to f_1, V_{out} increases to 7.07 V, and I_T decreases.

(b) As the frequency increases from f_1 to f_r, V_{out} increases from 7.07 V to 10 V, and I_T decreases to its minimum value.

(c) As the frequency increases from f_r to f_2, V_{out} decreases from 10 V to 7.07 V, and I_T increases from its minimum.

(d) As the frequency increases above f_2, V_{out} decreases below 7.07 V, and I_T continues to increase.

FIGURE 14–53
Example of the response of a parallel resonant band-pass filter with the input voltage at a constant 10 V rms.

EXAMPLE 14–20

A certain parallel resonant band-pass filter has a maximum output voltage of 4 V at f_r. What is the value of V_{out} at the cutoff frequencies?

Solution

V_{out} at the cutoff frequencies is 70.7% of maximum:

$$V_{out(1)} = V_{out(2)} = 0.707V_{out(max)} = 0.707(4 \text{ V}) = 2.828 \text{ V}$$

EXAMPLE 14–21   A parallel resonant circuit has a lower cutoff frequency of 3.5 kHz and an upper cutoff frequency of 6 kHz. What is the bandwidth?

Solution

$$BW = f_2 - f_1 = 6 \text{ kHz} - 3.5 \text{ kHz} = 2.5 \text{ kHz}$$

EXAMPLE 14–22   A certain parallel resonant band-pass filter has a resonant frequency of 12 kHz and a Q of 10. What is its bandwidth?

Solution

$$BW = \frac{f_r}{Q} = \frac{12 \text{ kHz}}{10} = 1.2 \text{ kHz}$$

How Loading Affects the Selectivity of a Parallel Resonant Band-Pass Filter

When a resistive load is connected across the output of a filter as shown in Figure 14–54(a), the Q of the filter is reduced. Since $BW = f_r/Q$, the bandwidth is increased,

(a) Tank circuit with load   (b) Equivalent circuit

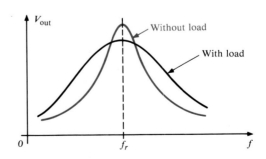

(c) Loading widens the bandwidth and reduces the output.

FIGURE 14–54
Effects of loading on a parallel resonant band-pass filter.

thus reducing the selectivity. Also, the impedance of the filter at resonance is decreased because R_L effectively appears in parallel with $R_{p(eq)}$. Thus, the maximum output voltage is reduced by the voltage divider effect of R_{pT} and the internal source resistance R_s, as illustrated in Figure 14–54(b). Part (c) of the figure shows the general effect on the filter response curve.

EXAMPLE 14–23

(a) Determine f_r, BW, and V_{out} at resonance for the unloaded filter in Figure 14–55. The source resistance is 600 Ω.

(b) Repeat part (a) when the filter is loaded with a 50-kΩ resistance, and compare the results.

FIGURE 14–55

Solution

(a) $f_r = \dfrac{\sqrt{1 - (R_W^2 C/L)}}{2\pi\sqrt{LC}} = 1.59 \text{ kHz}$

At resonance,

$$X_L = 2\pi f_r L = 2\pi(1.59 \text{ kHz})(100 \text{ mH}) = 999 \ \Omega$$

$$Q = \frac{X_L}{R_W} = \frac{999 \ \Omega}{50 \ \Omega} = 19.98$$

$$BW = \frac{f_r}{Q} = \frac{1.59 \text{ kHz}}{19.98} = 79.58 \text{ Hz}$$

$$R_{p(eq)} = R_W(Q^2 + 1) = 50 \ \Omega(19.98^2 + 1) = 20.01 \text{ k}\Omega$$

$$V_{out} = \left(\frac{R_{p(eq)}}{R_{p(eq)} + R_s}\right)V_s = \left(\frac{20.01 \text{ k}\Omega}{20.61 \text{ k}\Omega}\right)5 \text{ V} = 4.85 \text{ V}$$

(b) When a 50-kΩ load resistance is connected, the following values are obtained:

$$R_{pT} = R_{p(eq)} \| R_L = 20.01 \text{ k}\Omega \| 50 \text{ k}\Omega = 14.34 \text{ k}\Omega$$

$$Q_O = \frac{R_{pT}}{X_L} = \frac{14.34 \text{ k}\Omega}{999 \ \Omega} = 14.35$$

$$BW = \frac{f_r}{Q_O} = \frac{1.59 \text{ kHz}}{14.35} = 110.8 \text{ Hz}$$

$$V_{out} = \left(\frac{R_{pT}}{R_{pT} + R_s}\right)V_s = \left(\frac{14.34 \text{ k}\Omega}{14.94 \text{ k}\Omega}\right)5 \text{ V} = 4.799 \text{ V}$$

The result of adding the load resistance is an increased bandwidth and a decreased output voltage at resonance.

The Band-Stop Filter

A basic parallel resonant band-stop filter is shown in Figure 14–56. The output is taken across a load resistor in series with the tank circuit.

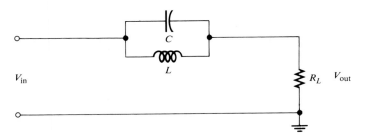

V_{in}

C

L

R_L V_{out}

FIGURE 14–56
A basic parallel resonant band-stop filter.

The variation of the tank circuit impedance with frequency produces the familiar current response that has been previously discussed; that is, the current is minimum at resonance and increases on both sides of resonance. Since the output voltage is across the series load resistor, it follows the current, thus creating the band-stop response characteristic as indicated in Figure 14–57.

Actually, the band-stop filter in Figure 14–56 can be viewed as a voltage divider created by Z_r of the tank and the load resistance. Thus, the output voltage at f_r is

$$V_{out} = \left(\frac{R_L}{R_L + Z_r}\right)V_{in}$$

FIGURE 14–57
Band-stop filter response.

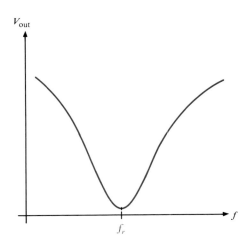

V_{out}

f_r

f

EXAMPLE 14–24

Find f_r and the output voltage at resonance for the band-stop filter in Figure 14–58.

FIGURE 14–58

Solution

$$f_r = \frac{\sqrt{1 - (R_W^2 C/L)}}{2\pi\sqrt{LC}} = 249.4 \text{ Hz}$$

$$X_L = 2\pi f_r L = 517 \ \Omega$$

$$Q = \frac{X_L}{R_W} = \frac{517 \ \Omega}{250 \ \Omega} = 2.07$$

$$Z_r = R_W(Q^2 + 1) = 1.32 \text{ k}\Omega$$

At resonance,

$$V_{out} = \left(\frac{R_L}{R_L + Z_r}\right)V_s = \left(\frac{1 \text{ k}\Omega}{2.32 \text{ k}\Omega}\right)12 \text{ V} = 5.17 \text{ V}$$

Basic filter configurations have been introduced in this section and in Section 14–4. These basic configurations are sometimes used in combination to increase filter selectivity. For example, a series resonant and a parallel resonant band-pass filter can be used together, with the series circuit connected between the input and output and the parallel circuit connected across the output.

Variable capacitors or inductors are sometimes used in certain filter applications so that the resonant circuits can be *tuned* over a range of resonant frequencies.

SECTION REVIEW 14–7

1. How can the bandwidth of a parallel resonant filter be increased?
2. The resonant frequency of a certain high-Q ($Q > 10$) filter is 5 kHz. If the Q is lowered to 2, does f_r change and, if so, to what value?
3. What is the impedance of a tank circuit at its resonant frequency if $R_W = 75 \ \Omega$ and $Q = 25$?

14–8

SYSTEM APPLICATIONS

Resonant filters are used in a wide variety of applications, particularly in communication systems. In this section, we will look briefly at a few common communication system applications. The purpose of this coverage is not to explain the systems that are used as examples but rather to show how important filters are in electronic communication.

Tuned Amplifiers

A tuned amplifier is a circuit that amplifies signals within a specified band. Typically, a parallel resonant circuit is used in conjunction with an amplifier to achieve the selectivity. In terms of the general operation, input signals with frequencies that range over a wide band are accepted on the amplifier's input and are amplified. The function of the resonant circuit is to allow only a relatively narrow band of those frequencies to be passed on. The variable capacitor allows tuning over the range of input frequencies so that a desired frequency can be selected, as indicated in Figure 14–59.

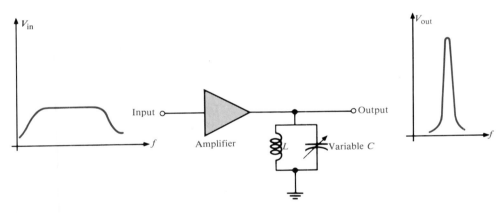

FIGURE 14–59
A basic tuned band-pass amplifier.

Antenna Input to a Receiver

Radio signals are sent out from a transmitter via electromagnetic waves which propagate through the atmosphere. When the electromagnetic waves cut across the receiving antenna, small voltages are induced. Out of all the wide range of electromagnetic frequencies, only one frequency or a limited band of frequencies must be extracted. Figure 14–60 shows a typical arrangement of an antenna coupled to the receiver input by a transformer. A variable capacitor is connected across the transformer secondary to form a parallel resonant circuit.

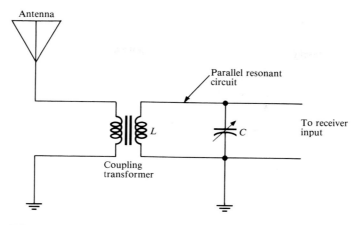

FIGURE 14–60
Resonant coupling from an antenna.

Double-Tuned Transformer Coupling in a Receiver

In some types of communication receivers, tuned amplifiers are transformer-coupled together to increase the amplification. Capacitors can be placed in parallel with the primary and secondary of the transformer, effectively creating two parallel resonant band-pass filters that are coupled together. This technique, illustrated in Figure 14–61, can result in a wider bandwidth and steeper slopes on the response curve, thus increasing the selectivity for a desired band of frequencies.

FIGURE 14–61
Double-tuned amplifiers.

Signal Reception and Separation in a TV Receiver

A television receiver must handle both video (picture) signals and audio (sound) signals. Each TV transmitting station is allotted a 6-MHz bandwidth. Channel 2 is allotted a band from 54 MHz through 59 MHz, channel 3 is allotted a band from 60 MHz through 65 MHz, on up to channel 13, which has a band from 210 MHz through 215 MHz. You can tune the front end of the TV receiver to select any one of these channels by using tuned amplifiers. The signal output of the front end of the receiver has a bandwidth from 41 MHz through 46 MHz, regardless of the channel that is tuned in. This band, called the *intermediate frequency* (IF) band, contains both video and audio. Amplifiers tuned to the IF band boost the signal and feed it to the video amplifier.

Before the output of the video amplifier is applied to the picture tube, the audio signal is removed by a 4.5-MHz band-stop filter (called a *wave trap*), as shown in Figure 14–62. This trap keeps the sound signal from interfering with the picture. The video amplifier output is also applied to band-pass circuits that are tuned to the sound carrier frequency of 4.5 MHz. The sound signal is then processed and applied to the speaker as indicated in Figure 14–62.

FIGURE 14–62
A simplified portion of a TV receiver showing filter usage.

Superheterodyne Receiver

Another good example of filter applications is in the common AM (amplitude modulation) receiver. The AM broadcast band ranges from 535 kHz to 1605 kHz. Each AM station is assigned a certain narrow bandwidth within that range. A simplified block diagram of a superheterodyne AM receiver is shown in Figure 14–63.

There are basically three parallel resonant band-pass filters in the front end of the receiver. Each of these filters is gang-tuned by capacitors; that is, the capacitors are mechanically or electronically linked together so that they change together as a station is selected. The front end is tuned to receive a desired station, for example, one that transmits at 600 kHz. The input filter from the antenna and the RF (radio frequency) amplifier filter select only a frequency of 600 kHz out of all the frequencies crossing the antenna. The actual audio (sound) signal is carried by the 600-kHz carrier frequency by modulating the amplitude of the carrier so that it follows the audio signal as indi-

FIGURE 14-63
A simplified diagram of a superheterodyne AM radio broadcast receiver showing the application of tuned resonant circuits.

cated. The variation in the amplitude of the carrier corresponding to the audio signal is called the *envelope*. The 600 kHz is then applied to a circuit called the *mixer*. The *local oscillator* (LO) is tuned to a frequency that is 455 kHz above the selected frequency (1055 kHz, in this case). By a process called *heterodyning* or *beating*, the AM signal and the local oscillator signal are mixed together, and the 600-kHz AM signal is converted to a 455-kHz AM signal (1055 kHz − 600 kHz = 455 kHz). The 455 kHz is the intermediate frequency (IF) for standard AM receivers. No matter which station within the broadcast band is selected, its frequency is always converted to the 455-kHz IF. The amplitude-modulated IF is applied to an *audio detector* which removes the IF, leaving only the envelope or audio signal. The audio signal is then amplified and applied to the speaker.

**SECTION REVIEW
14-8**

1. Generally, why is a tuned filter necessary when a signal is coupled from an antenna to the input of a receiver?
2. What is a *wave trap?*
3. What is meant by *ganged tuning?*

SUMMARY

At series resonance:

☐ The reactances are equal.

☐ The impedance is minimum and equal to the resistance.

☐ The current is maximum.

☐ The phase angle is zero.

☐ The voltages across L and C are equal in magnitude and, as always, $180°$ out of phase with each other and thus they cancel.

At parallel resonance:

☐ The reactances are approximately equal for $Q \geq 10$.

☐ The impedance is maximum.

☐ The current is minimum and, ideally, equal to zero.

☐ The phase angle is zero.

☐ The currents in the L and C branches are equal in magnitude and, as always, $180°$ out of phase with each other.

GLOSSARY

Band-pass filter   A filter that passes frequencies within a certain range and blocks all others.

Band-stop filter   A filter that blocks frequencies within a certain range.

Bandwidth (*BW*)   The characteristic of certain electronic circuits that specifies the usable range of frequencies that pass from input to output.

Cutoff frequencies   The frequencies at which the output of a filter is 70.7% of maximum.

Decibel (dB)   The unit of the logarithmic expression of a ratio, such as power or voltage ratios.

Half-power frequencies   The frequencies at which the output of a filter is 70.7% of maximum.

Pass band   The range of frequencies passed by a filter.

Quality factor (Q)   The ratio of true power to reactive power in a coil or a resonant circuit.

Resonance   In an *LC* circuit, the condition where the impedance is minimum (series) or maximum (parallel).

Selectivity   A measure of how well a filter passes certain frequencies and rejects others. The narrower the bandwidth, the greater the selectivity.

Tank   A parallel resonant circuit.

FORMULAS

$$X_T = |X_L - X_C| \qquad \text{Total series reactance (absolute value)} \tag{14–1}$$

$$Z = \sqrt{R^2 + (X_L - X_C)^2} \qquad \text{Series } RLC \text{ impedance} \tag{14–2}$$

$$\theta = \tan^{-1}\left(\frac{X_T}{R}\right) \qquad \text{Series } RLC \text{ phase angle} \tag{14–3}$$

$$X_L = X_C \qquad \text{Condition for series resonance} \tag{14–4}$$

$$Z_r = R \qquad \text{Series resonant impedance} \tag{14–5}$$

$$f_r = \frac{1}{2\pi\sqrt{LC}} \qquad \text{Series resonant frequency} \tag{14–6}$$

$$BW = f_2 - f_1 \qquad \text{Bandwidth} \tag{14–7}$$

$$dB = 20 \log\left(\frac{V_{out}}{V_{in}}\right) \qquad \text{Decibel formula for voltage ratio} \tag{14–8}$$

$$dB = 10 \log\left(\frac{P_{out}}{P_{in}}\right) \qquad \text{Decibel formula for power ratio} \tag{14–9}$$

$$Q = \frac{X_L}{R} \qquad \text{Series resonant } Q \tag{14–10}$$

$$BW \cong \frac{f_r}{Q} \qquad \text{Bandwidth} \tag{14–11}$$

$$Y = \sqrt{G^2 + (B_C - B_L)^2} \qquad \text{Parallel } RLC \text{ admittance} \tag{14–12}$$

$$\theta = \tan^{-1}\left(\frac{B_T}{G}\right) \qquad \text{Parallel } RLC \text{ phase angle} \tag{14–13}$$

$$I_T = \sqrt{I_R^2 + (I_C - I_L)^2} \qquad \text{Total parallel } RLC \text{ current} \tag{14–14}$$

$$\theta = \tan^{-1}\left(\frac{I_{CL}}{I_R}\right) \qquad \text{Parallel } RLC \text{ phase angle} \tag{14–15}$$

$$L_{eq} = L\left(\frac{Q^2 + 1}{Q^2}\right) \qquad \text{Equivalent parallel inductance} \tag{14–16}$$

$$R_{p(eq)} = R_W(Q^2 + 1) \qquad \text{Equivalent parallel resistance} \tag{14–17}$$

$$I_T = |I_C - I_L| \qquad \text{Total parallel } LC \text{ current (absolute value)} \tag{14–18}$$

$$Z_r = R_W(Q^2 + 1) \qquad \text{Impedance at parallel resonance} \tag{14–19}$$

$$I_T = \frac{V_s}{Z_r} \qquad \text{Total current at parallel resonance} \tag{14–20}$$

$$f_r = \frac{1}{2\pi\sqrt{LC}}\sqrt{\frac{Q^2}{Q^2 + 1}} \qquad \text{Parallel resonant frequency (exact)} \tag{14–21}$$

$$f_r = \frac{\sqrt{1 - (R_W^2 C/L)}}{2\pi\sqrt{LC}} \qquad \text{Parallel resonant frequency (exact)} \tag{14–22}$$

$$Q_O = \frac{R_{pT}}{X_{L(eq)}} \qquad \text{Parallel } RLC \text{ } Q \tag{14–23}$$

SELF-TEST

1. The total reactance of a series *RLC* circuit at resonance is
   (a) zero **(b)** equal to the resistance **(c)** infinity **(d)** capacitive

2. The phase angle of a series *RLC* circuit at resonance is
   (a) −90° **(b)** +90° **(c)** 0°
   (d) is dependent on the reactance

3. The impedance at the resonant frequency of a series RLC circuit with $L = 15$ mH, $C = 0.015$ μF, and $R_W = 80$ Ω is
    (a) 15 kΩ (b) 80 Ω (c) 30 Ω (d) 0 Ω

4. In a series RLC circuit that is operating below the resonant frequency, the current
    (a) is in phase with the applied voltage
    (b) lags the applied voltage
    (c) leads the applied voltage

5. If the value of C in a series RLC circuit is increased, the resonant frequency
    (a) is not affected   (b) increases   (c) remains the same   (d) decreases

6. In a certain series resonant circuit, $V_C = 150$ V, $V_L = 150$ V, and $V_R = 50$ V. The value of the source voltage is
    (a) 150 V   (b) 300 V   (c) 50 V   (d) 350 V

7. A certain series resonant band-pass filter has a bandwidth of 1 kHz. If the existing coil is replaced with one having a lower value of Q, the bandwidth will
    (a) increase   (b) decrease   (c) remain the same   (d) be more selective

8. At frequencies below resonance in a parallel RLC circuit, the current
    (a) leads the source voltage   (b) lags the source voltage
    (c) is in phase with the source voltage

9. The total current into the L and C branches of a parallel circuit at resonance is ideally
    (a) maximum   (b) low   (c) high   (d) zero

10. To tune a parallel resonant circuit to a lower frequency, the capacitance should be
    (a) increased   (b) decreased   (c) left alone   (d) replaced with inductance

11. The resonant frequency of a parallel circuit is the same as a series circuit when
    (a) the Q is very low   (b) the Q is very high   (c) there is no resistance

12. If the resistance in parallel with a parallel resonant filter is reduced, the bandwidth
    (a) disappears   (b) decreases   (c) becomes sharper   (d) increases

---

PROBLEM SET A

Section 14–1   Impedance and Phase Angle of Series RLC Circuits

1. A certain series RLC circuit has the following values: $R = 10$ Ω, $C = 0.05$ μF, and $L = 5$ mH. Determine the impedance and phase angle. What is the total reactance? $f = 5$ kHz.

2. Find the impedance in Figure 14–64.

3. If the frequency of the source voltage in Figure 14–64 is doubled from the value that produces the indicated reactances, how does the impedance change?

FIGURE 14–64

Section 14–2   Analysis of Series RLC Circuits

4. For the circuit in Figure 14–64, find I_T, V_R, V_L, and V_C.

5. Sketch the voltage phasor diagram for the circuit in Figure 14–64.

6. Analyze the circuit in Figure 14–65 for the following ($f = 25$ kHz):
    (a) I_T (b) P_{true} (c) P_r (d) P_a

FIGURE 14–65

Section 14–3   Series Resonance

7. Find X_L, X_C, Z, and I at the resonant frequency in Figure 14–66.

8. A certain series resonant circuit has a maximum current of 50 mA and a V_L of 100 V. The applied voltage is 10 V. What is Z? What are X_L and X_C?

FIGURE 14–66

9. For the RLC circuit in Figure 14–67, determine the resonant frequency and the cutoff frequencies.

10. What is the value of the current at the half-power points in Figure 14–67?

FIGURE 14–67

Section 14–4   Series Resonant Filters

11. Determine the resonant frequency for each filter in Figure 14–68. Are these filters bandpass or band-stop types?

(a)

(b)

FIGURE 14–68

12. Assuming that the coils in Figure 14–68 have a winding resistance of 10 Ω, find the bandwidth for each filter.

13. Determine the resonant frequency and bandwidth for each filter in Figure 14–69.

FIGURE 14–69

(a)   (b)

Section 14–5   Parallel *RLC* Circuits

14. Find the total impedance of the circuit in Figure 14–70.
15. Is the circuit in Figure 14–70 capacitive or inductive? Explain.
16. For the circuit in Figure 14–70, find all the currents and voltages.
17. Find the total impedance for the circuit in Figure 14–71.

FIGURE 14–70   **FIGURE 14–71**

Section 14–6   Parallel Resonance

18. What is the impedance of an ideal parallel resonant circuit (no resistance in either branch)?
19. Find Z at resonance and f_r for the tank circuit in Figure 14–72.
20. How much current is drawn from the source in Figure 14–72 at resonance? What are the inductive current and the capacitive current at the resonant frequency?

FIGURE 14–72

Section 14–7   Parallel Resonant Filters

21. At resonance, $X_L = 2$ kΩ and $R_W = 25$ Ω in a parallel resonant band-pass filter. The resonant frequency is 5 kHz. Determine the bandwidth.
22. If the lower cutoff frequency is 2400 Hz and the upper cutoff frequency is 2800 Hz, what is the bandwidth?
23. In a certain resonant circuit, the power at resonance is 2.75 W. What is the power at the lower and upper cutoff frequencies?
24. What values of L and C should be used in a tank circuit to obtain a resonant frequency of 8 kHz? The bandwidth must be 800 Hz. The winding resistance of the coil is 10 Ω.
25. A parallel resonant circuit has a Q of 50 and a BW of 400 Hz. If Q is doubled, what is the bandwidth for the same f_r?

PROBLEM SET B

26. For each following case, express the voltage ratio in decibels:
    (a) $V_{in} = 1$ V, $V_{out} = 1$ V   (b) $V_{in} = 5$ V, $V_{out} = 3$ V
    (c) $V_{in} = 10$ V, $V_{out} = 7.07$ V   (d) $V_{in} = 25$ V, $V_{out} = 5$ V
27. Find the current through each component in Figure 14–73. Find the voltage across each component.

FIGURE 14–73

28. Determine whether there is a value of C that will make $V_{ab} = 0$ V in Figure 14–74. If not, explain.

FIGURE 14–74

29. If the value of C is 0.2 μF, how much current flows through each branch in Figure 14–74? What is the total current?
30. Determine the resonant frequencies in Figure 14–75 and find V_{out} at each resonant frequency.

FIGURE 14–75

31. Design a band-pass filter using a parallel resonant circuit to meet the following specifications: $BW = 500$ Hz, $Q = 40$, $I_{C(max)} = 20$ mA, $V_{C(max)} = 2.5$ V.
32. Design a circuit in which the following series resonant frequencies are switch-selectable: 500 kHz, 1000 kHz, 1500 kHz, 2000 kHz.
33. Design a parallel-resonant network using a single coil and switch-selectable capacitors to produce the following resonant frequencies: 8 MHz, 9 MHz, 10 MHz, and 11 MHz. Assume a 10-μH coil with a winding resistance of 5 Ω.

ANSWERS TO
SECTION REVIEWS

Section 14–1
1. 70 Ω, capacitive.   **2.** 83.22 Ω, 57.26°, leading.

Section 14–2
1. 38.42 V.   **2.** Leads.   **3.** 6 Ω.

Section 14–3
1. $X_L = X_C$.   **2.** The impedance is minimum.   **3.** 159 kHz.   **4.** Capacitive.

Section 14–4
1. 10.61 V.   **2.** 10 kHz.   **3.** Maximum, minimum.

Section 14–5
1. I_R = 80 mA, I_C = 120 mA, I_L = 240 mA.   **2.** Inductive ($X_L < X_C$).
3. 20.13 mH, 1589 Ω.

Section 14–6
1. Maximum.   **2.** Minimum.   **3.** 22.51 kHz.   **4.** 20.9 kHz.   **5.** 8020 Ω.

Section 14–7
1. By reducing the parallel resistance.   **2.** Yes, 4.47 kHz.   **3.** 46.95 kΩ.

Section 14–8
1. To select a narrow band of frequencies.   **2.** A band-stop filter.
3. Several variable capacitors (or inductors) whose values can be varied simultaneously with a common control.

15 Pulse Response of *RC* and *RL* Circuits

In Chapters 12 and 13, the frequency (sinusoidal) response of *RC* and *RL* circuits was covered. In this chapter, the response of *RC* and *RL* filters to pulse inputs is examined. With pulse inputs, the *time* response of the circuits is of primary importance. In the areas of pulse and digital circuits, the technician is often concerned with how a circuit responds over an interval of time to rapid changes in voltage or current. The relationship of the circuit *time constant* to the input pulse characteristics, such as pulse width and period, determines the wave shapes of the circuit voltages.

Before starting this chapter, you should review the material in Sections 9–5 and 10–5. Understanding exponential changes in voltages and currents in inductors and capacitors is crucial to the study of pulse response. Exponential formulas that were given in Chapter 9 are used throughout this chapter.

Integrator and *differentiator,* terms used throughout this coverage, refer to mathematical functions that are approximated by these circuits under certain conditions. Integration is an averaging process, and differentiation is a process for establishing a rate of change of a function. We will not deal with the purely mathematical aspects at this level of coverage.

After completing this chapter, you will be able to

- ☐ Describe how *RC* and *RL* integrators work.
- ☐ Describe how *RC* and *RL* differentiators work.
- ☐ Define *transient time.*
- ☐ Define *steady state.*
- ☐ Determine the output voltage of an integrator or differentiator for a single input pulse or for a repetitive pulse input.
- ☐ Discuss how the shape of the output voltage of an integrator or differentiator is a function of the time constant.
- ☐ Discuss how the shape of the output voltage of an integrator or differentiator changes with frequency and pulse width.
- ☐ Troubleshoot integrators and differentiators.

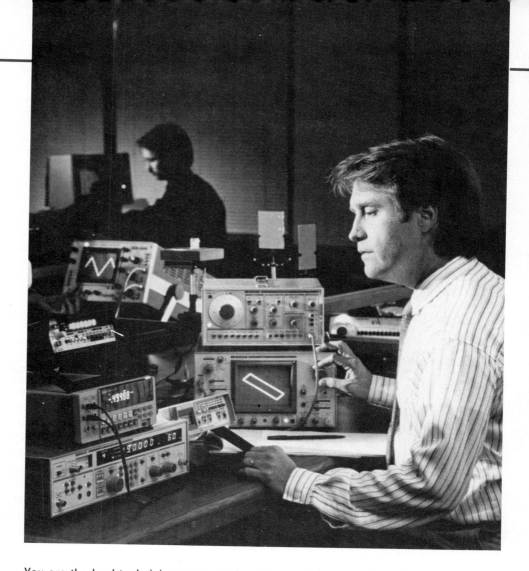

APPLICATION ASSIGNMENT

You are the lead technician on an engineering development project for a digital radar system. The design engineer has asked you to devise a simple circuit that will produce an alternating series of positive and negative voltage spikes (pulses of very short duration) derived from an available square wave source of variable amplitude and frequency. The spikes are to occur at each transition of the square wave and are to be used as timing pulses for testing a portion of the digital radar system. The spikes are to be no wider than 0.1 μs at their 37% amplitude points. The amplitude of each spike should be approximately the same as that of the square wave. This chapter will prepare you to approach this assignment.

15-1

THE *RC* INTEGRATOR

Figure 15–1 shows a series *RC* circuit with a pulse input. When the *output is taken across the capacitor,* this circuit is known as an *integrator* in terms of its pulse response. Recall that in terms of frequency response, it is a basic *low-pass* filter. The term *integrator* comes from the mathematical function called integration, which this circuit approximates under certain conditions.

FIGURE 15–1
An *RC* integrating circuit.

How the Capacitor Charges and Discharges with a Pulse Input

When a pulse generator is connected to the input, as shown in Figure 15–1, the capacitor will charge and discharge in response to the pulses. When the input goes from its low level to its high level, the capacitor *charges* toward the high level of the pulse through the resistor. This action is analogous to connecting a battery through a switch to the *RC* network, as illustrated in Figure 15–2(a). When the pulse goes from its high level back to its low level, the capacitor *discharges* back through the source. The resistance of the source is assumed to be negligible compared to *R*. This action is analogous to replacing the source with a closed switch, as illustrated in Figure 15–2(b).

The capacitor will charge and discharge following an *exponential curve.* Its rate of charging and discharging, of course, depends on the *RC* time constant ($\tau = RC$).

For an ideal pulse, each edge is considered to occur instantaneously. Two basic rules of capacitor behavior will aid in understanding *RC* circuit responses to pulse inputs:

(a) When the input pulse goes high, the source effectively acts as a battery in series with a switch closure, thereby charging the capacitor.

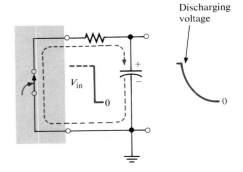

(b) When the input pulse goes back low, the source effectively acts as a closed switch, providing a discharge path for the capacitor.

FIGURE 15–2
The equivalent action when a pulse source charges and discharges the capacitor.

1. The capacitor appears as a *short* to an instantaneous change in current and as an open to dc.
2. The voltage across the capacitor cannot change instantaneously—it can change only exponentially.

The Capacitor Voltage

In an *RC* integrator, *the output is the capacitor voltage*. The capacitor charges during the time that the input pulse is high. If the pulse is at its high level long enough, the capacitor will fully charge to the voltage amplitude of the pulse, as illustrated in Figure 15–3. The capacitor discharges during the time that the pulse is low. If the low time between pulses is long enough, the capacitor will fully discharge to zero, as shown in the figure. Then when the next pulse occurs, it will charge again.

FIGURE 15–3
Illustration of a capacitor fully charging and discharging in response to a pulse input.

---

1. Define the term *integrator* in relation to an *RC* circuit.
2. What causes a capacitor in an *RC* circuit to charge and discharge?

15–2

RESPONSE OF AN *RC* INTEGRATOR TO A SINGLE PULSE

From the previous section, you have a general idea of how an *RC* integrator responds to a pulse input. In this section, we examine the response to a single pulse in detail. Then in the next section, we will expand this basic knowledge to include the general case of a series of pulses.

Two conditions of pulse response must be considered:

1. when the input pulse width (t_W) is equal to or greater than five time constants ($t_W \geq 5\tau$)
2. when the input pulse width is less than five time constants ($t_W < 5\tau$)

Recall that five time constants is accepted as the time a capacitor needs to fully charge or fully discharge; this time is often called the *transient time*.

When the Pulse Width Is Equal to or Greater than Five Time Constants

The capacitor will *fully charge* if the pulse width is equal to or greater than five time constants (5τ). This condition is expressed as $t_W \geq 5\tau$. At the end of the pulse, the capacitor *fully discharges* back through the source.

Figure 15–4 illustrates the output waveforms for various values of time constant and a fixed input pulse width. Notice that the shape of the output pulse approaches that of the input as the transient time is made small compared to the pulse width. In each case the output reaches the full amplitude of the input.

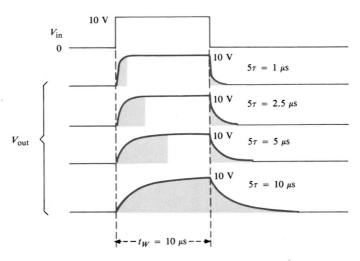

FIGURE 15–4
Variation of an integrator's output pulse shape with time constant. The color shaded areas indicate when the capacitor is charging and discharging.

Figure 15–5 shows how a fixed time constant and a variable input pulse width affect the integrator output. Notice that as the pulse width is increased, the shape of the output pulse approaches that of the input. Again, this means that the transient time is short compared to the pulse width.

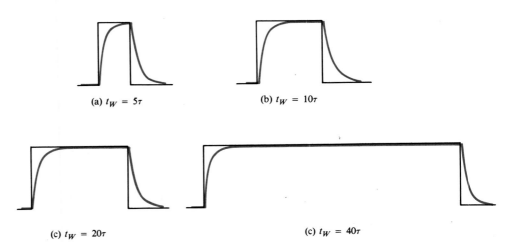

FIGURE 15–5
Variation of an integrator's output pulse shape with input pulse width (the time constant is fixed). Black is input and color is output.

When the Pulse Width Is Less than Five Time Constants

Now let us examine the case in which the width of the input pulse is less than five time constants of the integrator. This condition is expressed as $t_W < 5\tau$.

As before, the capacitor charges for the duration of the pulse. The pulse width is the time the capacitor has for charging. However, because the pulse width is less than the time the capacitor needs to fully charge (5τ), the output voltage will *not* reach the full input voltage before the end of the pulse. *The capacitor only partially charges,* as illustrated in Figure 15–6 for several values of *RC* time constants. Notice that for longer time constants, the output reaches a lower voltage because the capacitor cannot charge as much during the pulse width. Of course, in our examples with a single pulse input, the capacitor fully discharges after the pulse ends.

FIGURE 15–6
Capacitor voltage for various time constants which are longer than the input pulse width. Black is input and color is output.

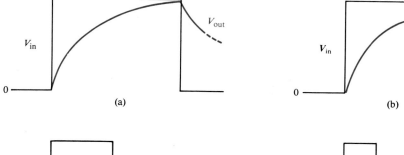

When the time constant is much greater than the input pulse width, the capacitor charges very little, and, as a result, the output voltage becomes almost negligible, as indicated in Figure 15–6.

Figure 15–7 illustrates the effect of reducing the input pulse width for a *fixed* time constant value. As the width is reduced, the output voltage becomes smaller because

FIGURE 15–7
The capacitor charges less and less as the input pulse width is reduced. The time constant is fixed.

the capacitor has less time to charge. However, it takes the capacitor the same length of time (5τ) to discharge back to zero for each condition after the pulse is removed.

EXAMPLE 15–1

A single 10-V pulse with a width of 100 μs is applied to the integrator in Figure 15–8.
(a)   To what voltage will the capacitor charge?
(b)   How long will it take the capacitor to discharge?
(c)   Sketch the output voltage.

FIGURE 15–8

Solution
(a)   The circuit time constant is

$$\tau = RC = (100 \text{ k}\Omega)(0.001 \text{ }\mu\text{F}) = 100 \text{ }\mu s$$

Notice that the pulse width is exactly equal to one time constant. Thus, the capacitor will charge 63% of the full input amplitude in one time constant, so the output will reach a maximum voltage of 6.3 V.
(b)   The capacitor discharges back through the source when the pulse ends. The total discharge time is

$$5\tau = 5(100 \text{ }\mu s) = 500 \text{ }\mu s$$

(c)   The output charging and discharging curve is shown in Figure 15–9.

FIGURE 15–9

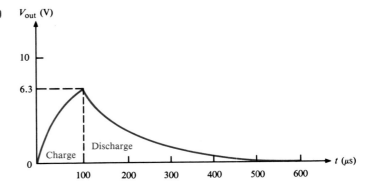

EXAMPLE 15–2

Determine how much the capacitor in Figure 15–10 will charge when the single pulse is applied to the input.

FIGURE 15–10

Solution

Calculate the time constant:

$$\tau = RC = (2.2 \text{ k}\Omega)(1 \text{ } \mu\text{F}) = 2.2 \text{ ms}$$

Because the pulse width is 5 ms, the capacitor charges for approximately 2.27 time constants (5 ms/2.2 ms = 2.27). We use the exponential formula from Chapter 9 to find the voltage to which the capacitor will charge. The calculation is done as follows:

$$
\begin{aligned}
v &= V_F(1 - e^{-t/RC}) \qquad \text{where } V_F = 25 \text{ V} \quad \text{and} \quad t = 5 \text{ ms} \\
&= (25 \text{ V})(1 - e^{-5/2.2}) = (25 \text{ V})(1 - e^{-2.273}) \\
&= (25 \text{ V})(1 - 0.103) = (25 \text{ V})(0.897) \\
&= 22.43 \text{ V}
\end{aligned}
$$

These calculations show that the capacitor charges to 22.43 V during the 5-ms duration of the input pulse. It will discharge back to zero when the pulse goes back to zero.

The calculator sequence for *v* is

$$\boxed{5} \; \boxed{\div} \; \boxed{2} \; \boxed{\cdot} \; \boxed{2} \; \boxed{=} \; \boxed{+/-} \; \boxed{\text{2nd}} \; \boxed{\ln x} \; \boxed{+/-} \; \boxed{+} \; \boxed{1} \; \boxed{=} \; \boxed{\times} \; \boxed{2} \; \boxed{5} \; \boxed{=}$$

SECTION REVIEW
15–2

1. When an input pulse is applied to an *RC* integrator, what condition must exist in order for the output voltage to reach full amplitude?
2. For the circuit in Figure 15–11, which has a single input pulse, find the maximum output voltage and determine how long the capacitor will discharge.
3. For Figure 15–11, sketch the approximate shape of the output voltage with respect to the input pulse.
4. If the integrator time constant equals the input pulse width, will the capacitor fully charge?
5. Describe the condition under which the output voltage has the approximate shape of a rectangular input pulse.

FIGURE 15–11

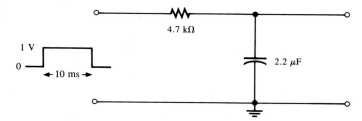

15–3

RESPONSE OF AN *RC* INTEGRATOR TO REPETITIVE PULSES

In the last section you learned how an *RC* integrator responds to a *single* pulse input. These basic ideas will be extended in this section to include the integrator response to *repetitive* pulses. In electronic systems, you will encounter repetitive pulse waveforms much more often than single pulses. However, an understanding of the *RC* integrator response to single pulses is essential to learning how these circuits act with repeated pulses.

If a periodic pulse waveform is applied to an *RC* integrator, as shown in Figure 15–12, *the output waveshape depends on the relationship of the circuit time constant and the frequency (period) of the input pulses.* The capacitor, of course, charges and discharges in response to a pulse input. *The amount of charge and discharge of the capacitor depends both on the circuit time constant and on the input frequency,* as mentioned.

If the pulse width and the time between pulses are each equal to or greater than five time constants, the capacitor will fully charge and fully discharge during each period of the input waveform. This case is shown in Figure 15–12.

$$\tau = RC$$

FIGURE 15–12
RC integrator with a repetitive pulse waveform input.

When the Capacitor Does Not Fully Charge and Discharge

When the pulse width and the time between pulses are *shorter than 5 time constants,* as illustrated in Figure 15–13 for a square wave, the capacitor will *not* completely charge or discharge. We will now examine the effects of this situation on the output voltage of the *RC* integrator.

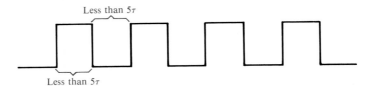

FIGURE 15–13
Waveform that does not allow full charge or discharge of integrator capacitor.

For convenience, let us take an example of an *RC* integrator with a charging and discharging time constant equal to the pulse width of a 10-V square wave input, as in

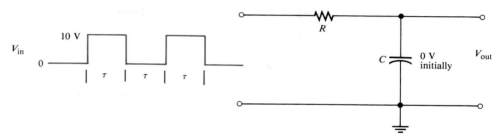

FIGURE 15–14

Integrator with a square wave input having a period equal to two time constants ($T = 2\tau$).

Figure 15–14. This choice will simplify the analysis and will demonstrate the basic action of the integrator under these conditions. At this point, we really do not care what the exact time constant value is because we know that an *RC* circuit charges 63% during one time constant interval.

We will assume that the capacitor in Figure 15–14 begins initially uncharged and examine the output voltage on a pulse-by-pulse basis. The results of this analysis are shown in Figure 15–15.

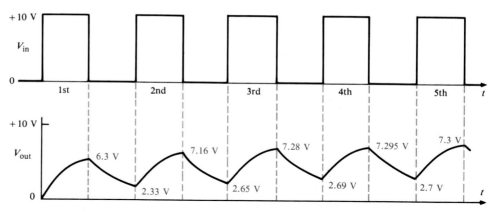

FIGURE 15–15

Input and output for the initially uncharged integrator in Figure 15–14.

First Pulse

During the first pulse, the capacitor charges. The output voltage reaches 6.3 V (63% of 10 V), as shown in Figure 15–15.

Between First and Second Pulses

The capacitor discharges, and the voltage decreases to 37% of the voltage at the beginning of this interval: 0.37(6.3 V) = 2.33 V.

Second Pulse

The capacitor voltage begins at 2.33 V and increases 63% of the way to 10 V. This calculation is as follows: The total charging range is 10 V − 2.33 V = 7.67 V. The capacitor voltage will increase an additional 63% of 7.67 V, which is 4.83 V. Thus, at the end of the second pulse, the output voltage is 2.33 V + 4.83 V = 7.16 V, as indicated in Figure 15–15. Notice that the *average* is building up.

Between Second and Third Pulses

The capacitor discharges during this time, and therefore the voltage decreases to 37% of the voltage by the end of the second pulse: 0.37(7.16 V) = 2.65 V.

Third Pulse

At the start of the third pulse, the capacitor voltage begins at 2.65 V. The capacitor charges 63% of the way from 2.65 V to 10 V: 0.63(10 V − 2.65 V) = 4.63 V. Therefore, the voltage at the end of the third pulse is 2.65 V + 4.63 V = 7.28 V.

Between Third and Fourth Pulses

The voltage during this interval decreases due to capacitor discharge. It will decrease to 37% of its value by the end of the third pulse. The final voltage in this interval is 0.37(7.28 V) = 2.69 V.

Fourth Pulse

At the start of the fourth pulse, the capacitor voltage is 2.69 V. The voltage increases by 0.63(10 V − 2.69 V) = 4.605 V. Therefore, at the end of the fourth pulse, the capacitor voltage is 2.69 V + 4.61 V = 7.295 V. Notice that the values are leveling off as the pulses continue.

Between Fourth and Fifth Pulses

Between these pulses, the capacitor voltage drops to 0.37(7.295 V) = 2.7 V.

Fifth Pulse

During the fifth pulse, the capacitor charges 0.63(10 V − 2.7 V) = 4.6 V. Since it started at 2.7 V, the voltage at the end of the pulse is 2.7 V + 4.6 V = 7.3 V.

Steady-State Response

In the preceding discussion, the output voltage gradually built up and then began leveling off. It takes approximately 5τ for the output voltage to build up to a constant *average* value, regardless of the number of pulses that may occur during that time interval. This interval is the *transient time* of the circuit. Once the output voltage reaches *the average value of the input voltage, a steady-state condition* is reached, which continues as long as the periodic input continues. This state is illustrated in Figure 15–16 based on the values obtained in the preceding discussion.

The transient time for our example circuit is the time from the beginning of the first pulse to the end of the third pulse. The reason for this interval is that the capacitor voltage at the end of the third pulse is 7.28 V, which is about 99% of the final voltage.

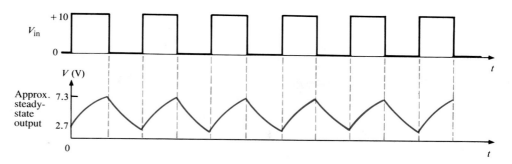

FIGURE 15–16
Output reaches steady state after 5τ.

The Effect of an Increase in Time Constant

What happens to the output voltage if the *RC* time constant of the integrator is increased with a variable resistor, as indicated in Figure 15–17? As the time constant is increased, the capacitor charges *less* during a pulse and discharges *less* between pulses. The result is a *smaller* fluctuation in the output voltage for increasing values of time constant, as shown in Figure 15–18.

FIGURE 15–17
Integrator with a variable time constant.

FIGURE 15–18
Effect of longer time constants on the output of an integrator $(\tau_3 > \tau_2 > \tau_1)$.

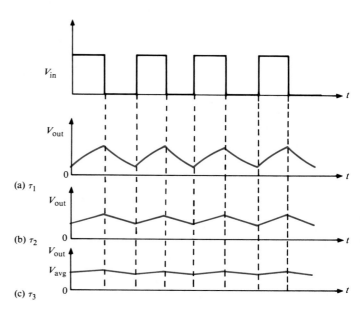

As the time constant becomes extremely long compared to the pulse width, the output voltage approaches a constant dc voltage, as shown in Figure 15–18(c). This value is the *average value* of the input. For a square wave, it is one-half the amplitude.

EXAMPLE 15–3

Determine the output voltage waveform for the first *two* pulses applied to the integrator circuit in Figure 15–19. Assume that the capacitor is initially uncharged.

FIGURE 15–19

Solution
First calculate the circuit time constant:

$$\tau = RC = (4.7 \text{ k}\Omega)(0.01 \text{ }\mu\text{F}) = 47 \text{ }\mu\text{s}$$

Obviously, the time constant is much longer than the input pulse width or the interval between pulses (notice that the input is not a square wave). Thus, in this case, the exponential formulas must be applied, and the analysis is relatively difficult. Follow the solution carefully.

Step 1: *Calculation for first pulse:* Use Equation (9–16) because C is charging. Note that V_F is 5 V, and t equals the pulse width of 10 μs. Therefore,

$$v_C = V_F(1 - e^{-t/RC}) = (5 \text{ V})(1 - e^{-10 \text{ }\mu\text{s}/47 \text{ }\mu\text{s}})$$
$$= (5 \text{ V})(1 - 0.808) = 0.96 \text{ V}$$

This result is plotted in Figure 15–20(a).

Step 2: *Calculation for interval between first and second pulse:* Use Equation (9–17) because C is discharging. Note that V_i is 0.96 V because C begins to discharge from this value at the end of the first pulse. The discharge time is 15 μs. Therefore,

$$v_C = V_i e^{-t/RC} = 0.96e^{-15 \text{ }\mu\text{s}/47 \text{ }\mu\text{s}}$$
$$= 0.96(0.727) = 0.698 \text{ V}$$

This result is shown in Figure 15–20(b).

Step 3: *Calculation for second pulse:* At the beginning of the second pulse, the output voltage is 0.698 V. During the second pulse, the capacitor will again charge. In this case it does not begin at zero volts. It already has 0.698 V from the

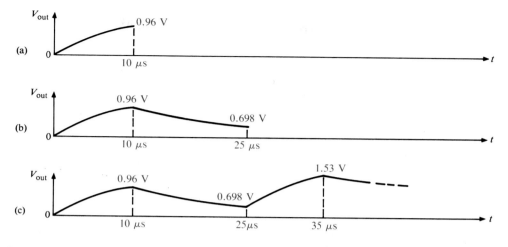

FIGURE 15–20

previous charge and discharge. To handle this situation, we must use the general formula of Equation (9–14):

$$v = V_F + (V_i - V_F)e^{-t/\tau}$$

Using this equation, we can calculate the voltage across the capacitor at the end of the second pulse as follows:

$$v_C = V_F + (V_i - V_F)e^{-t/RC}$$
$$= 5 \text{ V} + (0.698 \text{ V} - 5 \text{ V})e^{-10 \text{ } \mu s/47 \text{ } \mu s}$$
$$= 5 \text{ V} + (-4.3 \text{ V})(0.808) = 5 \text{ V} - 3.47 \text{ V}$$
$$= 1.53 \text{ V}$$

This result is shown in Figure 15–20(c).

Notice that the output waveform builds up on successive input pulses. After approximately 5τ, it will reach its steady state and will fluctuate between a constant maximum and a constant minimum, with an average equal to the average value of the input. We could demonstrate this pattern by carrying the analysis in this example further.

**SECTION REVIEW
15–3**

1. What conditions allow an *RC* integrator capacitor to fully charge and discharge when a periodic pulse waveform is applied to the input?
2. What will the output waveform look like if the circuit time constant is extremely small compared to the pulse width of a square wave input?
3. When 5τ is greater than the pulse width of an input square wave, the time required for the output voltage to build up to a constant average value is called _____ .
4. Define *steady-state response*.
5. What does the average value of the output voltage of an integrator equal during steady state?

15–4

RESPONSE OF AN *RC* DIFFERENTIATOR TO A SINGLE PULSE

Figure 15–21 shows a series *RC* circuit. Notice that the *output is taken across the resistor,* thereby distinguishing the *RC* differentiator from the *RC* integrator. This circuit is called a *differentiator* in terms of its pulse response. Recall that in terms of its frequency response, it is a basic *high-pass* filter. The term differentiator comes from the mathematical function called *differentiation.*

FIGURE 15–21
An *RC* differentiating circuit.

The same action occurs in the differentiator as in the integrator, except *the output voltage is taken across the resistor rather than the capacitor.* The capacitor charges exponentially at a rate depending on the *RC* time constant. The shape of the differentiator's resistor voltage is determined by the charging and discharging action of the capacitor.

To understand how the output voltage is shaped by a differentiator, we must consider the following:

1. the response to the rising pulse edge
2. the response between the rising and falling edges
3. the response to the falling pulse edge

Response to the Rising Edge of the Input Pulse

Assume that the capacitor is initially uncharged prior to the rising pulse edge. Prior to the pulse, the input is zero volts. Thus, there are zero volts across the capacitor and also zero volts across the resistor, as indicated in Figure 15–22(a).

Now assume that a 10-V pulse is applied to the input. When the rising edge occurs, point *A* goes to +10 V. Recall that *the voltage across a capacitor cannot change instantaneously,* and thus the capacitor appears instantaneously as a short. Therefore, if point *A* goes instantly to +10 V, then point *B* *must* also go instantly to +10 V, keeping the capacitor voltage zero for the instant of the rising edge. The capacitor voltage is the voltage from point *A* to point *B*.

The voltage at point *B* with respect to ground is the voltage across the resistor (and the output voltage). Thus, the output voltage suddenly goes to +10 V in response to the rising pulse edge, as indicated in Figure 15–22(b).

Response during Pulse When $t_W \geq 5\tau$

While the pulse is at its high level between the rising edge and the falling edge, the capacitor is charging. When the pulse width is equal to or greater than five time constants ($t_W \geq 5\tau$), the capacitor has time to *fully charge.*

As the voltage across the capacitor builds up exponentially, the voltage across the resistor *decreases* exponentially until it reaches zero volts at the time the capacitor reaches full charge ($+10$ V in this case). This decrease in the resistor voltage occurs because the sum of the capacitor voltage and the resistor voltage at any instant must be equal to the applied voltage, in compliance with Kirchhoff's voltage law ($v_C + v_R = v_{in}$). This part of the response is illustrated in Figure 15–22(c).

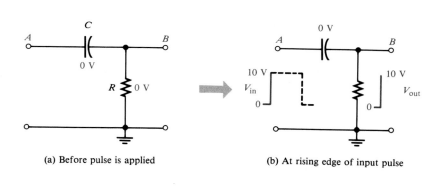

(a) Before pulse is applied   (b) At rising edge of input pulse

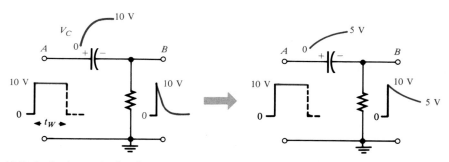

(c) During level part of pulse when $t_W \geqslant 5\tau$ (d) During level part of pulse when $t_W < 5\tau$

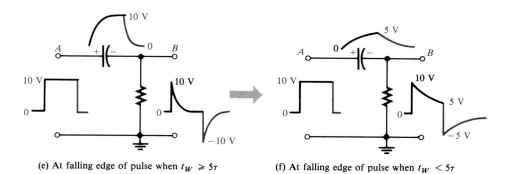

(e) At falling edge of pulse when $t_W \geqslant 5\tau$ (f) At falling edge of pulse when $t_W < 5\tau$

FIGURE 15–22

Examples of the response of a differentiator to a single input pulse under two conditions: $t_W \geq 5\tau$ and $t_W < 5\tau$.

Response during Pulse When $t_W < 5\tau$

When the pulse width is less than five time constants ($t_W < 5\tau$), *the capacitor does not have time to fully charge*. Its partial charge depends on the relation of the time constant and the pulse width.

Because the capacitor does not reach the full $+10$ V, *the resistor voltage will not reach zero volts* by the end of the pulse. For example, if the capacitor charges to $+5$ V during the pulse interval, the resistor voltage will decrease to $+5$ V, as illustrated in Figure 15–22(d).

Response to Falling Edge When $t_W \geq 5\tau$

Let us first examine the case in which the capacitor is *fully charged* at the end of the pulse ($t_W \geq 5\tau$). Refer to Figure 15–22(e). On the falling edge, the input pulse suddenly goes from $+10$ V back to zero. An instant before the falling edge, the capacitor is charged to 10 V, so point A is $+10$ V and point B is 0 V. Since the voltage across a capacitor cannot change instantaneously, when point A makes a transition from $+10$ V to zero on the falling edge, point B *must* also make a 10-V transition from zero to -10 V. This keeps the voltage across the capacitor at 10 V for the instant of the falling edge.

The capacitor now begins to *discharge* exponentially. As a result, the resistor voltage goes from -10 V to zero in an exponential curve, as indicated in Figure 15–22(e).

Response to Falling Edge When $t_W < 5\tau$

Next, let us examine the case in which the capacitor is only partially charged at the end of the pulse ($t_W < 5\tau$). For example, if the capacitor charges to $+5$ V, the resistor voltage at the instant before the falling edge is also $+5$ V, because the capacitor voltage plus the resistor voltage must add up to $+10$ V, as illustrated in Figure 15–22(d).

When the falling edge occurs, point A goes from $+10$ V to zero. As a result, point B goes from $+5$ V to -5 V, as illustrated in Figure 15–22(f). This decrease occurs, of course, because the capacitor voltage cannot change at the instant of the falling edge. Immediately after the falling edge, the capacitor begins to discharge to zero. As a result, the resistor voltage goes from -5 V to zero, as shown.

Summary of Differentiator Response to a Single Pulse

Perhaps a good way to summarize this section is to look at the general output waveforms of a differentiator as the time constant is varied from one extreme, when 5τ is much less than the pulse width, to the other extreme, when 5τ is much greater than the pulse width. These situations are illustrated in Figure 15–23. In part (a) of the figure, the output consists of narrow positive and negative "spikes." In part (f), the output approaches the shape of the input. Various conditions between these extremes are illustrated in parts (b) through (e).

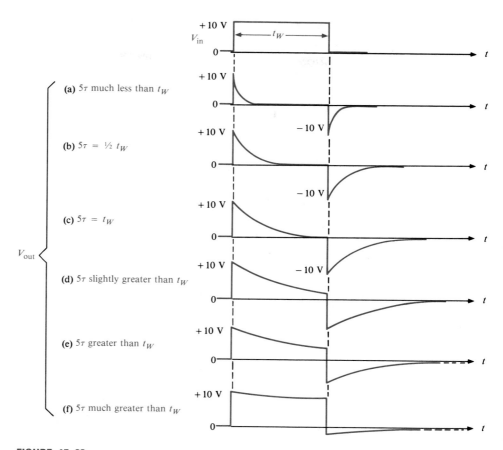

FIGURE 15–23
Effects of a change in time constant on the shape of the output voltage of a differentiator.

EXAMPLE 15–4

Sketch the output voltage for the circuit in Figure 15–24.

FIGURE 15–24

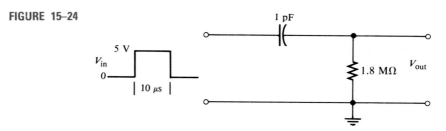

Solution
First calculate the time constant:

$$\tau = RC = (1.8 \text{ M}\Omega)(1 \text{ pF}) = 1.8 \text{ } \mu\text{s}$$

In this case, $t_W > 5\tau$, so the capacitor reaches full charge before the end of the pulse.

On the rising edge, the resistor voltage jumps to +5 V and then decreases exponentially to zero before the end of the pulse. On the falling edge, the resistor voltage jumps to −5 V and then goes back to zero exponentially. The resistor voltage is, of course, the output, and its shape is shown in Figure 15–25.

FIGURE 15-25

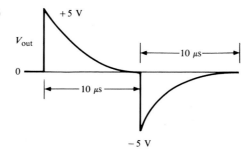

EXAMPLE 15–5

Determine the output voltage waveform for the differentiator in Figure 15–26.

FIGURE 15-26

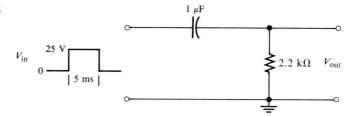

Solution
First calculate the time constant:

$$\tau = (2.2 \ k\Omega)(1 \ \mu F) = 2.2 \ ms$$

On the rising edge, the resistor voltage immediately jumps to +25 V. Because the pulse width is 5 ms, the capacitor charges for approximately 2.27 time constants during the pulse, and therefore does not reach full charge. Thus, we must use the formula for a decreasing exponential, Equation (9–17), in order to calculate to what voltage the output decreases by the end of the pulse:

$$V_{out} = V_i e^{-t/RC} \qquad \text{where} \quad V_i = 25 \ V \quad \text{and} \quad t = 5 \ ms$$
$$= 25e^{-5 \ ms/2.2 \ ms} = 25(0.103)$$
$$= 2.58 \ V$$

This calculation gives us the resistor voltage at the end of the 5-ms pulse width interval.
On the falling edge, the resistor voltage immediately jumps from +2.58 V down to −22.42 V (a 25-V transition). The resulting waveform of the output voltage is shown in Figure 15–27.

FIGURE 15–27

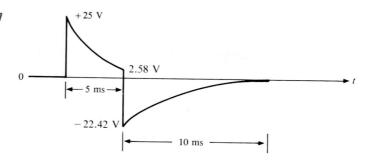

SECTION REVIEW
15–4

1. Sketch the output of a differentiator for a 10-V input pulse when $5\tau = 0.5t_W$.
2. Under what condition does the output pulse shape most closely resemble the input pulse for a differentiator?
3. What does the differentiator output look like when 5τ is much less than the pulse width of the input?
4. If the resistor voltage in a differentiating circuit is down to $+5$ V at the end of a 15-V input pulse, to what negative value will the resistor voltage go in response to the *falling* edge of the input?

15–5

RESPONSE OF AN *RC* DIFFERENTIATOR TO REPETITIVE PULSES

The differentiator response to a single pulse, covered in the preceding section, is extended in this section to repetitive pulses.

If a periodic pulse waveform is applied to an *RC* differentiating circuit, two conditions again are possible: $t_W \geq 5\tau$ or $t_W < 5\tau$. Figure 15–28 shows the output when $t_W = 5\tau$. As the time constant is reduced, both the positive and the negative portions of the output become narrower. Notice that the *average* value of the output is *zero*.

FIGURE 15–28
Example of differentiator response when $t_W = 5\tau$.

Figure 15–29 shows the *steady-state* output when $t_W < 5\tau$. As the time constant is increased, the positively and negatively sloping portions become flatter. For a very long time constant, the output approaches the shape of the input, but with an average value of zero. The average value of a waveform is its *dc* component. Because a capacitor blocks dc, the dc component of the input is prevented from passing through to the output.

FIGURE 15-29
Example of differentiator response when $t_W < 5\tau$.

Like the integrator, the differentiator output takes time (5τ) to reach steady state. To illustrate the response, let us take an example in which the time constant *equals* the input pulse width.

At this point, we do not care what the circuit time constant is, because we know that the resistor voltage will decrease to 37% of its maximum value during one pulse (1τ). We will assume that the capacitor in Figure 15–30 begins initially uncharged, and then we will examine the output voltage on a pulse-by-pulse basis. The results of the analysis to follow are shown in Figure 15–31.

FIGURE 15-30
RC differentiator with $\tau = t_W$.

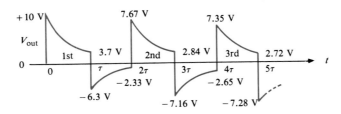

FIGURE 15-31
Differentiator output waveform during transient time for the circuit in Figure 15-30.

First Pulse

On the rising edge, the output instantaneously jumps to $+10$ V. Then the capacitor partially charges to 63% of 10 V, which is 6.3 V. Thus, the output voltage must decrease to 3.7 V, as shown in Figure 15–31.

On the falling edge, the output instantaneously makes a negative-going 10-V transition to -6.3 V (-10 V $+ 3.7$ V $= -6.3$ V).

Between First and Second Pulses

The capacitor discharges to 37% of 6.3 V, which is 2.33 V. Thus, the resistor voltage, which starts at -6.3 V, must increase to -2.33 V. Why? Because at *the instant prior to* the next pulse, the input voltage is zero. Therefore, the sum of v_C and v_R must be zero ($+2.33$ V $-$ 2.33 V $= 0$). Remember that $v_C + v_R = v_{in}$ at all times.

Second Pulse

On the rising edge, the output makes an instantaneous, positive-going, 10-V transition from -2.33 V to 7.67 V. Then the capacitor charges $0.63 \times (10$ V $- 2.33$ V$) = 4.83$ V by the end of the pulse. Thus, the capacitor voltage increases from 2.33 V to 2.33 V $+ 4.83$ V $= 7.16$ V. The output voltage drops to $0.37 \times (7.67$ V$) = 2.84$ V.

On the falling edge, the output instantaneously makes a negative-going transition from 2.84 V to -7.16 V, as shown in Figure 15–31.

Between Second and Third Pulses

The capacitor discharges to 37% of 7.16 V, which is 2.65 V. Thus, the output voltage starts at -7.16 V and increases to -2.65 V, because the capacitor voltage and the resistor voltage must add up to zero at the instant prior to the third pulse (the input is zero).

Third Pulse

On the rising edge, the output makes an instantaneous 10-V transition from -2.65 V to $+7.35$ V. Then the capacitor charges $0.63 \times (10$ V $- 2.65$ V$) = 4.63$ V to 2.65 V $+ 4.63$ V $= +7.28$ V. As a result, the output voltage drops to 0.37×7.35 V $= 2.72$ V.

On the falling edge, the output instantly goes from $+2.72$ V down to -7.28 V.

After the third pulse, five time constants have elapsed, and the output voltage is close to its steady state. Thus, it will continue to vary from a positive maximum of about $+7.3$ V to a negative maximum of about -7.3 V, with an average value of zero.

**SECTION REVIEW
15–5**

1. What conditions allow an *RC* differentiator to fully charge and discharge when a periodic pulse waveform is applied to the input?
2. What will the output waveform look like if the circuit time constant is extremely small compared to the pulse width of a square wave input?
3. What does the average value of the differentiator output voltage equal during steady state?

15-6

THE *RL* INTEGRATOR

Figure 15–32 shows a series *RL* circuit with *the output taken across the resistor*. This circuit is also known as an *integrator*. Its output waveform under equivalent conditions is the same as that for the *RC* integrator. Recall that in the *RC* case, the output was across the capacitor.

FIGURE 15–32
An *RL* integrating circuit.

As you know, both edges of an ideal pulse are considered to occur instantaneously. Two basic rules for inductor behavior will aid in analyzing *RL* circuit responses to pulse inputs:

1. The inductor appears as an open to an instantaneous change in current and as a short (ideally) to dc.
2. The current in an inductor cannot change instantaneously—it can change only exponentially.

Response of the Integrator to a Single Pulse

When a pulse generator is connected to the input of the integrator and the voltage pulse goes from its low level to its high level, the inductor prevents a sudden change in current. As a result, the inductor acts as an open, and all the input voltage is across it at the instant of the rising pulse edge. This situation is indicated in Figure 15–33(a).

After the rising edge, the current builds up, and the output voltage follows the current as it increases exponentially, as shown in Figure 15–33(b). The current can reach a maximum of V_p/R if the transient time is shorter than the pulse width ($V_p = 10$ V in this example).

When the pulse goes from its high level to its low level, an induced voltage with reversed polarity is created across the coil in an effort to keep the current equal to V_p/R. The output voltage begins to decrease exponentially, as shown in Figure 15–33(c).

The exact shape of the output depends on the L/R time constant as summarized in Figure 15–34 for various relationships between the time constant and the pulse width. You should note that the response of this *RL* circuit *in terms of the shape of the output* is identical to that of the *RC* integrator. The relationship of the L/R time constant to the input pulse width has the same effect as the *RC* time constant that we discussed earlier in this chapter. For example, when $t_W < 5\tau$, the output voltage will not reach its maximum possible value.

(a) At rising edge of pulse ($i = 0$)

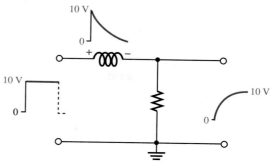

(b) During flat portion of pulse

(c) At falling edge of pulse and after

FIGURE 15–33
Illustration of the pulse response of an *RL* integrator ($t_W > 5\tau$).

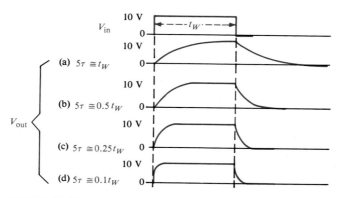

FIGURE 15–34
Illustration of the variation in integrator output pulse shape with time constant.

EXAMPLE 15-6   Determine the maximum output voltage for the integrator in Figure 15-35 when a single pulse is applied as shown.

FIGURE 15-35

Solution
Calculate the time constant:

$$\tau = \frac{L}{R} = \frac{100 \text{ mH}}{47 \text{ }\Omega} = 2.13 \text{ ms}$$

Because the pulse width is 5 ms, the inductor charges for approximately 2.35τ (5 ms/2.13 ms = 2.35). We must use the exponential formula to calculate the voltage, as follows:

$$v_{out} = V_F(1 - e^{-t/\tau}) = 25(1 - e^{-5/2.13})$$
$$= 25(1 - e^{-2.35}) = 25(1 - 0.095)$$
$$= 25(0.905) = 22.63 \text{ V}$$

EXAMPLE 15-7   A pulse is applied to the RL integrator circuit in Figure 15-36. Determine the complete waveshapes and the values for I, V_R, and V_L.

FIGURE 15-36

Solution
The circuit time constant is

$$\tau = \frac{L}{R} = \frac{5 \text{ mH}}{1.5 \text{ k}\Omega} = 3.33 \text{ }\mu s$$

Since $5\tau = 16.65$ μs is less than t_W, the current will reach its maximum value and remain there until the end of the pulse.

At the rising edge of the pulse:

$$i = 0 \text{ A}$$
$$v_R = 0 \text{ V}$$
$$v_L = 10 \text{ V}$$

Since the inductor is initially an open, all of the input voltage appears across L. During the pulse:

i increases exponentially to $V_p/R = $ 10 V/1.5 kΩ
$$= 6.67 \text{ mA in } 16.65 \text{ } \mu s$$
v_R increases exponentially to 10 V in 16.65 μs
v_L decreases exponentially to zero in 16.65 μs

At the falling edge of the pulse:

$$i = 6.67 \text{ mA}$$
$$v_R = 10 \text{ V}$$
$$v_L = -10 \text{ V}$$

After the pulse:

i decreases exponentially to zero in 16.65 μs
v_R decreases exponentially to zero in 16.65 μs
v_L decreases exponentially to zero in 16.65 μs

The waveforms are shown in Figure 15–37.

FIGURE 15–37

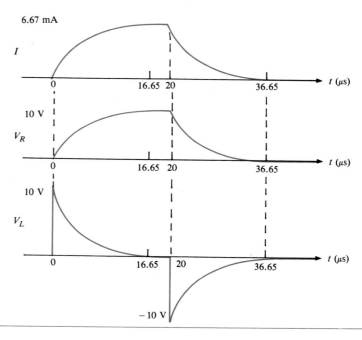

EXAMPLE 15–8

A 10-V pulse with a width of 1 ms is applied to the integrator in Figure 15–38. Determine the voltage level that the output will reach during the pulse. If the source has an internal resistance of 30 Ω, how long will it take the output to decay to zero? Sketch the output voltage waveform.

FIGURE 15–38

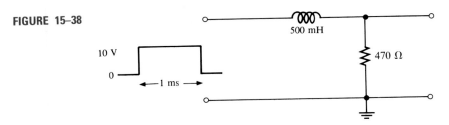

Solution

The coil charges through the 30-Ω source resistance plus the 470-Ω external resistor. The time constant is

$$\tau = \frac{L}{R_T} = \frac{500 \text{ mH}}{470 \text{ Ω} + 30 \text{ Ω}} = \frac{500 \text{ mH}}{0.5 \text{ kΩ}} = 1 \text{ ms}$$

Notice that in this case the pulse width is exactly equal to τ. Thus, the output V_R will reach 63% of the full input amplitude in 1τ. Therefore, the output voltage gets to 6.3 V at the end of the pulse.

After the pulse is gone, the inductor discharges back through the 30-Ω source and the 470-Ω resistor. The source takes 5τ to completely discharge; 5τ = 5(1 ms) = 5 ms.

The output voltage is shown in Figure 15–39.

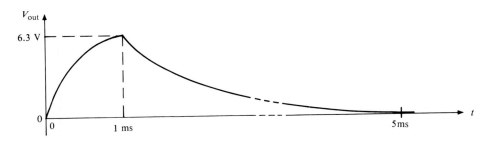

FIGURE 15–39

SECTION REVIEW
15–6

1. In an *RL* integrator, across which component is the output voltage taken?
2. When a pulse is applied to an *RL* integrator, what condition must exist in order for the output voltage to reach the amplitude of the input?
3. Under what condition will the output voltage have the approximate shape of the input pulse?

15–7

THE *RL* DIFFERENTIATOR

Figure 15–40 shows a series *RL* circuit with *the output across the inductor.* Compare this circuit to the *RC* differentiator in which the output is the resistor voltage.

FIGURE 15–40
An *RL* differentiating circuit.

Response of the Differentiator to a Single Pulse

A pulse generator is connected to the input of the differentiator. Initially, before the pulse, there is no current in the circuit. When the input pulse goes from its low level to its high level, the inductor prevents a sudden change in current. It does this, as you know, with an induced voltage equal and opposite to the input. As a result, *L* looks like an open, and *all* of the input voltage appears across it at the instant of the rising edge, as shown in Figure 15–41(a) with a 10-V pulse.

During the pulse, the current exponentially builds up. As a result, the inductor voltage decreases, as shown in Figure 15–41(b). The rate of decrease, as you know,

(a) At rising edge of pulse

(b) During flat portion of pulse

(c) At falling edge when $t_W < 5\tau$

(d) At falling edge when $t_W \geq 5\tau$.

FIGURE 15–41
Illustration of the response of an *RL* differentiator for both time constant conditions.

depends on the L/R time constant. When the falling edge of the input appears, the inductor reacts to keep the current as is by creating an induced voltage in a direction as indicated in part (c). This reaction is seen as a sudden negative-going transition of the inductor voltage, as indicated in parts (c) and (d).

Two conditions are possible, as indicated in parts (c) and (d). In part (c), 5τ is greater than the input pulse width, and the output voltage does not have time to decay to zero. In part (d), 5τ is less than the pulse width, and so the output decays to zero before the end of the pulse. In this case a full, negative, 10-V transition occurs at the trailing edge.

Keep in mind that as far as the input and output waveforms are concerned, the RL integrator and differentiator perform the same as their RC counterparts.

A summary of the RL differentiator response for relationships of various time constants and pulse widths is shown in Figure 15–42.

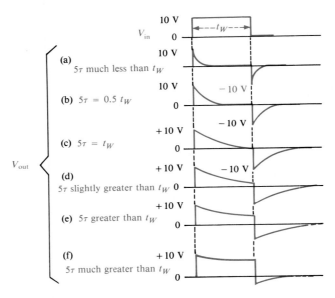

FIGURE 15–42
Illustration of the variation in output pulse shape with the time constant.

EXAMPLE 15–9   Sketch the output voltage for the circuit in Figure 15–43.

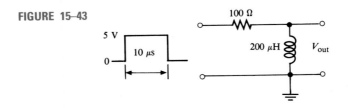

FIGURE 15–43

Solution
First calculate the time constant:

$$\tau = \frac{L}{R} = \frac{200\ \mu H}{100\ \Omega} = 2\ \mu s$$

In this case, $t_W = 5\tau$, so the output will decay to zero at the end of the pulse.

On the rising edge, the inductor voltage jumps to $+5$ V and then decays exponentially to zero. It reaches approximately zero at the instant of the falling edge. On the falling edge of the input, the inductor voltage jumps to -5 V and then goes back to zero. The output waveform is shown in Figure 15–44.

FIGURE 15–44

EXAMPLE 15–10   Determine the output voltage waveform for the differentiator in Figure 15–45.

FIGURE 15–45

Solution
First calculate the time constant:

$$\tau = \frac{L}{R} = \frac{20\ mH}{10\ \Omega} = 2\ ms$$

On the rising edge, the inductor voltage immediately jumps to $+25$ V. Because the pulse width is 5 ms, the inductor charges for only 2.5τ, so we must use the formula for a decreasing exponential:

$$
\begin{aligned}
v_L &= V_i e^{-t/\tau} = 25e^{-5/2} \\
&= 25e^{-2.5} = 25(0.082) \\
&= 2.05\ V
\end{aligned}
$$

This result is the inductor voltage at the end of the 5-ms input pulse.

On the falling edge, the output immediately jumps from $+2.05$ V down to -22.95 V (a 25-V negative-going transition). The complete output waveform is sketched in Figure 15–46.

FIGURE 15–46

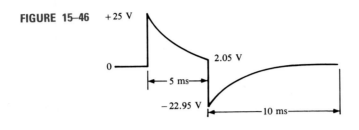

**SECTION REVIEW
15–7**

1. In an *RL* differentiator, across which component is the output taken?
2. Under what condition does the output pulse shape most closely resemble the input pulse?
3. If the inductor voltage in an *RL* differentiator is down to $+2$ V at the end of a $+10$-V input pulse, to what negative voltage will the output voltage go in response to the falling edge of the input?

15–8

APPLICATIONS

Although integrators and differentiators are found in many different kinds of applications, only three basic ones are discussed in this section to give you a general idea of how these circuits are used.

Timing Circuits

RC integrators can be used in timing circuits to set a specified time interval for various purposes. By changing the time constant, the interval can be adjusted. For example, the *RC* integrator circuit in Figure 15–47(a) is used to provide a delay between the time a power-on switch is closed and the time a certain event is activated by a *threshold* circuit. The threshold circuit is designed to respond to the input voltage only when the input reaches a certain specified level. Don't worry how the threshold circuit works at this point; you will learn about it in a later course.

When the switch is thrown, the capacitor begins to charge at a rate set by the *RC* time constant. After a certain time, the capacitor voltage reaches the threshold value and triggers (turns on) the circuit to activate a device such as a motor, relay, lamp, etc., depending on the particular application. This action is illustrated by the waveforms in Figure 15–47(b).

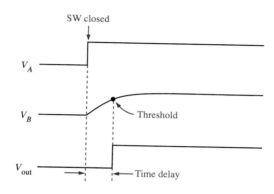

(a) Basic delay circuit

(b) Timing diagram

FIGURE 15–47
A basic time delay application of an *RC* integrator.

EXAMPLE 15–11

What is the time delay in the circuit of Figure 15–47(a) with V_{in} = 9 V, R = 10 MΩ and C = 0.47 μF? Assume the threshold voltage is 5 V.

Solution

Using the exponential formula from Chapter 9 and solving for time gives

$$v = V_F(1 - e^{-t/RC})$$

$$v = V_F - V_F e^{-t/RC}$$

$$V_F - v = V_F e^{-t/RC}$$

$$e^{-t/RC} = \frac{V_F - v}{V_F}$$

Taking the natural log (ln) of both sides of the equation yields

$$-\frac{t}{RC} = \ln\left(\frac{V_F - v}{V_F}\right)$$

$$t = -RC \ln\left(\frac{V_F - v}{V_F}\right)$$

AC CIRCUITS

V_F is the final voltage to which C will charge and equals V_{in}. Substituting, we solve for t:

$$t = -(10 \text{ M}\Omega)(0.47 \ \mu\text{F})\ln\left(\frac{9 \text{ V} - 5 \text{ V}}{9 \text{ V}}\right)$$

$$= (10 \text{ M}\Omega)(0.47 \ \mu\text{F})\ln\left(\frac{4 \text{ V}}{9 \text{ V}}\right)$$

$$= 3.8 \text{ s}$$

A Pulse Waveform-to-dc Converter

An integrating circuit can be used to convert a pulse waveform to a constant dc value equal to the average of the waveform. This is accomplished by using a time constant that is extremely large compared to the period of the pulse waveform, as illustrated in Figure 15–48(a). Actually, there will be a slight ripple on the output as the capacitor

(a)

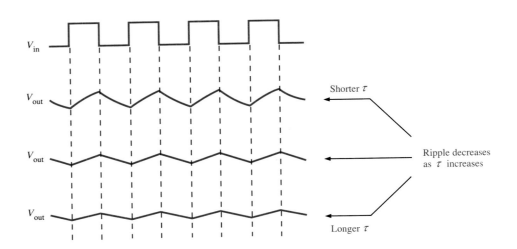

(b)

FIGURE 15–48
A long-time constant integrator acts as a pulse-to-dc converter.

charges and discharges a little. This ripple is less the larger the time constant and can approach a constant level, as shown in part (b).

A Positive and Negative Trigger Pulse Generator

A differentiator circuit can be used to produce short duration pulses with both positive and negative polarities, as illustrated in Figure 15–49(a). When the time constant is small compared to the pulse width of the input, the differentiator produces a positive spike on each positive-going edge of the input and a negative spike on each negative-going edge.

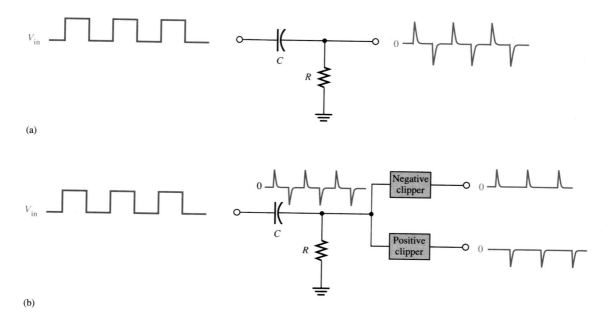

(a)

(b)

FIGURE 15–49
A very short time constant differentiator generates positive and negative spikes.

To separate the positive and negative spikes, two *clipping* circuits are used, as shown in part (b). Each clipping circuit consists of a *diode*, a semiconductor device that you will study later. The negative clipper removes the negative spike and passes the positive spike. The positive clipper does just the opposite.

**SECTION REVIEW
15–8**

1. How can the delay time in the circuit of Figure 15–47 be increased?
2. Reducing the value of *R* in Figure 15–48 will (increase, decrease) the amount of ripple on the output.

15–9   TROUBLESHOOTING

In this section we will use *RC* circuits with pulse inputs to demonstrate the effects of component failure in selected cases.

Open Capacitors

If the capacitor in an *integrator* opens, the output has the same waveshape as the input, as shown in Figure 15–50(a). If the capacitor in a *differentiator* opens, the output is zero because it is held at ground through the resistor, as illustrated in part (b).

(a) Integrator

(b) Differentiator

FIGURE 15–50
Examples of the effect of an open capacitor.

Leaky Capacitor

If the capacitor in an *integrator* becomes leaky, the time constant will be effectively reduced by the leakage resistance (when Thevenized, looking from C it appears in parallel with R) and the waveshape of the output voltage (across C) is altered from normal by a shorter charging time. Also, the amplitude of the output is reduced because R and R_{leak} effectively act as a voltage divider. These effects are illustrated in Figure 15–51(a).

If the capacitor in a *differentiator* becomes leaky, the time constant is reduced, just as in the integrator (they are both simply series *RC* circuits). When the capacitor reaches full charge, the output voltage (across R) is set by the effective voltage divider action of R and R_{leak}, as shown in Figure 15–51(b).

Looking from C, the time constant is $\tau = R_{\text{th}}C$

(a)

FIGURE 15–51
Examples of the effect of a leaky capacitor.

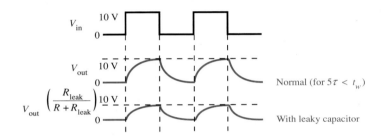

(b)

Shorted Capacitor

If the capacitor in an *integrator* shorts, the output is at ground as shown in Figure 15–52(a). If the capacitor in a *differentiator* shorts, the output is the same as the input, as shown in part (b).

(a)

(b)

FIGURE 15–52
Examples of the effect of a shorted capacitor.

Open Resistor

If the resistor in an integrator opens, the capacitor has no discharge path, and, ideally, it will hold its change. In an actual situation, the charge will gradually leak off or the capacitor will discharge slowly through a measuring instrument connected to the output. This is illustrated in Figure 15–53(a).

(a)

(b)

FIGURE 15–53
Examples of the effects of an open resistor.

If the resistor in a differentiator opens, the output looks like the input except for the dc level because the capacitor now must charge and discharge through the extremely high resistance of the oscilloscope, as shown in part (b).

SECTION REVIEW
15–9

1. An *RC* integrator has a zero output with a square wave input. What are the possible causes of this problem?
2. If the capacitor in a differentiator is shorted, what is the output for a square wave input?

SUMMARY

☐ In an *RC* integrating circuit, the output voltage is taken across the capacitor.
☐ In an *RC* differentiating circuit, the output voltage is taken across the resistor.
☐ In an *RL* integrating circuit, the output voltage is taken across the resistor.
☐ In an *RL* differentiating circuit, the output voltage is taken across the inductor.
☐ In an integrator, when the pulse width (t_W) of the input is much less than the transient time, the output voltage approaches a constant level equal to the average value of the input.
☐ In an integrator, when the pulse width of the input is much greater than the transient time, the output voltage approaches the shape of the input.
☐ In a differentiator, when the pulse width of the input is much less than the transient time, the output voltage approaches the shape of the input but with an average value of zero.
☐ In a differentiator, when the pulse width of the input is much greater than the transient time, the output voltage consists of narrow, positive- and negative-going spikes occurring on the leading and trailing edges of the input pulses.

GLOSSARY

dc component   The average value of a pulse waveform.

Derivative   The instantaneous rate of change of a function, determined mathematically.

Differentiator   A circuit producing an output that approaches the mathematical derivative of the input.

Integral   The area under the curve of a function, determined mathematically.

Integrator   A circuit producing an output that approaches the mathematical derivative of the input.

Periodic   Characterized by a repetition at fixed intervals.

Response   In electric circuits, the reaction of a circuit to a given input.

Steady state   The equilibrium condition of a circuit that occurs after an initial transient time.

Time constant   A fixed time interval, set by *R*, *C*, and *L* values, that determines the time response of a circuit.

Transient   A temporary passing condition in a circuit. A sudden and/or temporary change in circuit conditions.

Transient time   An interval equal to approximately five time constants.

SELF-TEST

1. The output of an RC integrator is taken across the
   (a) resistor   (b) capacitor   (c) source   (d) coil

2. When a 10-V input pulse with a width equal to one time constant is applied to an RC integrator, the capacitor charges to
   (a) 10 V   (b) 5 V   (c) 6.3 V   (d) 3.7 V

3. When a 10-V pulse with a width equal to one time constant is applied to an RC differentiator, the capacitor charges to
   (a) 6.3 V   (b) 10 V   (c) 0 V   (d) 3.7 V

4. In an RC integrator, the output pulse closely resembles the input pulse when
   (a) τ is much larger than the pulse width   (b) τ is equal to the pulse width
   (c) τ is less than the pulse width   (d) τ is much less than the pulse width

5. In an RC differentiator, the output pulse closely resembles the input pulse when
   (a) τ is much larger than the pulse width   (b) τ is equal to the pulse width
   (c) τ is less than the pulse width   (d) τ is much less than the pulse width

6. The positive and negative portions of a differentiator's output voltage are equal when
   (a) $5\tau < t_W$ (b) $5\tau > t_W$ (c) $5\tau = t_W$ (d) $5\tau > 0$

7. The output of an RL integrator is taken across the
   (a) resistor   (b) coil   (c) source   (d) capacitor

8. The maximum current in an RL differentiator is
   (a) $I = \dfrac{V_p}{X_L}$ (b) $I = \dfrac{V_p}{Z}$ (c) $I = \dfrac{V_p}{R}$

9. The current in an RL differentiator reaches its maximum possible value when
   (a) $5\tau = t_W$ (b) $5\tau < t_W$ (c) $5\tau > t_W$ (d) $\tau = 0.5t_W$

10. If you have an RC and an RL differentiator with equal time constants sitting side-by-side and you apply the same input pulse to both,
    (a) the RC has the widest output pulse
    (b) the RL has the most narrow spikes on the output
    (c) the output of one is an increasing exponential and the output of the other is a decreasing exponential
    (d) you can't tell the difference by observing the output waveforms

---

PROBLEM SET A

Section 15–1   The RC Integrator

1. An integrating circuit has $R = 2$ kΩ in series with $C = 0.05$ μF. What is the time constant?

2. Determine how long it takes the capacitor in an integrating circuit to reach full charge for each of the following series RC combinations:
   (a) $R = 50$ Ω, $C = 50$ μF   (b) $R = 3300$ Ω, $C = 0.015$ μF
   (c) $R = 22$ kΩ, $C = 100$ pF   (d) $R = 5$ MΩ, $C = 10$ pF

Section 15–2   Response of an RC Integrator to a Single Pulse

3. A 20-V pulse is applied to an RC integrator. The pulse width equals *one* time constant. To what voltage does the capacitor charge during the pulse? Assume that it is initially uncharged.

4. Repeat Problem 3 for the following values of t_W:
   (a) 2τ (b) 3τ (c) 4τ (d) 5τ

5. Sketch the approximate shape of an integrator output voltage where 5τ is much less than the pulse width of a 10-V square wave input. Repeat for the case in which 5τ is much larger than the pulse width.

6. Determine the output voltage for an integrator with a single input pulse, as shown in Figure 15–54. For repetitive pulses, how long will it take this circuit to reach steady state?

FIGURE 15–54

Section 15–3   Response of an *RC* Integrator to Repetitive Pulses

7. Sketch the integrator output in Figure 15–55, showing maximum voltages.

8. A 1-V, 10-kHz pulse waveform with a duty cycle of 25% is applied to an integrator with $\tau = 25$ μs. Graph the output voltage for three initial pulses. *C* is initially uncharged.

FIGURE 15–55

9. What is the steady-state output voltage of the integrator with a square wave input shown in Figure 15–56?

FIGURE 15–56

Section 15–4   Response of an *RC* Differentiator to a Single Pulse

10. Repeat Problem 5 for an *RC* differentiator.

11. Redraw the circuit in Figure 15–54 to make it a differentiator, and repeat Problem 6.

Section 15–5   Response of an *RC* Differentiator to Repetitive Pulses

12. Sketch the differentiator output in Figure 15–57, showing maximum voltages.

FIGURE 15–57

FIGURE 15–58

13. What is the steady-state output voltage of the differentiator with the square wave input shown in Figure 15–58?

Section 15–6   The *RL* Integrator

14. Determine the output voltage for the circuit in Figure 15–59. A single input pulse is applied as shown.

15. Sketch the integrator output in Figure 15–60, showing maximum voltages.

FIGURE 15–59   **FIGURE 15–60**

Section 15–7   The *RL* Differentiator

16. (a) What is τ in Figure 15–61?
    (b) Sketch the output voltage.

17. Draw the output waveform if a periodic pulse waveform with $t_w = 25\ \mu s$ and $T = 60\ \mu s$ is applied to the circuit in Figure 15–61.

FIGURE 15–61

Section 15–8   Applications

18. What is the instantaneous voltage at point *B* in Figure 15–47 440 μs after the switch is closed for $R = 22\ k\Omega$ and $C = 0.001\ \mu F$? $V_{IN} = 10$ V

19. Ideally, what is the output of an *RC* integrator when the input is a square wave with an amplitude of 12 V? Assume that the time constant is much greater than the period of the input signal?

Section 15–9   Troubleshooting

20. Determine the most likely fault(s), in the circuit of Figure 15–62(a) for each set of waveforms in parts (b) through (d). V_{in} is a square wave with a period of 8 ms.

21. Determine the most likely fault(s), if any, in the circuit of Figure 15–63(a) for each set of waveforms in parts (b) through (d). V_{in} is a square wave with a period of 8 ms.

(a)

(b)   (c)   (d)

FIGURE 15–62

(a)

(b)   (c)   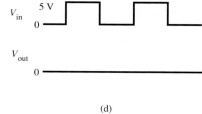(d)

FIGURE 15–63

PROBLEM SET B

22. (a) What is τ in Figure 15–64?
(b) Sketch the output voltage.

FIGURE 15–64

23. (a) What is τ in Figure 15–65?
(b) Sketch the output voltage.

FIGURE 15–65

24. Determine the time constant in Figure 15–66. Is this circuit an integrator or a differentiator?

FIGURE 15–66

25. In a delay circuit like that of Figure 15–47, what time constant will produce a 1-s delay if the threshold of the circuit is 2.5 V and the amplitude of the input is 5 V?

26. Draw the schematic for the circuit in Figure 15–67 and determine if the oscilloscope presentation is correct (a full-color photo appears following page 416).

ANSWERS TO SECTION REVIEWS

Section 15–1

1. A series RC circuit in which the output is across the capacitor.

2. A voltage applied to the input causes the capacitor to charge. A short across the input causes the capacitor to discharge.

Section 15–2

1. $5\tau \le t_W$. **2.** 0.62 V, 51.7 ms. **3.** See Figure 15–68. **4.** No.
5. $5\tau \ll t_W$ (5τ much less than t_W).

FIGURE 15–68

0.62 V

0

0   10 ms   61.7 ms

0

Section 15–3

1. $5\tau \leq t_w$, and $5\tau \leq$ time between pulses.   **2.** Like the input.

3. *transient time*.   **4.** The response after the transient time has passed.

5. The average value of the input voltage.

Section 15–4

1. See Figure 15–69.   **2.** $5\tau \gg t_w$.   **3.** Positive and negative spikes.

4. -10 V.

FIGURE 15–69

$+10$ V

-10 V

Section 15–5

1. $5\tau \leq t_w$ and $5\tau \leq$ time between pulses.   **2.** Positive and negative spikes.

3. 0 V.

Section 15–6

1. Resistor.   **2.** $5\tau \leq t_w$.   **3.** $5\tau \ll t_w$.

Section 15–7

1. Inductor.   **2.** $5\tau \gg t_w$.   **3.** -8 V.

Section 15–8

1. Increase the time constant.

2. Increase.

Section 15–9

1. Shorted capacitor, open resistor.

2. Same as input.

PART THREE

DEVICES

651

16

Introduction to Semiconductor Devices

To acquire a basic understanding of semiconductors, you must have some knowledge of atomic theory and the structure of semiconductor materials. In this chapter, you will learn about atomic theory. We will discuss the basic materials used in manufacturing diodes, transistors, and other semiconductor devices and will present PN junctions, an important concept essential for the understanding of diode and transistor operation. Also, other diode characteristics are introduced, and you will learn how to properly use a diode in a circuit.

After completing this chapter, you will be able to

- ☐ Discuss the significance of electron shells.
- ☐ Describe *ionization*.
- ☐ Explain the basic structure of silicon and germanium.
- ☐ Discuss how atoms bond together to form crystals.
- ☐ Explain how current in a material is related to the energy levels within an atom's structure.
- ☐ Define *P-type* and *N-type* semiconductors.
- ☐ Describe how a PN junction is formed and how it is used.
- ☐ Explain what a *diode* is.
- ☐ Describe forward and reverse bias of a diode.
- ☐ Interpret a diode characteristic curve.
- ☐ Define *anode* and *cathode.*
- ☐ Identify the terminals of a diode.

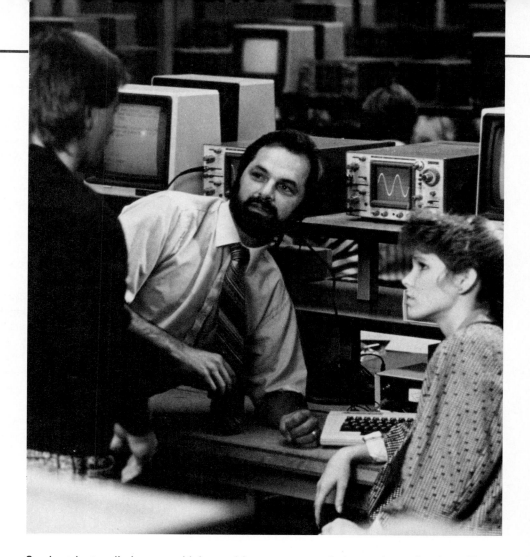

APPLICATION ASSIGNMENT

Semiconductor diodes are widely used in many types of electronic applications. They are of great importance in power supplies, where they are used as *rectifiers* to convert ac to dc. Your supervisor has removed the diodes from one particular power supply that has malfunctioned and has asked you to check the diodes to see which ones, if any, have failed. After completing this chapter, you will know how to carry out this assignment.

16–1   THE ATOMIC STRUCTURE OF SEMICONDUCTORS

The basic concept of the structure of an atom was studied in Chapter 2. In this chapter, atomic theory is extended to semiconductor materials that are used in electronic devices such as diodes and transistors. This coverage lays the foundation for a good understanding of how semiconductor devices function.

Electron Shells and Orbits

Electrons orbit the nucleus at certain distances from the nucleus. Electrons near the nucleus have less energy than those in more distant orbits. Only *discrete* (separate and distinct) values of electron energies exist within atomic structures. Therefore, electrons must orbit only at discrete distances from the nucleus.

Each discrete distance (orbit) from the nucleus corresponds to a certain energy level. In an atom, orbits are grouped into energy bands known as *shells*. A given atom has a fixed number of shells. Each shell has a fixed maximum number of electrons at permissible energy levels (orbits). The differences in energy levels within a shell are much smaller than the difference in energy between shells. The shells are designated K, L, M, N, and so on, with K being closest to the nucleus. This concept is illustrated in Figure 16–1.

$W \equiv$ energy
$r \equiv$ distance from nucleus

FIGURE 16–1
Energy levels increase as distance from the nucleus of the atom increases.

Valence Electrons

Electrons in orbits farther from the nucleus are less tightly bound to the atom than those closer to the nucleus, because the force of attraction between the positively charged nucleus and the negatively charged electron increases with decreasing distance.

Electrons with the highest energy levels exist in the outermost shell of an atom and are relatively loosely bound to the atom. These *valence electrons* contribute to chemical reactions and bonding within the structure of a material. The *valence* of an atom is the number of electrons in its outermost shell.

Ionization

When an atom absorbs energy from a heat source or from light, for example, the energy levels of the electrons are raised. When an electron gains energy, it moves to an orbit farther from the nucleus. Since the valence electrons possess more energy and are more loosely bound to the atom than inner electrons, they can jump to higher orbits more easily when external energy is absorbed.

If a valence electron acquires a sufficient amount of energy, it can be completely removed from the outer shell and the atom's influence. The departure of a valence electron leaves a previously neutral atom with an excess of positive charge (more protons than electrons). The process of losing a valence electron is known as *ionization,* and the resulting positively charged atom is called a *positive ion.* For example, the chemical symbol for hydrogen is H. When it loses its valence electron and becomes a positive ion, it is designated H^+. The escaped valence electron is called a *free electron.*

When a free electron falls into the outer shell of a neutral hydrogen atom, the atom becomes negatively charged (more electrons than protons) and is called a *negative ion,* designated H^-.

Silicon and Germanium Atoms

Two types of widely used semiconductor materials are *silicon* and *germanium.* Both the silicon and the germanium atoms have four valence electrons. They differ in that silicon has 14 protons in its nucleus and germanium has 32. Figure 16–2 shows a representation of the atomic structure for both materials.

FIGURE 16–2
Silicon and germanium atoms.

(a) Silicon   (b) Germanium

**SECTION REVIEW
16–1**

1. Define *atom.*
2. Name the components of an atom.
3. What is a valence electron?
4. What is a free electron?

16–2

ATOMIC BONDING

When silicon atoms combine into molecules to form a solid material, they arrange themselves in a fixed pattern called a *crystal*. The atoms within the crystal structure are held together by *convalent bonds*, which are created by interaction of the valence electrons of each atom.

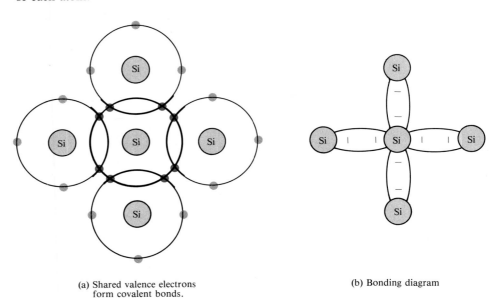

(a) Shared valence electrons   (b) Bonding diagram
form covalent bonds.

FIGURE 16–3
Covalent bonds in silicon.

Figure 16–3 shows how each silicon atom positions itself with four adjacent atoms. A silicon atom with its four valence electrons shares an electron with each of its four neighbors. This sharing of valence electrons produces the covalent bonds that hold the atoms together, because each shared electron is attracted equally by the two adjacent atoms which share it. Covalent bonding of a pure (intrinsic) silicon crystal is shown in Figure 16–4. Bonding for germanium is similar because it also has four valence electrons.

FIGURE 16–4
Covalent bonds in a pure
silicon crystal.

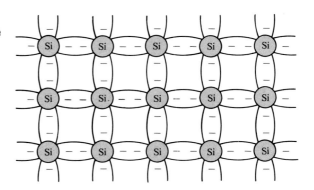

SECTION REVIEW
16–2

1. Name two semiconductor materials.
2. What is a covalent bond?

16–3

CONDUCTION IN SEMICONDUCTOR MATERIALS

As you have seen, the electrons of an atom can exist only within prescribed energy bands. Each shell around the nucleus corresponds to a certain energy band and is separated from adjacent shells by energy gaps, in which no electrons can exist. This condition, shown in Figure 16–5 for an unexcited silicon atom (no external energy), occurs only at absolute zero temperature.

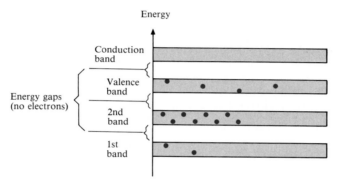

FIGURE 16–5
Energy band diagram for *unexcited* silicon atom.

Conduction Electrons and Holes

A pure silicon crystal at room temperature derives heat (thermal) energy from the surrounding air, causing some valence electrons to gain sufficient energy to jump the gap from the *valence band* into the *conduction band,* becoming free electrons. This situation is illustrated in the energy diagram of Figure 16–6(a) and in the bonding diagram of Figure 16–6(b).

When an electron jumps to the conduction band, a vacancy is left in the valence band. This vacancy is called a *hole*. For every electron raised to the conduction band by thermal or light energy, there is one hole left in the valence band, creating what is called an *electron-hole pair. Recombination* occurs when a conduction band electron loses energy and falls back into a hole in the valence band.

In summary, a piece of pure silicon at room temperature has, at any instant, a number of conduction band (free) electrons that are unattached to any atom and are essentially drifting randomly throughout the material. Also, an equal number of holes are created in the valence band when these electrons jump into the conduction band, as illustrated in Figure 16–7.

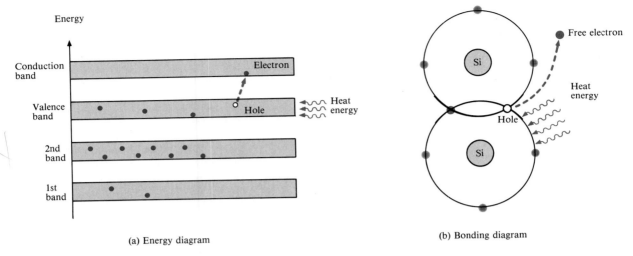

(a) Energy diagram

(b) Bonding diagram

FIGURE 16–6
Creation of an electron-hole pair in an *excited* silicon atom.

FIGURE 16–7
Electron-hole pairs in
a silicon crystal.

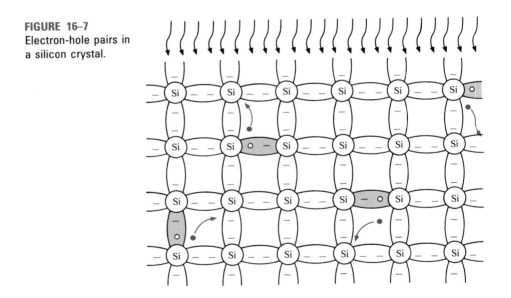

Germanium Versus Silicon

The situation in a germanium crystal is similar to that in silicon except that, because of its atomic structure, pure germanium has more free electrons than silicon and therefore a higher conductivity. Silicon, however, is the favored semiconductor material and is used far more widely than germanium. One reason for its wide usage is that silicon can be used at a much higher temperature than germanium.

Electron and Hole Current

When a voltage is applied across a piece of silicon, as shown in Figure 16–8, the free electrons in the conduction band are easily attracted toward the positive end. This move-

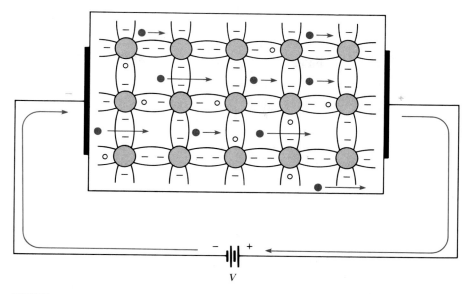

FIGURE 16–8
Free electron current in silicon.

ment of free electrons is one type of current in a semiconductor material, called *electron current.*

Another current mechanism occurs at the valence level, where the holes created by the free electrons exist. Electrons remaining in the valence band are still attached to their atoms and are not free to move randomly in the crystal structure. However, a valence electron can "fall" into a nearby hole, with little change in its energy level, thus leaving another hole where it came from. Effectively the hole has moved from one place to another in the crystal structure, as illustrated in Figure 16–9. This current is called *hole current.*

FIGURE 16–9
Hole current in silicon.

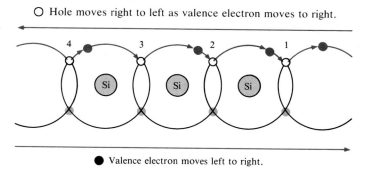

○ Hole moves right to left as valence electron moves to right.

● Valence electron moves left to right.

Semiconductors, Conductors, and Insulators

In a pure (intrinsic) semiconductor, there are relatively few free electrons, so neither silicon nor germanium is very useful in its intrinsic state. They are neither insulators nor good conductors because current in a material depends directly on the number of free electrons.

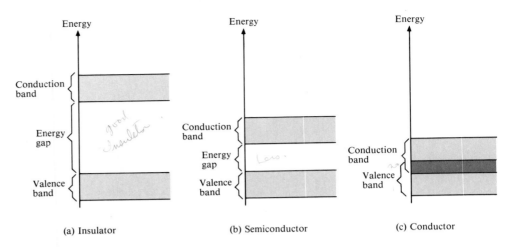

FIGURE 16–10
Energy diagrams for three categories of materials.

A comparison of the energy bands in Figure 16–10 for the three types of materials shows the essential differences among them regarding conduction. The energy gap for an insulator is so wide that hardly any electrons acquire enough energy to jump into the conduction band.

The valence band and the conduction band in a conductor (such as copper) overlap so that there are always many conduction electrons, even without the application of external energy. A semiconductor, as Figure 16–10 shows, has an energy gap that is much narrower than that in an insulator.

SECTION REVIEW 16–3

1. In the atomic structure of a semiconductor, within which energy band do free electrons exist? Valence electrons?
2. How are holes created in an intrinsic semiconductor?
3. Why is current established more easily in a semiconductor than in an insulator?

16–4

N-TYPE AND P-TYPE SEMICONDUCTORS

Intrinsic semiconductor materials do not conduct current very well because of the limited number of free electrons in the conduction band. Thus, the resistivity of a semiconductor is much greater than that of a conductor. For example, a one-cubic-centimeter sample of silver has a resistivity of 10^{-6} $\Omega \cdot$cm, whereas the resistivity is about 45 $\Omega \cdot$cm for pure germanium and several thousand $\Omega \cdot$cm for pure silicon.

Doping

The resistivities of silicon and germanium can be drastically reduced and controlled by the addition of *impurities* to the pure semiconductor material. This process, called *doping*, increases the number of current carriers (electrons or holes), thus increasing the conductivity and decreasing the resistivity. The two categories of impurities are *N-type* and *P-type*.

N-Type Semiconductor

To increase the number of conduction band electrons in pure silicon, *pentavalent* impurity atoms are added. These are atoms with five valence electrons, such as *arsenic, phosphorus,* and *antimony.*

As illustrated in Figure 16–11, each pentavalent atom (antimony, in this case) forms covalent bonds with four adjacent silicon atoms. Four of the antimony atom's valence electrons are used to form the covalent bonds, leaving *one extra electron.* This extra electron becomes a conduction electron because it is not attached to any atom. The number of conduction electrons can be controlled by the amount of impurity added to the silicon.

FIGURE 16–11
Pentavalent impurity atom in a silicon crystal. An antimony (Sb) impurity atom is shown in the center.

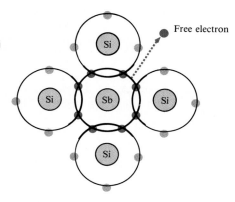

Since most of the current carriers are *electrons,* silicon (or germanium) doped in this way is an *N-type* semiconductor material where the *N* stands for the *negative* charge on an electron. The *electrons* are called the *majority carriers* in N-type material. Although the great majority of current carriers in N-type material are electrons, there are some holes. *Holes* in an N-type material are called *minority carriers.*

P-Type Semiconductor

To increase the number of holes in pure silicon, *trivalent* impurity atoms are added. These are atoms with three valence electrons, such as *aluminum, boron,* and *gallium.*

As illustrated in Figure 16–12, each trivalent atom (boron, in this case) forms covalent bonds with four adjacent silicon atoms. All three of the boron atom's valence

FIGURE 16–12
Trivalent impurity atom in a silicon crystal. A boron (B) impurity atom is shown in the center.

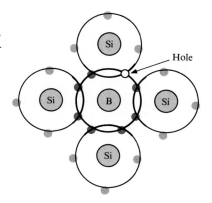

electrons are used in the covalent bonds; and, since four electrons are required, a *hole* is formed with each trivalent atom. The number of holes can be controlled by the amount of trivalent impurity added to the silicon.

Since most of the current carriers are *holes,* silicon (or germanium) doped in this way is called a *P-type* semiconductor material because holes can be thought of as positive charges. The *holes* are the *majority carriers* in P-type material. Although the great majority of current carriers in P-type material are holes, there are some electrons. *Electrons* in P-type material are called *minority carriers.*

SECTION REVIEW
16–4

1. How is an N-type semiconductor formed?
2. How is a P-type semiconductor formed?
3. What is a majority carrier?

16–5

PN JUNCTIONS

When a piece of silicon is doped so that half is N-type and the other half is P-type, a *PN junction* is formed between the two regions, as shown in Figure 16–13(a). This device is known as a *semiconductor diode*. The N region has many conduction electrons, and the P region has many holes, as shown in Figure 16–13(b). The PN junction is fundamental to the operation not only of diodes but also of transistors and other solid state devices.

FIGURE 16–13
Basic PN structure in a diode at the instant of junction formation.

The Depletion Layer

With no external voltage, the conduction electrons in the N region are aimlessly drifting in all directions. At the instant of junction formation, some of the electrons near the junction *drift* across into the P region and recombine with holes near the junction.

For each electron that crosses the junction and recombines with a hole, a pentavalent atom is left with a net positive charge in the N region near the junction, making it a *positive ion*. Also, when the electron recombines with a hole in the P region, a trivalent atom acquires net negative charge, making it a *negative ion*.

As a result of this recombination process, a large number of positive and negative ions builds up near the PN junction. As this buildup occurs, the electrons in the N region must overcome both the attraction of the positive ions and the repulsion of the negative ions in order to migrate into the P region. Thus, as the ion layers build up, the area on both sides of the junction becomes essentially depleted of any conduction electrons or holes and is known as the *depletion layer*. This condition is illustrated in Figure 16–14. When an equilibrium condition is reached, the depletion layer has widened to a point where no more electrons can cross the PN junction.

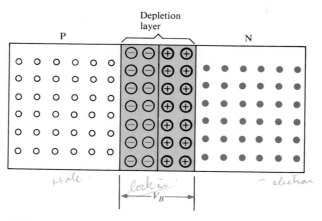

FIGURE 16–14
PN junction equilibrium condition in a diode.

The existence of the positive and negative ions on opposite sides of the junction creates a *barrier potential* (V_B) across the depletion layer, as indicated in Figure 16–14. At 25°C, the barrier potential is approximately 0.7 V for silicon and 0.3 V for germanium. As the junction temperature *increases*, the barrier potential *decreases*, and vice versa.

Energy Diagram of the PN Junction

Now, we will look at the operation of the PN junction in terms of its energy level. First consider the PN junction at the instant of its formation. The energy bands of the trivalent impurity atoms in the P-type material are at a slightly higher level than those of the pentavalent impurity atom in the N-type material, as shown in the graph of Figure 16–15. They are higher because the core attraction for the valence electrons (+3) in the trivalent atom is less than the core attraction for the valence electrons (+5) in the pentavalent atom. Thus, the trivalent valence electrons are in a slightly higher orbit and, thus, at a higher energy level.

Notice in Figure 16–15 that there is some overlap of the conduction bands in the P and N regions and also some overlap of the valence bands in the P and N regions. This overlap permits the electrons of higher energy near the top of the N-region conduction band to begin diffusing across the junction into the lower part of the P-region conduction band. As soon as an electron diffuses across the junction, it recombines with a hole in the valence band. As diffusion continues, the depletion layer begins to form. Also, the energy bands in the N region "shift" down as the electrons of higher energy are lost to diffusion. When the top of the N-region conduction band reaches the same level as the bottom of the P-region conduction band, diffusion ceases and the equilib-

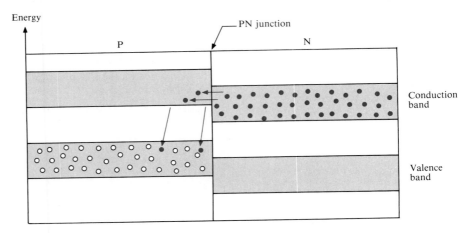

FIGURE 16–15
Graph of PN junction energy levels as diffusion begins.

rium condition is reached. This condition is shown in terms of energy levels in Figure 16–16. There is an *energy gradient* across the depletion layer rather than an abrupt change in energy level.

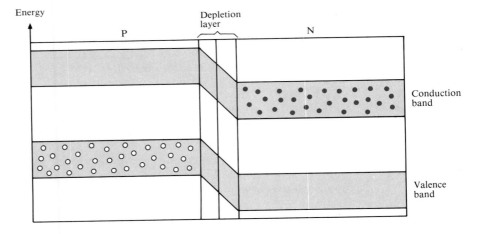

FIGURE 16–16
Graph of energy levels at equilibrium. The N-region bands are shifted down from their original positions when diffusion begins.

SECTION REVIEW 16–5

1. What is a PN junction?
2. When P and N regions are joined, a depletion layer forms. Describe the characteristics of the depletion layer.
3. The barrier potential for silicon is greater than for germanium (T or F).
4. What is the barrier potential for silicon at 25°C?

16–6

BIASING THE DIODE

As you have seen, there is no current across a PN junction at equilibrium. The primary usefulness of the PN junction diode is its ability to allow current in only *one* direction and to prevent current in the other direction as determined by the *bias*. There are two bias conditions for a PN junction: *forward* and *reverse*. Either of these conditions is created by application of an external voltage of the proper polarity.

Forward Bias

The term *bias* in electronics normally refers to a fixed dc voltage that sets the operating conditions for a semiconductor device. *Forward bias* is the condition that permits current across a PN junction. Figure 16–17 shows a dc voltage connected in a direction to forward-bias the diode. Notice that the negative terminal of the battery is connected to the N region (called the *cathode*), and the positive terminal is connected to the P region (called the *anode*).

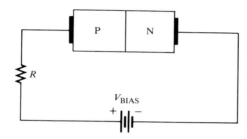

FIGURE 16–17
Forward-bias connection. The resistor limits the forward current in order to prevent damage to the diode.

A discussion of the basic operation of forward bias follows: The negative terminal of the battery pushes the conduction band electrons in the N region toward the junction, while the positive terminal pushes the holes in the P region also toward the junction. Recall from Chapter 2 that like charges repel each other.

When it overcomes the barrier potential, the external voltage source provides the N-region electrons with enough energy to penetrate the depletion layer and cross the junction, where they combine with the P-region holes. As electrons leave the N region, more flow in from the negative terminal of the battery. Thus, current through the N region is formed by the movement of conduction electrons (majority carriers) toward the junction.

Once the conduction electrons enter the P region and combine with holes, they become valence electrons. Then they move as valence electrons from hole to hole toward the positive connection of the battery. The movement of these valence electrons is the same as the movement of holes in the opposite direction. Thus, current in the P region is formed by the movement of holes (majority carriers) toward the junction. Figure 16–18 illustrates current in a forward-biased diode.

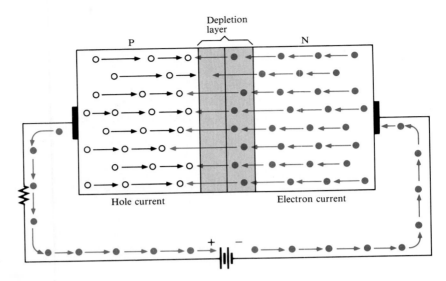

FIGURE 16–18
How current flows in a forward-biased diode.

Effect of Barrier Potential on Forward Bias

The barrier potential of the depletion layer can be thought of acting as a small battery that *opposes* bias, as illustrated in Figure 16–19. The resistances R_P and R_N represent the *bulk resistances* of the P and N materials.

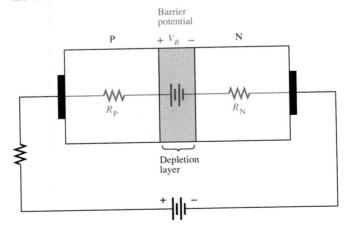

FIGURE 16–19
Barrier potential and bulk resistance equivalent for a PN junction diode.

The external bias voltage must overcome the barrier potential before the diode conducts, as illustrated in Figure 16–20. Conduction occurs at approximately 0.7 V for silicon and 0.3 V for germanium. Once the diode is conducting in the forward direction, the voltage drop across it remains at approximately the barrier potential and changes very little with changes in forward current (I_F) except for bulk resistance effects, as illustrated in Figure 16–20. The bulk resistances are usually only a few ohms and result in only a small voltage drop when the diode conducts. Often this drop can be neglected.

(a) When $V_{BIAS} < V_B$, the diode is essentially nonconducting.

(b) When $V_{BIAS} = V_B$, the diode begins to conduct.

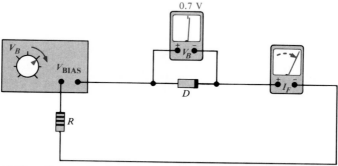

(c) When $V_{BIAS} > V_B$, the diode drop remains at approximately 0.7 V (silicon), but the forward current I_F continues to increase.

FIGURE 16–20
Illustration of diode operation under forward-bias conditions.

Energy Diagram for Forward Bias

Forward bias raises the energy levels of the conduction electrons in the N region, allowing them to move into the P region and combine with holes in the valence band. This condition is shown in Figure 16–21.

Energy

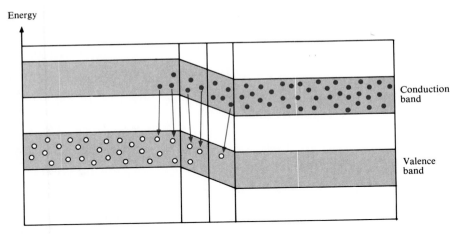

FIGURE 16–21
Energy diagram for forward bias, showing recombination in the depletion layer and in the P region as conduction electrons move across the junction.

Reverse Bias

Reverse bias is the condition that *prevents* current across the PN junction. Figure 16–22 shows a dc voltage source connected to reverse-bias the diode. Notice that the negative terminal of the battery is connected to the P region, and the positive terminal to the N region.

FIGURE 16–22
Reverse-bias connection.

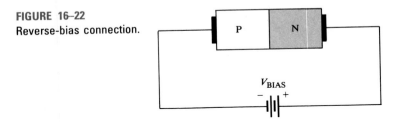

A discussion of the basic operation for reverse bias follows: The negative terminal of the battery attracts holes in the P region away from the PN junction, while the positive terminal also attracts electrons away from the junction. As electrons and holes move away from the junction, the depletion layer widens; more positive ions are created in the N region, and more negative ions are created in the P region, as shown in Figure 16–23(a).

The depletion layer widens until the potential difference across it equals the external bias voltage. At this point, the holes and electrons stop moving away from the junction, and majority current ceases, as indicated in Figure 16–23(b).

The initial flow of majority carriers away from the junction is called *transient current* and lasts only for a very short time upon application of reverse bias.

When the diode is reverse-biased, the depletion layer effectively acts as an insulator between the layers of oppositely charged ions, forming an effective capacitance, as illustrated in Figure 16–23(c). Since the depletion layer widens with increased

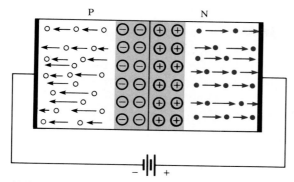

(a) Transient current as depletion layer widens

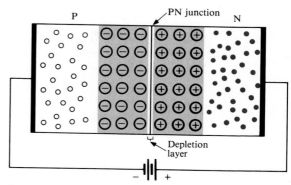

(b) Current ceases when barrier potential equals bias voltage

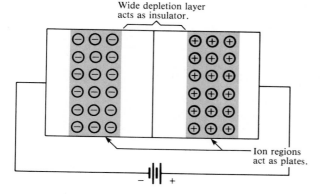

(c) Depletion layer widens as reverse bias increases

(d) Equivalent circuit with junction capacitance

FIGURE 16–23
Reverse bias.

reverse-biased voltage, the capacitance decreases, and vice versa. This internal capacitance is called the *depletion-layer capacitance* and can be represented by an equivalent circuit as shown in Figure 16–23(d).

Reverse Current

As you have learned, *majority current* very quickly becomes zero when reverse bias is applied. There is, however, a very small *current* produced by minority carriers during reverse bias. Germanium, as a rule, has a greater reverse current than silicon. This current is typically in the μA or nA range. A relatively small number of thermally produced electron-hole pairs exist in the depletion layer. Under the influence of the external voltage, some electrons manage to diffuse across the PN junction before recombination. This process establishes a small minority carrier current throughout the material.

The reverse current is dependent primarily on the junction temperature and not on the amount of reverse-biased voltage. A temperature increase causes an increase in reverse current.

Reverse Breakdown

If the external reverse-bias voltage is increased to a large enough value, *avalanche breakdown* occurs. The following describes what happens: Assume that one minority conduction band electron acquires enough energy from the external source to accelerate it toward the positive end of the diode. During its travel, it collides with an atom and imparts enough energy to knock a valence electron into the conduction band. There are now two conduction band electrons. Each will collide with an atom, knocking two more valence electrons into the conduction band. There are now four conduction band electrons which, in turn, knock four more into the conduction band. This rapid multiplication of conduction band electrons, known as an *avalanche effect,* results in a rapid build-up of reverse current.

Most diodes normally are not operated in reverse breakdown and can be damaged if they are. However, a particular type of diode (to be studied later), known as a *zener diode,* is specially designed for reverse-breakdown operation.

**SECTION REVIEW
16–6**

1. Name the two bias conditions.
2. Which bias condition produces majority carrier current?
3. Which bias condition produces a widening of the depletion region?
4. Minority carriers produce the current during avalanche breakdown (T or F).

16–7   DIODE CHARACTERISTICS

Diode Characteristic Curve

As you learned in the last section, a diode conducts current when it is forward-biased if the bias voltage exceeds the barrier potential, and the diode prevents current when it is reverse-biased at less than the breakdown voltage.

Figure 16–24 is a graph of diode current versus voltage. The upper right quadrant of the graph represents the *forward-biased condition.* As you can see, there is very little

FIGURE 16–24
Diode characteristic curve
(silicon).

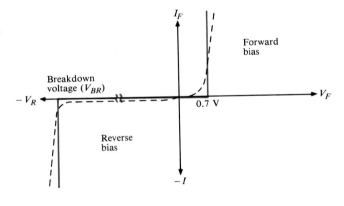

forward current (I_F) for forward voltages (V_F) below the barrier potential. As the forward voltage approaches the value of the barrier potential (0.7 V for silicon and 0.3 V for germanium), the current begins to increase. Once the forward voltage reaches the barrier potential, the current increases drastically and must be limited by a series resistor. *The voltage across the forward-biased diode remains approximately equal to the barrier potential.*

The lower left quadrant of the graph represents the *reverse-biased condition.* As the reverse voltage increases to the left, the current remains near zero until the breakdown voltage is reached. When breakdown occurs, there is a large reverse current which, if not limited, can destroy the diode. Typically, the breakdown voltage is greater than 50 V for most rectifier diodes. It should be noted here that most diodes should not be operated in reverse breakdown.

The solid curve in Figure 16–24 represents an *ideal diode.* In a practical diode, some forward conduction occurs before 0.7 V is reached. Also, there is a very small amount of reverse current before breakdown, as indicated by the dashed curve in the figure. In many applications, these nonideal characteristics can be neglected.

Diode Symbol

Figure 16–25(a) is the standard schematic symbol for a general-purpose diode. The arrow points in the direction opposite the electron flow. The two terminals of the diode are the *anode* and *cathode*.

When the anode is positive with respect to the cathode, the diode is *forward-biased* and current is from cathode to anode, as shown in Figure 16–25(b). Remember that when the diode is forward-biased, the barrier potential always appears between anode and cathode, as indicated in the figure. When the anode is negative with respect to the cathode, the diode is *reverse-biased,* as shown in Figure 16–25(c).

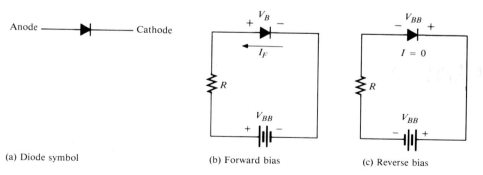

(a) Diode symbol   (b) Forward bias   (c) Reverse bias

FIGURE 16–25

Diode schematic symbol and bias circuits. V_{BB} is the bias battery voltage, and V_B is the barrier potential.

Some typical diodes are shown in Figure 16–26(a) to illustrate the variety of physical structures. Part (b) illustrates terminal identification.

(a) A variety of package types

Band
Cathode

Anode

Anode   Cathode

Anode   Cathode

Anode

Cathode
(case)

Anode

Anode

Cathode

(b) Examples of terminal identification

FIGURE 16–26
Typical diodes.

Testing the Diode with an Ohmmeter

The internal battery in certain ohmmeters will forward-bias or reverse-bias a diode, permitting a quick and simple check for proper functioning. This check works for 1.5-V ohmmeters such as VOMs but not for some of the digital multimeters. Many digital multimeters, however, have a diode test position.

To check the diode in the forward direction, connect the positive meter lead to the anode and the negative lead to the cathode, as shown in Figure 16–27(a). When a diode is forward-biased, its internal resistance is low (typically less than 100 Ω).

When the meter leads are reversed, as shown in Figure 16–27(b), the internal ohmmeter battery reverse-biases the diode, and a very large resistance value (ideally, infinite) is indicated. The PN junction is *shorted* if a low resistance is indicated in both bias conditions; it is *open* if a very high resistance is read for both checks.

Diode Approximations

The simplest way to visualize diode operation is to think of it as a *switch*. When forward-biased, the diode ideally acts as a closed (on) switch, and when reverse-biased,

(a) Typical foward check   (b) Typical reverse check

FIGURE 16–27
Checking a semiconductor diode with an ohmmeter.

it acts as an open (off) switch, as shown in Figure 16–28. The characteristic curve for this approximation is also shown. Note that the forward voltage and the reverse current are always zero in the ideal case.

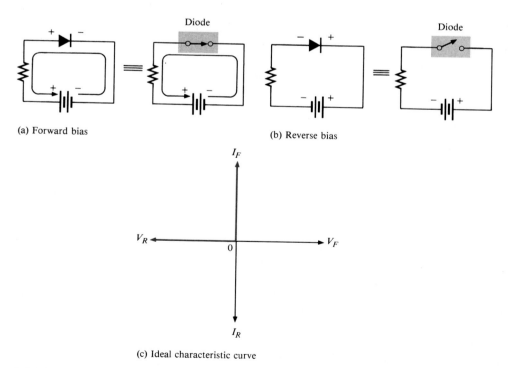

(a) Forward bias   (b) Reverse bias

(c) Ideal characteristic curve

FIGURE 16–28
Ideal approximation of the diode as a switch.

This ideal model, of course, neglects the effect of the barrier potential, the internal resistances, and other parameters. However, in many cases it is accurate enough.

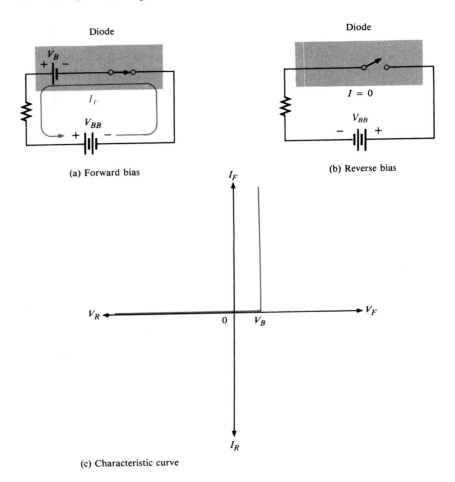

(a) Forward bias

(b) Reverse bias

(c) Characteristic curve

FIGURE 16–29
Diode approximation including barrier potential.

The next higher level of accuracy is the barrier potential model. In this approximation, the forward-biased diode is represented as a closed switch in series with a small "battery" equal to the barrier potential V_B (0.7 V for Si and 0.3 V for Ge), as shown in Figure 16–29(a). The positive end of the battery is toward the anode. Keep in mind that the barrier potential is not a voltage source and cannot be measured with a voltmeter; rather it only has the *effect* of a battery when forward bias is applied because the forward bias voltage, V_{BB}, must overcome the barrier potential before the diode begins to conduct current. The reverse-biased diode is represented by an open switch, as in the ideal case, because the barrier potential does not affect reverse bias, as shown in Figure 16–29(b). The characteristic curve for this ideal model is shown in part (c). In this book, the barrier potential is included in analysis unless otherwise stated.

One more level of accuracy will be considered at this point. Figure 16–30(a) shows the forward-biased diode model with both the *barrier potential* and the low *forward (bulk) resistance*. Figure 16–30(b) shows how the high *reverse resistance* affects the reverse-biased model. The characteristic curve is shown in part (c).

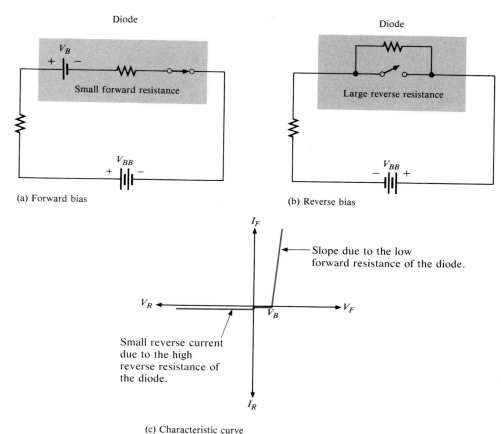

(a) Forward bias

(b) Reverse bias

(c) Characteristic curve

FIGURE 16–30
Diode approximation including barrier potential, forward resistance, and reverse leakage resistance.

Other parameters such as junction capacitance and breakdown voltage become important only under certain operating conditions and will be considered only where appropriate.

**SECTION REVIEW
16–7**

1. Sketch a rectifier diode symbol and label the terminals.
2. For a normal diode, the forward resistance is quite low and the reverse resistance is very high (T or F).
3. An open switch ideally represents a _____ -biased diode. A closed switch ideally represents a _____ -biased diode.

SUMMARY

- ☐ An atom is described as consisting of a nucleus containing protons and neutrons orbited by electrons.
- ☐ Protons are positive, neutrons are neutral, and electrons are negative.
- ☐ Atomic shells are energy bands.
- ☐ The outermost shell containing electrons is the valence shell.
- ☐ Silicon and germanium are semiconductor materials. Silicon is the most predominant.
- ☐ Atoms within a crystal structure are held together with covalent bonds.
- ☐ Electron-hole pairs are thermally produced.
- ☐ The process of adding impurities to an intrinsic (pure) semiconductor to increase and control conductivity is called *doping*.
- ☐ A P-type semiconductor is doped with trivalent impurity atoms.
- ☐ An N-type semiconductor is doped with pentavalent impurity atoms.
- ☐ The depletion layer is a region adjacent to the PN junction containing no majority carriers.
- ☐ Forward bias permits majority carrier current through the PN junction.
- ☐ Reverse bias prevents majority carrier current.
- ☐ A PN structure is called a *diode*.
- ☐ Reverse leakage current is due to thermally produced electron-hole pairs.
- ☐ Reverse breakdown occurs when the reverse-biased voltage exceeds a specified value.

GLOSSARY

Anode   The more positive terminal of a diode or other electronic device.

Atom   The smallest particle of an element possessing the unique characteristics of that element.

Barrier potential   The inherent voltage across the depletion layer of a PN junction.

Bias   The application of dc voltage to a diode or other electronic device to produce a desired mode of operation.

Cathode   The more negative terminal of a diode or other electronic device.

Covalent   Related to the bonding of two or more atoms by the interaction of their valence electrons.

Crystal   The pattern or arrangement of atoms forming a solid material.

Diode   An electronic device that permits current in only one direction.

Discrete device   An individual electrical or electronic component that must be used in combination with other components to form a complete functional circuit.

Doping   The process of imparting impurities to an intrinsic semiconductor material in order to control its conduction characteristics.

Electron   The basic particle of negative electrical charge in matter.

Energy   The ability to do work.

Free electron   A valence electron that has broken away from its parent atom and is free to move from atom to atom within the atomic structure of a material.

Germanium   A semiconductor material.

Hole   The absence of an electron.

Intrinsic semiconductor   A pure material with relatively few free electrons.

Ion   An atom that has gained or lost a valence electron resulting in a net positive or negative charge.

Ionization   The removal or addition of an electron from or to an atom so that the resulting atom (ion) has a net positive or negative.

PN junction The boundary between N-type and P-type materials.

Recombination The process of a free electron falling into a hole in the valence band of an atom.

Semiconductor A material that has a conductance value between that of a conductor and that of an insulator. Silicon and germanium are examples.

Silicon A semiconductor material used in diodes and transistors.

Terminal An external contact point on an electronic device.

Valence Related to the outer shell or orbit of an atom.

SELF-TEST

1. The charge on electrons is
   (a) positive   (b) negative   (c) neutral   (d) variable

2. The number of valence electrons in both silicon and germanium is
   (a) two   (b) eight   (c) four   (d) eighteen

3. When an atom loses or gains a valence electron, the atom becomes
   (a) covalent   (b) neutral   (c) a crystal   (d) an ion

4. Atoms within a crystal are held together by
   (a) atomic glue   (b) subatomic particles   (c) covalent bonds
   (d) the valence band

5. Free electrons exist in the
   (a) valence band   (b) conduction band   (c) lowest band
   (d) recombination band

6. A hole is
   (a) a vacancy in the valence band left by an electron
   (b) a vacancy in the conduction band
   (c) a positive electron
   (d) a conduction band electron

7. The widest energy gap between the valence band and the conduction band occurs in
   (a) semiconductors   (b) insulators   (c) conductors   (d) a vacuum

8. The process of adding impurity atoms to a pure semiconductor material is called
   (a) recombination   (b) crystalization   (c) bonding   (d) doping

9. The two types of current in a semiconductor material are called
   (a) positive current and negative current
   (b) electron current and conventional current
   (c) electron current and hole current
   (d) forward current and reverse current

10. The majority carriers in an N-type semiconductor are
    (a) electrons   (b) holes   (c) positive ions   (d) negative ions

11. The PN junction is found in
    (a) semiconductor diodes
    (b) transistors
    (c) all semiconductor materials
    (d) a and b

12. In a semiconductor diode, the region near the PN junction consisting of positive and negative ions is called the
    (a) neutral zone   (b) recombination region   (c) depletion layer
    (d) diffusion area

13. A fixed dc voltage that sets the operating condition of a semiconductor device is called the
    (a) bias   (b) depletion voltage   (c) battery   (d) barrier potential
14. In a semiconductor diode, the two bias conditions are
    (a) positive and negative   (b) blocking and nonblocking
    (c) open and closed   (d) forward and reverse
15. When a diode is forward-biased, it is
    (a) blocking current   (b) conducting current
    (c) similar to an open switch
    (d) similar to a closed switch
    (e) a and c   (f) b and d
16. The voltage across a forward-biased silicon diode is approximately
    (a) 0.7 V   (b) 0.3 V   (c) 0 V
    (d) dependent on the bias voltage
17. In Figure 16–31, identify the forward-biased diode(s).
    (a) D_1 (b) D_2 (c) D_3
    (d) D_1 and D_3 (e) D_2 and D_3

(a)   (b)   (c)

FIGURE 16–31

18. When the positive lead of an ohmmeter is connected to the cathode of a diode and the negative lead is connected to the anode, the meter reads
    (a) a very low resistance   (b) an infinitely high resistance
    (c) a high resistance initially, decreasing to about 100 Ω
    (d) a gradually increasing resistance

ANSWERS TO SECTION REVIEWS

Section 16–1
1. The smallest particle that retains the characteristics of its element.
2. Electrons, protons, neutrons.
3. An electron in the outermost shell (valence band).
4. An electron in the conduction band.

Section 16–2
1. Silicon, germanium.   2. The sharing of electrons with neighboring atoms.

Section 16–3
1. Conduction band, valence band.
2. An electron is thermally raised to the conduction band, leaving a hole in the valence band.
3. The energy gap between the valence band and the conduction band is narrower for a semiconductor.

Section 16–4
1. By the addition of pentavalent atoms to the intrinsic semiconductor material.
2. By the addition of trivalent atoms to the intrinsic semiconductor material.
3. The particle in greatest abundance: electrons in N-type material and holes in P-type material.

Section 16–5
1. The boundary between N-type and P-type materials.
2. Devoid of majority carriers, contains only positive and negative ions.   3. T.
4. 0.7 V.

Section 16–6
1. Forward, reverse.   2. Forward.   3. Reverse.   4. T.

Section 16–7
1. See Figure 16–32.   2. T.   3. Reverse, forward.

FIGURE 16–32   Anode   Cathode

17 Diodes and Applications

This chapter discusses the applications of diodes in converting ac to dc by the process known as *rectification*. Half-wave and full-wave rectification are introduced, and the basic circuits are studied. The limitations of diodes used in rectifier applications are examined, and diode clipping circuits and dc restoring (clamping) circuits are studied.

In addition to rectifier diodes, zener diodes and their applications in voltage regulation are introduced. Varactor diodes, light-emitting diodes, and photodiodes and their applications also are discussed.

After completing this chapter, you will be able to

☐ Recognize a half-wave rectifier circuit and explain how it works.
☐ Determine the average value of a half-wave rectified signal.
☐ Recognize a center-tapped full-wave rectifier circuit and explain how it works.
☐ Recognize a full-wave bridge rectifier and explain how it works.
☐ Determine the average value of a full-wave rectified signal.
☐ Determine the PIV (peak inverse voltage) across the diodes in full-wave rectifiers.
☐ Use a capacitor-input filter to smooth out a rectified voltage.
☐ Define *ripple voltage* and explain what causes it.
☐ Select a filter capacitor to minimize ripple voltage.
☐ Recognize other types of power supply filters.
☐ Use the zener diode as a voltage regulator.
☐ Explain *line regulation* and *load regulation*.
☐ Explain how a varactor diode is used as a variable capacitor.
☐ Discuss the basic principles of light-emitting diodes (LEDs) and photodiodes.

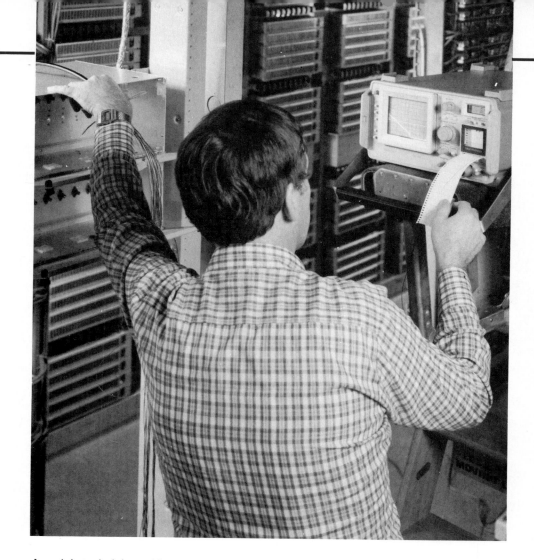

APPLICATION ASSIGNMENT

As a lab technician with a manufacturer of electronic test equipment, you are assigned to a group that is responsible for the development of a new signal generator. Today, your boss asks you to build a simple circuit that will regulate a dc voltage that can vary from 8 V to 12 V to maintain a nominal 6.8-V output. The 6.8 V is to be used as a reference voltage, so the current requirement is less than 1 mA. He tells you no more. After studying this chapter, you will know what kind of device to use and how to build the circuit. You will also know what additional information you may need.

17-1

HALF-WAVE RECTIFIERS

Because of their unique ability to conduct current in only one direction, diodes are used in rectifier circuits. *Rectification* is the process of converting ac to pulsating dc.

In Figure 17–1(a), an ac source is connected to a load resistor through a diode. Let's examine what happens during one cycle of the input voltage.

When the sine wave input voltage goes *positive*, the diode is *forward-biased* and conducts current to the load resistor, as shown in part (b). The current produces a voltage across the load which has the same shape as the positive half-cycle of the input voltage. When the input voltage goes *negative* during the second half of its cycle, the diode is *reversed-biased*. There is no current, so the voltage across the load resistor is 0, as shown in part (c). The net result is that only the positive half-cycles of the ac input voltage appear across the load, making the output a *pulsating dc voltage*, as shown in part (d). This process is called *half-wave rectification*.

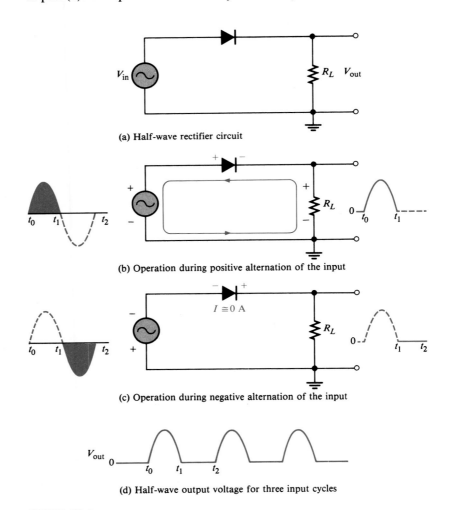

(a) Half-wave rectifier circuit

(b) Operation during positive alternation of the input

(c) Operation during negative alternation of the input

(d) Half-wave output voltage for three input cycles

FIGURE 17–1
Operation of half-wave rectifier.

Average Value of Half-Wave Output

The average value is the value that would be indicated by a dc voltmeter. It can be calculated with the following equation where V_p is the peak value:

$$V_{AVG} = \frac{V_p}{\pi} \qquad (17\text{--}1)$$

Figure 17–2 shows the half-wave voltage with its average value indicated by the colored dashed line.

FIGURE 17–2
Average value of half-wave rectified signal.

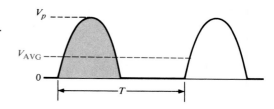

EXAMPLE 17–1

What is the average (dc) value of the half-wave rectified voltage waveform in Figure 17–3?

FIGURE 17–3

Solution

$$V_{AVG} = \frac{V_p}{\pi} = \frac{100 \text{ V}}{\pi} = 31.83 \text{ V}$$

Effect of Barrier Potential on Half-Wave Rectifier Output

In the previous discussion, the diode was considered ideal. When the diode barrier potential is taken into account, here is what happens: During the positive half-cycle, the input voltage must overcome the barrier potential before the diode becomes forward-biased. For a silicon diode this results in a half-wave output with a peak value that is 0.7 V less than the peak value of the input (0.3 V less for a germanium diode), as shown in Figure 17–4. The expression for *peak* output voltage is

$$V_{p(out)} = V_{p(in)} - 0.7 \text{ V} \qquad (17\text{--}2)$$

In working with diode circuits, you will find that it is often practical to neglect the effect of barrier potential when the peak value of the applied voltage is much greater (at least ten times) than the barrier potential. As mentioned before, we will always use silicon diodes and consider the barrier potential unless stated otherwise.

FIGURE 17–4
Effect of barrier potential on half-wave rectified output voltage (silicon diode shown).

EXAMPLE 17–2

Determine the peak output voltage of the silicon rectifier circuit in Figure 17–5 for the indicated input voltage.

FIGURE 17–5

Solution
The peak half-wave output voltage is

$$V_p = 5 \text{ V} - 0.7 \text{ V} = 4.3 \text{ V}$$

Peak Inverse Voltage (PIV)

The maximum value of reverse voltage, sometimes designated as *peak inverse voltage* (PIV), occurs at the peak of the *negative alternation* of the input cycle when the diode is *reverse-biased*. This condition is illustrated in Figure 17–6. The PIV equals the peak value of the input voltage, and the diode must be capable of withstanding this amount of repetitive reverse voltage.

FIGURE 17–6
The PIV occurs at the peak of the half-cycle when the diode is reverse-biased. In this circuit, the PIV occurs at the peak of the negative half-cycle.

1. At what point on the input cycle does the PIV occur?
2. For a half-wave rectifier, there is current through the load for approximately what percentage of the input cycle?
3. What is the average value of the voltage shown in Figure 17–7?

FIGURE 17–7   10 V

17–2

FULL-WAVE RECTIFIERS

The difference between full-wave and half-wave rectification is that a full-wave rectifier allows unidirectional current to the load during the entire input cycle, and the half-wave rectifier allows this only during one half of the cycle. The result of full-wave rectification is a dc output voltage that pulsates every half-cycle of the input, as shown in Figure 17–8.

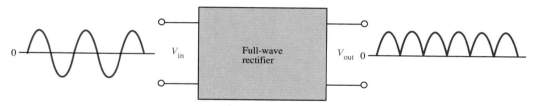

FIGURE 17–8
Full-wave rectification.

The average value for a full-wave rectified voltage is twice that of the half-wave, expressed as follows:

$$V_{avg} = \frac{2V_p}{\pi}$$

(17–3)

EXAMPLE 17–3

Find the average value of the full-wave rectified voltage in Figure 17–9.

FIGURE 17–9   15 V

Solution

$$V_{avg} = \frac{2V_p}{\pi} = \frac{2(15 \text{ V})}{\pi} = 9.55 \text{ V}$$

DEVICES

Center-Tapped Full-Wave Rectifier

This type of full-wave rectifier circuit uses two diodes connected to the secondary of a center-tapped transformer, as shown in Figure 17–10. The input signal is coupled through the transformer to the secondary. Half of the secondary voltage appears between the center tap and each end of the secondary winding as shown.

FIGURE 17–10
A center-tapped full-wave rectifier.

For a positive half-cycle of the input voltage, the polarities of the secondary voltages are as shown in Figure 17–11(a). *This condition forward-biases the upper diode D_1 and reverse-biases the lower diode D_2.* The current path is through D_1 and the load resistor, as indicated.

(a) During positive half-cycles, D_1 is forward-biased and D_2 is reverse-biased.

(b) During negative half-cycles, D_2 is forward-biased and D_1 is reverse biased.

FIGURE 17–11
Basic operation of a center-tapped full-wave rectifier. Note that the current through the load resistor is in the same direction during the entire input cycle.

For a negative half-cycle of the input voltage, the voltage polarities on the secondary are as shown in Figure 17–11(b). This condition *reverse-biases D_1 and forward-biases D_2*. The current path is through D_2 and the load resistor, as indicated.

Because the current during both the positive and the negative portions of the input cycle is in the *same direction* through the load, the output voltage developed across the load is a full-wave rectified dc voltage.

Effect of the Turns Ratio on Full-Wave Output Voltage

If the turns ratio of a transformer is 1, the peak value of the rectified output voltage equals half the peak value of the primary input voltage less the barrier potential (diode drop). This value occurs because half of the input voltage appears across each half of the secondary winding.

In order to obtain an output voltage equal to the input (less the barrier potential, you must use a step-up transformer with a turns ratio of 1 to 2 (1:2). In this case, the total secondary voltage is twice the primary voltage, so the voltage across each half of the secondary is equal to the input.

Peak Inverse Voltage (PIV)

Each diode in the full-wave rectifier is alternately forward-biased and then reverse-biased. The maximum reverse voltage that each diode must withstand is the peak value of the total secondary voltage (V_s), as illustrated in Figure 17–12. When the total secondary voltage V_s has the polarity shown, the anode of D_1 is $+V_s/2$ and the anode of D_2 is $-V_s/2$. Since D_1 is forward-biased, its cathode is at the same voltage as its anode ($+V_s/2$, neglecting the barrier potential); this is also the voltage on the cathode of D_2. The total reverse voltage across D_2 therefore is $+V_s/2 - (-V_s/2) = V_s/2 + V_s/2 = V_s$. Since $V_s = 2V_{out}$, the *peak inverse voltage* across either diode in the center-tapped full-wave rectifier is

$$PIV = 2V_{p(out)} \tag{17–4}$$

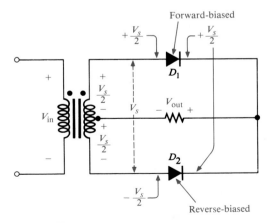

FIGURE 17–12
PIV across a diode is twice the peak value of the output voltage.

EXAMPLE 17–4

(a) Show the voltage waveforms across the secondary winding and across R_L when a 25-V peak sine wave is applied to the primary winding in Figure 17–13. This circuit is the same as the one in Figure 17–10, except the center tap and one end of R_L are connected to ground.

(b) What minimum PIV rating must the diodes have?

FIGURE 17–13

Solution

(a) The waveforms are shown in Figure 17–14.

(b) The total peak secondary voltage is

$$V_s = \left(\frac{N_2}{N_1}\right)V_{in} = (2)25 \text{ V} = 50 \text{ V}$$

FIGURE 17–14

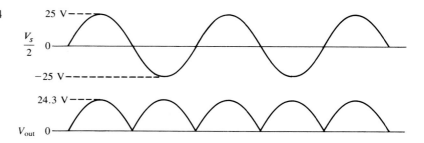

There is a 25-V peak across each half of the secondary. The output load voltage has a peak value of 25 V, less the 0.7-V drop across the diode. Each diode must have a minimum PIV rating of 50 V − 0.7 V = 49.3 V.

Full-Wave Bridge Rectifier

The full-wave bridge rectifier uses four diodes, as shown in Figure 17–15. When the input cycle is positive as in part (a), diodes D_1 and D_2 are forward-biased and conduct current in the direction shown. A voltage is developed across R_L which looks like the positive half of the input cycle. During this time, diodes D_3 and D_4 are reverse-biased.

When the input cycle is negative, as in part (b), diodes D_3 and D_4 are forward-biased and conduct current in the same direction through R_L as during the positive half-cycle. During the negative half-cycle, D_1 and D_2 are reverse-biased. A full-wave rectified output voltage appears across R_L as a result of this action.

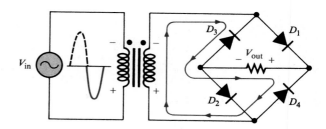

(a) During positive half-cycle of the input, D_1 and D_2 are forward-biased and conduct current. D_3 and D_4 are reverse-biased.

(b) During negative half-cycle, D_3 and D_4 are forward-biased and conduct current. D_1 and D_2 are reverse-biased.

FIGURE 17–15
Operation of full-wave bridge rectifier.

Bridge Output Voltage

During the positive half-cycle of the secondary voltage, diodes D_1 and D_2 are forward-biased. Neglecting the diode drops, the total secondary voltage V_s appears across the load resistor. The same is true when D_3 and D_4 are forward-biased during the negative half-cycle. Thus,

$$V_{\text{out}} = V_s \qquad \text{(17–5)}$$

As you can see in Figure 17–15, *two* diodes are always in series with the load resistor during both the positive and the negative half-cycles. If these diode drops are taken into account, the output voltage (with silicon diodes) is

$$V_{\text{out}} = V_s - 1.4 \text{ V} \qquad \text{(17–6)}$$

Peak Inverse Voltage (PIV)

Assuming that D_1 and D_2 are forward-biased, let us examine the reverse voltage across D_3 and D_4. Visualizing D_1 and D_2 as shorts (ideally), as in Figure 17–16, you can see that D_3 and D_4 have a peak inverse voltage equal to the peak secondary voltage. Since the output voltage (neglecting diode drops) is equal to the secondary voltage, we have

$$\text{PIV} \cong V_{p(\text{out})} \qquad \text{(17–7)}$$

The PIV rating of the bridge diodes is half that required for the center-tapped configuration.

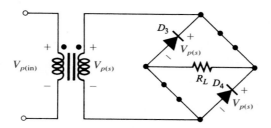

FIGURE 17–16
PIV in a bridge rectifier during the positive half-cycle of the input voltage.

EXAMPLE 17–5

(a) Determine the output voltage for the bridge rectifier in Figure 17–17.
(b) What minimum PIV rating is required for the silicon diodes?

FIGURE 17–17

Solution

(a) The peak output voltage is (taking into account the two diode drops)

$$V_{p(out)} = V_s - 1.4 \text{ V} = \left(\frac{N_2}{N_1}\right)V_{p(in)} - 1.4 \text{ V} = \left(\frac{1}{2}\right)24 \text{ V} - 1.4 \text{ V}$$
$$= 12 \text{ V} - 1.4 \text{ V} = 10.6 \text{ V}$$

(b) The PIV for each diode is

$$\text{PIV} \cong V_{p(out)} = 10.6 \text{ V}$$

SECTION REVIEW 17–2

1. What is the average value of a full-wave rectified voltage with a peak value of 60 V?
2. Which type of full-wave rectifier has the greater output voltage for the same input voltage and transformer turns ratio?
3. For a given output voltage, the PIV for bridge rectifier diodes is less than for center-tapped rectifier diodes (T or F).

17–3

RECTIFIER FILTERS

In most power supply applications, the standard 60-Hz ac power line voltage must be converted to a sufficiently constant dc voltage. The 60-Hz pulsating dc output of a half-wave rectifier or the 120-Hz pulsating output of a full-wave rectifier must be *filtered* to virtually eliminate the large voltage variations. Figure 17–18 illustrates the filtering

FIGURE 17–18
Basic block diagram of a power supply that converts 60-Hz ac to dc.

concept showing a smooth dc output voltage. The full-wave rectifier voltage is applied to the filter's input, and, ideally, a constant dc level appears on the output.

Capacitor-Input Filter

A half-wave rectifier with a capacitor-input filter is shown in Figure 17–19. We will use the half-wave rectifier to illustrate the filtering principle; then we will expand the concept to the full-wave rectifier.

During the positive first quarter-cycle of the input, the diode is forward-biased, allowing the capacitor to charge to within a diode drop of the input peak, as illustrated in Figure 17–19(a). When the input begins to decrease below its peak, as shown in part (b), the capacitor retains its charge and the diode becomes reverse-biased. During the remaining part of the cycle, the capacitor can discharge only through the load resistance at a rate determined by the R_LC time constant. The larger the time constant, the less the capacitor will discharge.

(a) Initial charging of capacitor (diode is forward-biased).

(b) Capacitor discharges through R_L after peak of the positive alternation (diode is reverse-biased).

(c) Capacitor charges back to the peak (diode is forward-biased).

FIGURE 17–19
Operation of a half-wave rectifier with a capacitor-input filter.

During the first quarter of the next cycle, the diode again will become forward-biased when the input voltage exceeds the capacitor voltage by approximately a diode drop, as illustrated in part (c).

Ripple

As you have seen, the capacitor quickly charges at the beginning of a cycle and slowly discharges after the positive peak (when the diode is reverse-biased). The variation in the output voltage due to the charging and discharging is called the *ripple*. The smaller the ripple, the better the filtering action, as illustrated in Figure 17–20.

FIGURE 17–20
Half-wave ripple voltage (colored line).

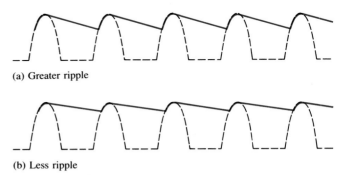

(a) Greater ripple

(b) Less ripple

For a given input frequency, the output frequency of a full-wave rectifier is twice that of a half-wave rectifier. As a result, a full-wave rectifier is easier to filter. When filtered, the full-wave rectified voltage has less ripple than does a half-wave signal for the *same* load resistance and capacitor values. Less ripple occurs because the capacitor discharges less during the shorter interval between full-wave pulses, as shown in Figure 17–21. A good rule of thumb for effective filtering is to make $R_L C \geq 10T$, where T is the period of the rectified voltage.

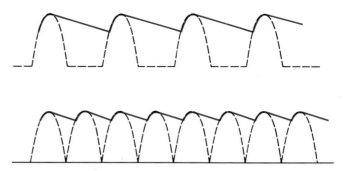

FIGURE 17–21
Comparison of ripple voltages for half-wave and full-wave signals with same filter and same input frequency.

The *ripple factor (r)* is an indication of the effectiveness of the filter and is defined as the ratio of the ripple voltage (V_r) to the dc (average) value of the filter output voltage (V_{dc}).

FIGURE 17–22
V_r and V_{dc} determine the ripple factor.

These factors are illustrated in Figure 17–22.

$$r = \left(\frac{V_r}{V_{dc}}\right)100\% \qquad (17\text{–}8)$$

The *lower* the ripple factor, the *better* the filter. The ripple factor can be decreased by increasing the value of the filter capacitor.

Surge Current in the Capacitor Filter

Before the switch in Figure 17–23(a) is closed, the filter capacitor is uncharged. At the instant the switch is closed, voltage is connected to the bridge and the capacitor appears as a short, as shown. An initial surge of current is produced through the two forward-biased diodes. The worst-case situation occurs when the switch is closed at a peak of the secondary voltage and a maximum surge current $I_{S(max)}$ is produced, as illustrated in the figure.

It is possible that the surge current could destroy the diodes, and for this reason a surge-limiting resistor is sometimes connected, as shown in Figure 17–23(b). The value

FIGURE 17–23
Surge current in a capacitor-input filter.

of this resistor must be small compared to R_L. Also, the diodes must have a forward current rating such that they can withstand the momentary surge of current.

Inductor-Input Filter

When a choke is added to the filter input, as in Figure 17–24, a reduction in the ripple voltage is achieved. The choke has a high reactance at the ripple frequency, and the capacitive reactance is low compared to both X_L and R_L. The two reactances form an ac voltage divider that tends to significantly reduce the ripple from that of a straight capacitor-input filter.

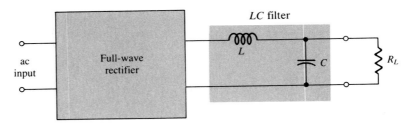

FIGURE 17–24
Rectifier with an *LC* filter.

It should be noted at this time that an inductor-input filter produces an output with a dc value approximately equal to the *average* value of the rectified input. The capacitor-input filter, however, produces an output with a dc value approximately equal to the *peak* value of the input.

Another point of comparison is that the amount of ripple voltage in the capacitor-input filter varies inversely with the load resistance. Ripple voltage in the *LC* filter is essentially independent of the load resistance and depends only on X_L and X_C, as long as X_C is sufficiently less than R_L.

π-Type and T-type Filter

A one-section π-type filter is shown in Figure 17–25(a). It can be thought of as a capacitor filter followed by an *LC* filter. The T-type filter in part (b) is basically an *LC*

FIGURE 17–25
π-type and T-type *LC* filters.

(a) π-type

(b) T-type

filter followed by an inductor filter. They combine the peak filtering action of the single-capacitor or inductor filter with the reduced ripple and load independence of the *LC* filter.

1. A 60-Hz sine wave is applied to a half-wave rectifier. What is the output frequency? What is the output frequency for a full-wave rectifier?
2. What causes the ripple voltage on the output of a capacitor-input filter?
3. The load resistance of a capacitor-filtered full-wave rectifier is reduced. What effect does this reduction have on the ripple voltage?
4. Name one advantage of an *LC* filter over a capacitor filter. Name one disadvantage.

17–4

TROUBLESHOOTING RECTIFIER CIRCUITS

Several types of failures can occur in power supply rectifiers. In this section we will examine some possible failures and the effects they would have on a circuit's operation.

Open Diode

A half-wave rectifier with a diode that has opened (a common failure mode) is shown in Figure 17–26. In this case, you would measure 0 V dc across the load resistor, as depicted.

FIGURE 17–26
Test for an open diode in a half-wave rectifier with capacitor-input filter.

Now consider the full wave, center-tapped rectifier in Figure 17–27. Assume that diode D_1 has failed open. With an oscilloscope connected to the output, as shown in part (a), you would observe the following: You would see a larger-than-normal ripple voltage at a frequency of 60 Hz rather than 120 Hz. Disconnecting the filter capacitor, you would observe a *half-wave* rectified voltage, as in part (b). Now let's examine the reason for these observations. If diode D_1 is open, there will be current through R_L only during the negative half-cycle of the input signal. During the positive half-cycle, an open path prevents current through R_L. The result is a half-wave voltage, as illustrated.

With the filter capacitor in the circuit, the half-wave signal will allow it to discharge more than it would with a normal full-wave signal, resulting in a larger rip-

(a) Ripple should be less and have a frequency of 120 Hz.

(b) With C removed, output should be a full-wave 120-Hz signal.

FIGURE 17–27
Symptoms of an open diode in a full-wave, center-tapped rectifier.

ple voltage. Basically, the same observations would be made for an open failure of diode D_2.

An open diode in a bridge rectifier would create symptoms identical to those just discussed for the center-tapped rectifier. As illustrated in Figure 17–28, the open diode

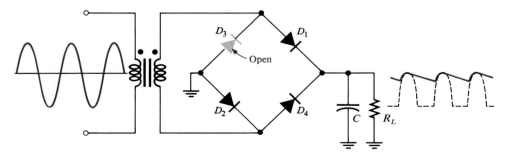

FIGURE 17–28
Effects of an open diode in a bridge rectifier.

would prevent current through R_L during half of the input cycle (in this case, the negative half). As a result, there would be a half-wave output and an increased ripple voltage at 60 Hz, as discussed before.

Shorted Diode

A shorted diode is one that has failed such that it has a very low resistance in both directions. If a diode suddenly became shorted in a bridge rectifier, it is likely that a sufficiently high current would exist during one half of the input cycle such that the shorted diode itself would burn open or the other diode in series with it would open. The transformer could also be damaged, as illustrated in Figure 17–29 with D_1 shorted.

(a) Positive half-cycle: The shorted diode acts as a forward-biased diode, so the load current is normal.

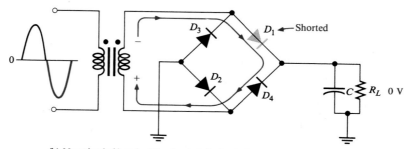

(b) Negative half-cycle: The shorted diode produces a short circuit across the source. As a result D_1, D_4, or the transformer secondary will probably burn open.

FIGURE 17–29
Effects of a shorted diode in a bridge rectifier.

In part (a) of the figure, current is supplied to the load through the shorted diode during the first positive half-cycle, just as though it were forward-biased. During the negative half-cycle, the current is shorted through D_1 and D_4, as shown in part (b). Again, damage to the transformer is possible. It is likely that this excessive current would burn either or both of the diodes open. If only one of the diodes opened, you would still observe a half-wave voltage on the output. If both diodes opened, there would be no voltage developed across the load. These conditions are illustrated in Figure 17–30.

(a) One open diode produces a half-wave output.

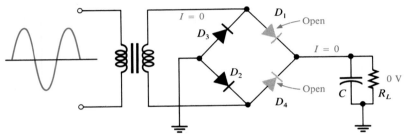

(b) Two open diodes produce a 0-V output.

FIGURE 17–30
Effects of open diodes.

Shorted or Leaky Filter Capacitor

A *shorted capacitor* would most likely cause some or all of the diodes in a full-wave rectifier to open due to excessive current. In any event, there would be no dc voltage on the output.

A *leaky capacitor* can be represented by a leakage resistance in parallel with the capacitor, as shown in Figure 17–31(a). The effect of the leakage resistance is to reduce the discharging time constant, causing an increase in ripple voltage on the output, as shown in Figure 17–31(b).

FIGURE 17–31
Effects of a leaky filter capacitor on the output of a full-wave rectifier.

1. What effect would an open D_2 produce in the rectifier of Figure 17–27?
2. You are checking a 60-Hz full-wave bridge rectifier and observe that the output has a 60-Hz ripple. What failure(s) do you suspect?
3. You observe that the output ripple of a full-wave rectifier is much greater than normal but its frequency is still 120 Hz. What component do you suspect?

17–5

DIODE CLIPPING AND CLAMPING CIRCUITS

Diode circuits are sometimes used to clip off portions of signal voltages above or below certain levels; these circuits are called *clippers* or *limiters*. Another type of diode circuit is used to restore a dc level to an electrical signal; these are called *clampers*. Both diode circuits will be examined in this section.

Diode Clippers

Figure 17–32(a) shows a diode circuit that clips off the positive part of the input signal. As the input signal goes positive, the diode becomes forward-biased. Since the cathode is at ground potential (0 V), the anode cannot exceed 0.7 V (assuming silicon). Thus, point A is clipped at $+0.7$ V when the input exceeds this value.

(a) Clipping of the positive alternation

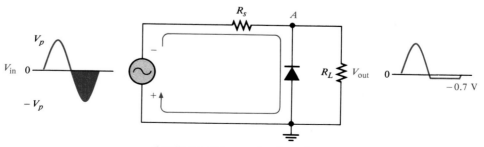

(b) Clipping of the negative alternation

FIGURE 17–32
Diode clipping circuits.

DEVICES

When the input goes back below 0.7 V, the diode reverse-biases and appears as an open. The output voltage looks like the negative part of the input, but with a magnitude determined by the R_s and R_L voltage divider as follows:

$$V_{out} = \left(\frac{R_L}{R_s + R_L}\right) V_{in}$$

If R_s is small compared to R_L, then $V_{out} \cong V_{in}$.

Turn the diode around, as in Figure 17–32(b), and the *negative* part of the input is clipped off. When the diode is forward-biased during the negative part of the input, point A is held at -0.7 V by the diode drop. When the input goes above -0.7 V, the diode is no longer forward-biased and a voltage appears across R_L proportional to the input.

EXAMPLE 17–6

What would you expect to see displayed on an oscilloscope connected as shown in Figure 17–33? The time base on the scope is set to show one and one-half cycles.

FIGURE 17–33

Solution
The diode conducts when the input voltage goes below -0.7 V. Thus, we have a negative clipper with a peak output voltage determined by the following equation:

$$V_{p(out)} = \left(\frac{R_L}{R_s + R_L}\right) V_{p(in)} = \left(\frac{1 \text{ k}\Omega}{1.1 \text{ k}\Omega}\right) 10 \text{ V}$$
$$= 9.09 \text{ V}$$

The scope will display an output waveform as shown in Figure 17–34.

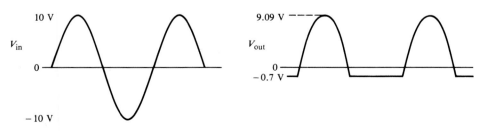

FIGURE 17–34
Waveforms for Figure 17–33.

Adjustment of the Clipping Level

To adjust the level at which a signal voltage is clipped, add a *bias* voltage in series with the diode, as shown in Figure 17–35. The voltage at point A must equal $V_{BB} + 0.7$ V before the diode will conduct. Once the diode begins to conduct, the voltage at point A is limited to $V_{BB} + 0.7$ V so that all input voltage above this level is clipped off, as shown in the figure.

FIGURE 17–35
Positively biased clipper.

If the bias voltage is varied up or down, the clipping level changes correspondingly, as shown in Figure 17–36. If the polarity of the bias voltage is reversed, as in Figure 17–37, voltages above $-V_{BB} + 0.7$ V are clipped, resulting in an output waveform as shown. The diode is reverse-biased only when the voltage at point A goes below $-V_{BB} + 0.7$ V.

FIGURE 17–36
Positive clipper with variable bias.

FIGURE 17–37

DEVICES

If it is necessary to clip off voltage below a specified negative level, then the diode and bias battery must be connected as in Figure 17–38. In this case, the voltage at point A must go below $-V_{BB} - 0.7$ V to forward-bias the diode and initiate clipping action, as shown.

FIGURE 17–38

EXAMPLE 17–7

Figure 17–39 shows a circuit combining a positive-biased clipper with a negative-biased clipper. Determine the output waveform.

FIGURE 17–39

Solution
When the voltage at point A reaches $+7.7$ V, diode D_1 conducts and clips the waveform at $+7.7$ V. Diode D_2 does not conduct until the voltage reaches -7.7 V. Therefore, positive voltages above $+7.7$ V and negative voltages below -7.7 V are clipped off. The resulting output waveform is shown in Figure 17–40.

FIGURE 17–40
Output waveform for Figure 17–39.

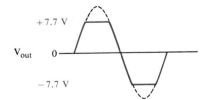

Diode Clampers

The purpose of a clamper is to add a dc level to an ac signal. Clampers are sometimes known as *dc restorers*.

Figure 17–41 shows a diode clamper that inserts a *positive* dc level. To understand the operation of this circuit, consider the first negative half-cycle of the input

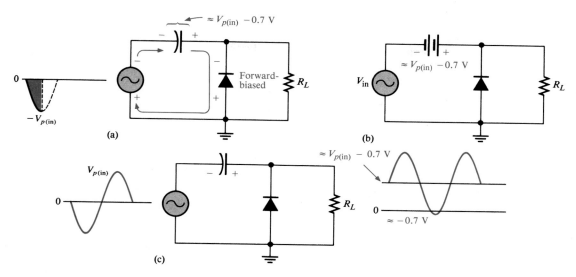

FIGURE 17–41
Positive clamping operation.

voltage. When the input initially goes negative, the diode is forward-biased, allowing the capacitor to charge to near the peak of the input ($V_{p(in)}$ − 0.7 V), as shown in Figure 17–41(a). Just past the negative peak, the diode becomes reverse-biased because the cathode is held near $V_{p(in)}$ by the charge on the capacitor.

The capacitor can discharge only through the high resistance of R_L. Thus, from the peak of one negative half-cycle to the next, the capacitor discharges very little The amount that is discharged, of course, depends on the value of R_L. For good clamping action, the RC time constant should be at least ten times the period of the input frequency.

The net effect of the clamping action is that the capacitor retains a charge approximately equal to the peak value of the input less the diode drop. The capacitor voltage acts essentially as a battery in series with the input signal, as shown in Figure 17–41(b). The dc voltage of the capacitor adds to the input voltage by superposition, as shown in Figure 17–41(c).

If the diode is turned around, a *negative* dc voltage is added to the input signal, as shown in Figure 17–42. If necessary, the diode can be biased to adjust the clamping level.

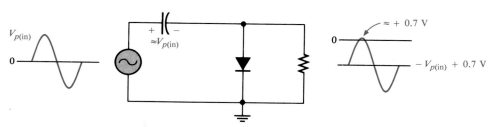

FIGURE 17–42
Negative clamping.

A Clamper Application

A clamping circuit is often used in television receivers as a dc restorer. The incoming composite video signal is normally processed through capacitively coupled amplifiers which eliminate the dc component, thus losing the black and white reference levels and the blanking level. Before being applied to the picture tube, these reference levels must be restored. Figure 17–43 illustrates this process in a general way.

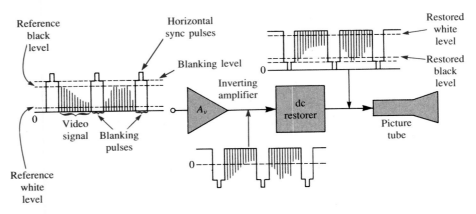

FIGURE 17–43
Clamping circuit (dc restorer) in a TV receiver.

EXAMPLE 17–8

What is the output voltage that you would expect to observe across R_L in the clamping circuit of Figure 17–44? Assume that RC is large enough to prevent significant capacitor discharge.

FIGURE 17–44

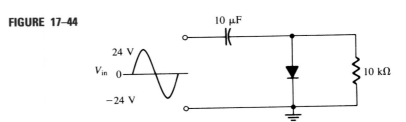

Solution
Ideally, a negative dc value equal to the input peak less the diode drop is inserted by the clamping circuit:

$$V_{dc} \cong V_{p(in)} - 0.7 \text{ V} = 24 \text{ V} - 0.7 \text{ V} = 23.3 \text{ V}$$

Actually, the capacitor will discharge slightly between peaks, and, as a result, the output voltage will have an average value of slightly less than that calculated above.

The output waveform goes to approximately 0.7 V above ground, as shown in Figure 17–45.

FIGURE 17–45
Output waveform for Figure 17–44.

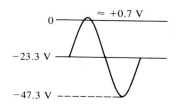

**SECTION REVIEW
17–5**

1. Determine the output waveform for the circuit of Figure 17–46.
2. What is the output voltage in Figure 17–47?
3. Sketch the approximate output waveform for a positive clamping circuit having a sine wave input with a peak of 50 V.

FIGURE 17–46

FIGURE 17–47

17–6

ZENER DIODES

FIGURE 17–48
Zener diode symbol.

The *zener diode* is used for voltage regulation and, like the general-purpose rectifier diode, is important in many power supply applications. The schematic symbol is shown in Figure 17–48.

The zener is a silicon PN junction device that differs from the rectifier diode in that it is designed for operation in the *reverse breakdown region*. The breakdown voltage of a zener diode is set by carefully controlling the doping level during manufacture. From the discussion of the diode characteristic curve in the last chapter, recall that when a diode reaches reverse breakdown, its voltage remains almost constant even though the current may change drastically. This volt-ampere characteristic is shown again in Figure 17–49.

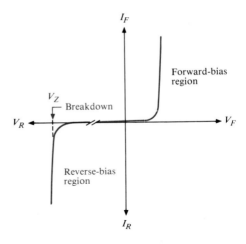

FIGURE 17–49
General diode characteristic.

Zener Breakdown

There are two types of reverse breakdown in a zener diode. One is the *avalanche* breakdown that was discussed in Chapter 16; this also occurs in rectifier diodes at a sufficiently high reverse voltage. The other type is *zener* breakdown which occurs in a zener diode at low reverse voltages. A zener diode is heavily doped to reduce the breakdown voltage, causing a very narrow depletion layer. As a result, an intense electric field exists within the depletion layer. Near the breakdown voltage (V_Z), the field is intense enough to pull electrons from their valence bands and create current.

Zener diodes with breakdown voltages of less than approximately 5 V operate predominantly in zener breakdown. Those with breakdown voltages greater than approximately 5 V operate predominantly in avalanche breakdown. Both types, however, are called zener diodes. Zeners with breakdown voltages of 1.8 V to 200 V are commercially available.

Breakdown Characteristics

Figure 17–50 shows the reverse portion of the characteristic curve of a zener diode. Notice that as the reverse voltage (V_R) is increased, the reverse current (I_R) remains

FIGURE 17–50
Reverse characteristic of a zener diode.

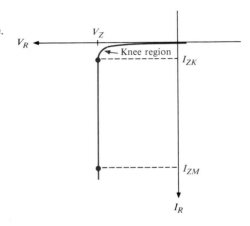

extremely small up to the "knee" of the curve. At this point, the breakdown effect begins; the zener resistance (R_Z) begins to decrease as the current (I_Z) increases rapidly. From the bottom of the knee, the breakdown voltage (V_Z) remains essentially constant. This *regulating* ability is the key feature of the zener diode: *It maintains an essentially constant voltage across its terminals over a specified range of reverse current values.*

A *minimum* value of reverse current, I_{ZK}, must be maintained in order to keep the diode in regulation. You can see on the curve that when the reverse current is reduced below the knee of the curve, the voltage changes drastically and regulation is lost. Also, there is a *maximum* current, I_{ZM}, above which the diode may be damaged.

Thus, basically, the zener diode maintains a constant voltage across its terminals for values of reverse current ranging from I_{ZK} to I_{ZM}.

Zener Equivalent Circuit

Figure 17–51(a) shows the ideal approximation of a zener diode in reverse breakdown. It *acts* simply as a battery having a value equal to the zener voltage. Part (b) represents the practical equivalent of a zener, where the zener resistance R_Z is included. Since the voltage curve is not ideally vertical, a change in reverse current (ΔI_Z) produces a small change in zener voltage (ΔV_Z), as illustrated in part (c).

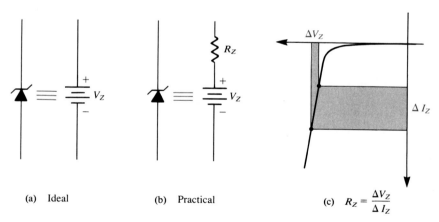

(a) Ideal   (b) Practical   (c) $R_Z = \dfrac{\Delta V_Z}{\Delta I_Z}$

FIGURE 17–51
Zener equivalent circuits.

The ratio of ΔV_Z to ΔI_Z is the resistance, as expressed in the following equation:

$$R_Z = \frac{\Delta V_Z}{\Delta I_Z} \tag{17–9}$$

Normally, R_Z is specified at a particular value of reverse current, I_{ZT}, called the *zener test current*. In most cases, this value of R_Z is approximately constant over the full range of reverse-current values.

EXAMPLE 17–9

A certain zener diode exhibits a 50-mV change in V_Z for a 2-mA change in I_Z. What is the zener resistance?

Solution

$$R_Z = \frac{\Delta V_Z}{\Delta I_Z} = \frac{50 \text{ mV}}{2 \text{ mA}} = 25 \ \Omega$$

EXAMPLE 17–10

A 6.8-V zener diode has a resistance of 5 Ω. What is the actual voltage across its terminals when the current is 20 mA?

Solution

Figure 17–52 represents the diode. The 20-mA current causes a voltage across R_Z as follows:

$$I_Z R_Z = (20 \text{ mA})(5 \ \Omega) = 100 \text{ mV}$$

FIGURE 17–52

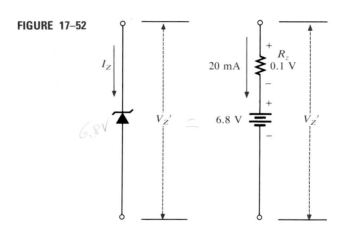

The polarity is in the same direction as the zener voltage, as shown. Therefore, it adds to V_Z, giving a total terminal voltage of

$$V_Z' = V_Z + I_Z R_Z = 6.8 \text{ V} + 100 \text{ mV} = 6.9 \text{ V}$$

Thus, for this particular current value, the terminal voltage is 0.1 V greater than the specified normal zener voltage of 6.8 V. This value, of course, changes with current.

Zener Voltage Regulation

Zener diodes are widely used for voltage regulation. Figure 17–53 illustrates how a zener diode can be used to regulate a varying dc voltage to keep it at a constant level. This process is called *input* or *line regulation*.

As the input voltage varies (within limits), the zener diode maintains an essentially constant voltage across the output terminals. However, as V_{IN} changes, I_Z will change proportionally, and therefore the limitations on the input variation are set by the mini-

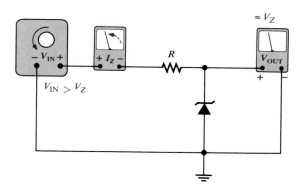

(a) As input voltage increases, V_{OUT} remains constant.

(b) As input voltage decreases, V_{OUT} remains constant.

FIGURE 17–53
Zener regulation of a varying input voltage.

mum and maximum current values with which the zener can operate and on the condition that $V_{IN} > V_Z$. R is the series current-limiting resistor.

For example, suppose that the zener diode in Figure 17–54 can maintain regulation over a range of current values from 4 mA to 40 mA. For the *minimum* current, the voltage across the 1-kΩ resistor is $V_R = (4 \text{ mA})(1 \text{ k}\Omega) = 4$ V. Since $V_R = V_{IN} - V_Z$, then $V_{IN} = V_R + V_Z = 4 \text{ V} + 10 \text{ V} = 14$ V. For the *maximum* current, the voltage across the 1-kΩ resistor is $V_R = (40 \text{ mA})(1 \text{ k}\Omega) = 40$ V. Therefore, $V_{IN} = 40 \text{ V} + 10 \text{ V} = 50$ V.

As you can see, this zener diode can regulate an input voltage from 14 V to 50 V and maintain approximately a 10-V output. (The output will vary slightly because of the zener resistance.)

FIGURE 17–54

R
1 kΩ

V_{IN}

+

10 V

+

−

EXAMPLE 17–11
Determine the minimum and the maximum input voltages that can be regulated by the zener diode in Figure 17–55. Assume that $I_{ZK} = 1$ mA, $I_{ZM} = 15$ mA, $V_Z = 5.1$ V, and $R_Z = 10 \ \Omega$.

FIGURE 17–55

R
680 Ω

V_{IN}

+

V_{OUT}

710

DEVICES

Solution

The equivalent circuit is shown in Figure 17–56. At $I_{ZK} = 1$ mA, the output voltage is

$$V_{OUT} = V_Z + I_{ZK}R_Z = 5.1 \text{ V} + (1 \text{ mA})(10 \text{ }\Omega)$$
$$= 5.1 \text{ V} + 0.010 \text{ V} = 5.11 \text{ V}$$

FIGURE 17–56
Equivalent of the circuit in Figure 17–55.

Therefore,

$$V_{IN(min)} = I_{ZK}R + V_{OUT} = (1 \text{ mA})(680 \text{ }\Omega) + 5.11 \text{ V}$$
$$= 5.79 \text{ V}$$

At $I_{ZM} = 15$ mA, the output voltage is

$$V_{OUT} = V_Z + I_{ZM}R_Z = 5.1 \text{ V} + (15 \text{ mA})(10 \text{ }\Omega)$$
$$= 5.1 \text{ V} + 0.15 \text{ V} = 5.25 \text{ V}$$

Therefore,

$$V_{IN(max)} = I_{ZM}R + V_{OUT} = (15 \text{ mA})(680 \text{ }\Omega) + 5.25 \text{ V}$$
$$= 15.45 \text{ V}$$

Regulation with a Varying Load

Figure 17–57 shows a zener regulator with a variable load resistor across the terminals. The zener diode maintains a constant voltage across R_L as long as the zener current is greater than I_{ZK} and less than I_{ZM}. This process is called *load regulation*.

FIGURE 17–57
Zener regulation with a variable load.

No Load to Full Load

When the output terminals are open ($R_L = \infty$), the load current is zero and all of the current is through the zener. When a load resistor is connected, part of the total current is through the zener and part through R_L.

As R_L is decreased, I_L goes up and I_Z goes down. The zener diode continues to regulate until I_Z reaches its minimum value, I_{ZK}. At this point the load current is maximum. The following example illustrates.

EXAMPLE 17–12

Determine the minimum and the maximum load curents for which the zener diode in Figure 17–58 will maintain regulation. What is the minimum R_L that can be used? $V_Z = 12$ V, $I_{ZK} = 3$ mA, and $I_{ZM} = 90$ mA. Assume that $R_Z = 0$ Ω.

FIGURE 17–58

Solution

When $I_L = 0$ A, I_Z is maximum and equal to the total circuit current I_T:

$$I_Z = \frac{V_{IN} - V_Z}{R} = \frac{24 \text{ V} - 12 \text{ V}}{470 \text{ }\Omega} = 25.53 \text{ mA}$$

Since this is much less than I_{ZM}, 0 A is an acceptable minimum for I_L. That is,

$$I_{L(min)} = 0 \text{ A}$$

The maximum value of I_L occurs when I_Z is minimum, so we can solve for $I_{L(max)}$ as follows:

$$I_{L(max)} = I_T - I_{Z(min)} = 25.53 \text{ mA} - 3 \text{ mA} = 22.53 \text{ mA}$$

The minimum value of R_L is

$$R_{L(min)} = \frac{V_Z}{I_{L(max)}} = \frac{12 \text{ V}}{22.53 \text{ mA}} = 533 \text{ }\Omega$$

Percent Regulation

The percent regulation is a figure of merit used to specify the performance of a voltage regulator. It can be in terms of input (line) regulation or load regulation.

The *percent input regulation* specifies how much change occurs in the output voltage for a given change in input voltage. It is usually expressed as a percent change in V_{OUT} for a 1-V change in V_{IN} (%/V).

The *percent load regulation* specifies how much change occurs in the output voltage over a certain range of load current values, usually from minimum current (no load)

DEVICES

to maximum current (full load). It is normally expressed as a percentage and can be calculated with the following formula:

$$\text{Percent load regulation} = \frac{V_{NL} - V_{FL}}{V_{FL}} \times 100\% \qquad (17-10)$$

where V_{NL} is the output voltage with no load, and V_{FL} is the output voltage with full load.

EXAMPLE 17–13

A certain regulator has a no-load output voltage of 6 V and a full-load output of 5.82 V. What is the percent load regulation?

Solution

$$\text{Percent load regulation} = \frac{V_{NL} - V_{FL}}{V_{FL}} \times 100\%$$

$$= \frac{6\text{ V} - 5.82\text{ V}}{5.82\text{ V}} \times 100\% = 3.09\%$$

SECTION REVIEW 17–6

1. Zener diodes are normally operated in the breakdown region (T or F).
2. A certain 10-V zener diode has a resistance of 8 Ω at 30 mA. What is the terminal voltage?
3. Explain the difference between input (line) regulation and load regulation.
4. In a zener diode regulator, for what value of load resistance is the zener current a maximum?
5. A zener regulator has an output voltage of 12 V with no load and 11.9 V with full load. What is the percent load regulation?

17–7

VARACTOR DIODES

Varactor diodes are used as *voltage-variable capacitors*. A varactor is basically a *reverse-biased* PN junction that utilizes the inherent capacitance of the depletion layer. The depletion layer, created by the reverse bias, acts as a capacitor *dielectric* because of its nonconductive characteristic. The P and N regions are conductive and act as the capacitor *plates*, as illustrated in Figure 17–59.

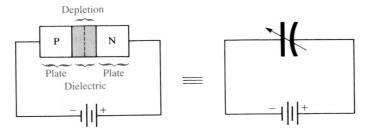

FIGURE 17–59
The reverse-biased varactor diode acts as a variable capacitor.

When the reverse-bias voltage increases, the depletion layer widens, effectively increasing the dielectric thickness *(d)* and thus *decreasing* the capacitance. When the reverse-bias voltage decreases, the depletion layer narrows, thus *increasing* the capacitance. This action is shown in Figures 17–60(a) and 17–60(b). A general curve of capacitance versus voltage is shown in Figure 17–60(c).

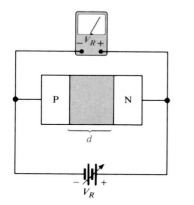

(a) Greater V_R, less capacitance

(b) Less V_R, greater capacitance

(c) Typical graph of capacitance versus reverse voltage

FIGURE 17–60
Varactor diode capacitance varies with reverse voltage.

Recall that capacitance is determined by the *plate area (A)*, *dielectric constant (ε)*, and *dielectric thickness*, as expressed in the following formula:

$$C = \frac{A\varepsilon}{d}$$

In a varactor diode, the capacitance parameters are controlled by the method of doping in the depletion layer and the size and geometry of the diode's construction. Varactor capacitances typically range from a few picofarads to a few hundred picofarads.

Figure 17–61(a) shows a common symbol for a varactor, and part (b) shows a simplified equivalent circuit. R_S is the reverse series resistance, and C_V is the variable capacitance.

FIGURE 17–61
Varactor diode.

(a) Symbol

(b) Equivalent circuit

Applications

A major application of varactors is in tuning circuits. For example, electronic tuners in TV and other commercial receivers utilize varactors as one of their elements.

When used in a resonant circuit, the varactor acts as a variable capacitor, thus allowing the resonant frequency to be adjusted by a variable voltage level, as illustrated in Figure 17–62 where two varactor diodes provide the total variable capacitance in a parallel resonant (tank) circuit. V_C is a variable dc voltage that controls the reverse bias and therefore the capacitance of the diodes.

FIGURE 17–62
Varactors in a resonant circuit.

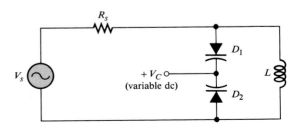

Recall that the resonant frequency of the tank circuit is

$$f_r \cong \frac{1}{2\pi\sqrt{LC}}$$

This approximation is valid for $Q > 10$.

EXAMPLE 17–14

The capacitance of a certain varactor can be varied from 5 pF to 50 pF. The diode is used in a tuned circuit similar to that shown in Figure 17–62. Determine the tuning range for the circuit if $L = 10$ mH.

Solution
The equivalent circuit is shown in Figure 17–63. Notice that the varactor capacitances are in series.

FIGURE 17–63

The *minimum* total capacitance is

$$C_{T(min)} = \frac{C_{1(min)}C_{2(min)}}{C_{1(min)} + C_{2(min)}} = \frac{(5 \text{ pF})(5 \text{ pF})}{10 \text{ pF}} = 2.5 \text{ pF}$$

The *maximum* resonant frequency therefore is

$$f_{r(max)} = \frac{1}{2\pi\sqrt{LC_{T(min)}}} = \frac{1}{2\pi\sqrt{(10 \text{ mH})(2.5 \text{ pF})}} \cong 1 \text{ MHz}$$

The *maximum* total capacitance is

$$C_{T(max)} = \frac{C_{1(max)}C_{2(max)}}{C_{1(max)} + C_{2(max)}} = \frac{(50 \text{ pF})(50 \text{ pF})}{100 \text{ pF}} = 25 \text{ pF}$$

The *minimum* resonant frequency therefore is

$$f_{r(min)} = \frac{1}{2\pi\sqrt{LC_{T(max)}}} = \frac{1}{2\pi\sqrt{(10 \text{ mH})(25 \text{ pF})}} \cong 318 \text{ kHz}$$

**SECTION REVIEW
17–7**

1. What is the purpose of a varactor diode?
2. Based on the general curve in Figure 17–60(c), what happens to the diode capacitance when the reverse voltage is increased?

17–8 LEDs AND PHOTODIODES

The Light-Emitting Diode (LED)

The basic operation of the LED is as follows: When the device is forward-biased, electrons cross the PN junction from the N-type material and *recombine* with holes in the P-type material. Recall that these free electrons are in the conduction band and at a higher energy level than the holes in the valence band. When recombination takes place, the recombining electrons release energy in the form of *heat* and *light*. A large exposed surface area on one layer of the semiconductor material permits the photons to be emitted as visible light. Figure 17–64 illustrates this process which is called *electroluminescence*.

FIGURE 17–64
Electroluminescence in an LED.

FIGURE 17–65
Symbol for an LED.

The semiconductor materials used in LEDs are *gallium arsenide* (GaAs), *gallium arsenide phosphide* (GaAsP), and *gallium phosphide* (GaP). Silicon and germanium are not used because they are essentially heat-producing materials and are very poor at producing light. GaAs LEDs emit *infrared* (IR) radiation, GaAsP produces either *red* or *yellow* light, and GaP emits *red* or *green* light. The symbol for an LED is shown in Figure 17–65.

The LED *emits* light in response to a sufficient *forward* current, as shown in Figure 17–66(a). The amount of light output is directly proportional to the forward current, as indicated in part (b). Typical LEDs are shown in part (c).

(a) Forward-biased operation

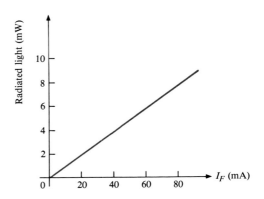

(b) Typical light output versus forward current

(c)

FIGURE 17–66
Light-emitting diodes (LEDs).

Applications

LEDs are commonly used for indicator lamps and readout displays on a wide variety of instruments, ranging from consumer appliances to scientific apparatus. A very common type of display device using LEDs is the seven-segment display. Combinations of the segments form the ten decimal digits. Also, IR-emitting diodes are employed in optical coupling applications, often in conjunction with fiber optics.

The Photodiode

The photodiode is a PN junction device that operates in reverse bias, as shown in Figure 17–67(a). Note the schematic symbol for the photodiode. The photodiode has a small transparent window that allows light to strike the PN junction. Typical photodiodes are shown in part (b).

Recall that when reverse-biased, a rectifier diode has a very small reverse leakage current. The same is true for the photodiode. The reverse-biased current is produced by thermally generated electron hole pairs in the depletion layer, which are swept across the junction by the electric field created by the reverse voltage. In a rectifier diode, the

FIGURE 17–67
Photodiode.

light — ... *diverted —*

(a) Reverse-bias operation   (b) Typical devices

reverse current increases with temperature due to an increase in the number of electron hole pairs.

In a photodiode, the reverse current increases with the light intensity at the PN junction. When there is no incident light, the reverse current I_λ is almost negligible and is called the *dark current*. An increase in the amount of light energy (measured in lumens per square meter, lm/m^2) produces an increase in the reverse current, as shown by the graph in figure 17–68(a). For a given value of reverse-bias voltage, part (b) shows a set of characteristic curves for a typical photodiode.

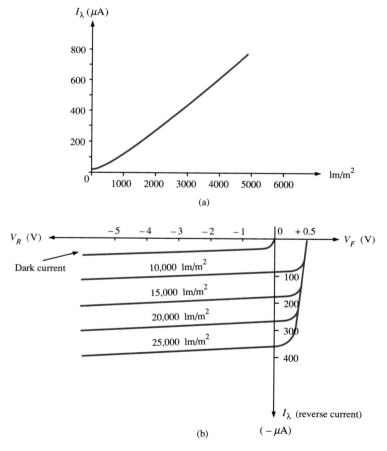

FIGURE 17–68
Typical photodiode characteristics.

From the characteristic curve in Figure 17–68(b), the dark current for this particular device is approximately 25 μA at a reverse bias voltage of 3 V. Therefore, the resistance of the device with no incident light is

$$R_R = \frac{V_R}{I_\lambda} = \frac{3 \text{ V}}{25 \text{ μA}} = 120 \text{ k}\Omega$$

At 25,000 lm/m², the current is approximately 375 μA at −3 V. The resistance under this condition is

$$R_R = \frac{V_R}{I_\lambda} = \frac{3 \text{ V}}{375 \text{ μA}} = 8 \text{ k}\Omega$$

These calculations show that the photodiode can be used as a variable-resistance device controlled by light intensity.

Figure 17–69 illustrates that the photodiode allows essentially no reverse current (except for a very small dark current) when there is no incident light. When a light beam strikes the photodiode, it conducts an amount of reverse current that is proportional to the light intensity.

FIGURE 17–69
Operation of a photodiode.

(a) No light, no current

(b) When there is incident light, reverse current flows.

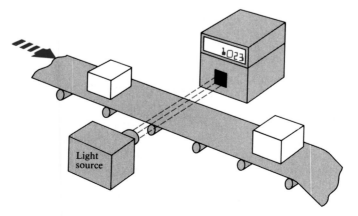

FIGURE 17–70
A photodiode circuit used in a system that counts objects as they pass on a conveyor belt.

A simple photodiode application is depicted in Figure 17–70. Here a beam of light continuously passes across a conveyor belt and into a transparent window behind which is a photodiode circuit. When the light beam is interrupted by an object passing by on the conveyor belt, the sudden reduction in diode current activates a control circuit that advances a counter by one. The total count of objects that have passed that point is displayed by the counter. This basic concept can be extended and used for production control, shipping, and monitoring of activity on production lines.

**SECTION REVIEW
17–8**

1. How does an LED differ from a photodiode?
2. List the semiconductor materials used in LEDs.
3. There is a very small reverse current in a photodiode under no-light conditions. What is this current called?

17–9

INTERPRETING DIODE DATA SHEETS

A manufacturer's data sheet gives detailed information on a device so that it can be used properly in a given application. A typical data sheet provides *maximum ratings, electrical characteristics, mechanical data,* and *graphs of various parameters.* We use a specific example to illustrate a typical data sheet. Table 17–1 shows the *maximum ratings* for a certain series of rectifier diodes (1N4001 through 1N4007). These are the absolute maximum values under which the diode can be operated without damage to the device. For greatest reliability and longer life, the diode should always be operated well under these maximums. Generally, the maximum ratings are specified at 25°C and must be adjusted downward for greater temperatures.

TABLE 17–1
Maximum ratings

Rating	Symbol	1N4001	1N4002	1N4003	1N4004	1N4005	1N4006	1N4007	Unit
Peak repetitive reverse voltage Working peak reverse voltage dc blocking voltage	V_{RRM} V_{RWM} V_R	50	100	200	400	600	800	1000	V
Nonrepetitive peak reverse voltage	V_{RSM}	60	120	240	480	720	1000	1200	V
rms reverse voltage	$V_{R(rms)}$	35	70	140	280	420	560	700	V
Average rectified forward current (single-phase, resistive load, 60 Hz, $T_A = 75°C$)	I_0	1.0							A
Nonrepetitive peak surge current (surge applied at rated load conditions)	I_{FSM}	30 (for 1 cycle)							A
Operating and storage junction temperature range	T_J, T_{stg}	−65 to +175							°C

An explanation of some of the parameters from Table 17–1 is as follows:

V_{RRM} The maximum reverse peak voltage that can be applied repetitively across the diode. Notice that in this case, it is 50 V for the 1N4001 and 1 kV for the 1N4007.

V_R The maximum reverse dc voltage that can be applied across the diode.

V_{RSM} The maximum reverse peak value of nonrepetitive (one-cycle) voltage that can be applied across the diode.

I_0 The maximum average value of a 60-Hz full-wave rectified forward current.

I_{FSM} The maximum peak value of nonrepetitive (one-cycle) forward current. The graph in Figure 17–71 expands on this parameter to show values for more than one cycle at temperatures of 25°C and 175°C. The dashed lines represent values where typical failures occur.

FIGURE 17–71
Nonrepetitive surge capability.

Notice what happens on the lower solid line when ten cycles of I_{FSM} are applied. The limit is 15 A rather than the one-cycle value of 30 A.

Table 17–2 lists typical and maximum values of certain electrical characteristics. These items differ from the maximum ratings in that they are not selected by design but are the result of operating the diode under specified conditions. A brief explanation of these parameters is as follows:

TABLE 17–2
Electrical characteristics

Characteristic and conditions	Symbol	Typical	Maximum	Unit
Maximum instantaneous forward voltage drop ($I_F = 1$ A, $T_J = 25°C$)	v_F	0.93	1.1	V
Maximum full-cycle average forward voltage drop ($I_0 = 1$ A, $T_L = 75°C$, 1-in. leads)	$V_{F(avg)}$	–	0.8	V
Maximum reverse current (rated dc voltage) $T_J = 25°C$ $T_J = 100°C$	I_R	0.05 1.0	10.0 50.0	μA
Maximum full-cycle average reverse current ($I_0 = 1$ A, $T_L = 75°C$, 1-in. leads)	$I_{R(avg)}$	–	30.0	μA

v_F The instantaneous voltage across the forward-biased diode when the forward current is 1 A at 25°C. Figure 17–72 shows how the forward voltages vary with forward current.

FIGURE 17–72
Forward voltage.

i_F, instantaneous forward current (amp)

v_F, instantaneous forward voltage (volts)

$V_{F(avg)}$ The maximum forward voltage drop averaged over a full cycle.
I_R The maximum current when the diode is reverse-biased with a dc voltage.
$I_{R(avg)}$ The maximum reverse current averaged over one cycle (when reverse-biased with an ac voltage).

The mechanical data for these particular diodes as they appear on a typical data sheet are shown in Figure 17–73.

FIGURE 17–73
Mechanical data.

DIM	Millimeters		Inches	
	Min	Max	Min	Max
A	5.97	6.60	0.235	0.260
B	2.79	3.05	0.110	0.120
D	0.76	0.86	0.030	0.034
K	27.94	–	1.100	–

Mechanical characteristics

Case: Transfer Molded Plastic
Maximum lead temperature for soldering purposes: 350°C, ⅜″ from case for 10 seconds at 5 lbs. tension
Finish: All external surfaces are corrosion-resistant, leads are readily solderable
Polarity: Cathode indicated by color band
Weight: 0.40 Grams (approximately)

**SECTION REVIEW
17–9**

1. List the three diode ratings categories.
2. Identify each of the following parameters: **(a)** V_F **(b)** I_R **(c)** I_0

SUMMARY

☐ The single diode in a half-wave rectifier conducts for half of the input cycle.
☐ The output frequency of a half-wave rectifier equals the input frequency.
☐ The average (dc) value of a half-wave rectified signal is 0.318 ($1/\pi$) times its peak value.
☐ The PIV is the maximum voltage appearing across the diode in reverse bias.
☐ Each diode in a full-wave rectifier conducts for half of the input cycle.
☐ The output frequency of a full-wave rectifier is twice the input frequency.
☐ The basic types of full-wave rectifier are center-tapped and bridge.
☐ The output voltage of a center-tapped, full-wave rectifier is approximately one-half of the total secondary voltage.
☐ The PIV for each diode in a center-tapped, full-wave rectifier is twice the output voltage.
☐ The output voltage of a bridge rectifier equals the total secondary voltage.
☐ The PIV for each diode in a bridge rectifier is half that required for the center-tapped configuration and is approximately equal to the peak output voltage.
☐ A capacitor-input filter provides a dc output approximately equal to the peak of the input.

☐ Ripple voltage is caused by the charging and discharging of the filter capacitor.

☐ The smaller the ripple, the better the filter.

☐ An *LC* filter provides improved ripple reduction over the capacitor-input filter.

☐ An inductor-input filter produces a dc output voltage approximately equal to the average value of the rectified input.

☐ Diode clippers cut off voltage above or below specified levels.

☐ Diode clampers add a dc level to an ac signal.

☐ The zener diode operates in reverse breakdown.

☐ There are two breakdown mechanisms in a zener diode; *avalanche* breakdown and *zener* breakdown.

☐ When $V_Z < 5$ V, *zener* breakdown is predominant.

☐ When $V_Z > 5$ V, *avalanche* breakdown is predominant.

☐ A zener diode maintains an essentially constant voltage across its terminals over a specified range of zener currents.

☐ Zener diodes are used as shunt voltage regulators.

☐ Regulation of output voltage over a range of input voltages is called *input* or *line* regulation.

☐ Regulation of output voltage over a range of load currents is called *load* regulation.

☐ The *smaller* the percent regulation, the better.

☐ A *varactor* diode acts as a variable capacitor under reverse-biased conditions.

☐ The capacitance of a varactor varies inversely with reverse-biased voltage.

☐ Diode symbols are shown in Figure 17–74.

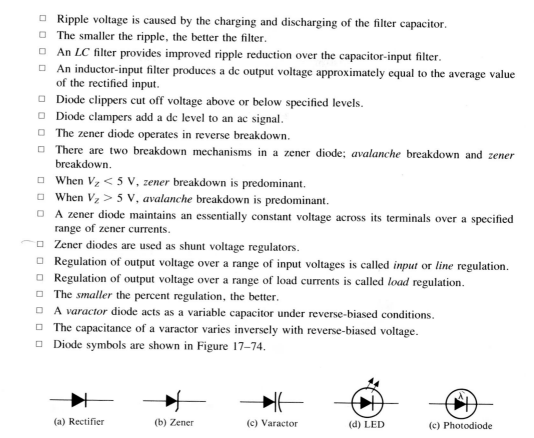

(a) Rectifier (b) Zener (c) Varactor (d) LED (c) Photodiode

FIGURE 17–74
Diode symbols.

GLOSSARY

Center-tap A connection at the midpoint of the secondary of a transformer.

Clamper A circuit that adds a dc level to an ac signal. A dc restorer.

Clipper A circuit that removes part of a waveform above or below a specified level. A limiter.

Filter A type of electrical circuit that passes certain frequencies and rejects all others.

Full-wave rectifier A circuit that converts an alternating sine wave into a pulsating dc consisting of both halves of a sine wave for each input cycle.

Half-wave rectifier A circuit that converts an alternating sine wave into a pulsating dc consisting of one-half of a sine wave for each input cycle.

Light-emitting diode (LED) A type of diode that emits light when forward current flows.

Line regulation The percent change in output voltage for a given change in line (input) voltage.

Load regulation The percent change in output voltage for a given change in load current.

Photodiode A diode whose reverse resistance changes with incident light.

Power supply An electronic instrument that produces voltage, current, and power from the ac power line or batteries in a form suitable for use in various applications to power electronic equipment.

Rectifier An electronic circuit that converts ac into pulsating dc.

Regulator   An electronic circuit that maintains an essentially constant output voltage with a changing input voltage or load.

Ripple   The variation in the dc voltage on the output of a filtered rectifier caused by the slight charging and discharging action of the filter capacitor.

Varactor   A diode that is used as a voltage-variable capacitor.

Voltage regulation   The process of maintaining an essentially constant output voltage over variations in input voltage or load.

Zener diode   A type of diode that operates in reverse breakdown (called zener breakdown) to provide voltage regulation.

FORMULAS

$$V_{avg} = \frac{V_p}{\pi} \qquad \text{Half-wave average value} \tag{17--1}$$

$$V_{p(out)} = V_{p(in)} - 0.7 \text{ V} \qquad \text{Peak half-wave rectifier output} \tag{17--2}$$

$$V_{avg} = \frac{2V_p}{\pi} \qquad \text{Full-wave average value} \tag{17--3}$$

$$PIV = 2V_{p(out)} \qquad \begin{array}{l}\text{Diode peak inverse voltage, center-tapped rectifier} \\ \text{(neglecting } V_B) \end{array} \tag{17--4}$$

$$V_{out} = V_s \qquad \text{Bridge full-wave output (neglecting diode drops)} \tag{17--5}$$

$$V_{out} = V_s - 1.4 \text{ V} \qquad \text{Bridge full-wave output (including diode drops)} \tag{17--6}$$

$$PIV \cong V_{p(out)} \qquad \text{Diode peak inverse voltage, bridge rectifier} \tag{17--7}$$

$$r = \left(\frac{V_r}{V_{dc}}\right)100\% \qquad \text{Ripple factor} \tag{17--8}$$

$$R_Z = \frac{\Delta V_Z}{\Delta I_Z} \qquad \text{Zener resistance} \tag{17--9}$$

$$\text{Percent load regulation} = \frac{V_{NL} - V_{FL}}{V_{FL}} \times 100\% \qquad \text{Load regulation} \tag{17--10}$$

SELF-TEST

1. The process of converting ac to pulsating dc is called
   (a) clipping   (b) charging   (c) rectification   (d) filtering

2. The output frequency of a half-wave rectifier with a 60-Hz sinusoidal input is
   (a) 30 Hz   (b) 60 Hz   (c) 120 Hz   (d) 0 Hz

3. The number of diodes used in a half-wave rectifier is
   (a) one   (b) two   (c) three   (d) four

4. If a 75-V peak sine wave is applied to a half-wave rectifier, the peak inverse voltage across the diode is
   (a) 75 V   (b) 150 V   (c) 37.5 V   (d) 0 V

5. The output frequency of a full-wave rectifier with a 60-Hz sinusoidal input is
   (a) 30 Hz   (b) 60 Hz   (c) 120 Hz   (d) 0 Hz

6. Two types of full-wave rectifier are
  (a) single diode and dual diode   **(b)** primary and secondary
  (c) forward-biased and reverse-biased **(d)** center-tapped and bridge

7. When a diode in a center-tapped rectifier opens, the output is
  (a) 0 V   **(b)** half-wave rectified
  (c) reduced in amplitude **(d)** unaffected

8. During the positive half-cycle of the input voltage in a bridge rectifier,
  (a) one diode is forward-biased **(b)** all diodes are forward-biased
  (c) all diodes are reverse-biased **(d)** two diodes are forward-biased

9. The process of changing a half-wave or a full-wave rectified voltage to a constant dc voltage is called
  (a) filtering   **(b)** ac to dc conversion
  (c) damping   **(d)** ripple suppression

10. The small variation in the output voltage of a dc power supply is called
  (a) average voltage **(b)** surge voltage
  (c) residual voltage **(d)** ripple voltage

11. One measure of the effectiveness of a power supply filter is the
  (a) surge current **(b)** ripple factor
  (c) peak value of the output **(d)** average value of the output

12. A diode clipping circuit
  (a) removes part of a waveform **(b)** inserts a dc level
  (c) produces an output equal to the average value of the input
  (d) increases the peak value of the input

13. A clamping circuit is also known as a(an)
  (a) averaging circuit **(b)** inverter
  (c) dc restorer **(d)** ac restorer

14. The zener diode is designed for operation in
  (a) zener breakdown **(b)** forward bias
  (c) reverse bias **(d)** avalanche breakdown

15. Zener diodes are widely used as
  (a) current limiters **(b)** power distributors
  (c) voltage regulators **(d)** variable resistors

16. Varactor diodes are used as
  (a) variable resistors **(b)** variable current sources
  (c) variable inductors **(d)** variable capacitors

17. LEDs are based on the principle of
  (a) forward bias **(b)** electroluminescence
  (c) photon sensitivity **(d)** electron-hole recombination

18. In a photodiode, light produces
  (a) reverse current **(b)** forward current
  (c) electroluminescence **(d)** dark current

PROBLEMS

Section 17–1   Half-wave Rectifiers

1. Calculate the average value of a half-wave rectified voltage with a peak value of 200 V.

2. Sketch the waveforms for the load current and voltage for Figure 17–75. Show the peak values. (The diode is silicon.)

3. Can a diode with a PIV rating of 5 V be used in the circuit of Figure 17–75?

FIGURE 17–75

4. Determine the peak power delivered to R_L in Figure 17–76.

FIGURE 17–76

Section 17–2   Full-wave Rectifiers

5. Calculate the average value of a full-wave rectified voltage with a peak value of 75 V.

6. Consider the circuit in Figure 17–77.
    (a) What type of circuit is this?
    (b) What is the total peak secondary voltage?
    (c) Find the peak voltage across each half of the secondary.
    (d) Sketch the voltage waveform across R_L.
    (e) What is the peak current through each diode?
    (f) What is the PIV for each diode?

FIGURE 17–77

7. Calculate the peak voltage rating of each half of a center-tapped transformer used in a full-wave rectifier that has an average output voltage of 110 V.

8. Show how to connect the diodes in a center-tapped rectifier in order to produce a *negative-going* full-wave voltage across the load resistor.

9. What PIV rating is required for the diodes in a bridge rectifier that produces an average output voltage of 50 V?

Section 17–3   Rectifier Filters

10. The *ideal* dc output voltage of a capacitor-input filter is the (peak, average) value of the rectified input.

11. Refer to Figure 17–78 and sketch the following voltage waveforms in relationship to the input waveform: V_{AB}, V_{AD}, V_{BD}, and V_{CD}.

FIGURE 17–78

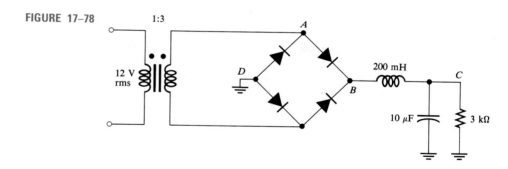

Section 17–4   Troubleshooting Rectifier Circuits

TBLS
+ −

12. From the meter readings in Figure 17–79, determine if the rectifier circuit is functioning properly. If it is not, determine the most likely failure(s).

TBLS
+ −

13. Each part of Figure 17–80 shows oscilloscope displays of rectifier output voltages. In each case, determine whether or not the rectifier is functioning properly and, if it is not, what the most likely failure(s) is(are).

FIGURE 17–79

(a)

(b)

(c)

(d)

FIGURE 17–80

Section 17-5   Diode Clipping and Clamping Circuits

14. Sketch the output waveforms for each circuit in Figure 17–81.

15. Describe the output waveform of each circuit in Figure 17–82. Assume that the *RC* time constant is much greater than the period of the input.

FIGURE 17–81

FIGURE 17–82

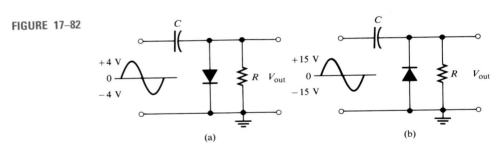

Section 17-6   Zener Diodes

16. A certain zener diode has a $V_Z = 7.5$ V and an $R_Z = 0.5$ Ω. Sketch the equivalent circuit.

17. Determine the *minimum* input voltage required for regulation to be established in Figure 17–83. Assume an ideal zener diode with $I_{ZK} = 1.5$ mA and $V_Z = 14$ V.

FIGURE 17–83

18. To what value must R be adjusted in Figure 17–84 to make $I_Z = 40$ mA? Assume that $V_Z = 12$ V and $R_Z = 30$ Ω.

19. A loaded zener regulator is shown in Figure 17–85. For $V_Z = 5.1$ V, $I_{ZK} = 5$ mA, $I_{ZM} = 70$ mA, and $R_Z = 12$ Ω, determine the minimum and maximum load currents.

20. Find the percent load regulation in Problem 19.

FIGURE 17–84

FIGURE 17–85

21. For the circuit of Problem 19, assume that the input voltage is varied from 6 V to 12 V. Determine the percent input regulation with no load and with maximum load.

22. The no-load output voltage of a certain zener regulator is 8.23 V, and the full-load output is 7.98 V. Calculate the percent load regulation.

Section 17–7 Varactor Diodes

23. Figure 17–86 is a curve of reverse voltage versus capacitance for a certain varactor. Determine the change in capacitance if V_R varies from 5 V to 20 V.

24. Refer to Figure 17–86 and determine the value of V_R that produces 25 pF.

FIGURE 17–86

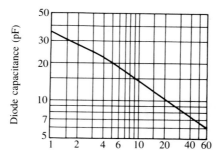

V_R, reverse voltage (V)

25. What capacitance value is required for each of the varactors in Figure 17–87 to produce a resonant frequency of 1 MHz?

26. At what value must the control voltage be set in Problem 25 if the varactors have the characteristic curve in Figure 17–86?

FIGURE 17–87

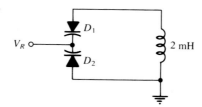

Section 17–8   LEDs and Photodiodes

27. When the switch in Figure 17–88 is closed, will the microammeter reading increase or decrease? Assume that D_1 and D_2 are optically coupled.

28. With no incident light, a certain amount of reverse current flows in a photodiode. What is this current called?

FIGURE 17–88

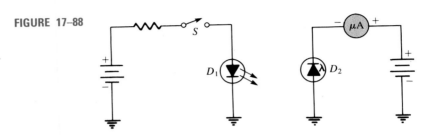

ANSWERS TO SECTION REVIEWS

Section 17–1
1. 270°.   **2.** 50%.   **3.** 3.18 V.

Section 17–2
1. 38.2 V.   **2.** Bridge.   **3.** True.

Section 17–3
1. 60 Hz, 120 Hz.   **2.** Capacitor charging and discharging slightly.
3. It increases ripple.   **4.** Reduced ripple, less dc voltage output.

Section 17–4
1. Half-wave output.   **2.** Open diode.   **3.** Leaky filter capacitor.

Section 17–5
1. See Figure 17–89.   **2.** See Figure 17–90.   **3.** See Figure 17–91.

FIGURE 17–89   **FIGURE 17–90**

FIGURE 17–91

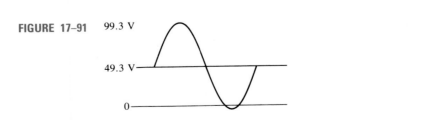

Section 17–6
1. True.   **2.**   10.24 V.

3. *Input regulation:* Constant output voltage for varying input voltage. *Load regulation:* Constant output voltage for varying load current.

4. Maximum R_L.   **5.**   0.833%.

Section 17–7
1. It is a variable capacitor.   **2.**   It decreases.

Section 17–8
1. LEDs give off light when forward-biased; photodiodes respond to light when reverse-biased.

2. Gallium arsenide, gallium arsenide phosphide, gallium phosphide.

3. Dark current.

Section 17–9
1. Forward current and voltage; reverse current and voltage; thermal.

2. **(a)** Maximum dc forward voltage drop;   **(b)**   maximum reverse direct current;   **(c)**   average maximum forward current.

18 Transistors and Thyristors

Two basic types of transistors will be discussed in this chapter: the *bipolar junction transistor (BJT)* and the *field-effect transistor (FET)*. The two major application areas of *amplification* and *switching* are introduced. Also, the *unijunction transistor (UJT)* will be introduced and common types of thyristors presented. After completing this chapter, you will be able to

- □ Describe the basic construction of bipolar junction transistors (BJTs).
- □ Distinguish between NPN and PNP transistors.
- □ Bias a BJT with dc voltages so that it can be operated as an amplifier.
- □ Define the transistor currents and explain how they are related.
- □ Explain how voltage-divider bias works in a transistor circuit.
- □ Interpret the characteristic curves of a transistor.
- □ Define the terms *cutoff* and *saturation*.
- □ Explain how a transistor produces voltage gain.
- □ Explain how a transistor can be used as a switch.
- □ Identify the significance of certain transistor parameters.
- □ Test a transistor.
- □ Describe the basic construction and operation of junction field-effect transistors (JFETs).
- □ Describe the basic construction and operation of metal oxide semiconductor field-effect transistors.
- □ Explain how a MOSFET functions in the depletion and the enhancement modes.
- □ Properly bias an FET.
- □ Describe how a unijunction transistor (UJT) works and how it is used in a basic oscillator.
- □ Explain the basic operation of SCRs, triacs, and diacs and how they are used as control devices.
- □ Recognize various types of transistor packages and identify the transistor leads.

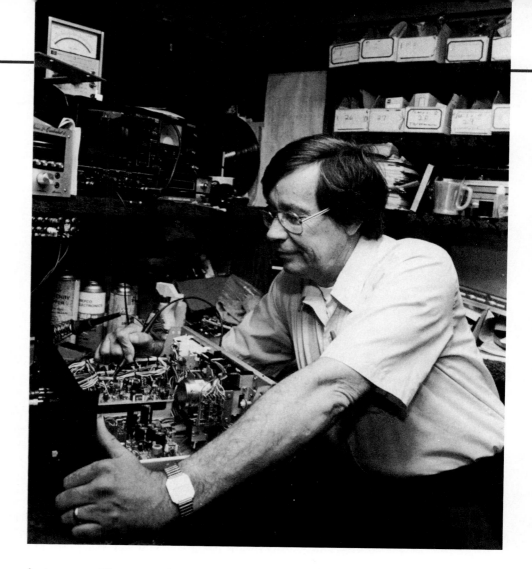

APPLICATION ASSIGNMENT

A stereo amplifier comes into your shop with one of the channels dead. There is no output signal to the speaker. You suspect that one of the bipolar power transistors has completely failed and thus must be removed and checked. How do you proceed, and what instrument(s) do you need? After completing this chapter, you will be able to carry out this assignment.

18-1

BIPOLAR JUNCTION TRANSISTORS (BJTs)

The bipolar junction transistor is constructed with *three* doped semiconductor regions separated by *two* PN junctions. The three regions are called *emitter, base,* and *collector.* The two types of bipolar transistors are shown in Figure 18–1. One type consists of two N regions separated by a P region (NPN), and the other consists of two P regions separated by an N region (PNP).

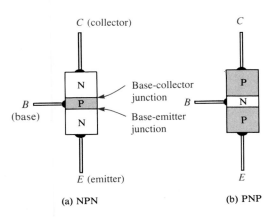

FIGURE 18–1
Construction of bipolar transistors.

The PN junction joining the base region and the emitter region is called the *base-emitter* junction. The junction joining the base region and the collector region is called the *base-collector* junction, as indicated. A wire lead connects to each of the three regions, as shown. These leads are labeled *E, B,* and *C* for emitter, base, and collector, respectively. The base material is lightly doped and very narrow compared to the heavily doped emitter and collector materials. The reason for this is discussed in the next section.

Figure 18–2 shows the schematic symbols for the NPN and PNP bipolar transistors. The term *bipolar* refers to the use of both *holes* and *electrons* as carriers in the transistor structure.

FIGURE 18–2
Transistor symbols.

(a) NPN   (b) PNP

Transistor Biasing

In order for the transistor to operate properly as an amplifier, the two PN junctions must be correctly *biased* with external voltages. We will use the NPN transistor to illustrate transistor biasing. The operation of the PNP is the same as for the NPN except that the roles of the electrons and holes, the bias voltage polarities, and the current directions are all reversed. Figure 18–3 shows the proper bias arrangement for both NPN and PNP transistors. Notice that in both cases the *base-emitter (BE) junction is forward-biased and the base-collector (BC) junction is reverse-biased*. This is called *forward-reverse* bias.

(a) NPN

(b) PNP

FIGURE 18–3
Forward-reverse bias of a bipolar transistor.

Now, let's examine what happens inside the transistor when it is forward-reverse-biased. The forward bias from base to emitter *narrows* the *BE* depletion layer, and the reverse bias from base to collector *widens* the *BC* depletion layer, as depicted in Figure 18–4(a). The N-type emitter region is teeming with conduction band (free) electrons which easily diffuse across the *BE* junction into the P-type base region, just as in a forward-biased diode.

The base region is lightly doped and very thin so that it has a very limited number of holes. Thus, only a small percentage of all the electrons flowing across the *BE* junction combine with the available holes. These relatively few recombined electrons flow out of the base lead as valence electrons, forming the small base current, I_B, as shown in Figure 18–4(b).

Most of the electrons flowing from the emitter into the base region diffuse into the *BC* depletion layer. Once in this layer, they are pulled across the *BC* junction by the depletion layer field set up by the force of attraction between the positive and negative ions. Actually, you can think of the electrons as being pulled across the reverse-biased *BC* junction by the attraction of the positive ions on the other side, as illustrated in part (c). The electrons now move through the collector region, out through the collector lead, and into the positive terminal of the external dc source, thereby forming the collector current, I_C, as shown. The amount of collector current depends directly on the amount of base current and is essentially independent of the dc collector voltage.

Transistor Currents

The directions of current in an NPN and a PNP transistor are as shown in Figures 18–5(a) and 18–5(b), respectively. An examination of these diagrams shows that the emitter current is the *sum* of the collector and base currents, expressed as follows:

$$I_E = I_C + I_B \tag{18–1}$$

As mentioned before, I_B is very small compared to I_E or I_C. The capital-letter subscripts indicate dc values.

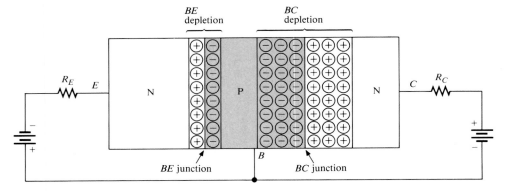

(a) Internal effects of forward-reverse bias

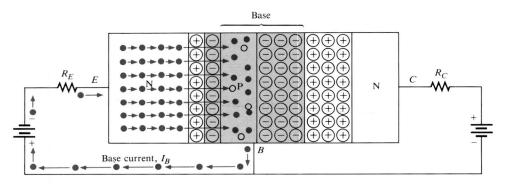

(b) Electron flow across the emitter-base junction

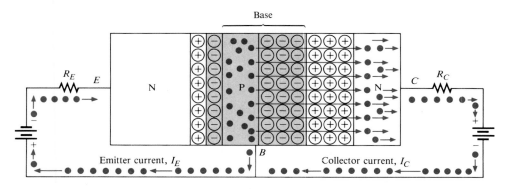

(c) Electron flow across the base-collector junction

FIGURE 18–4
Transistor action.

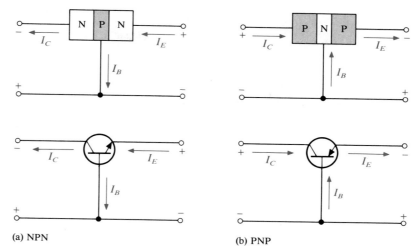

(a) NPN   (b) PNP

FIGURE 18–5
Current directions in a transistor.

These direct currents are also related by two parameters: the dc alpha (α_{dc}), which is the ratio I_C/I_E and the dc beta (β_{dc}), which is the ratio I_C/I_B. β_{dc} *is the direct current gain* and is usually designated as h_{FE} on the transistor data sheets.

The collector current is equal to α_{dc} times the emitter current:

$$I_C = \alpha_{dc}I_E \tag{18–2}$$

where α_{dc} typically has a value between 0.95 and 0.99.

The collector current is related to the base current by β_{dc}:

$$I_C = \beta_{dc}I_B \tag{18–3}$$

where β_{dc} typically has a value between 20 and 200.

Transistor Voltages

The three dc voltages for the biased transistor in Figure 18–6 are the emitter voltage V_E, the collector voltage V_C, and the base voltage V_B. These voltages are with respect

FIGURE 18–6
Bias voltages.

to ground. The collector voltage is equal to the dc supply voltage, V_{CC}, less the drop across R_C:

$$V_C = V_{CC} - I_C R_C \qquad (18\text{--}4)$$

The base voltage is equal to the emitter voltage plus the base-emitter junction barrier potential (V_{BE}), which is about 0.7 V for a silicon transistor:

$$V_B = V_E + V_{BE} \qquad (18\text{--}5)$$

In the configuration of Figure 18–6, *the emitter is the common terminal,* so $V_E = 0$ V.

EXAMPLE 18–1

Find I_B, I_C, I_E, V_B, and V_C in Figure 18–7, where β_{dc} is 50.

FIGURE 18–7

Solution

Since V_E is ground, $V_B = 0.7$ V. The drop across R_B is $V_{BB} - V_B$, so I_B is calculated as follows:

$$I_B = \frac{V_{BB} - V_B}{R_B} = \frac{3\text{ V} - 0.7\text{ V}}{10\text{ k}\Omega} = 0.23\text{ mA}$$

Now we can find I_C, I_E, and V_C.

$$I_C = \beta_{dc} I_B = 50(0.23\text{ mA}) = 11.5\text{ mA}$$

$$I_E = I_C + I_B = 11.5\text{ mA} + 0.23\text{ mA} = 11.73\text{ mA}$$

$$V_C = V_{CC} - I_C R_C = 20\text{ V} - (11.5\text{ mA})(1\text{ k}\Omega) = 8.5\text{ V}$$

SECTION REVIEW 18–1

1. Name the three transistor terminals.
2. Define *forward-reverse bias*.
3. What is β_{dc}?
4. If I_B is 10 μA and β_{dc} is 100, what is the collector current?

18-2

VOLTAGE DIVIDER BIAS

One of the most common bias arrangements in transistor circuits is the voltage divider bias. This configuration uses only a single dc source to provide forward-reverse bias to the transistor, as shown in Figure 18–8. Resistors R_1 and R_2 form a voltage divider that provides the base bias voltage. Resistor R_E allows the emitter to rise above ground potential.

FIGURE 18-8
Voltage divider bias.

The voltage divider is loaded by the resistance as viewed from the base of the transistor. In some cases, this loading effect is significant in determination of the base bias voltage. We will now examine this arrangement in more detail.

Input Resistance at the Base

FIGURE 18-9
Circuit for deriving input resistance.

The approximate input resistance of the transistor, viewed from the base of the transistor in Figure 18–9, is derived as follows:

$$R_{\text{IN}} = \frac{V_B}{I_B}$$

Neglecting the V_{BE} of 0.7 V, we have

$$V_B \cong V_E = I_E R_E$$

Since $I_E \cong I_C$ when $\alpha_{\text{dc}} \cong 1$, then

$$I_E \cong \beta_{\text{dc}} I_B$$

Substituting yields

$$V_B \cong \beta_{\text{dc}} I_B R_E$$

$$R_{\text{IN}} \cong \frac{\beta_{\text{dc}} I_B R_E}{I_B}$$

The I_B terms cancel, leaving

$$R_{\text{IN}} \cong \beta_{\text{dc}} R_E \qquad (18\text{-}6)$$

DEVICES

Base Voltage

Now, using the voltage divider formula, the following equation gives the base voltage for the circuit in Figure 18–8:

$$V_B = \left(\frac{R_2 \| R_{IN}}{R_1 + R_2 \| R_{IN}} \right) V_{CC} \qquad (18\text{–}7)$$

Generally, if R_{IN} is at least ten times greater than R_2, then Equation (18–7) can be simplified as follows:

$$V_B \cong \left(\frac{R_2}{R_1 + R_2} \right) V_{CC} \qquad (18\text{–}8)$$

Once you have determined the base voltage, you can determine the emitter voltage V_E (for an NPN transistor) as follows:

$$V_E = V_B - 0.7 \text{ V}$$

EXAMPLE 18–2

Determine V_B, V_E, V_C, V_{CE}, I_B, I_E, and I_C in Figure 18–10.

FIGURE 18–10

Solution

The input resistance at the base is

$$R_{IN} \cong \beta_{dc} R_E = 100(1 \text{ k}\Omega) = 100 \text{ k}\Omega$$

Since R_{IN} is ten times greater than R_2, the base voltage is approximately

$$V_B \cong \left(\frac{R_2}{R_1 + R_2} \right) V_{CC} = \left(\frac{10 \text{ k}\Omega}{30 \text{ k}\Omega} \right) 30 \text{ V} = 10 \text{ V}$$

Therefore, $V_E = V_B - 0.7 \text{ V} \cong 9.3 \text{ V}$.

Now that we know V_E, we can find I_E by Ohm's law:

$$I_E = \frac{V_E}{R_E} = \frac{9.3 \text{ V}}{1 \text{ k}\Omega} = 0.0093 \text{ A} = 9.3 \text{ mA}$$

Since α_{dc} is so close to 1 for most transistors, it is a good approximation to assume that $I_C \cong I_E$. Thus,

$$I_C \cong 9.3 \text{ mA}$$

Using $I_C = \beta_{dc}I_B$ and solving for I_B, we get

$$I_B = \frac{I_C}{\beta_{dc}} \cong \frac{9.3 \text{ mA}}{100} = 0.093 \text{ mA}$$

Since we know I_C, we can find V_C:

$$V_C = V_{CC} - I_C R_C = 30 \text{ V} - (9.3 \text{ mA})(1 \text{ k}\Omega)$$
$$= 30 \text{ V} - 9.3 \text{ V} = 20.7 \text{ V}$$

Since V_{CE} is the collector-to-emitter voltage, it is the difference of V_C and V_E:

$$V_{CE} = V_C - V_E = 20.7 \text{ V} - 9.3 \text{ V}$$
$$= 11.5 \text{ V}$$

SECTION REVIEW 18–2

1. How many dc voltage sources are required for voltage divider bias?
2. In Figure 18–10, if R_2 is 5 kΩ, what is the value of V_B?

18–3

THE BIPOLAR TRANSISTOR AS AN AMPLIFIER

The purpose of dc bias is to allow a transistor to operate as an amplifier. Thus, a transistor can be used to produce a larger signal using a smaller signal as a "pattern." In this section we will discuss how a transistor acts as an amplifier. The subject of amplifiers is covered in more detail in the next chapter.

Collector Curves

With a circuit such as that in Figure 18–11(a), a set of curves can be generated showing how I_C varies with V_{CE} for various values of I_B. These curves are the *collector characteristic* curves.

Notice that both V_{BB} and V_{CC} are adjustable. If V_{BB} is set to produce a specific value of I_B and V_{CC} is zero, then $I_C = 0$ and $V_{CE} = 0$. Now, as V_{CC} is gradually increased, V_{CE} will increase and so will I_C, as indicated on the portion of the curve between points A and B in Figure 18–11(b).

When V_{CE} reaches approximately 0.7 V, the base-collector junction becomes reverse-biased and I_C reaches its full value determined by the relationship $I_C = \beta_{dc}I_B$. At this point, I_C levels off to an almost constant value as V_{CE} continues to increase. This action appears to the right of point B on the curve. Actually, I_C increases slightly as V_{CE} increases due to widening of the base-collector depletion layer which results in fewer holes for recombination in the base region.

By using other values of I_B, we can produce additional I_C versus V_{CE} curves, as shown in Figure 18–11(c). These curves constitute a *family* of collector curves for a given transistor.

(a) Circuit

(b) I_C curve for one value of I_B with V_{CE} changing

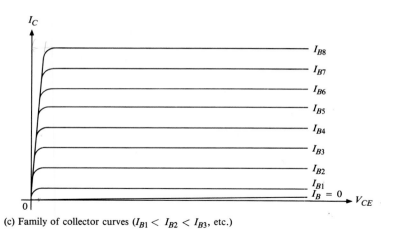

(c) Family of collector curves ($I_{B1} < I_{B2} < I_{B3}$, etc.)

FIGURE 18–11
Collector characteristic curves.

EXAMPLE 18–3

Sketch the family of collector curves for the circuit in Figure 18–12 for $I_B = 5\ \mu A$ to $25\ \mu A$ in 5-μA increments. Assume that $\beta_{dc} = 100$.

FIGURE 18–12

Solution

Using the relationship $I_C = \beta_{dc}I_B$, values of I_C are calculated and tabulated as follows:

When I_B =	Then I_C =
5 μA	0.5 mA
10 μA	1.0 mA
15 μA	1.5 mA
20 μA	2.0 mA
25 μA	2.5 mA

The resulting curves are plotted in Figure 18–13.

FIGURE 18–13

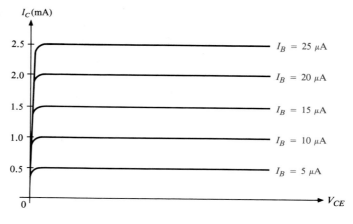

Cutoff and Saturation

When $I_B = 0$, the transistor is in *cutoff*. Under this condition, there is a very small amount of collector leakage current, I_{CEO}, due mainly to thermally produced carriers. In cutoff, both the base-emitter and the base-collector junctions are *reverse-biased*.

    Now let's consider the condition known as *saturation*. When the base current is increased, the collector current also increases and V_{CE} decreases as a result of more drop across R_C.

    When V_{CE} reaches a value called $V_{CE(sat)}$, the base-collector junction becomes *forward-biased* and I_C can increase no further even with a continued increase in I_B. At the point of saturation, the relation $I_C = \beta_{dc}I_B$ is no longer valid.

    For a transistor, $V_{CE(sat)}$ occurs somewhere below the knee of the collector curves and is usually only a few tenths of a volt for silicon transistors.

Load Line Operation

A straight line drawn on the collector curves between the cutoff and saturation points of the transistor is called the *load line*. Once set up, the transistor always operates along this line. Thus, any value of I_C and the corresponding V_{CE} will fall on this line.

Now we will set up a load line for the circuit in Figure 18–14, so that you can learn what it tells us about the transistor operation. First the *cutoff point* on the load line is determined as follows: When the transistor is cut off, there is essentially no collector current. Thus, the collector-emitter voltage V_{CE} is equal to V_{CC}. In this case, $V_{CE} = 30$ V.

FIGURE 18–14

Next the *saturation point* on the load line is determined. When the transistor is saturated, V_{CE} is approximately zero. (Actually, it is usually a few tenths of a volt, but zero is a good approximation.) Therefore, all the V_{CC} voltage is dropped across $R_C + R_E$. From this we can determine the saturation value of collector current, $I_{C(sat)}$. This value is the *maximum value* for I_C. We cannot possibly increase it further without changing V_{CC}, R_C, or R_E. In Figure 18–14, the value of $I_{C(sat)}$ is $V_{CC}/(R_C + R_E)$, which is 93.75 mA.

Next the cutoff and saturation points are plotted on the assumed curves in Figure 18–15, and a straight line, which is the *load line,* is drawn between them.

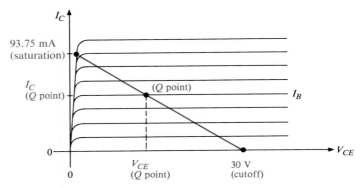

FIGURE 18–15
Load line for the circuit in Figure 18–14.

Q Point

The base current I_B is established by the base bias. The point at which the base current curve intersects the load line is the *quiescent* or *Q point* for the circuit. The coordinates of the Q point are the values for I_C and V_{CE} at that point, as illustrated in Figure 18–15.

Now we have completely described the *dc* operating conditions for the circuit. In the following paragraphs we discuss *ac* or signal conditions.

Signal Operation

The purpose of the circuit in Figure 18–16 is to produce an output signal with the same waveform as the input signal but with a greater amplitude. This increase is called *am-plification*. The figure shows an input signal, V_{in}, capacitively coupled to the base. The collector voltage is the *output signal*, as indicated. The input signal voltage causes the base current to vary at the same frequency above and below its dc value. This variation in base current produces a corresponding variation in collector current. However, the variation in collector current is much larger than the variation in base current because of the *current gain* through the transistor. The ratio of the collector current (I_c) to the base current (I_b) is designated β_{ac} (the ac beta) or h_{fe}:

$$\beta_{ac} = \frac{I_c}{I_b} \qquad (18-9)$$

The value of β_{ac} normally differs slightly from that of β_{dc} for a given transistor. Remember that lowercase subscripts distinguish ac currents and voltages from dc currents and voltages.

FIGURE 18–16
Voltage divider biased amplifier with capacitively coupled input signal. V_{in} and V_{out} are with respect to ground.

Signal Voltage Gain

Now let us take the amplifier in Figure 18–16 and examine its *voltage gain* with a signal input. The output voltage is the collector voltage. The variation in collector current produces a variation in the voltage across R_C and a resulting variation in the collector voltage, as shown in Figure 18–17.

FIGURE 18–17
Voltage amplification.

As the collector current increases, the I_cR_C drop increases. This increase produces a *decrease* in collector voltage because $V_c = V_{CC} - I_cR_C$. Likewise, as the collector current decreases, the I_cR_C drop decreases and produces an *increase* in collector voltage. Therefore, there is a 180° phase difference between the collector current and the collector voltage. The base voltage and collector voltage are also 180° out of phase, as indicated in the figure. This 180° phase difference between input and output is called an *inversion*.

The voltage gain A_v of the amplifier is V_{out}/V_{in}, where V_{out} is the signal voltage at the collector and V_{in} is the signal voltage at the base. Because the base-emitter junction is forward-biased, the signal voltage at the emitter is approximately equal to the signal voltage at the base. Thus, since $V_b \cong V_e$, the gain is approximately V_c/V_e or I_cR_C/I_eR_E. I_c and I_e are very close to the same value because the α_{ac} is close to 1. Therefore, they cancel, giving the voltage gain formula:

$$A_v \cong \frac{R_C}{R_E} \qquad (18\text{--}10)$$

A negative sign on A_v is often used to indicate inversion.

EXAMPLE 18–4

In Figure 18–18, a signal voltage of 50 mV rms is applied to the base.
(a) Determine the output voltage for the amplifier.
(b) Find the dc collector voltage on which the output signal voltage is riding.
(c) Sketch the output waveform.

FIGURE 18–18

Solution
(a) The signal voltage gain is

$$A_v \cong \frac{R_C}{R_E} = \frac{10\ \text{k}\Omega}{1\ \text{k}\Omega} = 10$$

The output signal voltage is the input signal voltage times the voltage gain:

$$V_{out} = A_vV_{in} = (10)(50\ \text{mV}) = 0.5\ \text{V rms}$$

(b) Next we find the dc collector voltage:

$$R_{IN} \cong \beta_{dc}R_E = (150)(1\ \text{k}\Omega) = 150\ \text{k}\Omega$$

R_{IN} can be neglected since it is more than ten times R_2. Thus,

$$V_B \cong \left(\frac{R_2}{R_1 + R_2}\right) V_{CC} = \left(\frac{4.7 \text{ k}\Omega}{51.7 \text{ k}\Omega}\right) 25 \text{ V} = 2.27 \text{ V}$$

$$I_C \cong I_E = \frac{V_E}{R_E} = \frac{V_B - 0.7 \text{ V}}{1 \text{ k}\Omega} = 1.57 \text{ mA}$$

$$V_C = V_{CC} - I_C R_C = 25 \text{ V} - (1.57 \text{ mA})(10 \text{ k}\Omega) = 9.3 \text{ V}$$

This value is the dc level of the output. The peak value of the output signal is

$$V_p = 1.414(0.5 \text{ V}) = 0.707 \text{ V}$$

(c) Figure 18–19 shows the waveform.

FIGURE 18–19

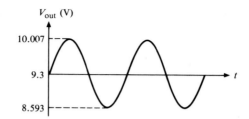

Signal Operation on the Load Line

We can obtain a graphical picture of an amplifier's operation by showing an example of signal variations on a set of collector curves with a load line, as shown in Figure 18–20. Let us assume that the dc Q-point values are as follows: $I_B = 40 \ \mu\text{A}$, $I_C = 4 \text{ mA}$, and $V_{CE} = 8 \text{ V}$. The input signal varies the base current from a maximum of 50 μA to a minimum of 30 μA. The resulting variations in collector current and V_{CE} are shown on the graph. The operation is linear as long as the variations do not reach cutoff at the lower end of the load line or saturation at the upper end.

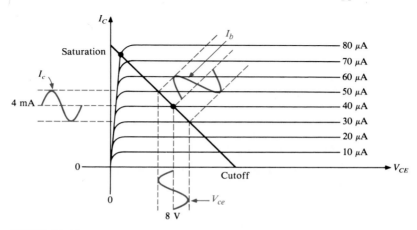

FIGURE 18–20
Signal operation on the load line.

SECTION REVIEW
18–3

1. What does the term *amplification* mean?
2. What is the Q point?
3. A certain transistor circuit has $R_C = 47 \text{ k}\Omega$ and $R_E = 2.2 \text{ k}\Omega$. What is the approximate voltage gain?

THE BIPOLAR TRANSISTOR AS A SWITCH

In the previous section, we discussed the transistor as a linear amplifier. The second major application area is switching applications. When used as an electronic switch, a transistor normally is operated alternately in cutoff and saturation.

Figure 18–21 illustrates the basic operation of a transistor as a switching device. In part (a), the transistor is cut off because the base-emitter junction is *not* forward-biased. In this condition, there is, ideally, an *open* between collector and emitter. In part (b), the transistor is saturated because the base-emitter junction is forward-biased and the base current is large enough to cause the collector current to reach its saturated value. In this condition there is, ideally, a *short* between collector and emitter. Actually, a voltage drop of a few tenths of a volt normally occurs.

(a) Cutoff   (b) Saturation

FIGURE 18–21
Ideal switching action of a transistor.

Conditions in Cutoff

As mentioned before, a transistor is in cutoff when the base-emitter junction is *not* forward-biased. All of the currents are approximately zero, and V_{CE} is approximately equal to V_{CC}:

$$V_{CE(\text{cutoff})} \cong V_{CC} \qquad (18\text{–}11)$$

Conditions in Saturation

When the emitter junction is forward-biased and there is enough base current to produce a maximum collector current, the transistor is saturated. Since V_{CE} is approximately zero at saturation, the collector current is

$$I_{C(\text{sat})} \cong \frac{V_{CC}}{R_C} \qquad (18\text{–}12)$$

The value of base current needed to produce saturation is

$$I_{B(min)} = \frac{I_{C(sat)}}{\beta_{dc}}$$ (18–13)

EXAMPLE 18–5

(a) For the transistor switching circuit in Figure 18–22, what is V_{CE} when V_{IN} = 0 V?

(b) What minimum value of I_B is required to saturate this transistor if the β_{dc} is 200?

(c) Calculate the maximum value of R_B when V_{IN} = 5 V.

FIGURE 18–22

Solution

(a) When V_{IN} = 0 V, the transistor if *off* and V_{CE} = V_{CC} = 10 V.

(b) When the transistor is saturated, $V_{CE} \cong 0$ V, so

$$I_{C(sat)} \cong \frac{V_{CC}}{R_C} = \frac{10 \text{ V}}{1 \text{ k}\Omega} = 10 \text{ mA}$$

$$I_{B(min)} = \frac{I_{C(sat)}}{\beta_{dc}} = \frac{10 \text{ mA}}{200} = 0.05 \text{ mA}$$

This is the value of I_B necessary to drive the transistor to the point of saturation. *Any further increase in I_B will drive the transistor deeper into saturation but will not increase I_C.*

(c) When the transistor is saturated, V_{BE} = 0.7 V. The voltage across R_B is V_{IN} − 0.7 V = 4.3 V. The maximum value of R_B needed to allow a minimum I_B of 0.05 mA is calculated by Ohm's law as follows:

$$R_B = \frac{V_{IN} - 0.7 \text{ V}}{I_B} = \frac{4.3 \text{ V}}{0.05 \text{ mA}} = 86 \text{ k}\Omega$$

SECTION REVIEW
18–4

1. When a transistor is used as a switching device, it is operated in either _____ or _____ .

2. When does the collector current reach its maximum value?

3. When is the collector current approximately zero?

4. Name the two conditions that produce saturation.

5. When is V_{CE} equal to V_{CC}?

BJT PARAMETERS AND RATINGS

More about β_{dc}

Because the β_{dc} is a very important bipolar transistor parameter, we need to examine it further. β_{dc} varies with both collector current and temperature. Keeping the junction temperature constant and increasing I_C causes β_{dc} to increase to a maximum. A further increase in I_C beyond this maximum point causes β_{dc} to decrease.

If I_C is held constant and the temperature is varied, β_{dc} changes directly with the temperature. If the temperature goes up, β_{dc} goes up, and vice versa. Figure 18–23 shows the variation of β_{dc} with I_C and junction temperature (T_J) for a typical transistor.

FIGURE 18–23
Variation of β_{dc} with I_C for several temperatures.

A transistor data sheet usually specifies β_{dc} (h_{FE}) at specific I_C values. Even at fixed values of I_C and temperature, β_{dc} varies from device to device for a given transistor.

The β_{dc} specified at a certain value of I_C is usually the minimum value, $\beta_{dc(min)}$, although the maximum and typical values are also sometimes specified. A typical transistor data sheet is shown in Appendix E.

Maximum Ratings

Like any other electronic device, the transistor has limitations on its operation. These limitations are stated in the form of *maximum ratings* and are normally specified on the manufacturer's data sheet. Typically, maximum ratings are given for *collector-to-base voltage* V_{CB}, *collector-to-emitter voltage* V_{CE}, *emitter-to-base voltage* V_{EB}, *collector current* I_C, and *power dissipation* P_D.

The product of V_{CE} and I_C must not exceed the maximum power dissipation. Both V_{CE} and I_C cannot be maximum at the same time. If V_{CE} is maximum, I_C can be calculated as

$$I_C = \frac{P_{D(max)}}{V_{CE}} \qquad (18\text{–}14)$$

If I_C is maximum, we can calculate V_{CE} by rearranging Equation (18–14) as follows:

$$V_{CE} = \frac{P_{D(\text{max})}}{I_C} \qquad (18\text{–}15)$$

For a given transistor, a maximum power dissipation curve can be plotted on the collector curves, as shown in Figure 18–24(a). These values are tabulated in part (b). For this transistor, $P_{D(\text{max})}$ is 0.5 W, $V_{CE(\text{max})}$ is 20 V, and $I_{C(\text{max})}$ is 50 mA. The curve shows that this particular transistor cannot be operated in the shaded portion of the graph. $I_{C(\text{max})}$ is the limiting rating between points A and B, $P_{D(\text{max})}$ is the limiting rating between points B and C, and $V_{CE(\text{max})}$ is the limiting rating between points C and D.

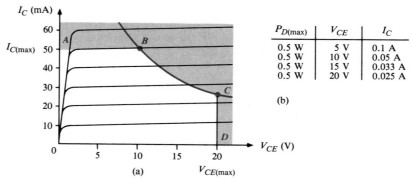

$P_{D(\text{max})}$	V_{CE}	I_C
0.5 W	5 V	0.1 A
0.5 W	10 V	0.05 A
0.5 W	15 V	0.033 A
0.5 W	20 V	0.025 A

(b)

(a)

FIGURE 18–24
Maximum power dissipation curve.

EXAMPLE 18–6

The silicon transistor in Figure 18–25 has the following maximum ratings: $P_{D(\text{max})} = 0.8$ W, $V_{CE(\text{max})} = 15$ V, $I_{C(\text{max})} = 100$ mA, $V_{CB(\text{max})} = 20$ V, and $V_{EB(\text{max})} = 10$ V. Determine the maximum value to which V_{CC} can be adjusted without exceeding a rating. Which rating would be exceeded first?

FIGURE 18–25

Solution
First find I_B so that I_C can be determined.

$$I_B = \frac{V_{BB} - V_{BE}}{R_B} = \frac{5\text{ V} - 0.7\text{ V}}{22\text{ k}\Omega} = 195.5\ \mu\text{A}$$

$$I_C = \beta_{\text{dc}}I_B = (100)(195.5\ \mu\text{A}) = 19.55\text{ mA}$$

I_C is much less than $I_{C(\text{max})}$ and will not change with V_{CC}. It is determined only by I_B and β_{dc}.

The voltage drop across R_C is

$$V_{R_C} = I_C R_C = (19.55 \text{ mA})(1 \text{ k}\Omega) = 19.55 \text{ V}$$

Now we can determine the maximum value of V_{CC} when $V_{CE} = V_{CE(\text{max})} = 15$ V:

$$V_{R_C} = V_{CC} - V_{CE}$$

Thus,

$$V_{CC(\text{max})} = V_{CE(\text{max})} + V_{R_C} = 15 \text{ V} + 19.55 \text{ V} = 34.55 \text{ V}$$

V_{CC} can be increased to 34.55 V, under the existing conditions, before $V_{CE(\text{max})}$ is exceeded. However, we do not know whether or not $P_{D(\text{max})}$ has been exceeded at this point. Let's find out.

$$P_D = V_{CE(\text{max})} I_C = (15 \text{ V})(19.55 \text{ mA}) = 0.293 \text{ W}$$

Since $P_{D(\text{max})}$ is 0.8 W, it is *not* exceeded when $V_{CE} = 34.55$ V. Thus, $V_{CE(\text{max})}$ is the limiting rating in this case.

It should be noted that if the base current is removed, causing the transistor to turn off, $V_{CE(\text{max})}$ will be exceeded because the entire supply voltage, V_{CC}, will be dropped across the transistor.

SECTION REVIEW
18–5

1. The β_{dc} of a transistor increases with temperature (T or F).
2. Generally, what effect does an increase in I_C have on the β_{dc}?
3. What is the allowable collector current in a transistor with $P_{D(\text{max})} = 0.32$ W when $V_{CE} = 8$ V?

18–6

TRANSISTOR TESTING

Ohmmeter Check of Transistor Junctions

The ohmmeter provides a simple test for open or shorted junctions. *The base-emitter and base-collector junctions are each treated as a diode for this test.* The junction should show a *low* resistance when forward-biased and a very *high* resistance when reverse-biased. The internal battery of the ohmmeter must provide the bias voltage.

Figure 18–26(a) shows a *forward-biased* check of an NPN transistor. Notice the polarities of the ohmmeter leads. Part (b) of the figure demonstrates a *reverse-biased* check. For a PNP transistor, the meter polarities are reversed from those shown in Figure 18–26.

(a) Forward-biased check   (b) Reverse-biased check

FIGURE 18–26
Ohmmeter check of NPN transistor junctions.

Leakage Measurement

Very small *leakage* currents exist in all transistors and in most cases are small enough to neglect.

When a transistor is connected as shown in Figure 18–27(a) with the base open ($I_B = 0$), it is in cutoff. Ideally $I_C = 0$, but actually there is a small current from collector to emitter, as mentioned earlier, called I_{CEO} (collector-to-emitter current with base open). This leakage current is usually in the nanoampere (nA) range for silicon. A faulty transistor will often have excessive leakage current and can be checked in a transistor tester which connects an ammeter as shown.

Another leakage current in transistors is the reverse collector-to-base current, I_{CBO}. This current is measured with the emitter open, as shown in part (b). If it is excessive, a shorted collector-base junction is likely.

FIGURE 18–27
Leakage-current test circuits.

(a) Circuit for I_{CEO} test   (b) Circuit for I_{CBO} test

Gain Measurement

In addition to leakage tests, the typical transistor tester also checks the β_{dc}. A known I_B is applied and the resulting I_C is measured. The reading will indicate the value of the I_C/I_B ratio, although in some units only a relative indication is given.

Most testers provide for an in-circuit β_{dc} check, so that a suspected device does not have to be removed from the circuit for testing. Figure 18–28 shows two typical transistor testers.

FIGURE 18–28
Transistor tester (courtesy of B&K Precision).

Curve Tracers

The *curve tracer* is an oscilloscope type of instrument that can display transistor characteristics such as a family of collector curves. In addition to measuring and displaying various transistor characteristics, the curve tracer can also display diode curves as well as the β_{dc}. A typical curve tracer is shown in Figure 18–29.

FIGURE 18–29
Curve tracer (courtesy Tektronix, Inc.).

**SECTION REVIEW
18–6**

1. The positive lead of an ohmmeter is on a transistor's base, the negative lead is on the emitter, and a low resistance reading is obtained. Assuming the transistor is good, is it a PNP or an NPN type?
2. Name two types of transistor leakage currents.

18–7   THE JUNCTION FIELD-EFFECT TRANSISTOR (JFET)

Depending on their structure, JFETs fall into either of two categories: *N-channel* or *P-channel*. Figure 18–30(a) shows the basic structure of an N-channel JFET. Wire leads are connected to each end of the N-channel; the *drain* is at the upper end and the *source* is at the lower end. Two P-type regions are diffused in the N-type material to form a channel, and *both* P-type regions are connected to the gate lead. In the remaining structure diagrams, the interconnection of *both* P-type regions is omitted for simplicity, with a connection to only one shown. A P-channel JFET is shown in Figure 18–30(b).

FIGURE 18–30
Structure of the two types of JFET.

Basic Operation

To illustrate the operation of a JFET, bias voltages are shown applied to an N-channel device in Figure 18–31(a). V_{DD} provides a drain-to-source voltage and supplies current from drain to source. V_{GG} sets the reverse-biased voltage between the gate and the source, as shown.

The JFET is always operated with the gate-to-source PN junction reverse-biased. Reverse biasing of the gate-source junction with a negative gate voltage produces a *depletion* region in the N channel and thus increases its resistance. The channel width can be controlled by varying the gate voltage, and thereby the amount of drain current, I_D, can also be controlled. This concept is illustrated in Figure 18–31(b) and (c).

The color shaded areas represent the depletion region created by the reverse bias. It is wider toward the drain end of the channel because the reverse-biased voltage between the gate and the drain is greater than that between the gate and the source.

JFET Symbols

The schematic symbols for both N-channel and P-channel JFETs are shown in Figure 18–32. Notice that the arrow on the gate points "in" for N-channel and "out" for P-channel.

(a) JFET biased for conduction

(b) Greater V_{GG} narrows the channel, thus decreasing I_D.

(c) Less V_{GG} widens channel and increases I_D.

FIGURE 18–31
Effects of V_{GG} on channel width and drain current ($V_{GG} = V_{GS}$).

FIGURE 18–32
JFET schematic symbols.

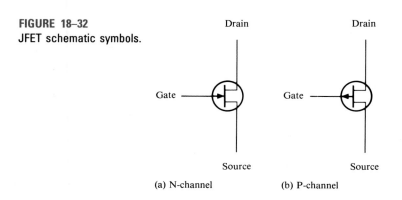

(a) N-channel   (b) P-channel

SECTION REVIEW
18–7

1. Name the three terminals of a JFET.
2. An N-channel JFET requires a (positive, negative, or 0) V_{GS}.
3. How is the drain current controlled in a JFET?

18–8

JFET CHARACTERISTICS

Drain Curves and Pinch-Off Voltage

First consider the case where the gate-to-source voltage is 0 ($V_{GS} = 0$ V). This voltage is produced by shorting the gate to the source, as in Figure 18–33(a). As V_{DD} (and thus V_{DS}) is increased from 0, I_D will increase proportionally, as shown in the graph of Figure 18–33(b) between points A and B. In this region, the channel resistance is essentially constant because the depletion region is not large enough to have significant effect. This region is called the *ohmic region* because V_{DS} and I_D are related by Ohm's law.

(a) JFET with $V_{GS} = 0$ V and a variable V_{DS}

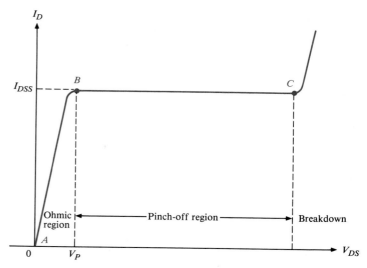

(b) Drain characteristic curve for $V_{GS} = 0$ V

FIGURE 18–33
Generation of drain characteristic curve for a JFET.

At point B, the curve levels off and I_D becomes a relatively constant value called I_{DSS}. It is at this point that the reverse-bias voltage across the *gate-to-drain* junction (V_{GD}) produces a depletion region sufficient to narrow the channel so that its resistance begins to *increase* significantly. The value of V_{GD} at this point is called the *pinch-off* voltage V_P. In the case where the gate bias voltage is zero, $-V_P = V_{DS}$ at the point of pinch-off, because V_{DS} and V_{DG} are equal. In general, however,

$$V_P = V_{GS} - V_{DS(P)} \qquad (18\text{–}16)$$

where $V_{DS(P)}$ is the pinch-off value of V_{DS} for a given value of V_{GS}.

V_P is a constant value for a given JFET and represents a fixed parameter. The pinch-off value of V_{DS} is a variable that depends on the gate-to-source bias voltage, V_{GS}.

As you can see from the curve for $V_{GS} = 0$ V in Figure 18–33(b), a continued increase in V_{DS} above point B produces an essentially constant I_D equal to a specific value called I_{DSS}. *I_{DSS} is the maximum value of I_D when $V_{GS} = 0$.* At point C, breakdown occurs and I_D increases rapidly with irreversible damage to the device very likely. JFETs are always operated below the breakdown point and within the pinch-off region (between points B and C).

Now consider a negative gate bias voltage, for example, $V_{GS} = -1$ V, as shown in Figure 18–34(a). As V_{DD} is increased from 0, pinch-off occurs at a lower value of V_{DS}, as shown in Figure 18–34(b). The reason for this is that the pinch-off voltage V_P is constant, and therefore, for a certain negative gate voltage, the drain voltage must only reach a value sufficient to make the gate-to-drain voltage equal to V_P in order to produce pinch-off.

As V_{GS} is set to increasingly negative values, a family of characteristic curves is produced as shown in Figure 18–34(b).

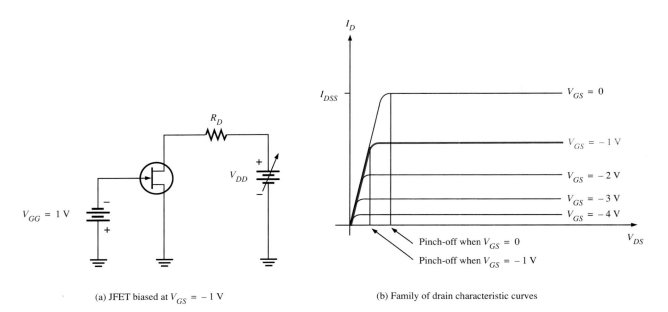

(a) JFET biased at $V_{GS} = -1$ V   (b) Family of drain characteristic curves

FIGURE 18–34
Pinch-off occurs at a lower V_{DS} when V_{GS} goes from 0 V to -1 V, and to successively more negative values.

Cutoff

As you know, for an N-channel JFET, the more negative V_{GS} is, the smaller I_D in the pinch-off region becomes. When V_{GS} is made sufficiently negative, I_D is reduced to 0 because the depletion region widens to a point where it completely closes the channel. The value of V_{GS} at the cutoff point is designated $V_{GS(off)}$.

For any given N-channel JFET, cutoff occurs when $V_{GS} = -V_P$. Since the pinch-off voltage V_P is a constant for a given JFET, when $V_{GS} = -V_P$ the drain-to-source voltage $V_{DS(P)}$ must be 0. Since there is no voltage drop between the drain and source, I_D *must* be 0. Even though V_{DS} may increase above 0 V, I_D remains essentially constant at near 0 A. Cutoff is illustrated in Figure 18–35.

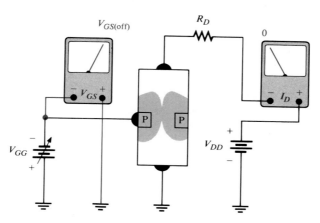

FIGURE 18–35
JFET at cutoff.

Do not confuse *cutoff* with *pinch-off*. The pinch-off voltage V_P is the value of the *gate-to-drain voltage* V_{GD} *at which the drain current reaches a constant value for a given value of* V_{GS}.

The cutoff voltage $V_{GS(off)}$ *is the value of* V_{GS} *at which the drain current is 0.* I_D is 0 only when the magnitude of V_{GS} is equal to or greater than the magnitude of V_P. I_D is nonzero for less negative values of V_{GS}.

EXAMPLE 18–7

For the JFET in Figure 18–36, $V_P = -8$ V and $I_{DSS} = 12$ mA.
(a) Determine the value of V_{DS} at which pinch-off begins.
(b) If the gate is grounded, what is the value of I_D for $V_{DD} = 12$ V when V_{DS} is above pinch-off?

FIGURE 18–36

Solution

(a) By rearranging Equation (18–16), we get

$$V_{DS(P)} = V_{GS} - V_P$$

Substituting $V_P = -8$ V and $V_{GS} = -5$ V yields

$$V_{DS(P)} = -5 \text{ V} - (-8 \text{ V}) = 3 \text{ V}$$

This is the value of V_{DS} at the beginning of pinch-off.

(b) When $V_{GS} = 0$ V, $I_D = I_{DSS} = 12$ mA for any value of V_{DS} between pinch-off (8 V) and breakdown.

Why V_P and $V_{GS(off)}$ Are Equivalent

I_D is 0 when $V_{GS} = V_P$. Because a zero drain current corresponds to an *off* condition, the magnitude of V_P is equivalent to the magnitude of $V_{GS(off)}$.

Most FET data sheets give a value only for $V_{GS(off)}$ and not for V_P. But once we know $V_{GS(off)}$, we also know V_P. For example, if $V_{GS(off)} = -5$ V, then $V_P = -5$ V.

JFET Input Resistance and Capacitance

A JFET operates with its gate-source junction reverse-biased. Therefore, the input resistance at the gate is very high. This high input resistance is one advantage of the JFET over the bipolar transistor. (Recall that a bipolar transistor operates with a forward-biased base-emitter junction.)

JFET data sheets often specify the input resistance by giving a value for the *gate reverse current* I_{GSS} at a certain gate-to-source voltage. The input resistance can then be determined using the following equation:

$$R_{IN} = \left| \frac{V_{GS}}{I_{GSS}} \right| \tag{18–17}$$

For example, the 2N3970 data sheet lists a maximum I_{GSS} of 250 pA for $V_{GS} = 20$ V at 25°C. I_{GSS} increases with temperature, so the input resistance decreases.

The input capacitance C_{iss} of a JFET is considerably greater than that of a bipolar transistor because the JFET operates with a reverse-biased PN junction. Recall that a reverse-biased PN junction acts as a capacitor whose capacitance depends on the amount of reverse voltage.

EXAMPLE 18–8

A certain JFET has an I_{GSS} of 1 nA for $V_{GS} = -20$ V. Determine the input resistance.

Solution

$$R_{IN} = \left| \frac{V_{GS}}{I_{GSS}} \right| = \frac{20 \text{ V}}{1 \text{ nA}} = 20,000 \text{ M}\Omega$$

SECTION REVIEW
18–8
1. The drain-to-source voltage at the pinch-off point of a particular JFET is 7 V. If the gate-to-source voltage is 0, what is V_P?

2. The V_{GS} of a certain N-channel JFET is increased negatively. Does the drain current increase or decrease?

3. What value must V_{GS} have to produce cutoff in a P-channel JFET with a $V_P = 3$ V?

18–9

THE METAL OXIDE SEMICONDUCTOR FET (MOSFET)

The MOSFET is the second category of field-effect transistor. It differs from the JFET in that it has no PN junction structure; instead, the gate of the MOSFET is *insulated* from the channel by a silicon dioxide (SiO_2) layer.

There are two basic types of MOSFETs: *depletion-enhancement* (DE) and *enhancement-only* (E).

Depletion-Enhancement (DE) MOSFET

Figure 18–37 illustrates the basic structure of DE MOSFETs. The drain and source are diffused into the substrate material and then connected by a narrow channel adjacent to the insulated gate. Both N-channel and P-channel devices are shown in the figure.

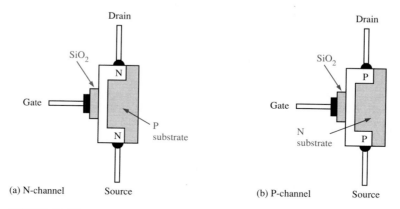

(a) N-channel

(b) P-channel

FIGURE 18–37
Basic structure of DE MOSFETs.

We will use the N-channel device to describe the basic operation. The P-channel operation is the same, except the voltage polarities are opposite those of the N-channel.

The DE MOSFET can be operated in either of two modes: the *depletion mode* or the *enhancement mode*. Since the gate is insulated from the channel, either a positive or a negative gate voltage can be applied. The MOSFET operates in the depletion mode when a *negative* gate-to-source voltage is applied and in the enhancement mode when a *positive* gate-to-source voltage is applied.

Depletion Mode

Visualize the gate as one plate of a parallel plate capacitor and the channel as the other plate. The silicon dioxide insulating layer is the dielectric. With a negative gate voltage, the negative charges on the gate repel conduction electrons from the channel, leaving positive ions in their place. Thereby, the N-channel is *depleted* of some of its electrons, thus decreasing the channel conductivity. The greater the negative voltage on the gate, the greater the depletion of N-channel electrons. At a sufficiently negative gate-to-source voltage, $V_{GS(off)}$, the channel is totally depleted and the drain current is 0. This depletion mode is illustrated in Figure 18–38(a).

(a) Depletion mode (V_{GS} negative and less than $V_{GS(off)}$)

(b) Enhancement mode (V_{GS} positive)

FIGURE 18–38
Operation of N-channel DE MOSFET.

Like the N-channel JFET, the N-channel DE MOSFET conducts drain current for gate-to-source voltages between $V_{GS(off)}$ and 0. In addition, the DE MOSFET conducts for values of V_{GS} above 0.

Enhancement Mode

With a positive gate voltage, more conduction electrons are attracted into the channel, thus increasing (enhancing) the channel conductivity, as illustrated in Figure 18–38(b).

DE MOSFET Symbols

The schematic symbols for both the N-channel and the P-channel depletion-enhancement MOSFETs are shown in Figure 18–39. The substrate, indicated by the arrow, is normally (but not always) connected internally to the source. An inward substrate arrow is for N-channel, and an outward arrow is for P-channel.

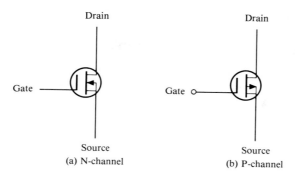

FIGURE 18–39
DE MOSFET schematic symbols.

Enhancement (E) MOSFET

This type of MOSFET operates *only* in the enhancement mode and has no depletion mode. It differs in construction from the DE MOSFET in that it has *no physical channel*. Notice in Figure 18–40(a) that the substrate extends completely to the SiO_2 layer.

For an N-channel device, a positive gate voltage above a threshold value, $V_{GS(th)}$, *induces* a channel by creating a thin layer of negative charges in the substrate region adjacent to the SiO_2 layer, as shown in Figure 18–40(b). The conductivity of the channel is enhanced by increasing the gate-to-source voltage, thus pulling more electrons into the channel. For any gate voltage below the threshold value, there is no channel.

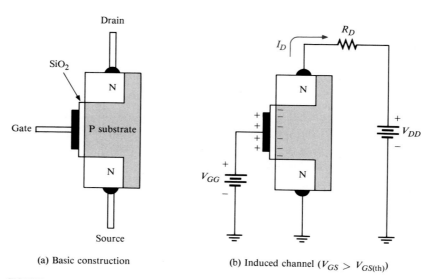

(a) Basic construction

(b) Induced channel ($V_{GS} > V_{GS(th)}$)

FIGURE 18–40
E MOSFET construction and operation.

The schematic symbols for the N-channel and P-channel E MOSFETs are shown in Figure 18–41. The broken lines symbolize the absence of a physical channel.

FIGURE 18–41
E MOSFET schematic symbols.

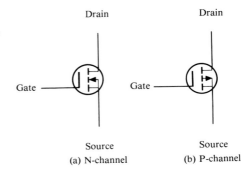

(a) N-channel   (b) P-channel

Handling Precautions

Because the gate of a MOSFET is insulated from the channel, the input resistance is extremely high (ideally infinite). The gate leakage current I_{GSS} for a typical MOSFET is in the pA range, whereas the gate reverse current for a typical JFET is in the nA range.

The input capacitance, of course, results from the insulated gate structure. Excess static charge can accumulate because the input capacitance combines with the very high input resistance and can result in damage to the device as a result of electrostatic discharge (ESD). To avoid ESD and possible damage, the following precautions should be taken:

1. MOS devices should be shipped and stored in conductive foam.
2. All instruments and metal benches used in assembly or test should be connected to earth ground (round prong of wall outlets).
3. The assembler's or handler's wrist should be connected to earth ground with a length of wire and a high-value series resistor.
4. Never remove a MOS device (or any other device, for that matter) from the circuit while the power is on.
5. Do not apply signals to a MOS device while the dc power supply is off.

**SECTION REVIEW
18–9**

1. Name two types of MOSFETs, and describe the major difference in construction.
2. If the gate-to-source voltage in a depletion-enhancement MOSFET is 0, is there current from drain to source?
3. If the gate-to-source voltage in an E MOSFET is 0, is there current from drain to source?

18–10

FET BIASING

Using some of the FET parameters discussed in the previous sections, we will now see how to dc-bias FETs. The purpose of biasing is to select a proper dc gate-to-source voltage to establish a desired value of drain current.

Self-Biasing a JFET

Recall that a JFET must be operated such that the gate-source junction is always reverse-biased. This condition requires a negative V_{GS} for an N-channel JFET and a positive

V_{GS} for a P-channel JFET. This can be achieved using the *self-bias* arrangements shown in Figure 18–42.

FIGURE 18–42
Self-biased JFETS ($I_S = I_D$ in all FETs).

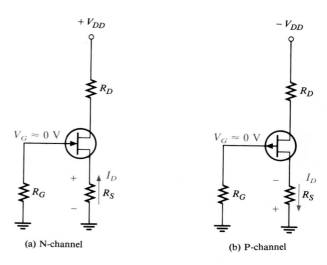

(a) N-channel

(b) P-channel

Notice that the gate is biased at approximately 0 V by resistor R_G connected to ground. The reverse leakage current I_{GSS} does produce a very small voltage across R_G, but this can be neglected in most cases, and it can be assumed that R_G has no voltage drop across it.

For the N-channel JFET in Figure 18–42(a), I_D produces a voltage drop across R_S and makes the source positive with respect to ground. Since $V_G = 0$ and $V_S = I_D R_S$, the gate-to-source voltage is

$$V_{GS} = V_G - V_S = 0 - I_D R_S$$

Thus,

$$V_{GS} = -I_D R_S \qquad (18–18)$$

For the P-channel JFET shown in Figure 18–42(b), the current through R_S produces a negative voltage at the source, and therefore

$$V_{GS} = +I_D R_S \qquad (18–19)$$

In the following analysis, the N-channel JFET is used for illustration. Keep in mind that analysis of the P-channel JFET is the same except for opposite polarity voltages.

The drain voltage with respect to ground is determined as follows:

$$V_D = V_{DD} - I_D R_D \qquad (18–20)$$

Since $V_S = I_D R_S$, the drain-to-source voltage is

$$V_{DS} = V_D - V_S$$
$$V_{DS} = V_{DD} - I_D(R_D + R_S) \qquad (18–21)$$

EXAMPLE 18–9

+10 V

R_D
1 kΩ

R_G

R_S
470 Ω

FIGURE 18–43

Find V_{DS} and V_{GS} in Figure 18–43, given that $I_D = 5$ mA.

Solution

$$V_S = I_D R_S = (5 \text{ mA})(470 \text{ } \Omega) = 2.35 \text{ V}$$

$$V_D = V_{DD} - I_D R_D = 10 \text{ V} - (5 \text{ mA})(1 \text{ k}\Omega)$$

$$= 10 \text{ V} - 5 \text{ V} = 5 \text{ V}$$

Therefore,

$$V_{DS} = V_D - V_S = 5 \text{ V} - 2.35 \text{ V} = 2.65 \text{ V}$$

Since $V_G = 0$ V,

$$V_{GS} = V_G - V_S = 0 \text{ V} - 2.35 \text{ V} = -2.35 \text{ V}$$

DE MOSFET Bias

Recall that depletion-enhancement MOSFETs can be operated with either positive or negative values of V_{GS}. A simple bias method is to set $V_{GS} = 0$ so that an ac signal at the gate varies the gate-to-source voltage above and below this bias point. A MOSFET with *0 bias* is shown in Figure 18–44.

FIGURE 18–44
A zero-biased DE MOSFET.

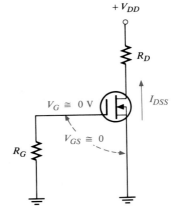

$+V_{DD}$

R_D

$V_G \cong 0$ V

I_{DSS}

$V_{GS} \cong 0$

R_G

Since $V_{GS} = 0$, $I_D = I_{DSS}$ as indicated. The drain-to-source voltage is expressed as follows:

$$V_{DS} = V_{DD} - I_{DSS}R_D \tag{18–22}$$

EXAMPLE 18–10

Determine the drain current in the circuit of Figure 18–45. The MOSFET data sheet gives $V_{GS(off)} = -8$ V and $I_{DSS} = 12$ mA.

FIGURE 18–45

+18 V

R_D
560 Ω

+
V_{DS}
−

R_G
10 MΩ

Solution

Since $I_D = I_{DSS} = 12$ mA, the drain-to-source voltage is calculated as follows:

$$V_{DS} = V_{DD} - I_{DSS}R_D = 18 \text{ V} - (12 \text{ mA})(560 \text{ }\Omega) = 11.28 \text{ V}$$

E MOSFET Bias

Recall that enhancement-only MOSFETs must have a V_{GS} greater than the threshold value, $V_{GS(th)}$. Figure 18–46 shows two ways to bias an E MOSFET. An N-channel device is used for illustration. In either bias arrangement, the purpose is to make the gate voltage more positive than the source by an amount exceeding $V_{GS(th)}$.

+V_{DD}

R_D

R_G

(a) Drain-feedback bias

+V_{DD}

R_1

R_D

R_2

(b) Voltage-divider bias

FIGURE 18–46
E MOSFET biasing arrangements.

In the drain feedback bias circuit in Figure 18–46(a), there is negligible gate current and, therefore, no voltage drop across R_G. As a result, $V_{GS} = V_{DS}$.

Equations for the voltage-divider bias in Figure 18–46(b) are as follows:

$$V_{GS} = \left(\frac{R_2}{R_1 + R_2}\right)V_{DD}$$

$$V_{DS} = V_{DD} - I_D R_D$$

EXAMPLE 18–11

Determine the amount of drain current in Figure 18–47. The MOSFET has a $V_{GS(th)}$ of 3 V.

FIGURE 18–47

Solution
The meter indicates that $V_{GS} = 8.5$ V. Since this is a drain feedback configuration, $V_{DS} = V_{GS} = 8.5$ V.

$$I_D = \frac{V_{DD} - V_{DS}}{R_D} = \frac{15 \text{ V} - 8.5 \text{ V}}{4.7 \text{ k}\Omega} = 1.38 \text{ mA}$$

**SECTION REVIEW
18–10**

1. A P-channel JFET must have a (positive, negative) V_{GS}.
2. In a certain self-biased N-channel JFET circuit, $I_D = 8$ mA and $R_S = 1$ kΩ. Determine V_{GS}.
3. For a DE MOSFET biased at $V_{GS} = 0$, is the drain current equal to 0, I_{GSS}, or I_{DSS}?
4. An N-channel E MOSFET with $V_{GS(th)} = 2$ V must have a V_{GS} in excess of _____ in order to conduct.

18–11

UNIJUNCTION TRANSISTORS (UJTs)

The unijunction transistor (UJT) consists of an emitter and two bases. It has no collector. The UJT is used mainly in switching and timing applications.

    Figure 18–48(a) shows the construction of a UJT. The base contacts are made to the N-type bar. The emitter lead is connected to the P-region. The UJT schematic symbol is shown in Figure 18–48(b).

FIGURE 18–48
Unijunction transistor (UJT).

(a)

(b)

UJT Operation

In normal UJT operation, base 2 (B_2) and the emitter are biased positive with respect to base 1 (B_1). Figure 18–49 will illustrate the operation. This equivalent circuit represents the internal UJT characteristics. The total resistance between the two bases, R_{BB}, is the resistance of the N-type material. R_{B1} is the resistance between point k and base 1. R_{B2} is the resistance between point k and base 2. The sum of these two resistances makes up the total resistance, R_{BB}. The diode represents the PN junction between the emitter and the N-type material.

FIGURE 18–49
Equivalent circuit for a UJT.

The ratio R_{B1}/R_{BB} is designated η (the Greek letter eta) and is defined as the *intrinsic standoff ratio*. It takes an emitter voltage of $V_B + \eta V_{BB}$ to turn the UJT on. This voltage is called the *peak voltage*. Once the device is on, resistance R_{B1} drops in value. Thus, as emitter current *increases*, emitter voltage *decreases* because of the decrease in R_{B1}. This characteristic is the *negative resistance* characteristic of the UJT.

As the emitter voltage decreases, it reaches a value called the *valley voltage*. At this point, the PN junction is no longer forward-biased, and the UJT turns off.

An Application

UJTs are commonly used in oscillator circuits. Figure 18–50 shows a typical circuit. Its operation is as follows: Initially the capacitor is uncharged and the UJT is off. When power is applied, the capacitor charges up exponentially. When it reaches the peak voltage, the UJT turns *on* and the capacitor begins to *discharge* as indicated. When the emitter reaches the valley voltage, the UJT turns off and the capacitor begins to charge again. The cycle repeats. Waveforms of the capacitor voltage and the R_1 voltage are shown in the diagram.

FIGURE 18–50
UJT oscillator circuit.

1. What does "UJT" stand for?
2. What are the terminals of a UJT?

18–12

THYRISTORS

Thyristors are a family of devices constructed of four layers of semiconductor material. The types of thyristors covered in this section are the *silicon-controlled rectifier (SCR)*, the *diac*, and the *triac*.

These thyristor devices share certain characteristics. They act as open circuits capable of withstanding a certain rated voltage until triggered. When turned on (triggered), they become low-resistance current paths and remain so, even after the trigger is removed, until the current is reduced to a certain level or until they are turned off, depending on the type of device.

Thyristors can be used to control the amount of ac power to a load and are used in lamp-dimming circuits, motor speed control, ignition systems, and charging circuits, to name a few. UJTs are often used as the trigger devices for thyristors.

Silicon-controlled Rectifiers (SCRs)

FIGURE 18–51
SCR symbol.

The silicon-controlled rectifier (SCR) has three terminals, as shown in the symbol in Figure 18–51. Like a rectifier diode, it is a unidirectional device, but the conduction of current is controlled by the gate, G. Current is from the cathode K to the anode A when the SCR is on. A positive voltage on the gate will turn the SCR on. The SCR will remain on as long as the current from the cathode to the anode is equal to or greater than a specified value called the *holding current*. When the current drops below the holding value, the SCR will turn off. It can be turned on again by a positive voltage on the gate.

SCR Construction

The SCR is a *four-layer* device. It has two P- and two N-regions, as shown in Figure 18–52(a). This construction can be thought of as an NPN and a PNP transistor interconnected as in part (b) of the figure. The upper N region is commonly shared by the base and the collector of the transistors.

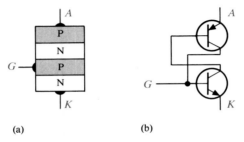

(a)   (b)

FIGURE 18–52
Construction of SCR and its transistor analogy.

An Application

Figure 18–53 shows an SCR used to rectify and control the average power delivered to a load. The SCR rectifies the 60-Hz ac just as a conventional rectifier diode. It can conduct only during the positive half-cycle. The purpose of the phase controller is to produce a trigger pulse at the SCR gate in order to turn it on during any portion of the positive half-cycle of the input. If the SCR is turned on earlier in the half-cycle, more average power is delivered to the load. If the SCR is fired later in the half-cycle, less average power is delivered.

FIGURE 18–53
SCR controls average power to load.

FIGURE 18–54
Triac symbol.

Triacs

The triac is also known as a *bidirectional triode thyristor*. It is equivalent to two SCRs connected to allow current in either direction. The SCRs share a common gate. The symbol for a triac is shown in Figure 18–54.

Applications

Like the SCR, triacs are also used to control average power to a load by the method of phase control. The triac can be triggered such that the ac power is supplied to the load for a controlled portion of each half-cycle. During each positive half-cycle of the ac, the triac is off for a certain interval, called the *delay angle* (measured in degrees). Then it is triggered on and conducts current through the load for the remaining portion of the positive half-cycle, called the *conduction angle*. Similar action occurs on the negative half-cycle except that, of course, current is conducted in the opposite direction through the load. Figure 18–55 illustrates this action.

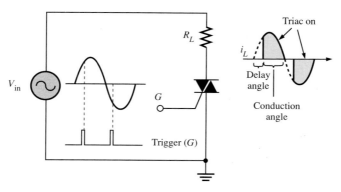

FIGURE 18–55
Basic triac phase control.

FIGURE 18–56
Diac symbol.

Diacs

The diac is a bidirectional device that does not have a gate. It conducts current in either direction when a sufficient voltage, called the *breakover potential,* is reached across the two terminals. The symbol for a diac is shown in Figure 18–56.

SECTION REVIEW
18–12

1. What does "SCR" stand for?
2. Explain how an SCR basically works.
3. Name a typical application of a triac.
4. A diac is gate-controlled (T or F).

18–13   TRANSISTOR PACKAGES AND TERMINAL IDENTIFICATION

Transistors are available in a wide range of package types for various applications. Those with mounting studs or heat sinks are power transistors. Low- and medium-power transistors are usually found in smaller metal or plastic cases. Still another package classification is for high-frequency devices.

As a technician, you often must check a transistor, either in the circuit or out of the circuit, for possible failure. First, of course, you must be able to recognize the transistor by its physical appearance. Also, you must be able to identify the base, collector, and emitter leads (or gate, drain, and source).

Transistors are found in a range of shapes and sizes. Figure 18–57 shows a few of the most typical case styles. Pin identification and construction diagrams are shown in Figure 18–58 for several types of transistor packages. Generally, the heat sink mounting or stud on power transistors is the collector terminal. On the metal "top hat" cases, the tab is closest to the emitter (or source) lead, and the collector (or drain) is often connected to the case.

FIGURE 18–57
Some common package configurations (courtesy of Motorola Semiconductor).

FIGURE 18–58
Typical transistor packages, construction views, and examples of pin arrangements. For FETs, replace base with gate, collector with drain, and emitter with source (courtesy of Motorola Semiconductor).

SECTION REVIEW 18–13

1. Identify the leads on the bipolar transistors in Figure 18–59.

FIGURE 18–59

(a)   (b)   (c)

SUMMARY

□ A bipolar junction transistor (BJT) consists of three regions: emitter, base, and collector.

□ The three regions of a BJT are separated by two PN junctions.

□ The two types of bipolar transistor are the NPN and the PNP.

□ The term *bipolar* refers to two types of current: electron current and hole current.

□ A field-effect transistor (FET) has three terminals: source, drain, and gate.

□ A junction field-effect transistor (JFET) operates with a reverse-biased gate-to-source PN junction.

□ JFETs have very high input resistance due to the reverse-biased gate-source junction.

□ JFET current flows between the drain and the source through a *channel* whose width is controlled by the amount of reverse bias on the gate-source junction.

□ The two types of JFETs are N-channel and P-channel.

□ Metal oxide semiconductor field-effect transistors (MOSFETs) differ from JFETs in that the gate of a MOSFET is insulated from the channel.

□ A depletion-enhancement MOSFET (DE MOSFET) can operate with a positive, negative, or zero gate-to-source voltage.

□ The DE MOSFET has a physical channel between the drain and the source.

□ An enhancement-only MOSFET (E MOSFET) can operate only when the gate-to-source voltage exceeds a threshold value.

□ The enhancement-only MOSFET has no physical channel.

□ Transistors are used as either amplifying devices or switching devices.

□ Transistor and thyristor symbols are shown in Figure 18–60.

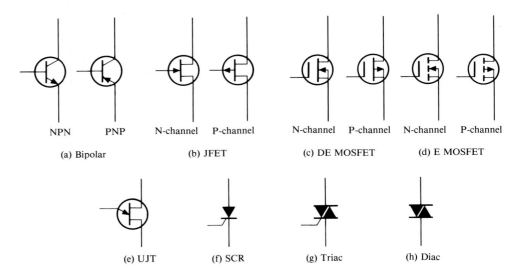

FIGURE 18–60
Transistor and thyristor symbols.

GLOSSARY

Alpha (α) The ratio of collector current to emitter in a bipolar junction transistor.

Amplification The process of producing a larger voltage, current, or power using a smaller input signal as a "pattern."

Amplifier An electronic circuit having the capability of amplification and designed specifically for that purpose.

Base One of the semiconductor regions in a bipolar junction transistor.

Beta (β) The ratio of collector current to base current in a bipolar junction transistor.

Bipolar Characterized by two PN junctions.

BJT Bipolar junction transistor.

Collector One of the semiconductor regions in a BJT.

Cutoff The nonconducting state of a transistor.

Diac A semiconductor device that can conduct current in either of two directions when properly activated.

Drain One of the three terminals of a field-effect transistor.

Emitter One of the three semiconductor regions in a BJT.

Field-effect transistor (FET) A type of transistor that uses an induced electric field to control current.

Gain The amount of amplification.

Gate One of the three terminals of an FET.

Junction field-effect transistor (JFET) A type of FET that operates with a reverse-biased junction to control current in a channel.

MOSFET Metal-oxide semiconductor field-effect transistor.

Q-point The dc operating (bias) point of an amplifier.

Saturation The state of a BJT in which the collector has reached a maximum and is independent of the base current.

Silicon-controlled rectifier (SCR) A device that can be triggered on to conduct current in one direction.

Switch An electrical or electronic device for opening and closing a current path.

Thyristor A class of four-layer semiconductor devices.

Transistor A semiconductor device used for amplification and switching applications in electronic circuits.

Triac A bidirectional thyristor that can be triggered into conduction.

Unijunction transistor (UJT) A type of transistor consisting of an emitter and two bases.

FORMULAS

$$I_E = I_C + I_B \qquad \text{Bipolar transistor currents} \tag{18--1}$$

$$I_C = \alpha_{dc} I_E \qquad \text{Relationship of collector and emitter currents} \tag{18--2}$$

$$I_C = \beta_{dc} I_B \qquad \text{Relationship of collector and base currents} \tag{18--3}$$

$$V_C = V_{CC} - I_C R_C \qquad \text{Collector voltage} \tag{18--4}$$

$$V_B = V_E + V_{BE} \qquad \text{Base voltage} \tag{18--5}$$

$$R_{IN} \cong \beta_{dc}R_E \qquad \text{Input resistance at base} \tag{18--6}$$

$$V_B = \left(\frac{R_2 \parallel R_{IN}}{R_1 + R_2 \parallel R_{IN}}\right)V_{CC} \qquad \text{Base voltage with voltage divider bias} \tag{18--7}$$

$$V_B \cong \left(\frac{R_2}{R_1 + R_2}\right)V_{CC} \qquad \text{Approximate base voltage } (R_{IN} \gg R_2) \tag{18--8}$$

$$\beta_{ac} = \frac{I_c}{I_b} \qquad \text{ac beta} \tag{18--9}$$

$$A_v \cong \frac{R_C}{R_E} \qquad \text{Voltage gain} \tag{18--10}$$

$$V_{CE(cutoff)} \cong V_{CE} \qquad V_{CE} \text{ at cutoff} \tag{18--11}$$

$$I_{C(sat)} \cong \frac{V_{CC}}{R_C} \qquad \text{Collector saturation current} \tag{18--12}$$

$$I_{B(min)} = \frac{I_{C(sat)}}{\beta_{dc}} \qquad \text{Minimum base current for saturation} \tag{18--13}$$

$$I_C = \frac{P_{D(max)}}{V_{CE}} \qquad I_C \text{ for maximum } V_{CE} \tag{18--14}$$

$$V_{CE} = \frac{P_{D(max)}}{I_C} \qquad V_{CE} \text{ for maximum } I_C \tag{18--15}$$

$$V_P = V_{GS} - V_{DS(P)} \qquad \text{Pinch-off voltage} \tag{18--16}$$

$$R_{IN} = \left|\frac{V_{GS}}{I_{GSS}}\right| \qquad \text{JFET input resistance} \tag{18--17}$$

$$V_{GS} = -I_D R_S \qquad \text{Self-bias voltage for N-channel JFET} \tag{18--18}$$

$$V_{GS} = +I_D R_S \qquad \text{Self-bias voltage for P-channel JFET} \tag{18--19}$$

$$V_D = V_{DD} - I_D R_D \qquad \text{Drain voltage} \tag{18--20}$$

$$V_{DS} = V_{DD} - I_D(R_D + R_S) \qquad \text{Drain-to-source voltage} \tag{18--21}$$

$$V_{DS} = V_{DD} - I_{DSS}R_D \qquad \text{Drain-to-source voltage DE MOSFET} \tag{18--22}$$

SELF-TEST

1. The N-type regions in an NPN bipolar junction transistor are
   (a) collector and base   (b) collector and emitter
   (c) base and emitter   (d) collector, base, and emitter

2. The N-region in a PNP transistor is the
   (a) base   (b) collector   (c) emitter   (d) case

3. For normal operation of a NPN transistor, the base must be
    (a) disconnected
    (b) negative with respect to the emitter
    (c) positive with respect to the emitter
    (d) positive with respect to the collector

4. The three currents in a BJT are
    (a) forward, reverse, neutral   (b) drain, source, gate
    (c) alpha, beta, and sigma   (d) base, emitter, and collector

5. Beta (β) is the ratio of
    (a) collector current to emitter current
    (b) collector current to base current
    (c) emitter current to base current
    (d) output voltage to input voltage

6. Alpha (α) is the ratio of
    (a) collector current to emitter current
    (b) collector current to base current
    (c) emitter current to base current
    (d) output voltage to input voltage

7. If the beta of a certain transistor is 30 and the base current is 1 mA, the collector current is
    (a) 0.33 mA   (b) 1 mA   (c) 30 mA   (d) unknown

8. If the base current is increased,
    (a) the collector current increases and the emitter current decreases
    (b) the collector current decreases and the emitter current decreases
    (c) the collector current increases and the emitter current does not change
    (d) the collector current increases and the emitter current increases

9. When an N-channel JFET is biased for conduction, the gate is
    (a) positive with respect to the source
    (b) negative with respect to the source
    (c) at the same voltage as the source
    (d) at the same voltage as the drain

10. When the gate-to-source voltage of an N-channel JFET is increased, the drain current
    (a) decreases   (b) increases   (c) stays constant   (d) become zero

11. When a negative gate-to-source voltage is applied to an N-channel MOSFET, it operates in the
    (a) cutoff state   (b) saturated state   (c) enhancement mode
    (d) depletion mode

12. A UJT consists of
    (a) two emitters   (b) a collector and two emitters   (c) an emitter and two bases
    (d) an emitter, base, and collector

13. An SCR is like a rectifier diode except that
    (a) it can be triggered into conduction by a voltage at the gate
    (b) it can conduct current in both directions
    (c) it can handle more power
    (d) it has four terminals

14. A triac is a device that
    (a) can be triggered into conduction
    (b) can conduct current in both directions
    (c) (a) and (b)
    (d) none of the above

PROBLEMS

Section 18–1　Bipolar Junction Transistors (BJTs)

1. What is the exact value of I_C for $I_E = 5.34$ mA and $I_B = 475$ μA?
2. What is the α_{dc} when $I_C = 8.23$ mA and $I_E = 8.69$ mA?
3. A certain transistor has an $I_C = 25$ mA and an $I_B = 200$ μA. Determine the β_{dc}.
4. In a certain transistor circuit, the base current is 2% of the 30-mA emitter current. Determine the approximate collector current.
5. Find I_B, I_E, and I_C in Figure 18–61 given that $\alpha_{dc} = 0.98$ and $\beta_{dc} = 49$.

FIGURE 18–61

6. Determine the terminal voltages of each transistor with respect to ground for each circuit in Figure 18–62. Also determine V_{CE}, V_{BE}, and V_{BC}. $\beta_{dc} = 25$.

(a)　　　　　　　　　　　　　　　　　(b)

FIGURE 18–62

Section 18–2　Voltage Divider Bias

7. Determine I_B, I_C, and V_C in Figure 18–63.

FIGURE 18–63

8. For the circuit in Figure 18–64, find V_B, V_E, I_E, I_C, and V_C.

9. In Figure 18–64, what is V_{CE}? What are the Q-point coordinates?

FIGURE 18–64

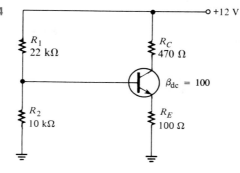

Section 18–3   The Bipolar Transistor as an Amplifier

10. A transistor amplifier has a voltage gain of 50. What is the output voltage when the input voltage is 100 mV?

11. To achieve an output of 10 V with an input of 300 mV, what voltage gain is required?

12. A 50-mV signal is applied to the base of a properly biased transistor with $R_E = 100\ \Omega$ and $R_C = 500\ \Omega$. Determine the signal voltage at the collector.

Section 18–4   The Bipolar Transistor as a Switch

13. Determine $I_{C(sat)}$ for the transistor in Figure 18–65. What is the value of I_B necessary to produce saturation? What minimum value of V_{IN} is necessary for saturation?

14. The transistor in Figure 18–66 has a β_{dc} of 50. Determine the value of R_B required to ensure saturation when V_{IN} is 5 V. What must V_{IN} be to cut off the transistor?

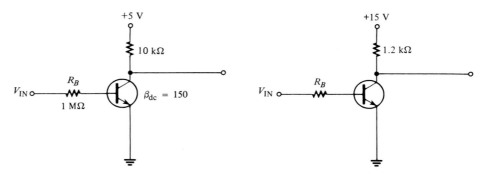

FIGURE 18–65   **FIGURE 18–66**

Section 18–5   BJT Parameters and Ratings

15. If the β_{dc} in Figure 18–62(a) changes from 100 to 150 due to a temperature increase, what is the change in collector current?

16. A certain transistor is to be operated at a collector current of 50 mA. How high can V_{CE} go without exceeding a $P_{D(max)}$ of 1.2 W?

Section 18–6   Transistor Testing

17. Which, if any, of the transistors in Figure 18–67 are bad?

18. Is the transistor in Figure 18–68 an NPN or a PNP device? Assume that it is good. (Refer to Section 18–13.)

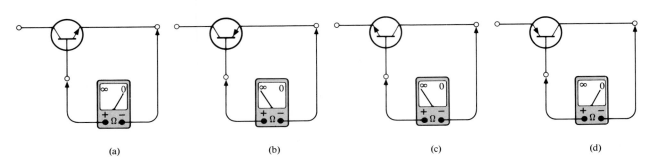

(a)   (b)   (c)   (d)

FIGURE 18–67

FIGURE 18–68

Section 18–7   The Junction Field-Effect Transistor (JFET)

19. The V_{GS} of a P-channel JFET is increased from 1 V to 3 V.

    (a) Does the depletion region narrow or widen?

    (b) Does the resistance of the channel increase or decrease?

20. Why must the gate-to-source voltage of an N-channel JFET always be either 0 or negative?

Section 18–8   JFET Characteristics

21. A JFET has a specified pinch-off voltage of -5 V. When $V_{GS} = 0$, what is V_{DS}?

22. A certain JFET is biased such that $V_{GS} = -2$ V. What is the value of V_{DS} if V_P is specified to be -6 V?

23. A certain JFET data sheet gives $V_{GS(off)} = -8$ V and $I_{DSS} = 10$ mA. Determine the value of V_{DS} at which pinch-off begins when $V_{GS} = -4$ V.

24. A certain P-channel JFET has a $V_{GS(off)} = 6$ V. What is I_D when $V_{GS} = 8$ V?

25. The JFET in Figure 18–69 has a $V_{GS(off)} = -4$ V. Assume that you increase the supply voltage, V_{DD}, beginning at 0 until the ammeter reaches a steady value. What does the voltmeter read at this point?

FIGURE 18–69

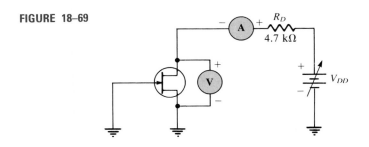

Section 18-9   The Metal-Oxide Semiconductor FET (MOSFET)

26. Sketch the schematic symbols for N-channel and P-channel DE MOSFETs and E MOS-FETs. Label the terminals.

27. Explain why both types of MOSFETs have an extremely high input resistance at the gate.

28. A N-channel DE MOSFET with a positive V_{GS} is operating in the _____ mode.

29. A certain E MOSFET has a $V_{GS(th)} = 3$ V. What is the minimum V_{GS} for the device to turn on?

Section 18-10   FET Biasing

30. For each circuit in Figure 18-70, determine V_{DS} and V_{GS}.

31. Determine in which mode (depletion or enhancement) each DE MOSFET in Figure 18-71 is biased.

FIGURE 18-70

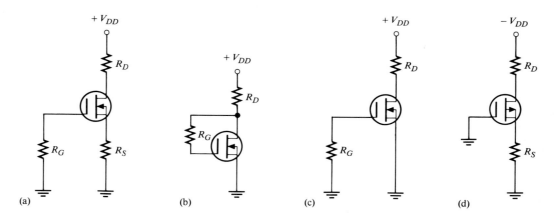

FIGURE 18-71

32. Each E MOSFET in Figure 18–72 has a $V_{GS(th)}$ of $+5$ V or -5 V, depending on whether it is an N-channel or a P-channel device. Determine whether each MOSFET is *on* or *off*.

FIGURE 18–72

Section 18–11 Unijunction Transistors (UJTs)

33. In a UJT, if R_{BB} is 5 kΩ and R_{B1} is 3 kΩ, what is η?

34. What is the charging time constant in the oscillator circuit of Figure 18–73?

35. How long will the capacitor in Figure 18–73 initially charge if the peak voltage of the UJT is 6.3 V?

36. If the valley voltage of the UJT in Figure 18–73 is 2.23 V, what is the peak-to-peak value of the voltage across the capacitor, if the peak voltage is the same as in Problem 35? Sketch its general shape.

FIGURE 18–73

Section 18–12 Thyristors

37. Assume that the holding current for a particular SCR is 10 mA. What is the maximum value of R in Figure 18–74 necessary to keep the SCR in conduction once it is turned on? Neglect the drop across the SCR.

FIGURE 18–74

38. Repeat Problem 37 for a holding current of 500 μA.

ANSWERS TO SECTION REVIEWS

Section 18–1

1. Emitter, base, collector.   **2.** The base-emitter junction is forward-biased and the base-collector junction is reverse-biased.

3. Direct current gain.   **4.** 1 mA.

Section 18–2

1. One.   **2.** 6 V.

Section 18–3

1. The process of increasing the amplitude of an electrical signal.

2. The dc operating point.   **3.** 21.36.

Section 18–4

1. Saturation, cutoff.   **2.** At saturation.   **3.** At cutoff.

4. The base-emitter is forward-biased, and there is sufficient base current.

5. At cutoff.

Section 18–5

1. True.   **2.** β_{dc} increases with I_C to a certain value, and then decreases.

3. 40 mA.

Section 18–6

1. NPN.   **2.** I_{CBO}, I_{CEO}.

Section 18–7

1. Drain, source, and gate.   **2.** Negative.   **3.** By V_{GS}.

Section 18–8

1. -7 V.   **2.** Decreases.   **3.** $+3$ V.

Section 18–9

1. Depletion-enhancement MOSFET; enhancement only MOSFET. The DE MOSFET has a physical channel; the E MOSFET does not.

2. Yes.   **3.** No.

Section 18–10

1. Positive.   **2.** -8 V.   **3.** I_{DSS} **4.** 2 V.

Section 18–11

1. Unijunction transistor.   **2.** Anode, cathode, gate.

Section 18–12

1. Silicon-controlled rectifier.   **2.** A positive voltage on the gate turns the SCR on, and the SCR conducts current from anode to cathode. If the current drops below the holding value, the SCR turns off.

3. Lamp dimmer.   **4.** False.

Section 18–13

1. See Figure 18–75.

FIGURE 18–75

19

Amplifiers and Oscillators

As you learned in the previous chapter, the *biasing* of a transistor is purely a dc operation. The purpose of biasing, however, is to establish an operating point (Q point) about which variations in current and voltage can occur in response to an ac input signal.

When very small signal voltages must be amplified, such as from an antenna in a receiver, variations about the Q point of an amplifier are relatively small.

Amplifiers designed to handle these small ac signals are called *small-signal* amplifiers. When large swings or variations in voltage and current about the Q point are required for power amplification, *large-signal* amplifiers are used. An example is the power amplifier that drives the speakers in a stereo system.

Regardless of whether an amplifier is in the small- or large-signal category, it will be in one of three configurations of amplifier circuits: *common-emitter, common-collector,* and *common-base.* For FETs the configurations are *common-source, common-drain,* and *common-gate.* Also, for any of the configurations there are three basic modes of operation possible—class A, class B, and class C—which are covered in this chapter.

Oscillators are also introduced in this chapter. An oscillator is a circuit that produces a sustained sine wave output without an input signal. It operates on the principle of positive feedback.

After completing this chapter, you will be able to

- ☐ Compare common-emitter (CE), common-collector (CC), and common-base (CB) amplifiers in terms of their configurations and operating characteristics.
- ☐ Determine how various circuit parameters affect the voltage gain of CE, CC, and CB amplifiers.
- ☐ Analyze BJT amplifiers for voltage gain, current gain, power gain, and input impedance.
- ☐ Explain the meaning of the term *emitter-follower* in relation to a common-collector amplifier.
- ☐ Describe what a *Darlington pair* is and explain its advantage.
- ☐ List the important characteristics of FET amplifiers.
- ☐ Show how multistage amplifiers are used to increase gain.
- ☐ Describe the differences that distinguish class A, class B, and class C amplifiers.
- ☐ Troubleshoot amplifiers.
- ☐ Explain basic oscillator operation.
- ☐ Describe how Colpitts, Hartley, Clapp, crystal, and *RC* oscillators differ.

**APPLICATION
ASSIGNMENT**

The sound is severely distorted in one of the channels of a stereo system. The power amplifiers that drive the speakers are class B push-pull circuits. The power amplifiers are located on a separate plug-in circuit board. You need to pull this board and perform a test to determine if one of the amplifiers has failed. After completing this chapter, you will be able to complete this assignment.

19–1   COMMON-EMITTER (CE) AMPLIFIERS

Figure 19–1 shows a *typical* common-emitter (CE) amplifier. The one shown has voltage divider bias, although other types of bias methods are possible. C_1 and C_2 are *coupling capacitors* used to *pass the signal* into and out of the amplifier such that the source or load will not affect the *dc* bias voltages. C_3 is a *bypass capacitor* that shorts the emitter signal voltage (ac) to ground without disturbing the dc emitter voltage. Because of the bypass capacitor, the emitter is at *signal ground* (but not dc ground), thus making the circuit a *common-emitter* amplifier. The purpose of the bypass capacitor is to increase the signal voltage gain. (The reason why this increase occurs will be discussed shortly.) Notice that the input signal is applied to the base, and the output signal is taken off the collector. All capacitors are assumed to have a reactance of approximately zero at the signal frequency.

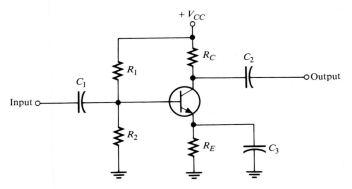

FIGURE 19–1
Typical common-emitter (CE) amplifier.

How a Bypass Capacitor Increases Voltage Gain

The bypass capacitor shorts the signal around the emitter resistor, R_E, in order to increase the voltage gain. To understand why, let us consider the amplifier *without* the bypass capacitor and see what the voltage gain is. The CE amplifier with the bypass capacitor removed is shown in Figure 19–2.

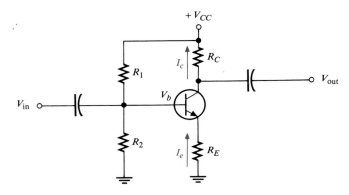

FIGURE 19–2
CE amplifier with bypass capacitor removed.

As before, lowercase subscripts indicate signal (ac) voltages and signal (alternating) currents. The voltage gain of the amplifier is V_{out}/V_{in}. The output *signal* voltage is

$$V_{out} = I_c R_C$$

The signal voltage at the base is approximately equal to

$$V_b \cong V_{in} \cong I_e(r_e + R_E)$$

where r_e is the *internal emitter resistance* of the transistor. The voltage gain A_v can now be expressed as

$$A_v = \frac{V_{out}}{V_{in}} = \frac{I_c R_C}{I_e(r_e + R_E)}$$

Since $I_c \cong I_e$, the currents cancel and the gain is the ratio of the resistances:

$$A_v = \frac{R_C}{r_e + R_E} \tag{19-1}$$

Keep in mind that this formula is for the CE *without* the bypass capacitor. If R_E is much greater than r_e, then $A_v \cong R_C/R_E$. This gain formula is similar to that presented in the last chapter, where r_e was neglected [Equation (18–10)].

If the bypass capacitor is connected across R_E, it effectively shorts the signal to ground, leaving only r_e in the emitter. Thus, the voltage gain of the CE amplifier with the bypass capacitor shorting R_E is

$$A_v = \frac{R_C}{r_e} \tag{19-2}$$

Now, r_e is a very important transistor parameter because it determines the voltage gain of a CE amplifier in conjunction with R_C. A formula for estimating r_e is given without derivation in the following equation:

$$r_e \cong \frac{25 \text{ mV}}{I_E} \tag{19-3}$$

EXAMPLE 19–1

Determine the voltage gain of the amplifier in Figure 19–3 both with and without a bypass capacitor. $\beta_{dc} = \beta_{ac} = 150$.

FIGURE 19-3

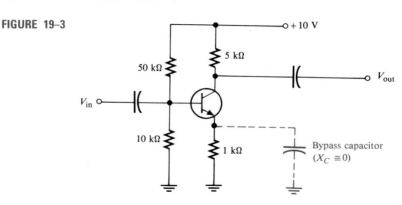

Solution

First we determine r_e. To do so, we need to find I_E. Thus,

$$V_B \cong \left(\frac{10 \text{ k}\Omega}{50 \text{ k}\Omega + 10 \text{ k}\Omega} \right) 10 \text{ V} = 1.67 \text{ V}$$

$$V_E = V_B - 0.7 \text{ V} = 0.97 \text{ V}$$

$$I_E = \frac{V_E}{R_E} = \frac{0.97 \text{ V}}{1 \text{ k}\Omega} = 0.97 \text{ mA}$$

$$r_e \cong \frac{25 \text{ mV}}{I_E} \cong \frac{25 \text{ mV}}{0.97 \text{ mA}} = 25.77 \text{ }\Omega$$

The voltage gain without a bypass capacitor is

$$A_v = \frac{R_C}{r_e + R_E} = \frac{5 \text{ k}\Omega}{1025.77 \text{ }\Omega} = 4.87$$

The voltage gain with the bypass capacitor installed is

$$A_v = \frac{R_C}{r_e} = \frac{5 \text{ k}\Omega}{25.77 \text{ }\Omega} = 194$$

As you can see, the voltage gain is greatly increased by the addition of the bypass capacitor. In terms of decibels (dB), the voltage gain is

$$A_v = 20 \log(194) = 45.76 \text{ dB}$$

Phase Inversion

As we discussed in the last chapter, the output voltage at the collector is 180° out of phase with the input voltage at the base. Therefore, the CE amplifier is characterized by a phase inversion between the input and the output. As mentioned before, this inversion is sometimes indicated by a negative voltage gain.

ac Input Resistance

In the last chapter the dc input resistance R_{IN}, viewed from the base of the transistor, was developed. The input resistance "seen" by the *signal* at the base is derived in a similar manner when the emitter resistor is bypassed to ground:

$$R_{in} = \frac{V_b}{I_b}$$

$$V_b = I_e r_e$$

$$I_e \cong \beta_{ac} I_b$$

$$R_{in} \cong \frac{\beta_{ac} I_b r_e}{I_b}$$

The I_b terms cancel, leaving

$$R_{in} \cong \beta_{ac} r_e \qquad (19\text{–}4)$$

Total Input Resistance to CE Amplifier

Viewed from the *base*, R_{in} is the ac resistance. The actual resistance seen by the source includes that of bias resistors. We will now develop an expression for the *total* input resistance.

The concept of ac ground was mentioned earlier. At this point it needs some additional explanation because it is important in the development of the formula for total input resistance, $R_{in(T)}$.

You have already seen that the bypass capacitor effectively makes the emitter appear as ground to the ac signal, because the X_C of the capacitor is nearly zero at the signal frequency. Of course, to a dc signal the capacitor looks like an open and thus does not affect the dc emitter voltage.

In addition to seeing ground through the bypass capacitor, the signal also sees ground through the dc supply voltage source, V_{CC}. It does so because there is *zero* signal voltage at the V_{CC} terminal. Thus, the $+V_{CC}$ terminal effectively acts as *ac ground*. As a result, the two bias resistors R_1 and R_2 appear in parallel to the ac input, because one end of R_2 goes to actual ground and one end of R_1 goes to ac ground (V_{CC} terminal). Also, R_{in} at the base appears in parallel with $R_1 \parallel R_2$. This situation is illustrated in Figure 19–4.

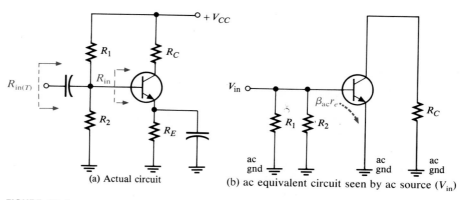

(a) Actual circuit   (b) ac equivalent circuit seen by ac source (V_{in})

FIGURE 19–4
Total input resistance.

The expression for the total input resistance to the CE amplifier as seen by the ac source is as follows:

$$R_{in(T)} = R_1 \parallel R_2 \parallel R_{in} \qquad (19\text{–}5)$$

R_C has no effect because of the reverse-biased, base-collector junction.

EXAMPLE 19–2

Determine the total input resistance seen by the signal source in the CE amplifier in Figure 19–5. $\beta_{ac} = 150$.

FIGURE 19–5

Solution

In Example 19–1, we found r_e for the same circuit. Thus,

$$R_{\text{in}} = \beta_{\text{ac}} r_e = 150(25.77 \ \Omega) = 3865.5 \ \Omega$$

$$R_{\text{in}(T)} = R_1 \| R_2 \| R_{\text{in}} = 50 \ \text{k}\Omega \| 10 \ \text{k}\Omega \| 3865.5 \ \Omega = 2.64 \ \text{k}\Omega$$

Current Gain

The signal current gain of a CE amplifier is

$$A_i = \frac{I_c}{I_s} \tag{19–6}$$

where I_s is the source current into the amplifier and is calculated by $V_s / R_{\text{in}(T)}$.

Power Gain

The power gain of a CE amplifier is the product of the voltage gain and the current gain:

$$A_p = A_v A_i \tag{19–7}$$

EXAMPLE 19–3

Determine the voltage gain, current gain, and power gain for the CE amplifier in Figure 19–6. $\beta_{\text{ac}} = 100$.

Solution

First we must find r_e. To do so, we must find I_E. We begin by calculating V_B. Since R_{IN} is ten times greater than R_2, it can be neglected. Thus,

$$V_B \cong \left(\frac{10 \ \text{k}\Omega}{110 \ \text{k}\Omega} \right) 30 \ \text{V} = 2.73 \ \text{V}$$

$$I_E = \frac{V_E}{R_E} = \frac{V_B - 0.7 \ \text{V}}{R_E} = \frac{2.03 \ \text{V}}{1 \ \text{k}\Omega} = 2.03 \ \text{mA}$$

$$r_e = \frac{25 \ \text{mV}}{I_E} = \frac{25 \ \text{mV}}{2.03 \ \text{mA}} = 12.32 \ \Omega$$

FIGURE 19–6

The ac voltage gain is

$$A_v = \frac{R_C}{r_e} = \frac{5 \text{ k}\Omega}{12.32 \text{ }\Omega} = 405.8$$

We determine the signal current gain by first finding $R_{in(T)}$ to get I_s:

$$R_{in(T)} = R_1 \| R_2 \| \beta_{ac}r_e = 100 \text{ k}\Omega \| 10 \text{ k}\Omega \| 1.232 \text{ k}\Omega$$
$$\cong 1.1 \text{ k}\Omega$$

$$I_s = \frac{V_s}{R_{in(T)}} = \frac{10 \text{ mV}}{1.1 \text{ k}\Omega} \cong 9.1 \text{ }\mu\text{A}$$

Next we must determine I_c:

$$I_c = \frac{V_{out}}{R_C} = \frac{A_v V_s}{R_C} = \frac{(405.8)(10 \text{ mV})}{5 \text{ k}\Omega} = 0.812 \text{ mA}$$

$$A_i = \frac{I_c}{I_s} = \frac{812 \text{ }\mu\text{A}}{9.1 \text{ }\mu\text{A}} \cong 89.2$$

The power gain is

$$A_p = A_v A_i = (405.8)(89.2) = 36,197$$

The voltage gain and the power gain in decibels are as follows:

$$A_v = 20 \log(405.8) = 52.17 \text{ dB}$$
$$A_p = 10 \log(36,197) = 45.59 \text{ dB}$$

**SECTION REVIEW
19–1**

1. What is the purpose of the bypass capacitor in a CE amplifier?
2. How is the CE voltage gain determined?
3. If A_v is 50 and A_i is 200, what is the power gain?

19–2 COMMON-COLLECTOR (CC) AMPLIFIERS

The common-collector (CC) amplifier is the second most important of the three basic amplifier configurations. It is commonly referred to as an *emitter-follower*. Figure 19–7 shows an emitter-follower with a voltage divider bias. Notice that the input is applied to the base and the output is taken off the emitter.

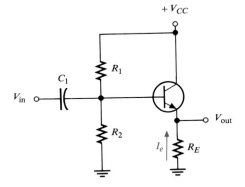

FIGURE 19–7
Typical emitter-follower (common-collector, CC) amplifier.

Voltage Gain

As in all amplifiers, the voltage gain in a CC amplifier is $A_v = V_{out}/V_{in}$. For the emitter-follower, V_{out} is $I_e R_E$, and V_{in} is $I_e(r_e + R_E)$. Therefore, the gain is $I_e R_E/I_e(r_e + R_E)$. The currents cancel, and the gain expression simplifies to

$$A_v = \frac{R_E}{r_e + R_E} \tag{19–8}$$

It is important to notice here that *the gain is always less than 1*. If r_e is much less than R_E, then a good approximation is $A_v \cong 1$.

Since the output voltage is the emitter voltage, it is *in phase* with the base or the input voltage. As a result, and because the voltage gain is close to 1, the output voltage *follows* the input voltage—thus the term *emitter-follower*.

Input Resistance

The emitter-follower is characterized by a high input resistance, which makes it a very useful circuit. Because of the high input resistance, the emitter follower can be used as a *buffer* to minimize loading effects when one circuit is driving another.

The derivation of the input resistance viewed from the base is similar to that for the CE amplifier. In this case, however, *the emitter resistor is not bypassed:*

$$R_{in} = \frac{V_b}{I_b} = \frac{I_e(r_e + R_E)}{I_b} \cong \frac{\beta_{ac} I_b(r_e + R_E)}{I_b} = \beta_{ac}(r_e + R_E)$$

If R_E is at least ten times larger than r_e, then the input resistance at the base is

$$R_{in} \cong \beta_{ac} R_E \tag{19–9}$$

In Figure 19–7, the bias resistors appear in parallel with R_{in} to the input signal, just as in the voltage divider biased CE amplifier. The total ac input resistance is

$$R_{in(T)} = R_1 \| R_2 \| R_{in} \qquad (19\text{–}10)$$

Since R_{in} can be made large with the proper selection of R_E, a much higher input resistance results for this configuration than for the CE circuit.

Current Gain

The signal current gain for the emitter-follower is I_e / I_s where I_s is the signal current and can be calculated as $V_s / R_{in(T)}$. If the bias resistors are large enough to be neglected so that $I_s = I_b$, then the current gain of the amplifier is equal to the current gain of the transistor, β_{ac}. Of course, the same was also true for the CE amplifier. β_{ac} is the maximum achievable current gain in both amplifiers.

$$A_i = \frac{I_e}{I_s} \qquad (19\text{–}11)$$

Power Gain

The power gain is the product of the voltage gain and the current gain. For the emitter-follower, the power gain is approximately equal to the current gain, because the voltage gain is approximately 1:

$$A_p \cong A_i \qquad (19\text{–}12)$$

EXAMPLE 19–4

Determine the input resistance of the emitter-follower in Figure 19–8. Also find the voltage gain, current gain, and power gain.

FIGURE 19–8

Solution

The approximate input resistance viewed from the base is

$$R_{in} \cong \beta_{ac} R_E = (175)(1 \text{ k}\Omega) = 175 \text{ k}\Omega$$

The total input resistance is

$$R_{\text{in}(T)} = R_1 \parallel R_2 \parallel R_{\text{in}} = 10 \text{ k}\Omega \parallel 10 \text{ k}\Omega \parallel 175 \text{ k}\Omega$$
$$= 4.86 \text{ k}\Omega$$

The voltage gain is, neglecting r_e,

$$A_v \cong 1$$

The current gain is

$$A_i = \frac{I_e}{I_s}$$

$$I_e = \frac{V_e}{R_E} = \frac{A_v V_b}{R_E} \cong \frac{1 \text{ V}}{1 \text{ k}\Omega} = 1 \text{ mA}$$

$$I_s = \frac{V_s}{R_{\text{in}(T)}} = \frac{1 \text{ V}}{4.86 \text{ k}\Omega} = 0.21 \text{ mA}$$

$$A_i = \frac{1 \text{ mA}}{0.21 \text{ mA}} = 4.76$$

The power gain is

$$A_p \cong A_i = 4.76$$

The Darlington Pair

As you have seen, β is a major factor in determining the input impedance. The β of the transistor limits the maximum achievable input impedance you can get from a given emitter-follower circuit.

One way to boost input impedance is to use a *Darlington pair*, as shown in Figure 19–9. The collectors of two transistors are connected, and the emitter of the first drives the base of the second. This configuration achieves β *multiplication* as shown in the following steps:

The emitter current of the first transistor is

$$I_{e_1} \cong \beta_1 I_{b_1}$$

This emitter current becomes the base current for the second transistor, producing a second emitter current of

$$I_{e_2} \cong \beta_2 I_{e_1}$$
$$I_{e_2} \cong \beta_1 \beta_2 I_{b_1}$$

Therefore, the effective current gain of the Darlington pair is

$$\beta = \beta_1 \beta_2 \qquad\qquad (19\text{–}13)$$

The input impedance is $\beta_1 \beta_2 R_E$.

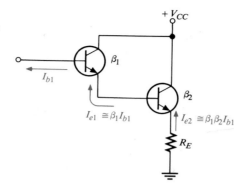

FIGURE 19–9
Darlington pair.

---

1. What is a common-collector amplifier called?
2. What is its ideal maximum voltage gain?
3. What is the most important characteristic of the CC amplifier?

19–3

COMMON-BASE (CB) AMPLIFIERS

The third basic amplifier configuration is the common-base (CB). A typical common-base circuit is pictured in Figure 19–10. The base is at signal (ac) ground, and the input is applied to the emitter. The output is taken off the collector.

FIGURE 19–10
Typical common-base (CB) amplifier.

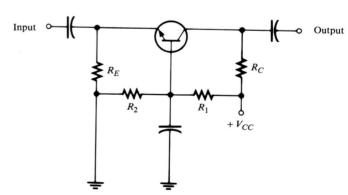

Voltage Gain

The input voltage is the emitter voltage V_e. The output voltage is the collector voltage V_c. With this in mind, we develop the voltage gain formula as follows:

$$A_v = \frac{V_c}{V_e} = \frac{I_c R_C}{I_e r_e} \cong \frac{I_e R_C}{I_e r_e}$$

$$A_v = \frac{R_C}{r_e} \tag{19–14}$$

Notice that the gain expression is the same as that for the CE amplifier (when R_E is bypassed).

Input Resistance

The resistance viewed from the emitter appears to the input signal as follows:

$$R_{in} = \frac{V_{in}}{I_{in}} = \frac{V_e}{I_e} = \frac{I_e r_e}{I_e}$$

$$R_{in} = r_e \qquad (19\text{--}15)$$

Viewed from the source, R_E appears in parallel with R_{in}. However, r_e is normally so small compared to R_E that Equation (19–15) is also valid for the total input resistance, $R_{in(T)}$.

Current Gain

The current gain is the output current I_c over the input current I_e. Since $I_c \cong I_e$, the signal current gain is approximately 1:

$$A_i \cong 1 \qquad (19\text{--}16)$$

Power Gain

Since the current gain is approximately 1 for the CB amplifier, the power gain is approximately equal to the voltage gain:

$$A_p \cong A_v \qquad (19\text{--}17)$$

EXAMPLE 19–5

Find the input resistance, voltage gain, current gain, and power gain for the CB amplifier in Figure 19–11.

FIGURE 19–11

Solution

First let us find I_E so that we can determine r_e. Then $R_{in} = r_e$. Thus,

$$V_B \cong \left(\frac{R_2}{R_1 + R_2}\right)V_{CC} = \left(\frac{20 \text{ k}\Omega}{120 \text{ k}\Omega}\right)10 \text{ V} = 1.67 \text{ V}$$

$$V_E = V_B - 0.7 \text{ V} = 1.67 \text{ V} - 0.7 \text{ V} = 0.97 \text{ V}$$

$$I_E = \frac{0.97 \text{ V}}{1 \text{ k}\Omega} = 0.97 \text{ mA}$$

$$R_{in} = r_e \cong \frac{25 \text{ mV}}{0.97 \text{ mA}} = 25.77 \text{ }\Omega$$

The signal voltage gain is

$$A_v = \frac{R_C}{r_e} = \frac{2 \text{ k}\Omega}{25.77 \text{ }\Omega} = 77.6$$

Thus,

$$A_i \cong 1$$

$$A_p \cong 77.6$$

Summary

Table 19–1 summarizes the important characteristics of each of the three amplifier configurations. Also, relative values are indicated for general comparison of the amplifiers.

TABLE 19–1

Comparison of amplifier configurations. The current gains and the input resistance are the maximum achievable values, with the bias resistors neglected.

	CE	CC	CB
Voltage gain, A_v	$\dfrac{R_C}{r_e}$ High	$\cong 1$ Low	$\dfrac{R_C}{r_e}$ High
Current gain, $A_{i(max)}$	β_{ac} High	β_{ac} High	$\cong 1$ Low
Power gain, A_p	$A_i A_v$ Very high	$\cong A_i$ High	$\cong A_v$ High
Input resistance, $R_{in(max)}$	$\beta_{ac} r_e$ Low	$\beta_{ac} R_E$ High	r_e Very low

1. The same voltage gain can be achieved with a CB as with a CE amplifier (T or F).
2. The CB amplifier has a very low input resistance (T or F).

19–4

FET AMPLIFIERS

Field-effect transistors, both JFETs and MOSFETs, can be used as amplifiers in any of three circuit configurations similar to those for the bipolar transistor. These configurations are *common-source, common-drain,* and *common-gate.*

Transconductance of an FET

Recall that in a bipolar transistor, the *base current* controls the collector current, and the relationship between these two currents is expressed by the parameter β ($I_C = \beta_{ac}I_b$). In an FET, the *gate voltage* controls the drain current. An important FET parameter is the *transconductance, g_m,* which is defined as

$$g_m = \frac{I_d}{V_{gs}} \tag{19–18}$$

The transconductance is one factor that determines the voltage gain of an FET amplifier. On data sheets, the transconductance is sometimes called the *forward transadmittance* and is designated y_{fs}.

Common-Source (CS) Amplifiers

A self-biased N-channel JFET with an ac source capacitively coupled to the gate is shown in Figure 19–12. The resistor, R_G, serves two purposes: (1) It keeps the gate at approximately 0 V dc (because I_{GSS} is extremely small), and (2) its large value (usually several megohms) prevents loading of the ac signal source. The bias voltage is created by the drop across R_S. The bypass capacitor, C_2, keeps the source of the FET effectively at ac ground.

The signal voltage causes the gate-to-source voltage to swing above and below its Q-point value, causing a swing in drain current. As the drain current increases, the

FIGURE 19–12
JFET common-source amplifier.

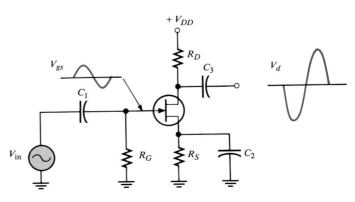

voltage drop across R_D also increases, causing the drain voltage (with respect to ground) to decrease.

The drain current swings above and below its Q-point value *in phase* with the gate-to-source voltage. The drain-to-source voltage swings above and below its Q-point value 180° *out of phase* with the gate-to-source voltage, as illustrated in the figure.

DE MOSFET

A zero-biased N-channel DE MOSFET with an ac source capacitively coupled to the gate is shown in Figure 19–13. The gate is at approximately 0 V dc and the source terminal is at ground, thus making $V_{GS} = 0$ V.

The signal voltage causes V_{gs} to swing above and below its 0 value, producing a swing in I_d. The negative swing in V_{gs} produces the depletion mode, and I_d decreases. The positive swing in V_{gs} produces the enhancement mode, and I_d increases.

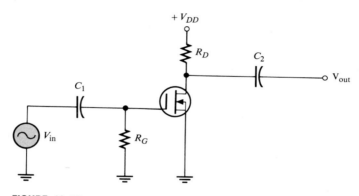

FIGURE 19–13
Zero-biased DE MOSFET common-source amplifier.

E MOSFET

A voltage divider-biased, N-channel E MOSFET with an ac signal source capacitively coupled to the gate is shown in Figure 19–14. The gate is biased with a positive voltage such that $V_{GS} > V_{GS(th)}$.

FIGURE 19–14
Common-source E MOSFET amplifier with voltage divider bias.

As with the JFET and DE MOSFET, the signal voltage produces a swing in V_{gs} above and below its Q-point value. This swing, in turn, causes a swing in I_d. Operation is entirely in the enhancement mode.

Voltage Gain of the CS Amplifier

Voltage gain, A_v, of an amplifier always equals V_{out}/V_{in}. In the case of the CS amplifier, V_{in} is equal to V_{gs}, and V_{out} is equal to the signal voltage developed across R_D which is $I_d R_D$. Thus,

$$A_v = \frac{I_d R_D}{V_{gs}}$$

Since $g_m = I_d/V_{gs}$, the common-source voltage gain is

$$A_v = g_m R_D \tag{19-19}$$

Input Impedance of the CS Amplifier

Because the input to a CS amplifier is at the gate, the input impedance is extremely high. Ideally, it approaches infinity and can be neglected. As you know, the high input impedance is produced by the reverse-biased PN junction in a JFET and by the insulated gate structure in a MOSFET.

The actual input impedance seen by the signal source is the gate-to-ground resistor R_G in parallel with the FET's input impedance, V_{GS}/I_{GSS}. The reverse leakage current I_{GSS} is typically given on the data sheet for a specific value of V_{GS} so that the input impedance of the device can be calculated.

EXAMPLE 19–6

(a) What is the total output voltage (dc + ac) of the amplifier in Figure 19–15? The g_m is 4500 μS, I_D is 2 mA, $V_{GS(off)}$ is -10 V, and I_{GSS} is 15 nA.

(b) What is the input impedance seen by the signal source?

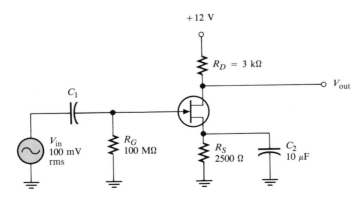

FIGURE 19–15

Solution

(a) First, find the dc output voltage:

$$V_D = V_{DD} - I_D R_D = 12 \text{ V} - (2 \text{ mA})(3 \text{ k}\Omega) = 6 \text{ V}$$

Next, find the ac output voltage by using the gain formula:

$$A_v = \frac{V_{out}}{V_{in}} = g_m R_D$$

$$V_{out} = g_m R_D V_{in} = (4500 \text{ }\mu\text{S})(3 \text{ k}\Omega)(100 \text{ mV})$$
$$= 1.35 \text{ V rms}$$

The total output voltage is an ac signal with a peak-to-peak value of 1.35 V × 2.828 = 3.82 V, riding on a dc level of 6 V.

(b) The input resistance is determined as follows (since $V_G = 0$ V):

$$V_{GS} = I_D R_S = (2 \text{ mA})(2500 \text{ }\Omega) = 5 \text{ V}$$

The input resistance at the gate of the JFET is

$$R_{IN(gate)} = \frac{V_{GS}}{I_{GSS}} = \frac{5 \text{ V}}{15 \text{ nA}} = 333 \text{ M}\Omega$$

The input impedance seen by the signal source is

$$R_{in} = R_G \| R_{IN(gate)} = 100 \text{ M}\Omega \| 333 \text{ M}\Omega = 76.9 \text{ M}\Omega$$

Common-Drain (CD) Amplifier

A common-drain JFET amplifier is shown in Figure 19–16 with voltages indicated. Self-biasing is used in this circuit. The input signal is applied to the gate through a

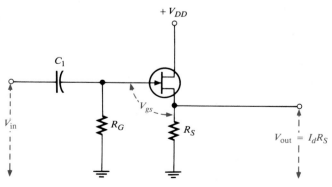

FIGURE 19–16
JFET common-drain amplifier (source-follower).

coupling capacitor, and the output is at the source terminal. There is no drain resistor. This circuit, of course, is analogous to the bipolar emitter-follower and is sometimes called a *source-follower*.

Voltage Gain of the CD Amplifier

As in all amplifiers, the voltage gain is $A_v = V_{out}/V_{in}$. For the source-follower, V_{out} is I_dR_S and V_{in} is $V_{gs} + I_dR_S$, as shown in Figure 19–16. Therefore, the gate-to-source voltage gain is $I_dR_S/(V_{gs} + I_dR_S)$. Substituting $I_d = g_mV_{gs}$ into the expression gives the following result:

$$A_v = \frac{g_mV_{gs}R_S}{V_{gs} + g_mV_{gs}R_S}$$

Canceling V_{gs}, we get

$$A_v \cong \frac{g_mR_S}{1 + g_mR_S} \tag{19–20}$$

Notice here that the gain is always slightly *less than 1*. If $g_mR_S \gg 1$, then a good approximation is $A_v \cong 1$.

Since the output voltage is at the source, it is in phase with the gate (input) voltage.

Input Impedance of the CD Amplifier

Because the input signal is applied to the gate, the input impedance seen by the input signal source is extremely high, just as in the CS amplifier configuration. The gate resistor R_G, in parallel with the input impedance looking in at the gate, is the total input impedance.

EXAMPLE 19–7

(a) Determine the voltage gain of the amplifier in Figure 19–17(a) using the data sheet information in Figure 19–17(b).

(b) Also determine the input impedance. Assume minimum data sheet values where available.

Solution

(a) From the data sheet, $g_m = y_{fs} = 8000\ \mu S$. The gain is

$$A_v \cong \frac{g_mR_S}{1 + g_mR_S} = \frac{(8000\ \mu S)(10\ k\Omega)}{1 + (8000\ \mu S)(10\ k\Omega)} = 0.988$$

(b) From the data sheet, $I_{GSS} = 10\ nA$ at $V_{GS} = 15\ V$. Therefore,

$$R_{IN(gate)} = \frac{15\ V}{10\ nA} = 1500\ M\Omega$$

$$R_{IN} = R_G \parallel R_{IN(gate)} = 100\ M\Omega \parallel 1500\ M\Omega = 93.75\ M\Omega$$

$+10$ V

V_{in} —| C |— ... 2N4360 ... V_{out}

100 MΩ 10 kΩ

(a)

*ELECTRICAL CHARACTERISTICS ($T_A = 25^\circ$C unless otherwise noted)

Characteristic	Symbol	Min	Max	Unit
OFF CHARACTERISTICS				
Gate-Source Breakdown Voltage ($I_G = 10\,\mu$Adc, $V_{DS} = 0$)	$V_{(BR)GSS}$	20	–	Vdc
Gate-Source Cutoff Voltage ($V_{DS} = -10$ Vdc, $I_D = 1.0\,\mu$Adc)	$V_{GS(off)}$	0.7	10	Vdc
Gate Reverse Current ($V_{GS} = 15$ Vdc, $V_{DS} = 0$) ($V_{GS} = 15$ Vdc, $V_{DS} = 0$, $T_A = 65^\circ$C)	I_{GSS}	– –	10 0.5	nAdc μAdc
ON CHARACTERISTICS				
Zero-Gate Voltage Drain Current (Note 1) ($V_{DS} = -10$ Vdc, $V_{GS} = 0$)	I_{DSS}	3.0	30	mAdc
Gate-Source Voltage ($V_{DS} = -10$ Vdc, $I_D = 0.3$ mAdc)	V_{GS}	0.4	9.0	Vdc
SMALL-SIGNAL CHARACTERISTICS				
Drain-Source "ON" Resistance ($V_{GS} = 0$, $I_D = 0$, $f = 1.0$ kHz)	$r_{ds(on)}$	–	700	Ohms
Forward Transadmittance (Note 1) ($V_{DS} = -10$ Vdc, $V_{GS} = 0$, $f = 1.0$ kHz)	$|y_{fs}|$	2000	8000	μmhos
Forward Transconductance ($V_{DS} = -10$ Vdc, $V_{GS} = 0$, $f = 1.0$ MHz)	$Re(y_{fs})$	1500	–	μmhos
Output Admittance ($V_{DS} = -10$ Vdc, $V_{GS} = 0$, $f = 1.0$ kHz)	$|y_{os}|$	–	100	μmhos
Input Capacitance ($V_{DS} = -10$ Vdc, $V_{GS} = 0$, $f = 1.0$ MHz)	C_{iss}	–	20	pF
Reverse Transfer Capacitance ($V_{DS} = -10$ Vdc, $V_{GS} = 0$, $f = 1.0$ MHz)	C_{rss}	–	5.0	pF
Common-Source Noise Figure ($V_{DS} = -10$ Vdc, $I_D = 1.0$ mAdc, $R_G = 1.0$ Megohm, $f = 100$ Hz)	NF	–	5.0	dB
Equivalent Short-Circuit Input Noise Voltage ($V_{DS} = -10$ Vdc, $I_D = 1.0$ mAdc, $f = 100$ Hz, BW = 15 Hz)	E_n	–	0.19	μV/$\sqrt{\text{Hz}}$

*Indicates JEDEC Registered Data.

Note 1: Pulse Test: Pulse Width $\leq$ 630 ms, Duty Cycle $\leq$ 10%.

(b)

FIGURE 19-17

Common-Gate (CG) Amplifier

A typical common-gate amplifier is shown in Figure 19–18. The gate is effectively at ac ground because of capacitor C_2. The input signal is applied at the source terminal through C_1. The output is coupled through C_3 from the drain terminal.

FIGURE 19–18
JFET Common-gate amplifier.

Voltage Gain of the CG Amplifier

The voltage gain from source to drain is developed as follows:

$$A_v = \frac{V_{\text{out}}}{V_{\text{in}}} = \frac{V_d}{V_{gs}} = \frac{I_d R_D}{V_{gs}} = \frac{g_m V_{gs}\, R_D}{V_{gs}}$$

$$A_v = g_m R_D \tag{19–21}$$

Notice that the gain expression is the same as for the CS JFET amplifier.

Input Impedance of the CG Amplifier

As you have seen, both the CS and the CD configuration have extremely high input impedances because the gate is the input terminal. In contrast, the common-gate configuration has a low input impedance, as shown in the following steps.

First, the input current (source current) is equal to the drain current:

$$I_{\text{in}} = I_d = g_m V_{gs}$$

The input voltage equals V_{gs}:

$$V_{\text{in}} = V_{gs}$$

The input impedance at the source terminal is, therefore,

$$R_{\text{in(source)}} = \frac{V_{\text{in}}}{I_{\text{in}}} = \frac{V_{gs}}{g_m V_{gs}}$$

$$R_{\text{in(source)}} = \frac{1}{g_m} \tag{19–22}$$

TABLE 19–2
Comparison of FET amplifier configurations.

	CS	CD	CG
Voltage gain, A_v	$g_m R_d$	$\dfrac{g_m R_S}{1 + g_m R_S}$	$g_m R_d$
Input impedance, R_{in}	$\left(\dfrac{V_{GS}}{I_{GSS}}\right) \parallel R_G$	$\left(\dfrac{V_{GS}}{I_{GSS}}\right) \parallel R_G$	$\left(\dfrac{1}{g_m}\right) \parallel R_G$

EXAMPLE 19–8

(a) Determine the voltage gain of the amplifier in Figure 19–19.
(b) Determine the input impedance.

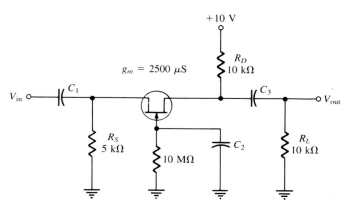

FIGURE 19–19

Solution

(a) This CG amplifier has a load resistor effectively in parallel with R_D, so the effective drain resistance is $R_D \| R_L$ and the gain is

$$A_v = g_m(R_D \| R_L) = (2500 \ \mu S)(10 \ k\Omega \| 10 \ k\Omega) = 12.5$$

(b) The input impedance at the source terminal is

$$R_{in(source)} = \frac{1}{g_m} = \frac{1}{2500 \ \mu S} = 400 \ \Omega$$

The signal source actually sees R_S in parallel with $R_{in(source)}$, so the total input impedance is

$$R_{in} = 400 \ \Omega \| 5 \ k\Omega = 370 \ \Omega$$

Summary

A summary of the gain and input impedance characteristics for the three FET amplifier configurations is given in Table 19–2.

SECTION REVIEW 19–4

1. What factors determine the voltage gain of a CS FET amplifier?
2. A certain CS amplifier has an $R_D = 1 \ k\Omega$. When a load resistance of 1 $k\Omega$ is capacitively coupled to the drain, how much does the gain change?
3. What is a major difference between a CG amplifier and the other two configurations?

19-5

MULTISTAGE AMPLIFIERS

Several amplifiers can be connected in a *cascaded* arrangement with the output of one amplifier driving the input of the next, as shown in Figure 19–20. Each amplifier in the cascaded arrangement is known as a *stage*. The purpose of a multistage arrangement is to increase the overall gain.

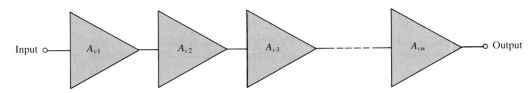

FIGURE 19–20
Cascaded amplifiers.

Multistage Gain

The overall gain A'_v of cascaded amplifiers is the product of the individual gains:

$$A'_v = A_{v1}A_{v2}A_{v3} \cdot \cdot \cdot A_{vn} \tag{19–23}$$

where n is the number of stages.

Decibel Voltage Gain

Amplifier voltage gain is often expressed in decibels (dB) as follows:

$$A_v \text{ (dB)} = 20 \log A_v \tag{19–24}$$

This formula is particularly useful in multistage systems because the overall dB voltage gain is the *sum* of the individual dB gains:

$$A'_v \text{ (dB)} = A_{v1} \text{ (dB)} + A_{v2} \text{ (dB)} + \cdot \cdot \cdot + A_{vn} \text{ (dB)} \tag{19–25}$$

EXAMPLE 19–9

A given cascaded amplifier arrangement has the following voltage gains: $A_{v1} = 10$, $A_{v2} = 15$, and $A_{v3} = 20$. What is the overall gain? Also express each gain in decibels and determine the total decibel voltage gain.

Solution

$$A'_v = A_{v1}A_{v2}A_{v3} = (10)(15)(20) = 3000$$
$$A_{v1} \text{ (dB)} = 20 \log 10 = 20 \text{ dB}$$
$$A_{v2} \text{ (dB)} = 20 \log 15 = 23.52 \text{ dB}$$
$$A_{v3} \text{ (dB)} = 20 \log 20 = 26.02 \text{ dB}$$
$$A'_v \text{ (dB)} = 20 \text{ dB} + 23.52 \text{ dB} + 26.02 \text{ dB} = 69.54 \text{ dB}$$

Multistage Analysis

The two-stage amplifier in Figure 19–21 illustrates multistage analysis. Notice that both stages are identical CE amplifiers with the output of the first stage *capacitively coupled* to the input of the second stage. Capacitive coupling prevents the dc bias of one stage from affecting that of the other. Also notice that the transistors are designated Q_1 and Q_2.

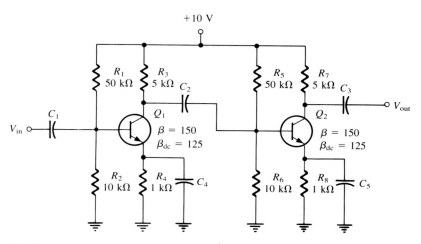

FIGURE 19–21
Two-stage amplifier.

Loading Effects

In determining the gain of the first stage, we must consider the loading effect of the second stage. Because the coupling capacitor C_2 appears as a short to the signal frequency, the total input impedance of the second stage presents an ac load to the first stage.

Looking from the collector of Q_1, the two biasing resistors, R_5 and R_6, appear in parallel with the input impedance at the base of Q_2. In other words, the signal at the collector of Q_1 "sees" R_3 and R_5, R_6, and $R_{in(base)}$ of the second stage all in parallel to ac ground. Thus, the *effective ac collector resistance* of Q_1 is the total of all these in parallel, as Figure 19–22 illustrates.

FIGURE 19–22
ac equivalent of first stage in Figure 19–21, showing loading from second stage.

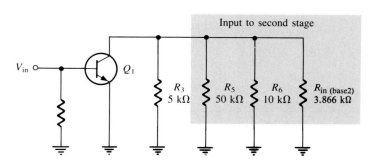

The voltage gain of the first stage is reduced by the loading of the second stage, because the effective ac collector resistance of the first stage is less than the actual value of its collector resistor, R_3. Remember that $A_v = R_C/r_e$ for an *unloaded* amplifier.

Voltage Gain of the First Stage

The ac collector resistance of the first stage is

$$R_c = R_3 \| R_5 \| R_6 \| R_{in(base2)}$$

Keep in mind that lowercase subscripts denote ac quantities such as for R_c.
You can verify that $I_E = 0.97$ mA, $r_e = 25.77$ Ω, and $R_{in(base2)} = 3.866$ kΩ. The effective ac collector resistance of the first stage is as follows:

$$R_c = 5 \text{ k}\Omega \| 50 \text{ k}\Omega \| 10 \text{ k}\Omega \| 3.866 \text{ k}\Omega = 1.73 \text{ k}\Omega$$

Therefore, the base-to-collector voltage gain of the first stage is

$$A_v = \frac{R_c}{r_e} = \frac{1.73 \text{ k}\Omega}{25.77 \text{ }\Omega} = 67$$

Voltage Gain of the Second Stage

The second stage has no load resistor, so the ac collector resistance is R_7, and the gain is

$$A_v = \frac{R_7}{r_e} = \frac{5 \text{ k}\Omega}{25.77 \text{ }\Omega} = 194$$

Compare this to the gain of the first stage, and notice how much the second-stage loading reduced the gain of the first stage.

Overall Voltage Gain

The overall amplifier gain is

$$A_v' = A_{v1}A_{v2} = (67)(194) \cong 13,000$$

If an input signal of, say, 100 μV, is applied to the first stage and if the attenuation of the input base circuit is neglected, an output from the second stage of $(100 \text{ }\mu\text{V})(13,000) = 1.3$ V will result. The overall gain can be expressed in decibels as follows:

$$A_v' \text{ (dB)} = 20 \log(13,000) = 82.28 \text{ dB}$$

SECTION REVIEW
19–5

1. What does the term *stage* mean?
2. How is the overall gain of a multistage amplifier determined?
3. Express a voltage gain of 500 in decibels.

19-6

CLASS A OPERATION

When an amplifier, whether it is one of the three configurations of bipolar or an FET type, is biased such that it always operates in the *linear* region where the output signal is an amplified replica of the input signal, it is a *class A amplifier,* as illustrated in Figure 19–23. The discussion and formulas in the previous sections apply to class A operation.

FIGURE 19–23
Class A operation.

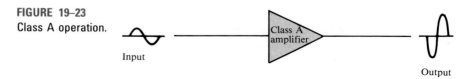

Input

Output

When the output signal takes up only a small percentage of the total load line excursion, the amplifier is a *small-signal* amplifier. When the output signal is larger and approaches the limits of the load line, the amplifier is a *large-signal* type. Amplifiers are typically operated as large-signal devices when power amplification is the major objective.

Why the *Q* Point Is Centered for Maximum Output Signal

When the dc operating point (*Q* point) is at the center of the load line, a maximum class A signal can be obtained. You can see this concept by examining the graph of the load line for a given amplifier in Figure 19–24(a). This graph shows the load line with the *Q* point at its center. The collector current can vary from its *Q*-point value, I_{CQ}, up to its saturation value, $I_{C(sat)}$, and down to its cutoff value of zero. Likewise, the collector-to-emitter voltage can swing from its *Q*-point value, V_{CEQ}, up to its cutoff value, $V_{CE(cutoff)}$, and down to its saturation value of near zero. This operation is indicated in Figure 19–24(b).

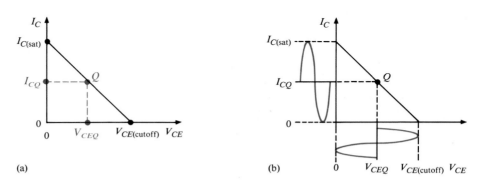

(a)

(b)

FIGURE 19–24
Maximum class A output occurs when the *Q* point is centered on the load line.

The *peak* value of the collector current is I_{CQ}, and the *peak* value of the collector-to-emitter voltage is V_{CEQ} in this case. This signal is the maximum that we can obtain from the class A amplifier. Actually, we cannot quite reach saturation or cutoff, so the practical maximum is slightly less.

How a Noncentered Q Point Limits Output Swing

If the Q point is not centered, the output signal is limited. Figure 19–25 shows a load line with the Q point moved away from center toward cutoff. The output variation is limited by cutoff in this case. The collector current can only swing down to near zero and an *equal amount* above I_{CQ}. The collector-to-emitter voltage can only swing up to its cutoff value and an *equal amount* below V_{CEQ}. This situation is illustrated in Figure 19–25(a). If the amplifier is driven any further than this, it will "clip" at cutoff, as shown in Figure 19–25(b).

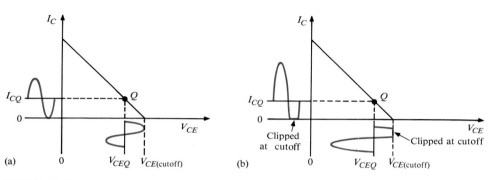

FIGURE 19–25
Q point closer to cutoff.

Figure 19–26 shows a load line with the Q point moved away from center toward saturation. In this case, the output variation is limited by saturation. The collector current can only swing up to near saturation and an *equal amount* below I_{CQ}. The collector-to-emitter voltage can only swing down to its saturation value and an *equal amount* above V_{CEQ}. This situation is illustrated in Figure 19–26(a). If the amplifier is driven any further, it will "clip" at saturation, as shown in Figure 19–26(b).

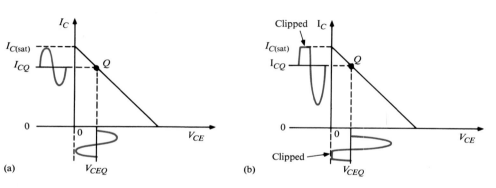

FIGURE 19–26
Q point closer to saturation.

Power Gain

The main purpose of a large-signal amplifier is to achieve power gain. If we assume that the large-signal current gain A_i is approximately equal to β_{dc}, then the power gain for a common-emitter amplifier is

$$A_p = A_i A_v = \beta_{dc} A_v$$

$$A_p = \beta_{dc} \left(\frac{R_c}{r_e} \right) \qquad (19\text{--}26)$$

dc Quiescent Power

The power dissipation of a transistor with no signal input is the product of its Q-point current and voltage:

$$P_{DQ} = I_{CQ} V_{CEQ} \qquad (19\text{--}27)$$

The quiescent power is the *maximum* power that the class A transistor must handle; therefore, its *power rating* should exceed this value.

Output Power

In general, for any Q-point location, the output power of a CE amplifier is the product of the *rms* collector current and the *rms* collector-to-emitter voltage:

$$P_{out} = V_{ce} I_c \qquad (19\text{--}28)$$

Q Point Centered

When the Q point is centered, the maximum collector current swing is I_{CQ}, and the maximum collector-to-emitter voltage swing is V_{CEQ}, as was shown in Figure 19–24(b). The output power therefore is

$$P_{out} = (0.707 V_{CEQ})(0.707 I_{CQ})$$

$$P_{out} = 0.5 V_{CEQ} I_{CQ} \qquad (19\text{--}29)$$

This is the *maximum* ac output power from a class A amplifier under signal conditions. Notice that it is one-half the quiescent power dissipation.

Efficiency

Efficiency of an amplifier is the ratio of ac output power to dc input power. The dc input power is the dc supply voltage times the current drawn from the supply.

$$P_{dc} = V_{CC} I_{CC}$$

The average supply current I_{CC} equals I_{CQ}, and the supply voltage V_{CC} is twice V_{CEQ} when the Q point is centered. Therefore, the maximum efficiency is

$$\text{eff}_{max} = \frac{P_{out}}{P_{dc}} = \frac{0.5 V_{CEQ} I_{CQ}}{V_{CC} I_{CC}} = \frac{0.5 V_{CEQ} I_{CQ}}{2 V_{CEQ} I_{CQ}} = \frac{0.5}{2}$$

$$\text{eff}_{max} = 0.25 \qquad (19\text{--}30)$$

Thus 25% is the highest possible efficiency available from a class A amplifier and is approached only when the Q point is at the center of the load line.

EXAMPLE 19–10

Determine the following values for a class A amplifier operated with a centered Q point with $I_{CQ} = 50$ mA and $V_{CEQ} = 7.5$ V:

(a) minimum transistor power rating
(b) ac output power
(c) efficiency

Solution
The quiescent dc values are

$$I_{CQ} = 50 \text{ mA}$$

$$V_{CEQ} = 7.5 \text{ V}$$

(a) The maximum power that the transistor must be able to handle is the minimum rating that you would use. Thus,

$$P_{DQ} = I_{CQ}V_{CEQ} = (50 \text{ mA})(7.5 \text{ V}) = 0.375 \text{ W}$$

(b) $P_{out} = 0.5V_{CEQ}I_{CQ} = 0.5(50 \text{ mA})(7.5 \text{ V}) = 0.1875 \text{ W}$
(c) Since the Q point is centered, the efficiency is at its maximum possible value of 25%.

**SECTION REVIEW
19–6**

1. What is the optimum Q-point location for class A amplifiers?
2. What is the maximum efficiency of a class A amplifier?
3. A certain amplifier has a centered Q point of $I_{CQ} = 10$ mA and $V_{CEQ} = 7$ V. What is the *maximum* ac output power?

19–7

CLASS B PUSH-PULL OPERATION

When an amplifier is biased such that it operates in the linear region for 180° of the input cycle and is in cutoff for 180°, it is a class B amplifier. This amplifier is illustrated in Figure 19–27, where the output waveform is shown relative to the input.

FIGURE 19–27
Class B amplifier (noninverting).

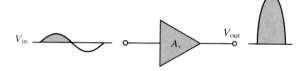

Why the Q Point Is at Cutoff

The class B amplifier is biased at cutoff so that $I_{CQ} = 0$ and $V_{CEQ} = V_{CE(\text{cutoff})}$. It is brought out of cutoff and operates in its linear region when the input signal drives it into conduction. This is illustrated in Figure 19–28 with an emitter-follower circuit.

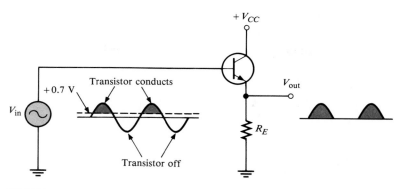

FIGURE 19–28
Common-collector class B amplifier.

Push-Pull Operation

Figure 19–29 shows one type of push-pull class B amplifier using two emitter-followers. This is a *complementary* amplifier, because one emitter-follower uses an NPN transistor and the other a PNP, which conducts on *opposite* alternations of the input cycle.

(a) Positive half-cycle   (b) Negative half-cycle

FIGURE 19–29
Class B push-pull operation.

Notice that there is no dc base bias voltage ($V_B = 0$). Thus only the signal voltage drives the transistors into conduction. Q_1 conducts during the positive half of the input cycle, and Q_2 conducts during the negative half.

Crossover Distortion

When the dc base voltage is zero, the input signal voltage must exceed V_{BE} before a transistor conducts. As a result, there is a time interval between the positive and negative alternations of the input when neither transistor is conducting, as shown in Figure

DEVICES

19–30. The resulting distortion in the output waveform is quite common and is called *crossover* distortion.

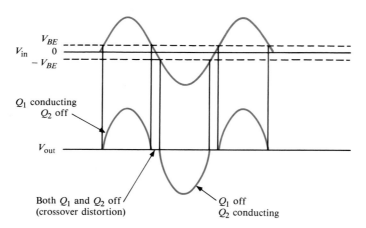

FIGURE 19–30
Illustration of crossover distortion in a class B push-pull amplifier.

Biasing the Push-Pull Amplifier

To eliminate crossover distortion, both transistors in the push-pull arrangement must be biased slightly above cutoff when there is no signal. This can be done with a voltage divider and diode arrangement, as shown in Figure 19–31. When the diode characteristics of D_1 and D_2 are closely matched to the characteristics of the transistor base-emitter junctions, a stable bias is maintained.

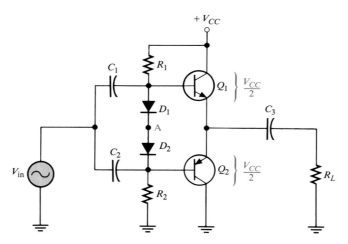

FIGURE 19–31
Biasing the push-pull amplifier to eliminate crossover distortion.

Since R_1 and R_2 are of equal value, the voltage with respect to ground at point A between the two diodes is $V_{CC}/2$. Assuming that both diodes and both transistors are identical, the drop across D_1 equals the V_{BE} of Q_1, and the drop across D_2 equals the V_{BE} of Q_2. As a result, the voltage at the emitters is also $V_{CC}/2$, therefore, $V_{CEQ1} = V_{CEQ2} = V_{CC}/2$, as indicated. Because both transistors are biased near cutoff, $I_{CQ} \cong 0$.

ac Operation

Under maximum conditions, transistors Q_1 and Q_2 are alternately driven from near cut-off to near saturation. During the positive alternation of the input signal, the Q_1 emitter is driven from its Q-point value of $V_{CC}/2$ to near V_{CC}, producing a positive peak voltage approximately equal to V_{CEQ}. At the same time, the Q_1 current swings from its Q-point value near zero to near-saturation value, as shown in Figure 19–32(a).

(a) Q_1 conducting with maximum signal output

(b) Q_2 conducting with maximum signal output

FIGURE 19–32
ac push-pull operation. Capacitors are assumed to be shorts at the signal frequency, and the dc source is at ac ground.

During the negative alternation of the input signal, the Q_2 emitter is driven from its Q-point value of $V_{CC}/2$ to near zero, producing a negative peak voltage approximately equal to V_{CEQ}. Also, the Q_2 current swings from near zero to near-saturation value, as shown in Figure 19–32(b).

Because the peak voltage across each transistor is V_{CEQ}, the ac saturation current is

$$I_{c(\text{sat})} = \frac{V_{CEQ}}{R_L} \qquad (19\text{–}31)$$

Since $I_e \cong I_c$ and the output current is the emitter current, the peak output current is also V_{CEQ}/R_L.

EXAMPLE 19–11

Determine the maximum peak values for the output voltage and current in Figure 19–33.

FIGURE 19–33

Solution

The maximum peak output voltage is

$$V_{p(\text{out})} \cong V_{CEQ} = \frac{V_{CC}}{2} = \frac{20 \text{ V}}{2} = 10 \text{ V}$$

The maximum peak output current is

$$I_{p(\text{out})} \cong I_{c(\text{sat})} = \frac{V_{CEQ}}{R} = \frac{10 \text{ V}}{5 \text{ }\Omega} = 2 \text{ A}$$

Maximum Output Power

It has been shown that the maximum peak output current is approximately $I_{c(\text{sat})}$, and the maximum peak output voltage is approximately V_{CEQ}. The maximum average output power therefore is

$$P_{\text{out}} = V_{\text{rms(out)}} I_{\text{rms(out)}}$$

Since

$$V_{\text{rms(out)}} = 0.707 V_{p(\text{out})} = 0.707 V_{CEQ}$$

and

$$I_{\text{rms(out)}} = 0.707 I_{p(\text{out})} = 0.707 I_{c(\text{sat})}$$

then

$$P_{\text{out}} = 0.5 V_{CEQ} I_{c(\text{sat})}$$

Substituting $V_{CC}/2$ for V_{CEQ}, we get

$$P_{\text{out}} = 0.25 V_{CC} I_{c(\text{sat})} \qquad (19\text{--}32)$$

Input Power

The input power comes from the V_{CC} supply and is

$$P_{\text{dc}} = V_{CC} I_{CC}$$

Since each transistor draws current for a half-cycle, the current is a *half-wave* signal with an average value of

$$I_{CC} = \frac{I_{c(\text{sat})}}{\pi}$$

Thus,

$$P_{\text{dc}} = \frac{V_{CC} I_{c(\text{sat})}}{\pi} \qquad (19\text{--}33)$$

Efficiency

The great advantage of push-pull class B amplifiers over class A is a much higher efficiency. This advantage usually overrides the difficulty of biasing the class B push-pull amplifier to eliminate crossover distortion.

The efficiency is again defined as the ratio of ac output power to dc input power:

$$\text{eff} = \frac{P_{\text{out}}}{P_{\text{dc}}}$$

The maximum efficiency for a class B amplifier is designated eff_{max} and is developed as follows, starting with Equation (19–32):

$$P_{\text{out}} = 0.25V_{CC}I_{c(\text{sat})}$$

$$\text{eff}_{\text{max}} = \frac{P_{\text{out}}}{P_{\text{dc}}} = \frac{0.25V_{CC}I_{c(\text{sat})}}{V_{CC}I_{c(\text{sat})}/\pi} = 0.25\pi$$

$$\text{eff}_{\text{max}} = 0.785 \tag{19–34}$$

Therefore, the maximum efficiency is 78.5%. Recall that the maximum efficiency for class A is 0.25 (25%).

EXAMPLE 19–12

Find the maximum ac output power and the dc input power of the amplifier in Figure 19–33 of Example 19–11.

Solution
In Example 19–11, $I_{c(\text{sat})}$ was found to be 2 A. Thus,

$$P_{\text{out}} = 0.25V_{CC}I_{c(\text{sat})} = 0.25(20 \text{ V})(2 \text{ A}) = 10 \text{ W}$$

$$P_{\text{dc}} = \frac{V_{CC}I_{c(\text{sat})}}{\pi} = \frac{(20 \text{ V})(2 \text{ A})}{\pi} = 12.73 \text{ W}$$

SECTION REVIEW 19–7

1. Where is the Q point for a class B amplifier?
2. What causes crossover distortion?
3. What is the maximum efficiency of a push-pull class B amplifier?
4. Explain the purpose of the push-pull configuration for class B.

19–8

CLASS C OPERATION

Class C amplifiers are biased so that conduction occurs for much less than 180°, as Figure 19–34 illustrates. Class C amplifiers are more efficient than either class A or push-pull class B. Thus, more output power can be obtained from class C operation. Because the output waveform is severely distorted, class C amplifiers are normally limited to applications as tuned amplifiers at radio frequencies (rf).

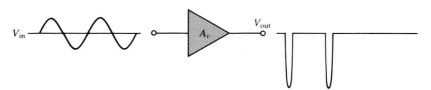

FIGURE 19–34
Class C amplifier (inverting).

Basic Operation

A basic common-emitter class C amplifier with a resistive load is shown in Figure 19–35(a). It is biased *below cutoff* with the $-V_{BB}$ supply. The ac source voltage has a peak value that is slightly greater than $V_{BB} + V_{BE}$ so that the base voltage exceeds the barrier potential of the base-emitter junction for a short time near the positive peak of each cycle, as illustrated in Figure 19–35(b). During this short interval, the transistor is turned on.

When the entire ac load line is used, the maximum collector current is approximately $I_{C(sat)}$, and the minimum collector voltage is approximately $V_{CE(sat)}$.

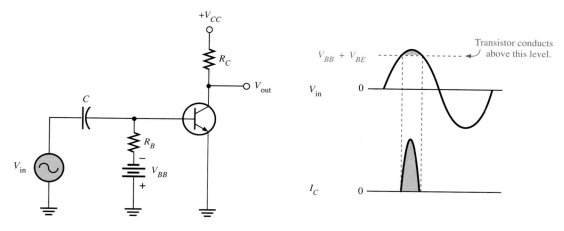

(a) Basic class C amplifier circuit

(b) Input voltage and output current waveforms

FIGURE 19–35
Class C operation.

Power Dissipation

The power dissipation of the transistor in a class C amplifier is low because it is on for only a small percentage of the input cycle. Figure 19–36 shows the collector current pulses. The time between the pulses is the *period (T)* of the ac input voltage.

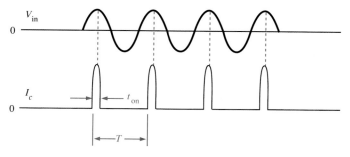

FIGURE 19–36
Collector current pulses in a class C amplifier.

The transistor is on for a short time, t_{on}, and *off* for the rest of the input cycle. Since the power dissipation averaged over the entire cycle depends on the ratio of t_{on} to T and on the power dissipation during t_{on}, it is typically very low.

Tuned Operation

Because the collector voltage (output) is not a replica of the input, the resistively loaded class C amplifier is of no value in linear applications. Therefore, it is necessary to use a class C amplifier with a parallel *resonant circuit* (tank), as shown in Figure 19–37(a). The resonant frequency of the tank circuit is determined by the formula $f_r = 1/(2 \pi \sqrt{LC})$.

(a) Basic circuit

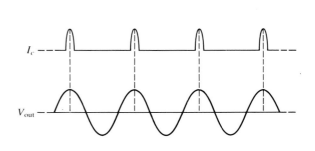

(b) Output waveforms

FIGURE 19–37
Tuned class C amplifier.

FIGURE 19–38
Tank circuit oscillations.

(a) Oscillation dies out due to energy loss.

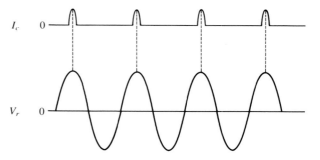

(b) Oscillation is sustained by short pulses of collector current.

The short pulse of collector current on each cycle of the input initiates and sustains the oscillation of the tank circuit so that an output sine wave voltage is produced, as illustrated in Figure 19–37(b).

The amplitude of each successive cycle of the oscillation would be less than that of the previous cycle because of energy loss in the resistance of the tank circuit, as shown in Figure 19–38(a), and the oscillation would eventually die out. However, the regular recurrences of the collector current pulse re-energizes the resonant circuit and sustains the oscillations at a constant amplitude. When the tank circuit is tuned to the frequency of the input signal, re-energizing occurs on each cycle of the tank voltage, as shown in Figure 19–38(b).

Maximum Output Power

Since the voltage developed across the tank circuit has a peak-to-peak value of approximately $2V_{CC}$, the maximum output power can be expressed as

$$P_{out} = \frac{V_{rms}^2}{R_c} = \frac{(0.707V_{CC})^2}{R_c}$$

$$P_{out} = \frac{0.5V_{CC}^2}{R_c} \tag{19–35}$$

where R_c is the equivalent parallel resistance of the collector tank circuit and represents the parallel combination of the coil resistance and the load resistance. It usually has a low value.

The total power that must be supplied to the amplifier is

$$P_T = P_{out} + P_{D(avg)}$$

Therefore, the efficiency is

$$\text{eff} = \frac{P_{out}}{P_{out} + P_{D(avg)}} \tag{19–36}$$

When $P_{out} \gg P_{D(avg)}$, the class C efficiency closely approaches 100%.

EXAMPLE 19–13

A certain class C amplifier has a $P_{D(avg)}$ of 2 mW, a V_{CC} equal to 24 V, and an R_c of 100 Ω. Determine the efficiency.

Solution

$$P_{out} = \frac{0.5V_{CC}^2}{R_c} = \frac{0.5(24 \text{ V})^2}{100 \text{ } \Omega} = 2.88 \text{ W}$$

Therefore,

$$\text{eff} = \frac{P_{out}}{P_{out} + P_{D(avg)}} = \frac{2.88 \text{ W}}{2.88 \text{ W} + 2 \text{ mW}} = 0.9993$$

or

$$\%\text{eff} = 99.93\%$$

SECTION REVIEW
19–8

1. A class C amplifier is normally biased in _____ .
2. What is the purpose of the tuned circuit in a class C amplifier?
3. A certain class C amplifier has a power dissipation of 100 mW and an output power of 1 W. What is its efficiency?

19–9

TROUBLESHOOTING AN AMPLIFIER

In working with any circuit, you must first know how it is supposed to work before you can troubleshoot it for a failure. For the multistage amplifier just discussed, since we know what to expect, we can identify and trace a failure.

Let's assume that transistor Q_2 in Figure 19–39 develops an open between the collector and emitter. You, as a technician, do not know what the trouble is, but the absence of a signal at the output indicates that the amplifier has malfunctioned.

To locate and identify the failure, *trace* the signal from the input to the point where it disappears. This procedure is illustrated in Figure 19–39. With a 100-μV rms test signal applied to the input, an inverted 6.7-mV signal should be at the collector of Q_1. The presence of this signal indicates that the first stage is operating as expected. Next, check the signal at the base of Q_2. If the coupling capacitor C_2 is good, this signal

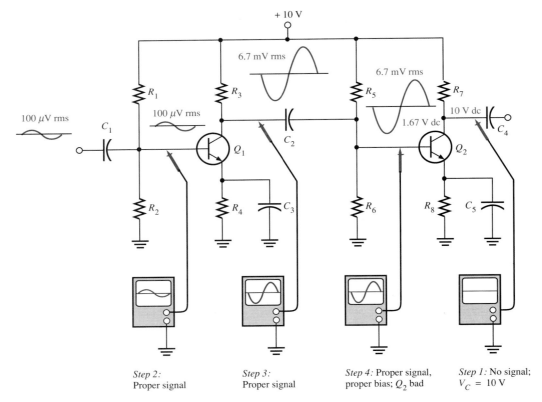

Step 2:
Proper signal

Step 3:
Proper signal

Step 4: Proper signal,
proper bias; Q_2 bad

Step 1: No signal;
V_C = 10 V

FIGURE 19–39
Using an oscilloscope to troubleshoot an amplifier.

should be the same as the Q_1 collector signal. Also, you should check the dc bias level at the base of Q_2 to verify correct bias. For this particular amplifier, it should be about 1.67 V.

So far you have found that there is a proper signal and bias level at the base of Q_2, but there is no collector signal. A check of the dc level at the Q_2 collector shows 10 V. Since the signal is not getting through Q_2 and since $V_{C2} = V_{CC}$, a collector-to-emitter open circuit should be suspected. The last step is to replace transistor Q_2 and check the amplifier's performance again. This general signal-tracing method of trouble-shooting can be used to isolate most problems in an electronic circuit.

As another example of isolating a component failure in a circuit, we will use a class A amplifier with the output monitored by an oscilloscope, as shown in Figure 19–40. As shown, the amplifier has a normal sine wave output when a sinusoidal input signal is applied.

FIGURE 19–40
Class A amplifier with proper output display.

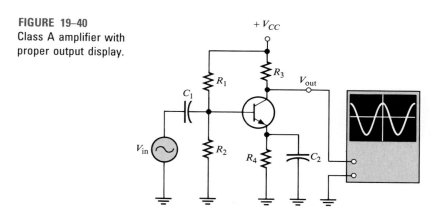

Now, several incorrect output waveforms will be considered and the most likely causes discussed. In Figure 19–41(a), the scope displays a dc level equal to the dc supply voltage, indicating that the transistor is in cutoff. The two possible causes of this

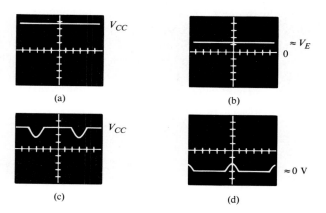

FIGURE 19–41
Oscilloscope displays of output voltage for the amplifier in Figure 19–40, illustrating several types of failures.

condition are (1) the transistor is open from collector to emitter, or (2) R_4 is open, preventing collector and emitter current.

In Figure 19–41(b), the scope displays a dc level at the collector approximately equal to the emitter voltage. The two possible causes of this indication are (1) the transistor is shorted from collector to emitter, or (2) R_2 is open, causing the transistor to be biased in saturation. In the second case, a sufficiently large input signal can bring the transistor out of saturation on its negative peaks, resulting in short pulses on the output.

In Figure 19–41(c), the scope displays an output waveform that is clipped at cutoff. Possible causes of this indication are (1) the Q point has shifted down due to a drastic out-of-tolerance change in a resistor value, or (2) R_1 is open, biasing the transistor in cutoff. In the second case, the input signal is sufficient to bring it out of cutoff for a small portion of the cycle.

In Figure 19–41(d), the scope displays an output waveform that is clipped at saturation. Again, it is possible that a resistance change has caused a drastic shift in the Q point up toward saturation, or R_2 is open, causing the transistor to be biased in saturation, and the input signal is bringing it out of saturation for a small portion of the cycle.

**SECTION REVIEW
19–9**

1. Assume that the base-emitter junction of Q_2 in Figure 19–39 shorts.
    (a) Will the ac signal at the base of Q_2 change? In what way?
    (b) Will the dc level at the base of Q_2 change? In what way?
2. What would you check for if you noticed clipping at both peaks of the output waveform?
3. A significant loss of gain in the amplifier of Figure 19–40 would most likely be caused by what type of failure?

19–10   OSCILLATORS

Oscillators are circuits that generate an output signal without having an externally applied input signal. They are used as *signal sources* in many applications. The oscillator is essentially an amplifier in which a portion of the output is fed back to the input. Its operation is based on the principle of *positive feedback*.

The block diagram in Figure 19–42 shows an amplifier with gain A. The output drives a feedback circuit with gain B. The output of the feedback circuit provides the input to the amplifier. This is the basic form of an oscillator. If conditions are proper, the oscillator will continuously produce an output signal by amplifying the feedback signal.

FIGURE 19–42
Block diagram of basic oscillator.

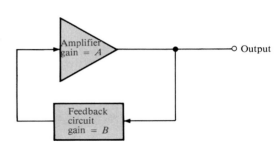

Conditions for Oscillation

There are two conditions that must be met in order for a circuit to oscillate. First, the phase shift through the amplifier and the feedback circuit must be 0°. There must be no phase shift so that the feedback signal will tend to *reinforce* itself rather than cancel. Second, the gain through the amplifier and the feedback circuit must be equal to or greater than 1 ($AB \geq 1$). This gain is called the *loop gain*. If the gain were less than 1, the output signal would decrease and die out.

The *RC* Oscillator

The basic *RC* oscillator shown in Figure 19–43 uses an *RC* network as its feedback circuit. In this case, three *RC* lag networks have a total phase shift of 180°. The common-emitter transistor contributes a 180° phase shift. The total phase shift through the amplifier and feedback circuit therefore is 360°, which is effectively 0° (no phase shift). The attenuation of the *RC* network and the gain of the amplifier must be such that the overall gain around the feedback loop is equal to 1 at the frequency of oscillation. This circuit will produce a *continuous sine wave output*.

FIGURE 19–43
Basic *RC* oscillator.

The Colpitts Oscillator

One basic type of *tuned* oscillator is the Colpitts, named after its inventor. As shown in Figure 19–44, this type of oscillator uses an *LC* circuit in the feedback loop to provide the necessary phase shift and to act as a filter that passes only the specified frequency of oscillation. The approximate frequency of oscillation is established by the values of C_1, C_2, and L according to the following familiar formula:

$$f_r \cong \frac{1}{2\pi\sqrt{LC_T}}$$

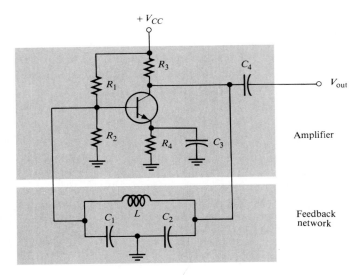

FIGURE 19–44
A basic Colpitts oscillator.

Because the capacitors effectively appear in series around the tank circuit, the total capacitance is

$$C_T = \frac{C_1 C_2}{C_1 + C_2}$$

The Hartley Oscillator

Another basic type of oscillator circuit is the Hartley, which is similar to the Colpitts except that the feedback network consists of two inductors and one capacitor, as shown in Figure 19–45.

FIGURE 19–45
A basic Hartley oscillator.

The frequency of oscillation is

$$f_r \cong \frac{1}{2\pi\sqrt{L_T C}}$$

The total inductance is the series combination of L_1 and L_2.

The Clapp Oscillator

The Clapp oscillator is similar to the Colpitts except that there is an additional capacitor in series with the inductor, as shown in Figure 19–46. C_1 and C_2 can be selected for optimum feedback, and C_3 can be adjusted to obtain the desired frequency of oscillation. Also, a capacitor having a negative temperature coefficient can be used for C_3 to stabilize the frequency of oscillation when there are temperature changes.

FIGURE 19–46
A basic Clapp oscillator.

The Crystal Oscillator

A crystal oscillator is essentially a tuned-circuit oscillator that uses a *quartz crystal* as the resonant tank circuit. Other types of crystals can be used, but quartz is the most prevalent. Crystal oscillators offer greater frequency stability than other types.

Quartz is a substance found in nature that exhibits a property called the *piezoelectric effect*. When a changing mechanical stress is applied across the crystal to cause it to vibrate, a voltage is developed at the frequency of the mechanical vibration. Conversely, when an ac voltage is applied across the crystal, it vibrates at the frequency of the applied voltage.

The symbol for a crystal is shown in Figure 19–47(a), the electrical equivalent is shown in part (b), and a typical mounted crystal is shown in part (c). In construction, a slab of quartz is mounted as shown in part (d).

Series resonance occurs in the crystal when the reactances in the series branch are equal. Parallel resonance occurs, at a higher frequency, when the reactance of L_S equals the reactance of C_m.

828   DEVICES

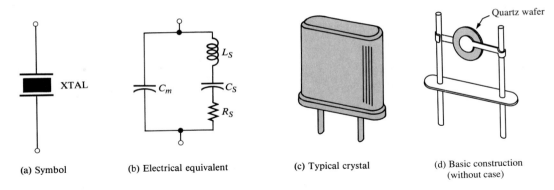

XTAL

(a) Symbol   (b) Electrical equivalent   (c) Typical crystal   (d) Basic construction
(without case)

FIGURE 19–47
Quartz crystal.

A crystal oscillator using the crystal as a series resonant tank circuit is shown in Figure 19–48(a). The impedance of the crystal is *minimum* at the series resonance, thus providing maximum feedback.

The capacitor C_c is a tuning capacitor used to fine-tune the frequency. A modified Colpitts configuration, shown in part (b), uses the crystal in its parallel resonant mode. The impedance of the crystal is maximum at parallel resonance, thus developing the maximum voltage across both C_1 and C_2. The voltage across C_2 is fed back to the input.

FIGURE 19–48
Basic crystal oscillators.

**SECTION REVIEW
19–10**

1. Name four types of oscillators.
2. Describe the basic difference between the Colpitts and Hartley oscillators.

SUMMARY

☐ The three bipolar transistor amplifier configurations are common-emitter (CE), common-collector (CC), and common-base (CB).

☐ The general characteristics of a common-emitter amplifier are high voltage gain, high current gain, very high power gain, and low input impedance.

☐ The general characteristics of a common-collector amplifier are low voltage gain (≤ 1), high current gain, high power gain, and high input impedance.

☐ The general characteristics of a common-base amplifier are high voltage gain, low current gain (≤ 1), high power gain, and very low input impedance.

☐ The three FET amplifier configurations are common-source (CS), common-drain (CD), and common-gate (CG).

☐ The overall voltage gain of a multistage amplifier is the product of the gains of all the individual stages.

☐ Any of the bipolar or FET configurations can be operated as class A, class B, or class C amplifiers.

☐ The class A amplifier conducts for the entire 360° of the input cycle.

☐ The class B amplifier conducts for 180° of the input cycle.

☐ The class C amplifier conducts for a small portion of the input cycle.

GLOSSARY

Attenuation   A gain less than unity.

Class A   A category of amplifier circuit that conducts for the entire input cycle and produces an output signal that is a replica of the input signal in terms of its waveshape.

Class B   A category of amplifier circuit that conducts for half of the input cycle.

Class C   A category of amplifier circuit that conducts for a veery small portion of the input cycle.

Common-base   A BJT amplifier configuration in which the base is the common (grounded) terminal.

Common-collector   A BJT amplifier configuration in which the collector is the common (grounded) terminal.

Common-emitter   A BJT amplifier configuration in which the emitter is the common (grounded) terminal.

Current gain   The ratio of output current to input current.

Darlington   A two-transistor arrangement that produces a multiplication of current gain.

Emitter-follower   A popular term for a common-collector amplifier.

Feedback   The process of returning a portion of a circuit's output signal to the input in such a way as to create certain specified operating conditions.

Gain   The amount by which an electrical signal is increased or decreased. The ratio of output to input.

Oscillator   An electronic circuit consisting of an amplifier and a phase-shift network connected in a feedback loop that produces a time-varying output signal without an external input signal using positive feedback.

Positive feedback   The return of a portion of the output signal to the input such that it is in-phase with the input signal.

Power gain   The ratio of output power to input power. The product of voltage gain and current gain.

Push-pull A type of class B amplifier in which one output transistor conducts for one half-cycle and the other conducts for the other half-cycle.

Transconductance, g_m The ratio of drain current to gate-to-source voltage in an FET.

Unity gain A gain of 1.

Voltage gain The ratio of output voltage to input voltage.

FORMULAS

$$A_v = \frac{R_C}{r_e + R_E}$$ CE voltage gain (unbypassed) (19–1)

$$A_v = \frac{R_C}{r_e}$$ CE voltage gain (bypassed) (19–2)

$$r_e \cong \frac{25 \text{ mV}}{I_E}$$ Internal emitter impedance (19–3)

$$R_{in} \cong \beta_{ac} r_e$$ CE input impedance (19–4)

$$R_{in(T)} = R_1 \| R_2 \| R_{in}$$ CE total input impedance (19–5)

$$A_i = \frac{I_c}{I_s}$$ CE current gain (19–6)

$$A_p = A_v A_i$$ CE power gain (19–7)

$$A_v = \frac{R_E}{r_e + R_E}$$ CC voltage gain (19–8)

$$R_{in} \cong \beta_{ac} R_E$$ CC input impedance (19–9)

$$R_{in(T)} = R_1 \| R_2 \| R_{in}$$ CC total input impedance (19–10)

$$A_i = \frac{I_e}{I_s}$$ CC current gain (19–11)

$$A_p \cong A_i$$ CC power gain (19–12)

$$\beta = \beta_1 \beta_2$$ Beta for a Darlington pair (19–13)

$$A_v = \frac{R_C}{r_e}$$ CB voltage gain (19–14)

$$R_{in} = r_e$$ CB input impedance (19–15)

$$A_i \cong 1$$ CB current gain (19–16)

$$A_p \cong A_v$$ CB power gain (19–17)

$$g_m = \frac{I_d}{V_{gs}}$$ FET transconductance (19–18)

$$A_v = g_m R_D \qquad \text{CS voltage gain} \tag{19-19}$$

$$A_v \cong \frac{g_m R_S}{1 + g_m R_S} \qquad \text{CD voltage gain} \tag{19-20}$$

$$A_v = g_m R_D \qquad \text{CG voltage gain} \tag{19-21}$$

$$R_{\text{in(source)}} = \frac{1}{g_m} \qquad \text{CG input impedance} \tag{19-22}$$

$$A_v' = A_{v1} A_{v2} A_{v3} \cdots A_{vn} \qquad \text{Multistage gain} \tag{19-23}$$

$$A_v \text{ (dB)} = 20 \log A_v \qquad \text{Voltage gain in dB} \tag{19-24}$$

$$A_v' \text{ (dB)} = A_{v1} \text{ (dB)} + A_{v2} \text{ (dB)} + \cdots + A_{vn} \text{ (dB)} \qquad \text{Multistage dB gain} \tag{19-25}$$

$$A_p = \beta_{dc} \left(\frac{R_C}{r_e} \right) \qquad \text{CE large-signal power gain} \tag{19-26}$$

$$P_{DQ} = I_{CQ} V_{CEQ} \qquad \text{Transistor power dissipation} \tag{19-27}$$

$$P_{\text{out}} = V_{ce} I_c \qquad \text{CE output power} \tag{19-28}$$

$$P_{\text{out}} = 0.5 V_{CEQ} I_{CQ} \qquad \text{CE output power with centered } Q \text{ point} \tag{19-29}$$

$$\text{eff}_{\text{max}} = 0.25 \qquad \text{Class A maximum efficiency} \tag{19-30}$$

$$I_{c(\text{sat})} = \frac{V_{CEQ}}{R_L} \qquad \text{Class B ac saturation current} \tag{19-31}$$

$$P_{\text{out}} = 0.25 V_{CC} I_{c(\text{sat})} \qquad \text{Class B output power (maximum)} \tag{19-32}$$

$$P_{dc} = \frac{V_{CC} I_{c(\text{sat})}}{\pi} \qquad \text{Class B input power} \tag{19-33}$$

$$\text{eff}_{\text{max}} = 0.785 \qquad \text{Class B maximum efficiency} \tag{19-34}$$

$$P_{\text{out}} = \frac{0.5 V_{CC}^2}{R_c} \qquad \text{Class C output power (maximum)} \tag{19-35}$$

$$\text{eff} = \frac{P_{\text{out}}}{P_{\text{out}} + P_{D(\text{avg})}} \qquad \text{Class C efficiency} \tag{19-36}$$

SELF-TEST

1. In a common-emitter (CE) amplifier, the capacitor from emitter to ground is called the
   (a) coupling capacitor (b) decoupling capacitor
   (c) bypass capacitor (d) tuning capacitor

2. If the capacitor from emitter to ground in a CE amplifier is removed, the voltage gain
   (a) increases (b) decreases (c) is not affected (d) becomes erratic

3. When the collector resistor in a CE amplifier is increased in value, the voltage gain
   (a) increases (b) decreases (c) is not affected (d) becomes erratic

4. The input resistance of a CE amplifier is affected by
    (a) α and r_e (b) β and r_e (c) R_c and r_e (D) R_e, r_e, and β

5. The output signal of a CE amplifier is always
    (a) in phase with the input signal   (b) out of phase with the input signal
    (c) larger than the input signal   (d) equal to the input signal

6. The output signal of a common-collector amplifier is always
    (a) in phase with the input signal   (b) out of phase with the input signal
    (c) larger than the input signal   (d) exactly equal to the input signal

7. The largest *theoretical* voltage gain obtainable with a CC amplifier is
    (a) 100   (b) 10   (c) 1   (d) dependent on β

8. Using a Darlington pair has the advantage of
    (a) increasing the overall voltage gain
    (b) less cost
    (c) decreasing the input impedance
    (d) increasing the overall β

9. Compared to CE and CC amplifiers, the common-base (CB) amplifier has
    (a) a lower input resistance   (b) a much larger voltage gain
    (c) a larger current gain   (d) a higher input resistance

10. In terms of higher power gain, the amplifier of choice is
    (a) CB   (b) CC   (c) CE   (d) any of these

11. The most efficient amplifier configuration is
    (a) class A   (b) class B   (c) class C

12. When an FET with a lower transconductance is substituted into an FET amplifier circuit,
    (a) the voltage gain increases   (b) the voltage gain decreases
    (c) the input resistance decreases   (d) nothing changes

13. When amplifiers are cascaded,
    (a) the gain of each amplifier is increased   (b) each amplifier has to work less
    (c) a lower supply voltage is required   (d) the overall gain is increased

14. If three amplifiers, each with a voltage gain of 30, are connected in a multistage arrangement, the overall voltage gain is
    (a) 27,000   (b) 90   (c) 10   (d) 30

15. In a class A amplifier, the output signal is
    (a) distorted   (b) clipped   (c) the same shape as the input
    (d) smaller in amplitude than the input

16. Oscillators operate on the principle of
    (a) signal feedthrough   (b) positive feedback   (c) negative feedback
    (d) attenuation

PROBLEMS   **Section 19–1   Common-Emitter (CE) Amplifiers**

1. Determine the voltage gain for Figure 19–49.

2. Determine each of the dc voltages, V_B, V_C, and V_E, with respect to ground in Figure 19–49.

3. Determine the following dc values for the amplifier in Figure 19–50:
    (a) V_B (b) V_E (c) I_E
    (d) I_C (e) V_C (f) V_{CE}

4. Determine the following ac values for the amplifier in Figure 19–50:
    (a) R_{in} (b) $R_{in(T)}$ (c) A_v
    (d) A_i (e) A_p

FIGURE 19–49

FIGURE 19–50

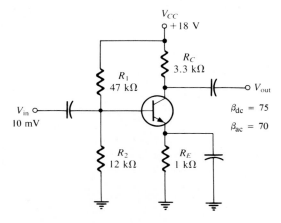

5. The amplifier in Figure 19–51 has a variable gain control, using a 100-Ω potentiometer for R_E with the wiper ac grounded. As the potentiometer is adjusted, more or less of R_E is bypassed to ground, thus varying the gain. The total R_E remains constant to dc, keeping the bias fixed. Determine the maximum and minimum gains for this amplifier.

6. If a load resistance of 600 Ω is placed on the output of the amplifier in Figure 19–51, what is the maximum gain?

FIGURE 19–51

Section 19-2   Common-Collector (CC) Amplifiers

7.   Determine the *exact* voltage gain for the emitter-follower in Figure 19–52.

8.   What is the total input impedance in Figure 19–52? What is the dc output voltage?

9.   A load resistance is capacitively coupled to the emitter in Figure 19–52. In terms of signal operation, the load appears in parallel with R_E and reduces the effective emitter resistance. How does this affect the voltage gain?

FIGURE 19–52

Section 19-3   Common-Base (CB) Amplifiers

10.   What is the main disadvantage of the CB amplifier compared to the CE and the emitter-follower?

11.   Find R_{in}, A_v, A_i, and A_p for the amplifier in Figure 19–53.

FIGURE 19–53

Section 19-4   FET Amplifiers

12.   Determine the voltage gain of each CS amplifier in Figure 19–54.

13.   Find the gain of each amplifier in Figure 19–55.

14.   Determine the gain of each amplifier in Figure 19–55 when a 10-kΩ load is capacitively coupled from source to ground.

Section 19-5   Multistage Amplifiers

15.   Each of three cascaded amplifier stages has a decibel voltage gain of 10. What is the overall decibel voltage gain? What is the actual overall voltage gain?

16.   For the two-stage, capacitively coupled amplifier in Figure 19–56, find the following values:
   (a)   Voltage gain of each stage
   (b)   Overall voltage gain
   (c)   Express the gains found above in decibels

FIGURE 19–54

FIGURE 19–55

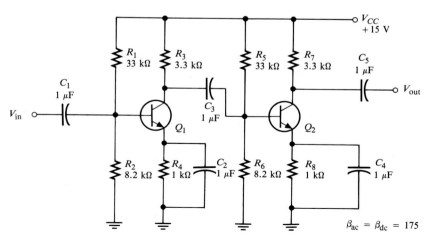

FIGURE 19–56

Section 19–6   Class A Operation

17. Determine the minimum power rating for each of the transistors in Figure 19–57.

(a) $\beta_{ac} = \beta_{dc} = 90$; $r_e = 10\ \Omega$ (b) $\beta_{ac} = \beta_{dc} = 175$; $r_e = 1.2\ \Omega$

FIGURE 19–57

Section 19–7   Class B Push-Pull Operation

18. Determine the dc voltages at the bases and emitters of Q_1 and Q_2 in Figure 19–58. Also determine V_{CEQ} for each transistor.

19. Determine the maximum peak output voltage and peak load current for the circuit in Figure 19–58.

FIGURE 19–58

20. The efficiency of a certain class B push-pull amplifier is 0.71, and the dc input power is 16.25 W. What is the ac output power?

Section 19–8   Class C Operation

21. What is the resonant frequency of the tank circuit in a class C amplifier with $L = 10$ mH and $C = 0.001$ μF?

22. Determine the efficiency of the class C amplifier when $P_{D(avg)} = 10$ mW, $V_{CC} = 15$ V, and the equivalent parallel resistance in the collector tank circuit is 50 Ω.

Section 19–9   Troubleshooting an Amplifier

23. Sketch the waveforms you would expect to see with a scope across R_L in Figure 19–58 if **(a)** Q_1 were open from collector to emitter and **(b)** Q_2 were open from collector to emitter.

24. What symptom(s) would indicate each of the following failures under signal conditions in Figure 19–59?

    (a) Q_1 open from drain to source   **(b)** R_3 open   **(c)** C_2 shorted

    (d) C_3 shorted   **(e)** Q_2 open from drain to source

FIGURE 19–59

Section 19–10   Oscillators

25. Calculate the frequency of oscillation for each circuit in Figure 19–60, and identify each type of oscillator.

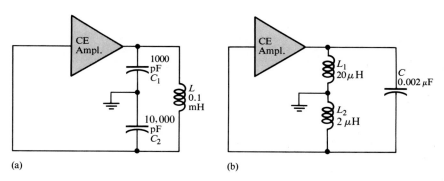

(a)   (b)

FIGURE 19–60

**ANSWERS TO
SECTION REVIEWS**

Section 19–1
1. To increase voltage gain.   **2.** Ratio of collector resistance to total emitter resistance.
3. 10,000.

Section 19–2
1. Emitter-follower.   **2.** 1.   **3.** High input impedance.

Section 19–3
1. True.   **2.** True.

Section 19–4
1. g_m and R_d.   **2.** It is halved.   **3.** Common-gate has low input impedance.

Section 19–5
1. One amplifier in a cascaded arrangement.   **2.** Product of individual gains.
3. 53.98 dB.

Section 19–6
1. Centered on the load line.   **2.** 25%.   **3.** 35 mW.

Section 19–7
1. At cutoff.   **2.** The barrier potential of the base-emitter junction.   **3.** 78.5%.
4. To reproduce both positive and negative alternations of the input signal with greater efficiency.

Section 19–8
1. Cutoff.   **2.** To produce a sine wave output.   **3.** 90.9%.

Section 19–9
1. **(a)** It will disappear because it is shorted to ground through the base-emitter junction and C_5.
    (b) Yes. It will decrease.
2. Excess input signal voltage.   **3.** Open bypass capacitor, C_2.

Section 19–10
1. *RC* phase-shift, Colpitts, Hartley, crystal.
2. Colpitts uses two capacitors, center-tapped to ground in parallel with an inductor. Hartley uses two coils, center-tapped to ground in parallel with a capacitor.

20

Operational Amplifiers (Op-Amps)

In the previous chapters of this book, a number of electronic devices were introduced. These devices, such as the diode and the transistor, are individually packaged and are connected in a circuit with other individual devices to form a functional unit. Individually packaged devices are referred to as *discrete components*.

We will now move into the area of *integrated circuits* (ICs), in which many transistors, diodes, resistors, and capacitors are fabricated on a single silicon chip and packaged in a single case to form a functional circuit. The manufacturing process for ICs is complex and beyond the scope of this coverage.

In our study of ICs, we will treat the entire circuit as a device. That is, we will be concerned with what the circuit does from an external point of view and will be less concerned about the internal, component-level operation.

After completing this chapter, you will be able to

☐ List the characteristics of an ideal operational amplifier (op-amp).
☐ List the characteristics of a practical op-amp and compare it to the ideal model.
☐ Explain the basic operation of a differential amplifier.
☐ Describe the various modes of differential amplifier operation.
☐ Define *common-mode gain.*
☐ Show how the differential amplifier is an essential part of the op-amp.
☐ Interpret an op-amp data sheet.
☐ Define *common-mode rejection ratio (CMRR).*
☐ Describe the application of negative feedback in op-amps.
☐ Distinguish various types of op-amp configurations such as *inverting, noninverting,* and *voltage-follower* and explain their operation.

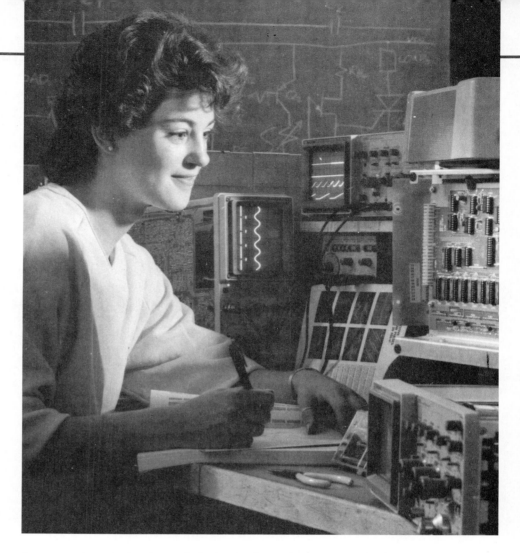

APPLICATION ASSIGNMENT

As a recent graduate of a technical school, you have accepted a job with a major electronics company involved in the design, development, and manufacturing of state-of-the-art radar systems for both commercial and military applications. Your first project assignment is with a design and development group working on an advanced terrain-following radar system for a military attack aircraft.

The design engineer asks you to build a high-input impedance, low-output impedance, noninverting amplifier that has a variable voltage gain from 1 to 50. This amplifier is needed to test a certain portion of the radar signal-processing circuitry. After completing this chapter, you will be able to carry out this assignment.

20–1

INTRODUCTION TO OPERATIONAL AMPLIFIERS

Early operational amplifiers (op-amps) were used primarily to perform *mathematical operations* such as addition, subtraction, integration, and differentiation—hence the term *operational*. These early devices were constructed with vacuum tubes and worked with high voltages.

Today's op-amps are *linear integrated circuits* (ICs) that use relatively low dc supply voltages and are very reliable and inexpensive.

Symbol and Terminals

The standard op-amp symbol is shown in Figure 20–1(a). It has *two input terminals,* called the *inverting input* ($-$) and the *noninverting input* ($+$), and one output terminal.

The typical op-amp operates with two dc supply voltages, one positive and the other negative, as shown in part (b). *Usually these dc voltage terminals are left off the schematic symbol for simplicity, but they are always understood to be there.*

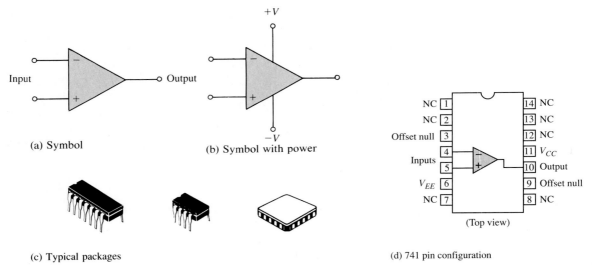

(a) Symbol

(b) Symbol with power

(c) Typical packages

(d) 741 pin configuration

FIGURE 20–1
Op-amp symbols and packages. (courtesy of Motorola Semiconductor)

The Ideal Op-Amp

To illustrate what an op-amp is, we will consider its *ideal* characteristics. A practical op-amp, of course, falls short of these ideal standards, but it is much easier to understand and analyze the device from an ideal point of view.

First, the ideal op-amp has *infinite voltage gain* and *infinite bandwidth*. Also, it has an *infinite input impedance* (open), so that it does not draw any power from the driving source. Finally, it has a *0 (zero) output impedance*. These characteristics are illustrated in Figure 20–2. The input voltage V_{in} appears between the two input terminals, and the output voltage is A_vV_{in}, as indicated by the symbol for internal voltage source.

FIGURE 20–2
Ideal op-amp representation.

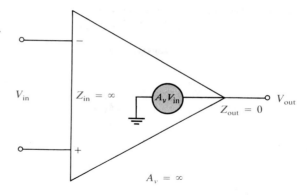

The concept of infinite input impedance is a particularly valuable analysis tool for the various op-amp configurations, which will be covered later.

The Practical Op-Amp

Although modern IC op-amps approach parameter values that can be treated as ideal in many cases, the ideal device has not been and probably will not be developed even though improvements continue to be made.

Any device has limitations, and the IC op-amp is no exception. Op-amps have both voltage and current limitations. Peak-to-peak output voltage, for example, is usually limited to slightly less than the two supply voltages. Output current is also limited by internal restrictions such as power dissipation and component ratings.

Characteristics of a practical op-amp are high voltage gain, high input impedance, low output impedance, and wide bandwidth, as illustrated in Figure 20–3.

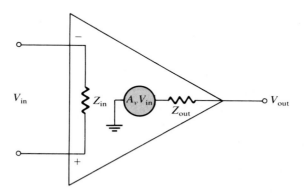

FIGURE 20–3
Practical op-amp representation.

**SECTION REVIEW
20–1**

1. Sketch the symbol for an op-amp.
2. Describe some of the characteristics of a *practical* op-amp.

20-2

THE DIFFERENTIAL AMPLIFIER

The op-amp, in its basic form, typically consists of two or more *differential amplifier* stages. Because the differential amplifier (diff-amp) is fundamental to the op-amp's internal operation, it is useful to spend some time in acquiring a basic understanding of this type of circuit. A basic diff-amp circuit is shown in Figure 20–4(a), and its block symbol in part (b).

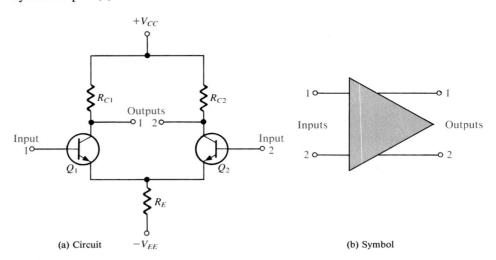

(a) Circuit   (b) Symbol

FIGURE 20–4
Basic differential amplifier.

The diff-amp stages that make up part of the op-amp provide *high voltage gain* and *common-mode rejection* (to be defined).

Basic Operation

Although an op-amp typically has more than one diff-amp stage, we will use a single diff-amp to illustrate the basic operation. The following discussion is in relation to Figure 20–5 and consists of a basic dc analysis of the diff-amp's operation.

First, when both inputs are grounded (0 V), the emitters are at −0.7 V, as indicated in Figure 20–5(a). It is assumed that the transistors are identically matched by careful process control during manufacturing so that their dc emitter currents are the same with no input signal. Thus,

$$I_{E1} = I_{E2}$$

Since both emitter currents combine through R_E,

$$I_{E1} = I_{E2} = \frac{I_{RE}}{2}$$

where

$$I_{RE} = \frac{V_E - V_{EE}}{R_E}$$

Based on the approximation that $I_C \cong I_E$, it can be stated that

$$I_{C1} = I_{C2} \cong \frac{I_{RE}}{2}$$

(a) Both inputs (bases) grounded

(b) Bias voltage on Q_1 base, Q_2 base grounded

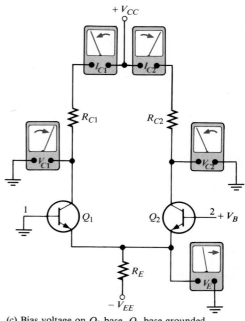

(c) Bias voltage on Q_2 base, Q_1 base grounded

FIGURE 20–5
Basic operation of a differential amplifier. Ground is 0 V.

DEVICES

Since both collector currents and both collector resistors are equal (when the input voltage is zero),

$$V_{C1} = V_{C2} = V_{CC} - I_{C1}R_{C1}$$

This condition is illustrated in Figure 20–5(a).

Next, input 2 is left grounded, and a positive bias voltage is applied to input 1, as shown in Figure 20–5(b). The positive voltage on the base of Q_1 increases I_{C1} and raises the emitter voltage. This action *reduces* the forward bias (V_{BE}) of Q_2 because its base is held at 0 V (ground), thus causing I_{C2} to *decrease,* as indicated in part (b) of the diagram. The net result is that the increase in I_{C1} causes a decrease in V_{C1}, and the decrease in I_{C2} causes an increase in V_{C2}, as shown.

Finally, input 1 is grounded and a positive bias voltage is applied to input 2, as shown in Figure 20–5(c). The positive bias voltage causes Q_2 to conduct more, thus increasing I_{C2}. Also, the emitter voltage is raised. This reduces the forward bias of Q_1, since its base is held at ground, and causes I_{C1} to decrease. The result is that the increase in I_{C2} produces a decrease in V_{C2}, and the decrease in I_{C1} causes V_{C1} to increase as shown.

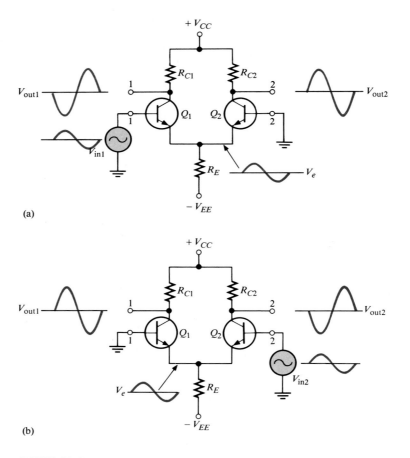

(a)

(b)

FIGURE 20–6
Single-ended input operation of a differential amplifier.

Modes of Signal Operation

Single-Ended Input

When a diff-amp is operated in this mode, one input is grounded and the signal voltage is applied only to the other input, as shown in Figure 20–6(a) and (b).

When the signal voltage is applied to input 1, as in part (a), an inverted, amplified signal voltage appears at output 1 as shown. Also, a signal voltage appears in phase at the emitter of Q_1. Since the emitters of Q_1 and Q_2 are common, this emitter signal becomes an input to Q_2 which functions as a common-base amplifier. The signal is amplified by Q_2 and appears, noninverted, at output 2.

When the signal is applied to input 2 with input 1 grounded, as in part (b), an inverted, amplified signal voltage appears at output 2. In this situation, Q_1 acts as a common-base amplifier, and a noninverted, amplified signal appears at output 1.

Differential Input

In this mode, two signals of *opposite polarity* (out of phase) are applied to the inputs, as shown in Figure 20–7(a). This type of operation is also referred to as *double-ended*. Each input affects the outputs, as you will see in the following discussion.

FIGURE 20–7
Differential operation of a differential amplifier.

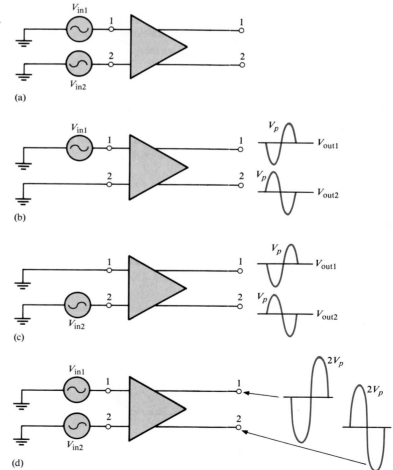

(a)

(b)

(c)

(d)

Figure 20–7(b) shows the output signals due to the signal on input 1 acting alone as a single-ended input. Figure 20–7(c) shows the output signals due to the signal on input 2 acting alone as a single-ended input. In parts (b) and (c), notice that the signals on output 1 are of the same polarity. The same is also true for output 2. By superimposing both output 1 signals and both output 2 signals, we get the total differential operation, as pictured in Figure 20–7(d).

Common-Mode Input

We will illustrate one of the most important aspects of the operation of a diff-amp by considering the case where two signal voltages of the *same phase, frequency, and amplitude* are applied to the two inputs, as shown in Figure 20–8(a). Again, by considering each input signal as acting alone, you can understand the basic operation.

Figure 20–8(b) shows the output signals due to the signal only on input 1, and part (c) shows the output signals due to the signal on only input 2. Notice that the component signals on output 1 are of the opposite polarity, and so are those on output

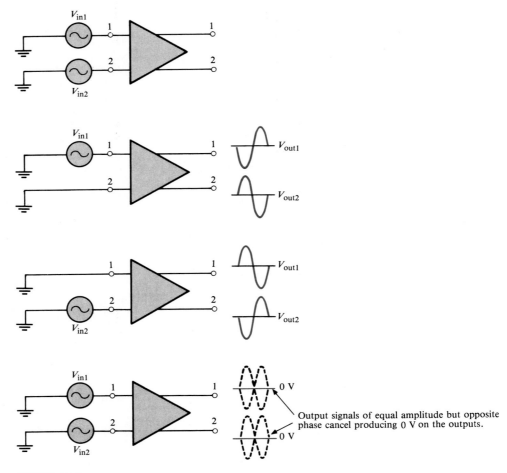

FIGURE 20–8
Common-mode operation of a differential amplifier.

2. When these are superimposed, they cancel, resulting in a near zero output voltage, as shown in Figure 20–8(d).

This action is called *common-mode rejection*. Its importance lies in the situation where an *unwanted* signal appears commonly on both diff-amp inputs. Common-mode rejection means that this unwanted signal will not appear on the outputs to distort the *desired* signal. Common-mode signals (noise) generally are the result of the pick-up of radiated energy on the input lines, or adjacent lines, or the 60-Hz power line, or other sources.

In summary, *desired* signals appear on only one input or with opposite polarities on both input lines. These desired signals are amplified and appear on the outputs as previously discussed. Unwanted signals (noise) appearing with the same polarity on both input lines are essentially canceled by the diff-amp and do not appear on the outputs. The *measure* of an amplifier's ability to reject common-mode signals is a parameter called the *common-mode rejection ratio* (CMRR) and is discussed in the following paragraphs.

Common-Mode Gain

Ideally, a diff-amp provides a very high gain for desired signals (single-ended or differential), and 0 (zero) gain for common-mode signals. Practical diff-amps, however, do exhibit a very small common-mode gain (usually much less than 1), while providing a high differential voltage gain (usually several thousand). The higher the differential gain with respect to the common-mode gain, the better the performance of the diff-amp in terms of rejection of common-mode signals. This suggests that a good measure of the diff-amp's performance in rejecting unwanted common-mode signals is the ratio of the differential gain $A_{v(d)}$ to the common-mode gain, A_{cm}. This ratio is called the *common-mode rejection ratio*, CMRR:

$$\text{CMRR} = \frac{A_{v(d)}}{A_{cm}} \tag{20–1}$$

The higher the CMRR, the better, as you can see. A very high value of CMRR means that the differential gain $A_{v(d)}$ is high and the common-mode gain A_{cm} is low.

The CMRR is often expressed in decibels (dB) as

$$\text{CMRR} = 20 \log \frac{A_{v(d)}}{A_{cm}} \tag{20–2}$$

EXAMPLE 20–1

A certain diff-amp has a differential voltage gain of 2000 and a common-mode gain of 0.2. Determine the CMRR and express it in decibels.

Solution

$$A_{v(d)} = 2000 \quad \text{and} \quad A_{cm} = 0.2$$

$$\text{CMRR} = \frac{A_{v(d)}}{A_{cm}} = \frac{2000}{0.2} = 10,000$$

In decibels,

$$\text{CMRR} = 20 \log(10,000) = 80 \text{ dB}$$

A CMRR of 10,000, for example, means that the desired input signal (differential) is amplified 10,000 times more than the unwanted noise (common-mode). So, as an example, if the amplitudes of the differential input signal and the common-mode noise are equal, the desired signal will appear on the output 10,000 times greater in amplitude than the noise. Thus, the noise or interference has been essentially eliminated.

The following example illustrates further the idea of *common-mode rejection* and the general signal operation of the diff-amp.

EXAMPLE 20–2

The diff-amp shown in Figure 20–9 has a differential voltage gain of 2500 and a CMRR of 30,000. In part (a) of the figure, a single-ended input signal of 500 μV rms is applied. At the same time a 1-V, 60-Hz interference signal appears on both inputs as a result of radiated pick-up from the ac power system.

In part (b) of the figure, differential input signals of 500 μV rms each are applied to the inputs. The common-mode interference is the same as in part (a).

(a) Determine the common-mode gain.
(b) Express the CMRR in dB.
(c) Determine the rms output signal for parts (a) and (b).
(d) Determine the rms interference voltage on the output.

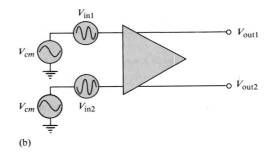

(a)   (b)

FIGURE 20–9

Solution

(a) $\text{CMRR} = \dfrac{A_{v(d)}}{A_{cm}}$

$A_{cm} = \dfrac{A_{v(d)}}{\text{CMRR}} = \dfrac{2500}{30{,}000} = 0.083$

(b) $\text{CMRR} = 20 \log(30{,}000) = 89.5 \text{ dB}$

(c)   In Figure 20–9(a), the differential input voltage is the *difference* between the voltage on input 1 and that on input 2. Since input 2 is grounded, its voltage is zero. Therefore,

$$V_{\text{in(diff)}} = V_{\text{in1}} - V_{\text{in2}} = 500 \ \mu\text{V} - 0 \text{ V} = 500 \ \mu\text{V}$$

The output signal voltage in this case is taken at output 1:

$$V_{\text{out1}} = A_{v(d)} V_{\text{in(diff)}} = (2500)(500 \ \mu\text{V}) = 1.25 \text{ V rms}$$

In Figure 20–9(b), the differential input voltage is the difference between the two opposite-polarity, 500-μV signals:

$$V_{in(diff)} = V_{in1} - V_{in2} = 500 \ \mu V - (-500 \ \mu V) = 1000 \ \mu V = 1 \ mV$$

The output signal voltage is

$$V_{out1} = A_{v(d)}V_{in(diff)} = (2500)(1 \ mV) = 2.5 \ V \ rms$$

This shows that a differential input (two opposite-polarity signals) results in a gain that is double that for a single-ended input.

(d) The common-mode input is 1 V rms. The common-mode gain A_{cm} is 0.083. The interference voltage on the output therefore is

$$A_{cm} = \frac{V_{out(cm)}}{V_{in(cm)}}$$

$$V_{out(cm)} = A_{cm}V_{in(cm)} = 0.083(1 \ V) = 0.083 \ V$$

Simple Op-Amp Arrangement

Figure 20–10 shows two diff-amp stages and an emitter-follower connected to form a very simple op-amp. The first stage can be used with a single-ended or a differential

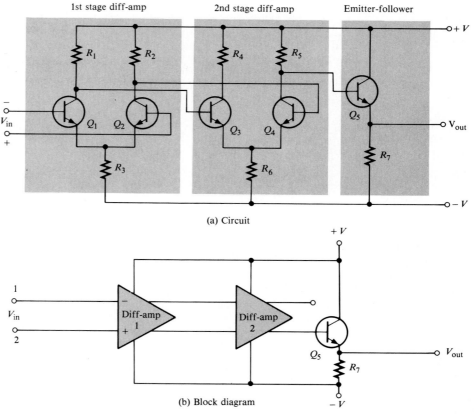

(a) Circuit

(b) Block diagram

FIGURE 20–10
Internal circuitry of a simple op-amp.

input. The differential outputs of the first stage feed into the differential inputs of the second stage. The output of the second stage is single-ended to drive an emitter-follower to achieve a relatively low output impedance. Both differential stages together provide a high voltage gain and a high CMRR.

**SECTION REVIEW
20–2**

1. Distinguish between differential and single-ended inputs.
2. Define *common-mode rejection*.

20–3

OP-AMP DATA SHEET PARAMETERS

In this section, several important op-amp parameters are defined, and four popular IC op-amps are compared in terms of their parameter values.

Input Offset Voltage

The ideal op-amp produces zero volts out for zero volts in. In a practical op-amp, however, a small dc voltage appears at the output when no differential input voltage is applied. Its primary cause is a slight mismatch of the base-emitter voltages of the differential input stage, as illustrated in Figure 20–11(a).

The output voltage of the differential input stage is expressed as

$$V_{OUT} = I_{C2}R_C - I_{C1}R_C \tag{20–3}$$

A small difference in the base-emitter voltages of Q_1 and Q_2 causes a small difference in the collector currents. This results in a nonzero value of V_{OUT}. (The collector resistors are equal.)

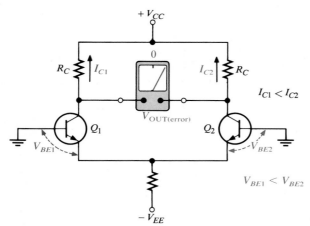

(a) A V_{BE} mismatch causes a small output error voltage.

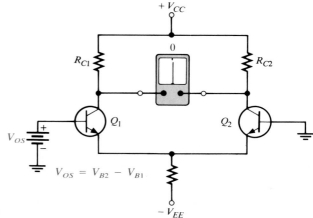

(b) The input offset voltage is the difference in the voltage between the inputs that is necessary to eliminate the output error voltage (makes $V_{OUT} = 0$)

FIGURE 20–11
Input offset voltage, V_{OS}.

As specified on an op-amp data sheet, *the input offset voltage V_{OS} is the differential dc voltage required between the inputs to force the differential output to zero volts,* as demonstrated in Figure 20–11(b).

Typical values of input offset voltage are in the range of 2 mV or less. In the ideal case, it is 0 V.

Input Offset Voltage Drift

The input offset voltage drift is a parameter related to V_{OS} that specifies how much change occurs in the input offset voltage for each degree change in temperature. Typical values range anywhere from about 5 μV per degree Celsius to about 50 μV per degree Celsius.

Usually, an op-amp with a higher nominal value of input offset voltage exhibits a higher drift.

Input Bias Current

You have seen that the input terminals of a diff-amp are the transistor bases and, therefore, the input currents are the base currents.

The input bias current is the direct current required by the inputs of the amplifier to properly operate the first stage. By definition, the input bias current is the *average* of both input currents and is calculated as follows:

$$I_{\text{BIAS}} = \frac{I_1 + I_2}{2} \tag{20–4}$$

The concept of input bias current is illustrated in Figure 20–12.

FIGURE 20–12
Input bias current is the average of the two op-amp input currents.

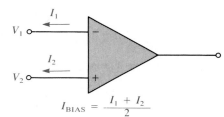

Input Impedance

There are two basic ways of specifying the impedance of an op-amp. The *differential input impedance* is the total resistance between the inverting and the noninverting inputs, as illustrated in Figure 20–13(a). Differential impedance is measured by determining the change in bias current for a given change in differential input voltage. The *common-mode input impedance* is measured from the inputs to ground and is depicted in Figure 20–13(b).

(a) Differential input impedance   (b) Common-mode input impedance

FIGURE 20–13
Op-amp input impedance.

Input Offset Current

Ideally, the two input bias currents are equal, and thus their difference is zero. In a practical op-amp, however, the bias currents are not exactly equal.

The *input offset current is the difference of the input bias currents,* expressed as

$$I_{OS} = |I_1 - I_2| \tag{20–5}$$

Actual magnitudes of offset current are usually at least an order of magnitude (ten times) less than the bias current.

In many applications the offset current can be neglected. However, high-gain, high-input impedance amplifiers should have as little I_{OS} as possible, because the difference in currents through large input resistances develops a substantial offset voltage, as shown in Figure 20–14.

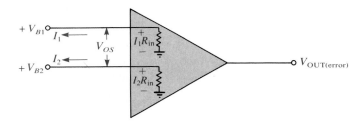

FIGURE 20–14
Effect of input offset current.

The offset voltage developed by the input offset current is

$$V_{OS} = I_1R_{in} - I_2R_{in} = (I_1 - I_2)R_{in}$$
$$V_{OS} = I_{OS}R_{in} \tag{20–6}$$

The error created by I_{OS} is amplified by the gain A_v of the op-amp and appears in the output as

$$V_{OUT(error)} = A_v I_{OS} R_{in} \tag{20–7}$$

The *change* in offset current with temperature is often an important consideration. Values of temperature coefficient in the range of 0.5 nA per degree Celsius are common.

Output Impedance

This parameter is the resistance viewed from the output terminal of the op-amp, as indicated in Figure 20–15.

FIGURE 20–15
Op-amp output impedance.

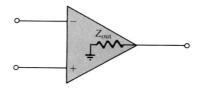

Common-Mode Range

All op-amps have limitations on the range of voltages over which they will operate. The *common-mode range* is the range of input voltages which, when applied to both inputs, will not cause clipping or other output distortion. Many op-amps have common-mode ranges of ±10 V with dc supply voltages of ±15 V.

Open-Loop Voltage Gain, A_{ol}

This is the gain of the op-amp without any external feedback from output to input. A good op-amp has a very high open-loop gain; 50,000 to 200,000 is typical.

Common-Mode Rejection Ratio

The *common-mode rejection ratio* (CMRR), as discussed in conjunction with the diff-amp, is a measure of an op-amp's ability to reject common-mode signals. An infinite value of CMRR means that the output is zero when the same signal is applied to both inputs (common-mode).

An infinite CMRR is never achieved in practice, but a good op-amp does have a very high value of CMRR. As previously mentioned, common-mode signals are undesired interference voltages such as 60-Hz power-supply ripple and noise voltages due to pickup of radiated energy. A high CMRR enables the op-amp to virtually eliminate these interference signals from the output.

The accepted definition of CMRR for an op-amp is the *open-loop gain* (A_{ol}) divided by the common-mode gain:

$$\text{CMRR} = \frac{A_{ol}}{A_{cm}} \tag{20–8}$$

It is commonly expressed in decibels as follows:

$$\text{CMRR} = 20 \log \frac{A_{ol}}{A_{cm}} \tag{20–9}$$

EXAMPLE 20–3

A certain op-amp has an open-loop gain of 100,000 and a common-mode gain of 0.25. Determine the CMRR and express it in decibels.

Solution

$$\text{CMRR} = \frac{A_{ol}}{A_{cm}} = \frac{100,000}{0.25} = 400,000$$

$$\text{CMRR} = 20 \log(400,000) = 112 \text{ dB}$$

Slew Rate

The maximum rate of change of the output voltage in response to a step input voltage is the *slew rate* of an op-amp. The slew rate is dependent upon the frequency response of the amplifier stages within the op-amp.

Slew rate is measured with an op-amp connected as shown in Figure 20–16(a). This particular op-amp connection is a unity-gain, noninverting configuration that will be discussed later. It gives a *worst-case* (slowest) slew rate.

As shown in part (b), a pulse is applied to the input, and the ideal output voltage is measured. The width of the input pulse must be sufficient to allow the output to "slew" from its lower limit to its upper limit, as shown.

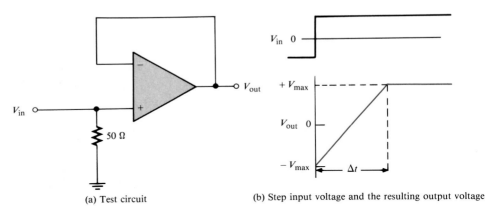

(a) Test circuit   (b) Step input voltage and the resulting output voltage

FIGURE 20–16
Measurement of slew rate.

As you can see, a certain time interval, Δt, is required for the output voltage to go from its lower limit $-V_{max}$ to its upper limit $+V_{max}$, once the input step is applied. The slew rate is expressed as

$$\text{Slew rate} = \frac{\Delta V_{out}}{\Delta t} \qquad (20\text{–}10)$$

where $\Delta V_{out} = +V_{max} - (-V_{max})$. The unit of slew rate is volts per microsecond (V/μs).

EXAMPLE 20–4

The output voltage of a certain op-amp appears as shown in Figure 20–17 in response to a step input. Determine the slew rate.

FIGURE 20–17

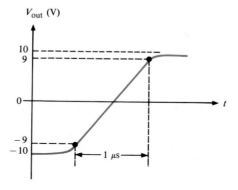

Solution

The output goes from the lower to the upper limit in 1 μs. Since this response is not ideal, the limits are taken at the 90% points, as indicated. Thus, the upper limit is $+9$ V and the lower limit is -9 V.

The slew rate is

$$\frac{\Delta V}{\Delta t} = \frac{+9 \text{ V} - (-9 \text{ V})}{1 \ \mu s} = 18 \text{ V}/\mu s$$

Frequency Response

The internal amplifier stages that make up an op-amp have voltage gains limited by junction capacitances. Although the diff-amps used in op-amps are somewhat different from the basic amplifiers discussed, the same principles apply. An op-amp has no internal coupling capacitors, however; therefore, the low frequency response extends down to dc. Frequency-related characteristics will be discussed in the next chapter.

Comparison of Op-Amp Parameters

Table 20–1 provides a comparison of values of some of the parameters just described for four popular IC op-amps. Any values not listed were not given on the manufacturer's data sheet. All values are typical at 25°C.

Other Features

Most available op-amps have two very important features: short-circuit protection and no latch-up. Short-circuit protection keeps the circuit from being damaged if the output becomes shorted, and the no-latch-up feature prevents the op-amp from hanging up in one output state (high or low voltage level) under certain input conditions.

TABLE 20-1

Parameter	Op-Amp Type			
	MC1741	LM101	LM108	LM118
Input offset voltage	1 mV	1 mV	0.7 mV	2 mV
Input bias current	80 nA	120 nA	0.8 nA	120 nA
Input offset current	20 nA	40 nA	0.05 nA	6 nA
Input impedance	2 MΩ	800 kΩ	70 MΩ	3 MΩ
Output impedance	75 Ω	—	—	—
Open-loop gain	200,000	160,000	300,000	200,000
Slew rate	0.5 V/μs	—	—	70 V/μs
CMRR	90 dB	90 dB	100 dB	100 dB

SECTION REVIEW 20-3

1. List ten or more op-amp parameters.
2. List two parameters, not including frequency response, that are frequency dependent.

20-4   OP-AMPS WITH NEGATIVE FEEDBACK

In this section we will discuss several basic ways in which an op-amp can be connected using *negative feedback* to stabilize the gain and increase frequency response.

The extremely high open-loop gain of an op-amp creates an unstable situation because a small noise voltage on the input can be amplified to a point where the amplifier is driven out of its linear region. Also, unwanted oscillations can occur. In addition, the open-loop gain parameter of an op-amp can vary greatly from one device to the next.

Negative feedback takes a portion of the output and applies it back *out of phase* with the input, creating an effective reduction in gain. This closed-loop gain is usually much less than the open-loop gain and independent of it.

Noninverting Amplifier

An op-amp connected as a *noninverting* amplifier with a controlled amount of voltage gain is shown in Figure 20-18. The input signal is applied to the noninverting input. The output is applied back to the inverting input through the feedback network formed by R_i and R_f, creating *negative feedback* as follows.

R_i and R_f form a voltage divider network that reduces the output V_{out} and connects the reduced voltage V_f to the inverting input. The feedback voltage is expressed as

$$V_f = \left(\frac{R_i}{R_i + R_f}\right)V_{out} \qquad (20\text{--}11)$$

The difference of the input voltage V_{in} and the feedback voltage V_f is the *differential input* to the op-amp, as shown in the figure. This differential voltage is amplified by the open-loop gain of the op-amp (A_{ol}) and produces an output voltage expressed as

$$V_{out} = A_{ol}(V_{in} - V_f) \qquad (20\text{--}12)$$

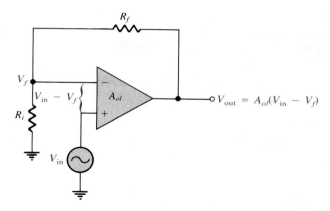

FIGURE 20–18
Noninverting amplifier.

Letting $R_i/(R_i + R_f) = B$ and then substituting BV_{out} for V_f in Equation (20–12), we get the following algebraic steps:

$$V_{out} = A_{ol}(V_{in} - BV_{out})$$
$$V_{out} = A_{ol}V_{in} - A_{ol}BV_{out}$$
$$V_{out} + A_{ol}BV_{out} = A_{ol}V_{in}$$
$$V_{out}(1 + A_{ol}B) = A_{ol}V_{in}$$

Since the total voltage gain of the amplifier in Figure 20–18 is V_{out}/V_{in}, it can be expressed as

$$\frac{V_{out}}{V_{in}} = \frac{A_{ol}}{1 + A_{ol}B}$$

The product $A_{ol}B$ is usually much greater than 1, so the equation simplifies to

$$\frac{V_{out}}{V_{in}} \cong \frac{A_{ol}}{A_{ol}B}$$

Since

$$A_{cl(NI)} = \frac{V_{out}}{V_{in}}$$

then

$$A_{cl(NI)} \cong \frac{1}{B} \qquad (20\text{–}13)$$

or

$$A_{cl(NI)} = 1 + \frac{R_f}{R_i} \qquad (20\text{–}14)$$

Equation (20–13) shows that the *closed-loop gain* $A_{cl(NI)}$ of the noninverting (NI) amplifier is the reciprocal of the attenuation (B) of the feedback network (voltage divider). It is very interesting to note that the closed-loop gain is not at all dependent on the op-amp's open-loop gain under the condition $A_{ol}B \gg 1$. The closed-loop gain can be set by selecting values of R_i and R_f.

EXAMPLE 20–5

Determine the gain of the amplifier in Figure 20–19. The open-loop voltage gain is 100,000.

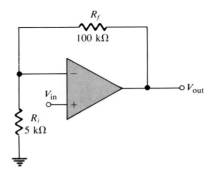

FIGURE 20–19

Solution

This is a noninverting op-amp configuration. The feedback gain is

$$B = \frac{R_i}{R_i + R_f} = \frac{5\ k\Omega}{105\ k\Omega} = 0.0476$$

Therefore,

$$A_{cl(NI)} \cong \frac{1}{B} = \frac{1}{0.0476} = 21$$

Voltage Follower

The voltage-follower configuration is a special case of the noninverting amplifier in which all of the output voltage is fed back to the inverting input, as shown in Figure 20–20. As you can see, the straight feedback connection has a voltage gain of approximately 1. The closed-loop voltage gain of a noninverting amplifier is $1/B$, as previously derived. Since $B = 1$, the closed-loop gain of the voltage follower (VF) is

$$A_{cl(VF)} \cong \frac{1}{B} = 1 \qquad \qquad (20–15)$$

The most important features of the voltage follower configuration are its *very high input impedance* and its *very low output impedance*. These features make it a nearly ideal buffer amplifier for interfacing high-impedance sources and low-impedance loads.

FIGURE 20–20
Op-amp voltage follower.

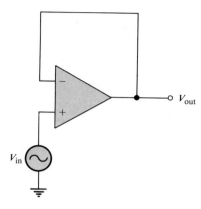

Inverting Amplifier

An op-amp connected as an *inverting* amplifier with a controlled amount of voltage gain is shown in Figure 20–21. The input signal is applied through a series input resistor R_i to the inverting input. Also, the output is fed back through R_f to the same input. The noninverting input is grounded.

FIGURE 20–21
Inverting amplifier.

At this point, the ideal op-amp parameters mentioned earlier are very useful in simplifying the analysis of this circuit. In particular, the concept of infinite input impedance is of great value. An infinite input impedance implies *zero current* to the inverting input. If there is zero current through the input impedance, then there must be *no* voltage drop between the inverting and noninverting inputs. That is, the voltage at the inverting $(-)$ input is *zero* because the other input $(+)$ is grounded. This zero voltage at the inverting input terminal is referred to as *virtual ground* and is illustrated in Figure 20–22(a).

Since there is no current into the inverting input, the current through R_i and the current through R_f are equal, as shown in Figure 20–22(b):

$$I_{in} = I_f$$

(a) Virtual ground

(b) $I_{in} = I_f$ and $I_1 = 0$

FIGURE 20–22
Virtual ground concept and development of closed-loop voltage gain for the inverting amplifier.

The voltage across R_i equals V_{in} because of virtual ground on the other side of the resistor. Therefore,

$$I_{in} = \frac{V_{in}}{R_i}$$

Also, the voltage across R_f equals $-V_{out}$ because of virtual ground, and therefore

$$I_f = \frac{-V_{out}}{R_f}$$

Since $I_f = I_{in}$,

$$\frac{-V_{out}}{R_f} = \frac{V_{in}}{R_i}$$

Rearranging the terms, we get

$$\frac{V_{out}}{V_{in}} = -\frac{R_f}{R_i}$$

Of course, you recognize V_{out}/V_{in} as the overall gain of the amplifier:

$$A_{cl(I)} = -\frac{R_f}{R_i} \qquad (20\text{–}16)$$

Equation (20–16) shows that the closed-loop voltage gain $A_{cl(I)}$ of the inverting (I) amplifier is the ratio of the feedback resistance R_f to the resistance R_i.

The *closed-loop gain* is independent of the op-amp's internal *open-loop gain*. Thus, the negative feedback stabilizes the voltage gain. The negative sign indicates inversion.

EXAMPLE 20–6

Given the op-amp configuration in Figure 20–23, determine the value of R_f required to produce a closed-loop voltage gain of 100.

FIGURE 20–23

Solution

$$R_i = 2 \text{ k}\Omega \quad \text{and} \quad A_{cl(I)} = 100$$

$$A_{cl(I)} = \left| \frac{R_f}{R_i} \right|$$

$$R_f = A_{cl(I)}R_i = (100)(2 \text{ k}\Omega) = 200 \text{ k}\Omega$$

SECTION REVIEW 20–4

1.　What is the main purpose of negative feedback?
2.　The closed-loop voltage gain of each of the op-amp configurations discussed is dependent on the internal open-loop voltage gain of the op-amp (T or F).
3.　The attenuation *(B)* of the negative feedback network of a noninverting op-amp configuration is 0.02. What is the closed-loop gain of the amplifier?

20–5

EFFECTS OF NEGATIVE FEEDBACK ON OP-AMP IMPEDANCES

Data sheets for IC op-amps usually give values for the input impedance (Z_{in}) and the output impedance (Z_{out}). These values are for the op-amp *without any external feedback connections*. Each of the negative feedback configurations discussed in the preceding section produces an overall input and output impedance that is different from the op-amp's internal Z_{in} and Z_{out} (with one exception).

Input and Output Impedances of the Noninverting Amplifier

The input impedance of the noninverting amplifier configuration, $Z_{in(NI)}$, shown in Figure 20–24, is greater than the internal input impedance of the op-amp itself (without feedback) by a factor of $1 + A_{ol}B$, as expressed by the following equation:

$$Z_{in(NI)} = (1 + A_{ol}B)Z_{in} \qquad (20\text{–}17)$$

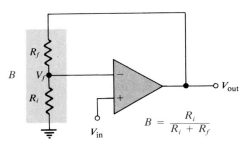

$$B = \frac{R_i}{R_i + R_f}$$

FIGURE 20–24
Noninverting amplifier.

The output impedance with the negative feedback, $Z_{out(NI)}$, is less than the op-amp output impedance by a factor of $1/(1 + A_{ol}B)$, expressed as follows:

$$Z_{out(NI)} = \frac{Z_{out}}{1 + A_{ol}B} \qquad (20\text{–}18)$$

In summary, the negative feedback in a noninverting configuration increases the input impedance and decreases the output impedance.

EXAMPLE 20–7

(a) Determine the input and output impedances of the amplifier in Figure 20–25. The op-amp data sheet gives $Z_{in} = 2 \text{ M}\Omega$, $Z_{out} = 75 \text{ }\Omega$, and $A_{ol} = 200,000$.

(b) Find the closed-loop voltage gain.

Solution

(a) The attenuation of the feedback network is

$$B = \frac{R_i}{R_i + R_f} = \frac{10 \text{ k}\Omega}{210 \text{ k}\Omega} = 0.048$$

$$Z_{in(NI)} = (1 + A_{ol}B)Z_{in} = [1 + (200,000)(0.048)]2 \text{ M}\Omega$$
$$= (1 + 9600)2 \text{ M}\Omega = 19,202 \text{ M}\Omega$$

$$Z_{out(NI)} = \frac{Z_{out}}{1 + A_{ol}B} = \frac{75 \text{ }\Omega}{1 + 9600} = 0.0078 \text{ }\Omega$$

(b) $A_{cl(NI)} = \dfrac{1}{B} = \dfrac{1}{0.048} = 20.83$

FIGURE 20–25

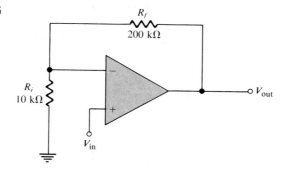

Voltage-Follower Impedances

Since the voltage-follower is a special case of the noninverting configuration, the same impedance formulas are used with $B = 1$:

$$Z_{in(VF)} = (1 + A_{ol})Z_{in} \qquad (20–19)$$

$$Z_{out(VF)} = \frac{Z_{out}}{1 + A_{ol}} \qquad (20–20)$$

As you can see, the voltage-follower input impedance is greater for a given A_{ol} and Z_{in} than for the noninverting configuration with the voltage divider feedback network. Also, its output impedance is much smaller because B is normally much less than 1 for a noninverting configuration.

EXAMPLE 20–8

The same op-amp in Example 20–7 is used in a voltage-follower configuration. Determine the input and output impedance.

Solution
Since $B = 1$,

$$Z_{in(VF)} = (1 + A_{ol})Z_{in} = (1 + 200,000)2 \text{ M}\Omega$$
$$\cong 400,000 \text{ M}\Omega$$
$$Z_{out(VF)} = \frac{Z_{out}}{1 + A_{ol}} = \frac{75 \ \Omega}{1 + 200,000}$$
$$= 0.00038 \ \Omega$$

Notice that $Z_{in(VF)}$ is much greater than $Z_{in(NI)}$, and $Z_{out(VF)}$ is much less than $Z_{out(NI)}$ from Example 20–7.

Input and Output Impedances of the Inverting Amplifier

For the inverting amplifier configuration shown in Figure 20–26, the input impedance, $Z_{in(I)}$, approximately equals the external input resistance, R_i:

$$Z_{in(I)} \cong R_i \qquad (20–21)$$

DEVICES

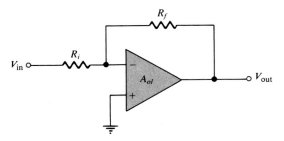

FIGURE 20–26
Inverting amplifier.

The output impedance, $Z_{out(I)}$, approximately equals the internal output impedance of the op-amp, Z_{out}:

$$Z_{out(I)} \cong Z_{out} \qquad (20\text{--}22)$$

EXAMPLE 20–9

The op-amp in Figure 20–27 has the following parameters: $A_{ol} = 50{,}000$, $Z_{in} = 4\ M\Omega$, and $Z_{out} = 50\ \Omega$.
(a) Find the values of the input and output impedances.
(b) Determine the closed-loop voltage gain.

FIGURE 20–27

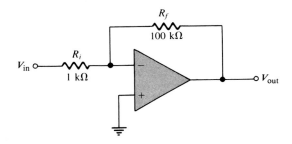

Solution
(a) $Z_{in(I)} \cong R_i = 1\ k\Omega$

$Z_{out(I)} \cong Z_{out} = 50\ \Omega$

(b) $A_{cl(I)} = -\dfrac{R_f}{R_i} = -\dfrac{100\ k\Omega}{1\ k\Omega} = -100$

**SECTION REVIEW
20–5**

1. How does the input impedance of a noninverting amplifier configuration compare to the input impedance of the op-amp itself?

2. Connecting an op-amp in a voltage-follower configuration (increases, decreases) the input impedance.

3. Given that $R_f = 100\ k\Omega$, $R_i = 2\ k\Omega$, $A_{ol} = 120{,}000$, $Z_{in} = 2\ M\Omega$, and $Z_{out} = 60\ \Omega$, determine $Z_{in(I)}$ and $Z_{out(I)}$ for an inverting amplifier configuration.

SUMMARY

☐ The op-amp has three terminals not including power and ground: inverting input ($-$), non-inverting input ($+$), and output.

☐ Most op-amps require both a positive and a negative dc supply voltage.

☐ The ideal (perfect) op-amp has infinite input impedance, zero output impedance, infinite open-loop voltage gain, infinite bandwidth, and infinite CMRR.

☐ A good practical op-amp has high input impedance, low output impedance, high open-loop voltage gain, and wide bandwidth.

☐ A diff-amp is normally used for the input stage of an op-amp.

☐ A differential input voltage appears between the inverting and noninverting inputs of a diff-amp.

☐ A single-ended input voltage appears between one input and ground (with the other inputs grounded).

☐ A differential output voltage appears between two output terminals of a diff-amp.

☐ A single-ended output voltage appears between the output and ground of a diff-amp.

☐ Common mode occurs when equal, in-phase voltages are applied to both input terminals.

☐ Input offset voltage produces an output error voltage (with no input voltage).

☐ Input bias current also produces an output error voltage (with no input voltage).

☐ Input offset current is the difference between the two bias currents.

☐ Open-loop voltage gain is the gain of the op-amp with no external feedback connections.

☐ Slew rate is the rate (in volts per microsecond) that the output voltage of an op-amp can change in response to a step input.

☐ There are three basic op-amp configurations: inverting, noninverting, and voltage-follower.

☐ All op-amp configurations (except comparators, to be covered in Chapter 21) employ negative feedback. Negative feedback occurs when a portion of the output voltage is connected back to the inverting input such that it subtracts from the input voltage, thus reducing the voltage gain but increasing the stability and bandwidth.

☐ A noninverting amplifier configuration has a higher input impedance and a lower output impedance than the op-amp itself (without feedback).

☐ An inverting amplifier configuration has an input impedance approximately equal to the input resistor R_i and an output impedance approximately equal to the internal output impedance of the op-amp itself.

☐ The voltage-follower has the highest input impedance and the lowest output impedance of the three configurations.

GLOSSARY

Closed loop   An op-amp configuration in which the output is connected back to the input.

Closed loop gain, A_{cl}   The overall voltage gain of an op-amp with feedback.

CMRR   Common-mode rejection ratio. A measure of an op-amp's ability to reject signals that appear the same on both inputs. The ratio of open-loop gain to common-mode gain.

Common-mode signals   Signals that appear the same on both inputs of an op-amp.

Differential amplifier   An amplifier that produces an output proportional to the difference of two inputs.

Negative feedback   The return of a portion of the output signal to the input such that it is out of phase with the input signal.

Open-loop gain, A_{ol}   The internal voltage gain of an op-amp without feedback.

Operational amplifier (op-amp)   A special type of amplifier exhibiting very high gain, very high input impedance, very low output impedance, and good rejection of common-mode signals.

FORMULAS

$$\text{CMRR} = \frac{A_{v(d)}}{A_{cm}} \qquad \text{Common-mode rejection ratio (diff-amp)} \qquad (20\text{--}1)$$

$$\text{CMRR} = 20 \log \frac{A_{v(d)}}{A_{cm}} \qquad \text{Common-mode rejection ratio (dB)} \qquad (20\text{--}2)$$

$$V_{\text{OUT}} = I_{C2}R_C - I_{C1}R_C \qquad \text{Differential output} \qquad (20\text{--}3)$$

$$I_{\text{BIAS}} = \frac{I_1 + I_2}{2} \qquad \text{Input bias current} \qquad (20\text{--}4)$$

$$I_{OS} = |I_1 - I_2| \qquad \text{Input offset current} \qquad (20\text{--}5)$$

$$V_{OS} = I_{OS}R_{\text{in}} \qquad \text{Offset voltage} \qquad (20\text{--}6)$$

$$V_{\text{OUT(error)}} = A_v I_{OS} R_{\text{in}} \qquad \text{Output error voltage} \qquad (20\text{--}7)$$

$$\text{CMRR} = \frac{A_{ol}}{A_{cm}} \qquad \text{Common-mode rejection ratio (op-amp)} \qquad (20\text{--}8)$$

$$\text{CMRR} = 20 \log \frac{A_{ol}}{A_{cm}} \qquad \text{Common-mode rejection ratio (dB)} \qquad (20\text{--}9)$$

$$\text{Slew rate} = \frac{\Delta V_{\text{out}}}{\Delta t} \qquad \text{Slew rate} \qquad (20\text{--}10)$$

$$V_f = \left(\frac{R_i}{R_i + R_f}\right)V_{\text{out}} \qquad \text{Feedback voltage (noninverting)} \qquad (20\text{--}11)$$

$$V_{\text{out}} = A_{ol}(V_{\text{in}} - V_f) \qquad \text{Output voltage (noninverting)} \qquad (20\text{--}12)$$

$$A_{cl(\text{NI})} \cong \frac{1}{B} \qquad \text{Voltage gain (noninverting)} \qquad (20\text{--}13)$$

$$A_{cl(\text{NI})} = 1 + \frac{R_f}{R_i} \qquad \text{Voltage gain (noninverting)} \qquad (20\text{--}14)$$

$$A_{cl(\text{VF})} \cong \frac{1}{B} = 1 \qquad \text{Voltage gain (voltage-follower)} \qquad (20\text{--}15)$$

$$A_{cl(\text{I})} = -\frac{R_f}{R_i} \qquad \text{Voltage gain (inverting)} \qquad (20\text{--}16)$$

$$Z_{\text{in(NI)}} = (1 + A_{ol}B)Z_{\text{in}} \qquad \text{Input impedance (noninverting)} \qquad (20\text{--}17)$$

$$Z_{\text{out(NI)}} = \frac{Z_{\text{out}}}{1 + A_{ol}B} \qquad \text{Output impedance (noninverting)} \qquad (20\text{--}18)$$

$$Z_{\text{in(VF)}} = (1 + A_{ol})Z_{\text{in}} \qquad \text{Input impedance (voltage-follower)} \qquad (20\text{--}19)$$

$$Z_{\text{out(VF)}} = \frac{Z_{\text{out}}}{1 + A_{ol}} \qquad \text{Output impedance (voltage-follower)} \qquad (20\text{--}20)$$

$Z_{\text{in(I)}} \cong R_i$ Input impedance (inverting)   (20–21)

$Z_{\text{out(I)}} \cong Z_{\text{out}}$ Output impedance (inverting)   (20–22)

SELF-TEST

1. Which characteristic does not necessarily apply to an op-amp:
   (a) high gain   **(b)** low power   **(c)** high input impedance
   (d) low output impedance

2. In selecting an op-amp suppose you have several choices. Of the CMRR values listed, the most desirable is
   (a) 10 dB   **(b)** 20 dB   **(c)** 50 dB   **(d)** 100 dB

3. The output voltage of a particular op-amp increases 8 V in 12 μs in response to a step voltage on the input. The slew rate is
   (a) 0.667 V/μs   **(b)** 1.5 V/μs   **(c)** 96 V/μs   **(d)** 0.75 V/μs

4. A noninverting op-amp configuration has an R_i of 1 kΩ and an R_f of 100 kΩ. If V_{out} is 5 V, the value of V_f is
   (a) 50 mV   **(b)** 49.5 mV   **(c)** 495 mV   **(d)** 500 mV

5. In the amplifier described in Problem 4, the value of B is
   (a) 0.01   **(b)** 0.1   **(c)** 0.0099   **(d)** 101

6. The closed loop gain of the amplifier in Problem 4 is
   (a) 0.0099   **(b)** 1   **(c)** 99   **(d)** 101

7. One characteristic of a voltage follower is
   (a) $A_{cl} > 1$ **(b)** inversion   **(c)** high Z_{out} **(d)** noninversion

8. An inverting amplifier has the following circuit values: $R_f = 220$ kΩ, $R_i = 2.2$ kΩ, and $A_{ol} = 25{,}000$. The closed-loop gain is
   (a) -100 **(b)** -0.01 **(c)** 100   **(d)** -250

9. If you know an op-amp's open-loop gain and nothing else, you can determine the closed-loop gain of
   (a) an inverting amplifier   **(b)** a noninverting amplifier
   (c) a voltage follower
   (d) none of the amplifier configurations without additional information

10. The feedback attenuation of a voltage-follower is
    (a) unity   **(b)** less than unity   **(c)** greater than unity   **(d)** variable

11. The value of B in a certain noninverting amplifier is 0.025. The closed-loop gain is
    (a) unity   **(b)** 40   **(c)** 0.025   **(d)** undeterminable

12. The highest possible input impedance is achieved with the
    (a) inverting amplifier   **(b)** noninverting amplifier
    (c) differential amplifier   **(d)** voltage-follower

PROBLEMS

Section 20–1   Introduction to Operational Amplifiers

1. Compare a practical op-amp to the ideal.

2. Two IC op-amps are available to you. Their characteristics are as follows:
   Op-amp 1: $Z_{\text{in}} = 5$ MΩ, $Z_{\text{out}} = 100$ Ω, $A_{ol} = 50{,}000$
   Op-amp 2: $Z_{\text{in}} = 10$ MΩ, $Z_{\text{out}} = 75$ Ω, $A_{ol} = 150{,}000$
   Choose the one you think is generally more desirable.

Section 20–2   The Differential Amplifier

3.   Identify the type of input and output configuration for each diff-amp in Figure 20–28.

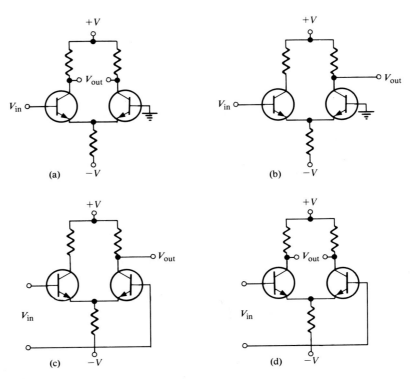

(a)   (b)   (c)   (d)

FIGURE 20–28

4.   The dc base voltages in Figure 20–29 are zero. Using your knowledge of transistor analysis, determine the dc differential output voltage. Assume that Q_1 has an $\alpha = 0.98$ and Q_2 has an $\alpha = 0.975$.

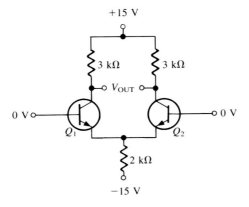

FIGURE 20–29

Section 20–3   Op-Amp Data Sheet Parameters

5. Determine the bias current, I_{BIAS}, given that the input currents to an op-amp are 8.3 μA and 7.9 μA.

6. Distinguish between *input bias current* and *input offset current*, and then calculate the input offset current in Problem 5.

7. A certain op-amp has a CMRR of 250,000. Convert this to decibels.

8. The open-loop gain of a certain op-amp is 175,000. Its common-mode gain is 0.18. Determine the CMRR in decibels.

9. An op-amp data sheet specifies a CMRR of 300,000 and an A_{ol} of 90,000. What is the common-mode gain?

10. Figure 20–30 shows the output voltage of an op-amp in response to a step input. What is the slew rate?

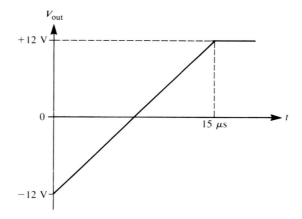

FIGURE 20–30

11. How long does it take the output voltage of an op-amp to go from -10 V to $+10$ V if the slew rate is 0.5 V/μs?

Section 20–4   Op-Amps with Negative Feedback

12. Identify each of the op-amp configurations in Figure 20–31.

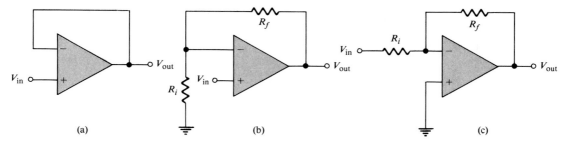

FIGURE 20–31

13. For the amplifier in Figure 20–32, determine the following:
    (a) $A_{cl(NI)}$ **(b)** V_{out} **(c)** V_f

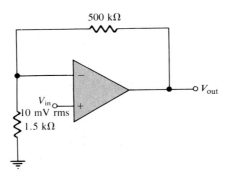

FIGURE 20–32

14. Determine the closed-loop gain of each amplifier in Figure 20–33.

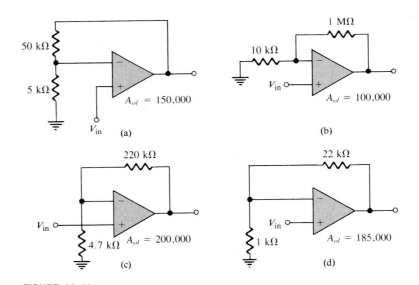

FIGURE 20–33

15. Find the value of R_f that will produce the indicated closed-loop gain in each amplifier in Figure 20–34.

16. Find the gain of each amplifier in Figure 20–35.

17. If a signal voltage of 10 mV rms is applied to each amplifier in Figure 20–35, what are the output voltages and what is their phase relationship with inputs?

FIGURE 20–34

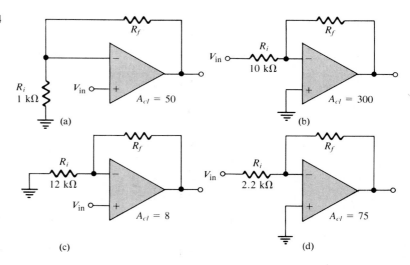

(a)

(b)

(c)

(d)

FIGURE 20–35

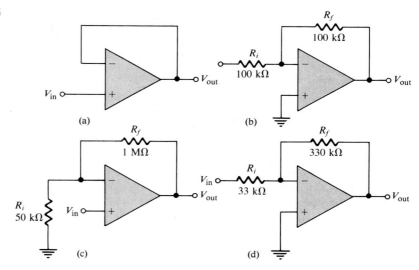

(a)

(b)

(c)

(d)

18. Determine the approximate values for each of the following quantities in Figure 20–36:
   (a) I_{in} **(b)** I_f
   (c) V_{out} **(d)** closed-loop gain

FIGURE 20–36

Section 20–5   Effects of Negative Feedback on Op-Amp Impedances

19. Determine the input and output impedances for each amplifier configuration in Figure 20–37.

FIGURE 20–37

20. Repeat Problem 19 for each circuit in Figure 20–38.

21. Repeat Problem 19 for each circuit in Figure 20–39.

FIGURE 20–38

FIGURE 20–39

ANSWERS TO SECTION REVIEWS

Section 20–1

1. See Figure 20–40.   2. High Z_{in}, low Z_{out}, high voltage gain, wide bandwidth.

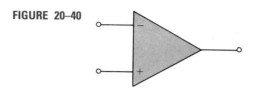

FIGURE 20–40

Section 20–2

1. Differential: between two input terminals. Single-ended: from one input terminal to ground (with the other inputs grounded).
2. The ability of an op-amp to reject common-mode signals.

Section 20–3

1. Input bias current, input offset voltage, drift, input offset current, input impedance, output impedance, common-mode range, CMRR, open-loop voltage gain, slew rate, frequency response.
2. Slew rate and gain.

Section 20–4

1. To stabilize gain.   2. False.   3. 50.

Section 20–5

1. Z_{in} of the noninverting amplifier is higher.   2. Increases.   3. 2 kΩ, 60 Ω.

21

Basic Applications of Op-Amps

Op-amps are used in such a wide variety of applications that it is impossible to cover all of them in one chapter or even in one book. Therefore, in this chapter we examine some of the more fundamental applications to illustrate the versatility of the op-amp.

After completing this chapter, you will be able to

☐ Use an op-amp as a voltage comparator.
☐ Distinguish between zero-level detection and nonzero-level detection in comparators.
☐ Use an op-amp to implement a summing amplifier.
☐ Adjust the gain of a summing amplifier.
☐ Make a summing amplifier produce an output that is the average of the inputs.
☐ Assign various weights to the inputs of a summing amplifier.
☐ Use an op-amp as an integrator or differentiator.
☐ Connect an op-amp as a sawtooth generator.
☐ Implement a Wien-bridge oscillator using an op-amp.
☐ Analyze an oscillator for conditions that assure proper start-up and sustained oscillation.
☐ Use an op-amp as the active element in low-pass, high-pass, and band-pass filters.
☐ Explain what a filter *pole* is.
☐ Define the term *roll-off* in relation to filters.
☐ Describe a three-terminal regulator.
☐ Explain how series and shunt regulators work.

APPLICATION ASSIGNMENT

As a manufacturing technician in a chemical plant, you are requested by your supervisor to devise an electronic method of determining when the *average* volume of a chemical solution contained in four different storage tanks falls below a specified level. A sensor in each of the storage tanks produces a dc voltage proportional to the remaining volume of solution. After completing this chapter, you will be able to complete this assignment.

21–1   COMPARATORS

Operational amplifiers are often used to *compare* the amplitude of one voltage with another. In this application the op-amp is used in the open-loop configuration, with the input voltage on one input and a *reference voltage* on the other.

Zero-Level Detection

A basic application of the op-amp as a comparator is in determining when an input voltage exceeds a certain level. Figure 21–1(a) shows a *zero-level detector*. Notice that the inverting ($-$) input is grounded and the input signal voltage is applied to the non-inverting ($+$) input. Because of the high open-loop voltage gain, a very small difference voltage between the two inputs drives the amplifier into saturation, causing the output voltage to go to its limit.

For example, consider an op-amp having A_{ol} = 100,000. A voltage difference of only 0.25 mV between the inputs could produce an output voltage of (0.25 mV)(100,000) = 25 V *if the op-amp were capable*. However, since most op-amps have output voltage limitations of less than ± 15 V, the device would be driven into saturation.

Figure 21–1(b) shows the result of a sine wave input voltage applied to the non-inverting input of the zero-level detector. When the sine wave is *negative*, the output is at its maximum *negative* level. When the sine wave crosses 0, the amplifier is driven to its opposite state and the output goes to its maximum *positive* level, as shown.

As you can see, the zero-level detector can be used as a *squaring circuit* to produce a square wave from a sine wave.

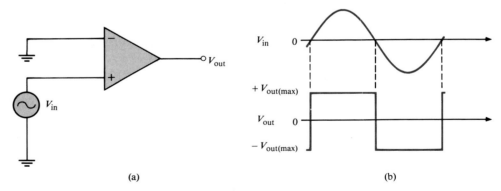

(a)   (b)

FIGURE 21–1
The op-amp as a zero-level detector.

Nonzero-Level Detection

The zero-level detector in Figure 21–1 can be modified to detect voltages other than zero by connecting a *fixed reference voltage*, as shown in Figure 21–2(a). A more practical arrangement is shown in part (b) using a voltage divider to set the reference voltage as follows:

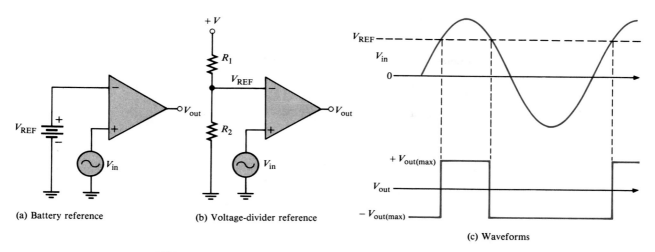

(a) Battery reference

(b) Voltage-divider reference

(c) Waveforms

FIGURE 21–2
Nonzero-level detectors.

$$V_{REF} = \frac{R_2}{R_1 + R_2}(+V) \qquad (21\text{–}1)$$

where $+V$ is the positive op-amp supply voltage.

As long as the input voltage V_{in} is less than V_{REF}, the output remains at the maximum negative level. When the input voltage exceeds the reference voltage, the output goes to its maximum positive state, as shown in Figure 21–2(c) with a sine wave input voltage.

EXAMPLE 21–1

The input signal in Figure 21–3(a) is applied to the comparator circuit in part (b). Make a sketch of the output showing its proper relationship to the input signal. Assume that the maximum output levels are ± 12 V.

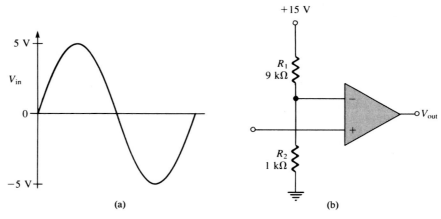

FIGURE 21–3

Solution

The reference voltage is set by R_1 and R_2 as follows:

$$V_{REF} = \frac{R_2}{R_1 + R_2}(+V) = \frac{1 \text{ k}\Omega}{9 \text{ k}\Omega + 1 \text{ k}\Omega}(+15 \text{ V}) = 1.5 \text{ V}$$

As shown in Figure 21–4, each time the input exceeds $+1.5$ V, the output voltage switches to its $+12$-V level, and each time the input goes below $+1.5$ V, the output switches back to its -12-V level.

FIGURE 21–4

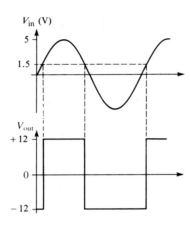

SECTION REVIEW 21–1

1. What is the reference voltage for the comparator in Figure 21–5?
2. Sketch the output waveform for Figure 21–5 when a sine wave with a 5-V peak is applied to the input.

FIGURE 21–5

21–2

SUMMING AMPLIFIERS

The summing amplifier (adder) is a variation of the *inverting* op-amp configuration covered in the last chapter. A two-input summing amplifier is shown in Figure 21–6, but any number of inputs can be used. The output voltage is proportional to the negative of the algebraic sum of the input voltages.

FIGURE 21–6
Two-input, inverting summing amplifier.

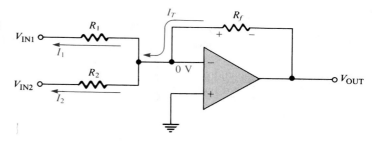

The operation of the circuit and derivation of the output expression are as follows: Two voltages, V_{IN1} and V_{IN2}, are applied to the inputs and produce currents I_1 and I_2, as shown in Figure 21–6.

Using the concepts of infinite input impedance and virtual ground, you can see that the inverting input of the op-amp is approximately zero volts, and there is no current into the input. That is, both input currents I_1 and I_2 come together at this summing point, and the total current is through R_f, as indicated:

$$I_T = I_1 + I_2$$

Since $V_{OUT} = -I_T R_f$, the following steps apply:

$$V_{OUT} = -(I_1 + I_2)R_f$$

$$= -\left(\frac{V_{IN1}}{R_1} + \frac{V_{IN2}}{R_2}\right)R_f$$

If all three of the resistors are *equal* ($R_1 = R_2 = R_f = R$), then

$$V_{OUT} = -\left(\frac{V_{IN1}}{R} + \frac{V_{IN2}}{R}\right)R$$

$$V_{OUT} = -(V_{IN1} + V_{IN2}) \qquad (21\text{–}2)$$

Equation (21–2) shows that the output voltage is the *sum* of the two input voltages. A general expression is given in Equation (21–3) for a summing amplifier with n inputs, as shown in Figure 21–7:

$$V_{OUT} = -(V_{IN1} + V_{IN2} + V_{IN3} + \cdots + V_{INn}) \qquad (21\text{–}3)$$

FIGURE 21–7
Summing amplifier with n inputs.

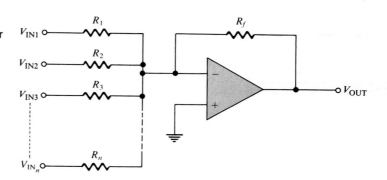

EXAMPLE 21–2   Determine the output voltage in Figure 21–8.

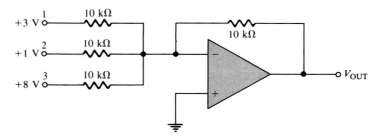

FIGURE 21–8

Solution

$$V_{OUT} = -(V_{IN1} + V_{IN2} + V_{IN3}) = -(3 \text{ V} + 1 \text{ V} + 8 \text{ V}) = -12 \text{ V}$$

Summing Amplifier with Gain Greater than Unity

When R_f is larger than the input resistors, the amplifier has a gain of R_f/R where R is the value of each input resistor. The general expression for the output is

$$V_{OUT} = -\frac{R_f}{R}(V_{IN1} + V_{IN2} + \cdots + V_{INn}) \tag{21–4}$$

As you can see, the output is the sum of all the input voltages *multiplied* by a constant determined by the ratio R_f/R.

EXAMPLE 21–3   Determine the output voltage for the summing amplifier in Figure 21–9.

FIGURE 21–9

Solution

$$R_f = 10 \text{ k}\Omega \quad \text{and} \quad R = 1 \text{ k}\Omega$$

$$V_{OUT} = -\frac{R_f}{R}(V_{IN1} + V_{IN2}) = -\frac{10 \text{ k}\Omega}{1 \text{ k}\Omega}(0.2 \text{ V} + 0.5 \text{ V})$$

$$= -10(0.7 \text{ V}) = -7 \text{ V}$$

Averaging Amplifier

A summing amplifier can produce the mathematical average of the input voltages. To do so, set the ratio R_f/R equal to *the reciprocal of the number of inputs*. You know that you obtain the average of several numbers by first adding the numbers and then dividing by the quantity of numbers you have. Examination of Equation (21–4) and a little thought will convince you that a summing amplifier will do the same. The next example will illustrate.

EXAMPLE 21–4

Show that the amplifier in Figure 21–10 produces an output whose magnitude is the average of the input voltages.

FIGURE 21–10

Solution
The output voltage is

$$V_{OUT} = -\frac{R_f}{R}(V_{IN1} + V_{IN2} + V_{IN3} + V_{IN4})$$

$$= -\frac{25\ k\Omega}{100\ k\Omega}(1\ V + 2\ V + 3\ V + 4\ V) = -\frac{1}{4}(10\ V)$$

$$= -2.5\ V$$

It can easily be shown that the average of the input values is the same as V_{OUT}:

$$\frac{1\ V + 2\ V + 3\ V + 4\ V}{4} = \frac{10\ V}{4} = 2.5\ V$$

Scaling Adder

A different *weight* can be assigned to each input of a summing amplifier by simply adjusting the values of the input resistors. As you have seen, the output voltage can be expressed as

$$V_{OUT} = -\left(\frac{R_f}{R_1}V_{IN1} + \frac{R_f}{R_2}V_{IN2} + \cdots + \frac{R_f}{R_n}V_{INn}\right) \qquad (21\text{–}5)$$

The weight of a particular input is set by the ratio of R_f to the input resistance R. For example, if an input voltage is to have a weight of 1, then $R_f = R$. Or, if a weight of 0.5 is required, $R_f = 0.5\,R$. The smaller the value of R, the greater the weight.

EXAMPLE 21–5

For the scaling adder in Figure 21–11 determine the weight of each input voltage and find the output voltage.

FIGURE 21–11

Solution

Weight of input 1: $\dfrac{R_f}{R_1} = \dfrac{10\ \text{k}\Omega}{10\ \text{k}\Omega} = 1$

Weight of input 2: $\dfrac{R_f}{R_2} = \dfrac{10\ \text{k}\Omega}{100\ \text{k}\Omega} = 0.1$

Weight of input 3: $\dfrac{R_f}{R_3} = \dfrac{10\ \text{k}\Omega}{50\ \text{k}\Omega} = 0.2$

The output voltage is

$$V_{OUT} = -\left(\frac{R_f}{R_1}V_{IN1} + \frac{R_f}{R_2}V_{IN2} + \frac{R_f}{R_3}V_{IN3}\right)$$
$$= -[1(3\ \text{V}) + 0.1(2\ \text{V}) + 0.2(8\ \text{V})]$$
$$= -(3\ \text{V} + 0.2\ \text{V} + 1.6\ \text{V})$$
$$= -4.8\ \text{V}$$

SECTION REVIEW 21–2

1. Define *summing point*.
2. What is the value of R_f/R for a five-input averaging amplifier?
3. A certain scaling adder has two inputs, one having twice the weight of the other. If the value of the *lowest* input resistor is 10 kΩ, what is the value of the other input resistor?

21–3

INTEGRATORS AND DIFFERENTIATORS

Integration and differentiation are presented in a way that will give you a very fundamental idea of what they are so you can understand the basic operation of the op-amp
integrator and differentiator circuits.

Integration

Basically, integration is the mathematical process of finding the *area under a curve.*
Therefore, the op-amp integrator shown in Figure 21–12 produces an output that is
proportional to the area under the curve of a graph of the input voltage.

FIGURE 21–12
Op-amp integrator.

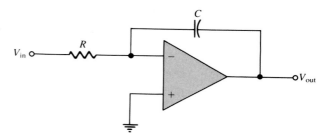

To aid in understanding the concept of integration, let's use a rectangular pulse of
voltage as the input, as shown in Figure 21–13(a) with amplitude V and width t_W. The
total area A under this curve is the height times the width, Vt_W. Now, visualize the total
area divided into, say, four equal smaller areas, as shown in part (b). Beginning at
$t = 0$ and going to $t = t_1$, the area encompassed up to this point is A_1. At $t = t_2$
the accumulated area is $A_1 + A_2$. At $t = t_3$, the accumulated area is $A_1 + A_2 + A_3$.
At $t = t_W$, the total accumulated area is $A_1 + A_2 + A_3 + A_4$. So, as you can see, you

FIGURE 21–13
Total area under rectangle.

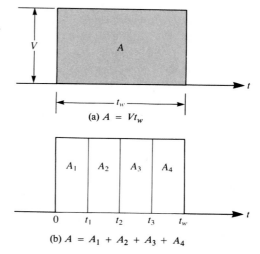

(a) $A = Vt_W$

(b) $A = A_1 + A_2 + A_3 + A_4$

can find the total area by summing all of the smaller areas. Integration is the process of summing infinitely small areas under a curve to obtain the total area. Mathematically, the *integral* of the expression for a curve represents the *area* under that curve.

The integral can be plotted graphically, as we will show using the rectangular pulse for simplicity. Starting at $t = 0$ and going to $t = t_1$, the area up to this point can be represented by a point on a graph having a vertical height KA_1, which is proportional to the area, as shown in Figure 21–14. (K is a constant of proportionality.) At each interval of time, another point is added to the graph (we are showing only four points for simplicity). Connecting the points, we get a sloping straight line called a *ramp*.

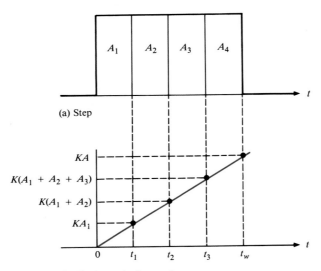

(a) Step

(b) The integral of a step is a ramp.

FIGURE 21–14
Integration concept.

Of course, integration is not limited to a rectangular input voltage. Any type of input voltage curve can be integrated, but the rectangular input is perhaps the easiest way to illustrate the principle.

Now let's see how the op-amp integrator in Figure 21–12 performs the integration function. First, remember that the inverting input is at virtual ground. When a positive step input voltage (beginning of a rectangular pulse) is applied to the input, the current through R is

$$I = \frac{V_{IN}}{R}$$

Since the input amplitude has a constant value V_{IN}, the voltage across R is constant, and therefore I is constant.

All of the input current is through the capacitor, charging it in the direction shown in Figure 21–15. Recall that the rate of change of voltage across a capacitor is proportional to the current. Since the current is constant, the capacitor charges at a fixed rate, producing a negative-going ramp because of the inversion.

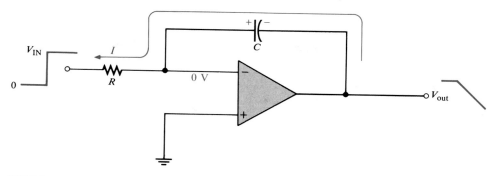

FIGURE 21–15
Inverting integrator with a step input voltage.

Rate of Change

The rate at which the capacitor charges, and therefore the slope of the output ramp, is set by the ratio I/C. Since $I = V_{IN}/R$, the rate of change of the output voltage is

$$\frac{\Delta V_{out}}{\Delta t} = -\frac{V_{IN}}{RC} \qquad (21\text{–}6)$$

where the symbol Δ means "a change in."

EXAMPLE 21–6

(a) Determine the rate of change of the output voltage in response to a single pulse input, as shown for the integrator in Figure 21–16(a). The output voltage is initially zero.

(b) Draw the output waveform.

FIGURE 21–16

(a)

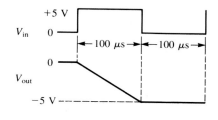

(b)

Solution

(a)   The rate of change of the output voltage is

$$\frac{\Delta V_{out}}{\Delta t} = \frac{-V_{IN}}{RC} = -\frac{5\text{ V}}{(10\text{ k}\Omega)(0.01\ \mu\text{F})}$$

$$= -50\text{ kV/s} = -50\text{ mV}/\mu\text{s}$$

(b)   The rate of change was found to be 50 mV/μs in part (a). When the input is at +5 V, the output is a negative-going ramp. When the input is at 0 V, the output is a constant level. In 100 μs the voltage decreases:

$$\Delta V_{out} = (50\text{ mV}/\mu\text{s})(100\ \mu\text{s}) = 5\text{ V}$$

Therefore, the negative-going ramp reaches -5 V at the end of the pulse. The output voltage then remains constant at -5 V for the time that the input is zero. The waveforms are shown in Figure 21–16(b).

Differentiator

An op-amp *differentiator* is shown in Figure 21–17. Notice the arrangement of the resistor and capacitor compared to the integrator.

FIGURE 21–17
Op-amp differentiator.

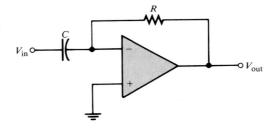

Differentiation is a process from calculus by which the *rate of change, $\Delta V/\Delta t$,* of a voltage curve at any given point can be determined. It is essentially the inverse of integration. The *derivative* of a curve at a point is its instantaneous rate of change. Therefore, the output of an op-amp differentiator is proportional to the derivative of the input voltage.

We will use two examples to illustrate differentiation.

EXAMPLE 21–7

What is the derivative of the triangular waveform in Figure 21–18?

Solution

The derivative is the rate of change. The triangle changes linearly from 0 V to 4 V during the time interval 0 to t_1. Thus,

$$\text{Rate of change} = \frac{\Delta V}{\Delta t} = \frac{4\text{ V}}{0.5\text{ ms}} = +8\text{ V/ms}$$

Therefore, during the interval of 0 to t_1, the derivative is a positive constant.

FIGURE 21–18

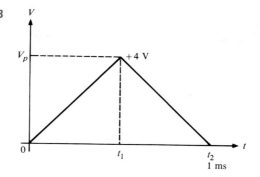

The triangle changes linearly from 4 V to 0 V during the time interval t_1 to t_2:

$$\text{Rate of change} = \frac{\Delta V}{\Delta t} = \frac{-4 \text{ V}}{0.5 \text{ ms}} = -8 \text{ V/ms}$$

The minus sign indicates a negative slope. Therefore, during the interval t_1 to t_2, the derivative is a negative constant. A graph of the derivative of the triangular wave is shown in Figure 21–19.

FIGURE 21–19
Derivative of a triangular waveform.

From this example, you can see that a differentiator circuit has a constant output proportional to the slope of a ramp input.

Operation of the Op-Amp Differentiator

First consider the performance of the op-amp differentiator with the application of a triangular input as shown in Figure 21–20. Again, remember that the inverting input is at virtual ground (0 V). When the triangular voltage is applied to the input, the voltage across the capacitor increases at a constant rate. Recall from circuit theory that the current in a capacitor is proportional to the rate of change of the voltage across the capacitor.

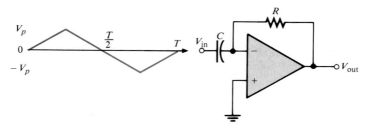

FIGURE 21–20
Differentiator with triangular input voltage.

Since the rate of change of the voltage ramp is constant, the current is also constant and equal to

$$I = \frac{CV_{pp}}{T/2} \tag{21-7}$$

Also, since the output voltage equals the voltage across R, and since I is constant, the output voltage is constant and equal to

$$V_{out} = \pm RC\left(\frac{V_{pp}}{T/2}\right) \tag{21-8}$$

A negative output with an amplitude given by Equation (21-8) occurs during the positive slope of the triangular input due to the inversion of this op-amp. Likewise, a positive output occurs during the negative slope of the input. V_{pp} is the peak-to-peak amplitude of the input, and T is its period.

The waveform diagrams of the input and output for the circuit in Figure 21-20 are shown in Figure 21-21. Of course, other input waveforms are also possible, but the triangular input has been used to illustrate the basic operation.

FIGURE 21-21
Input and output waveforms for the differentiator in Figure 21-20.

SECTION REVIEW 21-3

1. Sketch the circuit diagram for an op-amp integrator.
2. An integrator produces a _____ output in response to a step input.
3. A differentiator produces a _____ output in response to a ramp input.

21-4

SAWTOOTH GENERATOR

One way to build a sawtooth generator is with an op-amp integrator that uses a switching device (PUT) in parallel with the feedback capacitor to terminate each ramp at a prescribed level and effectively "reset" the circuit. Figure 21-22(a) shows the implementation.

The PUT is a programmable unijunction transistor with a structure similar to that of an SCR except for the placement of the gate. The gate is always biased positively with respect to the cathode. When the anode voltage exceeds the gate voltage by ap-

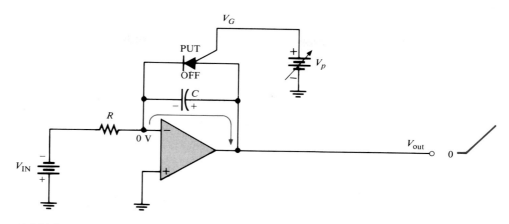

(a) Initially, the capacitor charges, the output ramp begins, and the PUT is off.

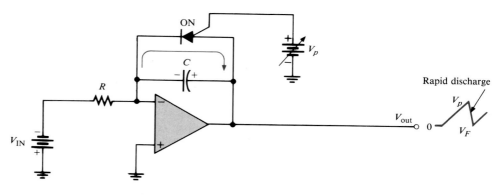

(b) The capacitor rapidly discharges when the PUT momentarily turns on.

FIGURE 21–22
Sawtooth generator operation.

proximately 0.7 V, the PUT turns on. When the anode voltage falls below this level, the PUT turns off. Also, the current must be above the *holding* value to maintain conduction.

The operation of the sawtooth generator begins when the negative dc input voltage, $-V_{IN}$, produces a positive-going ramp on the output. During the time that the ramp is increasing, the circuit acts as a regular integrator.

The PUT triggers *on* when the output ramp (at the anode) exceeds the gate voltage by 0.7 V. The gate is set to the approximate desired sawtooth peak voltage. When the PUT turns on, the capacitor rapidly discharges, as shown in Figure 21–22(b). The capacitor does not discharge completely to zero because of the PUT's forward voltage, V_F. Discharge continues until the PUT current falls below the holding value. At this point, the PUT turns *off* and the capacitor begins to charge again, thus generating a new output ramp. The cycle continually repeats, and the resulting output is a repetitive sawtooth waveform, as shown. The sawtooth amplitude and period can be adjusted by varying the PUT gate voltage.

The frequency is determined by the RC time constant of the integrator and the peak voltage set by the PUT. Recall that the charging rate of the capacitor is V_{IN}/RC. The time it takes the capacitor to charge from V_F to V_p is the period of the sawtooth (neglecting the rapid discharge time):

$$T = \frac{V_p - V_F}{|V_{IN}|/RC} \tag{21-9}$$

From $f = 1/T$, we get

$$f = \frac{|V_{IN}|}{RC}\left(\frac{1}{V_p - V_F}\right) \tag{21-10}$$

EXAMPLE 21-8

(a) Find the amplitude and frequency of the sawtooth output in Figure 21–23. Assume that the forward PUT voltage V_F is approximately 1 V.

(b) Sketch the output waveform.

FIGURE 21-23

Solution

(a) First, find the gate voltage in order to establish the approximate voltage at which the PUT turns on:

$$V_G = \frac{R_4}{R_3 + R_4}(+V) = \frac{10\text{ k}\Omega}{20\text{ k}\Omega}(15\text{ V}) = 7.5\text{ V}$$

This voltage sets the approximate maximum peak value of the sawtooth output (neglecting the 0.7 V):

$$V_p \cong 7.5\text{ V}$$

The minimum peak value (low point) is

$$V_F \cong 1\text{ V}$$

The period is determined as follows:

$$V_{IN} = \frac{R_2}{R_1 + R_2}(-V) = \frac{10 \text{ k}\Omega}{78 \text{ k}\Omega}(-15 \text{ V}) = -1.92 \text{ V}$$

$$T = \frac{V_p - V_F}{|V_{IN}|/RC}$$

$$= \frac{7.5 \text{ V} - 1 \text{ V}}{1.92 \text{ V}/(100 \text{ k}\Omega)(0.005 \ \mu\text{F})} = 1.69 \text{ ms}$$

$$f = \frac{1}{1.69 \text{ ms}} \cong 592 \text{ Hz}$$

(b) The output waveform is shown in Figure 21–24.

FIGURE 21–24
Output of circuit in Figure 21–23.

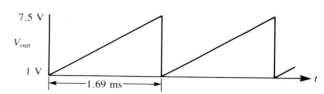

SECTION REVIEW
21–4

1. The rate of change of the ramp portion of a sawtooth is 50 V/ms, and the peak-to-peak value is 10 V. How long does it take to reach 10 V?

2. What is the frequency of the sawtooth described in Problem 1 if the discharge time is neglected?

21–5

WIEN-BRIDGE OSCILLATOR

One type of sine wave oscillator is called the *Wien-bridge* oscillator. A fundamental part of this oscillator is a *lead-lag* network such as that shown in Figure 21–25(a). R_1

(a) Network

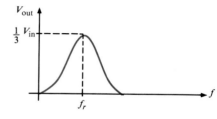

(b) Response curve

FIGURE 21–25
Lead-lag network and frequency response curve.

and C_1 together form the *lag* portion of the network; R_2 and C_2 form the *lead* portion. The operation of this circuit is as follows: At lower frequencies, the lead network dominates due to the high reactance of C_2. As the frequency increases, X_{C2} decreases, thus allowing the output voltage to increase. At some specified frequency, the response of the lag network takes over, and the decreasing value of X_{C1} causes the output voltage to decrease.

Thus, we have a response curve as shown in Figure 21–25(b) where the output voltage peaks at a frequency f_r. At this point, the attenuation of the network is $\frac{1}{3}$, expressed as follows:

$$\frac{V_{\text{out}}}{V_{\text{in}}} = \frac{1}{3} \tag{21–11}$$

The proof of this equation is beyond the scope of this book.

The resonant frequency of the lead-lag network is

$$f_r = \frac{1}{2\pi RC} \tag{21–12}$$

where $R_1 = R_2 = R$ and $C_1 = C_2 = \text{C}$.

To summarize, the lead-lag network has a *resonant frequency* f_r at which the phase shift through the network is $0°$ and the attenuation is $\frac{1}{3}$. Below f_r, the lead network dominates and the output leads the input. Above f_r, the lag network dominates and the output lags the input.

The lead-lag network is used in the *positive* feedback loop of an op-amp, as shown in Figure 21–26(a). A voltage divider is used in the *negative* feedback loop.

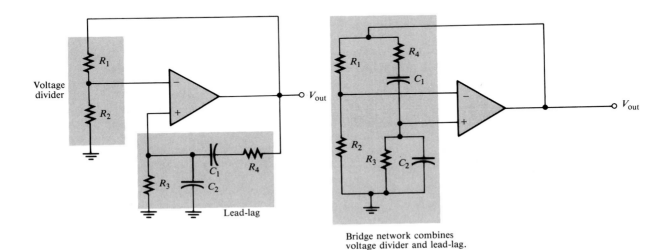

Bridge network combines
voltage divider and lead-lag.

FIGURE 21–26
Wien-bridge oscillator showing two ways to draw the schematic.

Basic Circuit

This oscillator circuit can be viewed as a *noninverting* amplifier configuration with the input signal fed back from the output through the lead-lag network. Recall that the *closed-loop* gain is determined by the voltage divider:

$$A_{cl} = \frac{1}{B} = \frac{1}{R_2/(R_1 + R_2)} = \frac{R_1 + R_2}{R_2}$$

The circuit has been redrawn in Figure 21–26(b) to show that the op-amp is connected *across* the Wien bridge. One leg of the bridge is the lead-lag network, and the other is the voltage divider.

The Positive Feedback Conditions for Oscillation

In order for the circuit to produce a sustained sine wave output (oscillate), the phase shift around the positive feedback loop must be 0°, and the gain around the loop must be at least unity (1).

The 0° phase-shift condition is met when the frequency is f_r, because the phase shift through the lead-lag network is 0° and there is no inversion from the noninverting input (+) of the op-amp to the output. This condition is shown in Figure 21–27(a).

(a) The phase shift around the loop is 0°.

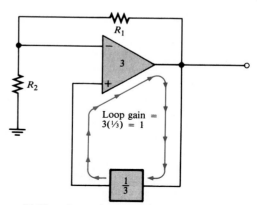

(b) The voltage gain around the loop is 1.

FIGURE 21–27
Conditions for oscillation.

The unity-gain condition is met when

$$A_{cl} = 3$$

This offsets the ⅓ attenuation of the lead-lag network, thus making the gain around the positive feedback equal to 1, as depicted in Figure 21–27(b).

To achieve a closed-loop gain of 3,

$$R_1 = 2R_2$$

Then

$$A_{cl} = \frac{1}{B} = \frac{R_1 + R_2}{R_2} = \frac{R_2 + 2R_2}{R_2} = \frac{3R_2}{R_2} = 3$$

Start-Up Conditions

The unity-gain condition must be met for oscillation to be *sustained*. For oscillation to *begin,* the voltage gain around the positive feedback loop must be *greater than unity* so that the amplitude of the output can build up. Thus, to start with, the loop gain must be more than 1 ($A_{cl} > 3$) until the output signal builds up to a desired level. The gain must then decrease to 1 so that the output signal stays at the desired level, as illustrated in Figure 21–28.

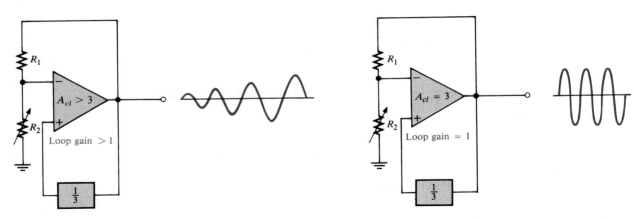

(a) Loop gain greater than 1 causes output to build up.   (b) Loop gain of 1 causes a sustained constant output.

FIGURE 21–28
Oscillator start-up conditions.

The circuit in Figure 21–29 illustrates a basic method for achieving the conditions described above. Notice that the voltage divider network has been modified to include an additional resistor R_3 in parallel with a back-to-back zener diode arrangement.

When dc power is first applied, both zener diodes appear open. This places R_3 in series with R_1, thus increasing the closed-loop gain as follows ($R_1 = 2R_2$):

$$A_{cl} = \frac{R_1 + R_2 + R_3}{R_2} = \frac{3R_2 + R_3}{R_2} = 3 + \frac{R_3}{R_2}$$

Initially, a small positive feedback signal develops from noise or turn-on transients. The lead-lag network permits only a signal with a frequency equal to f_r to appear *in phase* on the noninverting input. This feedback signal is amplified and continually reinforced, resulting in a buildup of the output voltage.

FIGURE 21–29
Self-starting Wien-bridge oscillator.

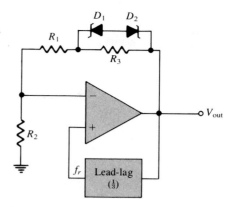

When the output signal reaches the zener breakdown voltage, the zeners conduct and effectively short out R_3. This lowers the closed-loop gain to 3. At this point the output signal levels off and the oscillation is sustained.

Incidentally, the frequency of oscillation can be adjusted by using gang-tuned capacitors in the lead-lag network.

EXAMPLE 21–9

Determine the frequency of the Wien-bridge oscillator in Figure 21–30. Also verify that oscillations will start and then continue when the output signal reaches 5.4 V.

FIGURE 21–30

Solution
The frequency is

$$f_r = \frac{1}{2\pi RC} = \frac{1}{2\pi(10 \text{ k}\Omega)(0.001 \text{ }\mu\text{F})} = 15.92 \text{ kHz}$$

Initially the closed-loop gain is

$$A_{cl} = \frac{R_1 + R_2 + R_3}{R_2} = \frac{40 \text{ k}\Omega}{10 \text{ k}\Omega} = 4$$

Since $A_{cl} > 3$, the start-up condition is met.

When the output reaches 5.4 V (4.7 V + 0.7 V), the zeners conduct (their forward resistance is assumed small, compared to 20 kΩ), and the closed-loop gain is reached. Thus oscillation is sustained.

$$A_{cl} = \frac{R_1 + R_2}{R_2} = \frac{30 \text{ k}\Omega}{10 \text{ k}\Omega} = 3$$

The Hartley, Colpitts, and Clapp oscillators, discussed in Chapter 19, can also be implemented with op-amps.

SECTION REVIEW 21–5

1. There are two feedback loops in the Wien-bridge oscillator. State the purpose of each.
2. A certain lead-lag network has $R_1 = R_2$ and $C_1 = C_2$. An input voltage of 5 V rms is applied. Its frequency equals the resonant frequency of the network. What is the rms output voltage?

21–6

ACTIVE FILTERS

In this section, we will examine selected low-pass, high-pass, and band-pass filters with op-amps as the active (gain) elements.

Low-Pass Active Filters

Figure 21–31 shows a basic active filter and its response curve. Notice that the input circuit is a single low-pass RC network, and unity gain is provided by the op-amp with a negative feedback loop. Simply stated, this is a voltage-follower with an RC filter between the input signal and the noninverting input.

(a)

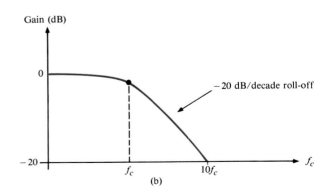
(b)

FIGURE 21–31
Single-pole, active low-pass filter and response curve.

The voltage at the noninverting input is as follows:

$$V+ = \left(\frac{X_C}{\sqrt{R^2 + X_C^2}}\right)V_{in}$$

Since the gain of the op-amp is 1, the output voltage is equal to $V+$:

$$V_{out} = \left(\frac{X_C}{\sqrt{R^2 + X_C^2}}\right)V_{in} \qquad \textbf{(21–13)}$$

A filter with *one RC* network that produces a -20-dB/decade roll-off beginning at f_c is said to be a *single-pole* or *first-order* filter. The term "-20 dB/decade" means that the voltage gain decreases by ten times (-20 dB) when the frequency increases by ten times (decade).

Low-Pass Two-pole Filters

There are several types of active filters and they can have varying numbers of poles, but we will use a two-pole filter to illustrate. Figure 21–32(a) shows a *two-pole (second-order)* low-pass filter. Since each *RC* network in a filter is considered to be *one-pole*, the *two-pole* filter uses *two RC* networks to produce a roll-off rate of -40 dB/decade, as indicated in Figure 21–32(b). The active filter in Figure 21–32 has unity gain up to near f_c because the op-amp is connected as a voltage-follower.

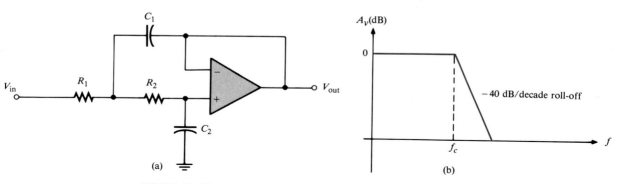

FIGURE 21–32
Two-pole, active low-pass filter and its ideal response curve.

One of the *RC* networks is formed by R_1 and C_1, and the other by R_2 and C_2. The critical frequency of this filter can be calculated using the following formula:

$$f_c = \frac{1}{2\pi\sqrt{R_1 R_2 C_1 C_2}} \qquad \textbf{(21–14)}$$

Figure 21–33(a) shows an example of a two-pole low-pass filter with values chosen to produce a response with a critical frequency of 1 kHz. Note that $C_1 = 2C_2$ and $R_1 = R_2$, because these relationships result in a gain of 0.707 (-3 dB) at f_c. For critical frequencies other than 1 kHz, the capacitance values can be *scaled* inversely with the frequency. For example, as shown in Figure 21–33(b) and (c), to get a 2-kHz filter, halve the values of C_1 and C_2; for a 500-Hz filter, double the values.

(a) $f_c = 1$ kHz

(b) $f_c = 2$ kHz

(c) $f_c = 500$ Hz

FIGURE 21–33
Examples of low-pass filters (two-pole).

EXAMPLE 21–10

Calculate the capacitance values required to produce a 3-kHz critical frequency in the filter of Figure 21–34.

FIGURE 21–34

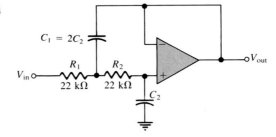

Solution
The resistor values have already been set at 22 kΩ each. Since these differ from the 1-kHz reference filter, we cannot use the scaling method to get the capacitance values. We will use the formula

$$f_c = \frac{1}{2\pi\sqrt{R_1 R_2 C_1 C_2}}$$

$$f_c^2 = \frac{1}{4\pi^2 R_1 R_2 C_1 C_2}$$

Since $C_1 = 2C_2$ and $R_1 = R_2 = R$,

$$f_c^2 = \frac{1}{4\pi^2 R^2 (2C_2^2)}$$

Solving for C_2, we obtain

$$C_2^2 = \frac{1}{8\pi^2 R^2 f_c^2}$$

$$C_2 = \frac{1}{\sqrt{2}\, 2\pi R f_c} = \frac{0.707}{2\pi R f_c} = \frac{0.707}{2\pi(22\ \text{k}\Omega)(3\ \text{kHz})}$$

$$= 0.0017\ \mu\text{F}$$

$$C_1 = 2C_2 = 2(0.0017\ \mu\text{F}) = 0.0034\ \mu\text{F}$$

High-Pass Active Filters

In Figure 21–35(a), a high-pass active filter with a 20 dB/decade roll-off is shown. Notice that the input circuit is a single high-pass RC network and that unity gain is provided by the op-amp with negative feedback. The response curve is shown in Figure 21–35(b).

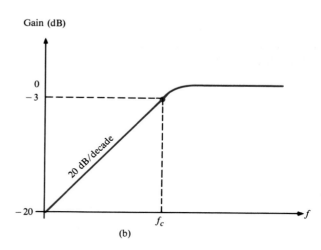

FIGURE 21–35
Single-pole, active high-pass filter and response curve.

Ideally, a high-pass filter passes all frequencies above f_c without limit, as indicated in Figure 21–36(a). In practice, of course, such is not the case. All op-amps inherently have internal RC networks that limit the amplifier's response at high frequencies. Such is the case with the active high-pass filter. There is an upper frequency limit to its response, which essentially makes this type of filter a wide band-pass filter rather than a true high-pass filter, as indicated in Figure 21–35(b). In many applications, the internal high-frequency cutoff is so much greater than the filter's f_c that the internal high-frequency cutoff can be neglected.

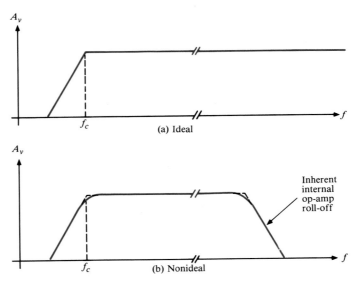

FIGURE 21–36
High-pass filter response.

The voltage at the noninverting input is as follows:

$$V+ = \left(\frac{R}{\sqrt{R^2 + X_C^2}}\right)V_{in}$$

Since the op-amp is connected as a voltage-follower with unity gain, the output voltage is the same as $V+$:

$$V_{out} = \left(\frac{R}{\sqrt{R^2 + X_C^2}}\right)V_{in} \qquad (21\text{--}15)$$

If the internal critical frequencies of the op-amp are assumed to be much greater than the desired f_c of the filter, the gain will roll off at 20 dB/decade as shown in Figure 21–36(b). This is a single-pole filter because it has one RC network.

High-Pass Two-Pole Filters

Figure 21–37 shows a *two-pole* active high-pass filter. Notice that it is identical to the corresponding low-pass type, except for the positions of the resistors and capacitors. This filter has a roll-off rate of 40 dB/decade below f_c, and the critical frequency is the same as for the low-pass filter given in Equation (21–14).

Figure 21–38 shows a two-pole high-pass filter with values chosen to produce a response with a critical frequency of 1 kHz. Note that $R_2 = 2R_1$ and $C_1 = C_2$ because these relationships result in a gain of 0.707 (-3 dB) at f_c. For frequencies other than

1 kHz, the resistance values can be *scaled* inversely, as was done with the capacitors in the low-pass case.

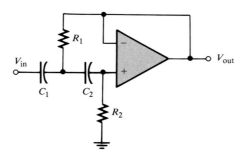

FIGURE 21–37
Two-pole, active high-pass filter.

FIGURE 21–38
Two-pole, high-pass filter (f_c = 1 kHz).

EXAMPLE 21–11

For the filter of Figure 21–39, calculate the resistance values required to produce a critical frequency of 5.5 kHz.

FIGURE 21–39

Solution

The capacitor values have been preselected to be 0.0022 μF each. Since these differ from the 1-kHz reference filter, we cannot use the scaling method to get the resistor values.

$$f_c = \frac{1}{2\pi\sqrt{R_1 R_2 C_1 C_2}}$$

$$f_c^2 = \frac{1}{4\pi^2 R_1 R_2 C_1 C_2}$$

Since $R_2 = 2R_1$ and $C_1 = C_2 = C$,

$$f_c^2 = \frac{1}{4\pi^2 (2R_1^2)C^2}$$

Solving for R_1, we have

$$R_1^2 = \frac{1}{8\pi^2 C^2 f_c^2}$$

$$R_1 = \frac{1}{\sqrt{2}\ 2\pi C f_c} = \frac{0.707}{2\pi C f_c}$$

$$= \frac{0.707}{2\pi(0.0022\ \mu\text{F})(5.5\ \text{kHz})} \cong 9.3\ \text{k}\Omega$$

$$R_2 = 2R_1 = 2(9.3\ \text{k}\Omega) = 18.6\ \text{k}\Omega$$

Band-Pass Filter Using High-Pass/Low-Pass Combination

One way to implement a band-pass filter is to use a cascaded arrangement of a high-pass filter followed by a low-pass filter, as shown in Figure 21–40(a). Each of the filters shown is a two-pole configuration so that the roll-off rates of the response curve are ± 40 dB/decade, as indicated in the composite response curve of part (b). The critical frequency of each filter is chosen so that the response curves overlap, as indicated. The critical frequency of the high-pass filter is lower than that of the low-pass filter.

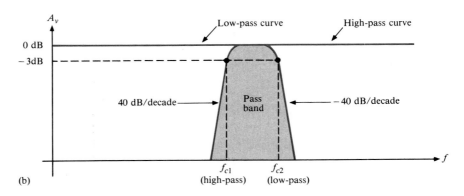

FIGURE 21–40
Band-pass filter formed by combining two-pole, high-pass filter with two-pole, low-pass filter. (The order in which the filters are cascaded does not matter.)

The lower frequency f_{c1} of the pass band is set by the critical frequency of the high-pass filter. The upper frequency f_{c2} of the pass band is the critical frequency of the low-pass filter. Ideally, the center frequency f_r of the pass band is the geometric average of f_{c1} and f_{c2}. The following formulas express the three frequencies of the band-pass filter in Figure 21–40:

$$f_{c1} = \frac{1}{2\pi\sqrt{R_1 R_2 C_1 C_2}} \qquad (21\text{--}16)$$

$$f_{c2} = \frac{1}{2\pi\sqrt{R_3 R_4 C_3 C_4}} \qquad (21\text{--}17)$$

$$f_r = \sqrt{f_{c1} f_{c2}} \qquad (21\text{--}18)$$

EXAMPLE 21–12

(a) Determine the bandwidth and center frequency for the filter in Figure 21–41.

(b) Sketch the response curve.

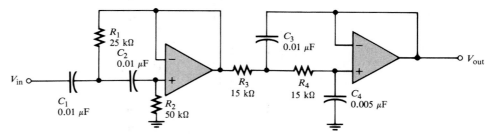

FIGURE 21–41

Solution

(a) The critical frequency of the high-pass filter is

$$f_{c1} = \frac{1}{2\pi\sqrt{R_1 R_2 C_1 C_2}}$$

$$= \frac{1}{2\pi\sqrt{(25\ \text{k}\Omega)(50\ \text{k}\Omega)(0.01\ \mu\text{F})(0.01\ \mu\text{F})}}$$

$$= 450\ \text{Hz}$$

The critical frequency of the low-pass filter is

$$f_{c2} = \frac{1}{2\pi\sqrt{R_3 R_4 C_3 C_4}}$$

$$= \frac{1}{2\pi\sqrt{(15\ \text{k}\Omega)(15\ \text{k}\Omega)(0.01\ \mu\text{F})(0.005\ \mu\text{F})}}$$

$$= 1.5\ \text{kHz}$$

$$BW = f_{c2} - f_{c1} = 1.5\ \text{kHz} - 450\ \text{Hz} = 1.05\ \text{kHz}$$

$$f_r = \sqrt{f_{c1} f_{c2}} = \sqrt{(1.5\ \text{kHz})(450\ \text{Hz})} = 822\ \text{Hz}$$

FIGURE 21–42

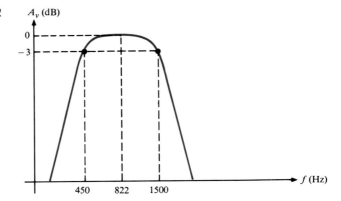

(b) The response curve is shown in Figure 21–42.

---

1. In terms of circuit components, what does the term *pole* refer to?
2. What type of response characterizes the single-pole low-pass filter?
3. How does a high-pass filter differ in its implementation from a corresponding low-pass version?
4. If the resistance values of a high-pass filter are doubled, what happens to the critical frequency?

21–7

THREE-TERMINAL REGULATORS

Three-terminal linear voltage regulators consist of an input, an output, and a ground terminal. There are two basic types of linear regulator: the *series regulator* and the *shunt regulator*.

Basic Series Regulator

A simple representation of a series linear regulator is shown in Figure 21–43(a), and the basic components are shown in the block diagram in part (b). Notice that the control element is in *series* with the load between input and output. The output sample circuit senses a change in the output voltage. The error detector compares the sample voltage with a reference voltage and causes the control element to compensate in order to maintain a constant output voltage.

Regulating Action

A basic op-amp series regulator circuit is shown in Figure 21–44. Its operation is as follows: The resistive voltage divider formed by R_2 and R_3 senses any change in the output voltage. When the output tries to decrease because of a decrease in V_{IN} or because of an increase in I_L, a *proportional* voltage decrease is applied to the op-amp's inverting input by the voltage divider. Since the zener diode holds the other op-amp

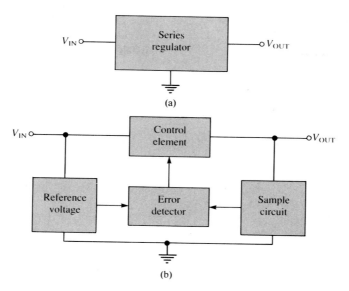

FIGURE 21–43
Block diagram of simple, three-terminal, series voltage regulator.

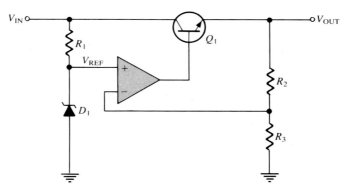

FIGURE 21–44
Basic op-amp series regulator.

input at a nearly *fixed* reference voltage V_{REF}, a small difference voltage (error voltage) is developed across the op-amp's inputs. This difference voltage is amplified, and the op-amp's output voltage increases. This increase is applied to the base of Q_1, causing the emitter voltage V_{OUT} to increase until the voltage to the inverting input again equals the reference (zener) voltage. This action offsets the attempted decrease in output voltage, thus keeping it nearly constant. Q_1 is a power transistor and is often used with a heat sink because it must handle all of the load current.

The opposite action occurs when the output tries to increase. Figure 21–45 illustrates the regulating action of the circuit. Percent regulation was discussed in Chapter 17.

(a) When V_{IN} or R_L increases, V_{OUT} attempts to increase.
The feedback voltage, V_{FB}, also attempts to increase,
and, as a result, the op-amp's output voltage, V_B,
applied to the base of the control transistor, Q_1, attempts to
decrease, thus compensating for the attempted increase in V_{OUT}
by decreasing the Q_1 emitter voltage.

(b) When V_{IN} (or R_L) stabilizes at its new higher value, the voltages return
to their original values, thus keeping V_{OUT} constant as a
result of the negative feedback.

FIGURE 21–45

Illustration of series regulator action that keeps V_{OUT} constant when V_{IN} or R_L changes.

(c) When V_{IN} or R_L decreases, V_{OUT} attempts to decrease, V_{FB} also attempts to decrease, and as a result, V_B attempts to increase, thus compensating for the attempted decrease in V_{OUT} by increasing the Q_1 emitter voltage.

(d) When V_{IN} (or R_L) stabilizes at its new higher value, the voltages return to their original values, thus keeping V_{OUT} constant as result of the negative feedback.

FIGURE 21–45
(Continued)

The op-amp in Figure 21–44 is actually connected as a noninverting amplifier in which the reference voltage V_{REF} is the input at the noninverting terminal, and the R_2/R_3 voltage divider forms the negative feedback network. The closed-loop voltage gain is

$$A_{cl} = 1 + \frac{R_2}{R_3} \tag{21–19}$$

Therefore, the regulated output voltage (neglecting the base-emitter voltage of Q_1) is

$$V_{OUT} \cong \left(1 + \frac{R_2}{R_3}\right)V_{REF} \tag{21–20}$$

From this analysis you can see that the output voltage is determined by the zener voltage and the resistors R_2 and R_3. It is relatively independent of the input voltage, and therefore, regulation is achieved (as long as the input voltage and load current are within specified limits).

EXAMPLE 21–13

Determine the output voltage for the regulator in Figure 21–46.

FIGURE 21–46

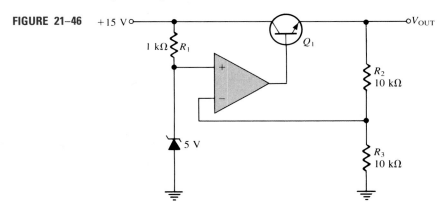

Solution

$$V_{REF} = 5 \text{ V}$$
$$V_{OUT} = \left(1 + \frac{R_2}{R_3}\right)V_{REF} = \left(1 + \frac{10 \text{ k}\Omega}{10 \text{ k}\Omega}\right)5 \text{ V}$$
$$= (2)5 \text{ V} = 10 \text{ V}$$

Short-Circuit or Overload Protection

If an excessive amount of load current is drawn, the series-pass transistor can be quickly damaged or destroyed. Most regulators employ some type of protection from excess current in the form of a current-limiting mechanism.

Figure 21–47 shows one method of current limiting to prevent overloads called *constant current limiting*. The current-limiting circuit consists of transistor Q_2 and resistor R_4.

FIGURE 21–47
Series regulator with constant current limiting.

The load current through R_4 creates a voltage from base to emitter of Q_2. When I_L reaches a predetermined maximum value, the voltage drop across R_4 is sufficient to forward-bias the base-emitter junction of Q_2, thus causing it to conduct. Enough Q_1 base current is diverted into the collector of Q_2 so that I_L is limited to its maximum value $I_{L(\text{max})}$. Since the base-to-emitter voltage of Q_2 cannot exceed about 0.7 V for a silicon transistor, the voltage across R_4 is held to this value, and the load current is limited to

$$I_{L(\text{max})} = \frac{0.7 \text{ V}}{R_4} \tag{21–21}$$

Basic Shunt Regulator

As you have seen, the control element in the series regulator is the series-pass transistor. A simple representation of a shunt type of linear regulator is shown in Figure 21–48(a), and the basic components are shown in the block diagram in part (b).

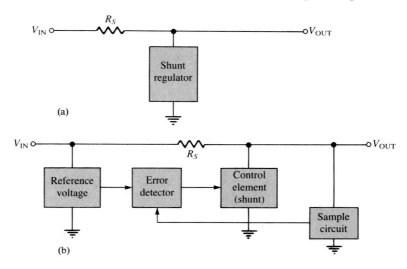

FIGURE 21–48
Block diagrams of simple, three-terminal shunt regulator.

FIGURE 21–49
Basic op-amp shunt
regulator.

(a) Initial response to a decrease in V_{IN} or R_L

FIGURE 21–50
Sequence of responses
when V_{IN} or R_L decreases.

(b) V_{OUT} held constant by the feedback action.

In the basic shunt regulator, the control element is a series resistor (R_1) and a transistor Q_1 in parallel with the load, as shown in Figure 21–49. The operation of the circuit is similar to that of the series regulator, except that regulation is achieved by controlling the current through the parallel transistor Q_1.

When the output voltage tries to decrease due to a change in input voltage, load current, or temperature, the attempted decrease is sensed by R_3 and R_4 and applied to the op-amp's noninverting input. The resulting difference in voltage reduces the op-amp's output, driving Q_1 less, thus reducing its collector current (shunt current) and increasing its effective collector-to-emitter resistance r_{ce}. Since r_{ce} acts as a voltage divider with R_1, this action offsets the attempted decrease in V_{OUT} and maintains it at an almost constant level. The opposite action occurs when the output tries to increase. This regulating action of the shunt element is illustrated in Figure 21–50.

With I_L and V_{OUT} constant, a change in the input voltage produces a change in shunt current (I_S) as follows:

$$\Delta I_S = \frac{\Delta V_{IN}}{R_1}$$

With a constant V_{IN} and V_{OUT}, a change in load current causes an opposite change in shunt current:

$$\Delta I_S = -\Delta I_L$$

This formula says that if I_L increases, I_S decreases, and vice versa.

The shunt regulator is less efficient than the series type but offers inherent short-circuit protection. If the output is shorted ($V_{OUT} = 0$), the load current is limited by the series resistor R_1 to a maximum value as follows ($I_S = 0$):

$$I_{L(max)} = \frac{V_{IN}}{R_1}$$

EXAMPLE 21–14

In Figure 21–51, what power rating must R_1 have if the maximum input voltage is 12.5 V?

FIGURE 21–51

Solution

The worst-case power dissipation in R_1 occurs when the output is short-circuited. $V_{OUT} = 0$, and when $V_{IN} = 12.5$ V, the voltage dropped across R_1 is $V_{IN} - V_{OUT} = 12.5$ V.

The power dissipation in R_1 is

$$P_{R1} = \frac{V_{R1}^2}{R_1} = \frac{(12.5 \text{ V})^2}{20 \text{ } \Omega} = 7.8 \text{ W}$$

Therefore, a resistor of at least 10 W should be used.

SECTION REVIEW
21–7

1. How does the control element in a shunt regulator differ from that in a series regulator?
2. Name one advantage and one disadvantage of a shunt regulator over a series type.

SUMMARY

☐ In an op-amp comparator, when the input voltage exceeds a specified reference voltage, the output changes state.

☐ The output voltage of a summing amplifier is proportional to the *sum* of the input voltages.

☐ An *averaging amplifier* is a summing amplifier with a closed-loop gain equal to the reciprocal of the number of inputs.

☐ In a *scaling adder,* a different weight can be assigned to each input, thus making the input contribute more or contribute less to the output.

☐ The integral of a step is a ramp.

☐ The derivative of a ramp is a step.

☐ In a Wien-bridge oscillator, the closed-loop gain must be equal to 3 in order to have unity gain around the positive feedback loop.

☐ In filter terminology, a single RC network is called a *pole*.

☐ Each pole in a filter causes the output to roll off (decrease) at a rate of 20 dB/decade.

☐ Two-pole filters roll off at a rate of 40 dB/decade.

☐ In a *series* voltage regulator, the control element is a transistor in series with the load.

☐ In a *shunt* voltage regulator, the control element is a transistor (or zener diode) in parallel with the load.

☐ The terminals on a three-terminal regulator are *input voltage, output voltage,* and *ground.*

GLOSSARY

Active filter   A frequency-selective circuit consisting of active devices such as transistors or op-amps.

Decade   A tenfold change in the value of a quantity. When a quantity becomes ten times less or ten times greater, it has changed a decade.

Decibel (dB)   The unit of logarithmic expression of a ratio, such as a power ratio or a voltage ratio.

Pole   In practical terms, a single RC network in a filter or amplifier that causes the response to change at a 20-dB/decade rate above or below a certain frequency.

Roll-off   The decrease in the response of a filter below or above a critical frequency.

FORMULAS

$$V_{REF} = \frac{R_2}{R_1 + R_2}(+V) \qquad \text{Comparator reference} \tag{21-1}$$

$$V_{OUT} = -(V_{IN1} + V_{IN2}) \qquad \text{Two-input adder} \tag{21-2}$$

$$V_{OUT} = -(V_{IN1} + V_{IN2} + V_{IN3} + \cdots + V_{INn}) \qquad n\text{-input adder} \tag{21-3}$$

$$V_{OUT} = -\frac{R_f}{R}(V_{IN1} + V_{IN2} + \cdots + V_{INn}) \qquad \text{Adder with gain} \tag{21-4}$$

$$V_{OUT} = -\left(\frac{R_f}{R_1}V_{IN1} + \frac{R_f}{R_2}V_{IN2} + \cdots + \frac{R_f}{R_n}V_{INn}\right) \qquad \text{Adder with gain} \tag{21-5}$$

$$\frac{\Delta V_{out}}{\Delta t} = -\frac{V_{IN}}{RC} \qquad \text{Rate of change in integrator} \tag{21-6}$$

$$I = \frac{CV_{pp}}{T/2} \qquad \text{Differentiator current with triangular input} \tag{21-7}$$

$$V_{out} = \pm RC\left(\frac{V_{pp}}{T/2}\right) \qquad \text{Differentiator output with triangular input} \tag{21-8}$$

$$T = \frac{V_p - V_F}{|V_{IN}|/RC} \qquad \text{Sawtooth period} \tag{21-9}$$

$$f = \frac{|V_{IN}|}{RC}\left(\frac{1}{V_p - V_F}\right) \qquad \text{Sawtooth frequency} \tag{21-10}$$

$$\frac{V_{out}}{V_{in}} = \frac{1}{3} \qquad \text{Lead-lag attenuation at } f_r \tag{21-11}$$

$$f_r = \frac{1}{2\pi RC} \qquad \text{Lead-lag resonant frequency} \tag{21-12}$$

$$V_{out} = \left(\frac{X_C}{\sqrt{R^2 + X_C^2}}\right)V_{in} \qquad \text{Output of one-pole filter (low-pass only)} \tag{21-13}$$

$$f_c = \frac{1}{2\pi\sqrt{R_1R_2C_1C_2}} \qquad \text{Critical frequency of two-pole filter (low-pass)} \tag{21-14}$$

$$V_{out} = \left(\frac{R}{\sqrt{R^2 + X_C^2}}\right)V_{in} \qquad \text{Output of one-pole filter (high-pass only)} \tag{21-15}$$

$$f_{c1} = \frac{1}{2\pi\sqrt{R_1R_2C_1C_2}} \qquad \text{Lower critical frequency} \tag{21-16}$$

$$f_{c2} = \frac{1}{2\pi\sqrt{R_3R_4C_3C_4}} \qquad \text{Upper critical frequency} \tag{21-17}$$

$$f_r = \sqrt{f_{c1}f_{c2}} \qquad \text{Center frequency} \tag{21-18}$$

$$A_{cl} = 1 + \frac{R_2}{R_3} \qquad \text{Closed-loop voltage gain} \tag{21-19}$$

$$V_{\text{OUT}} \cong \left(1 + \frac{R_2}{R_3} \right) V_{\text{REF}} \qquad \text{Regulator output} \qquad\qquad (21\text{–}20)$$

$$I_{L(\text{max})} = \frac{0.7 \text{ V}}{R_4} \qquad \text{For constant current limiting} \qquad\qquad (21\text{–}21)$$

SELF-TEST

1. The purpose of a comparator is to
    (a) amplify an input voltage
    (b) detect the occurrence of a changing input voltage
    (c) produce a change in output when an input voltage equals a reference voltage
    (d) maintain a constant output when the dc input voltage changes

2. To use a comparator for zero-level detection, the inverting input is connected to
    (a) ground   (b) the dc supply voltage
    (c) a positive reference voltage   (d) a negative reference voltage

3. In a 5-V level detector circuit,
    (a) the noninverting input is connected to +5 V
    (b) the inverting input is connected to +5 V
    (c) the input signal is limited to a 5-V peak value
    (d) the input signal must be riding on a +5 dc level

4. In a certain four-input summing amplifier, all the input resistors are 2.2 kΩ and the feedback resistor is 2.2 kΩ. If all the input voltages are 2 V, the output voltage is
    (a) 2 V   (b) 10 V   (c) 2.2 V   (d) 8 V

5. The gain of the amplifier in Problem 4 is
    (a) 1   (b) 2.2   (c) 4   (d) unknown

6. To convert a summing amplifier to an averaging amplifier,
    (a) all input resistors must be a different value
    (b) the ratio R_f/R must equal to the reciprocal of the number of inputs
    (c) the ratio R_f/R must equal to the number of inputs
    (d) a and b

7. In a scaling adder,
    (a) the input resistors are all the same value
    (b) the input resistors are all different values
    (c) the input resistors have values depending on the assigned weight of each input
    (d) the ratio R_f/R must be the same for each input

8. The feedback path in an op-amp integrator consists of
    (a) a resistor   (b) a capacitor   (c) a resistor and a capacitor in series
    (d) a resistor and a capacitor in parallel

9. The feedback path in an op-amp differentiator consists of
    (a) a resistor   (b) a capacitor   (c) a resistor and a capacitor in series
    (d) a resistor and a capacitor in parallel

10. The op-amp comparator circuit uses
    (a) positive feedback   (b) negative feedback   (c) regenerative feedback
    (d) no feedback

11. Unity gain and zero phase shift around the feedback loop are conditions that describe
    (a) an active filter   (b) a comparator   (c) an oscillator
    (d) an integrator or differentiator

12. The input frequency of a single-pole, low-pass active filter increases from 1.5 kHz to 150 kHz. If the critical frequency is 1.5 kHz, the gain decreases by
    (a) 3 dB   **(b)** 20 dB   **(c)** 40 dB   **(d)** 60 dB

PROBLEMS

Section 21–1   Comparators

1. Determine the output level (maximum positive or maximum negative) for each comparator in Figure 21–52.

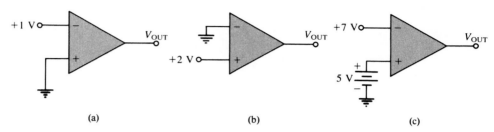

(a)   (b)   (c)

FIGURE 21–52

2. A certain op-amp has an open-loop gain of 80,000. The maximum saturated output levels of this particular device are ± 12 V when the dc supply voltages are ± 15 V. If a differential voltage of 0.15 mV rms is applied between the inputs, what is the peak-to-peak value of the output?

3. Sketch the output voltage waveform for each circuit in Figure 21–53 with respect to the input. Show voltage levels.

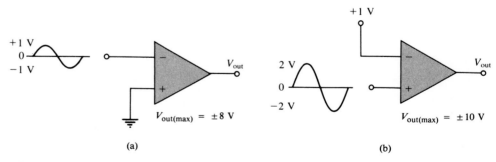

(a)   (b)

FIGURE 21–53

Section 21–2   Summing Amplifiers

4. Determine the output voltage for each circuit in Figure 21–54.

5. Determine the following in Figure 21–55:
    (a) V_{R1} and V_{R2} **(b)** current through R_f
    (c) V_{OUT}

DEVICES

(a)   (b)

FIGURE 21–54

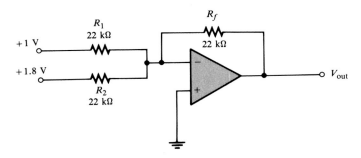

FIGURE 21–55

6. Find the value of R_f necessary to produce an output that is 5 times the sum of the inputs in Figure 21–55.

7. Find the output voltage when the input voltages shown in Figure 21–56 are applied to the scaling adder. What is the current through R_f?

FIGURE 21–56

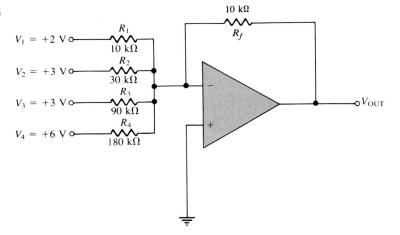

8. Determine the values of the input resistors required in a six-input scaling adder so that the lowest weighted input is 1 and each successive input has a weight *twice* the previous one. Use $R_f = 100$ kΩ.

Section 21–3   Integrators and Differentiators

9. Determine the rate of change of the output voltage in response to the step input to the integrator in Figure 21–57.

FIGURE 21–57

10. A triangular waveform is applied to the input of the circuit in Figure 21–58 as shown. Determine what the output should be, and sketch its waveform in relation to the input.

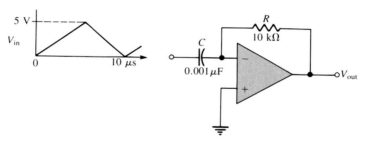

FIGURE 21–58

Section 21–4   Sawtooth Generator

11. Determine the amplitude and frequency of the output voltage in Figure 21–59. Use 1 V as the forward PUT voltage.

12. Modify the sawtooth generator in Figure 21–59 so that its peak-to-peak output is 4 V.

FIGURE 21–59

13. A certain sawtooth generator has the following parameter values: $V_{IN} = 3$ V, $R = 4.7$ kΩ, $C = 0.001$ μF, and V_F for the PUT is 1.2 V. Determine its peak-to-peak output voltage if the period is 10 μs.

Section 21–5   Wien-Bridge Oscillator

14. Determine the necessary value of R_2 in Figure 21–60 so that the circuit will oscillate. Neglect the forward resistance of the zener diodes.

15. Explain the purpose of R_3 in Figure 21–60.

16. What is the initial closed-loop gain in Figure 21–60? At what value of output voltage does A_{cl} change, and to what value does it change?

17. Find the frequency of oscillation for the Wien-bridge oscillator Figure 21–60.

FIGURE 21–60

Section 21–6   Active Filters

18. Determine the number of poles in each active filter in Figure 21–61, and identify its type.

19. Calculate the critical frequencies for the filters in Figure 21–61.

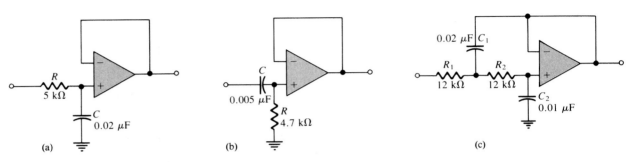

(a)   (b)   (c)

FIGURE 21–61

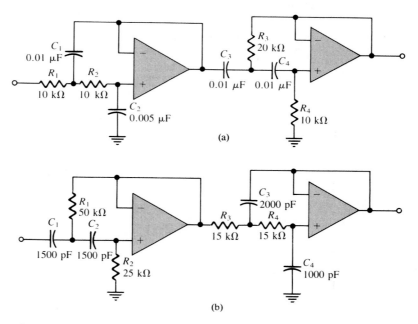

(a)

(b)

FIGURE 21-62

20. Determine the bandwidth and center frequency of each filter in Figure 21–62.

Section 21-7   Three-Terminal Regulators

21. Determine the output voltage for the series regulator in Figure 21–63.
22. If R_3 in Figure 21–63 is doubled, what happens to the output voltage?
23. If the zener voltage is 2.7 V instead of 2 V in Figure 21–63, what is the output voltage?

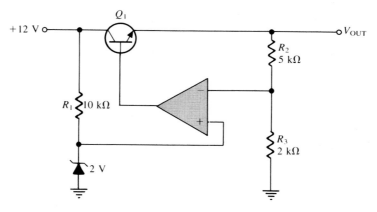

FIGURE 21-63

24. A series voltage regulator with constant current limiting is shown in Figure 21–64. Determine the value of R_4 if the load current is to be limited to a maximum value of 250 mA. What power rating must R_4 have?

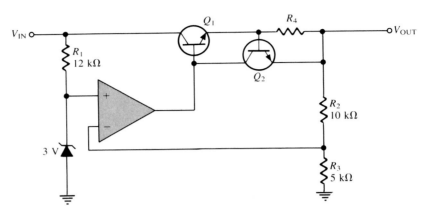

FIGURE 21–64

25. If R_4 (determined in Problem 24) is halved, what is the maximum load current?

26. In the shunt regulator of Figure 21–65, when the load current increases, does Q_1 conduct more or less? Why?

27. Assume that I_L remains constant and V_{IN} increases by 1 V in Figure 21–65. What is the change in the collector current of Q_1?

FIGURE 21–65

28. With a constant input voltage of 18 V, the load resistance in Figure 21–65 is varied from 1 kΩ to 1.2 kΩ. Neglecting any change in output voltage, how much does the shunt current through Q_1 change?

ANSWERS TO SECTION REVIEWS

Section 21–1

1. 1.36 V. **2.** See Figure 21–66.

FIGURE 21–66

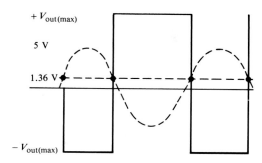

Section 21–2

1. The terminal of the op-amp where the input resistors are commonly connected.
2. $\frac{1}{5}$. **3.** 20 kΩ.

Section 21–3

1. See Figure 21–67. **2.** Ramp. **3.** Step.

FIGURE 21–67

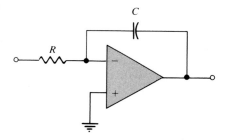

Section 21–4

1. 0.2 ms. **2.** 5 kHz.

Section 21–5

1. Negative feedback loop sets closed-loop gain; positive feedback loop sets frequency.
2. 1.67 V.

Section 21–6

1. A single *RC* network. **2.** Flat from dc to the critical frequency.
3. The *R* and *C* positions are interchanged. **4.** It is halved.

Section 21–7

1. In a shunt regulator, the control element is in parallel with the load rather than in series with the load.
2. A shunt regulator has inherent current limiting, but it is less efficient than a series regulator because part of the load current must be bypassed through the control element.

APPENDIX A
Circuit Breadboarding and Measurements

BREADBOARDING A CIRCUIT

The term *breadboarding* refers to the process of installing components on a circuit board and interconnecting them to form a specified circuit. One common type of circuit board used for constructing circuits in the laboratory is shown in Figure A–1. It consists of rows of small sockets into which component leads and jumper wires are inserted.

Top view

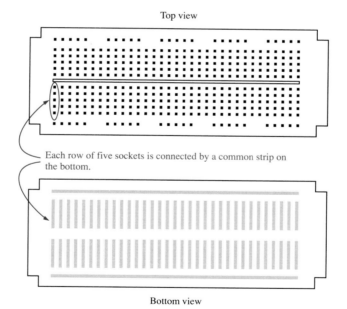

Each row of five sockets is connected by a common strip on the bottom.

Bottom view

FIGURE A–1
A typical circuit board used for breadboarding.

In this particular configuration, all five sockets in each row are connected together and are effectively one electrical point as shown in the bottom view. All the sockets arranged on the outer edges of the board are typically connected together as shown. These two strips of sockets are normally used to distribute the supply voltages and ground to the circuit.

As examples of breadboarding a circuit, Figure A–2 shows how both simple series and parallel resistive circuits might be connected on the circuit board.

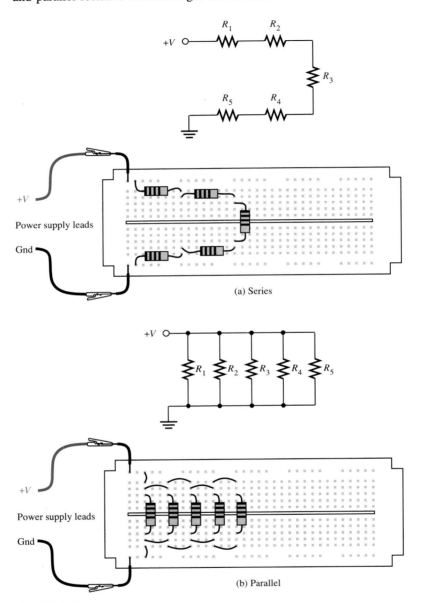

(a) Series

(b) Parallel

FIGURE A–2
Examples of breadboarding a circuit.

Voltage Measurement

To measure voltage in a circuit using a multimeter, the following general steps should be taken:

1. Insert the red meter lead into the connector labeled for *voltage* (V) on the meter and insert the black meter lead into the connector labeled *common* (COM).

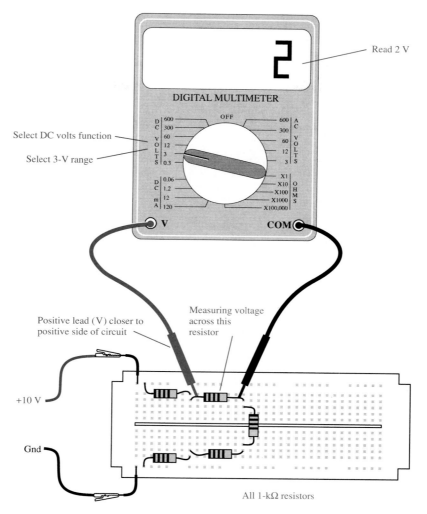

FIGURE A–3
Example of measuring voltage in a circuit.

2. Set the function switch to the *volts* position (select either dc or ac).
3. Set the range switch (the range switch and function switch are often the same) to the lowest setting that is higher than the voltage you expect to measure.
4. Select the two points in the circuit between which you want to measure the voltage.
5. Place the red meter lead on the point closest to the positive side of the circuit and the black lead on the point closest to the negative or ground side of the circuit as shown in Figure A–3.
6. Read the voltage.

Current Measurement

To measure current in a circuit using a multimeter, the following general steps are taken:

1. Insert the red meter lead into the connector labeled for *current* (A or mA) on the meter and insert the black meter lead into the connector labeled *common* (COM).
2. Set the function switch to the *amps* position (select either dc or ac).
3. Set the range switch to the lowest setting that is higher than the current you expect to measure.

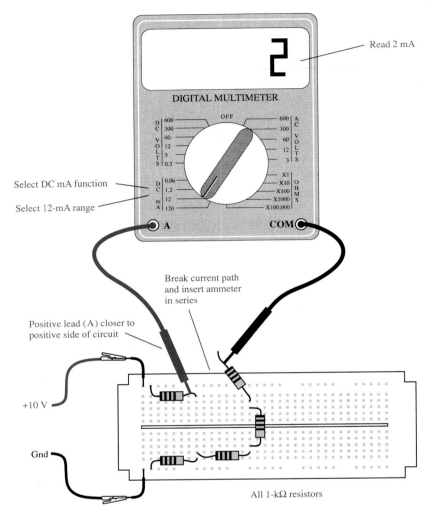

Read 2 mA

DIGITAL MULTIMETER

Select DC mA function

Select 12-mA range

Break current path
and insert ammeter
in series

Positive lead (A) closer to
positive side of circuit

+10 V

Gnd

All 1-kΩ resistors

FIGURE A–4
Example of measuring current in a circuit.

4. Select the point in the circuit at which you want to measure the current.
5. Turn off the power and break the circuit at the selected point by simply lifting a component lead or one end of a jumper wire from the socket. This permits the ammeter to be placed in series with the current path.
6. Insert the ammeter in the break by connecting the red lead to the point closest to the positive side of the circuit and the black lead to the point closest to the negative or ground side of the circuit as shown in Figure A–4.
7. Turn the power back on and read the current.

MEASURING RESISTANCE

To measure resistance in a circuit using a multimeter, the following general steps are taken:

1. Insert one meter lead into the connector labeled for *ohms* (Ω) on the meter and the other lead into the connector labeled *common* (COM). Zero the ohmmeter if necessary.

FIGURE A–5
Example of measuring resistance in a circuit.

2. Set the function switch to the *ohms* (Ω) position.
3. Select an appropriate multiplier setting (X1, X10, etc.).
4. Disconnect power from the circuit.
5. Remove the resistor to be measured or lift one lead so that it is disconnected from the rest of the circuit. If you wish to measure *total* resistance, none of the components has to be removed.
6. Connect the ohmmeter leads across the selected resistor as shown in Figure A–5. Polarity does not matter.
7. Read the resistance and reconnect the resistor and the power.

APPENDIX B
Batteries

Batteries are an important source of dc voltage. They are available in two basic categories: the wet cell and the dry cell. A battery is formed by two or more cells connected in series.

A cell consists basically of two electrodes immersed in an electrolyte. A voltage is developed between the electrodes as a result of the chemical action between the electrodes and the electrolyte. The electrodes typically are two dissimilar metals, and the electrolyte is a chemical solution or substance.

Wet Cells and Dry Cells

In a wet cell, the electrolyte is in liquid form, as you might guess. The automotive battery is a common type of wet-cell battery. In a dry cell, the electrolyte is typically in a pastelike form. There are many more varieties of dry-cell batteries than wet-cell batteries.

Primary and Secondary Cells

Primary cells cannot be recharged and thus are used once and then discarded. The 1.5-V carbon-zinc or alkaline flashlight battery is an example of a primary cell.

Secondary cells are rechargable and can be reused many times. Examples are nickel-cadmium (Ni-Cad) cells and lead-acid automotive cells.

Charging Rate

The charging rate for secondary cells specifies the amount of reverse current recommended over a given period of time. For example, the charging rate might be specified as 100 mA for 10 h. The same battery could be charged at a faster *equivalent* rate of, say, 200 mA for 5 h or at a slower *equivalent* rate of 50 mA for 20 h.

Charge/Discharge Cycle

When a battery charges and is then discharged to below its useful output voltage, it has gone through *one* charge/discharge cycle. The total number of charge/discharge cycles that a battery can go through before it becomes no longer usable is called the *cycle life*. The cycle life of a typical rechargeable battery can be several hundred cycles.

Capacity

The capacity of any type of battery specifies how much current the battery can deliver to a load over a certain period of time before its output voltage decreases to an unusable level. Capacity is expressed in ampere-hours (Ah). For example, if a certain battery has a capacity of 50 Ah, it can deliver 50 A for 1 h or 100 A for ½ h, 25 A for 2 h, and so on.

TYPES OF BATTERIES

Lead-Acid Cell

The positive electrode of a lead-acid cell is lead peroxide (PbO_2), and the negative electrode is spongy lead (Pb). The electrolyte is sulfuric acid (H_2SO_4) in water. Thus, the lead-acid cell is classified as a wet cell.

Two positive hydrogen ions ($2H^+$) and one negative sulfate ion (SO_4^{--}) are formed when the sulfuric acid ionizes in the water. Lead ions (Pb^{++}) from both electrodes displace the hydrogen ions in the electrolyte solution. When the lead ion from the spongy lead electrode enters the solution, it combines with a sulfate ion (SO_4^{--}) to form lead sulfate ($PbSO_4$), and it leaves two excess electrons on the electrode.

When a lead ion from the lead peroxide electrode enters the solution, it also leaves two excess electrons on the electrode and forms lead sulfate in the solution. However, because this electrode is lead peroxide, two free oxygen atoms are created when a lead atom leaves and enters the solution as a lead ion. These two oxygen atoms take four electrons from the lead peroxide electrode and become oxygen ions (O^{--}). This process creates a deficiency of two electrons on this electrode (there were initially two excess electrons).

The two oxygen ions ($2O^{--}$) combine in the solution with four hydrogen ions ($4H^+$) to produce two molecules of water ($2H_2O$). This process dilutes the electrolyte over a period of time. Also, there is a buildup of lead sulfate on the electrodes. These two factors result in a reduction in the voltage produced by the cell and necessitate periodic recharging.

As you have seen, for each departing lead ion, there is an excess of two electrons on the spongy lead electrode, and there is a deficiency of two electrons on the lead peroxide electrode. Therefore, the lead peroxide electrode is positive, and the spongy lead electrode is negative. This chemical reaction is pictured in Figure B–1.

FIGURE B–1
Chemical reaction in a discharging lead-acid cell.

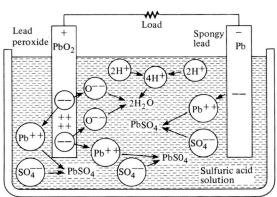

As mentioned, the dilution of the electrolyte by the formation of water and lead sulfate requires that the lead-acid cell be recharged to *reverse* the chemical process. The cell is recharged by connection of an external voltage source to the electrodes, as shown in Figure B–2. The formula for the chemical reaction in a lead-acid cell is as follows:

$$Pb + PbO_2 + 2H_2SO_4 \longrightarrow 2PbSO_4 + 2H_2O$$

FIGURE B–2
Recharging a lead-acid cell.

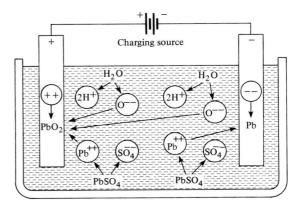

Carbon-Zinc Cell

A typical carbon-zinc dry cell is illustrated in Figure B–3. The zinc container or can is dissolved by the electrolyte. As a result of this reaction, an excess of electrons accumulates on the container, making it the negative electrode.

FIGURE B–3
Cutaway view of a typical carbon-zinc cell.

TABLE B–1

Type	Carbon-zinc	Alkaline	Mercury	Silver	Ni-cadmium
Type	Dry/primary	Dry/primary	Dry/primary	Dry/primary	Dry/primary
Volts/cell	1.5	1.5	1.35	1.5	1.2
Negative Electrode	Zinc	Zinc	Zinc	Zinc	Cadmium
Positive Electrode	Manganese dioxide	Manganese dioxide	Mercuric oxide	Silver oxide	Nickel hydroxide
Electrolyte	Zinc chloride	Potassium hydroxide	Potassium hydroxide	Potassium hydroxide	Potassium hydroxide
Cycle Life	1	1	1	1	Up to 1000
Capacity	Up to 18 Ah	Up to 45 Ah	Up to 14 Ah	Up to 210 mAh	Up to 4 Ah
Discharge Curve	Sloping	Sloping	Flat	Flat	Flat
Relative cost	Low	Medium	High	High	High

The hydrogen ions in the electrolyte take electrons from the carbon rod, making it the positive electrode. Hydrogen gas is formed near the carbon electrode, but this gas is eliminated by reaction with manganese dioxide (called a *depolarizing agent*). This depolarization prevents bursting of the container due to gas formation. Because the chemical reaction is not reversible, the carbon-zinc cell is a primary cell. Several common types of dry cells are summarized in Table B–1.

APPENDIX C
Norton's Theorem, Millman's Theorem, and Δ-Y/Y-Δ Network Conversions

NORTON'S THEOREM

Like Thevenin's theorem, Norton's theorem provides a method of reducing a more complex circuit to a simpler form. The basic difference is that Norton's theorem gives *an equivalent current source in parallel with an equivalent resistance*. The form of Norton's equivalent circuit is shown in Figure C–1, Regardless of how complex the original circuit is, it can always be reduced to this equivalent form. The equivalent current source is designated I_N, and the equivalent resistance, R_N.

FIGURE C–1
The form of Norton's equivalent circuit.

To apply Norton's theorem, you must know how to find the two quantities I_N and R_N. Once you know them for a given circuit, simply connect them in parallel to get the complete Norton circuit.

Norton's Equivalent Current (I_N)

As stated, I_N is one part of the complete Norton equivalent circuit; R_N is the other part. I_N is defined to be the *short circuit current* between two points in a circuit. Any component connected between these two points effectively "sees" a current source of value I_N in parallel with R_N.

To illustrate, suppose that a resistive circuit of some kind has a resistor connected between two points in the circuit, as shown in Figure C–2(a). We wish to find the Norton circuit that is equivalent to the one shown as "seen" by R_L. To find I_N, calculate the current between points A and B with these two points *shorted*, as shown in Figure C–2(b). Example C–1 demonstrates how to find I_N.

FIGURE C–2
Determining the Norton
equivalent current, I_N.

(a) Original circuit   (b) Short the terminals to get I_N.

EXAMPLE C–1

Determine I_N for the circuit within the shaded area in Figure C–3(a).

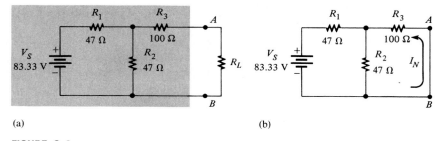

(a)   (b)

FIGURE C–3

Solution
Short terminals A and B are shown in Figure C–3(b). I_N is the current through the short and is calculated as follows: First, the total resistance seen by the voltage source is

$$R_T = R_1 + \frac{R_2 R_3}{R_2 + R_3}$$

$$= 47\ \Omega + \frac{(47\ \Omega)(100\ \Omega)}{147\ \Omega} = 78.97\ \Omega$$

The total current from the source is

$$I_T = \frac{V_S}{R_T} = \frac{83.33\ \text{V}}{78.97\ \Omega}$$

$$= 1.06\ \text{A}$$

Now apply the current divider formula to find I_N (the current through the short):

$$I_N = \left(\frac{R_2}{R_2 + R_3}\right)I_T = \left(\frac{47\ \Omega}{147\ \Omega}\right)1.06\ \text{A}$$

$$= 0.339\ \text{A}$$

This is the value for the equivalent Norton current source.

Norton's Equivalent Resistance (R_N)

We define R_N in the same way as R_{TH}: it is the total resistance appearing between two terminals in a given circuit with all sources replaced by their internal resistances. Example C–2 demonstrates how to find R_N.

EXAMPLE C–2

Find R_N for the circuit within the shaded area of Figure C–3 (see Example C–1).

Solution
First reduce V_S to zero by shorting it, as shown in Figure C–4.

FIGURE C–4

Looking in at terminals A and B, we see that the parallel combination of R_1 and R_2 is in series with R_3. Thus,

$$R_N = R_3 + \frac{R_1}{2} = 100\ \Omega + \frac{47\ \Omega}{2}$$

$$= 123.5\ \Omega$$

The last two examples have shown how to find the two equivalent components of a Norton equivalent circuit, I_N and R_N. Keep in mind that these values can be found for any linear circuit. Once these are known, they must be connected in *parallel* to form the Norton equivalent circuit, as illustrated in Example C–3.

EXAMPLE C–3

Draw the complete Norton circuit for the original circuit in Figure C–3 (Example C–1).

Solution
We found in Examples C–1 and C–2 that $I_N = 0.339$ A and $R_N = 123.5\ \Omega$. The Norton equivalent circuit is shown in Figure C–5.

FIGURE C–5

Summary of Norton's Theorem

Any load resistor connected between the terminals of a Norton equivalent circuit will have the same current through it and the same voltage across it as if it were connected to the terminals of the original circuit. A summary of steps for theoretically applying Norton's theorem is as follows:

1. Short the two terminals between which you want to find the Norton equivalent circuit.
2. Determine the current (I_N) through the shorted terminals.
3. Determine the resistance (R_N) between the two terminals (opened) with all voltage sources shorted and all current sources opened $(R_N = T_{TH})$.
4. Connect I_N and R_N in parallel to produce the complete Norton equivalent for the original circuit.

Norton's equivalent circuit can also be derived from Thevenin's equivalent circuit by use of the source conversion method.

MILLMAN'S THEOREM

Millman's theorem allows us to reduce *any number of parallel voltage sources to a single equivalent voltage source*. It simplifies finding the voltage across or current through a load. Millman's theorem gives the same results as Thevenin's theorem for the special case of parallel voltage sources. A conversion by Millman's theorem is illustrated in Figure C–6.

FIGURE C–6
Reduction of parallel voltage sources to a single equivalent voltage source.

Millman's Equivalent Voltage (V_{EQ}) and Equivalent Resistance (R_{EQ})

Millman's theorem gives us a formula for calculating the equivalent voltage, V_{EQ}. To find V_{EQ}, convert each of the parallel voltage sources into current sources, as shown in Figure C–7.

In Figure C–7(b), the total current from the parallel current sources is

$$I_T = I_1 + I_2 + I_3 + \cdots + I_n$$

The total conductance between terminals A and B is

$$G_T = G_1 + G_2 + G_3 + \cdots + G_n$$

FIGURE C–7
Parallel voltage sources converted to current sources.

where $G_T = 1/R_T$, $G_1 = 1/R_1$, and so on. Remember, the current sources are effectively open. Therefore, by Millman's theorem, the *equivalent resistance* is the total resistance R_T.

$$R_{EQ} = \frac{1}{G_T} = \frac{1}{(1/R_1) + (1/R_2) + (1/R_3) + \cdots + (1/R_n)} \qquad \text{(C–1)}$$

By Millman's theorem, the *equivalent voltage* is $I_T R_{EQ}$, where I_T is expressed as follows:

$$I_T = I_1 + I_2 + I_3 + \cdots + I_n$$

$$= \frac{V_1}{R_1} + \frac{V_2}{R_2} + \frac{V_3}{R_3} + \cdots + \frac{V_n}{R_n}$$

The following is the formula for the equivalent voltage:

$$V_{EQ} = \frac{(V_1/R_1) + (V_2/R_2) + (V_3/R_3) + \cdots + (V_n/R_n)}{(1/R_1) + (1/R_2) + (1/R_3) + \cdots + (1/R_n)} \qquad \text{(C–2)}$$

Equations (C–1) and (C–2) are the two Millman formulas. The equivalent voltage source has a polarity such that the total current through a load will flow in the same direction as in the original circuit.

EXAMPLE C–4

Use Millman's theorem to find the voltage across R_L and the current through R_L in Figure C–8.

FIGURE C–8

Solution

Apply Millman's theorem as follows:

$$R_{EQ} = \frac{1}{(1/R_1) + (1/R_2) + (1/R_3)}$$

$$= \frac{1}{(1/22\ \Omega) + (1/22\ \Omega) + (1/10\ \Omega)}$$

$$= \frac{1}{0.19} = 5.24\ \Omega$$

$$V_{EQ} = \frac{(V_1/R_2) + (V_2/R_2) + (V_3/R_3)}{(1/R_1) + (1/R_2) + (1/R_3)}$$

$$= \frac{(10\ V/22\ \Omega) + (5\ V/22\ \Omega) + (15\ V/10\ \Omega)}{(1/22\ \Omega) + (1/22\ \Omega) + (1/10\ \Omega)}$$

$$= \frac{2.18\ A}{0.19\ S} = 11.47\ V$$

The single equivalent voltage source is shown in Figure C–9.

FIGURE C–9

Now we calculate I_L and V_L for the load resistor.

$$I_L = \frac{V_{EQ}}{R_{EQ} + R_L} = \frac{11.47\ V}{52.24\ \Omega}$$

$$= 0.220\ A$$

$$V_L = I_L R_L = (0.220\ A)(47\ \Omega)$$

$$= 10.34\ V$$

DELTA-WYE (Δ-Y) AND WYE-DELTA (Y-Δ) NETWORK CONVERSIONS

A resistive delta (Δ) network has the form shown in Figure C–10(a). A wye (Y) network is shown in Figure C–10(b). Notice that letter subscripts are used to designate resistors in the delta network and that numerical subscripts are used to designate resistors in the wye.

Conversion between these two forms of circuits is sometimes helpful in areas such as bridge analysis and three-phase power systems. In this section, the conversion formulas and rules for remembering them are given.

FIGURE C–10
Delta and wye networks.

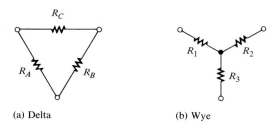

(a) Delta   (b) Wye

Δ-To-Y Conversion

It is convenient to think of the wye positioned within the delta, as shown in Figure C–11. To convert from delta to wye, we need R_1, R_2, and R_3 in terms of R_A, R_B, and R_C. The conversion rule is as follows: *Each resistor in the wye is equal to the product of the resistors in two adjacent delta branches, divided by the sum of all three delta resistors.*

FIGURE C–11
"Y within Δ" aid for conversion formulas.

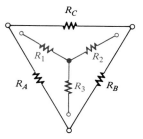

In Figure C–11, R_A and R_C are "adjacent" to R_1:

$$R_1 = \frac{R_A R_C}{R_A + R_B + R_C} \tag{C–3}$$

Also, R_B and R_C are "adjacent" to R_2:

$$R_2 = \frac{R_B R_C}{R_A + R_B + R_C} \tag{C–4}$$

and R_A and R_B are "adjacent" to R_3:

$$R_3 = \frac{R_A R_B}{R_A + R_B + R_C} \tag{C–5}$$

Y-To-Δ Conversion

To convert from wye to delta, we need R_A, R_B, and R_C in terms of R_1, R_2 and R_3. The conversion rule is as follows: *Each resistor in the delta is equal to the sum of all possible products of wye resistors taken two at a time, divided by the opposite wye resistor.*

In Figure C–11, R_2 is "opposite" to R_A:

$$R_A = \frac{R_1R_2 + R_1R_3 + R_2R_3}{R_2} \qquad \text{(C–6)}$$

Also, R_1 is "opposite" to R_B:

$$R_B = \frac{R_1R_2 + R_1R_3 + R_2R_3}{R_1} \qquad \text{(C–7)}$$

and R_3 is "opposite" to R_C:

$$R_C = \frac{R_1R_2 + R_1R_3 + R_2R_3}{R_3} \qquad \text{(C–8)}$$

The following two examples illustrate conversion between these two forms of circuits.

EXAMPLE C–5

Convert the delta network in Figure C–12 to a wye network.

FIGURE C–12

100 Ω

R_C

R_A R_B
220 Ω   560 Ω

Solution
Use Equations (C–3), (C–4), and (C–5):

$$R_1 = \frac{R_AR_C}{R_A + R_B + R_C} = \frac{(220\ \Omega)(100\ \Omega)}{220\ \Omega + 560\ \Omega + 100\ \Omega} = 25\ \Omega$$

$$R_2 = \frac{R_BR_C}{R_A + R_B + R_C} = \frac{(560\ \Omega)(100\ \Omega)}{880\ \Omega} = 63.64\ \Omega$$

$$R_3 = \frac{R_AR_B}{R_A + R_B + R_C} = \frac{(220\ \Omega)(560\ \Omega)}{880\ \Omega} = 140\ \Omega$$

The resulting wye network is shown in Figure C–13.

FIGURE C–13

R_1 R_2

25 Ω   63.64 Ω

R_3 140 Ω

EXAMPLE C–6   Convert the wye network in Figure C–14 to a delta network.

FIGURE C–14

Solution
Use Equations (C–6), (C–7), and (C–8):

$$R_A = \frac{R_1 R_2 + R_1 R_3 + R_2 R_3}{R_2}$$

$$= \frac{(1 \text{ k}\Omega)(2.2 \text{ k}\Omega) + (1 \text{ k}\Omega)(5.6 \text{ k}\Omega) + (2.2 \text{ k}\Omega)(5.6 \text{ k}\Omega)}{2.2 \text{ k}\Omega} = 9.15 \text{ k}\Omega$$

$$R_B = \frac{R_1 R_2 + R_1 R_3 + R_2 R_3}{R_1}$$

$$= \frac{(1 \text{ k}\Omega)(2.2 \text{ k}\Omega) + (1 \text{ k}\Omega)(5.6 \text{ k}\Omega) + (2.2 \text{ k}\Omega)(5.6 \text{ k}\Omega)}{1 \text{ k}\Omega} = 20.12 \text{ k}\Omega$$

$$R_C = \frac{R_1 R_2 + R_1 R_3 + R_2 R_3}{R_3}$$

$$= \frac{(1 \text{ k}\Omega)(2.2 \text{ k}\Omega) + (1 \text{ k}\Omega)(5.6 \text{ k}\Omega) + (2.2 \text{ k}\Omega)(5.6 \text{ k}\Omega)}{5.6 \text{ k}\Omega} = 3.59 \text{ k}\Omega$$

The resulting delta network is shown in Figure C–15.

FIGURE C–15

Application of Δ-Y Conversion to the Simplification of a Bridge Circuit

You have already seen how Thevenin's theorem can be used to simplify a bridge circuit. Now we will see how Δ-Y conversion can be used for conversion of a bridge circuit to a series parallel form for easier analysis.

Figure C–16 illustrates how the delta (Δ) formed by R_A, R_B, and R_C can be converted to a wye (Y), thus creating an equivalent series-parallel circuit. Equations (C–3), (C–4), and (C–5) are used in this conversion.

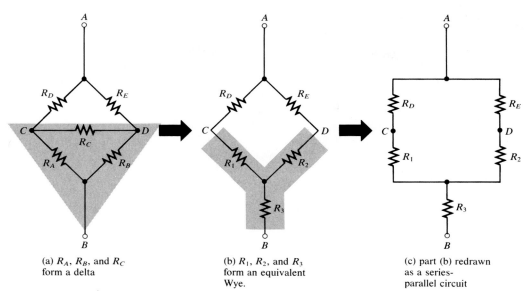

(a) R_A, R_B, and R_C form a delta

(b) R_1, R_2, and R_3 form an equivalent Wye.

(c) part (b) redrawn as a series-parallel circuit

FIGURE C–16
Conversion of a bridge circuit to a series-parallel configuration.

In a bridge circuit, the load is connected across points C and D. In Figure C–16(a), R_C represents the load resistor. When voltage is applied across points A and B, the voltage from C to D (V_{CD}) can be determined using the equivalent series-parallel circuit in Figure C–16(c) as follows.

The total resistance from point A to point B is

$$R_T = \frac{(R_1 + R_D)(R_2 + R_E)}{(R_1 + R_D) + (R_2 + R_E)} + R_3$$

Then

$$I_T = \frac{V_{AB}}{R_T}$$

The current through the left branch is

$$I_{AC} = \left(\frac{R_T}{R_1 + R_D}\right)I_T$$

The current through the right branch is

$$I_{AD} = \left(\frac{R_T}{R_2 + R_E}\right)I_t$$

The voltage at point C with respect to point A is

$$V_{CA} = V_A - I_{AC}R_D$$

The voltage at point D with respect to point A is

$$V_{DA} = V_A - I_{AD}R_E$$

The voltage from point C to point D is

$$V_{CD} = V_{CA} - V_{DA} = (V_A - I_{AC}R_D) - (V_A - I_{AD}R_E)$$
$$= I_{AD}R_E - I_{AC}R_D$$

V_{CD} is the voltage across the load (R_C) in the bridge circuit of Figure C–16(a). The current through the load can be found by Ohm's law:

$$I_C = \frac{V_{CD}}{R_C}$$

EXAMPLE C–7

Determine the load voltage and the load current in the bridge circuit in Figure C–17. Notice that the resistors are labeled for convenient conversion using the previous formulas. R_C is the load resistor.

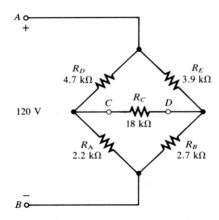

FIGURE C–17

Solution
First convert the delta formed by R_A, R_B, and R_C to a wye:

$$R_1 = \frac{R_A R_C}{R_A + R_B + R_C} = \frac{(2.2\ \text{k}\Omega)(18\ \text{k}\Omega)}{2.2\ \text{k}\Omega + 2.7\ \text{k}\Omega + 18\ \text{k}\Omega} = 1.73\ \text{k}\Omega$$

$$R_2 = \frac{R_B R_C}{R_A + R_B + R_C} = \frac{(2.7\ \text{k}\Omega)(18\ \text{k}\Omega)}{22.9\ \text{k}\Omega} = 2.12\ \text{k}\Omega$$

$$R_3 = \frac{R_A R_B}{R_A + R_B + R_C} = \frac{(2.2\ \text{k}\Omega)(2.7\ \text{k}\Omega)}{22.9\ \text{k}\Omega} = 259\ \Omega$$

The resulting equivalent series-parallel circuit is shown in Figure C–18.

FIGURE C–18

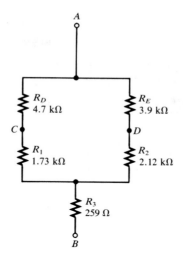

Now determine R_T and the branch currents in Figure C–18.

$$R_T = \frac{(R_1 + R_D)(R_2 + R_E)}{(R_1 + R_D) + (R_2 + R_E)} + R_3 = \frac{(6.43 \text{ k}\Omega)(6.02 \text{ k}\Omega)}{6.43 \text{ k}\Omega + 6.02 \text{ k}\Omega} + 259 \ \Omega = 3.37 \text{ k}\Omega$$

$$I_T = \frac{V_{AB}}{R_T} = \frac{120 \text{ V}}{3.37 \text{ k}\Omega} = 35.61 \text{ mA}$$

$$I_{AC} = \left(\frac{R_T}{R_1 + R_D}\right)I_T = \left(\frac{3.37 \text{ k}\Omega}{1.73 \text{ k}\Omega + 4.7 \text{ k}\Omega}\right)35.61 \text{ mA} = 18.66 \text{ mA}$$

$$I_{AD} = \left(\frac{R_T}{R_2 + R_E}\right)I_T = \left(\frac{3.37 \text{ k}\Omega}{2.12 \text{ k}\Omega + 3.9 \text{ k}\Omega}\right)35.61 \text{ mA} = 19.94 \text{ mA}$$

The voltage across the load is

$$\begin{aligned}
V_{CD} &= I_{AD}R_E - I_{AC}R_D \\
&= (19.94 \text{ mA})(3.9 \text{ k}\Omega) - (18.66 \text{ mA})(4.7 \text{ k}\Omega) \\
&= 77.77 \text{ V} - 87.70 \text{ V} = -9.93 \text{ V}
\end{aligned}$$

The load current is

$$I_C = \frac{V_{CD}}{R_C} = \frac{-9.93 \text{ V}}{18 \text{ k}\Omega} = -0.552 \text{ mA}$$

APPENDIX D
Standard Resistance Values

These values are generally available in multiples of 0.1, 1, 10, 100, 1 k, and 1 M.

Resistance Tolerance (±%)

0.1% 0.25% 0.5%	1%	2% 5%	10%	0.1% 0.25% 0.5%	1%	2% 5%	10%	0.1% 0.25% 0.5%	1%	2% 5%	10%	0.1% 0.25% 0.5%	1%	2% 5%	10%	0.1% 0.25% 0.5%	1%	2% 5%	10%	0.1% 0.25% 0.5%	1%	2% 5%	10%
10.0	10.0	10	10	14.7	14.7	—	—	21.5	21.5	—	—	31.6	31.6	—	—	46.4	46.4	—	—	68.1	68.1	68	68
10.1	—	—	—	14.9	—	—	—	21.8	—	—	—	32.0	—	—	—	47.0	—	—	—	69.0	—	—	—
10.2	10.2	—	—	15.0	15.0	15	15	22.1	22.1	22	22	32.4	32.4	—	—	47.5	47.5	47	47	69.8	69.8	—	—
10.4	—	—	—	15.2	—	—	—	22.3	—	—	—	32.8	—	—	—	48.1	—	—	—	70.6	—	—	—
10.5	10.5	—	—	15.4	15.4	—	—	22.6	22.6	—	—	33.2	33.2	33	33	48.7	48.7	—	—	71.5	71.5	—	—
10.6	—	—	—	15.6	—	—	—	22.9	—	—	—	33.6	—	—	—	49.3	—	—	—	72.3	—	—	—
10.7	10.7	—	—	15.8	15.8	—	—	23.2	23.2	—	—	34.0	34.0	—	—	49.9	49.9	—	—	73.2	73.2	—	—
10.9	—	—	—	16.0	—	—	—	23.4	—	—	—	34.4	—	—	—	50.5	—	—	—	74.1	—	—	—
11.0	11.0	11	—	16.2	16.2	16	—	23.7	23.7	—	—	34.8	34.8	—	—	51.1	51.1	51	—	75.0	75.0	75	—
11.1	—	—	—	16.4	—	—	—	24.0	—	—	—	35.2	—	—	—	51.7	—	—	—	75.9	—	—	—
11.3	11.3	—	—	16.5	16.5	—	—	24.3	24.3	24	—	35.7	35.7	—	—	52.3	52.3	—	—	76.8	76.8	—	—
11.4	—	—	—	16.7	—	—	—	24.6	—	—	—	36.1	—	—	—	53.0	—	—	—	77.7	—	—	—
11.5	11.5	—	—	16.9	16.9	—	—	24.9	24.9	—	—	36.5	36.5	36	—	53.6	53.6	—	—	78.7	78.7	—	—
11.7	—	—	—	17.2	—	—	—	25.2	—	—	—	37.0	—	—	—	54.2	—	—	—	79.6	—	—	—
11.8	11.8	—	—	17.4	17.4	—	—	25.5	25.5	—	—	37.4	37.4	—	—	54.9	54.9	—	—	80.6	80.6	—	—
12.0	—	—	—	17.6	—	—	—	25.8	—	—	—	37.9	—	—	—	55.6	—	—	—	81.6	—	—	—
12.1	12.1	12	12	17.8	17.8	—	—	26.1	26.1	—	—	38.3	38.3	—	—	56.2	56.2	56	56	82.5	82.5	82	82
12.3	—	—	—	18.0	—	—	—	26.4	—	—	—	38.8	—	—	—	56.9	—	—	—	83.5	—	—	—
12.4	12.4	—	—	18.2	18.2	18	18	26.7	26.7	—	—	39.2	39.2	39	39	57.6	57.6	—	—	84.5	84.5	—	—
12.6	—	—	—	18.4	—	—	—	27.1	—	—	—	39.7	—	—	—	58.3	—	—	—	85.6	—	—	—
12.7	12.7	—	—	18.7	18.7	—	—	27.4	27.4	27	27	40.2	40.2	—	—	59.0	59.0	—	—	86.6	86.6	—	—
12.9	—	—	—	18.9	—	—	—	27.7	—	—	—	40.7	—	—	—	59.7	—	—	—	87.6	—	—	—
13.0	13.0	13	—	19.1	19.1	—	—	28.0	28.0	—	—	41.2	41.2	—	—	60.4	60.4	—	—	88.7	88.7	—	—
13.2	—	—	—	19.3	—	—	—	28.4	—	—	—	41.7	—	—	—	61.2	—	—	—	89.8	—	—	—
13.3	13.3	—	—	19.6	19.6	—	—	28.7	28.7	—	—	42.2	42.2	—	—	61.9	61.9	62	—	90.9	90.9	91	—
13.5	—	—	—	19.8	—	—	—	29.1	—	—	—	42.7	—	—	—	62.6	—	—	—	92.0	—	—	—
13.7	13.7	—	—	20.0	20.0	20	—	29.4	29.4	—	—	43.2	43.2	43	—	63.4	63.4	—	—	93.1	93.1	—	—
13.8	—	—	—	20.3	—	—	—	29.8	—	—	—	43.7	—	—	—	64.2	—	—	—	94.2	—	—	—
14.0	14.0	—	—	20.5	20.5	—	—	30.1	30.1	30	—	44.2	44.2	—	—	64.9	64.9	—	—	95.3	95.3	—	—
14.2	—	—	—	20.8	—	—	—	30.5	—	—	—	44.8	—	—	—	65.7	—	—	—	96.5	—	—	—
14.3	14.3	—	—	21.0	21.0	—	—	30.9	30.9	—	—	45.3	45.3	—	—	66.5	66.5	—	—	97.6	97.6	—	—
14.5	—	—	—	21.3	—	—	—	31.2	—	—	—	45.9	—	—	—	67.3	—	—	—	98.8	—	—	—

APPENDIX E
Devices Data Sheets

MOTOROLA

1N4001 thru 1N4007

Designers Data Sheet

"SURMETIC" RECTIFIERS

. . . subminiature size, axial lead mounted rectifiers for general-purpose low-power applications.

Designers Data for "Worst Case" Conditions

The Designers Data Sheets permit the design of most circuits entirely from the information presented. Limit curves — representing boundaries on device characteristics — are given to facilitate "worst case" design.

LEAD MOUNTED SILICON RECTIFIERS

50-1000 VOLTS
DIFFUSED JUNCTION

*MAXIMUM RATINGS

Rating	Symbol	1N4001	1N4002	1N4003	1N4004	1N4005	1N4006	1N4007	Unit
Peak Repetitive Reverse Voltage Working Peak Reverse Voltage DC Blocking Voltage	V_{RRM} V_{RWM} V_R	50	100	200	400	600	800	1000	Volts
Non-Repetitive Peak Reverse Voltage (halfwave, single phase, 60 Hz)	V_{RSM}	60	120	240	480	720	1000	1200	Volts
RMS Reverse Voltage	$V_{R(RMS)}$	35	70	140	280	420	560	700	Volts
Average Rectified Forward Current (single phase, resistive load, 60 Hz, see Figure 8, T_A = 75°C)	I_O	1.0							Amp
Non-Repetitive Peak Surge Current (surge applied at rated load conditions, see Figure 2)	I_{FSM}	30 (for 1 cycle)							Amp
Operating and Storage Junction Temperature Range	T_J, T_{stg}	−65 to +175							°C

*ELECTRICAL CHARACTERISTICS

Characteristic and Conditions	Symbol	Typ	Max	Unit
Maximum Instantaneous Forward Voltage Drop (i_F = 1.0 Amp, T_J = 25°C) Figure 1	v_F	0.93	1.1	Volts
Maximum Full-Cycle Average Forward Voltage Drop (I_O = 1.0 Amp, T_L = 75°C, 1 inch leads)	$V_{F(AV)}$	—	0.8	Volts
Maximum Reverse Current (rated dc voltage) T_J = 25°C T_J = 100°C	I_R	0.05 1.0	10 50	μA
Maximum Full-Cycle Average Reverse Current (I_O = 1.0 Amp, T_L = 75°C, 1 inch leads	$I_{R(AV)}$	—	30	μA

*Indicates JEDEC Registered Data.

MECHANICAL CHARACTERISTICS

CASE: Transfer Molded Plastic
MAXIMUM LEAD TEMPERATURE FOR SOLDERING PURPOSES: 350°C, 3/8" from case for 10 seconds at 5 lbs. tension
FINISH: All external surfaces are corrosion-resistant, leads are readily solderable
POLARITY: Cathode indicated by color band
WEIGHT: 0.40 Grams (approximately)

▲Trademark of Motorola Inc.

CATHODE BAND

	MILLIMETERS		INCHES	
DIM	MIN	MAX	MIN	MAX
A	5.97	6.60	0.235	0.260
B	2.79	3.05	0.110	0.120
D	0.76	0.86	0.030	0.034
K	27.94	—	1.100	—

CASE 59-04
Does Not Conform to DO-41 Outline.

DS 6015 R3

MOTOROLA

1N4728, A
thru
1N4764, A

Designers▲ Data Sheet

ONE WATT HERMETICALLY SEALED GLASS SILICON ZENER DIODES

- Complete Voltage Range — 2.4 to 100 Volts
- DO-41 Package — Smaller than Conventional DO-7 Package
- Double Slug Type Construction
- Metallurgically Bonded Construction
- Nitride Passivated Die

Designer's Data for "Worst Case" Conditions

The Designers▲ Data sheets permit the design of most circuits entirely from the information presented. Limit curves — representing boundaries on device characteristics — are given to facilitate "worst case" design.

1.0 WATT
ZENER REGULATOR DIODES
3.3–100 VOLTS

*MAXIMUM RATINGS

Rating	Symbol	Value	Unit
DC Power Dissipation @ $T_A = 50^{\circ}C$ Derate above $50^{\circ}C$	P_D	1.0 6.67	Watt mW/$^{\circ}$C
Operating and Storage Junction Temperature Range	T_J, T_{stg}	–65 to +200	$^{\circ}$C

MECHANICAL CHARACTERISTICS

CASE: Double slug type, hermetically sealed glass

MAXIMUM LEAD TEMPERATURE FOR SOLDERING PURPOSES: 230°C, 1/16" from case for 10 seconds

FINISH: All external surfaces are corrosion resistant with readily solderable leads.

POLARITY: Cathode indicated by color band. When operated in zener mode, cathode will be positive with respect to anode.

MOUNTING POSITION: Any

FIGURE 1 — POWER TEMPERATURE DERATING CURVE

L = LEAD LENGTH TO HEAT SINK

L = 1"
L = 1/8"
L = 3/8"

P_D, MAXIMUM POWER DISSIPATION (WATTS)

T_L, LEAD TEMPERATURE ($^{\circ}$C)

NOTE:
1. POLARITY DENOTED BY CATHODE BAND
2. LEAD DIAMETER NOT CONTROLLED WITHIN "F" DIMENSION.

DIM	MILLIMETERS		INCHES	
	MIN	MAX	MIN	MAX
A	4.07	5.20	0.160	0.205
B	2.04	2.71	0.080	0.107
D	0.71	0.86	0.028	0.034
F	–	1.27	–	0.050
K	27.94	–	1.100	–

All JEDEC dimensions and notes apply.

CASE 59-03
(DO-41)

*Indicates JEDEC Registered Data
▲Trademark of Motorola Inc.

©MOTOROLA INC., 1978   DS 7039 R1

*ELECTRICAL CHARACTERISTICS (T_A = 25°C unless otherwise noted) V_F = 1.2 V max, I_F = 200 mA for all types.

JEDEC Type No. (Note 1)	Nominal Zener Voltage V_Z @ I_{ZT} Volts (Notes 2 and 3)	Test Current I_{ZT} mA	Maximum Zener Impedance (Note 4)			Leakage Current		Surge Current @ T_A = 25°C i_r – mA (Note 5)
			Z_{ZT} @ I_{ZT} Ohms	Z_{ZK} @ I_{ZK} Ohms	I_{ZK} mA	I_R µA Max	V_R Volts	
1N4728	3.3	76	10	400	1.0	100	1.0	1380
1N4729	3.6	69	10	400	1.0	100	1.0	1260
1N4730	3.9	64	9.0	400	1.0	50	1.0	1190
1N4731	4.3	58	9.0	400	1.0	10	1.0	1070
1N4732	4.7	53	8.0	500	1.0	10	1.0	970
1N4733	5.1	49	7.0	550	1.0	10	1.0	890
1N4734	5.6	45	5.0	600	1.0	10	2.0	810
1N4735	6.2	41	2.0	700	1.0	10	3.0	730
1N4736	6.8	37	3.5	700	1.0	10	4.0	660
1N4737	7.5	34	4.0	700	0.5	10	5.0	605
1N4738	8.2	31	4.5	700	0.5	10	6.0	550
1N4739	9.1	28	5.0	700	0.5	10	7.0	500
1N4740	10	25	7.0	700	0.25	10	7.6	454
1N4741	11	23	8.0	700	0.25	5.0	8.4	414
1N4742	12	21	9.0	700	0.25	5.0	9.1	380
1N4743	13	19	10	700	0.25	5.0	9.9	344
1N4744	15	17	14	700	0.25	5.0	11.4	304
1N4745	16	15.5	16	700	0.25	5.0	12.2	285
1N4746	18	14	20	750	0.25	5.0	13.7	250
1N4747	20	12.5	22	750	0.25	5.0	15.2	225
1N4748	22	11.5	23	750	0.25	5.0	16.7	205
1N4749	24	10.5	25	750	0.25	5.0	18.2	190
1N4750	27	9.5	35	750	0.25	5.0	20.6	170
1N4751	30	8.5	40	1000	0.25	5.0	22.8	150
1N4752	33	7.5	45	1000	0.25	5.0	25.1	135
1N4753	36	7.0	50	1000	0.25	5.0	27.4	125
1N4754	39	6.5	60	1000	0.25	5.0	29.7	115
1N4755	43	6.0	70	1500	0.25	5.0	32.7	110
1N4756	47	5.5	80	1500	0.25	5.0	35.8	95
1N4757	51	5.0	95	1500	0.25	5.0	38.8	90
1N4758	56	4.5	110	2000	0.25	5.0	42.6	80
1N4759	62	4.0	125	2000	0.25	5.0	47.1	70
1N4760	68	3.7	150	2000	0.25	5.0	51.7	65
1N4761	75	3.3	175	2000	0.25	5.0	56.0	60
1N4762	82	3.0	200	3000	0.25	5.0	62.2	55
1N4763	91	2.8	250	3000	0.25	5.0	69.2	50
1N4764	100	2.5	350	3000	0.25	5.0	76.0	45

* Indicates JEDEC Registered Data.

NOTE 1 — Tolerance and Type Number Designation. The JEDEC type numbers listed have a standard tolerance on the nominal zener voltage of ±10%. A standard tolerance of ±5% on individual units is also available and is indicated by suffixing "A" to the standard type number.

NOTE 2 — Specials Available Include:

A. Nominal zener voltages between the voltages shown and tighter voltage tolerances,
B. Matched sets.

For detailed information on price, availability, and delivery, contact your nearest Motorola representative.

NOTE 3 — Zener Voltage (V_Z) Measurement. Motorola guarantees the zener voltage when measured at 90 seconds while maintaining the lead temperature (T_L) at 30°C ± 1°C, 3/8" from the diode body.

NOTE 4 — Zener Impedance (Z_Z) Derivation. The zener impedance is derived from the 60 cycle ac voltage, which results when an ac current having an rms value equal to 10% of the dc zener current (I_{ZT} or I_{ZK}) is superimposed on I_{ZT} or I_{ZK}.

NOTE 5 — Surge Current (i_r) Non-Repetitive. The rating listed in the electrical characteristics table is maximum peak, non-repetitive, reverse surge current of 1/2 square wave or equivalent sine wave pulse of 1/120 second duration superimposed on the test current, I_{ZT}, per JEDEC registration; however, actual device capability is as described in Figures 4 and 5.

APPLICATION NOTE

Since the actual voltage available from a given zener diode is temperature dependent, it is necessary to determine junction temperature under any set of operating conditions in order to calculate its value. The following procedure is recommended:

Lead Temperature, T_L, should be determined from

$$T_L = \theta_{LA}P_D + T_A$$

θ_{LA} is the lead-to-ambient thermal resistance (°C/W) and P_D is the power dissipation. The value for θ_{LA} will vary and depends on the device mounting method. θ_{LA} is generally 30 to 40°C/W for the various clips and tie points in common use and for printed circuit board wiring.

The temperature of the lead can also be measured using a thermocouple placed on the lead as close as possible to the tie point. The thermal mass connected to the tie point is normally large enough so that it will not significantly respond to heat surges generated in the diode as a result of pulsed operation once steady-state conditions are achieved. Using the measured value of T_L, the junction temperature may be determined by:

$$T_J = T_L + \Delta T_{JL}.$$

ΔT_{JL} is the increase in junction temperature above the lead temperature and may be found as follows:

$$\Delta T_{JL} = \theta_{JL}P_D$$

θ_{JL} may be determined from Figure 3 for dc power conditions. For worst-case design, using expected limits of I_Z, limits of P_D and the extremes of $T_J(\Delta T_J)$ may be estimated. Changes in voltage, V_Z, can then be found from:

$$\Delta V = \theta_{VZ}\Delta T_J$$

θ_{VZ}, the zener voltage temperature coefficient, is found from Figure 2.

Under high power-pulse operation, the zener voltage will vary with time and may also be affected significantly by the zener resistance. For best regulation, keep current excursions as low as possible.

Surge limitations are given in Figure 5. They are lower than would be expected by considering only junction temperature, as current crowding effects cause temperatures to be extremely high in small spots resulting in device degradation should the limits of Figure 5 be exceeded.

 MOTOROLA *Semiconductor Products Inc.* ——

MOTOROLA

2N4877

MEDIUM-POWER NPN SILICON TRANSISTOR

. . . designed for switching and wide band amplifier applications.

- Low Collector-Emitter Saturation Voltage — $V_{CE(sat)}$ = 1.0 Vdc (Max) @ I_C = 4.0 Amp

- DC Current Gain Specified to 4 Amperes

- Excellent Safe Operating Area

- Packaged in the Compact TO-39 Case for Critical Space-Limited Applications.

**4 AMPERE
POWER TRANSISTOR**

**NPN SILICON
60 VOLTS
10 WATTS**

---

* MAXIMUM RATINGS

Rating	Symbol	Value	Unit
Collector-Emitter Voltage	V_{CEO}	60	Vdc
Collector-Base Voltage	V_{CB}	70	Vdc
Emitter-Base Voltage	V_{EB}	5.0	Vdc
Collector Current — Continuous	I_C	4.0	Adc
Base Current	I_B	1.0	Adc
Total Device Dissipation @ T_C = 25°C Derate above 25°C	P_D	10 57.2	Watts mW/°C
Operating and Storage Junction Temperature Range	T_J, T_{stg}	−65 to +200	°C

*Indicates JEDEC Registered Data

THERMAL CHARACTERISTICS

Characteristic	Symbol	Max	Unit
Thermal Resistance, Junction to Case	θ_{JC}	17.5	°C/W

FIGURE 1 — POWER–TEMPERATURE DERATING CURVE

Safe Area Curves are indicated by Figure 2. All limits are applicable and must be observed.

Pin 1. Emitter
2. Base
3. Collector

To convert inches to millimeters multiply by 25.4.
All JEDEC TO-39 dimensions and notes apply.

**CASE 79 (1)
TO-39**

© MOTOROLA INC., 1971

Characteristic	Symbol	Min	Max	Unit
OFF CHARACTERISTICS				
Collector-Emitter Sustaining Voltage (1) (I_C = 200 mAdc, I_B = 0)	$V_{CEO(sus)}$	60	—	Vdc
Collector Cutoff Current (V_{CE} = 70 Vdc, $V_{EB(off)}$ = 1.5 Vdc) (V_{CE} = 70 Vdc, $V_{EB(off)}$ = 1.5 Vdc, T_C = 100°C)	I_{CEX}	— —	100 1.0	μAdc mAdc
Collector Cutoff Current (V_{CB} = 70 Vdc, I_E = 0)	I_{CBO}	—	100	μAdc
Emitter Cutoff Current (V_{BE} = 5.0 Vdc, I_C = 0)	I_{EBO}	—	100	μAdc
ON CHARACTERISTICS(1)				
DC Current Gain (I_C = 1.0 Adc, V_{CE} = 2.0 Vdc) (I_C = 4.0 Adc, V_{CE} = 2.0 Vdc)	h_{FE}	30 20	— 100	—
Collector-Emitter Saturation Voltage (I_C = 4.0 Adc, I_B = 0.4 Adc)	$V_{CE(sat)}$	—	1.0	Vdc
Base-Emitter Saturation Voltage (I_C = 4.0 Adc, I_B = 0.4 Adc)	$V_{BE(sat)}$	—	1.8	Vdc
DYNAMIC CHARACTERISTICS				
Current-Gain-Bandwidth Product (I_C = 0.25 Adc, V_{CE} = 10 Vdc, f = 1.0 MHz) (I_C = 0.25 Adc, V_{CE} = 10 Vdc, f = 10 MHz)**	f_T	4.0 30	— —	MHz
SWITCHING CHARACTERISTICS				
Rise Time (V_{CC} = 25 Vdc, I_C = 4.0 Adc, I_{B1} = 0.4 Adc)	t_r	—	100	ns
Storage Time (V_{CC} = 25 Vdc, I_C = 4.0 Adc,	t_s	—	1.5	μs
Fall Time I_{B1} = I_{B2} = 0.4 Adc)	t_f	—	500	ns

*Indicates JEDEC Registered Data.
**Motorola guarantees this value in addition to JEDEC Registered Data.
Note 1: Pulse Test: Pulse Width $\leq$ 300 μs, Duty Cycle $\leq$ 2.0%.

FIGURE 2 — ACTIVE-REGION SAFE OPERATING AREA

FIGURE 3 — SWITCHING TIME TEST CIRCUIT

There are two limitations on the power handling ability of a transistor: average junction temperature and second breakdown. Safe operating area curves indicate I_C–V_{CE} limits of the transistor that must be observed for reliable operation; i.e., the transistor must not be subjected to greater dissipation than the curves indicate.

The data of Figure 2 is based on $T_{J(pk)}$ = 200°C; T_C is variable depending on conditions. Second breakdown pulse limits are valid for duty cycles to 10% provided $T_{J(pk)} \leq$ 200°C. At high case temperatures, thermal limitations will reduce the power that can be handled to values less than the limitations imposed by second breakdown. (See AN-415)

MOTOROLA *Semiconductor Products Inc.*

BOX 20912 • PHOENIX, ARIZONA 85036 • A SUBSIDIARY OF MOTOROLA INC.
6025-2 PRINTED IN USA 3-71 IMPERIAL LITHO 821668   10M

DS 3189

2N1595
thru
2N1599

REVERSE BLOCKING TRIODE THYRISTORS

These devices are glassivated planar construction designed for gating operation in mA/μA signal or detection circuits.

- Low-Level Gate Characteristics —
  I_{GT} = 10 mA (Max) @ 25°C
- Low Holding Current —
  I_H = 5.0 mA (Typ) @ 25°C
- Glass-to-Metal Bond for Maximum Hermetic Seal

SILICON CONTROLLED
RECTIFIERS

1.6 AMPERE RMS
50 thru 400 VOLTS

*MAXIMUM RATINGS (T_J = 125°C unless otherwise noted).

Rating	Symbol	Value	Unit
Repetitive Peak Reverse Blocking Voltage	V_{RRM}		Volts
2N1595		50	
2N1596		100	
2N1597		200	
2N1598		300	
2N1599		400	
Repetitive Peak Forward Blocking Voltage	V_{DRM}		Volts
2N1595		50	
2N1596		100	
2N1597		200	
2N1598		300	
2N1599		400	
RMS On-State Current (All Conduction Angles)	$I_{T(RMS)}$	1.6	Amps
Peak Non-Repetitive Surge Current (One Cycle, 60 Hz, T_J = –65 to +125°C)	I_{TSM}	15	Amps
Peak Gate Power	P_{GM}	0.1	Watt
Average Gate Power	$P_{G(AV)}$	0.01	Watt
Peak Gate Current	I_{GM}	0.1	Amp
Peak Gate Voltage — Forward	V_{GFM}	10	Volts
Reverse	V_{GRM}	10	
Operating Junction Temperature Range	T_J	–65 to +125	°C
Storage Temperature Range	T_{stg}	–65 to +150	°C

*Indicates JEDEC Registered Data.

STYLE 3:
PIN 1. CATHODE
2. GATE
3. ANODE (CONNECTED TO CASE)

DIM	MILLIMETERS		INCHES	
	MIN	MAX	MIN	MAX
A	8.89	9.40	0.350	0.370
B	8.00	8.51	0.315	0.335
C	6.10	6.60	0.240	0.260
D	0.406	0.533	0.016	0.021
E	0.229	3.18	0.009	0.125
F	0.406	0.483	0.016	0.019
G	4.83	5.33	0.190	0.210
H	0.711	0.864	0.028	0.034
J	0.737	1.02	0.029	0.040
K	12.70	–	0.500	–
L	6.35	–	0.250	–
M	45° NOM		45° NOM	
P	–	1.27	–	0.050
Q	90° NOM		90° NOM	
R	2.54	–	0.100	–

All JEDEC dimensions and notes apply.
CASE 79-02
TO-39

DS 6503 R1

ELECTRICAL CHARACTERISTICS (T$_C$ = 25°C unless otherwise noted).

Characteristic	Symbol	Min	Typ	Max	Unit
*Peak Reverse Blocking Current (Rated V$_{RRM}$, T$_J$ = 125°C)	I$_{RRM}$	—	—	1000	μA
*Peak Forward Blocking Current (Rated V$_{DRM}$, T$_J$ = 125°C)	I$_{DRM}$	—	—	1000	μA
*Peak On-State Voltage (I$_F$ = 1.0 Aac, Pulsed, 1.0 ms (Max), Duty Cycle ≤ 1%)	V$_{TM}$	—	1.1	2.0	Volts
*Gate Trigger Current (V$_{AK}$ = 6.0 V, R$_L$ = 12 Ohms)	I$_{GT}$	—	2.0	10	mA
*Gate Trigger Voltage (V$_{AK}$ = 6.0 V, R$_L$ = 12 Ohms) (V$_{AK}$ = 6.0 V, R$_L$ = 12 Ohms, T$_J$ = 125°C)	V$_{GT}$	— 0.2	0.7 —	3.0 —	Volts
Reverse Gate Current (V$_{GK}$ = 10 V)	I$_{GR}$	—	17	—	mA
Holding Current (V$_{AK}$ = 12 V)	I$_H$	—	5.0	—	mA
Turn-On Time (I$_{GT}$ = 10 mA, I$_F$ = 1.0 A) (I$_{GT}$ = 20 mA, I$_F$ = 1.0 A)	t$_{gt}$	— —	0.8 0.6	— —	μs
Turn-Off Time (I$_F$ = 1.0 A, I$_R$ = 1.0 A, dv/dt = 20 V/μs, T$_J$ = 125°C)	t$_q$	—	10	—	μs

*Indicates JEDEC Registered Data.

CURRENT DERATING

FIGURE 1 – CASE TEMPERATURE REFERENCE

NOTES:
(1) dc, 1ɸ, 3ɸ, 6ɸ, CIRCUIT – RESISTIVE OR INDUCTIVE LOAD, 50 to 400 Hz
(2) CASE TEMPERATURE MEASURED AT CENTER OF BOTTOM OF CASE
(3) 125°C JUNCTION TEMPERATURE

FIGURE 2 – AMBIENT TEMPERATURE REFERENCE

NOTES:
(1) dc, 1ɸ, 3ɸ, 6ɸ, CIRCUIT – RESISTIVE OR INDUCTIVE LOAD, 50 TO 400 Hz
(2) 125°C JUNCTION TEMPERATURE

MOTOROLA *Semiconductor Products Inc.*

BOX 20912 • PHOENIX, ARIZONA 85036 • A SUBSIDIARY OF MOTOROLA INC

10137 PRINTED IN USA (8/77) MPS 10M

MOTOROLA
Semiconductors
BOX 20912 • PHOENIX, ARIZONA 85036

INTERNALLY COMPENSATED, HIGH PERFORMANCE OPERATIONAL AMPLIFIERS

. . . designed for use as a summing amplifier, integrator, or amplifier with operating characteristics as a function of the external feedback components.

- No Frequency Compensation Required
- Short-Circuit Protection
- Offset Voltage Null Capability
- Wide Common-Mode and Differential Voltage Ranges
- Low-Power Consumption
- No Latch Up
- Low Noise Selections Offered — N Suffix

OPERATIONAL AMPLIFIER
SILICON MONOLITHIC
INTEGRATED CIRCUIT

G SUFFIX
METAL PACKAGE
CASE 601

P1 SUFFIX
PLASTIC PACKAGE
CASE 626
(MC1741C, MC1741NC)

U SUFFIX
CERAMIC PACKAGE
CASE 693

L SUFFIX
CERAMIC PACKAGE
CASE 632
TO-116

P2 SUFFIX
PLASTIC PACKAGE
CASE 646
(MC1741C, MC1741NC)

F SUFFIX
CERAMIC PACKAGE
CASE 606-04
TO-91

MAXIMUM RATINGS (T$_A$ = +25°C unless otherwise noted)

Rating	Symbol	MC1741C	MC1741	Unit
Power Supply Voltage	V$_{CC}$	+18	+22	Vdc
	V$_{EE}$	-18	-22	Vdc
Input Differential Voltage	V$_{ID}$	±30		Volts
Input Common Mode Voltage (Note 1)	V$_{ICM}$	±15		Volts
Output Short Circuit Duration (Note 2)	t$_S$	Continuous		
Operating Ambient Temperature Range	T$_A$	0 to +70	-55 to +125	^{o}C
Storage Temperature Range	T$_{stg}$			^{o}C
Metal, Flat and Ceramic Packages		-65 to +150		
Plastic Packages		-55 to +125		

Note 1.   For supply voltages less than + 15 V, the absolute maximum input voltage is equal to the supply voltage.

Note 2.   Supply voltage equal to or less than 15 V.

EQUIVALENT CIRCUIT SCHEMATIC

DS 9123 R3

ELECTRICAL CHARACTERISTICS (V_{CC} = 15 V, V_{EE} = 15 V, T_A = 25°C unless otherwise noted).

Characteristic	Symbol	MC1741 Min	MC1741 Typ	MC1741 Max	MC1741C Min	MC1741C Typ	MC1741C Max	Unit
Input Offset Voltage ($R_S \leqslant$ 10 k)	V_{IO}	—	1.0	5.0	—	2.0	6.0	mV
Input Offset Current	I_{IO}	—	20	200	—	20	200	nA
Input Bias Current	I_{IB}	—	80	500	—	80	500	nA
Input Resistance	r_i	0.3	2.0	—	0.3	2.0	—	MΩ
Input Capacitance	C_i	—	1.4	—	—	1.4	—	pF
Offset Voltage Adjustment Range	V_{IOR}	—	±15	—	—	±15	—	mV
Common Mode Input Voltage Range	V_{ICR}	±12	±13	—	±12	±13	—	V
Large Signal Voltage Gain (V_O = ±10 V, $R_L \geqslant$ 2.0 k)	A_v	50	200	—	20	200	—	V/mV
Output Resistance	r_o	—	75	—	—	75	—	Ω
Common Mode Rejection Ratio ($R_S \leqslant$ 10 k)	CMRR	70	90	—	70	90	—	dB
Supply Voltage Rejection Ratio ($R_S \leqslant$ 10 k)	PSRR	—	30	150	—	30	150	μV/V
Output Voltage Swing	V_O							V
($R_L \geqslant$ 10 k)		±12	±14	—	±12	±14	—	
($R_L \geqslant$ 2 k)		±10	±13	—	±10	±13	—	
Output Short-Circuit Current	I_{os}	—	20	—	—	20	—	mA
Supply Current	I_D	—	1.7	2.8	—	1.7	2.8	mA
Power Consumption	P_C	—	50	85	—	50	85	mW
Transient Response (Unity Gain — Non-Inverting)								
(V_I = 20 mV, $R_L \geqslant$ 2 k, $C_L \leqslant$ 100 pF) Rise Time	t_{TLH}	—	0.3	—	—	0.3	—	μs
(V_I = 20 mV, $R_L \geqslant$ 2 k, $C_L \leqslant$ 100 pF) Overshoot	os	—	15	—	—	15	—	%
(V_I = 10 V, $R_L \geqslant$ 2 k, $C_L \leqslant$ 100 pF) Slew Rate	SR	—	0.5	—	—	0.5	—	V/μs

ELECTRICAL CHARACTERISTICS (V_{CC} = 15 V, V_{EE} = 15 V, T_A = *T_{high} to T_{low} unless otherwise noted.)

Characteristic	Symbol	MC1741 Min	MC1741 Typ	MC1741 Max	MC1741C Min	MC1741C Typ	MC1741C Max	Unit
Input Offset Voltage ($R_S \leqslant$ 10 kΩ)	V_{IO}	—	1.0	6.0	—	—	7.5	mV
Input Offset Current	I_{IO}							nA
(T_A = 125°C)		—	7.0	200	—	—	—	
(T_A = -55°C)		—	85	500	—	—	—	
(T_A = 0°C to +70°C)		—	—	—	—	—	300	
Input Bias Current	I_{IB}							nA
(T_A = 125°C)		—	30	500	—	—	—	
(T_A = -55°C)		—	300	1500	—	—	—	
(T_A = 0°C to +70°C)		—	—	—	—	—	800	
Common Mode Input Voltage Range	V_{ICR}	±12	±13	—	—	—	—	V
Common Mode Rejection Ratio ($R_S \leqslant$ 10 k)	CMRR	70	90	—	—	—	—	dB
Supply Voltage Rejection Ratio ($R_S \leqslant$ 10 k)	PSRR	—	30	150	—	—	—	μV/V
Output Voltage Swing	V_O							V
($R_L \geqslant$ 10 k)		±12	±14	—	—	—	—	
($R_L \geqslant$ 2 k)		±10	±13	—	±10	±13	—	
Large Signal Voltage Gain ($R_L \geqslant$ 2 k, V_{out} = ±10 V)	A_v	25	—	—	15	—	—	V/mV
Supply Currents	I_D							mA
(T_A = 125°C)		—	1.5	2.5	—	—	—	
(T_A = -55°C)		—	2.0	3.3	—	—	—	
Power Consumption (T_A = +125°C)	P_C	—	45	75	—	—	—	mW
(T_A = -55°C)		—	60	100	—	—	—	

*T_{high} = 125°C for MC1741 and 70°C for MC1741C
T_{low} = -55°C for MC1741 and 0°C for MC1741C

MOTOROLA *Semiconductor Products Inc.*

Answers to Self-Tests

Chapter 1

1.	(c)	2.	(c)	3.	(c)	4.	(d)	5.	(b)
6.	(b)	7.	(a)	8.	(d)	9.	(d)	10.	(d)
11.	(c)	12.	(c)						

Chapter 2

1.	(b)	2.	(a)	3.	(c)	4.	(c)	5.	(b)
6.	(b)	7.	(c)	8.	(d)	9.	(b)	10.	(b)
11.	(e)	12.	(b)	13.	(b)	14.	(c)		

Chapter 3

1.	(b)	2.	(c)	3.	(b)	4.	(d)	5.	(a)
6.	(d)	7.	(b)	8.	(c)	9.	(d)	10.	(a)
11.	(a)	12.	(d)	13.	(c)	14.	(a)	15.	(c)

Chapter 4

1.	(a)	2.	(d)	3.	(b)	4.	(d)	5.	(d)
6.	(a)	7.	(b)	8.	(c)	9.	(b)	10.	(c)
11.	(a)	12.	(d)	13.	(d)	14.	(d)		

Chapter 5

1.	(b)	2.	(c)	3.	(b)	4.	(d)	5.	(a)
6.	(c)	7.	(c)	8.	(c)	9.	(d)	10.	(b)
11.	(a)	12.	(d)	13.	(c)	14.	(b)		

Chapter 6

1.	(e)	2.	(c)	3.	(c)	4.	(c)	5.	(b)
6.	(a)	7.	(b)	8.	(b)	9.	(d)	10.	(b)
11.	(b)	12.	(c)	13.	(d)	14.	(a)	15.	(d)

Chapter 7

1.	(b)	2.	(c)	3.	(a)	4.	(d)	5.	(b)
6.	(c)	7.	(a)	8.	(d)	9.	(c)	10.	(c)
11.	(b)	12.	(c)						

Chapter 8

1.	(b)	2.	(b)	3.	(c)	4.	(b)	5.	(d)
6.	(a)	7.	(a)	8.	(a)	9.	(c)	10.	(b)
11.	(a)	12.	(d)	13.	(c)	14.	(b)	15.	(d)

Chapter 9

1.	(g)	2.	(b)	3.	(c)	4.	(d)	5.	(a)
6.	(d)	7.	(a)	8.	(f)	9.	(c)	10.	(b)
11.	(d)	12.	(a)	13.	(b)	14.	(a)	15.	(b)
16.	(d)								

Chapter 10

1.	(c)	2.	(d)	3.	(c)	4.	(b)	5.	(d)
6.	(a)	7.	(d)	8.	(c)	9.	(b)	10.	(a)
11.	(d)	12.	(b)						

Chapter 11

1.	(b)	2.	(c)	3.	(d)	4.	(a)	5.	(b)
6.	(c)	7.	(d)	8.	(b)	9.	(a)	10.	(c)
11.	(d)	12.	(c)	13.	(a)	14.	(c)		

Chapter 12

1.	(c)	2.	(b)	3.	(b)	4.	(a)	5.	(d)
6.	(b)	7.	(a)	8.	(d)	9.	(c)	10.	(b)
11.	(d)	12.	(d)	13.	(c)	14.	(b)	15.	(d)

Chapter 13

1.	(f)	2.	(b)	3.	(a)	4.	(d)	5.	(a)
6.	(d)	7.	(b)	8.	(b)	9.	(a)	10.	(d)
11.	(c)	12.	(d)	13.	(b)	14.	(c)	15.	(a)

Chapter 14

1.	(a)	2.	(c)	3.	(b)	4.	(c)	5.	(d)
6.	(c)	7.	(a)	8.	(b)	9.	(d)	10.	(a)
11.	(b)	12.	(c)						

Chapter 15

| 1. | (b) | 2. | (c) | 3. | (a) | 4. | (d) | 5. | (a) |
| 6. | (a) | 7. | (a) | 8. | (c) | 9. | (b) | 10. | (d) |

Chapter 16

1.	(b)	2.	(c)	3.	(d)	4.	(c)	5.	(b)
6.	(a)	7.	(b)	8.	(d)	9.	(c)	10.	(a)
11.	(d)	12.	(c)	13.	(a)	14.	(d)	15.	(f)
16.	(a)	17.	(d)	18.	(b)				

Chapter 17

1.	(c)	2.	(b)	3.	(a)	4.	(a)	5.	(c)
6.	(d)	7.	(b)	8.	(d)	9.	(a)	10.	(d)
11.	(b)	12.	(a)	13.	(c)	14.	(a)	15.	(c)
16.	(d)	17.	(b)	18.	(a)				

Chapter 18

1.	(b)	2.	(a)	3.	(c)	4.	(d)	5.	(b)
6.	(a)	7.	(c)	8.	(d)	9.	(b)	10.	(a)
11.	(d)	12.	(c)	13.	(a)	14.	(c)		

Chapter 19

1.	(c)	2.	(b)	3.	(a)	4.	(d)	5.	(b)
6.	(a)	7.	(c)	8.	(d)	9.	(a)	10.	(c)
11.	(c)	12.	(b)	13.	(d)	14.	(a)	15.	(c)
16.	(b)								

Chapter 20

1.	(b)	2.	(d)	3.	(a)	4.	(b)	5.	(c)
7.	(d)	7.	(d)	8.	(a)	9.	(c)	10.	(a)
11.	(b)	12.	(d)						

Chapter 21

1.	(c)	2.	(a)	3.	(b)	4.	(d)	5.	(a)
6.	(b)	7.	(c)	8.	(b)	9.	(a)	10.	(d)
11.	(c)	12.	(c)						

Answers to Selected Odd-Numbered Problems

Chapter 1

1. **(a)** 3×10^3 **(b)** 75×10^3
   (c) 2×10^6.
3. **(a)** $8.4 \times 10^3 = 0.84 \times 10^4 = 0.084 \times 10^5$
   (b) $99 \times 10^3 = 9.9 \times 10^4 = 0.99 \times 10^5$
   (c) $200 \times 10^3 = 20 \times 10^4 = 2 \times 10^5$.
5. **(a)** 0.0000025   **(b)** 5000   **(c)** 0.39.
7. **(a)** 126×10^6 **(b)** 5.00085×10^3
   (c) 6060×10^{-9}.
9. **(a)** 20×10^8 **(b)** 36×10^{14}
   (c) 15.4×10^{-15}.
11. **(a)** 31 mA   **(b)** 5.5 kV   **(c)** 200 pF.
13. **(a)** 5×10^{-6} A   **(b)** 43×10^{-3} V
    (c) 275×10^3 Ω **(d)** 10×10^6 W.

Chapter 2

1. 80×10^{12} C.
3. **(a)** 10 V   **(b)** 2.5 V   **(c)** 4 V.
5. 20 V.
7. **(a)** 75 A   **(b)** 20 A   **(c)** 2.5 A.
9. 2 s.
11. **(a)** 6800 Ω ± 10%   **(b)** 33 Ω ± 10%
    (c) 47,000 Ω ± 20%.
13. **(a)** Red, violet, brown
    (b) 330 Ω: orange, orange, brown
    2.2 kΩ: red, red, red
    56 kΩ: green, blue, orange
    100 kΩ: brown, black, yellow
    39 kΩ: orange, white, orange.
15. Current flows through lamp 2.
17. Ammeter in series with resistors, with its negative terminal to the negative terminal of source and its positive terminal to one side of R_1. Voltmeter placed across (in parallel with) the source (negative to negative, positive to positive).
19. Position 1: V1 = 0, V2 = V_S
    Position 2: V1 = V_S, V2 = 0.
21. **(a)** 0.25 V   **(b)** 250 V.

23. **(a)** 200 Ω **(b)** 150 MΩ **(c)** 4500 Ω.
25. 33.33 V.
27. AWG #27.
29. (b).
31. One ammeter in series with battery. One ammeter in series with each resistor (seven total).
33. See Figure P–1.

FIGURE P–1

Chapter 3

1. **(a)** 3 A   **(b)** 0.2 A   **(c)** 1.5 A.
3. 15 mA.
5. **(a)** 3.33 mA   **(b)** 0.55 mA.
   (c) 0.588 mA   **(d)** 0.5 A   **(e)** 6.6 mA.
7. **(a)** 2.5 mA   **(b)** 2.27 μA
   (c) 8.33 mA.
9. **(a)** 10 mV   **(b)** 1.65 V   **(c)** 15 kV
   (d) 3.52 V   **(e)** 0.25 V   **(f)** 750 kV
   (g) 8.5 kV   **(h)** 3.75 mV.
11. **(a)** 81 V   **(b)** 500 V   **(c)** 125 V.
13. **(a)** 2 kΩ **(b)** 3.5 kΩ **(c)** 2 kΩ
    (d) 100 kΩ **(e)** 1 MΩ.
15. **(a)** 4 Ω **(b)** 3 kΩ **(c)** 200 kΩ.
17. 0.417 W.
19. **(a)** 1 MW   **(b)** 3 MW   **(c)** 150 MW
    (d) 8.7 MW.
21. **(a)** 2,000,000 μW   **(b)** 500 μW
    (c) 250 μW   **(d)** 6.67 μW.
23. 16.5 mW.
25. 1.175 kW.

27. 6 W.

29. 25 Ω.

31. 0.00186 kWh.

33. 1 W.

35. 150 Ω.

37. $V = 0$ V, $I = 0$ A; $V = 10$ V, $I = 0.1$ A;
$V = 20$ V, $I = 0.2$ A; $V = 30$ V, $I = 0.3$ A;
$V = 40$ V, $I = 0.4$ A; $V = 50$ V, $I = 0.5$ A;
$V = 60$ V, $I = 0.6$ A; $V = 70$ V, $I = 0.7$ A;
$V = 80$ V, $I = 0.8$ A; $V = 90$ V, $I = 0.9$ A;
$V = 100$ V, $I = 1$ A.
The graph is a straight line. There is a linear relationship between V and I.

39. $R_1 = 0.5$ Ω, $R_2 = 1$ Ω, $R_3 = 2$ Ω.

41. 10 V, 30 V.

43. $I_{MAX} = 3.83$ mA, $I_{MIN} = 3.46$ mA.

45. 216 kWh

47. 12 W.

49. 2 A.

27. Measure V_A and V_B with respect to ground; then
$V_{R2} = V_A - V_B$.

29. (a) R_4 open   (b) R_4 and R_5 shorted.

31. 780 Ω

33. $V_{AF} = 100$ V, $V_{BF} = 77.23$ V, $V_{CF} = 66.88$ V, $V_{DF} = 18.23$ V, $V_{EF} = 5.81$ V, $V_F = 0$ V.

35. 500 Ω.

37. (a) 19.09 mA   (b) 45.82 V
(c) $R(\frac{1}{8}$ W$) = 343$ Ω, $R(\frac{1}{4}$ W$) = 686$ Ω, $R(\frac{1}{2}$ W$) = 1372$ Ω.

39. See Figure P–3.

FIGURE P–3

Chapter 4

1. See Figure P–2.

FIGURE P–2

3. 0.1 A.

5. 138 Ω.

7. (a) 7.9 kΩ   (b) 33 Ω   (c) 13.24 MΩ
The series circuit is disconnected from the source, and the ohmmeter is connected across the circuit terminals.

9. 1126 Ω.

11. (a) 0.625 mA   (b) 4.26 μA
The ammeter is connected in series.

13. 0.355 A.

15. 14 V.

17. 26 V.

19. (a) $V_2 = 6.8$ V
(b) $V_R = 8$ V, $V_{2R} = 16$ V, $V_{3R} = 24$ V, $V_{4R} = 32$ V
The voltmeter is connected across (in parallel with) each resistor for which the voltage is unknown.

21. (a) 3.84 V   (b) 6.77 V.

23. $V_{5.6kΩ} = 10$ V, $V_{1kΩ} = 1.79$ V, $V_{560Ω} = 1$ V, $V_{10kΩ} = 17.9$ V.

25. 55 mW.

41. R_1, R_7, R_8, and R_{10}; R_2, R_4, R_6, and R_{11}; R_3, R_5, R_9, and R_{12}.

43. Yes, R_3 and R_5 are shorted.

45. (a) R_{11} has burned open due to excessive power dissipation.
(b) Replace R_{11} (10 kΩ)
(c) 337.5 V.

Chapter 5

1. See Figure P–4.

FIGURE P–4

3. 12 V, 5 mA.
5. 1350 mA.
7. $I_2 = I_3 = 7.5$ mA. An ammeter in series with each resistor in each branch.
9. (a) 359 Ω   (b) 25.6 Ω   (c) 819 Ω
   (d) 996 Ω.
11. 2 kΩ.
13. (a) 0.909 A   (b) 76 mA.
15. Circuit (a).

43. (a) 940 Ω
   (b) 518 Ω
   (c) 518 Ω
   (d) 422 Ω.

Chapter 6

1. R_2, R_3, and R_4 are in parallel, and this parallel combination is in series with both R_1 and R_5.
3. See Figure P–5.

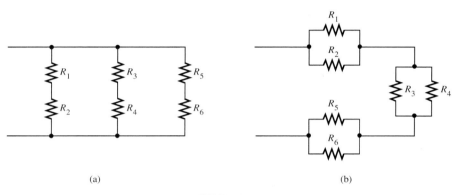

(a)   (b)

FIGURE P–5

17. $I_1 = 2.19$ A, $I_2 = 0.81$ A.
19. 200 mW.
21. 0.682 A, 4.092 A.
23. The 1 kΩ resistor is open.
25. $R_2 = 25$ Ω, $R_3 = 100$ Ω, $R_4 = 12.5$ Ω.
27. $I_R = 4.8$ A, $I_{2R} = 2.4$ A, $I_{3R} = 1.6$ A, $I_{4R} = 1.2$ A.
29. (a) $R_1 = 100$ Ω, $R_2 = 200$ Ω, $I_2 = 50$ mA
   (b) $I_1 = 125.09$ mA, $I_2 = 74.91$ mA,
      $R_1 = 79.9$ Ω, $R_2 = 133.5$ mA, $V_s = 10$ V
   (c) $I_1 = 0.253$ A, $I_2 = 0.147$ A, $I_3 = 0.1$ A,
      $R_1 = 395.26$ Ω.
31. 53.68 Ω
33. 1.084 A
35. R_1, R_2, R_5, R_9, R_{10}, and R_{12};
   R_4, R_6, R_7, and R_8;
   R_3 and R_{11}.
37. $R_2 = 750$ Ω, $R_4 = 423$ Ω
39. The 4.7-kΩ resistor is open.
41. (a) One of the resistors has opened due to excessive power dissipation.
   (b) 30 V
   (c) Replace the 1.8-kΩ resistor.

5. 2003 Ω.
7. (a) 128 Ω   (b) 790.5 Ω.
9. (a) $I_1 = I_4 = 11.7$ mA
      $I_2 = I_3 = 5.85$ mA
      $V_1 = 0.655$ V
      $V_2 = V_3 = 0.585$ V
      $V_4 = 0.257$ V
   (b) $I_1 = 3.8$ mA
      $I_2 = 0.618$ mA
      $I_3 = 1.27$ mA
      $I_4 = 1.91$ mA
      $V_1 = 2.58$ V
      $V_2 = V_3 = V_4 = 0.42$ V.
11. 7.5 V unloaded, 7.29 V loaded.
13. 56-kΩ load.
15. 1000 V.
17. (a) 0.5-V range   (b) Approximately 1 mV.
19. 360 Ω.
21. 20.99 mA.
23. $R_{TH} = 18$ kΩ, $V_{TH} = 2.7$ V.
25. No, the meter should read 4.39 V. The 680-Ω is open.

27. The 7.5-V and 5-V readings are incorrect, indicating that the 3-kΩ resistor is open.

29. See Figure P–6.

15. Material A.

17. 1 mA.

19. To electrically connect the loop to the external circuit.

FIGURE P–6

These resistors do not affect the circuit because they are shorted and can be deleted.

31. $R_T = 5.76$ kΩ, $V_A = 3.3$ V, $V_B = 1.7$ V, $V_C = 0.85$ V.

33. $V_1 = 1.61$ V, $V_2 = 6.7$ V, $V_3 = 1.77$ V, $V_4 = 3.4$ V, $V_5 = 0.387$ V, $V_6 = 2.63$ V, $V_7 = 0.387$ V, $V_8 = 1.77$ V, $V_9 = 1.61$ V.

35. 109.9 Ω.

37. $R_1 = 180$ Ω, $R_2 = 60$ Ω.

39. 0.844 mA.

41. 11.73 V.

43. See Figure P–7.

Chapter 8

1. (a) 1 Hz (b) 5 Hz (c) 20 Hz (d) 1000 Hz (e) 2 kHz (f) 100 kHz.

3. 2 μs.

5. (a) 8.48 V (b) 24 V (c) 7.64 V.

7. $V_p = 25$ V, $V_{pp} = 50$ V, $V_{rms} = 17.68$ V, $V_{avg} = 15.93$ V.

9. 120 Hz.

FIGURE P–7

45. One of the 12-kΩ resistors is open.

47. The 2.2-kΩ resistor is open.

Chapter 7

1. Decrease.

3. 37.5 μWb.

5. 596.8

7. 150 At.

9. (a) Electromagnetic (b) Spring.

11. Electromagnetic force.

13. Change the current.

11. 15°, A leads.

13. See Figure P–8.

FIGURE P–8

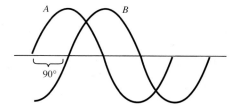

15. (a) 22.5° (b) 60° (c) 90°
  (d) 108° (e) 216° (f) 324°.
17. (a) 57.36 mA (b) 99.62 mA (c) −17.36 mA
  (d) −57.36 mA (e) −99.62 mA (f) 0 mA.
19. 30°: 12.99 V
  45°: 14.49 V
  90°: 12.99 V
  180°: −7.5 V
  200°: −11.49 V
  300°: −7.5 V.
21. See Figure P–9.

FIGURE P–9

23. (a) 7.07 mA (b) 6.37 mA
  (c) 10 mA (d) 20 mA
  (e) 10 mA.
25. 7.38 V.
27. 4.242 V.
29. $t_r \cong 3.5$ ms, $t_f \cong 3.5$ ms, $t_w \cong 12.5$ ms
31. (a) −0.375 V (b) 3.01 V.
33. (a) 50 kHz (b) 10 Hz.
35. 25 kHz.
37. 1.56 V, 735.3 kHz
39. 23.08 V.
41. 3 V
45. $V_p = 4.71$ V, $T = 2$ ms

Chapter 9
1. (a) 5 μF (b) 1 μC
  (c) 10 V.
3. (a) 0.001 μF (b) 0.0035 μF
  (c) 0.00025 μF.
5. 2 μF.
7. 88.5 pF.
9. A 12.5-pF increase.
11. Ceramic.
13. (a) 0.02 μF (b) 0.047 μF
  (c) 0.001 μF (d) 220 pF.
15. (a) Encapsulation
  (b) Dielectric (ceramic disk)
  (c) Plate (metal disk)
  (d) Leads.

17. (a) 0.67 μF (b) 68.97 pF (c) 2.7 μF.
19. $V_1 = 2$ V, $V_2 = 10$ V, $V_3 = 5$ V, $V_4 = 1$ V.
21. (a) 2.5 μF (b) 716.66 pF (c) 1.6 μF.
23. (a) 100 μs (b) 500 μs
  (c) 23.5 μs (d) 15 ms.
25. (a) 9.48 V (b) 12.97 V (c) 14.25 V
  (d) 14.73 V (e) 14.899 V.
27. (a) 2.72 V (b) 5.90 V (c) 11.65 V
29. (a) 338.6 kΩ (b) 13.55 kΩ
  (c) 677 Ω (d) 33.86 Ω.
31. (a) 31.83 Ω (b) 111 kΩ (c) 49.7 Ω.
33. 200 Ω.
35. $P_{\text{true}} = 0$ W, $P_r = 3.39$ mVAR.
37. 0 Ω.
39. Shorted.
41. 3.18 ms.
43. 2.94 μs.
45. (a) Charges to 3.2 V in 10 ms, then discharges to
  0 V in 215 ms.
  (b) Charges to 3.2 V in 10 ms, then discharges to
  2.85 V in 5 ms, then charges toward 20 V.
47. 0.0052 μF.
49. $V_1 = 7.25$ V, $V_2 = 2.76$ V, $V_3 = 0.79$ V,
  $V_4 = 1.98$ V, $V_5 = 1.19$ V, $V_6 = 0.78$ V.

Chapter 10
1. (a) 1000 mH (b) 0.25 mH
  (c) 0.01 mH (d) 0.5 mH.
3. 3536 turns.
5. 0.05 J.
7. 155 μH.
9. 7.14 μH.
11. (a) 4.33 H (b) 50 mH
  (c) 0.57 μH.
13. (a) 1 μs (b) 2.13 μs (c) 2 μs.
15. (a) 5.52 V (b) 2.03 V (c) 0.75 V
  (d) 0.27 V (e) 0.1 V.
17. (a) 136 kΩ (b) 1.57 kΩ
  (c) 0.0179 Ω.
19. $I_T = 10.05$ A, $I_{L2} = 6.7$ A, $I_{L3} = 3.35$ A.
21. 100.5 VAR.
23. (a) Infinite resistance
  (b) Zero resistance
  (c) Lower R_W.
25. 2.86 mA.
27. 26.1 mA.

Chapter 11

1. 3

3. (a) Positive at top
(b) Positive at bottom
(c) Positive at top
(d) Positive at bottom.

5. 50 turns.

7. (a) Same polarity, 100 V rms
(b) Opposite polarity, 100 V rms.

9. 240 V.

11. 33.33 mA.

13. 27.3 Ω.

15. 5.75 mA.

17. 0.5.

19. 38.93 W.

21. No.

23. 94.5 W.

25. 0.98

27. 25 kVA.

29. Section 1: 2
Section 2: 0.5
Section 3: 0.25.

31. (a) 48 V   (b) 25 V.

33. Excessive primary current is drawn, potentially burning out the source and/or the transformer, unless the primary is protected by a fuse.

35. (a) V_{L1} = 35 V, I_{L1} = 2.92 A, V_{L2} = 15 V, I_{L2} = 1.5 A   (b) 29.2 Ω.

37. (a) 20 V   (b) 10 V

39. 0.0145.

Chapter 12

1. 8 kHz, 8 kHz.

3. (a) 269.3 Ω.   (b) 1166 Ω.

5. (a) 37.1 mA   (b) 4.29 mA.

7. I_T = 12 mA, $V_{C(0.1)}$ = 1.27 V, $V_{C(0.2)}$ = 0.637 V, V_R = 0.6 V, θ = 72.48° (V_s lagging I_T).

9. 807.8 Ω, −36.12°

11. 353.4 Ω, 61.99°

13. 261 Ω, −79.83°

15. I_{C1} = 118.2 mA, I_{C2} = 50.3 mA, I_{R1} = 36.4 mA, I_{R2} = 44.4 mA, I_T = 186.9 mA, θ = 64.4° (V_s lagging I_T).

17. (a) 3.9 kΩ (b) 20 μA   (c) 15.7 μA
(d) 25.6 μA   (e) 38.1° (V_s lagging I_T).

19. V_{C1} = 8.77 V, V_{C2} = 3.24 V, V_{C3} = 3.24 V
V_{R1} = 2.1 V, V_{R2} = 1.14 V.

21. I_T = 82.7 mA, I_{C2} = 15.3 mA, I_{C3} = 67.5 mA, I_{R1} = I_{R2} = 6.35 mA.

23. 4.03 VA.

25. 0.91

27. (a) 0.06°   (b) 5.74°   (c) 45.1°
(d) 84.32°.

29. (a) 89.96°   (b) 86.4°   (c) 57.86°
(d) 9.04°.

31. See Figure P–10.

(a) For Figure 12-73

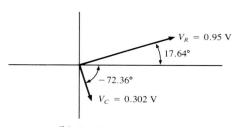

(b) For Figure 12-74

FIGURE P–10

33. Figure 12–73: 995 Hz
Figure 12–74: 1.59 kHz.

35. No leakage: V_{out} = 3.21 V, θ = −18.71°
Leakage: V_{out} = 2.83 V, θ = −33.34°.

37. (a) 0 V   (b) 0.3 V   (c) 0.5 V   (d) 0 V.

39. (a) I_{LA} = 4.4 A, I_{LB} = 3.06 A
(b) P_{rA} = 509.2 VAR, P_{rB} = 210.5 VAR
(c) $P_{true(A)}$ = 822.8 W, $P_{true(B)}$ = 640.5 W
(d) P_{aA} = 968 VA, P_{aB} = 674.2 VA.

41. 11.43 kΩ.

43. P_r = 1.32 kVAR, P_a = 2 kVA.

45. 0.1 μF

Chapter 13

1. 15 kHz.

3. (a) 111.8 Ω (b) 1.8 kΩ.

5. (a) 17.38 Ω (b) 63.97 Ω
(c) 126.23 Ω (d) 251.61 Ω.

7. (a) 89.4 mA   (b) 2.78 mA.

9. 37°.

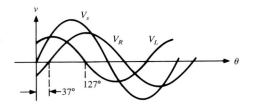

FIGURE P–11

11. See Figure P–11.
13. 7.69 Ω.
15. 2387 Hz.
17. **(a)** 274 Ω   **(b)** 89.3 mA
    (c) 159 mA   **(d)** 182.5 mA
    (e) 60.7° (I_T lagging V_s).
19. V_{R1} = 7.92 V, V_{R2} = V_L = 20.84 V.
21. I_T = 36 mA, I_L = 33.2 mA, I_{R2} = 13.9 mA.
23. 1.28 W, 0.96 VAR.
25. PF = 0.386, P_{true} = 0.347 W, P_r = 0.692 VAR, P_a = 0.9 VA.
27. See Figure P–12.
29. See Figure P–13.

FIGURE P–12

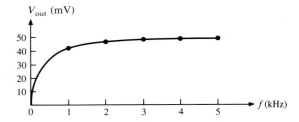

FIGURE P–13

31. V_{R1} = V_{L1} = 18 V; V_{R2} = V_{R3} = V_{L2} = 0 V.
33. 5.83 V.
35. 92.9 mA.
37. **(a)** 394 mA   **(b)** 221.6 mA
    (c) 333.33 mA   **(d)** 325.8 mA.

41. See Figure P–14.

FIGURE P–14

Chapter 14

1. 479.6 Ω, 88.8° (V_s lagging I), 479.5 Ω capacitive.
3. Impedance increases.
5. See Figure P–15.

FIGURE P–15

7. X_L = 4.472 kΩ, X_C = 4.472 kΩ, Z = 20 Ω, I = 0.6 A.
9. f_r = 459.44 kHz, f_{C1} = 359.995 kHz, f_{C2} = 558.885 kHz.
11. **(a)** 14.53 kHz   **(b)** 25.16 kHz.
13. **(a)** f_r = 339 kHz, BW = 238.732 kHz
    (b) f_r = 10.4 kHz, BW = 2613 Hz.
15. Capacitive, $X_C < X_L$.
17. 757.6 Ω.
19. 532 MΩ, 103.82 kHz.
21. 62.5 Hz.
23. 1.375 W.
25. 200 Hz.
27. I_{R1} = I_C = 2.11 mA,
I_{L1} = 1.33 mA,
I_{L2} = 0.667 mA,
I_{R2} = 0.667 mA,
V_{R1} = 6.96 V,
V_C = 2.11 V,
V_{L1} = V_{L2} = V_{R2} = 6.67 V.
29. I_{R1} = I_{L1} = 41.5 mA,
I_C = I_{L2} = 105.3 mA,
I_T = 77.24 mA.

31. $L = 989\ \mu H,\ C = 0.064\ \mu F.$
33. 8 MHz: $C = 40$ pF
9 MHz: $C = 31$ pF
10 MHz: $C = 25$ pF
11 MHz: $C = 21$ pF.

Chapter 15
1. 100 $\mu s.$
3. 12.6 V.
5. See Figure P–16.

FIGURE P–16

7. See Figure P–17.

FIGURE P–17

9. 15-V dc level with a very small charge/discharge fluctuation.
11. Exchange positions of R and C. See Figure P–18.
5 ms.

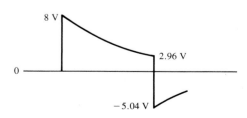

FIGURE P–18

13. Approximately the same shape as input, but with an average value of 0 V.
15. See Figure P–19.

FIGURE P–19

17. See Figure P–20.

FIGURE P–20

19. A 6-V dc level.
21. **(b)** No fault   **(c)** Leaky capacitor   **(d)** Open capacitor or shorted resistor.
23. **(a)** 25 ms   **(b)** See Figure P–21.
25. 1.44 s.

FIGURE P–21

Chapter 16
No problem set.

Chapter 17
1. 63.66 V.
3. Yes.
5. 47.75 V.
7. 172.79 V.
9. 78.54 V.
11. See Figure P–22.

FIGURE P–22

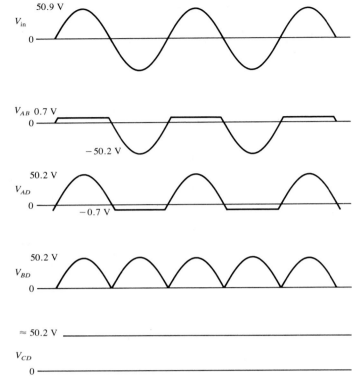

13. **(a)** Good   **(b)** Bad, open diode
    (c) Good   **(d)** Bad, open diode.
15. **(a)** A sine wave with a positive peak at $+0.7$ V, a negative peak at -7.3 V, and a dc value of -3.3 V.
    (b) A sine wave with a positive peak at $+29.3$ V, a negative peak at -0.7 V, and a dc value of $+14.3$ V.
17. 15.02 V.
19. $I_{L(min)} = 0$ A, $I_{L(max)} = 20.9$ mA.
21. 10.7%, 8.7%.
23. A decrease of 10 pF.
25. 25.34 pF each.
27. Increase.

Chapter 18
1. 4.865 mA.
3. 125.
5. $I_B = 26$ μA, $I_E = 1.3$ mA, $I_C = 1.274$ mA.
7. $I_B = 13.64$ μA, $I_C = 682$ μA, $V_C = 9.32$ V.
9. $V_{CE} = 3.45$ V. Q point: $I_C = 15$ mA, $V_{CE} = 3.45$ V.
11. 33.33
13. 0.5 mA, 3.33 μA, 4.03 V.
15. 0.0031 mA.
17. **(a)** Good   **(b)** Good
    (c) Bad   **(d)** Bad.

19. **(a)** Narrows   **(b)** Increases.
21. 5 V.
23. 4 V
25. 4 V.
27. The gate is insulated from the channel by an SiO_2 layer.
29. 3 V.
31. **(a)** Depletion   **(b)** Enhancement
    (c) Zero bias   **(d)** Depletion.
33. 0.6
35. 1 ms.
37. 2.4 kΩ.

Chapter 19
1. 204.1
3. **(a)** 3.25 V   **(b)** 2.55 V
   (c) 2.55 mA   **(d)** $\cong 2.55$ mA
   (e) 9.59 V   **(f)** 7.04 V.
5. $A_{v(max)} = 92.3$, $A_{v(min)} = 2.91$
7. 0.978
9. A_v is reduced slightly.
11. $R_{in} = 2.36$ Ω, $A_v = 508.5$, $A_i \cong 1$, $A_p = 508.5$
13. **(a)** 0.938   **(b)** 0.301.
15. 30 dB, 31.62.

17. **(a)** 7.71 mW   **(b)** 53.56 mW.
19. 10 V, 625 mA.
21. 50.33 kHz.
23. See Figure P–23.
25. **(a)** 527.9 kHz   **(b)** 758.7 kHz.

FIGURE P–23

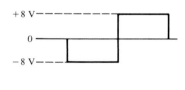

(a) Q_1 open

21. **(a)** $Z_{in(I)} = 10 \text{ k}\Omega$
           $Z_{out(I)} = 40 \text{ }\Omega$
       (b) $Z_{in(I)} = 100 \text{ k}\Omega$
           $Z_{out(I)} = 50 \text{ }\Omega$
       (c) $Z_{in(I)} = 500 \text{ }\Omega$
           $Z_{out(I)} = 70 \text{ }\Omega$

(b) Q_2 open

Chapter 20

1. *Practical op-amp:* High open-loop gain, high input impedance, low output impedance, large bandwidth, high CMRR.
   Ideal op-amp: Infinite open-loop gain, infinite input impedance, zero output impedance, infinite bandwidth, infinite CMRR.

(a)

3. **(a)** Single-ended input, differential output
   (b) Single-ended input, single-ended output
   (c) Differential input, single-ended output
   (d) Differential input, differential output.
5. 8.1 μA.
7. 107.96 dB.
9. 0.3.
11. 40 μs.
13. **(a)** 334.45   **(b)** 3.35 V   **(c)** 10 mV
15. **(a)** 49 kΩ **(b)** 3 MΩ **(c)** 84 kΩ
   (d) 165 kΩ
17. **(a)** 10 mV, in-phase
   (b) -10 mV, 180° out-of-phase
   (c) 210 mV, in-phase
   (d) -100 mV, 180° out-of-phase
19. **(a)** $Z_{in(NI)} = 8725 \text{ M}\Omega$
       $Z_{out(NI)} = 0.086 \text{ }\Omega$
   (b) $Z_{in(NI)} = 6201 \text{ M}\Omega$
       $Z_{out(NI)} = 0.004 \text{ }\Omega$
   (c) $Z_{in(NI)} = 4762 \text{ M}\Omega$
       $Z_{out(NI)} = 0.021 \text{ }\Omega$

Chapter 21

1. **(a)** Maximum negative
   (b) Maximum positive
   (c) Maximum negative.
3. See Figure P–24.

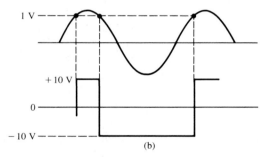

FIGURE P–24

5. **(a)** $V_{R1} = 1$ V, $V_{R2} = 1.8$ V
   (b) 127.27 μA   **(c)** -2.8 V.
7. -3.66 V, 366 μA.
9. -5 mV/μs.
11. 3 V peak-to-peak, 3.33 kHz.
13. 6.38 V.
15. When dc power is first applied, both zener diodes appear as opens. As a result, R_3 is placed in series with R_1, thus increasing the closed-loop gain to a value greater than unity to assure that oscillation will begin.
17. 884.2 Hz.
19. **(a)** 1.59 kHz   **(b)** 6772.6 Hz
   (c) 937.83 Hz.
21. 7 V.
23. 9.45 V.
25. 500 mA.
27. 10 mA.

Glossary

Active filter A frequency-selective circuit consisting of active devices such as transistors or op-amps.

Admittance A measure of the ability of a reactive circuit to permit current. The reciprocal of impedance. The unit is siemens (S).

Alpha (α) The ratio of collector current to emitter in a bipolar junction transistor.

Alternating current (ac) Current that reverses direction in response to a change in source voltage polarity.

Ammeter An electrical instrument used to measure current.

Ampere (A or amp) The unit of electrical current.

Ampere-turn The unit of magnetomotive force (mmf).

Amplification The process of producing a larger voltage, current, or power using a smaller input signal as a pattern.

Amplifier An electronic circuit having the capability of amplification and designed specifically for that purpose.

Anode The more positive terminal of a diode or other electronic device.

Apparent power The product of the voltage times the current, expressed in volt-amperes (VA). In a purely resistive circuit it is the same as the true power.

Atom The smallest particle of an element possessing the unique characteristics of that element.

Atomic number The number of electrons in a neutral atom.

Atomic weight The number of protons and neutrons in the nucleus of an atom.

Attenuation A gain less than unity.

Autotransformer A transformer in which the primary and secondary are in a single winding.

Average value The average of a sine wave over one half-cycle. It is 0.637 times the peak value.

AWG American wire gage. A standardization based on wire diameter.

Band-pass filter A filter that passes frequencies within a certain range and blocks all others.

Band-stop filter A filter that blocks frequencies within a certain range.

Bandwidth *(BW)* The characteristic of certain electronic circuits that specifies the usable range of frequencies that pass from input to output.

Barrier potential The inherent voltage across the depletion layer of a PN junction.

Base One of the semiconductor regions in a bipolar junction transistor.

Baseline The normal level of a pulse waveform. The voltage level in the absence of a pulse.

Battery An energy source that uses a chemical reaction to convert chemical energy into electrical energy.

Beta (β) The ratio of collector current to base current in a bipolar junction transistor.

Bias The application of dc voltage to a diode or other electronic device to produce a desired mode of operation.

Bipolar Characterized by two PN junctions.

BJT Bipolar junction transistor.

Bleeder current. The current left after the total load current is subtracted from the total current into the circuit.

Branch One current path in a parallel circuit.

Capacitance The ability of a capacitor to store electrical charge.

Capacitive reactance The opposition of a capacitor to sinusoidal current. The unit is the ohm.

Capacitive susceptance The ability of a capacitor to permit current.

Capacitor An electrical device consisting of two conductive plates separated by an insulating material and possessing the property of capacitance.

Cathode The more negative terminal of a diode or other electronic device.

Center tap A connection at the midpoint of the secondary of a transformer.

Charge An electrical property of matter that exists because of an excess or a deficiency of electrons. Charge can be either positive or negative.

Choke An inductor. The term is used more commonly in connection with inductors used to block or choke off high frequencies.

Circuit An interconnection of electrical components designed to produce a desired result. A basic circuit consists of a source, a load, and an interconnecting current path.

Circuit breaker A resettable protective device used for interrupting excessive current in an electric circuit.

Circular mil (CM) The unit of the cross-sectional area of a wire.

Clamper A circuit that adds a dc level to an ac signal. A dc restorer.

Class A A category of amplifier circuit that conducts for the entire input cycle and produces an output signal that is a replica of the input signal in terms of its waveshape.

Class B A category of amplifier circuit that conducts for half of the input cycle.

Class C A category of amplifier circuit that conducts for a very small portion of the input cycle.

Clipper A circuit that removes part of a waveform above or below a specified level.

Closed circuit A circuit with a complete current path.

Closed loop An op-amp configuration in which the output is connected back to the input.

Closed loop gain, A_{cl} The overall voltage gain of an op-amp with feedback.

CMRR Common-mode rejection ratio. A measure of an op-amp's ability to reject signals that appear the same on both inputs. The ratio of open-loop gain to common-mode gain.

Coefficient of coupling A constant associated with transformers that is the ratio of secondary flux to primary flux. The ideal value of 1 indicates that all the flux in the primary is coupled into the secondary.

Coil A common term for an inductor.

Collector One of the semiconductor regions in a BJT.

Common base A BJT amplifier configuration in which the base is the common (grounded) terminal.

Common-collector A BJT amplifier configuration in which the collector is the common (grounded) terminal.

Common-emitter A BJT amplifier configuration in which the emitter is the common (grounded) terminal.

Common-mode signals   Signals that appear the same on both inputs of an op-amp.

Conductance   The ability of a circuit to allow current. The unit is the siemen (S).

Conductor   A material in which electrical current can flow with relative ease. An example is copper.

Core   The structure around which the winding of an inductor is formed. The core material influences the electromagnetic characteristics of the inductor.

Coulomb (C)   The unit of electrical charge.

Coulomb's law   A physical law that states a force exists between two charged bodies that is directly proportional to the product of the two charges and inversely proportional to the square of the distance between them.

Covalent   Related to the bonding of two or more atoms by the interaction of their valence electrons.

CRT   Cathode-ray tube.

Crystal   The pattern or arrangement of atoms forming a solid material.

Current   The rate of flow of electrons.

Current gain   The ratio of output current to input current.

Cutoff   The nonconducting state of a transistor.

Cutoff frequency   The frequency at which the output voltage of a filter is 70.7% of the maximum output voltage.

Cycle   One repetition of a periodic waveform.

Darlington   A two-transistor arrangement that produces a multiplication of current gain.

DC component   The average value of a pulse waveform.

Decade   A tenfold change in the value of a quantity. When a quantity becomes ten times less or ten times greater, it has changed a decade.

Decibel (dB)   The unit of logarithmic expression of a ratio, such as a power ratio or a voltage ratio.

Degree   The unit of angular measure corresponding to 1/360 of a complete revolution.

Derivative   The instantaneous rate of change of a function determined mathematically.

Diac   A semiconductor device that can conduct current in either of two directions when properly activated.

Dielectric   The insulating material between the plates of a capacitor.

Dielectric constant   A measure of the ability of a dielectric material to establish an electric field.

Dielectric strength   A measure of the ability of a dielectric material to withstand voltage.

Differential amplifier   An amplifier that produces an output proportional to the difference of two inputs.

Differentiator   A circuit producing an output that approaches the mathematical derivative of the input.

Diode   An electronic device that permits current in only one direction.

Discrete device   An individual electrical or electronic component that must be used in combination with other components to form a complete functional circuit.

Doping   The process of imparting impurities to an intrinsic semiconductor material in order to control its conduction characteristics.

Drain   One of the three terminals of a field-effect transistor.

Duty cycle   A characteristic of a pulse waveform that indicates the percentage of time that a pulse is present during a cycle. The ratio of pulse width to period.

Effective value   A measure of the heating effect of a sine wave. Also known as the rms (root mean square) value.

Electrical   Related to the use of electrical voltage and current to achieve desired results.

Electromagnetic field   A formation of a group of magnetic lines of force surrounding a conductor created by electrical current in the conductor.

Electromagnetic induction   The phenomenon or process by which a voltage is produced in a conductor when there is relative motion between the conductor and a magnetic or electromagnetic field.

Electron   The basic particle of negative electrical charge in matter.

Electronic   Related to the movement and control of free electrons in semiconductors or vacuum devices.

Element   One of the unique substances that make up the known universe. Each element is characterized by a unique atomic structure.

Emitter   One of the three semiconductor regions in a BJT.

Emitter-follower   A popular term for a common-collector amplifier.

Energy   The ability to do work.

Falling edge   The negative-going transition of a pulse.

Fall time   The time interval required for a pulse to change from 90% to 10% of its full amplitude.

Farad (F)   The unit of capacitance.

Faraday's law   A law stating that the voltage induced across a coil of wire equals the number of turns in the coil times the rate of change of the magnetic flux.

Feedback   The process of returning a portion of a circuit's output signal to the input in such a way as to create certain specified operating conditions.

Field-effect transistor (FET)   A type of transistor that uses an induced electric field to control current.

Filter   A type of electrical circuit that passes certain frequencies and rejects all others.

Flux density   The number of lines of force per unit area in a magnetic field.

Free electron   A valence electron that has broken away from its parent atom and is free to move from atom to atom within the atomic structure of a material.

Frequency   A measure of the rate of change of a periodic function. The number of cycles completed in 1 s. The unit of frequency is the hertz.

Full-wave rectifier   A circuit that converts an alternating sine wave into a pulsating dc consisting of both halves of a sine wave for each input cycle.

Fuse   A protective device that burns open when excessive current flows in a circuit.

Gain   The amount by which an electrical signal is increased or decreased. The ratio of output to input. The amount of amplification.

Gate   One of the three terminals of an FET.

Generator   An energy source that produces electrical signals.

Germanium   A semiconductor material.

Ground   In electric circuits, the common or reference point.

Half-power frequencies   The frequencies at which the output of a filter is 70.7% of maximum.

Half-wave rectifier   A circuit that converts an alternating sine wave into a pulsating dc consisting of one-half of a sine wave for each input cycle.

Harmonics   The frequencies contained in a composite waveform, which are integer multiples of the repetition frequency (fundamental).

Henry (H)   The unit of inductance.

Hertz (Hz)   The unit of frequency. One hertz equals one cycle per second.

High-pass   The characteristic of a certain type of filter whereby higher frequencies are passed and lower frequencies are rejected.

Hole   The absence of an electron.

Hypotenuse   The longest side of a right triangle.

Hysteresis   A lagging in the values of resulting magnetization in a magnetic material due to a changing magnetizing force.

Impedance   The total opposition to sinusoidal current expressed in ohms.

Induced voltage   Voltage produced as a result of a changing magnetic field.

Inductance   The property of an inductor that produces an opposition to any *change* in current.

Inductive reactance   The opposition of an inductor to sinusoidal current. The unit is ohms.

Inductive susceptance   The reciprocal of inductive reactance. The unit is siemens.

Inductor   An electrical device formed by a wire wound around a core having the property of inductance and the capability to store energy in its electromagnetic field. Also known as a coil or a choke.

Input   The voltage, current, or power applied to an electrical circuit to produce a desired result.

Instantaneous value   The voltage or current value of a waveform at a given instant in time.

Insulator   A material that does not allow current under normal conditions.

Integral   The area under the curve of a function, determined mathematically.

Integrator   A circuit producing an output that approaches the mathematical derivative of the input.

Intrinsic semiconductor   A pure material with relatively few free electrons.

Ion   An atom that has gained or lost a valence electron resulting in a net positive or negative charge.

Ionization   The removal or addition of an electron from or to an atom so that the resulting atom (ion) has a net positive or negative.

Joule (J)   The unit of energy.

Junction   A point at which two or more branches are connected together.

Junction field-effect transistor (JFET)   A type of FET that operates with a reverse-biased junction to control current in a channel.

Kilowatthour (kWh)   A common unit of energy used mainly by utility companies.

Kirchhoff's current law   A law stating that the total current into a junction equals the total current out of the junction.

Kirchhoff's voltage law   A law stating that (1) the sum of the voltage drops around a closed loop equals the source voltage or (2) the sum of all the voltages (drops and sources) around a closed loop is zero.

Lag   A condition of the phase or time relationship of waveforms in which one waveform is behind the other in phase or time.

Lead   A condition of the phase or time relationship of waveforms in which one waveform is ahead of the other in phase or time. Also, a wire or cable connection to a device or instrument.

Leading edge   The first step or transition of a pulse.

Lenz's law A physical law that states when the current through a coil changes, an induced voltage is created in a direction to oppose the change in current. The current cannot change instantaneously.

Light-emitting diode (LED) A type of diode that emits light when forward current flows.

Linear Characterized by a straight-line relationship.

Line regulation The percent change in output voltage for a given change in line (input) voltage.

Load An element (resistor or other component) connected across the output terminals of a circuit that draws current from the circuit.

Load regulation The percent change in output voltage for a given change in load current.

Low-pass The characteristic of a certain type of filter whereby lower frequencies are passed and higher frequencies are rejected.

Magnetic field intensity The amount of mmf per unit length of magnetic material.

Magnetic flux The lines of force between the north and south poles of a permanent magnet or an electromagnet.

Magnitude The value of a quantity, such as the number of volts of voltage or the number of amperes of current.

Maximum power transfer theorem A theorem that states the maximum power is transferred from a source to a load when the load resistance equals the source resistance.

MOSFET Metal-oxide semiconductor field-effect transistor.

Mutual inductance The inductance between two separate coils, such as a transformer.

Negative feedback The return of a portion of the output signal to the input such that it is out of phase with the input signal.

Neutron An atomic particle having no electrical charge.

Ohm (Ω) The unit of resistance.

Ohmmeter An instrument for measuring resistance.

Ohm's law A law stating that current is directly proportional to voltage and inversely proportional to resistance.

Open circuit A circuit in which the current path is broken.

Open-loop gain, A_{ol} The internal voltage gain of an op-amp without feedback.

Operational amplifier (op-amp) A special type of amplifier exhibiting very high open-loop gain, very high input impedance, very low output impedance, and good rejection of common-mode signals.

Oscillator An electronic circuit consisting of an amplifier and a phase-shift network connected in a positive feedback loop that produces a time-varying output signal without an external input signal using positive feedback.

Oscilloscope A measurement instrument that displays signal waveforms on a screen.

Parallel The relationship in electric circuits in which two or more current paths are connected between the same two points.

Pass band The range of frequencies passed by a filter.

Peak value The voltage or current value of a waveform at its maximum positive or negative points.

Peak-to-peak value The voltage or current value of a waveform measured from its minimum to its maximum points.

Period The time interval of one complete cycle of a periodic waveform.

Periodic Characterized by a repetition at fixed time intervals.

Permeability   The measure of ease with which a magnetic field can be established in a material.

Phase   The relative displacement of a time-varying waveform in terms of its occurrence with respect to a reference.

Phasor   A representation of a sine wave in terms of both magnitude and phase angle.

Photodiode   A diode whose reverse resistance changes with incident light.

PN junction   The boundary between N-type and P-type materials.

Pole   In practical terms, a single *RC* network in a filter or amplifier that causes the response to change at a 20-dB/decade rate above or below a certain frequency.

Positive feedback   The return of a portion of the output signal to the input such that it is inphase with the input signal.

Power   The rate of energy consumption.

Power factor   The relationship between volt-amperes and true power or watts. Volt-amperes multiplied by the power factor equals true power.

Power gain   The ratio of output power to input power. The product of voltage gain and current gain.

Power supply   An electronic instrument that produces voltage, current, and power from the ac power line or batteries in a form suitable for use in various applications to power electronic equipment.

Primary   The input winding of a transformer.

Pulse   A type of waveform that consists of two equal and opposite steps in voltage or current separated by a time interval.

Pulse width   The time interval between the opposite steps of an ideal pulse. For a nonideal pulse, the time between the 50% points on the leading and trailing edges.

Push-pull   A type of class B amplifier in which one output transistor conducts for one half-cycle and the other conducts for the other half-cycle.

Q-point   The dc operating (bias) point of an amplifier.

Quality factor *(Q)*   The ratio of true power to reactive power in a coil or a resonant circuit.

Radian   A unit of angular measurement. There are 2π radians in one complete revolution. One radian equals 57.3°.

Ramp   A type of waveform characterized by a linear increase or decrease in voltage or current.

Reactive power   The rate at which energy is stored and alternately returned to the source by a capacitor or inductor. The unit is the VAR.

Recombination   The process of a free electron falling into a hole in the valence band of an atom.

Rectifier   An electronic circuit that converts ac into pulsating dc.

Reflected load   The secondary load as it appears to the source in the primary.

Regulator   An electronic circuit that maintains an essentially constant output voltage with a changing input voltage or load.

Relay   An electromagnetically controlled mechanical device in which electrical contacts are opened or closed by a magnetizing current.

Reluctance   The opposition to the establishment of a magnetic field.

Resistance   Opposition to current. The unit is the ohm (Ω).

Resistor   An electrical component designed specifically to provide resistance.

Resonance   In an *LC* circuit, the condition where the impedance is minimum (series) or maximum (parallel).

Response   In electrical circuits, the reaction of a circuit to a given input.

Retentivity   The ability of a material, once magnetized, to maintain a magnetized state without the presence of a magnetizing force.

Rheostat   A two-terminal variable resistor.

Right angle   A 90° angle.

Ripple   The variation in the dc voltage on the output of a filtered rectifier caused by the slight charging and discharging action of the filter capacitor.

Rise time   The time interval required for a pulse to change from 10% to 90% of its amplitude.

Rising edge   The positive-going transition of a pulse.

Roll-off   The decrease in the response of a filter below or above a critical frequency.

Root mean square (rms)   The value of a sine wave that indicates its heating effect, also known as the effective value. It is equal to 0.707 times the peak value.

Saturation   The state of a BJT in which the collector current has reached a maximum and is independent of the base current.

Sawtooth   A type of electrical waveform composed of ramps. A special case of a triangular wave.

Schematic   A symbolized diagram of an electrical or electronic circuit.

Secondary   The output winding of a transformer.

Selectivity   A measure of how well a filter passes certain frequencies and rejects others. The narrower the bandwidth, the greater the selectivity.

Semiconductor   A material that has a conductance value between that of a conductor and that of an insulator. Silicon and germanium are examples.

Sensitivity factor   The ohms-per-volt rating of a voltmeter.

Series   In an electrical circuit, a relationship of components in which the components are connected such that they provide a single current path between two points.

Short circuit   A zero or abnormally low resistance between two points. Usually an inadvertent condition.

Silicon   A semiconductor material used in diodes and transistors.

Silicon-controlled rectifier (SCR)   A device that can be triggered on to conduct current in one direction.

Solenoid   An electromagnetically controlled device in which the mechanical movement of a shaft or plunger is activated by a magnetizing current.

Source   Any device that produces energy.

Steady state   The equilibrium condition of a circuit that occurs after an initial transient time.

Step-down transformer   A transformer in which the secondary voltage is less than the primary voltage.

Step-up transformer   A transformer in which the secondary voltage is greater than the primary voltage.

Superposition   A method for analyzing circuits with two or more sources by examining the effects of each source by itself and then combining the effects.

Switch   An electrical or electronic device for opening and closing a current path.

Tank   A parallel resonant circuit.

Tapered   Nonlinear, such as a tapered potentiometer.

Temperature coefficient   A constant specifying the amount of change in the value of a quantity for a given change in temperature.

Terminal   An external contact point on an electronic device.

Terminal equivalency   A condition that occurs when two circuits produce the same load voltage and load current where the same value of load resistance is connected to either circuit.

Tesla   The unit of flux density.

Thevenin's theorem   A circuit theorem that provides for reducing any resistive circuit to a single equivalent voltage source in series with an equivalent resistance.

Thyristor   A class of four-layer semiconductor devices.

Time constant   A fixed time interval set by $R,C,$ and/or L values that determines the time response of a circuit.

Tolerance   The limits of variation in the value of a component.

Trailing edge   The second step or transition of a pulse.

Transconductance, g_m   The ratio of drain current to gate-to-source voltage in an FET.

Transformer   A device formed by two or more windings that are magnetically coupled to each other and providing a transfer of power electromagnetically from one winding to another.

Transient   A temporary passing condition in a circuit. A sudden or temporary change in circuit conditions.

Transient time   An interval equal to approximately five time constants.

Transistor   A semiconductor device used for amplification and switching applications in electronic circuits.

Triac   A bidirectional thyristor that can be triggered into conduction.

Triangular wave   A type of electrical waveform that consists of two ramps.

Trigger   The activating unit of some electronic devices or instruments.

Trimmer   A small variable capacitor.

Turns ratio   The ratio of secondary turns to primary turns.

Unijunction transistor (UJT)   A type of transistor consisting of an emitter and two bases.

Unit gain   A gain of 1.

Valence   Related to the outer shell or orbit of an atom.

VAR   Volt-ampere reactive. The unit of reactive power.

Varactor   A diode that is used as a voltage-variable capacitor.

Volt   The unit of voltage or electromotive force.

Voltage   The amount of energy available to move a certain number of electrons from one point to another in an electrical circuit.

Voltage gain   The ratio of output voltage to input voltage.

Voltage regulation   The process of maintaining an essentially constant output voltage over variations in input voltage or load.

Watt (W)   The unit of power.

Watt's law   A law that states the relationships of power to current, voltage, and resistance.

Waveform   The pattern of variations of a voltage or current showing how the quantity changes with time.

Weber   The unit of magnetic flux.

Winding   The loops of wire in an inductor.

Wiper   The sliding contact in a potentiometer.

Zener diode   A type of diode that operates in reverse breakdown (called zener breakdown) to provide voltage regulation.

INDEX